FINDING AUSTRALIAN BIRDS

A Field Guide to Birding Locations

Tim Dolby and Rohan Clarke

CSIRO
PUBLISHING

Reprinted with corrections in 2015

National Library of Australia Cataloguing-in-Publication entry

Dolby, Tim, author.

Finding Australian Birds : a field guide to birding locations / Tim Dolby and Rohan Clarke.

9780643097667 (paperback)
9781486300846 (epdf)
9781486300853 (epub)

Includes bibliographical references and index.

Bird watching – Australia – Guidebooks.
Birds – Australia.
Australia – Guidebooks.
Clarke, Rohan, author.
598.0723494

Published by
CSIRO Publishing
Locked Bag 10
Clayton South VIC 3169
Australia

Telephone: +61 3 9545 8400
Email: publishing.sales@csiro.au
Website: www.publish.csiro.au

Front cover: Turquoise Parrot. Photo by Rohan Clarke.

Back cover: (Clockwise from left) Plains Wanderer, Blue-faced Honeyeater, Southern Emu-wren, Eclectus Parrot. Photos by Rohan Clarke.

Set in 9/11 Stempel Schneidler
Edited by Peter Storer Editorial Services
Cover and text design by James Kelly

Typeset by Thomson Digital
Printed in China by 1010 Printing International Ltd

Contents

Contents

Acknowledgements

We've received much assistance from experienced birdwatchers throughout Australia. We are especially grateful to the following people (in alphabetically order) who joined us in the field at various times, provided tips on great birding locations, feedback on chapter drafts or assisted with the project in other ways: Richard and Margaret Alcorn, Tim Bawden, Rebecca Boulton, Adrian Boyle, Martin Cachard, Mike Carter, Mark Carter, Martin Cake, Greg Clancy, Stuart Dashper, Jeff Davies, Paul Dodd, John Ewen, Tim Faulkner, Keith Fisher, Lindsay Fisher, Dougald Frederick, Alan Gillanders, Denise Lawungkurr Goodfellow, John Graff, John Harris, Brian Hawkins, Ashley Herrod, Dean Ingwersen, Nick Leseberg, Murray Lord, Duncan Mackenzie, Braden McDonald, Ross McDonald, James Melville, Peter Menkhorst, Bob Moffatt, Alan Morris, Richard Noske, Greg Oakley, Kevin Parker, Fiona Parkin, Carol Probets, Michael Ramsey, Del Richards, Greg Roberts, Mick Roderick, Martin Schulz, Peter Shute, Jen Spry, Simon Starr, Alistair Stewart, George Swann, Tom Tarrant, Brian Thomas, John Tongue, Andrew Turbill, Chris Tzaros, Peter Waanders, Paul Walbridge, Chris Watson, John Weigel, Kailash Willis and Ruth Woodrow. While we are truly thankful for the exceptional assistance we have received from others, any omissions or errors remain our own. We are grateful to the CSIRO Publishing staff, in particular John Manger, Briana Melideo, Helena Clements and Peter Storer, for their efforts in seeing this book through to fruition. Finally, we would like to thank our families for their support throughout this project, for sharing the many family trips that have involved at least some element of birdwatching and for tolerating our birdwatching absences at other times: to Tanya Bramley, Barbara, Rhys and James Dolby and Kate and Aidan Clarke – thank you.

Photo credits

Adrian Boyle: 36, 97, 103, 106, 107, 157r, 313, 319, 331, 370, 382, 390, 410, 429, 449, 451, 455, 456l, 458, 460, 501, 503, 505

Rebecca Boulton: 335, 338

Martin Cake: 14, 28, 56, 401u, 407, 414, 419, 427, 434, 438, 454

John Ewen: 4, 18, 23

Ashley Herrod: 113, 362

Alan Morris: 118

Kevin Parker: 25

Mick Roderick: 196, 198

Martin Schulz: 178, 218, 220, 346

Brian Thomas: 134, 135, 137

Chris Tzaros: 139, 222, 345, 348u, 364, 366, 368, 371, 380

John Weigel: 32, 75, 77, 128, 456u, 461

Kailash Willis: 326

Tim Dolby: xii, 125, 129, 186, 191, 227, 254, 261, 303, 308, 309, 311, 320, 322, 324, 327, 329, 360, 404, 412

Rohan Clarke: ix, xiv, xvii, 5, 6, 10, 15, 22, 24, 30, 33, 42, 45, 47, 50, 51, 52, 53, 54, 58, 59, 61, 62, 64, 69, 71, 72, 78, 79, 81, 82, 87, 89, 91, 94, 95, 100, 109, 111, 115, 122, 124, 131, 140, 141, 144, 145, 147, 148, 152, 153, 157l, 159, 160, 164, 169, 170, 172, 179, 181, 185, 189, 201, 203, 206, 207, 209, 213, 216, 219, 228, 231, 232, 234, 235, 238, 243, 244, 245, 247, 248, 250, 252, 259u, 259l, 260, 263, 265, 267, 269, 272, 276, 278, 280, 282, 283, 287, 288, 290, 292, 293, 294, 296u, 296l, 300, 305, 314, 316, 318, 323, 330, 337, 340, 348l, 349, 351, 352, 353, 357, 367, 374, 375, 376, 378, 393, 396, 401l, 406, 411, 417, 419, 432, 442, 444, 446, 447, 452, 466, 468, 470, 471, 473, 475, 476, 479, 482, 484, 486, 487, 488, 490, 493, 495, 496, 498, 499

Introduction

How to use this book

Finding Australian Birds aims to assist readers in locating Australia's wonderful birdlife. It is designed to accommodate a broad spectrum of interests and skill. At one end, it provides information to birdwatchers regarding the birds they may expect to see in a diversity of oft-visited locations, such as remnants of habitat in and around our major cities and popular national parks. At the other end, it enables birders with particular targets to seek those specific species, including those that are rare and often hard to observe. Where possible, this book also seeks to highlight sites that may offer alternative natural attractions for non-birding family members and other travelling companions. The book is divided into two core sections. The first describes the best birdwatching sites in Australia, conveniently divided into the six states – Queensland, New South Wales, Victoria, Tasmania, South Australia, and Western Australia – and the Northern Territory, and offshore islands. Owing to its relatively small size, the Australian Capital Territory section is included in the account for NSW. The state and territory chapters are then further divided into 'birding regions' to encapsulate major geographic or biogeographic areas. For example, the Northern Territory is broken into two regions – the first focussed on birding opportunities across the Top End and the second providing coverage of birding opportunities in the Red Centre.

At the beginning of each chapter, the reader will find a range of information under several headings:

- The **regional overview** provides: an outline of what to expect in the region; information on the climate and conditions, including the best season(s) for birdwatchers to visit; a summary of travel options that may include insight regarding the quality of the roads, optimal flights, and so on; and recommended birdwatching itineraries. Because Australia is a vast land mass, and the distances between sites are often great, the itineraries should prove especially useful to first-time international visitors: they recommend the minimum amount of time needed per site to ensure travel plans aren't too ambitious.
- **Birding highlights** provides a summary of the birding specialties in the region. This includes information on endemic and other speciality species and key subspecies within the region, as well as discussion of the common species that typify the bird communities that visitors can expect to experience.

Each birding region is further divided into sites. These start with the site name, followed by a concise list of key species to be found at the site – key species are a combination of locally significant species that are readily seen at the site and nationally or regional significant species that occur at the site. More details for each are provided in the subsequent account. A concise list of **other species** highlights further species of interest that may not meet the above criteria. This is followed by an introductory paragraph detailing how to get there, where to stay (if appropriate) and other useful tips that may assist the birdwatcher at that location. Subsequent text provides detailed description(s) of the site, including specific tips on just how to locate those difficult-to-see and/or sought-after species.

The second core section of the book is a bird-finding guide. This acts as a cross-referenced tool, linking individual species with specific sites mentioned in the text. Species summaries also include insight regarding the likelihood of encounter and typical habitats. Where appropriate, subspecific information is also summarised. Importantly, this is not a bird field guide and readers are urged to use this book in conjunction with one of the excellent Australian bird field guides (see *Further reading* p. xix).

Throughout this book, the taxonomy of the IOC World Bird List (v 3.4) is followed (www.worldbirdnames.org). The cut-off date for updates and changes to this list was July 2013.

Australia: an overview

Australia provides opportunities for some of the world's most spectacular birding. A continent in its own right, with a landmass of 7.7 million km², it is the world's largest island and the sixth largest country. Aside from the mainland states and territories, Australia also administers the islands of Christmas, Cocos (Keeling), Ashmore, Norfolk, Lord Howe, Heard and Macquarie, so the coverage of this book spans a quarter of the Earth's circumference.

Australia hosts some remarkable geographical features that provide a spectacular backdrop (and, of course, travel highlights in their own right) while exploring the continent's birdlife. The Great Dividing Range stretches 3500 km from the north-east tip of Queensland, down through eastern New South Wales to the Grampians in western Victoria. These ranges include, among other habitats, tropical rainforest, vast eucalypt forests and woodlands and the only alpine areas on the mainland. Off the south-eastern coast lies the island state of Tasmania, with its spectacular mountain ranges, extensive areas of native vegetation (42% of the state is either reserved or exposed to minimal use) and pristine coastlines, including the entire south-west coast for which there is no road access. The Great Barrier Reef, located off the coast of Queensland, is the world's largest coral reef, spanning over 2000 km of coastline. Covering most of continental Australia, the 'outback' is predominantly semi-arid rangelands and desert, and includes some of the world's largest deserts (such as the Great Victoria, Great Sandy, Tanami, Simpson and Gibson Deserts), the vast Nullarbor Plain, spanning the South Australian and Western Australian border region on the southern coast of Australia, and the vast dry mulga woodlands (encompassing 20% of the Australian continent). Across northern Australia, from west to east, are the Kimberley, the Top End, Gulf Country and Cape York – all have a tropical climate with a predominantly summer monsoon season, with large tracks of woodland, grassland and escarpment. Although Australia is a relatively dry continent, some of its most arid lands are crossed by remarkable river systems. For example, the Cooper Creek system begins in the Channel Country of south-western Queensland and, after ~1300 km, drains into the largest Australian lake bed, Lake Eyre. The largest, and perhaps the most iconic river systems, are the Murray and Darling Rivers that collectively drain the Murray–Darling Basin – an area of over 1 million km^2 in the eastern half of the continent, or one-seventh of the land mass of Australia. The continent is bounded by the Indian, Pacific and Southern Oceans, and separated from land masses to the north by the Arafura and Timor Seas. Other notable marine features include: Bass Strait, which separates Tasmania from Victoria; Torres Strait, which separates Queensland from Papua New Guinea; and the Great Australia Bight bounding the southern coastline of Western Australia and the coastline of South Australia.

The National Reserve System is Australia's network of protected areas and includes more than 9300 separate reserves covering ~14% of the country (almost 106 million ha).The reserve system consists of Commonwealth, state and territory reserves, Indigenous lands, protected areas that are run by non-government organisations, and ecosystems protected by farmers on their private working properties. Many of these places are specifically covered in this book. Australia currently has 65 Ramsar-listed wetlands covering 7.5 million ha. Recently, 314 Important Bird Areas (IBAs) have been identified across Australia (see www.birdata.com.au/iba.vm). These IBAs are considered to be of international significance for bird conservation. Not surprisingly, many IBAs are captured within the National Reserve System, but there are also many that are not.

Birdlife Australia (www.birdlife.org.au) is Australia's premier birding organisation, and provides invaluable assistance for both local and international birders. It is well worth joining Birdlife Australia just to participate in the many monthly activities that take place. An award-winning quarterly magazine forms part of the

Subtropical rainforests in northern NSW support over 100 species of bird including sought-after species such as Wompoo Fruit-Dove, Marbled Frogmouth, Regent Bowerbird and Paradise Riflebird.

membership package, while your membership also provides a voice for the birds, because BirdLife Australia is the leading advocacy group for birds and bird conservation in Australia. BirdLife Australia consists of numerous state and regional groups, so it is also possible to 'tap into' local activities and knowledge through this network. Contact details for BirdLife Australia and some of the more active state and regional bird groups are listed on pp. xxi–xxiii.

Finding Australian birds

Due to the great age of the Australian continent, its variable weather, extended periods of geographic isolation and the vast areas involved, much of the birdlife is unique. At last count, 898 bird species have been recorded in Australia and its offshore islands and territories. Although ~165 of these are considered vagrant or accidental visitors, more than 45% of the remainder are Australian endemics: found nowhere else on Earth. Some species are extraordinarily specialised with very restricted distributions, endemic to a localised area (e.g. Golden Bowerbird), while others are widespread and occur over most of the continent (e.g. Brown Falcon).

Australian birds include such recognisable species as the world's second largest bird and Australia's unofficial bird emblem, the iconic Emu, one of the world's great songsters and mimics, the Lyrebird and, with its distinctive chorus of laughter, the Kookaburra. Australia is also home to over 50 species of brightly coloured parrots and cockatoos, with many of the most colourful species (e.g. rosellas, lorikeets and galahs), easily seen in urban environments. The diversity of finches is also high, and includes the stunningly beautiful Gouldian Finch. Owing to Australia's Gondwanan origins (the southernmost of two ancient continents) many of the most well-represented passerine groups so characteristic of Australia also have ancient origins. For example the honeyeaters, fairy-wrens, grasswrens, thornbills, gerygones, scrubwrens, treecreepers and the Australo-papuan robins can all be traced back to ancient songbirds with local origins. Finally, being an island continent, the surrounding oceans hold a truly remarkable diversity of seabirds: if you participate in a pelagic boat trip off the southern coastline, on a good day it is possible to see as many as eight species of albatross, including the majestic Wandering Albatross.

Ten tips for finding birds in Australia

1. Know your birds by knowing what to look for – in conjunction with this book, a good field guide is a must.
2. Many species in Australia are small, skulking or both. When combined with the challenges posed by dense and sometimes tall vegetation, good birdwatching optics, especially binoculars, will enhance the views obtained and provide more opportunity for a positive experience.
3. Most birds display distinct habitat preferences. Species-specific pointers and images of habitat in this book, alongside habitat descriptions presented in field guides, will help to narrow the search.
4. Many birds are most easily located by call. A growing familiarity with the calls of common species allows a birdwatcher to focus more on the unfamiliar. When an unknown songster is tracked, there is a higher probability of it being a less common species.
5. Many small birds associate with other species in mixed flocks. Looking for birds 'where the birds are' enhances the chance of encountering many of the more sought-after species – for

example, the elusive Grey Honeyeater is most frequently found in the company of thornbills.

6. Australian birds display a remarkable diversity of movement patterns, with some key strategies being north–south migration, altitudinal migration, nomadism, and population irruptions and dispersal in response to boom and bust cycles driven by above average rainfall and drought years. Knowledge of seasonality (as presented in this book) and other less-well-defined movement strategies can be the key to locating some species.
7. Australia experiences a climate of extremes. On hot days, birds are generally most active in the morning, especially just after dawn, but become progressively harder to find through the heat of the day. Most species show increased activity again in the late afternoon, but, in exceptionally hot weather (>38°C), this second activity peak may not occur. Conversely, in cold weather (<5°C) the dawn chorus may initially be more muted until birds have a chance to soak up the first rays of sunlight.
8. Many bird species drink daily from freshwater sources, especially in warmer weather. Along with areas with concentrated food resources, such as eucalypts heavy with blossom (honeyeaters and lorikeets), caterpillar-infested shrubs and trees (cuckoos) or drying wetlands (herons, egrets and other piscivores), these locations warrant extra time.
9. Good field craft isn't essential but it certainly helps! Wear muted colours, avoid sharp noises and use slow steady movements (or better remain still) while observing – with these techniques birds will be less disturbed by your presence and as an added bonus you'll also get better views.
10. To continue honing your bird finding skills, take note of where and how you find birds. Some places are simply better for birds than others, with different species responding differently to the available opportunities – open areas, sheltered areas, sites with water, the presence of a specific plant species or updrafts to assist flight are just some elements that create the right opportunities for particular species.

Seasonal variation that affects birdwatching opportunities in Australia

Although there are good opportunities for birdwatching in most areas of Australia at any time of year, seasonal variation means that for many sites, specific times of year can be better than others. Where appropriate, seasonal occurrences are specified for mobile species within the accounts for each site. Nevertheless, there are some broad principles that may be useful for travel-planning purposes. Because Australia is such a large country (the continent spans 33 degrees of latitude), the seasonal patterns differ greatly between the tropical northern regions and the temperate southern regions. Here, a broad zone centred on the Tropic of Capricorn provides a convenient distinction between the tropical north and the temperate south.

September to November (spring in the south, late dry season/build-up in the north)

This can be one of the best seasons to birdwatch in Australia. Across much of southern Australia, bird activity increases with the arrival of the breeding season. It is also a period of great movement, with large numbers of birds migrating south, especially along the Great Dividing Range and eastern seaboard, but there are also migrants passing through most other areas of Australia. Many of the migrant land birds spend the 'winter' in central and northern Queensland and New Guinea. Also included here are as many as 90 species (mostly shorebirds and seabirds) that are returning to continental Australia and its waters. This timing means that it is possible to see most of the migrants that visit northern Australia during the wet season before the wet season actually commences. In this regard, November generally provides the best pre-wet season opportunities for species such as Oriental Pratincole, Eastern Yellow Wagtail and Barn Swallow in the Kimberley and Top End regions. The September–October period is also a great time for birdwatchers to head inland: with daytime temperatures that are generally mild and nighttime temperatures not too cold, it is a perfect time to visit fabulous inland birding locations such as Alice Springs (NT), Gluepot Reserve (SA), Sturt National Park (NSW), Bowra Station (Qld), Wyperfeld and Hattah-Kulkyne National Parks (Vic).

December to February (summer in the south, wet season in the north)

Between December and February, temperatures can be very high. In the centre and south of Australia, this period is also characterised by reduced rainfall, whereas the northern tropics and subtropics generally receive most of their annual rainfall at this time. For birdwatching, the temperate coastal areas of the east and south coasts (including Tasmania) provide the most pleasant weather conditions. The same period in the tropical north can be more challenging because it is usually hot and humid, and high rainfall events may mean some areas are inaccessible. An average of 11 tropical cyclones form each year in the Australian region, with almost all occurring

Coastal heathlands in Tasmania and eastern Australia provide important habitat for Eastern Ground Parrot. Other species that frequent this habitat include Southern Emu-wren, Striated Fieldwren, Tawny-crowned Honeyeater and Beautiful Firetail.

in the period from December to April. Many tropical migrants are, however, present at this time so if you can tolerate the humidity, heat and the chance of a cyclone, birdwatching in the wet season can have its rewards. If possible, it is best to avoid Australia's inland areas during this period because of the extreme heat. Each summer, Australia provides critical non-breeding habitat for millions of migratory shorebirds. At this time of year, it is worth visiting coastal mudflats and wetland systems such as the Cairns Esplanade and Toorbul Shorebird Roost (Qld), Western Treatment Plant (Vic) and Roebuck Bay and Eighty mile Beach (WA). Summer is also a good time to head to offshore islands, such as Christmas Island and the Cocos Islands, for a chance of vagrant species (in addition to all of the endemics) or Macquarie Island because this is the period when most expeditions visit. The summer school holidays in Australia begin in mid-December and end in late January (actual dates vary between states and year to year). As a result, coastal accommodation in non-tropical areas is often heavily booked during this period, such that advance bookings are essential.

March to May (autumn in the south, late wet/early dry season in the north)

As in spring, the mild days and nights mean that autumn is a very comfortable time to visit inland birdwatching sites. During autumn, various groups of eucalypts open up their flowers to a host of bird species that rely on nectar and pollen for their survival. It's a good time to visit the box-ironbark forests, such as Chiltern and Greater Bendigo National Parks (Vic), the Barraba Region and Capertee Valley (NSW), that line the western slopes of the Great Dividing Range. In Western Australia, it is an excellent time to visit Dryandra Forest, and then continue south to sites around Albany, such as the Stirling Range, Two Peoples Bay and Cheynes Beach. In the tropical north, water levels are usually at their annual peak by the late wet season, meaning waterbirds (and many seed-eating species that must drink daily) are widely dispersed and more difficult to observe. The northern country is, however, at its finest, with most vegetation being lush and green by the end of the 'growing season'. It is also a period of northern migration – migratory shorebirds depart for the northern hemisphere via the East Asian–Australasian Flyway and many acquire splendid breeding plumages just before departure. Other species, such as Yellow-faced Honeyeater, Red Wattlebird, Noisy Friarbird, Spotted and Striated Pardalote, and Black-faced Cuckoo-shrike participate in the autumn migration within Australia, and some can occasionally be seen moving north in large flocks.

June to August (winter in the south, dry season in the north)

Southern Australia sustains rich evergreen flora and abundant birdlife throughout the year because its winter climate is relatively mild. Some species, such as Orange-bellied Parrot and Swift Parrot, migrate north from Tasmania across Bass Strait to winter on the Australian mainland, while Double-banded Plover and White-fronted Tern undertake an east–west journey from New Zealand to spend the winter in Australia. In winter, the southern oceans host a great diversity of cold-water seabirds such as albatrosses, petrels and prions, so it is an excellent time to take part in a pelagic seabird excursion. Alternatively, sea-watching from coastal headlands can also be good at this time of year as winter storms bring some species within sight from land. Good sites include Mistral and Magic Points, Cape St George and Green Cape Lighthouses (NSW), Point Addis, Point Lonsdale and Cape Nelson (Vic), Cape Bruny and Cape Tourville Lighthouses (Tas), Newland Heads and Cape De Couedic (SA), and The Gap and Cape Leeuwin (WA). The dry season is the most pleasant time to visit the tropical north, given the low rainfall, mild to warm temperatures and low humidity. Notably, vehicular access to some tropical areas such as Cape York and the much of the Kimberley is only possible once the roads re-open in the early dry season through until the first rains of the wet season.

Nomadism

The movements of many of the birds in Australia's arid and semi-arid interior are largely nomadic, and often made in response to periods of above average rainfall or extended drought. These events lead to large population fluctuations as species both breed up and decline, and depart and return, in direct response to local or regional conditions. These fluctuations are exemplified by a range of bird species, including Freckled Duck, Budgerigar, Flock Bronzewing, Black Honeyeater, Pied Honeyeater, Crimson Chat and White-browed and Masked Woodswallow. These, and other species, may appear in an area where they have not been seen for several years, responding to favourable conditions over vast distances.

Climatic conditions

Due to its large size, Australia displays a great diversity of climatic conditions. The tropical regions in the north have high temperatures and humidity, and distinct wet and dry seasons. The arid expanses of the interior have high daytime temperatures especially in summer, when temperatures can exceed 40°C and low annual

The Blue Mountains in NSW are part of the Great Dividing Range, which stretches 3500 km from the north-east tip of Queensland, down through eastern New South Wales to the Grampians in western Victoria.

Ethical birdwatching

All birders have a responsibility to adhere to ethical birdwatching principles. BirdLife Australia has developed a policy on ethical birdwatching that can be found at www.birdlife.org.au. The guidelines are especially appropriate at sites such as those outlined in this book because many of these will be visited over time by relatively large numbers of birdwatchers. Even a very small impact on the birds (and landholders and other community members) at such sites by individual birdwatchers needs to be viewed in the context of cumulative impacts from all visitors. For this reason, we ask that birders strive to adhere to the following principles at all times.

Promote the welfare of birds and their environment

- Avoid stressing birds or exposing birds to danger.
- Avoid using methods such as flushing and call playback, particularly during nesting season when the consequences of even minor disturbance may be severe (e.g. nest failure).
- Remain on roads, trails and paths where they exist – this avoids habitat disturbance.
- Do not handle birds.
- Report rare bird sightings to conservation authorities but consider the wellbeing of the bird or population before making this knowledge more publicly available.
- Support the protection of birds and their habitat.

Respect the law and the rights of others

- Do not enter private property without the owner's explicit permission.
- Follow all laws, rules and regulations governing use of roads and public areas.
- Consider and respect the rights of landholders.
- Practise common courtesy in interactions with other people.

rainfall, with much of it falling in short intense rainfall events. The temperate regions in Australia's south-east and south-west corners have moderate rainfall and generally moderate temperatures though a full range of temperatures can be experienced (sub-zero in winter at alpine sites to 40+°C in summer).

The Australian landmass is relatively arid, with 80% experiencing annual rainfall of less than 600 mm and 50% experiencing less than 300 mm; it is the second driest continent on Earth (after Antarctica). The climate is influenced by southern hemisphere perturbations including the Indian Ocean Dipole and the El Niño-Southern Oscillation, with particular components of these (and others) correlating with periods of above average rainfall followed by periodic drought. There are also seasonal tropical low-pressure systems producing cyclones across northern Australia. These factors produce rainfall that varies markedly from year to year – as part of this much of the continent is prone to periodic flood events.

If you are birdwatching in the hotter times of the year, it is important to be suitably attired to ensure protection from the elements. A wide-brimmed hat protects the head from heat exposure, keeps the sun out of the eyes and helps avoid sunburn; long sleeved shirts and long pants provide further protection from the sun. Long pants and solid footwear also provide protection from prickly plants, scratches, insects and the very rare incidence of snake bite. Australia's sunshine is particularly unforgiving and a good broad spectrum sunscreen should be applied whenever bare skin is exposed to the sun. Generally, in hot conditions restrict your birdwatching activities to the cooler periods in the early morning and early evening. Not only is it safer and more pleasant for the observer, but birds are much more active at these times also. Of course, Australia is a climate of extremes and although the risks of heat and sun exposure should be carefully managed, so too should exposure to cold and wet conditions – for birdwatchers, this especially includes travel in temperate and southern mountain areas and southern ocean pelagic trips.

Getting around Australia

If you are planning to travel in the Australian bush, or through the outback, it is important to remember that these areas are vast, and there are often large to very large distances between towns and facilities. To assist travellers with planning, recommended itineraries are provided in the state and regional sections of this book. Appropriate preparedness and safe travel techniques cannot be emphasised enough. There have been several instances of bird-watchers dying in Australia's outback (as well as several other close calls) and safe travel strategies are something that all bird-watchers must consider. There are entire books focussed on remote area travel and remote area survival in Australia, so it is beyond the scope of this introduction to fully cover the issues. Nevertheless some of the key principles you should adhere to when travelling in remote areas include:

1. letting a reliable person or persons know of your plans and your return date so that they can raise the alarm with appropriate authorities should you fail to return
2. carrying appropriate maps and navigational aids and knowing how to use them (although online resources are great for trip planning, they should not be relied upon during remote-area travel)
3. ensuring your vehicle is in good working order and you know how to complete basic maintenance tasks (for example, in hire cars it is advisable to check both the location of the wheel-changing equipment and that it is all present and working properly at the time you take possession of the car)
4. carrying extra water, food, fuel and other spares such as tyres to ensure self-sufficiency for a period beyond the duration of your intended trip
5. carrying appropriate communications equipment (e.g. satellite phone and recharge options) to facilitate emergency communication
6. carrying appropriate first aid equipment
7. avoiding the hottest periods of the year (e.g. Dec–Mar) in remote inland semi-arid and desert areas, because the consequences of become lost or breaking down at these times can be considerably greater.

Canning Stock Route, WA. Some areas in Australia are very remote: the nearest petrol supply and store was ~500 km away!

In Australia, road conditions vary greatly: excellent sealed highways provide relatively straightforward access around the continent and through the centre. Highway 1, the world's longest road, with a length of ~14 500 km circles the entire Australian land mass. A 2WD vehicle is suitable for the majority of birding sites in the county, but, if you plan to explore more remote locations, a 4WD may be essential. The need for 4WD includes some sites covered in this book such as the desert areas of South Australia, Cape York in Queensland and the Kimberley in Western Australia. Of particular note, intense rainfall events in Australia are often unpredictable. This is especially so in desert regions where both anomalous southern low pressure systems can cause heavy winter falls and occasional tropical cyclones can lead to heavy summer falls far from the coast. Many roads (including occasionally our major highways) and most inland tracks may be impassable after such events. It therefore pays to monitor both the weather (especially in remote areas) and the road conditions (state road authority websites) so that you can plan accordingly. Kangaroos, other wildlife and livestock all stray onto outback roads (and indeed many less remote roads). Most roads where this occurs frequently include signage that warns of the threat. This threat is greatest at dawn and dusk when animals move between watering points, feeding areas and daytime resting areas. For this reason, if nighttime travel on outback roads is necessary, a good travel strategy is to take travel breaks that span the twilight periods. With careful planning, this, of course, also means maximising opportunities to birdwatch at the best times of day rather than drive.

For air travel, daily scheduled flights operate from every Australian capital city, and there is a large network of regional airports. It is possible when booking your flights to arrange to have a rental car awaiting your arrival at most of the larger airports. The major hire car companies in Australia are AVIS, Europcar, Budget, Thifty and Hertz, but there are also smaller local providers at many sites. It is worth checking the terms and conditions of any hire car companies because the hire of 2WDs is often restricted to sealed roads and hire of 4WDs many include clauses that limit or exclude liability and cover when travelling off gazetted roads – the latter would not be a problem when visiting most sites within this book, but it is a case of 'renter beware'. Rental cars are usually readily available but at certain times of the year, and at certain sites, 4WD can be in high demand. Most major cities are also accessible by rail, though unlike some other regions (Europe and the USA) there is not a strong culture of rail travel in Australia. See Rail Australia (www.railaustralia.com.au) for further details.

Northern Territory

The Northern Territory (NT) is a federal territory extending from the arid-lands in the centre of the continent, as a rectangular block, north to the coast and humid tropics. The Territory shares borders with Western Australia, South Australia and Queensland. The Capital City is Darwin, but there are only two other large population centres in the region, Katherine and Alice Springs. Despite its large area (~1.35 million km^2), it is sparsely populated, with ~230 000 residents. The north has a tropical climate with high humidity and is characterised by a wet (Nov–Apr) and a dry season (May–Oct), whereas central Australia is arid and experiences cool to mild winters (sometimes below zero at night) and hot dry summers. The Wedge-tailed Eagle is the bird emblem of the NT.

The Top End

Regional overview

The Top End in the NT provides some of Australia's most spectacular birding. Covering 400 000 km^2, it stretches from Darwin and Kakadu in the north, south to Katherine and Mataranka, east to the Barkley Tableland and the Gulf Region and west to the Victoria River and Gregory National Park. Habitats are diverse: they range from marine environments and river systems, floodplains and wetlands, monsoon forests, tropical woodlands and, in the south of the region, grasslands and shrublands. This diversity of habitats and the vast areas that remain uncleared means the Top End is a haven for birdlife and other wildlife.

On the edge of a harbour seven times bigger than Sydney Harbour, Darwin provides a gateway to many of the Top End birding sites. Although isolated by distance, the city is well serviced by air, road and rail links to other Australian cities. There are several excellent birdwatching locations in and around Darwin, with all sites easily reached by car.

A suggested 3-day itinerary with Darwin as a focal point is as follows:

- **Day 1.** Darwin Botanical Gardens, Lee Point, Buffalo Creek and Holmes Jungle (Rufous Owl, Chestnut Rail, Rainbow Pitta, Lesser Crested Tern, Arafura Fantail, Red-backed Button-quail, shorebirds).
- **Day 2.** East Point, Leanyer Treatment Plant and Howard Springs (Rainbow Pitta, Collared Kingfisher, Green-backed Gerygone, Beach Stone-curlew, Little Kingfisher, Black Bittern and, in summer, Eastern Yellow Wagtail, Little Ringed Plover, Swinhoe's Snipe).
- **Day 3.** Fogg Dam and Adelaide River (White-browed Crake, Little Kingfisher, Broad-billed Flycatcher, Rose-crowned Fruit-Dove, Mangrove Golden Whistler, waterbirds).

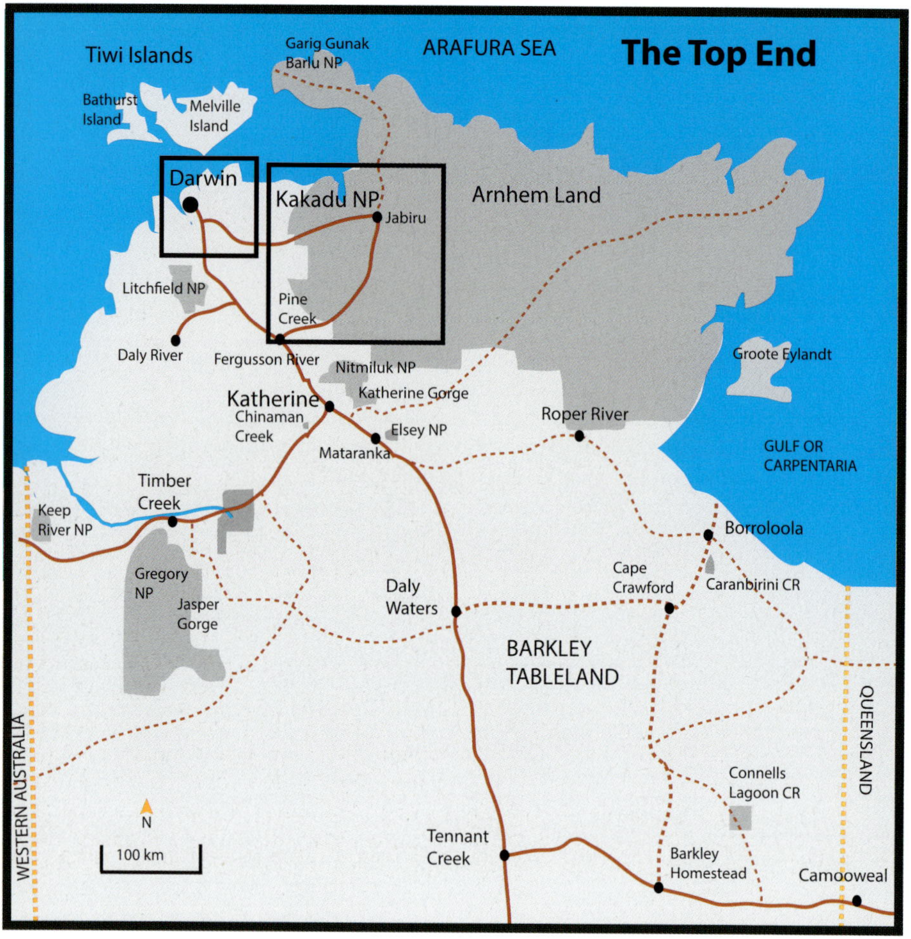

A suggested 10-day itinerary that connects Kakadu with both Pine Creek and the Katherine region involves:

- **Days 1–3.** Darwin to Kakadu along the Arnhem Hwy, but also visiting Mamukala Wetlands and concluding in Kakadu National Park on day 3, ~250 km (Gouldian, Masked and Long-tailed Finch, Black-tailed Treecreeper, Buff-sided Robin, Paperbark Flycatcher, waterbirds).
- **Days 4–5.** Kakadu National Park: Nourlangie and Yellow Water, 80 km (Partridge Pigeon, White-lined Honeyeater, Banded Fruit-Dove, Sandstone Shrike-thrush, Great-billed Heron, Little kingfisher, Red Goshawk, waterbirds and rock art).
- **Day 6.** Kakadu National Park: Gunlom, 120 km (White-throated Grasswren, White-lined Honeyeater, Sandstone Shrike-thrush, Banded Fruit-Dove, Chestnut-quilled Rock-Pigeon, and waterfalls and swimming holes).
- **Days 7–9.** Nitmiluk National Park, Chinaman Creek, Pine Creek, Fergusson River, Mataranka, 200 km (Red

Goshawk, Gouldian Finch, Hooded Parrot, Chestnut-backed Button-quail, and spectacular gorge country and more swimming holes).
- **Day 10.** Elsey National Park (Hooded Parrot, Gouldian Finch and swimming holes) and return to Darwin, 400 km.

From Katherine, plan side trips to the Victoria River Region (Purple-crowned Fairy-wren, White-quilled Rock-Pigeon, Gouldian and Star Finch, Pictorella and Yellow-rumped Mannikin) in the west, allowing 3 days, or east to the Gulf Region and Barkly Tableland (Flock Bronzewing, Yellow Chat, Pictorella Mannikin, Oriental Plover (summer), Grey Falcon, Letter-winged Kite) – again allow 3 days. An increasingly popular route for birdwatchers (with time on their hands) is to drive the Savannah Way. This outback tourist route links Darwin to both Cairns on the east coast and Broome on the west coast, a total distance of 3500 km, linking in most of the areas mentioned in the text; allow at least 14 days to explore the NT section (or 30 days for the entire journey).

Listed as a World Heritage Area, Kakadu National Park covers almost 2 million ha. The park entrance is 152 km east of Darwin, but there is little here and the first fuel stop, store and small information centre is another ~60 km further east along the Arnhem Hwy just before the South Alligator River. There is a diverse range of habitats within Kakadu, including easily accessible areas of floodplain (typically inundated by water for 2–6 months), monsoon forest, tropical woodlands and spectacular sandstone escarpments. The bird list for the park is impressive, boasting almost all of the Top End specialties and attracting a range of rarities. Sought-after species include the Top End endemic White-throated Grasswren, White-lined Honeyeater, Banded Fruit-Dove and Chestnut-quilled Rock-Pigeon. Kakadu also provides good opportunities to see Sandstone Shrike-thrush, Partridge Pigeon, Red Goshawk and Chestnut-backed Button-quail. The waterbirds, especially when massed in dry season concentrations, are a spectacle to behold; even non-birdwatchers flock to Kakadu to witness these events. Ideally, aim to devote at least 3 days within Kakadu and 4 if you plan to seek the White-throated Grasswren at Gunlom. Kakadu National Park is well serviced by camping areas, all with various levels of facilities. Some are run by the national park, whereas others function as independent commercial operations. Remote camping requires a permit from park management. For information on camping and accommodation options visit www.environment.gov.au/parks/kakadu. An entry fee of $25.00 provides access for a 14-day period. Information is also available from the Bowali Visitor Centre, 5 km west of Jabiru (open 8 a.m. to 5 p.m. daily).

The Top End has a tropical monsoonal climate with two major weather extremes: heavy rain during the humid wet season (Nov–Apr) and cooler less humid conditions with little rain during the dry season (May–Sept). The most pleasant times to visit are during the June to August period when the humidity is relatively low (40–60%), the average maximum daily temperature is 30–31°C and the average minimum night-time temperature is 19–21°C. October and November, known as the 'build-up', can be extremely hot and humid. If you can handle the conditions then, this period can be very rewarding, with many of the summer migrants, such as waders and larger cuckoos, arriving back in the Top End. Generally, the summer migrants arrive between August and November and leave between March and May. Visiting the Top End during the wet summer months can be stifling (temperatures can reach into the high 30s°C and the humidity is regularly above 70%), but again it can be rewarding, with an increased potential to find

Escarpment and floodplain, Kakadu National Park. Chestnut-quilled Rock-Pigeon and Sandstone Shrike-thrush occur in sandstone country visible in the foreground, Rainbow Pitta, Yellow Oriole and other specialists frequent the gallery forest visible in the background, while waterbirds congregate on the floodplains in season.

rarities that may visit from south-east Asia (e.g. Garganey). There are a few species (such as Spotted Nightjar, White-breasted, White-browed and Little Woodswallow, Australian Pratincole, Tree and Fairy Martin and many birds of prey) that are present in the Top End during the dry season but move to more southerly areas of Australia during the wet season.

Excellent sealed highways provide easy access across the Top End: the Stuart Hwy from Adelaide, SA, the Barkly Hwy from Mount Isa, Qld and the Victoria Hwy from Kununurra, WA. Darwin is well serviced by national coach companies and the iconic Ghan railway service links Adelaide and Darwin via Alice Springs. A 2WD vehicle will get you to the majority of birding sites in the region, but, if you intend to explore remote locations (including Jim Jim and Twin Falls in Kakadu), a 4WD is essential. Daily scheduled flights to Darwin operate from every mainland Australian capital city. Darwin, being the closest Australian capital city to Asia, also has regular international air services from Singapore, Malaysia, Indonesia and Brunei.

Birding highlights

The Top End supports four range-restricted endemic bird species found nowhere else. The Chestnut-quilled Rock-Pigeon, White-throated Grasswren and White-lined Honeyeater are restricted to the Arnhem Land escarpment and plateau, while the Hooded Parrot occurs in the southern section of the Top End north to areas including Gimbat. Within Australia, the Banded Fruit-Dove is also restricted to the Arnhem Land escarpment though a closely related taxon (sometimes treated as a full species) also occurs in Indonesia. The South Alligator floodplain subspecies of the Yellow Chat is a poorly known endemic form that is also worth looking for.

Among birdwatchers, the Top End is well known for opportunities to observe many sought-after species. These include Pied Heron, Lesser Crested Tern, Chestnut Rail, Chestnut-backed and Red-backed Button-Quail, Rufous Owl, Partridge Pigeon, Red-collared Lorikeet, Varied Lorikeet, Northern Rosella, Rainbow Pitta, Arafura Fantail, Paperbark Flycatcher, Buff-sided Robin, Mangrove Golden Whistler, Sandstone Shrike-thrush, Green-backed Gerygone, Rufous-banded Honeyeater and Gouldian Finch. In the far west of the Top End, you can also expect to see White-quilled Rock-Pigeon, Purple-crowned Fairy-wren and Yellow-rumped Mannikin. In the Gulf country and on the Barkly Tableland, you can find Flock Bronzewing, Letter-winged Kite, Purple-crowned Fairy-wren, Carpentarian Grasswren (now very rare in the NT), Yellow Chat and Pictorella Mannikin. Wet season migrants include Little Curlew, Swinhoe's Snipe, Oriental Plover and Oriental Pratincole. A wide range of rarities that originated in south-east Asia or further afield have been recorded, including Eurasian Little Grebe, Garganey, Northern Pintail, Spotted Whistling-Duck, Green and Stilt Sandpiper, Little Stint, Pin-tailed Snipe, Little-ringed Plover (annual at Leanyer Sewage Ponds, but very rare elsewhere), Caspian, Kentish and Ringed Plover, Oriental Honey Buzzard, Elegant Imperial-Pigeon, Black-headed, Franklin's, Sabine's and Black-tailed Gull, Oriental Reed-Warbler, House Swift, Barn and Red-rumped Swallow, White and Grey Wagtail.

In addition to the Buff-sided Robin, White-lined Honeyeater, Helmeted Friarbird and Arafura Fantail which have been elevated to full species status in recent checklists; some authorities also consider the 'Silver-backed Butcherbird' – currently a subspecies of Grey Butcherbird – to be a full species. Some other distinct or range-restricted subspecies found in the Top End include White-quilled Rock-Pigeon (cinnamon brown ssp. *boothi* in the western Top End), Partridge Pigeon (red-eyed ssp. *smithii*), Variegated Fairy-wren ('Lavender-flanked Fairy-wren' ssp. *rogersi* and *dulcis*), Black-chinned Honeyeater ('Golden-backed Honeyeater' ssp. *laetior*), Grey Whistler ('Brown Whistler' ssp. *simplex*), Crested Shrike-tit ('Northern Shrike-tit' ssp. *whitei*), Crimson Finch (black-bellied ssp. *phaeton*) and Yellow Chat (South Alligator floodplain ssp. *tunneyi*). The Tiwi Islands, isolated from mainland Australia since the

A female Variegated Wren, ssp. *dulcis*: one of a number of distinctive range-restricted subspecies that can be found in the Top End. For this form, look especially in woodland in close proximity to sandstone escarpment.

last ice age, display considerable endemism, with eight unique subspecies, including distinct forms of Masked Owl, Helmeted Friarbird and Hooded Robin (sadly, the last is possibly extinct).

Darwin Botanical Gardens

Key species: Rufous Owl, Barking Owl, Magpie Goose, Orange-footed Scrubfowl, Oriental Cuckoo (wet season)
Other species: Little Curlew (wet season), Cicadabird, Pale-vented Bush-hen

Just 1.5 km from Darwin's CBD (access via Gardens Rd or Gilruth Ave), the Darwin Botanical Gardens (42 ha) has in the past been one of the most reliable places in Australia to see Rufous Owl. A pair frequented the gardens with their favourite roosting site being the trees immediately adjacent to the main toilet block; alternative locations worth checking included further along the path near the playground and the larger open trees along the rainforest walk, especially near the pond in the north-east section or in the taller trees along the small path at the bottom of the rainforest habitat. In recent years, this pair has proved more difficult to find, but the above locations are still worth checking as they provide ideal roosting sites for the species.

While looking for Rufous Owl, keep your eyes open for roosting Barking Owl, particularly in the larger trees between the fountain and the rainforest walk. Pigeons and cuckoos are also often well represented in the gardens, with some likelihood of Rose-crowned Fruit-Dove, Pied Imperial-Pigeon, Little Bronze-Cuckoo and Pheasant Coucal. Wet season migrants include Eastern Koel, Channel-billed Cuckoo and Oriental Cuckoo. Orange-footed Scrubfowl are common within the gardens and relatively tame. Occasionally, especially in the late dry season and into the wet season large flocks of Magpie Geese frequent the lawns. Other birds to look for include Forest Kingfisher, Pale-vented Bush-hen (uncommon), Radjah Shelduck, Little Curlew (wet season), Varied Triller, Cicadabird and Spangled Drongo.

A Rufous Owl in the Darwin Botanic Gardens. Check the taller trees with a dense canopy, especially those in the rainforest section.

Charles Darwin University

Key species: Red-headed Honeyeater, Masked Finch, Long-tailed Finch, Green-backed Gerygone
Other species: Eastern Osprey, Yellow White-eye, Northern Fantail, Pied Imperial-Pigeon

Located 12 km from Darwin's CBD, the Casuarina campus of Charles Darwin

Darwin Harbour

Make time for a walk around the harbour district, visiting Stokes Hill, Fishermans and Fort Hill wharves. Out over the harbour, look for Lesser Frigatebird and Brown Booby, while White-breasted Woodswallow (dry season) are often overhead. Darwin Harbour has played host to several rare gulls; since 1998 Black-headed Gull have been a sporadic visitor to the wharf area with most records between December and March. In April 2008, a Franklin's Gull visited Fort Hill Wharf and there are also historic records of both Black-tailed and Sabine's Gull from within the harbour.

University is set on 56 ha of urban parkland bordering mangroves and beaches. To get there from the Stuart Hwy, take Bagot Rd onto Trower Rd and turn left into Lakeside Drive. At the roundabout, turn left on to Dripstone Rd and then turn left into University Drive South.

On the west side of University Drive South, riverine monsoon forest borders Rapid Creek. Here you can see most of Darwin's common honeyeaters such as Red-headed, Dusky, White-gaped, White-throated and Bar-breasted, as well as Varied Lorikeet and Red-winged Parrot, Pied Imperial-Pigeon, Little Bronze-Cuckoo, Grey (Brown) Whistler and raptors such as Collared Sparrowhawk, Grey Goshawk, Pacific Baza, Brahminy Kite and Eastern Osprey (the latter nest at the university). Further along University Drive South (at the intersection of University Drive West) is a pathway that leads to Casuarina Beach. The track winds through melaleuca and mangrove forest; expect to see Yellow White-eye, Large-billed and Green-backed Gerygone, particularly near the footbridge 100 m from the road. On rare occasions, Chestnut Rail has been reported feeding openly in the grassy areas of the university.

Further along the walk, search the forest fringe for Orange-footed Scrubfowl, White-bellied Cuckoo-shrike, Northern Fantail, Lemon-bellied Flycatcher and Channel-billed Cuckoo (wet season). Around the

Royal Darwin Hospital

Royal Darwin Hospital is located off Rocklands Drive. Open woodland around Royal Darwin Hospital holds similar species to those found at Charles Darwin University. The best birding is usually had by taking the walkway to the beach located west of the Paracelsus Rd carpark. At night, listen for the far carrying and haunting call of the Bush Stone-curlew – the species often frequents the carparks. During the dry season, the carpark can also be a good spot for Spotted Nightjar as they occasionally hawk for insects in the streetlights.

small grass-covered sand dunes you may encounter Masked, Long-tailed and Double-barred Finch, Chestnut-breasted Mannikin, Peaceful and Bar-shouldered Dove, Varied Triller, Red-backed Fairy-wren and Striated Pardalote.

Lee Point and Buffalo Creek

Key species: Chestnut Rail, Rainbow Pitta, Lesser Crested Tern, Arafura Fantail
Other species: Great-billed Heron, Red-headed Honeyeater, Large-tailed Nightjar, Mangrove Golden Whistler (rare), Mangrove Grey Fantail, Black-headed Gull (rare), waders, seabirds

At the northern end of Casuarina Coastal Reserve (1500 ha), 18 km from Darwin's CBD, the Lee Point beach and shore-point serves as a high tide roost for shorebirds. Buffalo Creek, a 2.5 km drive from Lee Point (reached via Buffalo Creek Rd, which runs east off Lee Point Rd), is an area of mangrove and coastal monsoon forests adjoining tidal sand banks. To get to either location from the Stuart Hwy, take Bagot Rd, turn right into McMillans Rd and left into Lee Point Rd. Lee Point is 7.5 km from the turn-off, while the turn-off to Buffalo Creek Rd is 6.5 km from McMillans Rd.

Around the Lee Point carpark, there is a good chance of species such as Spangled Drongo, Varied Triller, Leaden Flycatcher, Pied Imperial-Pigeon and Pacific Baza. Barn Swallow is also possible between October and March, though it is generally uncommon. At the point itself are several small reefs. If one continues along the beach towards Buffalo Creek, there is high-tide wader roost at a point where a small inlet forms a narrow sand spit (alternatively accessed from Buffalo Creek carpark). Look for Grey and Pacific Golden Plover, Grey-tailed Tattler, Terek Sandpiper, Ruddy Turnstone, Australian Pied and Sooty Oystercatcher, Eastern Reef Egret and Crested, Lesser Crested, Little, Roseate (uncommon) and Caspian Tern. Scan the waters out to sea for Lesser Frigatebird and Brown Booby. The beach also regularly hosts breeding pairs of Red-capped Plovers.

The Buffalo Creek mud flats can be a reliable place to see Chestnut Rail, including areas directly opposite the boat ramp. The best time to see this shy, chicken-sized bird is in the early morning or late evening on a low tide. Listen for their loud harsh '*karkkark*' call. Buffalo Creek is a very popular boating spot, so you may need to wait for quiet periods between boat launches. (If you miss seeing Chestnut Rail here, another good site is the Stuart Park mangroves, 3 km from Darwin City Centre on Tiger Brennan Drive.) Immediately to the right of the boat ramp, a small (often-muddy) trail leads into the mangroves. Look for Rufous-banded, White-gaped and Red-headed Honeyeater (particularly when Grey Mangrove is flowering), Rose-crowned Fruit-Dove, Green-backed and Large-billed Gerygone, Grey (Brown) Whistler, Yellow White-eye, Varied Triller, Arafura and Mangrove Grey Fantail and Mangrove Golden Whistler (uncommon). Along the creek's edge, there is the possibility of Little and Azure Kingfisher, Striated and Nankeen Night-Heron. Raptors that often patrol the creek include White-bellied Sea-Eagle and Brahminy and Whistling Kite. Renting a dinghy and outboard from a boat hire company in Darwin can be worthwhile, giving you access to the river and areas of mangroves. Hiring a boat also enhances ones chance of spotting Chestnut Rail, Little Kingfisher and Great-billed Heron. Take care, because Saltwater Crocodiles are regularly seen in this area.

The coastal monsoon forest between Buffalo Creek Rd and Casuarina Beach supports resident Rainbow Pitta. The best way to see them is to walk quietly through the middle of the forest from the eastern end of the carpark, then west for several hundred metres. Other birds here include Orange-footed Scrubfowl, Rose-crowned Fruit-Dove, Emerald Dove, Pied Imperial-Pigeon, Pheasant Coucal and Black Butcherbird. At night around the carpark, Large-tailed Nightjar can also be found.

The beach and extensive tidal flats adjacent to the mouth of Buffalo Creek hold large numbers of shorebirds, including Whimbrel, Eastern Curlew, Black-tailed and Bar-tailed Godwit, Common Greenshank, Terek Sandpiper, Sanderling, Great Knot, Grey and Pacific Golden Plover, Lesser Sand and Greater Sand Plover, and Little, Caspian, Crested and Lesser Crested Tern. Beach Stone-curlew also occurs in the area. The mud flats have good potential for rarities, such as the Kentish Plover, which was recorded here in November 1988. Birders should also scrutinise flocks of Silver Gulls carefully: although a vagrant to Australia, Black-headed Gull has been detected here (especially in December and January) with some frequency. As an added bonus, Dugongs are known to feed on the inundated sea-grass meadow just off Casuarina Beach.

Leanyer (Darwin) Sewage Ponds

Key species: Eastern Yellow Wagtail (wet season), Little Ringed Plover (Oct–Apr), Swinhoe's Snipe, Pale-vented Bush-hen, Yellow-rumped Mannikin, Garganey (rare during the build up and wet season), waterbirds

Other species: Little Curlew (wet season), Great-billed Heron, Little Kingfisher, Oriental Plover, Oriental Pratincole (both wet season), Flock Bronzewing, Long-toed Stint, White-winged Black Tern, Mangrove Gerygone, Broad-billed Flycatcher, Horsfield's Bushlark

Like many large sewage treatment plants that support numerous settling ponds, Leanyer Treatment Plant is an excellent birdwatching site. The best time to visit Leanyer is between September and March when migrant waders and rarities are a feature. Access to the ponds is restricted, so you will need to contact Northern Territory Power and Water to enquire about key collection arrangements and to complete the induction and indemnity process before entering the site. For details contact Customer Service on 1800 245 092

Charles Darwin National Park

Charles Darwin National Park (1300 ha) is the closest national park to Darwin and is a particularly good place to see woodland birds such as Varied Lorikeet, Northern Rosella, Red-tailed Black-Cockatoo, Brown Quail, Banded Honeyeater, Helmeted and Silver-crowned Friarbird and Great Bowerbird. At the picnic area at the end of the national park access road, a path leads down the slope to a large expanse of mangroves; look for Yellow White-eye, Red-headed Honeyeater, Chestnut Rail, Pale-vented Bush Hen and White-breasted Whistler. Charles Darwin National Park is 8 km from the CBD along Tiger Brennan Drive.

or customerservice@powerwater.com.au. Note that access arrangements periodically change and access is completely restricted at times (e.g. when construction works take place) so early contact is advised to learn of the current situation. If access is approved, from the Stuart Hwy take Bagot Rd (3.4 km), turn right into McMillans Rd (3 km), left into Lee Point Rd (3.5 km) and then right into Fitzmaurice Drive (1.4 km).

Just before the main entrance to the treatment ponds, a drain and track lead north for ~80 m. This area is good for finches such as Double-barred, Long-tailed and Crimson Finch and Chestnut-breasted Mannikin; be sure to look carefully because Yellow-rumped Mannikin (uncommon) have been seen. During the wet season, look for Eastern Yellow Wagtail and vagrant species such as Grey and White Wagtail.

Once inside Leanyer Treatment Plant, the best method for birding is to slowly drive the causeways between ponds. These hold large numbers of waterbirds, include Australasian Grebe, Great, Intermediate, Little and Cattle Egret, Pied Heron (also common at the Shoal Bay Waste Disposal Site near Karama off Vanderlin Drive) and ducks such as Wandering and Plumed Whistling-Duck, Radjah Shelduck, Hardhead and Pink-eared Duck. Waders include Pacific Golden Plover, Common, Wood and Sharp-tailed Sandpiper, Red-necked Stint, Whimbrel, Black-tailed Godwit and Black-winged Stilt. Also look for Gull-billed, Whiskered and White-winged Black Tern (wet season), White-breasted Woodswallow, Welcome Swallow (rare in the Top End), Brown Quail, Pale-vented Bush Hen (around the vegetated fringes during the wet), Horsfield's Bushlark and Flock Bronzewing (rare). Leanyer is the only site in Australia where Little Ringed Plover is known to occur annually – between two and five birds are usually present between October and April. The site regularly hosts a selection of other rare or difficult to find species including Swinhoe's Snipe, Long-toed Stint, Little Curlew, Oriental Plover and Oriental Pratincole. More infrequent rarities have included Little (Eurasian) Grebe, Red-necked Phalarope, Ruff, Garganey (odd birds are occasionally found in flocks of Grey Teal and Pacific Black Duck), Northern Pintail and Red-rumped Swallow. All of these species are more likely in the wet season and especially in late November and December when many migrants have arrived in Australia before significant wet season rainfall. This very dry period just before the wet season can result in huge concentrations of birds at key wetlands such as the sewage ponds.

The third area in which to concentrate your birding is the mangroves bordering the ponds on the north-east side. Look for Great-billed Heron, Little Kingfisher, Little Bronze-Cuckoo, Mangrove Gerygone (in areas of low-growing Grey Mangrove), Yellow White-eye, Dollarbird (wet season), Helmeted Friarbird, Red-headed Honeyeater, Cicadabird, Broad-billed, Paperback and Shining Flycatcher, Northern, Arafura and Mangrove Grey Fantail,

Wood Sandpipers are freshwater wetland specialists. A migrant from the northern hemisphere, the species can usually be found at Leanyer Sewage Ponds from August until mid-April.

Palmerston Sewage Works

Palmerston Sewage Works holds similar species to those found at Leanyer. Interesting species to look for are Pale-vented Bush-hen (wet season), White-browed Crake, Garganey (rare wet season visitor), Grey Goshawk and Yellow-rumped Mannikin (uncommon). In the mangroves to the north of the sewage ponds, look for Collared and Little Kingfisher, Chestnut Rail, Mangrove Grey Fantail, Mangrove Robin, Broad-billed Flycatcher, Cicadabird, Helmeted Friarbird and Mangrove Golden (uncommon) and White-breasted Whistler. To get to the Palmerston Sewage Works from the Stuart Hwy, take Roystonea Ave, then right into University Ave, left into Elrundie Ave and after 2.4 km turn right at Catalina Rd before driving to the end. Pale-vented Bush-hen is also possible on the small creek that crosses Wishart Rd (500 m east of Berrimah Rd), between Palmerston and Darwin. There is no direct access to the sewage works but most areas can be viewed from the boundary fence.

Mangrove and Green-backed Gerygone and Grey (Brown), Mangrove Golden and White-breasted (uncommon) Whistlers.

East Point Reserve

Key species: Rainbow Pitta, Collared Kingfisher, Green-backed Gerygone, Beach Stone-curlew, waders
Other species: Rose-crowned Fruit-Dove, Mangrove Golden Whistler, Rufous Owl, Large-tailed Nightjar

East Point Reserve (200 ha) has a range of habitats, including monsoon forest, mangroves, open parkland and beaches with rocky outcrops. This variety of habitats in a relatively small area provides excellent opportunities for birdwatching. To get there from Darwin, take Gardens Rd and continue on to Gilruth Ave and then East Point Rd, 3 km from the Darwin City Centre. Continue down East Point Rd and then Alec Fong Lim Drive for a further 3.8 km.

On the monsoon forest walk, starting just west of 'Pee-Wee's at the Point' restaurant, search the forest floor for Rainbow Pitta, Emerald Dove and Orange-footed Scrubfowl (there are several active nest-mounds along the walk). In the forest mid-storey, birds include Green-backed Gerygone, Grey (Brown) Whistler, Lemon-bellied Flycatcher, White-bellied Cuckoo-shrike, cuckoos such as Little and Horsfield's Bronze-Cuckoo, Oriental Cuckoo (uncommon during the wet) and Pheasant Coucal. In the canopy, look for Rose-crowned Fruit-Dove (in the dry season this species may perch on power lines at East Point), Yellow Oriole and Spangled Drongo. Honeyeaters include Brown, White-gaped, Banded (uncommon) and Bar-breasted, and Helmeted, Little and Silver-crowned Friarbird are also present at times. Resident night birds include Rufous Owl (uncommon), Large-tailed Nightjar, Australian Owlet-Nightjar and Bush Stone-curlew.

Mixed flocks of shorebirds roost on the reefs and beaches at the end of Alec Fong Lim Drive in the north-west section of East Point. Species that may be found here include Pacific Golden and Grey Plover, Lesser and Greater Sand Plover, Grey-tailed Tattler, Terek and Common Sandpiper, Great Knot, Eastern Curlew, Red-necked Stint, Whimbrel, Ruddy Turnstone, Bar-tailed and Black-tailed Godwit, while less frequent visitors include Pectoral Sandpiper, Asian Dowitcher and Common Redshank. Other coastal species include Intermediate and Eastern Reef Egret, Striated Heron, Beach Stone-curlew and terns such as Little, Gull-billed, Caspian, Common (wet season), Lesser Crested and Crested Tern. At low tide, when the mud flats are exposed, search the mangroves on the north-east shoreline of the reserve (between Ludmilla Creek and Colivas Rd) for Chestnut Rail, Collared Kingfisher, Green-backed and Mangrove Gerygone, Yellow White-eye, Black Butcherbird, Shining Flycatcher and Mangrove Golden Whistler (uncommon). Nearby there is an excellent mangrove boardwalk that starts at the Lake Alexander carpark in Bayview St (~1 km return). Coastal vine scrub near the start of the boardwalk can be good for White-gaped, Bar-breasted and Rufous-banded Honeyeater if there is nectar bearing trees in flower, while Long-tailed and Double-barred Finch and Varied Triller are usually common. The well-made aluminium boardwalk within the mangroves can be productive for Collared Kingfisher, Lemon-bellied and Broad-billed Flycatcher, Red-headed Honeyeater, and, occasionally, Mangrove Fantail and Mangrove Golden Whistler (both have bred nearby).

Holmes Jungle Nature Reserve

Key species: Red-backed Button-quail, Red-chested Button-quail, King Quail, Rainbow Pitta, Zitting Cisticola
Other species: Brolga, Black Bittern, Horsfield's Bushlark

Holmes Jungle Nature Reserve (250 ha), 20 km from Darwin's CBD, protects important areas of open grassland, monsoon forest and tropical woodland. Palm Creek winds its way through the centre of the reserve. To get there from the Stuart Hwy turn left into Vanderlin Drive (12.5 km from Darwin CBD). After 3.5 km turn right at the Shoal Bay Recycling Centre and take the second right on to Holmes Jungle Rd – signs show the way. From the Hilltop Picnic Area there is a good view of the reserve. Gates are open from 7 a.m. to 7 p.m.

From the upper carpark (the lower carparks are often closed in the wet season), walk down Holmes Jungle Rd to the bottom of the hill. Grasslands border the fence line on the northern side of the reserve. In this area, Red-backed and Red-chested Button-quail, Brown and King Quail have been recorded, with Red-backed Button-quail and Brown Quail being seen most frequently; the two button-quails display a preference for denser areas of grass. Another sought-after grassland species at Holmes Jungle is the Zitting Cisticola. Search around wetland edges and listen for the zit-zit-zit flight song. Take care with identification: the similar Golden-headed Cisticola can be common. Around marshy areas, also look for Pale-vented Bush-hen, Swinhoe's Snipe (wet season), White-browed Crake, Horsfield's Bushlark, Tawny Grassbird, Australian Reed-Warbler, Chestnut-breasted Mannikin (thousands sometimes roost in reedbeds here) and Eastern Grass Owl (rare). Oriental Reed-Warbler has been reported here in the wet season, but is best considered a rare vagrant. Towards the end of the dry season, waterbirds gather in large numbers on the north side of Holmes Jungle. At this time, species may include Brolga, Sarus Crane (exceptional), Magpie Goose, Pied Heron, Black Bittern, White-browed Crake and shorebirds such Sharp-tailed, Marsh and Wood Sandpiper, Black-tailed Godwit, Little Curlew and Oriental Pratincole. The rainforest walk winds along Palm Creek and through monsoon forest. Possibilities here include Black Bittern, Pacific Baza, Orange-footed Scrubfowl, Rainbow Pitta, Rose-crowned Fruit-Dove, Pied Imperial-Pigeon, Grey (Brown) Whistler, Spangled Drongo, Northern Fantail and Red-backed Fairy-wren; every so often, exotic species such as Helmeted Guineafowl and Indian Peafowl turn up. Night birds to look out for include Rufous Owl and Australian Owlet-Nightjar, though note that gates are closed at night.

Howard Springs Nature Park

Key species: Rainbow Pitta, Little Kingfisher, Black Bittern
Other species: Emerald Dove, Little Shrike-thrush, Bar-breasted Honeyeater

The Howard Springs Nature Park (283 ha) protects an area of monsoon forest and extensive swamp habitats. This reserve is one of the most reliable places in the Darwin region for Rainbow Pitta. Howard Springs Nature Park is a ~34 km drive south of Darwin; from the Stuart Hwy, turn east into Howard Springs Rd ~28 km south of the Darwin CBD and follow this road for another 6 km. Gates are open from 8 a.m. to 8 p.m. daily.

A 1.8 km walking track from the carpark winds its way along the creek line that feeds the spring. Rainbow Pitta is reasonably common. Listen carefully for rustling leaf-litter as they forage on the ground. Alternatively, their loud '*we-wik-to-wik*' call, particularly from September to February, and often given from an elevated perch within the monsoon forest, is a sure indication of their presence. An especially good area for the species is approximately halfway down the creek-line track near to the footbridge. Look for Little Kingfisher along the stream; although the species may be seen flying through any part of the monsoon forest, they typically forage over clear water pools. Other birds along the track include Emerald and Bar-shouldered Dove, Rose-crowned Fruit-Dove, White-winged and Varied Triller, Little Shrike-thrush, Grey (Brown) Whistler, Spangled Drongo, Shining Flycatcher, Arafura and Northern Fantail, Forest and Azure Kingfisher. Honeyeaters here include Brown, Dusky, White-gaped, White-throated, Rufous-banded and Bar-breasted (uncommon) Honeyeater, Silver-crowned Friarbird and Yellow-throated Miner. Finches can be common in the grassy northern section of the reserve, with Crimson, Masked, Double-barred and Long-tailed Finch and Chestnut-breasted Mannikin regularly seen. At the spring itself, Black Bittern occasionally feed on the edge of the pool, especially at dawn and dusk. Other species recorded include Nankeen Night-Heron, Green Pygmy-Goose, Magpie Goose, Wandering Whistling-Duck, Radjah Shelduck, Orange-footed Scrubfowl, Yellow Oriole, Great Bowerbird and Little Corella.

Fogg Dam Conservation Reserve

Key species: White-browed Crake, Comb-crested Jacana, Magpie Goose, Brolga, Little Kingfisher, Broad-billed Flycatcher, Rainbow Pitta, huge concentrations of waterbirds
Other species: Black-necked Stork, Tawny Grassbird

With over 230 bird species recorded here, Fogg Dam Conservation Reserve (1569 ha) is a must for birdwatchers visiting the Top End. Initially built to provide water for rice production, the dam (68 km east of Darwin) attracts large numbers of local and migratory waterbirds, especially during the late dry season. Given its location, Fogg Dam is an ideal day trip from Darwin or alternatively it involves a short detour (14 km round trip) while en route to Kakadu National Park. To get there from Darwin, take the Stuart Hwy south (35 km) to the Arnhem Hwy, a further

Howard Springs is not only a popular swimming hole, the monsoon forest that surrounds the springs supports sought-after species such as Rose-crowned Fruit-Dove, Rainbow Pitta, Shining Flycatcher, Arafura Fantail and Northern Fantail.

25 km along the Arnhem Hwy turn left into Anzac Pde. Fogg Dam Rd is a left turn 6 km down Anzac Pde, with the route well signposted.

There are two excellent walks, both leaving from the first carpark. The 2.5 km-return Monsoon Forest Walk winds its way through a variety of habitats including monsoon and paperbark forests and then on to the edge of the floodplain. The denser forest near the start of the walk can be good for Rainbow Pitta. Other birds to look for include Rose-crowned Fruit-Dove and Emerald Dove, Orange-footed Scrubfowl, Blue-winged Kookaburra, Sacred Kingfisher, Olive-backed and Yellow Oriole and Brown Goshawk. The 2 km-return Woodlands to Waterlily Walk leads you through forests that fringe the floodplains and then to a boardwalk on the edge of the dam. Birds along here include Little and Azure Kingfisher, Little Bronze-Cuckoo, Green-backed Gerygone, Lemon-bellied and Broad-billed Flycatcher, Grey (Brown) Whistler, Dusky and Red-headed Honeyeater, Northern Fantail, White-bellied Cuckoo-shrike, Varied Triller and Yellow White-eye. The grassy fringes of the floodplain can be especially good for Tawny Grassbird, Golden-headed Cisticola, Australian Reed-Warbler, Crimson Finch and Chestnut-breasted and Yellow-rumped Mannikin (though the latter is uncommon).

In the morning, the trees around the first carpark can be alive with bird song.

Knuckey Lagoons Conservation Reserve

When Knuckey Lagoons hold water, they can be excellent sites for species such as Magpie Goose, Green Pygmy Goose, Cattle, Little, Intermediate and Great Egret, Pied Heron and Marsh and Wood Sandpiper. Garganey has occasionally been recorded, especially during December before the onset of the wet season when wetlands are at a premium and many waterfowl are concentrated at the few remaining sites that hold water. The grasslands bordering the lagoon are also well known for rarities during the wet season, with Swinhoe's Snipe, Oriental Pratincole, Oriental Plover, Little Curlew, Yellow Chat, Yellow-rumped Mannikin and Zitting Cisticola all recorded. Knuckey Lagoons is located 12 km south-east of Darwin CBD. There are two good access points: from Darwin's CBD, drive east along the Stuart Hwy and after ~14 km turn left on to Lagoon Rd. After another 2.3 km turn right on to Randall Rd – the lagoon is another 200 m; alternatively, after 14.8 km from Darwin's CBD, from the Stuart Hwy turn left on to McMillans Rd and then after 2.2 km turn left on to Fiddlers Lane, with the lagoon another 200 m.

Knuckey Lagoons are open year round, but access is easiest and waterbirds are usually most abundant as the water levels recede through October and November.

Expect to see Paperback Flycatcher, Forest and Sacred Kingfisher, Rainbow Bee-eater, Red-backed Fairy-wren, Leaden and Broad-billed Flycatcher (quite common), Brush Cuckoo, White-gaped, Bar-breasted and Rufous-banded Honeyeaters, and Black-faced Woodswallow (look also for this last species on power lines on the drive in). Birding along the dam wall is best done from a vehicle, stopping at each viewing platform, as large crocodiles frequent the adjacent water bodies. The sealed road follows the causeway along Fogg Dam Rd and provides convenient viewing across the wetlands. Keep an eye open for White-browed and Baillon's Crake and Buff-banded Rail along the marshy fringes near the central viewing platform, Comb-crested Jacana (common) walking on lily pads and for terns such as Gull-billed, Whiskered and White-winged Black Tern (wet season). The Pandanus Lookout on the west side of Fogg Dam provides clear views over the northern section of the wetlands and floodplain. From this vantage point, look for larger waterbirds such as Magpie Goose, Wandering Whistling-Duck, Radjah Shelduck, Green Pygmy-Goose, Royal Spoonbill, Black-necked Stork, and Glossy and Straw-necked Ibis. Egrets and herons include Little, Intermediate and Great Egret, and White-faced, White-necked and Pied Heron. Depending on the water levels, many of these larger waterbirds may also be easily observed from the dam wall. Just west of the Pandanus Lookout is a causeway that on occasions can teem with birdlife, especially fish-eating species that congregate for the easy pickings; if conditions are appropriate, great flocks of Pied Heron, Intermediate and Great Egret may occur. South of the lookout (across a small bridge) in the first section of forest, Nankeen Night-Heron and, occasionally, Black Bittern roost. Birds of prey hunting over the wetlands include Swamp Harrier, Whistling and Black Kite and White-bellied Sea-Eagle, while Pacific Baza is occasionally seen over the adjacent forest. Letter-winged Kite has roosted on the power lines on the drive into Fogg Dam and Red Goshawk has also been recorded, but both are very rare. At night in the dry season, Spotted Nightjar feed along the causeway, while Large-tailed Nightjar is present year round.

Adelaide River

Continuing down the Arnhem Hwy, 7 km east of the Fogg Dam turn-off, brings you to the Adelaide River. This is a reliable site for Mangrove Golden Whistler. Just before the bridge, park your car in a small pull-in on the north-west side of the road. Mangrove Golden Whistler occur in the riparian forest bordering the river. Other birds here include Little Bronze-'Cuckoo' and Oriental Cuckoo (wet season, especially around bamboo thickets), Shining, Paperbark and Broad-billed Flycatcher, Yellow Oriole, Arafura Fantail and Yellow White-eye and Brahminy Kite. Great-billed Heron and Black Bittern are occasionally seen on the Adelaide River boat cruises (which are also renowned for their Saltwater Crocodile experiences).

Arnhem Hwy and entry to Kakadu National Park

Key species: Gouldian Finch, Masked Finch, Long-tailed Finch
Other species: Northern Rosella, Varied Lorikeet, Black-tailed Treecreeper, Chestnut-backed Button-quail, waterbirds

The drive east from Darwin (via the Stuart Hwy) along the Arnhem Hwy to Kakadu National Park can be very productive for birding. Along the Arnhem Hwy, check the numerous wetlands (many ephemeral) for Plumed and Wandering Whistling-Duck, Radjah Shelduck, Pacific Black Duck, Australian Pelican, White-necked Heron, Great Egret, Intermediate Egret, Glossy Ibis, Australian White Ibis, Straw-necked Ibis and Royal Spoonbill. In the floodplain areas, particularly between the Adelaide River and the South Alligator River, including Marrakai and the Mary River

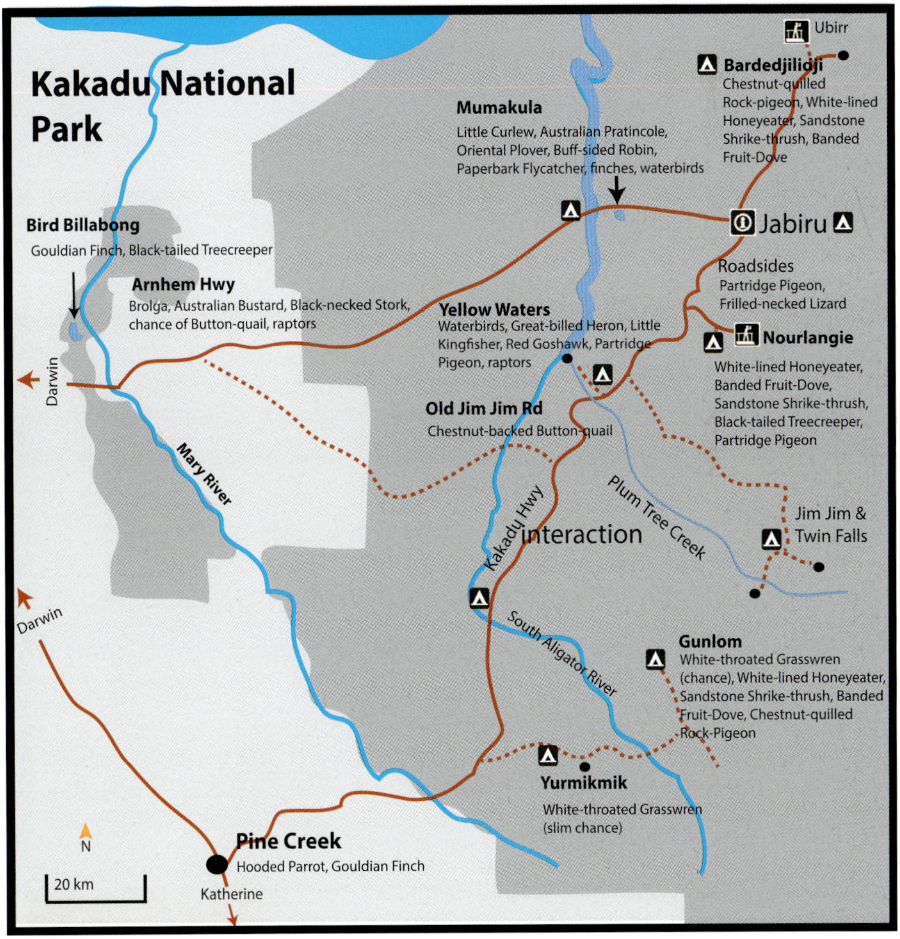

floodplain, you may see Brolga, Australian Bustard, Black-necked Stork, Blue-winged Kookaburra and Brown Songlark. At night, remain alert for Red-backed and Red-chested Button-quail, which are occasionally encountered sitting on the road.

It is worth stopping in roadside areas that support monsoon forest with long grasses because finches favour these areas. Crimson Finch shows a distinct preference for these habitats, but there are also regular reports of Gouldian Finch in adjacent areas of drier woodland. One such location can be found 20.6 km from the Stuart Hwy turn-off along the Arnhem Hwy where a small unnamed road heads south. In addition to the above finches, Masked, Long-tailed and Double-barred Finch, Chestnut-breasted Mannikin, Northern Rosella, White-bellied Cuckoo-shrike, Varied Lorikeet and Black-tailed Treecreeper all occur. The first 300 m is a public road easement, but beyond this a permit is required.

Further along the Arnhem Hwy, 68.6 km east of the Stuart Hwy turn-off (or 9.7 km west of the Mary River) there is rough pull-off on the south side of the road that provides access to a gravel extraction area with two small waterholes adjacent to woodlands and a low range of hills. (Another landmark to help find this site is an information bay for Mary River National Park, which is situated 850 m west of the gravel pits.) In recent years, especially when conditions are dry, Gouldian Finch have been observed drinking here (first hour or two after dawn is best), along with Crimson, Masked, Double-barred and Long-tailed Finch. If both pools contain water, then the eastern waterhole with a few more surrounding shrubs is often the more productive of the two.

Huge paperbarks line the upper reaches of the East Alligator River (accessed via the road to Gunlom). During the dry season, large numbers of honeyeaters and other species frequent these habitats: when in flower, look especially for Bar-breasted and Banded Honeyeater.

A further 7 km east of the extraction area (~2.5 km west of Mary River) Gouldian Finch have been seen feeding around the Bird Billabong carpark and along the walk to the billabong. Look also for Black-tailed Treecreeper in the tropical woodlands here. Nearby, another good area for Gouldian Finch is around the intersection of the Arnhem Hwy and the access road to Bird Billabong. Here flocks of up to 100 Gouldian Finches have occurred. When flowering, the surrounding woodlands in this area can also be good for Varied Lorikeet. While driving the Arnhem Hwy, maintain a sharp lookout for raptors such as Brown Falcon, Spotted Harrier, Wedge-tailed Eagle, Black-breasted Buzzard, Black and Whistling Kite and Brown Goshawk. The rare Red Goshawk is also occasionally observed in this area.

Mamukala Wetlands

Key species: Little Curlew (wet season), Australian Pratincole, Oriental Plover (wet season), Buff-sided Robin, Paperbark Flycatcher
Other species: White-winged Black Tern (wet season), waterbirds

One of Kakadu's finest wetlands, Mamukala is a large billabong that can hold thousands of waterbirds, particularly in the late dry season. As the wetland dries, Mamukala can also be an excellent site for grassland-frequenting shorebirds such as Little Curlew, Australian Pratincole and Oriental Plover. To get there turn south off the Arnhem Hwy 7 km east of the South Alligator River.

Depending on the season and water level, there may be quite an assemblage of birds at Mamukala. When there is water, from the bird hide expect to see waterfowl such as Magpie Goose, Radjah Shelduck, Hardhead, Plumed and Wandering Whistling-Duck, Green Pygmy-Goose, Pacific Black Duck, Magpie Goose and Australian Wood Duck (uncommon). Egrets and herons include Little, Intermediate and Great Egret and Nankeen Night-Heron and White-necked and Pied Heron. Other waterbirds regularly recorded include all three Australian ibis species, Royal Spoonbill, Black-necked Stork, Comb-crested Jacana, Australasian Grebe and Australian Pelican. In the rushes and reeds beside the wetland, look for Tawny Grassbird, Golden-headed Cisticola and Australian Reed-Warbler. The extremely rare Oriental Reed-Warbler is a potential vagrant during the wet season. Terns such as Whiskered, White-winged Black (wet season) and Gull-billed Tern may also be hawking for insects over the water.

As Mamukala starts drying out (usually best from mid-October), the wetland takes on a different character, attracting a range of grassland birds and waders. Check the fringes and the area in the front of the hide for Australian Pratincole, while Black-winged Stilt, Red-kneed and Black-fronted Dotterel may all frequent the muddy shorelines. From late August each year, Pacific Golden Plover, Wood, Marsh and Sharp-tailed Sandpiper and Bar-tailed Godwit may also frequent the wetland. Two additional, and often sought-after, migratory shorebirds, the Little Curlew and Oriental Plover, don't usually arrive in numbers until early November. In the taller grass and sedge tussocks, look for finches such as Crimson, Doubled-barred, Long-tailed, Masked and Star (rare) Finch. The bush surrounding the path to the hide supports a good selection of woodland species. Buff-sided Robin, a strikingly coloured robin species (recently split from White-browed Robin), is resident; listen for its distinctive piping call. Other woodland birds include Brush Cuckoo, Lemon-bellied, Paperbark and Leaden Flycatcher, White-winged and Varied Triller, Rufous Whistler, White-throated Gerygone, White-throated and Banded Honeyeater, Red-winged Parrot and Red-winged Fairy-wren. The 3 km track that skirts the wetlands (only accessible during the drier months) has many of the birds mentioned above, as well as Blue-winged Kookaburra, Peaceful and Bar-shouldered Dove, Little Friarbird, Blue-faced Honeyeater, White-bellied Cuckoo-shrike and, at night, Australian Owlet-Nightjar and Barking Owl.

Yellow Water and Mardugal

Key species: Red Goshawk, Great-billed Heron, Little Kingfisher
Other species: Buff-sided Robin, Partridge Pigeon, waterbirds

Within Kakadu National Park at the confluence of Jim Jim Creek and South Alligator River, Yellow Water consists of a series of billabongs surrounded by floodplains and extensive woodlands. The wetlands are among the most spectacular in the world, acting as a haven for waterbirds (especially, late in the dry season). The turn-off to Yellow Water on the Kakadu Hwy is 50 km south of Jabiru and 160 km north of Pine Creek. From the turn-off, it is a 4.3 km drive to the Warradjan Aboriginal Cultural Centre and a further 3 km drive to Yellow Water. On the drive in, remain alert for Partridge Pigeon along the roadside.

Yellow Water is the main access point for boat cruises that explore Jim Jim Creek and South Alligator River. Most cruises tend to concentrate their efforts on viewing crocodiles; however, birdwatchers will find them worthwhile due to the multitude of waterbirds. The cruises operate regularly during the day, although those in the early morning are generally best for birding. The list of waterbirds is impressive. Little and Azure Kingfisher are found at various places along the river, becoming more numerous where there is extensive riparian vegetation together with clear water for fishing; typically they are found perched on exposed branches just above the water. Other kingfishers include Forest and Sacred Kingfishers and Blue-winged Kookaburra. Australia's largest heron, the impressive Great-billed Heron, is occasionally seen along the shady waterways. Birds commonly encountered on the cruises include Magpie Goose, Radjah Shelduck, Grey Teal, Plumed and Wandering Whistling-Duck, Green Pygmy-Goose, Pacific Black Duck, White-necked, White-faced and Pied Heron, Nankeen Night-Heron, Little, Intermediate, Great and Cattle Egret, Australasian Grebe, Australian Pelican, Australasian Darter, Glossy, Straw-necked and Australian White Ibis, Royal Spoonbill, Black-necked Stork, Brolga, Black-winged Stilt and Comb-crested Jacana. In the late dry season and through the wet season, expect to see Rainbow Bee-eater and Dollarbird hawking for insects. Emu, which are generally rare to uncommon in the Top End, can occasionally be seen feeding in open grassy areas. Yellow Water is also an excellent location for birds of prey: the most commonly encountered species include White-bellied Sea-Eagle, Black and Whistling Kite, Brown Goshawk, Australian Hobby and Brown Falcon. Keep an eye open for less common species such Eastern Osprey, Pacific Baza, Brahminy Kite, Square-tailed Kite and the rare Red Goshawk. The areas around Yellow Water, and indeed many of the paperbark-lined watercourses within Kakadu, are a stronghold for this last species. Night birds that have been encountered here include Barking Owl, Large-tailed Nightjar and Tawny Frogmouth. Owing to the frequent sightings of crocodiles and the numerous waterbirds, these cruises are also very appealing to non-birding family members and companions.

Just west of the Kakadu Highway, 2 km south of the turn-off to Yellow Water and Cooinda, is the Mardugal Park campground – a well-serviced camping area with shower and toilet facilities. There are a several excellent walks: the 2 km Gun-gardun Walk can be productive for forest birds such as Leaden, Paperbark and Shining

Kakadu Hwy

While driving in Kakadu, scan the roadsides for Partridge Pigeon, particularly in the drier, red soil area south of Jabiru on the Kakadu Hwy. Frilled-necked Lizard (mostly wet season) and Antilopine Wallaroo are also commonly seen in this stretch of the Hwy. On power lines, look for kingfishers such as Red-backed (mainly in the dry season) and Forest Kingfisher. There is also a good variety of cockatoos and parrots in the adjacent roadside vegetation: species include Sulphur-crested and Red-tailed Black-Cockatoo, Little Corella, Galah, Red-collared and Varied Lorikeet, Northern Rosella and Red-winged Parrot. Also remain alert for Red Goshawk, especially around the turn-off to Yellow Water and the forested area ~20 km south of the turn-off, as birds have been seen here previously. The Old Jim Jim Rd (also known as the Old Darwin Rd) leaves the Kakadu Hwy 8.5 km south of the Cooinda turn-off. Chestnut-backed Button-quail occur in the woodlands around the intersection, with the west side of the Kakadu Hwy 300 to 500 m north of the turn-off being a regular site.

Flycatcher, Northern and Arafura Fantail, Lemon-bellied Flycatcher, Rufous Whistler, Little Shrike-thrush, Grey-crowned Babbler and Weebill. In the monsoon thickets, look for Rainbow Pitta, while Black-tailed Treecreeper prefer woodland areas dominated by Darwin Stringybark. Nectar feeders that may be present include Brown, White-gaped, Bar-breasted, Dusky, White-throated and Banded Honeyeater, as well as Silver-crowned Friarbird. The circular Home Billabong walk at Yellow Waters (an easy 1 km loop accessible in the dry season only), features a boardwalk adjacent to the wetlands that leads to a viewing platform at Home Billabong. Here there is a chance of Little and Azure Kingfisher, Great-billed Heron, Striated and Nankeen Night-Heron and Black Bittern. As this walk also passes through floodplain forest, expect to see a variety of birds that seek this habitat including Pheasant Coucal, Brush and Little Bronze-Cuckoo, Bar-breasted, Banded and Rufous-throated Honeyeater, Olive-backed and Yellow Oriole, Buff-sided Robin and a range of flycatchers such as Broad-billed, Leaden, Paperbark and Shining Flycatcher. Two ground foraging birds, the Partridge Pigeon and Pheasant Coucal, may be encountered around the camping area and along the nearby roadsides.

Nourlangie

Key species: White-lined Honeyeater, Banded Fruit-Dove, Sandstone Shrike-thrush
Other species: Black-tailed Treecreeper, Partridge Pigeon, Helmeted Friarbird

Nourlangie is an extensive sandstone rock escarpment with pockets of monsoon forest. A very popular tourist attraction, it contains some of Australia's best-known Aboriginal rock paintings. Nourlangie can be reached by turning east off the Kakadu

Hwy 24 km south of Jabiru and then travelling a further 12 km to the parking area. Nourlangie is open from 7.00 a.m. to sunset daily. Camping is not permitted.

On the road into Nourlangie, look for Partridge Pigeon, particularly in the early morning and late afternoon. When seeking the escarpment endemics, it is best to start the 1.5 km circular walk early (it can get very busy with other visitors). Look along here for White-lined Honeyeater – during the late dry to early wet season they feed in eucalypts at the base of Nourlangie Rock (Anbangbang). This can also be a good area for the Banded Fruit-Dove, which can be found in fruiting trees such as figs, especially in the cooler canyons around the rock. One such site can be found behind the main rock shelter at the lower paintings; again, the best time of year is the late dry and early wet season. Another bird to look for here is Helmeted Friarbird.

The circular walk ends with a moderately steep climb to Gunwarddehwardde Lookout, providing views along the escarpment and Anbangbang. This area can be a good place to see Chestnut-quilled Rock-Pigeon and Sandstone Shrike-thrush; the latter often broadcasts its song from ledges along the cliff tops. Banded Fruit-Dove also occurs in association with fruiting native figs, especially along the gully west of the lookout. The Barrk Sandstone Bushwalk (12 km circuit) commences near the Gunwarddehwardde Lookout. After 2 km, a plateau is reached where there is a good chance of Chestnut-quilled Rock-Pigeon. Considered one of the park's best walks, the Barrk Sandstone Bushwalk has no formed path and is rated as a strenuous walk, so should only be attempted by experienced and well-prepared bushwalkers. Other birds that occur regularly at Nourlangie include Cicadabird, Leaden and Paperbark Flycatcher, Emerald Dove, Olive-backed Oriole, Black-tailed Treecreeper, Partridge Pigeon, Great Bowerbird, Pied Butcherbird, Little Woodswallow, Weebill, Mistletoebird, Brown, Dusky, White-throated and Banded Honeyeaters, and Little and Silver-crowned Friarbirds.

If you miss the Sandstone Shrike-thrush at Nourlangie, they also occur along the Nawurlandja Lookout Walk (600 m return), which begins 1 km north of the Nourlangie carpark. Look especially along the rocky ridge on the left side of the track before you reach the lookout. Nearby, waterbirds congregate at the Anbangbang Billabong. Along the Anbangbang Billabong Walk (2.5 km return), expect many of the familiar Kakadu wetland and floodplain forest species such as Magpie Goose, Green Pygmy-Goose, Wandering and Plumed Whistling-Duck, Radjah Shelduck, Black-necked Stork, Australasian Darter, Blue-winged Kookaburra and Paperbark Flycatcher.

White-lined Honeyeaters are restricted to the Arnhem sandstone escarpments, which are most easily accessed in Kakadu National Park. Most walking trails in this habitat provide a chance to encounter the species: look especially in flowering trees in close proximity to steep rock faces.

Gunlom (Waterfall Creek)

Key species: White-throated Grasswren, White-lined Honeyeater, Sandstone Shrike-thrush, Banded Fruit-Dove, Chestnut-quilled Rock-Pigeon

Other species: Banded Honeyeater, Helmeted Friarbird, Variegated Fairy-wren, Buff-sided Robin, Grey (Silver-backed) Butcherbird

Located near the southern border of Kakadu, Gunlom (also known as Waterfall Creek) is the most accessible site where White-throated Grasswren is a possibility. All

Bardedjilidji

The 2.5 km Bardedjilidji Walk leads through sandstone outcrops adjacent to the East Alligator River (just south from Ubirr). There are pockets of monsoon vine forest, paperbark woodlands and some wetland areas. It is good for the 'sandstone endemics' such as Chestnut-quilled Rock-pigeon, White-lined Honeyeater, Sandstone Shrike-thrush and Banded Fruit-Dove, as well as White-breasted and Little Woodswallow, Red-backed Fairy-wren, Grey-crowned Babbler, White-throated and Large-billed Gerygone, Spangled Drongo and Mistletoebird. Rainbow Pitta can be common in the narrow strip of riparian vegetation on the dry season extension to this walk in the vicinity of the East Alligator River. Mammals here include Black Wallaroo and Short-eared Rock Wallaby. The walk starts at the 'upstream' East Alligator carpark adjacent to the Oenpelli Rd, 40 km north of Jabiru.

Beehive rock formations along the Bardedjilidji walk. Chestnut-quilled Rock-Pigeon, White-lined Honeyeater and Sandstone Shrike-thrush are regular here, while Banded Fruit-Dove is also a chance if fig trees are in fruit.

the other sandstone endemics (e.g. White-lined Honeyeater, Sandstone Shrike-thrush, Chestnut-quilled Rock-Pigeon and Banded Fruit-Dove) are also possible. A gravel access road (30 km) runs off the Arnhem Hwy some 90 km south of the Yellow Water turn-off (70 km north of Pine Creek). During the dry season, Gunlom is accessible to conventional 2WD (though the road is usually heavily corrugated); during the wet season, access is restricted and the road is often closed, even to 4WD. There is an excellent campground at Gunlom Falls, while the nearest hotels and caravan parks are located in Pine Creek and near Yellow Water.

The steep Gunlom Lookout Walk (2 km return) leads to the escarpment at the top of the falls. From here, there are spectacular views from cliff-top swimming holes over the southern portion of the Kakadu National Park. Once it is warm enough for thermals to form, watch for birds of prey such as Wedge-tailed and Little Eagle, Brown Goshawk and Black-breasted Buzzard, circling over the surrounding woodlands – the second mainland record of a vagrant Oriental Honey-buzzard occurred here in May 2004. Little Woodswallow also frequent the airspace along the escarpment edge and over the falls.

From the top, the best way to find the escarpment-specialities is to continue climbing (clambering) up the sandstone boulders north-east of the falls. There are no clear paths and it can be tough going, so you will need a reasonable level of fitness. Also, take plenty of water, as it can be especially hot in this terrain. Look for an easy access point and then make your way east by climbing higher; a good access place is 100 m along the walk track that follows the creek line. White-throated Grasswren inhabit the higher sandstone plateau areas, with their optimum habitat being flat, sparsely vegetated escarpment with mature Spinifex. One such area is located ~500 m east from the track. Listen for their distinctive buzzing call: much harsher than, but still reminiscent of, fairy-wrens. The Grasswren can be a notoriously difficult bird to find, so allow at least half a day for your search. Although they are most active in the cooler parts of the day, birders have had success throughout daylight hours. White-throated Grasswren appear to run 'hot and cold' at this site, with periods where birders report finding them with relative ease, followed by months (or even years) without sightings. This, in part, appears to be dictated by just how recently the Spinifex habitat has been burnt (the Grasswren shows a distinct preference for mature Spinifex). The campground manager may be able to offer insight regarding recent sightings. If you do not feel like clambering over the rocky areas north-east of the falls, White-throated Grasswren have been recorded in a large open plateau ~300 m along the creek line. In the centre of the plateau is a rocky outcrop and they have been recorded on the outcrop and in the area between it and the escarpment to the north-east.

While searching for White-throated Grasswren, look for the other sandstone

An Arnhem sandstone escarpment endemic, the Chestnut-quilled Rock-pigeon. This species can be very confiding along some of the more popular walking tracks: early morning is usually best.

specialists, including Banded Fruit-Dove, Chestnut-quilled Rock-Pigeon (seen most often on rocky outcrops), Sandstone Shrike-thrush, Helmeted Friarbird, White-lined and Banded Honeyeater, Black-tailed Treecreeper and Variegated (Lavender-flanked) Fairy-wren. Banded Fruit-Dove feed and roost in fig-lined gorges and gullies and can sometimes be found sitting quietly in dense foliage; their presence may be given away by falling fruit. Another endemic to the Arnhem Land escarpments is the Black Wallaroo and this species is encountered regularly in the more elevated areas around Gunlom.

Back at the campground, a short walk leads to the plunge pool and superb opportunities for a cooling swim at the base of the falls. The short path also passes through some good areas of monsoon forest. Black Bittern can occur along Waterfall Creek (just below the pool). At night, this same area can be good for Rufous Owl. The Murill Billabong Walk is a 2 km return walk that commences from the west side of the campground, travels through woodlands to the seasonal Murill Billabong and then further west to the South Alligator River. Woodland species that are possible include Partridge Pigeon, Red-tailed Black-Cockatoo, Northern Rosella, Brown Goshawk, White-breasted and Little Woodswallow, Dollarbird (wet season), Leaden Flycatcher, Little Shrike-thrush, Grey (Silver-backed) Butcherbird, Red-backed Fairy-wren, Striated Pardalote and finches including Crimson, Masked, Long-tailed and Double-barred. At the billabong, expect to see Plumed Whistling-Duck, Radjah Shelduck, Pacific Black

The creekline behind Gunlom Falls, Kakadu National Park. White-throated Grasswren used to be frequently sighted on the Spinifex-clad slopes to either side of the creek. In recent years, the frequency of fire has rendered much of the habitat unsuitable. Although the grasswren can be a 'long-shot', all other sandstone endemics can still be found in the area.

Yurmikmik and Plum Tree Creek

There are couple of other sites near Gunlom that present opportunities for the elusive White-throated Grasswren. Yurmikmik, on Sleisbeck Rd 16 km before Gunlom (14 km from the Kakadu Hwy) presents the easiest access. Search in the Spinifex sandstone areas on the higher section of the 5 km Yurmikmik Lookout Walk. Banded Fruit-Dove, White-lined Honeyeater, Helmeted Friarbird are found in the rainforest on the Boulder Creek Walk (2 km) and Banded Fruit-Dove use the gorge at Yurmikmik as a flight corridor in the morning – look for them from the base of the waterfall. Another historic location for White-throated Grasswren is on escarpment to the east of Plum Tree Creek, 1 km east of Yurmikmik carpark. Try in areas with Spinifex and darker conglomerate rock 1 km east of where the road intersects with the creek, though it should be recognised that any sighting would be very fortuitous.

Duck, Great and Intermediate Egret, Azure and Forest Kingfisher, and Blue-winged Kookaburra. A good variety of birds can also be found In the vicinity of the campsite including Red-tailed Black-Cockatoo, Little Woodswallow, Varied Lorikeet, Northern Rosella, Great Bowerbird, Bush Stone-curlew, Blue-faced Honeyeater, Pheasant Coucal, Blue-winged Kookaburra, White-gaped and White-throated Honeyeater, Silver-crowned Friarbird and Olive-backed Oriole. Night birds include Barking Owl, Southern Boobook and Australian Owlet-Nightjar – while Spotted Nightjar can be found in the dry season perched on the road into Gunlom at night.

Pine Creek

Key species: Hooded Parrot, Gouldian Finch
Other species: Red-backed Kingfisher, Grey Falcon, King Quail, Brown Quail, Partridge Pigeon, Blue-faced Honeyeater, Grey (Silver-backed) Butcherbird

Pine Creek is a reliable site for Hooded Parrot. The town is 226 km south of Darwin and 90 km north of Katherine and, although small, offers a range of accommodation options, including caravan parks and cabins. The Water Gardens in the middle of town (between Main Terrace and Railway Terrace) supports a series of man-made pools shaded by ornamental trees and shrubs. Here the much sought-after Hooded Parrot roosts during the heat of the day along with more common species such as Red-winged Parrot and Great Bowerbird. If unsuccessful, other sites worth a visit include the trees around Mine Lookout and the nearby water tower, at the southern end of Moule St. They also regularly roost at the Pussy Cat Flats Racecourse, 1 km from town on the northern side of the Kakadu Hwy. It is also worth trying for Hooded Parrot and Gouldian, Long-tailed and Masked Finch at the Pine Creek Cemetery, especially along the fence line that leads right from the cemetery gate. The road to the cemetery leaves the Stuart Hwy on the by-pass around Pine Creek.

Pine Creek Treatment Plant, located on Sewer Ponds-Pine Creek Rd, and the surrounding woodland, can also be productive for Hooded Parrot, Gouldian Finch, Partridge Pigeon and Chestnut-backed Button-quail. Other birds here include King (rare) and Brown Quail (both species have bred within the treatment plant), Pink-eared Duck, Radjah Shelduck and Common Sandpiper (wet season). Another good spot for birdwatching in Pine Creek is the Miners Park in the north part of town and along the creek to the north of the Miners Park, which often contains a series of waterholes. Species to look for include the Hooded Parrot, Gouldian, Masked and Long-tailed Finch, Olive-backed Oriole and Northern Rosella. Around the township of Pine Creek, expect to see Varied Lorikeet, Northern Rosella, Jacky Winter, Black-chinned (Golden-backed) and Bar-breasted Honeyeater, Little and Silver-crowned Friarbird, Grey-crowned Babbler, White-throated Gerygone and Little, Masked and Black-faced Woodswallow, and, on power lines, White-breasted Woodswallow and Red-backed Kingfisher.

Copperfield Dam is accessed from the Stuart Hwy, via Umbrawarra Gorge Rd, 4 km south of Pine Creek. Gouldian Finch, Hooded Parrot, Northern Rosella, Chestnut-backed Button-quail and Partridge Pigeon occur in the woodlands around the day visitor area. Umbrawarra Gorge Nature Park is also worth visiting. Here you may see Grey (Silver-backed) Butcherbird, Hooded Parrot, Grey Falcon (rare) and Buff-sided Robin. To get there, continue along Umbrawarra Gorge Rd (from Copperfield Dam) – the Nature Park is 20 km from the Stuart Hwy turn-off. It is also worth visiting Copperfield Creek, particularly at the end of the dry season, when Gouldian Finch and Hooded Parrot are known to drink at the waterholes along the creek. Copperfield Creek can be reached from the Stuart Hwy, via Dorisvale Rd, 23 km south of Pine Creek. The creek is 1 km from the Stuart Hwy turn-off. Once you reach the creek, you will see a track leading to the waterholes on the west side of the causeway.

Fergusson River

Key species: Gouldian Finch, Hooded Parrot
Other species: Northern Rosella, Varied Lorikeet, Black Bittern, Black-tailed Treecreeper, Varied Sittella

The Fergusson River passes under the Stuart Hwy 33 km south of Pine Creek (57 km north of Katherine). The grassy fringe and waterholes along the river can be a good site for Gouldian Finch and Hooded Parrot. Park near the bridge and walk either east or west along the river. The prime time to visit is during the late dry season, when water is limited and even the Fergusson River is a series of sandy pools. A wide range of birds uses these ponds for drinking, especially at dawn.

To the west of the bridge, the best birding is along the northern edge of the river ~200 m downstream. In the morning, Gouldian, Masked, Long-tailed Finch and Hooded Parrot, Northern Rosella and Cockatiel may drink here. Don't overlook the paperbark trees bordering the river for woodland birds. Among others, nectar feeders found along the river include Yellow-tinted, Rufous-throated, Blue-faced and, occasionally, Banded Honeyeater. The largest trees overhanging the water sometimes shelter roosting Black Bittern and Nankeen Night-Heron. To the east of the bridge (past several older bridges), is another site where finches and parrots regularly congregate to drink.

Another productive location for finches is a rest stop 3 km south of the Fergusson River Bridge on Stuart Hwy. Here Gouldian Finch has been observed feeding among the roadside grasses, as have Masked,

Savannah woodlands of the Top End support a wide range of birds. Some of the more distinctive species include Partridge Pigeon, Peaceful Dove, Red-tailed Black-Cockatoo, Northern Rosella, Blue-winged Kookaburra and Black-tailed Treecreeper.

Long-tailed, Double-barred and Crimson Finch. Hooded Parrot, Varied Lorikeet and Northern Rosella also frequent this area. Other species occupying the savannah woodlands include Australian Bustard (foraging in the roadside grasslands), Crested (Northern) Shrike-tit, Black-tailed Treecreeper, Hooded Robin, Varied Sittella, White-winged Triller and Yellow-tinted Honeyeater.

Nitmiluk National Park (Katherine Gorge)

Key species: Gouldian Finch, Hooded Parrot, Chestnut-quilled Rock-Pigeon, Sandstone Shrike-thrush
Other species: Varied Lorikeet, Northern Rosella, Black Bittern, Great-billed Heron, Channel-billed Cuckoo, Variegated Fairy-wren, Crested Shrike-tit, Grey (Silver-backed) Butcherbird, Masked Finch

Nitmiluk National Park (292 800 ha) features spectacular sandstone valleys and gorges, extensive riverine habitat and expansive tropical savannahs. Like other sites in the area, the key species to look for are Gouldian Finch and Hooded Parrot. There are two main sections of the park: Leliyn (also known as Edith Falls) in the north and Katherine Gorge in the south. Access to Leliyn can be gained by turning off the Stuart Hwy 48 km south of Pine Creek (42 km north of Katherine), with the falls a further 19 km down Jatbula Rd. The park has good camping facilities. The

Katherine Gorge section is located 32 km north-east of Katherine and is accessed via Gorge Rd.

Before heading into Leliyn (Edith Falls), it is worth stopping at the Edith River crossing, located immediately north of the turn-off on the Stuart Hwy. This can be an excellent site for finches, with the potential for Gouldian, Long-tailed and Masked Finch in the grasses beside the road in the vicinity of the bridge. During the day, Hooded Parrot feed in the woodlands adjacent to the river. As the species can be very unobtrusive when feeding, a careful search may be required to locate them. Once on Jatbula Rd, stop at the creek 5.6 km from the Stuart Hwy. Hooded Parrot and Gouldian Finch drink at the isolated waterholes along the creek in the late dry season, particular north of the road. Look for Black Bittern feeding along the creek and Bar-breasted, Banded, Rufous-throated Honeyeater. For easy access, park in the circular turn-in that travels north for ~100 m. Once at Edith Falls, the Leliyn Trail (2.6 km circuit), ascends to the head of the falls before descending to a river crossing and popular swimming area. Species that have been seen along the trail include the Gouldian Finch, Hooded Parrot, Crested (Northern) Shrike-tit, Black-tailed Treecreeper, Northern Rosella, Banded and Bar-breasted Honeyeater, Silver-crowned Friarbird, Black-faced and Little Woodswallow, Pied Butcherbird and Great Bowerbird. Although rare, the Grey Falcon has also been reported in this area.

On route to Katherine Gorge, stop at Donkey Camp Weir. Located on a natural rock bar, the site can be good for Great-billed Heron and Black Bittern. The weir entrance is 10 km from Katherine, 500 m past the Kumbidgee Tea Rooms. Park at the Pumping Station and then walk downstream of the weir, searching sand bars and rocky areas for Great-billed Heron and clumps of Pandanus for Black Bittern. Other birds that occur here include Nankeen Night-Heron, White-bellied Sea-Eagle and Radjah Shelduck.

Once at Katherine Gorge, the area around the picnic areas and campgrounds can be productive. Species include Northern Rosella, Grey (Silver-backed) Butcherbird and Banded Honeyeater. The Baruwei Lookout Walk (1.8 km) traverses the escarpment and provides spectacular views of the Katherine River before continuing as the Baruwei Loop Walk (3.7 km). Along both walks, Gouldian Finch and Hooded Parrot are possible, as are escarpment specialists such as Chestnut-quilled Rock-Pigeon and Sandstone Shrike-thrush. In the riverine areas where there are huge paperbarks, species include Bar-breasted Honeyeater. Other bush birds that may be encountered include Red-winged Parrot, Varied Lorikeet, Channel-billed Cuckoo, Yellow Oriole, Australasian Figbird, Black-tailed Treecreeper, Striated Pardalote, Northern Fantail, Little Woodswallow, and Long-tailed and Masked Finch. The campgrounds and picnic areas are good places to find Apostlebird, Blue-faced Honeyeater and Great Bowerbirds, often tame from feeding, they are at times a nuisance if your food is left unattended.

Chinaman Creek and Horse Hole

Key species: Gouldian Finch, Hooded Parrot, Chestnut-backed Button-quail
Other species: Black Bittern, Crested (Northern) Shrike-tit

The Chinaman Creek crossing, situated on the Victoria Hwy 16 km south-west of Katherine, can be a reliable site for Hooded Parrot, Gouldian Finch and other seed-eaters that congregate to drink in the early mornings and late afternoons. The surrounding woodlands are also a good site for the sought-after and sometimes difficult to find Chestnut-backed Button-quail.

The best birding is in the dry season when the creek is reduced to a series of pools. To reach these pools, just east of the Chinaman Creek crossing, take an indistinct vehicle track north of the highway for 100 m and park adjacent to the old Victoria Hwy. From here, walk a further 50 m north-west to a series of small waterholes (known as Horse Hole) along the creek line. A wide range of seed-eaters such as Hooded Parrot, Cockatiel, Budgerigar Gouldian, Long-tailed and Masked Finches, and Yellow-rumped Mannikin use the waterholes to drink, particularly early in the morning. After drinking, Hooded Parrots move to feeding areas in the surrounding woodlands; they can occasionally be found between the creek and a truck stop 2.5 km west of here. One approach is to walk along the Old Victoria Hwy (running parallel to the new Victoria Hwy) searching for Hooded Parrots on either side of the road. Chestnut-backed Button-quail frequent the grassy savannah woodland either side of the creek, typically preferring areas with some bare stony ground. Here you may also see Crested (Northern) Shrike-tit, Grey-crowned Babbler, Red-backed Fairy-wren, Weebill, White-throated Gerygone, Varied Sittella, Black-tailed Treecreeper and Apostlebird. The savannah woodlands can also be good for raptors such as Spotted Harrier, Australian Hobby and Brown Falcon, while Wedge-tailed Eagle and Black-breasted Buzzard often patrol the highway for carrion.

If you do not see Chestnut-backed Button-quail at Chinaman Creek, another location worth trying is at the crossing for the Chainman Creek, 4 km west of Chinaman Creek on the Victoria Hwy. Again, concentrate your search in grassy savannah woodland on the northern side of the Hwy, particularly between the new and the Old Victoria Hwy and north of the Old Victoria Hwy in woodlands with areas of bare stony ground. Other birds to look for here include Varied Lorikeet, Rufous-throated Honeyeater, Silver-crowned Friarbird and Crested (Northern) Shrike-tit.

Hooded Parrots are a regional endemic, restricted to the humid woodlands in the southern half of the Top End. During the dry season, the species is most easily found when it visits waterholes to drink in the early morning (e.g. Ferguson River, Chinaman Creek, waterholes on Jatbula Rd).

Elsey National Park and Mataranka

Key species: Red Goshawk, Crested (Northern) Shrike-tit

Other species: Grey Goshawk, Flock Pigeon, Black Bittern, Square-tailed Kite, Grey Falcon

Elsey National Park (13 840 ha), near Mataranka, is 106 km south-east of Katherine on the Stuart Hwy. Dominated by tropical woodland and dense stands of palm and paperbark along the rivers and streams, the park provides some excellent inland birding opportunities. It is particularly good for birds of prey, with 17 species recorded, including the rare Red Goshawk.

There is an established campground at 12 Mile Yards and camping and accommodation at Mataranka Cabins (near Bitter Springs) and Mataranka Homestead (~9 km south-east of Mataranka via the Stuart Hwy and Homestead Rd).

Bitter Springs is in the northern section of the park, accessed via Martin Rd – a sealed road that leads directly east from Mataranka. Red Goshawk (rare) occurs around Bitter Springs, with a preference for watercourses and tall riverine forest; investigate the area along the north side of Little Roper River just before Bitter Springs. Other birds of prey found along the Little Roper River include Square-tailed Kite, Grey and Brown Goshawk, Collared Sparrowhawk and Pacific Baza. Look also for Azure Kingfisher, Australasian Darter, Black Bittern, Nankeen Night-Heron, Great-billed Heron and Dollarbird (wet season).

In the southern section of the park, rewarding birding can be had along the Roper River, running parallel to John Hauser Drive, which begins at the Mataranka Homestead and heads east along the river to the Korowan (Mataranka Falls), a total distance of ~12 km. There are several good access points to the river including the 1.5 km Botanical Walk, 4 Mile Picnic Area, Wabalarr, 12 Mile Yards and Korowan (Mataranka Falls). Along the river, expect similar species to those found at Bitter Springs. In open woodland areas, look for mixed flocks of White-throated and Western Gerygone, Paperbark Flycatcher, Varied Sittella, Lemon-bellied Flycatcher, Buff-sided Robin and, particularly when the gums are flowering, Banded, Yellow-tinted, Bar-breasted, Grey-headed and White-fronted Honeyeater. The rare Crested (Northern) Shrike-tit is also a possibility. The park also supports a long list of finch species including Gouldian, Star, Long-tailed, Masked, Crimson, Double-barred and Zebra Finch and Pictorella and Chestnut-breasted Mannikin. Other birds to look for include Flock Bronzewing (scarce), Australian Bustard, Brown Quail, Ground Cuckoo-shrike, White-breasted, Black-faced and Little Woodswallow and Apostlebird. Night birds here include Rufous Owl, Southern Boobook, Tawny Frogmouth, Australian Owlet-Nightjar and Spotted Nightjar.

Victoria River, Gregory National Park and Timber Creek

Key species: Purple-crowned Fairy-wren, Grey Falcon, White-quilled Rock-Pigeon, Sandstone Shrike-thrush, Yellow-rumped Mannikin
Other species: Gouldian Finch, Star Finch, Pictorella Mannikin, Pale-vented Bush-hen

Gregory National Park, ~160 km west of Katherine on the Victoria Hwy, covers an area of ~1.3 million ha divided into two sections. The park features open woodland, dissected by rugged escarpments and deep gorges, carved by rivers that weave through the western section of the Top End. The small township of Timber Creek is situated near the northern boundary of the park between the eastern and western sections, while the Victoria River Roadhouse is located in the park's eastern section. There is camping at Sullivan Creek, located on the banks of a permanent waterhole, 17 km east of the Victoria River Roadhouse. The Victoria River Roadhouse also has camping and accommodation.

The Victoria River crossing is one of the most accessible sites for Purple-crowned Fairy-wren in the NT. Living in small family groups, they display a distinct preference for the narrow band of Pandanus and cane grass along the river banks. Concentrate your search in the vicinity of the boat ramp (500 m west of Victoria River Roadhouse turn south to the boat ramp) searching the cane grass around both the carpark and the road between the carpark and boat ramp. Remain alert along the river for the possibility of uncommon species such as Pale-vented

Bush-hen (in grassy fringes), Black Bittern, Star Finch and Yellow-rumped Mannikin. Expect to see Australasian Darter, Nankeen Night-Heron, Radjah Shelduck, Little Corella, Blue-winged Kookaburra, Pheasant Coucal, Yellow-tinted, Blue-faced, Banded and Bar-breasted Honeyeater, Golden-headed Cisticola, Variegated Fairy-wren and Crimson Finch. At night Barking Owl occasionally feed around the lights near the roadhouse. Raptors in the area include White-bellied Sea-Eagle, Brown Goshawk, Brown and Black Falcon. Both Victoria River and Gregory National Park also seem to generate more than their fair share of Grey Falcon sightings.

Two escarpment specialties to look for in Gregory National Park are the White-quilled Rock-Pigeon and Sandstone Shrike-thrush. Both can be found on the Escarpment Walk (3 km return), accessed from the Victoria Hwy 2 km west of Victoria River Roadhouse. The walk is of moderate difficulty, with some steep sections; concentrate your search in the escarpment areas. Look also for Grey (Silver-backed) Butcherbird, White-throated Gerygone, Silver-crowned Friarbird, Yellow-tinted, Banded and Blue-faced Honeyeater, and Black-faced and Little Woodswallow. This area of escarpment can also be a good place to observe Short-eared Rock Wallaby. Before heading up to the escarpment, it can be rewarding to birdwatch around the carpark and the creek at the start of the walk; uncommon species that have been recorded here include Yellow-rumped Mannikin, Star and Gouldian

Escarpment in Gregory National Park provides habitat for the eastern form of the White-quilled Rock-Pigeon, Sandstone Shrike-thrush and Little Woodswallow, while the woodland (in the foreground) provides opportunities for White-throated Gerygone, Silver-crowned Friarbird, Yellow-tinted Honeyeater, Banded Honeyeater and Grey (Silver-backed) Butcherbird.

Finch, Purple-crowned Fairy-wren, Ground Cuckoo-shrike, as well as Crimson Finch, Chestnut-breasted Mannikin and Golden-headed Cisticola. The Rock-Pigeon and Shrike-thrush can also be found along the steep rocky slopes of the Nawulbinbin Walk (1.7 km loop), which starts from Joe's Creek Picnic Area, situated off the Victoria Hwy, 10 km west of Victoria River Roadhouse. On the road into Joe's Creek (2 km unsealed but usually suitable for 2WD), look for (white-bellied) Spinifex Pigeon along the roadside.

At the start of the Buchanan Hwy (the road down to Jasper Gorge – see below), there are several good sites for finches. The waterholes along a small, unnamed creek 4.5 km from the Victoria Hwy attract Pictorella Mannikin and Masked, Long-tailed, Double-barred and Gouldian Finch. Skull Creek, a further 9 km south of here, provides drinking opportunities for Pictorella Mannikin, Star and Gouldian Finch during drier periods when these species may also feed in the grassy areas alongside the creek.

Gouldian Finch occasionally feed on the lawn and drink at the sprinklers at Timber Creek and in the street trees Black-chinned (Golden-backed) Honeyeater is a possibility. Another good site for Gouldian, Star and Long-tailed Finch and Chestnut-breasted and Yellow-rumped Mannikin is the Timber Creek Airfield, 6 km north-west of the township. Flocks of finches often congregate along the fence lines of the airfield and feed on the short grass areas of the airfield. The airfield can also be a good site for Oriental Plover during the wet season.

West of Timber Creek, Gouldian, Masked and Long-tailed Finch are regularly noted along the track to Policeman's Point. This track proceeds north off the Victoria Hwy 4 km west of the township. A good birding area that provides opportunities for Buff-sided Robin, Purple-crowned Fairy-wren, Yellow-rumped Mannikin and Crimson and Star Finch can be found a few hundred metres west of Policeman's Point. For many of the above-listed species, pay particular attention to the grassy areas along the Victoria River. To get there from the Policeman's Point track, take the first track on your left after you leave the Hwy. In the surrounding woodland, look for Red-winged Parrot, Great Bowerbird, Red-backed Fairy-wren, Black-faced Woodswallow, Grey Shrike-thrush and Yellow-tinted, White-gaped, Rufous-throated and Banded Honeyeater. If you have time, the stoney hills immediately west of Timber Creek has produced sightings of the sought-after Chestnut-backed Button-Quail.

Crimson Finch more closely associate with tall grass and Pandanus habitats along drainage lines and rivers than any other finch. They are common along the banks of the Victoria River.

Barkly Tableland and the Tablelands Hwy

Key species: Flock Bronzewing, Yellow Chat, Pictorella Mannikin, Oriental Plover, Letter-winged Kite, Grey Falcon
Other species: Black Falcon, Australian Bustard, Australian Pratincole, Ground Cuckoo-shrike

Jasper Gorge

As a side excursion, it is worth visiting Jasper Gorge, 107 km west of Victoria River. To get there from the Victoria River Roadhouse, travel west along the Victoria Hwy for 64 km, and then travel south along the Buchanan Hwy for 43 km. Along the gorge, there are permanent waterholes and areas of spectacular sandstone escarpment. The gorge area and drainage lines support good numbers of finches, with a remarkable 10 species recorded. These include such sought-after species as Painted, Gouldian and Star Finch, and Yellow-rumped and Pictorella Mannikin. Also look for Sandstone Shrike-thrush, Spinifex Pigeon and White-quilled Rock-Pigeon – the last species at its southern limit here. There are unconfirmed historic reports of White-lined Honeyeater at Jasper Gorge, which is intriguing given this location is situated roughly halfway between the known range of the White-lined Honeyeater of the Arnhem Land escarpment and the Kimberley Honeyeater of the Mitchell Plateau region of WA (see p. 454). Other birds you may see include Red-backed, Variegated and Purple-crowned Fairy-wren (in areas of Pandanus along the creek line), Northern Rosella, Red-winged Parrot, Little Button-quail, Yellow-tinted and Banded Honeyeater, Grey-crowned Babbler, Pied Butcherbird and Crimson Chat (a scarce nomad to the region). Nightbirds here include Spotted Nightjar and Tawny Frogmouth, and there have been several records of Masked Owl. The grassy woodlands around the gorge also support Northern Nail-tailed Wallaby. There are camping facilities at Jasper Gorge.

Covering one-fifth of the NT, the Barkly Tableland is recognised as a stronghold for the nomadic and irruptive Flock Bronzewing. Owing to its boom and bust life cycle, the species can often prove difficult to track down within the vast arid interior of Australia. Even when scarce, there are usually a few to be found on the tablelands, because they show a preference for the dominant habitat: black soil plains with Mitchell Grass. One of the best strategies to find the species is to traverse the highways that crisscross the tablelands (particularly the Tablelands Hwy that runs north off the Barkly Hwy) and focus your efforts around freshwater bores (also known as turkey nests, because they resemble the mounded nest of a Brush Turkey). These dams, with raised earthen walls and an adjacent windmill are dotted across the landscape and are easily visible from roads – here, the bronzewings may fly into drink of an early morning or evening. In March 2010, in an area just south of Kennedy Creek (from ~25 km north of Alroy Downs Station to Brunette Creek 45 km to the north) – an enormous flock of 6000 Flock Bronzewing was present. Another potentially good location is at a bore 21 km north of the Playford River on the Tablelands Hwy. Freshwater bores, especially those with emergent vegetation, represent small oases in otherwise expansive black soil plains and thus serve to attract other special birds including Pictorella Mannikin, Ground Cuckoo-shrike and Yellow Chat. Another suitable site for Yellow Chat is Kennedy Creek where it crosses the Tablelands Hwy 80 km north of the Barkly Hwy. Other species to be seen on the Barkly Tableland include Plumed Whistling-Duck, Magpie Goose, Pacific Black Duck, Hardhead, Red-kneed Dotterel, Red-backed Kingfisher, Diamond Dove, Budgerigar, Zebra, Long-tailed and Masked Finch, Variegated Fairy-wren, and Black-chinned, Grey-fronted and Singing Honeyeater. In the roadside grasslands, Australian Bustard, Australian Pratincole, Oriental Plover (wet season) and Horsfield's Bushlark also occur. Overhead look for Fork-tailed Swift (wet season), especially ahead of storms. The highly sought-after Grey Falcon and Letter-winged Kite are two species that are encountered with some frequency within the tablelands. Because Grey Falcon frequent areas where prey congregates and Letter-winged Kite roost by day in isolated trees, freshwater bores and their associated vegetation are an excellent place to start the search. Other birds of prey that are often common on the tablelands include Black-breasted Buzzard, Spotted Harrier, Black and Whistling Kite, Wedge-tailed Eagle, Nankeen Kestrel and Black and Brown Falcon.

Connells Lagoon Conservation Reserve

Key species: Flock Bronzewing, Pictorella Mannikin, Red-chested Button-quail, Yellow Chat, Letter-winged Kite, Grey Falcon, Little Curlew
Other species: Australian Bustard, Australian Pratincole, Oriental Plover, Eastern Grass Owl

Connells Lagoon Conservation Reserve (1600 ha) lies in the heart of the Barkly Tableland. Dominated by perennial Mitchell Grass, the reserve at times supports large numbers of Flock Bronzewing, Pictorella Mannikin, Red-chested Button-quail and Australian Bustard. There is no campground (though one is planned), so overnight visitors will need to bush camp. The reserve can be reached by conventional vehicle via Ranken Rd (a gravel 2WD road between Brunette Downs and Alexandria Stations), but travel within Connells Lagoon is 4WD. It is 130 km north of the Barkly Hwy and 40 km east of the Tableland Hwy.

The two main bores (Middle and West) and Hidden and Pictorella Swamps attract sought-after species such as Flock Bronzewing, Pictorella Mannikin and Yellow Chat. Birds found in the grassy areas include Red-chested Button-quail, Eastern Grass Owl (mainly wet season), Australian Pratincole, Little Curlew (wet season), Budgerigar, Brolga and Sarus Crane (uncommon), Australian Bustard, large numbers of Horsfield's Bushlark, Zebra Finch and Australasian Pipit. Red-capped Robin inhabits the reserve, a northerly record for this woodland species. Keep an eye open for rare and uncommon raptors such as Letter-winged Kite, Black and Grey Falcon (seen mainly during the dry), as well as more common species such as Nankeen Kestrel and Brown Falcon. Iconic mammals recorded within the reserve include the Red Kangaroo and Dingo.

Caranbirini Conservation Reserve

Key species: Sandstone Shrike-thrush, Purple-crowned Fairy-wren, Carpentarian Grasswren, Gouldian Finch
Other species: Red Goshawk, Masked Owl, Spinifex Pigeon, Little Woodswallow, Yellow-tinted Honeyeater, Banded Honeyeater, Grey-headed Honeyeater, Jacky Winter.

Caranbirini Conservation Reserve (1200 ha) is on the Carpentaria Hwy, 46 km south of Borroloola. Best known for its sandstone spires that resemble a mini version of the Bungle Bungles (p. 463), the reserve protects a surprising diversity of habitats for such a small area, including sandstone escarpments, a semi-permanent waterhole surrounded by riverine vegetation and open woodland. At Caranbirini, look for Sandstone Shrike-thrush in the escarpment country and Purple-crowned Fairy-wren in the riparian vegetation around water. Carpentarian Grasswren has been recorded at Caranbirini in the sandstone escarpment 1 km east of Caranbirini Waterhole (and nearby on the south side of the Carpentarian Hwy 40 km south of Borroloola); however, there have been no recent confirmed records and it may be that changing fire regimes in this area have led to the decline of this species here.

From the carpark, a 300-m-return walk leads to a shelter where visitors can view the birds at the semi-permanent Caranbirini Waterhole. Purple-crowned Fairy-wren can be found here, as can the Gouldian Finch, with small numbers drinking at the waterhole during the height of the dry season. Keep an eye open for Red Goshawk (rare), as well as Azure Kingfisher, Leaden and Paperbark Flycatcher, Jacky Winter, Crested (Northern) Shrike-tit, Little Shrike-thrush and Northern Fantail. Brown, Yellow-tinted and White-throated Honeyeater can be numerous, but check also for the less common Grey-headed, Bar-breasted and Banded Honeyeater. Night forays may produce sightings of Tawny Frogmouth, Barking and Masked Owl (rare).

The Barrawulla Walk (2 km loop) explores the sandstone formations that the park is so well known for. Search here for Sandstone Shrike-thrush, particularly atop the rock formations. Other birds include Peregrine Falcon, Little Woodswallow, Rainbow Bee-eater, Spinifex Pigeon, Northern Rosella, Red-backed Fairy-wren and Banded, Yellow-tinted and Rufous-throated Honeyeater, and Silver-crowned Friarbird. The Carpentarian Grasswren has been reported on the southern section of the Jagududgu Walk (5 km loop), 400 m from its intersection with the Barrawulla Walk (last recorded 2010). Two large species that frequent grasslands and tropical savannahs, Emu and Australian Bustard, can occasionally be found in the surrounding countryside.

Despite their rarity in southern Australia, Bustards are a regular feature of the humid grasslands and savannahs across northern Australia.

Borroloola

Key species: Purple-crowned Fairy-wren, Pictorella Mannikin, Mangrove Golden Whistler, White-breasted Whistler
Other species: Buff-sided Robin, Grey-headed Honeyeater, Red-headed Honeyeater, Beach Stone-curlew

The small town of Borroloola is located on the McArthur River 50 km upstream from the river mouth on the coastal plain between the Barkly Tableland and the Gulf of Carpentaria. Borroloola has pub-style accommodation, a guesthouse and a good caravan park. Purple-crowned Fairy-wren can be found along the McArthur River at Borroloola, especially near the boat ramp at the south end of town, which is accessed via a dirt track off Robinson Rd. Buff-sided Robin is resident in the vegetation along the river bank around the boat ramp and Buff-banded Rail feed along the edge of the creek. If you do not see Purple-crowned Fairy-wren here, Borroloola's northern boat ramp and the rubbish dump are further locations worth searching.

In the woodlands surrounding Borroloola, look for Black-tailed Treecreeper, Northern Rosella, Red-winged Parrot and Grey-headed, Grey-fronted, Yellow-tinted and Bar-breasted Honeyeater, while in open grassy areas look for Black-necked Stork, Brolga, Oriental Plover (wet season) and Red-backed Button-quail. At the Borroloola water treatment plant (west of the township and accessed via a track on the south side of the showgrounds), look for finches such as Star, Masked and Long-tailed Finch and Pictorella Mannikin. Orange Chat has also been recorded here, but the species is very rare this far north. The McArthur River attracts an array of raptors including Square-tailed Kite, Eastern Osprey, Collared Sparrowhawk and Peregrine Falcon.

Situated 57 km north of Borroloola on the Gulf of Carpentaria, Bing Bong is a small port for the mining industry. To access this area, take the sealed road from Borroloola to its end, turning east down an unsealed track for 2 km to reach the boat ramp on the McArthur River. Check the mangroves to the north of the boat ramp for White-breasted and Mangrove Golden Whistler, Red-headed Honeyeater, Arafura and Mangrove Grey Fantail and Buff-sided Robin. In grassy plains adjacent to the river, the patchily distributed Zitting Cisticola occurs. Another sought-after species, the Beach Stone-curlew, can occasionally be found on the beach near the Bing Bong Port.

Tiwi Islands

Key species: Red Goshawk, Masked Owl, Partridge Pigeon, Chestnut Rail
Other species: Little Kingfisher, Helmeted Friarbird, waders

Located some 100 km north of Darwin, Melville and Bathurst Islands are collectively known as the Tiwi Islands. After Tasmania, the Tiwi Islands are the largest landmass off the Australian coast (total land area of 8320 km^2). A permit is required to visit the Tiwi Islands and surrounding waters – before visiting see www.tiwilandcouncil.com. There are regular flights from Darwin, while the ferry to Bathurst Island takes approximately 2 hours. Most visitors stay in Darwin, visiting the islands on day tours. Due to the islands isolation from the mainland Australia (since the end of the last ice age), there is a high level of endemism among the flora and fauna (including the birds) at the subspecific level.

Birding can be rewarding in the islands' extensive woodland and tall eucalypt forests. Resident species include Red Goshawk (with the islands being a stronghold for this rare species – estimated at 80 to 160 adults and ~10% of the global population), Chestnut-backed Button-Quail, Partridge Pigeon, Varied Lorikeet, Northern Rosella, Striated Pardalote, Yellow-tinted Honeyeater, Yellow-throated Miner, and Helmeted and Silver-crowned Friarbird. Search grassland areas on the island for Australian Bustard and Red-backed Button-Quail, and finches such as Masked, Crimson, Double-barred and Long-tailed. At night, listen for Masked Owl (Tiwi Islands ssp. *melvillensis*). Pay special attention to any black and white robins because the endemic Hooded Robin ssp. *melvillensis*, has not been recorded in recent years and is now considered by some to be extinct.

Impressively, the islands support 18 000 ha of rainforest, mainly in coastal and riparian areas along watercourses and, as a consequence, there are resident populations of Rainbow Pitta, Orange-footed Scrubfowl, Emerald Dove, Rose-crowned Fruit-Dove and Australasian Figbird. In the mangrove-lined creeks and rivers, especially if one has boat access, look for Chestnut Rail, Little Kingfisher, Red-headed Honeyeater, Mangrove Golden Whistler, Shining Flycatcher and Yellow White-eye. With long beaches, rocky headlands and extensive areas of tidal mud flats (particularly on the south-east coast of Melville Island), the Tiwi Islands attract large numbers of waders including Great Knot, Red-necked Stint, Greater and Lesser Sand Plover, and Bar-tailed Godwit. Terns include Black-naped, Little, Lesser Crested and Roseate Tern. Chestnut Rail can be relatively tame on parts of the Tiwi Islands, such as the mud flats south-west of Nguiu. Scattered among the Tiwi Islands, there are also at least 19 seabird colonies, with Seagull Island (on the west coast of Melville Island) supporting the world's largest known breeding colonies of Crested Tern. The Tiwi group also support Australia's largest known colonies of Black-naped Tern.

Groote Eylandt

Key species: Chestnut Rail, Rainbow Pitta, Beach Stone-curlew, Great-billed Heron
Other species: Australian Magpie

Groote Eylandt (and the associated archipelago of 40 islands), is 650 km east of Darwin. When compared with other islands in the NT, the habitats on this archipelago are relatively pristine. This is partly because there are few feral animals (and those few that are present occur in low numbers), and the frequency of fires is much less than that on the adjacent mainland. To date, the number of species recorded on the island is 209 species, while there are at least 17 seabird-breeding colonies within the archipelago, with significant colonies of Roseate, Bridled and Black-naped Tern.

Only recently opened to tourism, Groote Eylandt is part of Arnhem Land and permits from the Anindilyakwa Land Council are mandatory for all visitors travelling off the Mining Leases. Short-term visitor permits can be obtained by contacting the Anindilyakwa Land Council, on (08) 8987 4002 or permits@anindilyakwa.org. There is accommodation at the Dugong Resort: a new resort commissioned by the Land Council.

A good site for seabirds is Alyangula's main wharf. Look here for Little, Crested and Gull-billed Tern, Lesser Frigatebird (uncommon and most frequently found during or after windy weather), Australasian Darter and White-breasted Woodswallow, and raptors such as Brahminy Kite and Eastern Osprey – the latter nests in the port stockpile area in easy view from the fuel station. Beach Stone-curlew can commonly be seen near the mouths of the Angurugu and Amagula Rivers, while Bush Stone-curlew occur at the Alyangula Golf Club.

Around Mission Landing at the mouth of the Angurugu Creek, there is a recreational area available for public use. Look for Beach Stone-curlew, Great-billed Heron and Chestnut Rail along the mud flats and adjacent mangroves. Other species in this area include Eastern Reef Egret, Pacific Golden Plover, Lesser Sand Plover, Grey-tailed Tattler, Forest Kingfisher, Mangrove Gerygone and Northern Fantail. There are also some productive monsoon thickets in this area that support species such Rainbow Pitta, Shining Flycatcher, Grey (Brown) Whistler, Green-backed Gerygone and Arafura and Northern Fantail. In the same area, the Angurugu Sewage Pond provides opportunities for waterbirds including Wandering Whistling-Duck and Radjah Shelduck; views of the ponds can be gained from the carpark. Within the monsoon forests bordering the beach around Alyangula and Angurugu, check for Pied Imperial-Pigeon, Rose-crowned Fruit-Dove and White-gaped Honeyeater. Other birds that may occur include Pheasant Coucal, and Brown, White-throated and Bar-breasted Honeyeater. Common in Alyangula, the Groote Eylandt form of the Australian Magpie (*eylandtensis*) is a locally endemic taxon that is smaller than its mainland counterpart. During a more extended visit, other sites that may be worth a visit include Salt Creek (in the south-west corner of the island), Amagula River and Leski Pools.

The Red Centre – Alice Springs and nearby arid lands

Regional overview

The 'Red Centre' provides opportunities for a true outback experience, with almost endless desert landscapes. Indeed, situated in the geographic centre of the continent, the region is bounded by some of Australia's most expansive arid lands, including the Simpson, Tamami, Great Sandy, Gibson and Great Victoria Deserts, to name a few. Do not be fooled into thinking that desert country is barren though, because the area supports a wide range of habitats and a great diversity of life that is particularly noticeable after rainfall events. Generally, the landscape is flat to gently undulating, with Mulga woodlands and sand dunes swathed in desert shrubs and Spinifex. However, rocky ranges and outcrops that reveal hidden gorges and picturesque waterholes bisect the central deserts. Indeed, the area is known globally for its outstanding geological features including the MacDonnell Ranges and the World Heritage listed Uluru (Ayer's Rock) and the Olgas (Kata Tjuta), all of which rise abruptly from their surrounds. Sandy river-beds lined with River Red Gums also criss-cross the region, providing further birding opportunities. These are invariably dry, but in any case, it is the larger trees along the sandy banks that provide the habitat here.

As the only major town in the region, Alice Springs provides a necessary and convenient base from which to explore. It is equipped with all the major services. Most site visits are day trips from the town, though, given the distances involved, the majority of travellers would stay overnight near Uluru and at Kings Canyon (440 km and 330 km distant from Alice Springs, respectively).

A recommended 3-day itinerary (with key target species bracketed) based in Alice Springs includes:

- **Day 1.** Old Telegraph Station Historic Reserve, Olive Pink Botanic Gardens, Alice Springs Waste Stabilisation Ponds, Alice Springs Desert Park (Red-browed Pardalote, Redthroat, Grey Honeyeater (spring), Western Bowerbird, Orange Chat, waterbirds, shorebirds).
- **Day 2.** Santa Teresa Rd, Kunoth Bore in afternoon/evening (Rufous-crowned Emu-wren, Dusky Grasswren, Grey-fronted, Black, Pied and Grey Honeyeater, Slaty-backed Thornbill, Bourke's Parrot (dusk)).
- **Day 3.** Mount Gillen and Simpson's Gap (Dusky Grasswren, Spinifex Pigeon, Painted Finch, Little Woodswallow, Grey-headed Honeyeater, Western Bowerbird).

An 8-day (or more) itinerary undertaking the Red Centre Way, offers an excellent option for adding to your bird list and seeing some of Australia's most spectacular scenery. The Red Centre Way connects

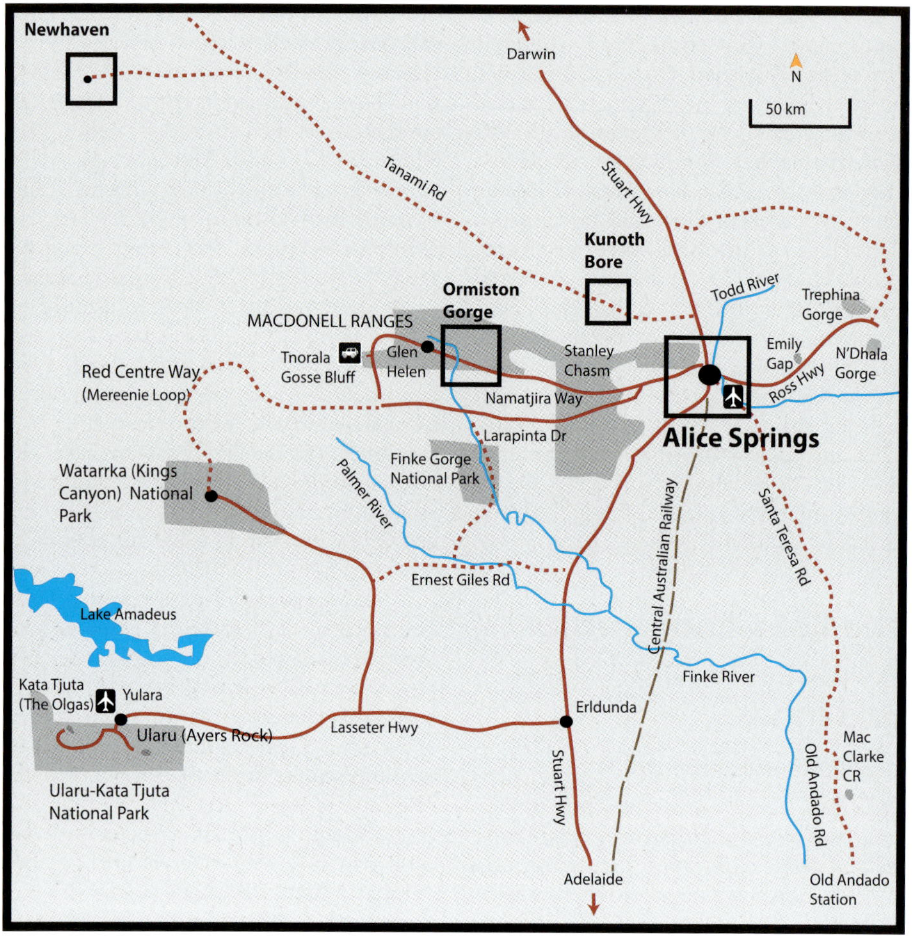

Alice Springs to Uluru through the West MacDonnell Ranges and the Mereenie and the Kings Canyon regions:

- **Days 1–2.** Alice Springs to Ormiston Gorge, 130 km (Dusky Grasswren, Rufous-crowned Emu-wren, Spinifexbird, Painted Finch, Spinifex Pigeon).
- **Days 3–4.** Ormiston Gorge to Kings Canyon, via the Mereenie Loop (4WD), 260 km (Dusky Grasswren, Spinifex Pigeon, Painted Finch, Western Bowerbird, Red-browed Pardalote, Little Woodswallow, Princess Parrot (rare)). Note: an alternative to the Mereenie Loop is to take the Stuart Hwy south of Alice Springs to Kings Canyon.
- **Days 5–6.** Kings Canyon to Uluru, 300 km (Striated (Sandhill) Grasswren, Banded Whiteface, Chiming Wedgebill, Black-eared Cuckoo, Black and Pied Honeyeater, Princess Parrot (rare)).
- **Day 7.** Uluru to Erldunda, 246 km (Banded Whiteface, Cinnamon Quail Thrush, Chiming Wedgebill, Pied Honeyeater).
- **Day 8.** Erldunda to Alice Springs, 200 km.

Extended side trips would include a 3-day trip to Newhaven 350 km (Dusky Grasswren, Rufous-crowned Emu-wren, Spinifexbird, Grey Falcon (rare)) – and/or a 4-day visit Old Andado Station 330 km (Eyrean Grasswren, Gibberbird, Cinnamon Quail-thrush, Banded Whiteface, Letter-winged Kite, Grey Falcon, Bourke's Parrot).

The cooler months, from April to September, are the most pleasant times to visit. In summer, temperatures reach an average maximum of 37.8°C during the day, with the best birding times from dawn to ~10 a.m., then later in the day from ~4:30 p.m. until dusk. Despite the heat, birding in summer has its advantages, because September to December is the peak breeding time for most species, meaning many are more active and easier to find (e.g. Grey Honeyeater), while occasional high rainfall arising from cyclonic events on the northern coast in January and February bring nomads such as Black and Pied Honeyeaters. By contrast, in winter, conditions for birding can be very comfortable throughout the day, but the nights can get very cold (average minimum 4.7°C).

Getting to Alice Springs, Uluru and the MacDonnell Ranges is relatively straightforward. There are regular daily flights to and from major Australian cities (Sydney, Melbourne, Adelaide, Darwin, Perth and Cairns) and to Uluru. Most flights from the coastal areas take around 2 hours. There are also two Adelaide–Alice Springs services of the iconic Ghan railway each week, while express coaches from Adelaide, Darwin and Brisbane depart daily. Alice Springs is located halfway between Darwin (to the north) and Adelaide (to the south) along the Stuart Hwy. Both major cities are ~1500 km and 2+ full days of driving. Although there are many dirt roads connecting Alice Springs to other settlements, the Stuart Hwy and the Alice Springs to Yulara Rd are the only fully sealed roads in the region.

There are numerous hotels, motels and caravan parks in Alice Springs. For information on available accommodation contact the Tourism Central Australia Visitor Information Centre (1800 645 199 or (08) 8952 5800, visinfo@tourismca.asn.au, www.centralaustraliantourism.com). The Centre is located on the corner of Gregory Terrace and Todd St, next to the town library. Many birders use Alice Springs as a base for visiting the surrounding areas, including the MacDonnell Ranges; however, it is well worth taking the opportunity to stay or camp in some of the more remote sites such as Ormiston Gorge in the West MacDonnell Ranges or Trephina Gorge in the East MacDonnell Ranges.

Birding highlights

The Red Centre includes a variety of sought-after bird species, including the only regional endemic to central Australia: the Dusky Grasswren. In the Mulga shrublands to the north of Alice Springs, target birds include Grey Honeyeater, Bourke's Parrot and Slaty-backed Thornbill. Around Alice Springs and the MacDonnell Ranges, you can see the Spinifex Pigeon (white-bellied ssp. *plumifera*), Spinifexbird, Rufous-crowned Emu-wren, Pied and Black Honeyeater, Crimson and Orange Chat, Painted Finch and Western Bowerbird. After significant rainfall events, large flocks of Budgerigar may congregate. To the south and south-east of Alice Springs, highlights include Eyrean Grasswren, Striated Grasswren (ssp. *oweni*, a subspecies that may be elevated to full species status, known as Sandhill Grasswren), Cinnamon Quail-thrush, Gibberbird and Banded Whiteface. Rarities such as Grey Falcon and Princess Parrot frequent the area from time to time and there was a report of the critically endangered Night Parrot in 1996 from Newhaven. In May 2013, the first mainland record of a Forest

Marble Gum *Eucalyptus gongylocarpa* woodland to the west of Kings Canyon, NT. Following above average rains, this particular habitat can be alive with birds. The numerous hollows provide breeding opportunities for parrots and Red-backed Kingfisher while the flowering shrubs and seeding grasses attract Budgerigar, Black Honeyeater, Pied Honeyeater, Masked Woodswallow and Zebra Finch.

Wagtail was obtained in a private garden in Alice Springs.

Populations of birds fluctuate from season to season in desert environments. Good rainfall can transform an apparently barren landscape into a thriving environment, with vigorous growth of seeding grasses and flowering of many local species of *Eremophila*, *Grevillea* and *Acacia*. Such events can also lead to large numbers of insects, which in turn feed increased populations of birds. Many of the desert birds are either nomadic or migratory and, after periods of good rain, they will move into the area in great numbers. Two species that are highly responsive to such conditions are the Budgerigar, which can be abundant with large flocks wheeling around open shrublands and gorges, their distinctive chattering filling the air, and the nomadic Crimson Chat, which can arrive in large numbers such that they are encountered in almost all areas of open country.

Alice Springs Waste Stabilisation Ponds

Key species: Waterfowl (including Freckled Duck), Black Falcon, Australian Spotted Crake, Spotless Crake, Orange Chat, shorebirds

Other species: Rufous Songlark, Golden-headed Cisticola, Zebra Finch, Crimson Chat, White-winged Fairy-wren, Variegated Fairy-wren, Little Crow, Major Mitchell's Cockatoo

The waste stabilisation ponds are 4 km south of Alice Springs CBD. To get there, drive south down the Stuart Hwy through Heavitree Gap and turn right into Commonage Rd. After a further 0.4 km,

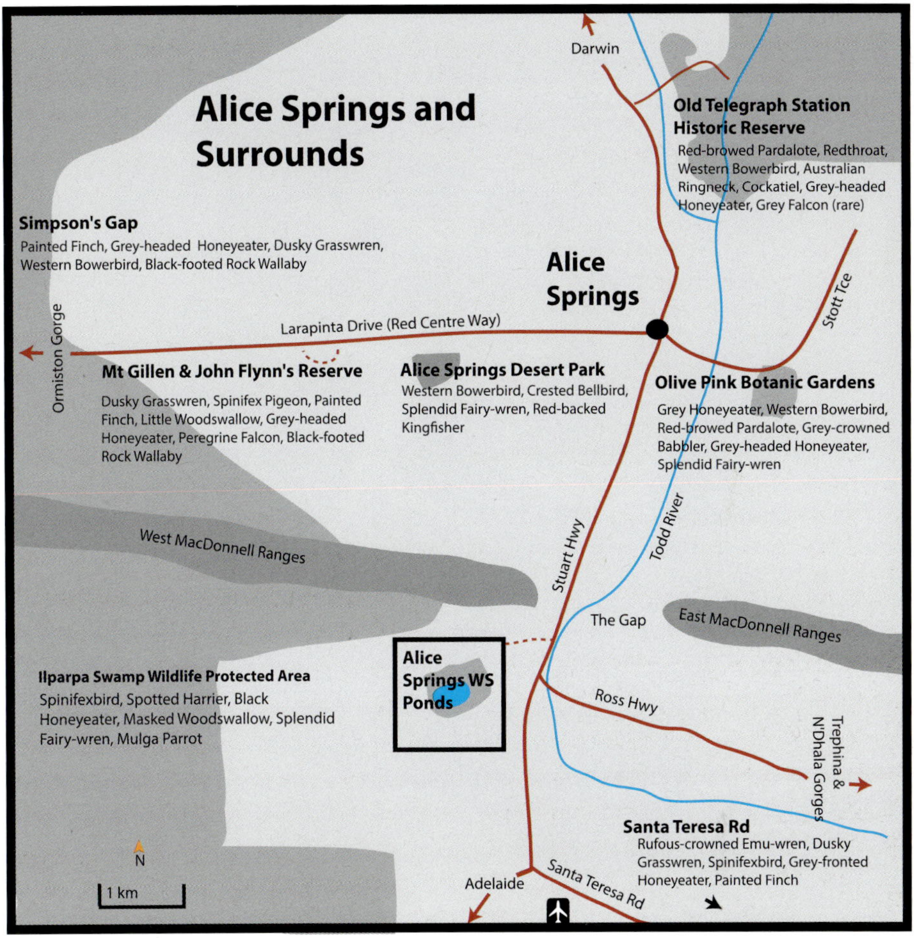

turn left into an unsealed lane (just before the rubbish dump gate); there is a birders' access gate several hundred metres down this lane. You will need to obtain a key ($50 deposit) from the Power and Water office in the Alice Plaza (36 Todd Mall, www.powerwater.com.au/sustainability_and_environment/birdwatching_ilparpa_swamp); while there, also collect their excellent birdwatchers' map and guide for the site.

With over a 140 bird species recorded, the ponds can support an abundance of birdlife, with a good mix of waterbirds and waders. The potential for this isolated body of water is evident from the number of rare birds reported in recent years. The ponds are divided into the 'A ponds' in the north and the 'EP ponds' in the south, with the EP ponds generally being better for birding. The EP10 pond has a bird hide and can be good for ducks such as Grey Teal, Hardhead, Pink-eared and Pacific Black Duck, as well as rarer species such as Blue-billed and Freckled Duck. During spring and summer, waders congregate on the northern edge of the EP10 pond. Species include Red-necked Stint, Black-tailed

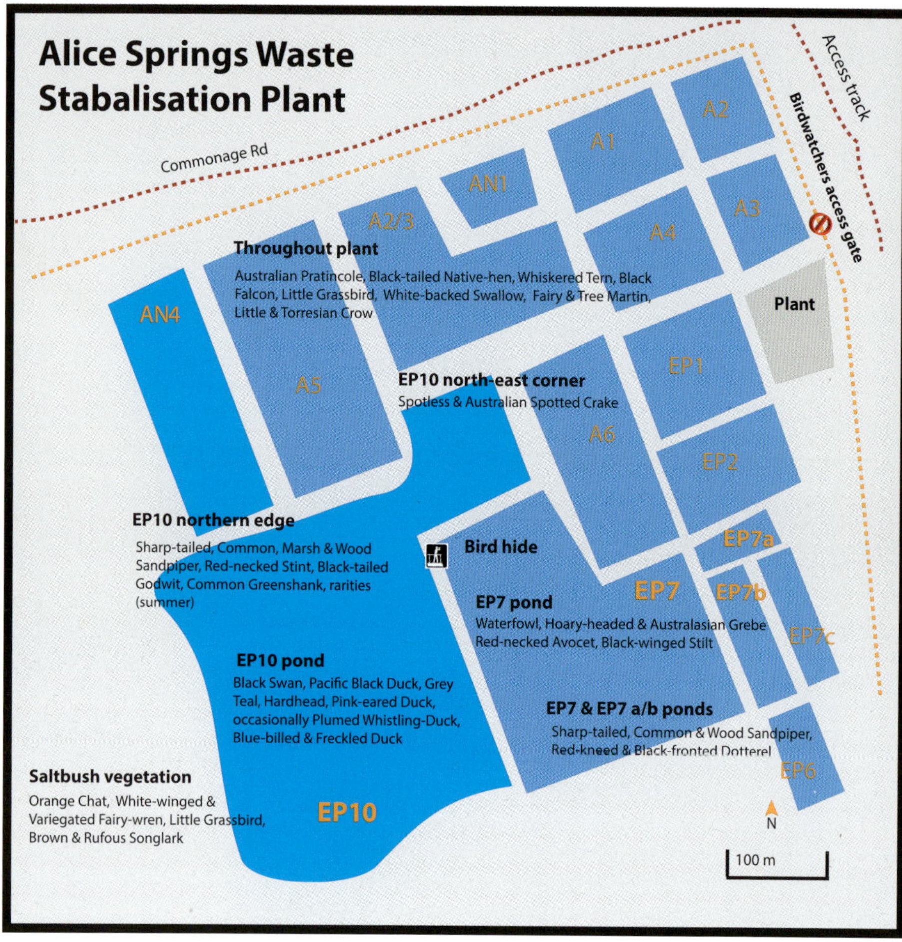

Godwit, Common Greenshank, and Sharp-tailed, Common, Marsh and Wood Sandpiper, with occasional rarities such as Long-toed Stint, Little Curlew, Pectoral Sandpiper, Ruff, Latham's and Swinhoe's Snipe and Oriental Plover turning up. The north-east corner of EP10 can be a good place for skulking species such as Spotless and Australian Spotted Crake.

There is a good area of saltbush vegetation to the south-west of the EP10 pond. A survey of this area from the viewpoint of the pond wall can reveal White-winged and Variegated Fairy-wren, Orange Chat and Rufous Songlark. Other good birding ponds are the EP7 and the smaller EP7a ponds, both of which can hold numbers of waders and waterbirds. Once again, look for waders, such as Sharp-tailed, Common and Wood Sandpiper that congregate around the edge of the pond, and other wetland species such as Black-tailed Native-hen, Hoary-headed and Australasian Grebe, and Red-kneed and Black-fronted Dotterel. The EP7 pond usually supports flocks of Red-necked Avocet and Black-winged Stilt.

Throughout the stabilisation ponds, check the reeds for Golden-headed Cisticola (rare) and Australia Reed-Warbler and on the paths between the ponds look for Australian

Alice Springs Waste Stabilisation Ponds and the view north to Heavitree Gap. The best site for waterbirds in the Alice Springs region the area can also be good for Black Falcon, Orange Chat and Little Crow.

Pipit, occasional flocks of Orange Chat and, in summer, Australian Pratincole. Species that hawk over the ponds from time to time include Gull-billed and Whiskered Tern, White-backed Swallow, and Fairy and Tree Martin. The common corvids here are Little and Torresian Crow. There is also a chance of seeing Major Mitchell's Cockatoo and Red-tailed Black-Cockatoo, especially in the surrounding trees. Although Black Falcon is generally uncommon, the species seems to appear at the ponds with some regularity.

Just west of the Alice Spring Stabilisation Ponds is the Ilparpa Swamp Wildlife Protected Area. The swamp only floods in wet years and, at these times, Baillon's, Spotted and Spotless Crakes can arrive to breed. Other species seen include Orange Chat, White-winged Fairy-wren and Brown Songlark, especially in the area adjacent to the swamp. Along the northern edge, there are also some areas of Spinifex that run uninterrupted to the peak of Mount Gillen. Spinifexbird is quite common here, while it is also worth keeping a sharp lookout for Rufous-crowned Emu-wren and Dusky Grasswren. In the areas of nearby Mulga, look for Spotted Harrier, Black Honeyeater, Masked Woodswallow, Pied Butcherbird, Splendid Fairy-wren and Mulga Parrot. In 2011, Bourke's Parrot was also present in this area.

Old Telegraph Station Historic Reserve

Key species: Red-browed Pardalote, Redthroat, Grey Falcon (rare)
Other species: Western Bowerbird, Australian Ringneck, Cockatiel, Grey-headed Honeyeater

Open between 8 a.m. and 9 p.m. year round, the Old Telegraph Station Historic Reserve (450 ha) is 3 km from the CBD. Drive north on the Stuart Hwy and look for the sign on the right ~2.2 km past the Wills Terrace traffic lights. The Old Telegraph Station can also be accessed with a leisurely 4 km walk or cycle from Todd Mall in the centre of Alice Springs: the walking track follows the west side of the Todd River for much of the route.

The Old Telegraph Station is a good site for Red-browed Pardalote. Try in the River Red Gums along the Todd River, particularly the area in front of the picnic ground and historic precinct, though the species also occurs in the trees in the main carpark from time to time. Weebill and Striated Pardalote also feed in this area. Around the picnic ground, there is generally a good mix of the larger passerine species, including Yellow-throated Miner, Pied Butcherbird, Grey-crowned Babbler and Western Bowerbird. Channel-billed Cuckoo also occurs in the area and young can occasionally be found being attended to by crows in spring and summer. The Bradshaw and Stuart Walks can be good for ground-foraging woodland species such as Diamond Dove, Red-capped Robin, Redthroat, Western Gerygone and Yellow-rumped Thornbill. Other birds to look for include Australian Ringneck, Cockatiel, and Singing, Brown, White-plumed and Grey-headed Honeyeaters. In spring, birds of prey such as Brown Falcon, and Black and Whistling Kite nest in the gums along the Riverside Walk. Also be alert to the possibility of the rare Grey Falcon because it has been recorded at the Telegraph station on several occasions. There are several colonies of Black-footed Rock Wallaby, with the best place to look being on the ridge immediately south-west of the junction of the Bradshaw and Riverside Walks.

Olive Pink Botanic Gardens

Key species: Grey Honeyeater, Western Bowerbird
Other species: Red-browed Pardalote, Grey-crowned Babbler, Grey-headed Honeyeater, Splendid Fairy-wren

Within walking distance (~1.4 km) of the Alice Springs CBD, the Olive Pink Botanic Gardens (16 ha) are located on Tuncks Rd, accessed via Stott Terrace, on the eastern bank of the Todd River. The gardens are open to the public daily from 8 a.m. to 6 p.m., with extended opening periods during peak visiting months (Apr–Oct). There is also a pleasant café.

Aside from the diverse and impressive collection of central Australian plants (over 250 species), the gardens are also excellent for birdwatching. In particular, Grey Honeyeater, which can prove one of the more challenging birds to see in Australia, appears regularly at this site. In recent years, the species has nested in the Mulga groves on the west side of the visitor centre; this is usually when rains have been favourable and the Mulga is in flower, around September. It is worth asking the staff at the centre if there are recent reports of Grey Honeyeater because that may assist to narrow the search. When the conditions are suitable, other places to look for Grey Honeyeater include the mallee and wattle sections of the gardens. The other special bird in the Olive Pink Botanic Gardens is the Western Bowerbird. There are several resident individuals and at least one active bower. A good spot to look is in the Wattle section south-east of the visitor centre; if you can't find a bower, ask the cafe staff where the current bower is. This is a popular site, so be respectful of the bird and his bower because the cumulative number of visitors is high. When walking around the gardens, look also for Red-capped Robin, Zebra Finch and Inland Thornbill within shrubby areas. The spectacular Splendid Fairy-wren (with males acquiring their iridescent blue plumage during spring) also frequents this habitat. Higher up in the mid and upper canopy, look out for Striated and Red-browed Pardalote (mainly winter), and Brown, White-plumed, Singing, Spiny-cheeked (summer) and Grey-headed

Western Bowerbird tending to its bower in Olive Pink Botanic Gardens. There is usually at least one active bower present in the gardens – if you can't find it, ask for directions at the cafe.

Honeyeater. Australian Ringneck and Rainbow Bee-eater (summer) also frequent the gardens, as do several groups of Grey-crowned Babbler. Common Wallaroo and, less commonly, Black-footed Rock Wallaby occur in the rocky habitat on Annie Meyers Hill.

Alice Springs Desert Park

Key species: Western Bowerbird, Crested Bellbird
Other species: Splendid Fairy-wren, Red-backed Kingfisher

Located on Larapinta Drive, the Alice Springs Desert Park (1300 ha) is ~6.6 km from the centre of Alice Springs. Designed to provide an insight into the ecology of central Australia's arid zone habitats, the park includes three desert types: sand country, woodland habitat and desert river systems.

The Desert Park is included here because of the wide selection of captive inland birds (some of which are held nowhere else in captivity) that often prove difficult to observe in the wild. Fauna on display are reviewed regularly and no species is necessarily always on display, but birds held often include, Inland Dotterel, Cinnamon Quail Thrush and Redthroat. There is also a good selection of rare (or indeed locally extinct) inland mammals and the park has an excellent nocturnal tour of a large free range enclosure in dense mulga woodland. Mammals seen usually include Bilby, Mala, Stick-nest Rats, Golden Bandicoot and others. Bush Stone-curlew are also commonly seen; bookings are essential. Aside from the captive species, the Desert Park is also a good site for a range of free-flying birds. Expect to see Crested Bellbird, Splendid Fairy-wren, Zebra Finch, Grey Shrike-thrush, Common Bronzewing, Peaceful Dove, Spinifex Pigeon, Australian Ringneck, Red-tailed Black-Cockatoo and Red-backed Kingfisher. Grey Honeyeater, Slaty-backed Thornbill and Redthroat breed on the greater site and are sometimes seen within the visitor area. If you did not manage to see Western Bowerbird at the Olive Pink Botanic Gardens they are also found at the Desert Park, as are Grey-crowned and White-browed Babbler. The Desert Park also runs bird walkabout tours. Conducted on Wednesday and Sunday mornings, these start early and take about 2 hours; bookings are required because these are run on demand only. See www.alicespringsdesertpark.com.au

Mount Gillen and John Flynn's Memorial Historical Reserve

Key species: Dusky Grasswren, Spinifex Pigeon
Other species: Painted Finch, Little Woodswallow, Peregrine Falcon

The resting place for the Reverend John Flynn, founder of the Flying Doctor Service, John Flynn's Memorial Historical Reserve is ~7 km west of the town centre on the Larapinta Drive, just past the entrance to the Alice Springs Desert Park. In the shrublands behind Flynn's Grave, you can usually find a good variety of birds including Yellow-rumped Thornbill, Variegated and Splendid Fairy-wren, Grey-headed Honeyeater, Grey Honeyeater (uncommon), Grey Shrike-thrush, Pied Butcherbird and, occasionally, Crimson Chat and Western Gerygone. For those with extra time, the walk to the summit of Mount Gillen can be rewarding. Requiring about 2 hours for the return journey, the walk involves a vertical ascent of ~300 m. This is a strenuous walk to the top, best attempted by fit and experienced bushwalkers only. Importantly, the walk is one of the most reliable sites in the MacDonnell Ranges to find the Dusky Grasswren: look especially in rocky Spinifex-clad areas near the summit. Another good place for the Grasswren is at the foot of the gully that leads to the summit where there are stands of native pines (*Callitris*). Painted Finch feed on grass seeds from the saddle up and in the same area, with a little luck, you may find Spinifex Pigeon, Variegated and Splendid Fairy-wren. Wedge-tailed Eagle, Peregrine Falcon and Little Woodswallow frequently soar around the cliffs in this area, and in spring Little Woodswallow nest on the adjacent cliffs. Common Wallaroo feed on the lower slopes of Mount Gillen, while Black-footed Rock Wallaby may be sighted perched on all-but-inaccessible ledges on the higher cliffs.

Kunoth Bore

Key species: Grey Honeyeater, Slaty-backed Thornbill, Bourke's Parrot
Other species: Black Honeyeater, Pied Honeyeater, Ground Cuckoo-shrike

Kunoth Bore is ~50 km from Alice Springs. To get there, take the Stuart Hwy north out of Alice Springs; after 20 km, turn west into the Tanami Rd. After a further ~30 km

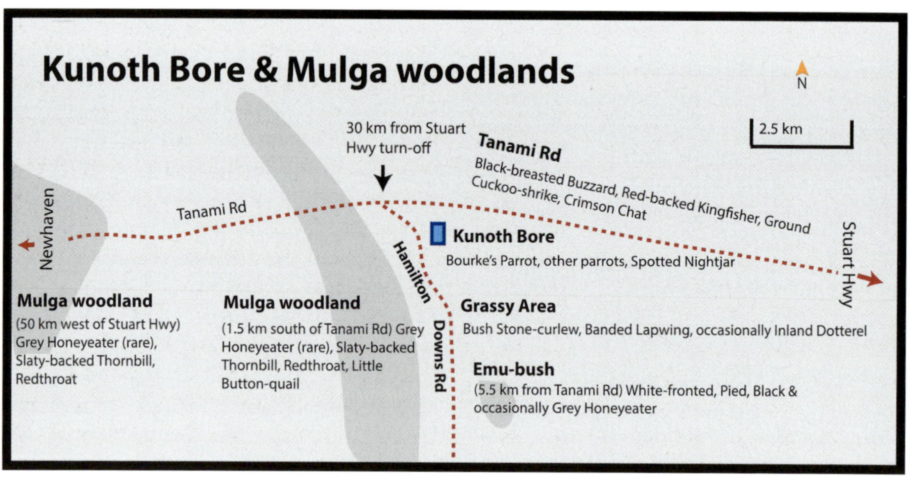

on the Tanami Rd, turn south into the Hamilton Downs Youth Camp Rd. This is a well-graded road suitable for all cars. The bore is on the east side of the road ~200 m down. This area is leased Aboriginal Land and, although there are currently no restrictions on entry, please try not to disturb the cattle. A visit to this site provides a good day trip from Alice Springs, with three main areas in which to concentrate your birding.

On the drive to Kunoth Bore, maintain a vigil for Spotted Harrier, Black-breasted Buzzard, Red-backed Kingfisher, Ground Cuckoo-shrike and Crimson Chat along the Tanami Rd. The most reliable area to search for Grey Honeyeater is in the Mulga woodlands (a small wattle tree with grey-green foliage) on the west side of the road 1.5 km south of Kunoth Bore. The Mulga in this area continues for ~4 km on the west side of the road and the honeyeater may occur anywhere along here. The Mulga in this area can also be a good location for the uncommon Slaty-backed Thornbill, which are best identified by their dark streaked crown and pale chestnut rump. Other thornbills found here include the similar Inland Thornbill (with a scaly crown), and the Chestnut-rumped and Yellow-rumped Thornbills. There are also a range of other small passerine species such as Southern Whiteface, Western Gerygone and Redthroat. On occasion, the female Redthroat may resemble the Grey Honeyeater, so you will need to be careful not to confuse the two. Redthroat tend to forage on the ground and, when flushed, typically reveal their largely black, white-tipped, tail. The Mulga in this area is very productive during the winter and spring periods. You can usually find Little Button-quail, Grey Fantail, Diamond Dove, Australian Owlet-Nightjar, Black-faced and Masked Woodswallow, Red-capped and Hooded Robin, Rufous Whistler, Crested Bellbird, White-browed Babbler, White-browed Treecreeper, Splendid Fairy-wren, Horsfield's Bronze-Cuckoo, Spiny-cheeked and Singing Honeyeater and Mulga Parrot. Mulga is also a favoured habitat of the rather cryptic Bourke's Parrot and the species is not uncommonly encountered during a walk in this area.

On the east side, 5.5 km down the Old Hamilton Downs Rd, there is a patch of Emu-bush (*Eremophila*) located between an area of Mulga and open grasslands. When in flower (usually Jul–Aug, but dependent on rain), these nectar-rich shrubs attract a variety of nomadic species including White-fronted, Pied and Black Honeyeater.

The final place to concentrate your birding is around Kunoth Bore itself. A good way to see Bourke's Parrot is to stake out Kunoth Bore in the evenings. Typically, in the half-light just after sunset, small flocks of Bourke's Parrot come in to drink at the bore, particularly around the small rise in the south-west corner of the dam. Many other birds are attracted to the bore also, so expect to see Galah, Cockatiel, Budgerigar, Australian Ringneck, Red-tailed Black-Cockatoo and White-backed Swallow. Check the nearby grassy areas for Brown Songlark, Bush Stone-curlew, Banded Lapwing and, occasionally, Inland Dotterel (rare). Along the roadside, look for Rufous Songlark, White-winged Triller and Crimson Chat. Raptors that regularly occur around Kunoth Bore include Black Kite, Wedge-tailed Eagle, Nankeen Kestrel, and Brown and Peregrine Falcon. At night, Spotted Nightjar sometimes drinks on the wing and hunts for insects over the dam.

If you do not find Grey Honeyeater and Slaty-backed Thornbill at Kunoth Bore, another spot to look is on the Tanami Rd, 50 km from Sturt Hwy turnoff – look for them in flowering patches of mulga, with the best times between April and September.

Mulga Woodland near Kunoth Bore: the preferred habitat of the elusive Grey Honeyeater. The species is often found in the company of thornbills and is more likely to be foraging among the Mulga stands than frequenting patches of flowering Emu Bush *Eremophila* sp.

Santa Teresa Rd

Key species: Rufous-crowned Emu-wren, Dusky Grasswren, Spinifexbird, Grey-fronted Honeyeater
Other species: Painted Finch, Bourke's Parrot

A very good area of Spinifex-covered limestone-ridge habitat near Alice Springs is located on the Santa Teresa Rd 31.5 km from the airport roundabout. This is an area where Spinifex occurs along the road on both sides; to further assist the visiting birdwatcher, there is a car tyre wedged in a metal power pole at this point!

Santa Teresa Rd has recently become popular among birdwatchers as an accessible location near Alice Springs where Rufous-crowned Emu-wren and Dusky Grasswren may be seen. Often the most reliable area is among the larger clumps of Spinifex on the rocky slopes and valleys on the west side of the road, because the hills to the east of the road were badly burnt in late 2011 and it will take several years before the Spinifex is again suitable for grasswrens and emu-wrens. Rufous-crowned Emu-wren seems to prefer the small valley areas – listen for their high pitched insect-like call – while the Dusky Grasswren tends to prefer the rocky escarpment and scree country. Dusky Grasswren have a trilling call that is noticeably more strident than that of the Rufous-crowned Emu-wren. Grasswrens seldom fly, preferring to bounce over boulders with their tails cocked. When disturbed, they will bound away, with their tails lowered to seek cover within or behind a well-placed Spinifex clump. Another sought-after species, the Spinifexbird, also occurs here. If you do not find Dusky Grasswren at the above site, another area worth a visit is the cliff-top edges of the flat-topped hills 26.5 km

Rufous-crowned Emu-wren atop Spinifex or hummock grass. The species prefers areas of long-unburnt Spinifex that has developed into large clumps. At the Santa Teresa Rd site, optimal habitat occurs along narrow drainage lines towards the foot of the slope. Notoriously elusive, a 'thin' insect-like contact call is the best means of detection.

from the airport roundabout: an area also good for Grey-fronted Honeyeater. While searching for the Grass wren and Emu-wren, there is also a good chance of encountering Red-backed Kingfisher (often sitting on electric power lines along the roadside), Singing, Grey-fronted and Pied Honeyeater, Chiming Wedgebill, Crested Bellbird, White-browed Babbler, Crimson Chat, Black-faced Woodswallow, Bourke's Parrot (uncommon and more frequent in the nearby patches of Mulga), and Painted and Zebra Finch. On the road between the airport and the emu-wren site, look for Black Falcon, Spotted Harrier and Red-backed Kingfisher, and more common species such as Australian Ringneck, Cockatiel, Budgerigar, Black-faced Woodswallow, Rufous Songlark, Diamond Dove and Zebra Finch.

Trephina Gorge Nature Park

Key species: Painted Finch, Little Woodswallow, Dusky Grasswren, Western Bowerbird
Other species: Budgerigar, Grey-headed Honeyeater, Grey-fronted Honeyeater

Trephina Gorge Nature Park (1771 ha) is in the East MacDonnell Ranges. Known for its imposing sheer quartzite cliffs and River Red Gum-lined watercourses, Trephina Gorge can be one of the most productive areas in the MacDonnell Ranges for birdwatching. Situated 85 km east of Alice Springs, the site is accessible to all vehicles via the sealed Ross Hwy, followed by a 9 km drive into Trephina Gorge, the last 5 km on gravel. Trephina boasts excellent camping facilities, gas barbecues, toilets and showers.

After rains and when the grasses, shrubs and eucalypts are blooming or seeding, Budgerigar can be very common at Trephina. There are several very pleasant walks, with the Trephina Gorge Walk being particularly good for birding, including the likelihood of observing Painted Finch. A loop walk, it commences at the creek's edge and provides scenic views of the sheer red cliffs towering over the tree-lined creek. Look up to enjoy Little Woodswallow as they circle the surrounding ridges. You can either walk directly down into the creek bed or alternatively walk up along the gorge rim, where Dusky Grasswren, Western Bowerbird, Spinifex Pigeon and Grey Shrike-thrush are found, as well as White-plumed, Grey-headed, Grey-fronted and, occasionally, White-fronted Honeyeater. Black-footed Rock Wallaby inhabit the rocky crags in this area. Birding around the picnic areas can be quite rewarding, with

The Dusky Grasswren is typically one the easier grasswren species to find. Look in areas of mature Spinifex on rocky slopes and especially along boulder strewn drainage lines at the base of these of slopes.

Variegated Fairy-wren, Hooded Robin, Crested Bellbird, Zebra and Painted Finch occurring, while at night Southern Boobook and Australian Owlet-Nightjar are easily heard and occasionally sighted.

It is worth making several brief stops on the road into the Trephina Gorge. Along here, expect to encounter a diversity of bush birds including Crimson Chat, Diamond and Peaceful Dove, Common Bronzewing, Black-faced Woodswallow, Zebra Finch, Spinifexbird, Rufous Songlark, Cockatiel, Mulga Parrot, Western Gerygone and Hooded Robin. Spotted Nightjar is also possible in this area. At night, look for the nightjar's orange-red eye shine by driving slowly along the access road; drive carefully though, because the species often sits in the middle of the road and at times may not take flight until the approaching vehicle is almost upon it!

N'Dhala Gorge Nature Park

Key species: Painted Finch, Grey Falcon
Other species: White-winged Fairy-wren, Splendid Fairy-wren, Variegated Fairy-wren

N'Dhala Gorge Nature Park (501 ha) is located in the Eastern MacDonnell Range. The reserve is accessed via a sealed road to Ross River Homestead (90 km east of Alice Springs), followed by 11 km of unsealed track. 4WD drive, or at least high clearance, is required on this last section as the track crosses the Ross River several times; indeed, it becomes impassable even to 4WDs after heavy rain. The N'Dhala Gorge Walk (1.5 km) takes ~1.5 h (return) to complete. There is a basic bush camping area at the entrance to the park.

N'Dhala Gorge is a reliable site for Painted Finch, with a good location being a rock pool about three-quarters of the way down the Gorge Walk. The pool has a large rock face on the south side and a sandy area on the north side. This is an ideal spot for birds because there are shrubs overhanging the water that provide birds with a perfect vantage point to scan for danger before drinking. Painted Finch, along with large numbers of Zebra Finch and Diamond Dove, visit the waterhole, especially during the heat of the day. Honeyeaters at N'Dhala include Grey-headed, Grey-fronted and White-fronted honeyeater, and three species of fairy-wren, White-winged, Splendid and Variegated, also occur. Other species that may be encountered include Mulga Parrot, Australian Ringneck and Red-tailed Black-Cockatoo, while Peregrine Falcon, hunts along the gorge, as does the rare Grey Falcon on occasion.

Simpson's Gap

Key species: Painted Finch, Grey-headed Honeyeater, Western Bowerbird
Other species: Peregrine Falcon, Zebra Finch

Open from 5 a.m. to 8 p.m. daily. The turn-off to Simpson's Gap is 18 km west of Alice Springs along Larapinta Drive; Simpson's Gap is a further 6 km via this access road.

Simpson's Gap is a spectacular narrow gorge with a permanent waterhole. When combined with its proximity to Alice Springs, the Gap is an extremely popular gorge for visiting tourists. Despite this, wildlife can be abundant and birdwatching at this site can be very rewarding, especially early in the morning before large numbers of tourists arrive.

There are several good birdwatching walks in the area. The Ghost Gum Walk (200 m, 15 min return) near the information centre is an excellent short walk, while the longer Cassia Hill Walk (1.5 km, 1 h return) can be very productive. On both walks, expect to see a range of nectar-feeding species such as Grey-headed, Brown, Spiny-cheeked (summer) and Singing Honeyeater. All of these species can be locally nomadic, with a preference for the River Red Gums and Ghost Gums that are found along the gorges and watercourses. With work, Dusky Grasswren can occasionally be located on the rock scree and Spinifex-covered slopes on the left as you approach Simpson's Gap. Parrots found here include Budgerigar and Major Mitchell's Cockatoo and, during the heat of the day, Zebra and Painted Finch may drink from the main pool. Little Woodswallow are also a common sight circling the ridges. Expect to see a good variety of other bush birds, including Chestnut-rumped and Inland Thornbill, Western Gerygone, Pied Butcherbird, Western Bowerbird and Collared Sparrowhawk. At night, Australian Owlet-Nightjar and Southern Boobook occur around the picnic area. Simpson's Gap is also one of the more reliable sites for Black-footed Rock Wallaby, especially along the short walking track

Low cloud on the West MacDonnell Ranges, with the sandy riverbed of Roe Creek in the foreground, Simpsons Gap, NT. The River Red Gum-lined creek and surrounding woodland supports a good range of birds including Little Corella, Brown Honeyeater, White-plumed Honeyeater, Pied Butcherbird and, occasionally, Major Mitchell's Cockatoo.

into the Gap and among the large boulder field on the east side of the gorge entrance. Nearby, Roe Creek between Simpsons Gap and Temple Bar Gap can be especially good for Major Mitchell's Cockatoo.

Ellery Creek Big Hole

Key species: Rufous-crowned Emu-wren, Spinifexbird
Other species: Little Button-quail, Grey-headed Honeyeater, Black Honeyeater, Red-browed Pardalote

Situated 90 km west of Alice Springs along Larapinta Drive, this site receives fewer tourists than Simpson's Gap, so, at busy times, can be a more peaceful birding retreat. Spectacular red cliffs, a large waterhole (Big Hole) and a sandy creek fringed by River Red Gums and Ghost Gums make this an appealing destination in the West MacDonnell Ranges.

The 3 km Dolomite Walk provides birders with an opportunity to explore a Spinifex-dominated habitat and is a well known site for hummock grass specialists such as Rufous-crowned Emu-wren and Spinifexbird. Extensive fires have affected large areas around Ellery Creek in recent years and, as a consequence, finding these species may require a bit of persistence. With time, and in the absence of further fires, the thicker long-unburnt Spinifex that is sought by these two species will recover. Grey-headed and White-plumed Honeyeater are both reasonably common here and the nomadic Black Honeyeater can be present when the *Eremophila* is flowering. Also, look for Little Button-quail, Australian Ringneck, Sacred and Red-backed Kingfisher, Red-browed Pardalote, Pied Butcherbird and White-winged Fairy-wren, and raptors such as Little Eagle and Peregrine Falcon. Big Hole also often supports small numbers of waterbirds such as Grey Teal, Hardhead, Dusky Moorhen, Pink-eared Duck, Eurasian Coot and Australasian Grebe.

The Grey-headed Honeyeater frequents woodlands and shrublands in association with rocky slopes and gorges. It can be encountered at most accessible sites dotted along the MacDonnell Ranges.

Ormiston Gorge

Key species: Dusky Grasswren, Rufous-crowned Emu-wren, Spinifexbird, Painted Finch, Spinifex Pigeon
Other species: Little Woodswallow, Red-browed Pardalote, Western Bowerbird, Black-chinned Honeyeater

Located 135 km west of Alice Springs on Namatjira Drive, Ormiston Gorge is one of the most interesting birding locations in the Red Centre. Situated in the West MacDonnell National Park (190 000 ha), the visitor centre is ~8 km from Namatjira Drive. A sealed road provides access to conventional vehicles year round. There is an excellent camping ground at Ormiston, commercial accommodation nearby at Glen Helen and bush camping along the Finke River. At Ormiston Gorge, there are barbecues, drinking water, toilets, showers and a canteen (with basic food and drinks). Ormiston Gorge waterhole is also ideal for swimming, especially in the warmer months.

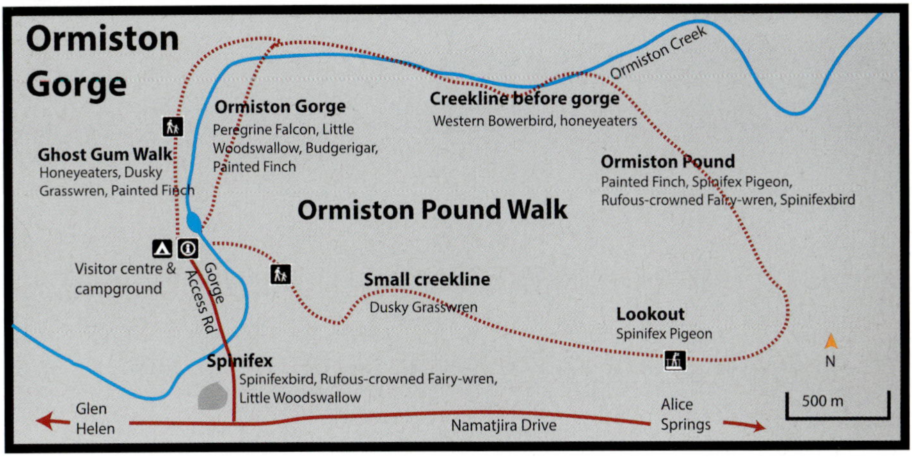

The first site to visit is a section of mature Spinifex just 50 m or so after the turn-off from Namatjira Drive into Ormiston Gorge Access Rd. On the west side of the road, there is a large patch of Spinifex, where, with a bit of careful searching and patience, birdwatchers may find Spinifexbird and (less commonly) Rufous-crowned Emu-wren. Keep an eye out for Little Woodswallow and White-backed Swallow overhead. Also look out for Black-breasted Buzzard and Wedge-tailed Eagle: the latter are a common sight along the roadsides where they feed on roadside carrion.

The best walk for birding is the wonderful Ormiston Pound Walk. Taking ~3–4 h, this is a 7.5 km loop that is best walked in an anti-clockwise direction. This walk starts at the roadside ~500 m east of the visitor centre. It begins by taking you over the Ormiston Creek, a tributary of the Finke River, and then up and along a picturesque rocky valley with a ridge of quartzite scattered with stately Ghost Gums. From here you drop down into the Ormiston Pound, a flat Spinifex-covered valley before re-joining the Ormiston Creek to follow it through the majestic Ormiston Gorge. Here, walkers pass a series of waterholes before returning to the visitor centre.

In the River Red Gums at the Ormiston Creek crossing, you can usually find Grey-headed and Black-chinned Honeyeater. Striated and Red-browed Pardalote are also often present and, if there is water in this part of the creek, Black-fronted Dotterel is a possibility on the sandy edges. From here the track follows a small creek line. About 1.5 km into the walk, there is an excellent area for Dusky Grasswren, where, at times, several family parties can be located in the general vicinity. Concentrate your efforts in the valley before reaching the 'saddle' area and lookout at the top. Around the saddle, Spinifex Pigeon feed and shelter among boulders and heavily fissured rock. After the lookout, the track descends into the pound itself: an area in excess of 1000 ha that supports extensive flats of Spinifex. When walking through the pound, be alert to the possibility of Rufous-crowned Emu-wren and Spinifexbird, particularly where there is a dense cover of Spinifex. Approximately halfway through the pound, there are several rocky outcrops. In these areas, look for Spinifex Pigeon and Painted Finch, especially if there are pools of water present. After leaving the pound, the track re-joins Ormiston Creek. Along the creek, Western Bowerbird can

The view along Ormiston Gorge. The Spinifex-clad slope on the left can be good for Dusky Grasswren, while the gorge walls are frequented by Peregrine Falcon, Nankeen Kestrel and Little Woodswallow. Budgerigar is often common and tame Spinifex Pigeon sometimes occur in the carpark.

be quite vocal and honeyeaters here can include Black-chinned, Grey-headed, White-plumed, Brown and Singing. The track then passes through Ormiston Gorge. If conditions are good, Budgerigar may be abundant. Small flocks will fly up and down at break-neck speed attempting to avoid the resident Peregrine Falcon that nest in the central part of the gorge. The falcons often sit high up on the cliff-face surveying their territory, before diving down into the gorge to catch their prey. At the gorge, a near-permanent waterhole, up to 14 m deep, provides drinking opportunities for species such as the Painted Finch. Black-footed Rock Wallaby frequent the gorge, especially near the main rock pool. There are several additional short walking tracks that lead out of the main picnic area; the short 20 min Waterhole Track, the 1 h Ghost Gum Lookout Walk and the longer 1.5 h Ghost Gum Walk. For casual birding, all can be productive. Some of the additional birds you can expect to see are Grey-crowned Babbler, Masked and Black-faced Woodswallow, Pied Butcherbird, Hooded and Red-capped Robin, Variegated Fairy-wren, and Red-browed and Striated Pardalote. Little Button-quail are also a possibility in grassy areas.

Watarrka (Kings Canyon) National Park

Key species: Dusky Grasswren, Spinifex Pigeon, Painted Finch, Western Bowerbird, Red-browed Pardalote
Other species: Little Woodswallow, Ground Cuckoo-shrike

Kings Canyon is an imposing landmark of spectacular red sandstone cliffs that tower

Tnorala Conservation Reserve and the Mereenie Loop Rd

Tnorala Conservation Reserve (4800 ha), also known as Gosse Bluff, is located on Larapinta Drive, 60 km west of Glen Helen. The main picnic area is located 10 km from the highway and 4WD is recommended. Birds you may see here include Red-browed Pardalote, Black Honeyeater, Little Button-quail, Crested Bellbird, Red-capped Robin and Splendid Fairy-wren. Near the first picnic table, there is a bower of the Western Bowerbird. On the short walk to a lookout, Dusky Grasswren, Little Woodswallow, Grey-headed Honeyeater, Tree Martin and Brown Goshawk may be encountered. The rare Grey Falcon has also been recorded on occasion. (Another, perhaps more regular site for Grey Falcon is Roma Gorge – the turn-off is located between Glen Helen Gorge and Red Bank Gorge and 4WD is necessary.)

The Mereenie Loop Rd (part of the Red Centre Way) connects the West MacDonnell Ranges with Kings Canyon. It passes through several areas of deep dunes that support Desert Oak (*Allocasuarina decaisneana*) in association with *Eremophila*, *Grevillea* and the Upside-down Plant (*Leptosema chambersii*). When flowering, these smaller shrubs and herbs attract honeyeaters such as Black, Pied, Grey-fronted and Spiny-cheeked. You may also encounter Emu, Major Mitchell's Cockatoo, Australian Ringneck, Mulga Parrot, Budgerigar, Diamond Dove, White-backed Swallow, Crested Bellbird, Black-faced and Masked Woodswallow, Crimson Chat, Western Gerygone and Zebra Finch. Princess Parrot, which is rare anywhere within its range, is a possibility here, particularly after periods of above-average rain. In order to travel along the Mereenie Loop Rd, you need to acquire a transit permit from the Central Land Council (www.clc.org.au). Permits are also available at Glen Helen, Hermannsburg and Kings Canyon Resort.

up to 100 m above the surrounding landscape. Situated in Watarrka National Park (71 000 ha) it is 330 km (via the Mereenie Loop Rd) south-west of Alice Springs. The park also encompasses the western end of the Gorge Gill Range and is an important conservation area. It can be reached by 2WD vehicle via Luritja Rd from Yulara and Lasseters Hwy (sealed road), or via the West MacDonnell Ranges via the Red Centre Way. Accommodation and camping is available at both Kings Canyon Resort and Kings Creek Station.

There are several good walks for birdwatching in the park, varying in difficulty and length. The Rim Walk (6 km loop) traces the top of the canyon and is regarded as one of the best walking tracks in central Australia for its spectacular and varied vistas. Along here, birds include Spinifex

Pigeon, Zebra Finch, Grey-headed and Grey Honeyeater (the latter is rare), Red-browed Pardalote, Australian Ringneck and, occasionally, Ground Cuckoo-shrike. Dusky Grasswren is reasonably common, particularly in the denser areas of mature Spinifex. The Kings Creek Walk (2.6 km return) traces the bottom of the gorge along Kings Creek, offering stunning views of the canyon walls above. The gully along the walk has a wetter microclimate than the surrounding plains and ridges so provides additional opportunities for plants and wildlife. Kings Creek is a very good location for Painted Finch, which drink from the small pools along the gorge. Dusky Grasswren often frequent the start of the walk, particularly where Spinifex meets the creek line, and Spinifex Pigeon is common along the walk (with some hand-tame individuals frequenting the picnic area and carpark). Other birds to look for include Red-browed Pardalote, usually high in the canopy of River Red Gums along the creek, Red-capped Robin, Rufous Songlark, Grey Shrike Thrush, Rufous Whistler, Splendid Fairy-wren (again at the start of the walk), Brown, Grey-headed, White-plumed and Spiny-cheeked Honeyeater, Inland Thornbill, Horsfield's Bronze-Cuckoo and Pallid Cuckoo. Overhead, Little Woodswallow routinely cruise the cliff-tops. Approximately halfway along the walk is the Garden of Eden, a delightful area of cool waterholes and riverine vegetation communities. Western Bowerbird occurs here, especially in association with the Desert Fig. At the end of the walk, there is a platform with spectacular views of the canyon walls. The Kathleen Springs Walk (2.6 km return) is perfect for families and visitors with limited mobility. It leads to a delightful spring-fed waterhole at the head

Desert Oak country with an understorey of Spinifex along the Mereenie Loop Rd. These areas can be good for a range of species including Little Button-Quail, Diamond Dove, Major Mitchell's Cockatoo, Budgerigar, Western Gerygone, Crimson Chat, Crested Bellbird, White-backed Swallow and Zebra Finch. Although an elusive species, this is also a favoured foraging and daytime roosting habitat for the Princess Parrot.

of Kathleen Gorge. Birds along here include Spinifex Pigeon, Red-browed Pardalote, Western Bowerbird, Grey Shrike-thrush, and Grey-headed and White-plumed Honeyeater.

A classic tree of the arid lands, the Desert Oak, lines Luritja Drive. This tree is also a favoured habitat (mostly as a daytime roost) for the rarely encountered Princess Parrot (e.g. in 2010 a flock of five birds were recorded here). Other birds to look for in this area include Crimson Chat, Galah, Little Corella and Major Mitchell's Cockatoo, Mulga Parrot, Australian Ringneck, Pied Butcherbird, Willie Wagtail, Crested Pigeon and Common Bronzewing, and raptors such as Nankeen Kestrel, Brown Falcon and Spotted Harrier.

Erldunda

Key species: Banded Whiteface, Cinnamon Quail-thrush, Chiming Wedgebill
Other species: Pied Honeyeater, Black Honeyeater, White-fronted Honeyeater, Bourke's Parrot

Erldunda is a fuel stop and roadhouse situated at the junction of the Stuart and Lasseter Hwys, 200 km south of Alice Springs. The roadhouse has good camping facilities, which includes gas barbecues, toilets and showers and hotel-style accommodation. At the roadhouse, expect to see birds such as White-plumed Honeyeater, Yellow-throated Miner, Black-faced Woodswallow, Welcome Swallow, Crested Pigeon, Galah, Australian Ringneck, Australian Magpie, Little Crow and Major Mitchell's Cockatoo. Raptors in the local area include Wedge-tailed Eagle, Spotted Harrier and Nankeen Kestrel.

An area of gibber plain and open shrublands north of Erldunda is a well-known and relatively reliable site for the uncommon Banded Whiteface. The site is 21 km north of the roadhouse, or 175 km south of Alice Springs (using The Gap as the start point). A useful landmark is a dune ridge that crosses the road on the southern boundary of the area. The best birding is usually had in the open areas up to 1 km to the west of the road, or 1 km north and south of this area. The Banded Whiteface displays a preference for stony 'gibber' ground and often occurs in mixed flocks with Southern Whiteface, Inland Thornbill and other small passerines.

Where Mulga and *Eremophila* merges with more open areas, usually on dune rises, listen for the high-pitched calls of Cinnamon Quail Thrush; this site is near to the northern limit for this species. Among the Mulga, Chiming Wedgebill and families of White-browed Babbler regularly occur, while nomadic Pied, White-fronted and Black Honeyeater feed in the *Eremophila* when flowering (usually winter and spring). Other birds that regularly occur in and around this site include Red-backed Kingfisher (often perched on power lines), Crimson and Orange Chat, Bourke's Parrot, Crested Bellbird, White-backed Swallow, Brown and Rufous Songlark, Western Gerygone, White-winged, Splendid and Variegated Fairy-wren, Zebra Finch, Banded Lapwing and Inland Dotterel (rare).

The Black Honeyeater is a truly nomadic species. Flowering Emu-bush *Eremophila* sp. are a preferred nectar source, so thickets of flower-laden shrubs should be checked carefully.

Uluru–Kata Tjuta National Park

Key species: Striated (Sandhill) Grasswren, Banded Whiteface, Princess Parrot

Other species: Chiming Wedgebill, Black-eared Cuckoo, Black Honeyeater, Pied Honeyeater

One of the world's most iconic locations, and a place of staggering beauty, Uluru–Kata Tjuta National Park (201 000 ha) is located in the 'very middle of the Red Centre'. The park features two of Australia's most recognisable natural landmarks: Uluru (previously named Ayers Rock), which stands 348 m above the surrounding plain and 40 km to its west, Kata Tjuta (formerly known as The Olgas), standing at 546 m. The area was declared a national park in 1950; the name was changed in 1995 from Ayers Rock–Mount Olga National Park to Uluru–Kata Tjuta National Park. The park is located 440 km south-west of Alice Springs accessed via the Lasseter Hwy. One can fly to Uluru (Connellan Airport) from Alice Springs connecting to most capital cities. Regular bus services also run from Alice Springs. There are picnic areas with gas barbecues at the cultural centre and at the sunset viewing areas at Kata Tjuta. A resort village located just outside the park at Yulara offers different levels of accommodation, from luxury to budget, as well as a caravan park with camping areas.

In and around Uluru–Kata Tjuta National Park, 178 bird species have been reported, 66 of these are resident, with the remainder either nomadic or migratory. Widespread species found within Uluru–Kata Tjuta National Park include Emu, Peregrine Falcon, Australian Bustard, Australian Ringneck, Mulga Parrot, Red-backed Kingfisher, Ground Cuckoo-shrike, Wedge-tailed Eagle, Nankeen Kestrel, Splendid and Variegated Fairy-wren, and Redthroat (regularly recorded near the Yulara rubbish dump). Nightbirds at Uluru include Southern Boobook, Eastern Barn Owl, Australian Owlet-Nightjar and Spotted Nightjar (often seen near the airport after sunset). Rarities include Grey Honeyeater and Bourke's, Scarlet-chested Parrot and Princess Parrot.

The sedentary Striated Grasswren (ssp. *oweni*, a subspecies that may be elevated to full species status, known as Sandhill Grasswren) is one of the more sought-after species to occur within the Park. The species favours mature Spinifex grasslands, so any areas with extensive habitat are worth a careful search. The best spot to look for them is in the large area of Spinifex west of the sunset viewing carpark on Uluru Rd. Grasswrens, in general, are a notoriously difficult genus to find, so you need to allow yourself a bit of time for any search. The highly nomadic Black and Pied Honeyeater may be present near the carpark when shrubs are flowering. Another spot to look for Striated (Sandhill) Grasswren is the larger areas of Spinifex behind Connellan Airport.

Another sought-after species at Uluru is the Banded Whiteface. Patchily distributed, the species prefers expansive areas of flat stony ground with a sparse cover of small shrubs. Research suggests that they prefer areas recovering from fire (usually about a year after fire). In such habitats, they occur in small parties of 5–10 birds, often in mixed flocks with other small passerines. Depending on the recent fire history, good sites may include areas on either side of the road to the airport (Coote Rd), along the road to the Yulara water treatment plant (Marla Rd) and in the roadside areas between Yulara and Uluru.

One excellent site for more general birdwatching in the Park is the Liru Walk (4 km return). This takes you through Mulga shrublands where you may see Black-eared Cuckoo, Horsfield's Bronze-Cuckoo, Western Gerygone, Inland Thornbill,

Red-capped and Hooded Robin, Rufous Whistler, Crested Bellbird and Chiming Wedgebill. Black and Pied Honeyeater, both highly nomadic species, may be seen here if there is flowering *Eremophila*. In good seasons, Budgerigar may occur in very large numbers. A longer walk is the Uluru Base Walk (9.4 km loop walk, 3–4 h), which circumnavigates the rock. This pleasant walk usually provides good opportunities to view Singing, Grey-headed, White-plumed and Spiny-cheeked Honeyeater, particularly along the Mutijulu section. Other common birds that may be encountered include Crested Pigeon, Galah, Black-faced Cuckoo-shrike, Willie-wagtail, Yellow-throated Miner, Black-faced Woodswallow, Little and Torresian Crow, Pied Butcherbird, Zebra Finch and, occasionally, Major Mitchell's Cockatoo.

From Uluru, it is a short drive (50 km) to Kata Tjuta (The Olgas).To get there, continue west along the Lesseter Hwy. There are several very enjoyable walks for birding at Kata Tjuta including the Walpa Gorge Walk (2. 6 km, 1 h) and the Valley of the Winds Walk (7.4 km, 3 h). Many of the birds that occur at Uluru can also be expected here; because it is often quieter (fewer tourists), it can be a much more pleasant birding experience.

Red desert dunes with a sparse cover of shrubs and large clumps of Spinifex: the favoured habitat of the Striated (Sandhill) Grasswren within Uluru–Kata Tjuta National Park. Other species found in this habitat include Black-eared Cuckoo, Inland Thornbill, Crested Bellbird and Black-faced Woodswallow.

Old Andado Station

Key species: Eyrean Grasswren, Gibberbird, Cinnamon Quail-thrush, Banded Whiteface, Letter-winged Kite, Grey Falcon
Other species: White-winged Fairy-wren, Bourke's Parrot

An extremely remote property, Old Andado Station is located in the south-east of the NT on the edge of the Simpson Desert. It is a good day's drive from Alice Springs. You can access the property several ways. From Alice Springs, head south-east 94 km along the Old Andado Track to Santa Teresa. Old Andado Station is a further 237 km. Alternatively, you can access the site via Stuart Hwy from Kulgera (257 km) through Finke, and from the south from Oodnadatta (240 km) and Birdsville (369 km). All access is 4WD only and, even then, roads may become impassable after heavy rain. Comprising ~1 million ha, Andado Station is one of the largest privately owned cattle stations in Australia. Accommodation includes a well-serviced campsite and hotel-type accommodation in the homestead. Details can be found at www.oldandado.com.

Located further into the Simpson Desert than any other property, Andado Station is an excellent site for Eyrean Grasswren. They live on the system of parallel cane grass-covered red sand dunes: a common feature of the property. Cinnamon Quail-thrush can also be quite common in the sand dune areas. Other species with a preference for this habitat type are White-winged Fairy-wren, White-backed Swallow, Nankeen Kestrel and Singing Honeyeater. Gibberbird (rare) and Banded Whiteface also occur at Old Andado, but both display a distinct preference for gibber plains – a desert surface covered with closely packed rock fragments. While at Old Andado, look carefully for Bourke's Parrot, Ground Cuckoo-shrike and Red-backed Kingfisher, which occur along the Mulga-lined watercourses. Also, look for Diamond Dove, Tree Martin, Brown Songlark, Crimson Chat, Pied Butcherbird, Budgerigar and Zebra Finch. The nomadic Australian Bustard occurs on the property too, regularly moving into the areas when conditions are favourable; one of Australia's heaviest flying birds, Bustards prefer open grasslands. Nectar-feeders at Andado include Singing, White-plumed and Grey Honeyeater (rare) and Yellow-throated Miner. With very few trees on the property, those that are present provide some of the few roosting sites for birds of prey. Look for Black Kite, Nankeen Kestrel, Brown Falcon, Wedge-tailed Eagle, Black-breasted Buzzard and Spotted Harrier and rarer raptors such as Letter-winged Kite (in La Nina years when native rodents are abundant), Black and Grey Falcon. Finally, the lowlands around the homestead become a large lake and swamp in La Nina years, attracting large numbers of Pink-eared Duck, Black-tailed Native Hen, crakes and other waterbirds. In such conditions, there are also records Grass Owl here hunting over dense grasses beside creek lines.

If you are travelling to Andado along the Santa Teresa Rd, it is worth stopping

Eremophila labtobei, a species of Emu Bush, in full flower. This particular thicket was situated in a Marble Gum woodland and was being frequented by Masked Woodswallow and Pied, Black, Grey-fronted, White-fronted and Spiny-cheeked Honeyeaters.

at the Mac Clark Conservation Reserve, about halfway between Andado and Alice Springs. The reserve protects a rare stand of Waddywood, the preferred timber for the Aboriginal waddy or club. The reserve protects ~1000 mature trees that grow to heights of 17 m and may live for 500 years. Interesting birds recorded within Mac Clark Conservation Reserve include Orange Chat, Banded Whiteface (check along the fence lines for perched birds) and Inland Dotterel, as well as raptors such Black Falcon, Letter-winged Kite and Grass Owl (the last two when native rodents are abundant).

Newhaven Sanctuary

Key species: Dusky Grasswren, Rufous-crowned Emu-wren, Princess Parrot (rare)
Other species: Spinifexbird, Grey Falcon, Night Parrot (reported once in 1996), Grey Honeyeater

Newhaven Sanctuary (262 000 ha) is ~350 km north-west of Alice Springs. Located at the junction of the Great Sandy Desert and MacDonnell Ranges, Newhaven is an extremely isolated reserve. Of note, Newhaven is also the site of a recent claimed sighting of the Night Parrot. Originally a pastoral property, it was purchased by

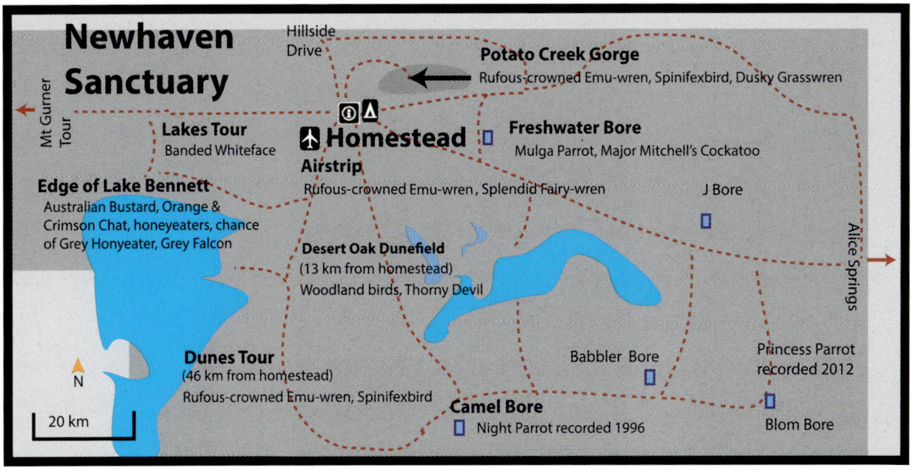

Birds Australia as a wildlife reserve before ownership passed to the Australian Wildlife Conservancy. From Alice Springs, head north along the Stuart Hwy to the Tanami Rd then drive 200 km along the Tanami Rd to the Tilmouth Well Roadhouse. Turn left onto the sign posted Newhaven Access Rd. Newhaven Reserve is a further 140 km from this turn-off. There is a well-equipped campground with toilets and showers. Remember to take ample supplies of food, water and fuel, because the nearest supply centre is the Tilmouth Roadhouse. The reserve is closed to visitors during the hotter months (Oct–Mar). Before travelling to Newhaven, it is essential to contact the Australian Wildlife Conservancy regarding current access arrangements (www.australianwildlife.org).

On the long drive into Newhaven, look for raptors such as Black-breasted Buzzard and Wedge-tailed Eagle and it is also worth stopping and searching areas of Mulga for Grey Honeyeater – in May 2012, they were recorded on the Tanami Rd 50 km west of the Stuart Hwy turn-off. Once at Newhaven, expect to find a good variety of birds around the homestead and campsite. Some of the common species include Australian Ringneck, Diamond Dove, Red-capped Robin, Crested Bellbird, Rufous Whistler, Black-faced and Ground Cuckoo-shrike, Crimson Chat and Western Gerygone, and honeyeaters such as Singing, Grey-fronted, Spiny-cheeked (summer) and Grey-headed. The Zebra Finch, which can sometimes be abundant, comes in to drink at a water tray near the campground.

One of the features of Newhaven is the self-guided driving tours, designed to take visitors around the property. Each has a slightly different feel. Fact sheets for each of these are available at the homestead. Along the Potato Creek Gorge Tour (15 km), the dominant tree species is the Ghost Gum and the area has a good ground cover of Spinifex. Rufous-crowned Emu-wren and Spinifexbird occur in the more extensive areas of Spinifex, while Dusky Grasswren inhabit the rocky escarpment areas in the Gorge. Honeyeaters such as Singing, Grey-headed and Grey-fronted are attracted to Holly Grevillea when flowering, while Zebra Finch, Painted Finch and Diamond Dove all drink at the pool in the gorge. On the Hillside Drive Tour (50 km), visit the Freshwater Bore, where you may find a good variety of small seed-eating birds, especially Zebra Finch and Budgerigar, as well as Major Mitchell's Cockatoo and Mulga Parrot. Keep an eye open for Grey Honeyeater, which have been record in Mulga just south of Honeyeater Hill. The Lakes Tour (50 km) passes the airstrip and a series of small wetlands, before continuing to the northern shore of Lake Bennett. Rainfall determines the state of these wetlands and the wildlife you may encounter here. Passing the airstrip, there are several good areas of Spinifex: potential sites for Rufous-crowned Emu-wren. Further along the tour, in areas with shrub-clad plains, a short walk may produce Banded Whiteface, among the more common Southern Whiteface. These same areas also support Crested Bellbird, Mulga Parrot, Red-capped and Hooded Robin, and Little and Black-faced Woodswallow. The area surrounding Lake Bennett can be particularly productive, with potential species including Crimson and Orange Chat and Black-faced Woodswallow. Where freshwater is present at the surface, Red-capped Plover may occur, while Zebra Finch and honeyeaters such as Pied, White-fronted and Brown are all attracted to drink. Australian Bustard occur throughout the area in low densities, and rarities such as Grey Honeyeater and Grey Falcon are always a possibility. The Dunes Tour (80 km) takes you south-west of the homestead. The first place to stop is the Desert Oak dune-fields ~13 km from the start of the tour. Deep red in colour, these dunes support mixed-species feeding flocks such as Yellow-rumped and Chestnut-rumped Thornbill, Weebill, Western Gerygone, Red-capped and Hooded Robin, Crested Bellbird and Rufous Whistler. This area is also a regular site for the intriguing Thorny Devil: look in the sand for the delicate tracks of a small reptile without any tail drag. Some 45 km into the tour are several mature stands of Spinifex,

Spinifex plains with termite mounds on the Tanami Rd: suitable habitat for Little Button-Quail, Rufous-crowned Emu-wren, Crimson Chat and Spinifexbird, among a host of other more common species.

which are a good site for Rufous-crowned Emu-wren and Spinifexbird. The final place to visit on the Dunes Tour is Camel Bore, ~52 km from the beginning. It was here that a sighting of the critically endangered Night Parrot was made in 1996.

Another rare and sought-after bird to look for at Newhaven is Princess Parrot – in 2012, as many as 50 birds were recorded. The best chance of seeing them is after periods of good rainfall; look in the south-east side of the reserve where, during the heat of the day, the most likely place to see them is roosting in the dense foliage of Desert Bloodwood (*Corymbia opaca*) and Desert Oak.

Barrow Creek

Striated (Sandhill) Grasswren has been found beside the Stuart Hwy 8 km south of Barrow Creek (278 km north of Alice Springs), an area characterised by low plateaus rising 50 m above the plain with extensive areas of Spinifex and a few small creek lines. Other birds to look for in the area include Rufous-crowned Emu-wren, Spinifexbird, Ground Cuckoo-shrike, Little Button-quail and Black-breasted Buzzard.

Queensland

Queensland (Qld) occupies the north-eastern section of mainland Australia, and is bordered by New South Wales (NSW) and South Australia (SA) to the south and south-west and Northern Territory (NT) to the west. The islands of Torres Strait, Boigu, Dauan and Saibai Islands, located just off the coast of Papua New Guinea, mark the northernmost point in the state (these islands are dealt with on pp. 477–484). Qld is Australia's second largest state (1 852 642 km^2) and the third most populous. The capital city is Brisbane. Due to its large size, there is considerable variation in the climate across the state. There is a monsoonal 'wet' season in the far north, warm temperate conditions along the east coast, and low rainfall and hot summer seasons in the arid south-west of the state. The Brolga is the bird emblem of Qld.

Cape York Peninsula

Regional overview

Cape York Peninsula is an excellent region for birding, with over 320 species recorded. From Rinyirru (Lakefield) National Park in the south, the Cape covers 137 000 km^2. Habitats range from tropical savannahs and grasslands scattered with termite mounds, extensive wetlands (which include expansive saline mangrove deltas and freshwater lagoon systems) to Australia's largest area of lowland rainforest at Kutini–Payamu (Iron Range) National Park.

In order to observe most of the Cape York specialties, a minimum itinerary of 9 days is recommended.

- **Days 1–2.** Cairns to Musgrave Roadhouse and Rinyirru (Lakefield) National Park, 444 km. Note that this trip can be shortened by commencing the journey from Kingfisher Lodge near Julatten. (Golden-shouldered Parrot, Black-backed Butcherbird, Red Goshawk, Sarus Crane, Star Finch).
- **Days 3–8.** Musgrave Roadhouse to Kutini–Payamu (Iron Range) National Park, 300 km (Red-cheeked and Eclectus Parrot, Palm Cockatoo, Yellow-billed Kingfisher, Chestnut-breasted Cuckoo, Red-bellied Pitta (summer), Fawn-breasted Bowerbird, Northern Scrub-robin, Trumpet Manucode, Magnificent Riflebird, Green-backed, White-streaked and Tawny-breasted Honeyeater, White-faced Robin, Frilled-necked and Black-winged Monarch (summer), Yellow-legged Flycatcher, Tropical Scrubwren).
- **Days 9–10.** Return to Cairns.

From the Iron Range (if you have time), travel across to Weipa and on to the tip of Cape York, allowing an additional 4+ days. Another alternative is to travel to Musgrave via Cooktown and Rinyirru (Lakefield) National Park, allowing an additional 2 days.

A 4WD is essential for travel in Cape York, but, even then, it is also important to

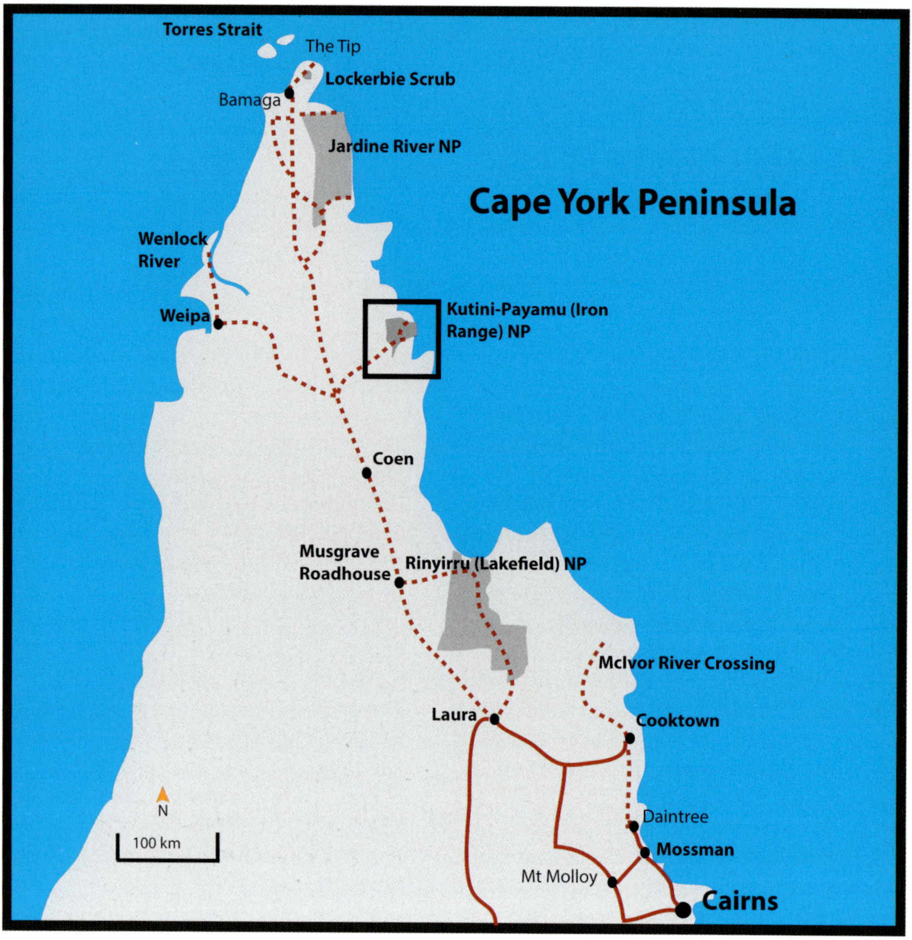

note that during the wet season (Dec–Apr) many of the roads on Cape York are regularly closed. There are also regular fights to Coen, Weipa, Lockhart River, Bamaga and Horn Island. Local 4WD car hire is available in most of these towns. Black-winged Monarch is a regular visitor to the Cooktown region, particularly McIvor River Crossing, which is generally accessible from Cairns by road during the wet season.

As with much of tropical Australia, the region is dominated by two seasons: the wet and the dry. The mean daily temperatures in the wet range from night-time lows of 23.5°C to daytime highs of 32°C, while those in the mid-dry season range from 18.5°C to 28°C. The mean monthly rainfall mid-wet season ranges between 350 and 450 mm; while in mid-dry season it is only 20 mm. The dry season is from May to November, with September being the month with the lowest average rainfall, while the wet season runs from December to April, with March being the month of highest average rainfall. Late September to early November (before the first rains of the wet season) is considered

Rainforest habitats support more species of birds endemic to Qld than any other habitat type. Here, a deciduous species sits among other rainforest trees in Lockerbie Scrub. These bare open branches are often favoured perching sites for starlings, pigeons and parrots.

an interesting time for birdwatching because some early migrating individuals of sought-after species may arrive from New Guinea during this period.

Extending 750 km up the centre of Cape York, the Peninsula Development Rd starts at Mareeba (west of Cairns) and ends at Weipa. From Weipa, to reach the top of Cape York, you then travel a further 315 km along the Telegraph Rd. Check www.racq.com.au for current local road conditions. Supplies are available from Laura, Coen, Cooktown and Weipa, although, if travelling up from Cairns, it is worth obtaining most major supplies before departure.

Birding highlights

Cape York supports two endemic species: Golden-shouldered Parrot and White-streaked Honeyeater. Other species whose Australian distribution is largely or entirely restricted to Cape York include Spotted Whistling-Duck, Red-cheeked and Eclectus Parrot (both also in northern Torres Strait), Palm Cockatoo, Yellow-billed Kingfisher, Chestnut-breasted Cuckoo (also present in very small numbers in the Wet Tropics), Red-bellied Pitta (summer), Black-backed Butcherbird, Green-backed Honeyeater, Pale White-eye (islands), Fawn-breasted Bowerbird, Northern Scrub-robin, White-faced Robin, Frilled-necked Monarch, Black-winged Monarch (summer), Yellow-legged Flycatcher, Tropical Scrubwren, Trumpet Manucode and Magnificent Riflebird. As noted earlier, Black-winged Monarch and Red-bellied Pitta are only present on Cape York during the wet season (the Monarch arrives as early as October, the Pitta as early as December; both depart by April, although very small numbers may overwinter on occasion). To have a realistic chance of finding these two species, you will need to visit during the wet season, with a good option being to fly to the Lockhart River (birding at Iron Range National Park) from Cairns.

Aside from these, other birding highlights on Cape York include Sooty, Bridled and Lesser Crested Tern, Lesser and Great Frigatebird, Sarus Crane, Beach Stone-curlew, Great-billed Heron, Black Bittern, Red Goshawk, Wompoo, Superb and Rose-crowned Fruit-Dove, White-eared Monarch, Yellow-tinted, Banded, Bar-breasted and Red-headed Honeyeater, Mangrove Golden Whistler, Lovely Fairy-wren and Star Finch. Night birds include Rufous and Eastern

Grass Owl, Papuan Frogmouth, Marbled Frogmouth (ssp. *marmoratus*) and Large-tailed Nightjar. Noteworthy subspecies that occur on the cape include Australian Brush-turkey (violet-wattled ssp. *purpureicollis*), Brown Goshawk (smaller, paler ssp. *didimus*), Rufous Owl (smaller darker ssp. *meesi*), Double-eyed Fig-Parrot ('Marshall's Fig-Parrot' ssp. *marshalli*), Pale-headed Rosella (ssp. *adscitus*), Crimson Finch (white-bellied ssp. *evangelinae*), Brown Treecreeper ('Black Treecreeper' ssp. *melanota*), Pied Currawong (large ssp. *magnirostris*), Little Bronze-Cuckoo (Gould's Bronze-Cuckoo ssp. *russatus*), Masked Finch ('White-eared Finch' ssp. *leucotis*), Black-throated Finch (black-rumped ssp. *atropygialis*) and Red-browed Finch (ssp. *minor*). Summer migrants include Oriental, Lesser Sand, Greater Sand and Pacific Golden Plover, Whimbrel, Eastern Curlew, Curlew and Sharp-tailed Sandpiper, Grey-tailed Tattler, Common Greenshank, Bar-tailed Godwit, Pied Imperial-Pigeon, Buff-breasted Paradise-Kingfisher and Eastern Koel. Vagrant species recorded to date have included Red-necked Phalarope, and Grey and White Wagtail.

Cooktown

Key species: Black-winged Monarch (Oct–Apr), Tropical Scrubwren, Chestnut-breasted Cuckoo, White-streaked Honeyeater
Other species: Oriental Cuckoo, Buff-breasted Paradise-Kingfisher, Lovely Fairy-wren, Barred Cuckoo-shrike

Cooktown is located in the far north of north Qld. There are two roads into Cooktown: either inland via the Peninsula Development Rd (326 km and suitable for 2WD), or via the Cape Tribulation Rd from Cairns (235 km, 4WD only, and often impassable after rain). Cooktown airport has regular flights from Cairns. The Cooktown Visitor Information Centre is located in the Botanic Gardens: www.naturespowerhouse.com.au, (07) 4069 6004.

The Webber Esplanade, with a rocky embankment, runs parallel to the estuary at the mouth of Endeavour River. Look along here for Common Sandpiper, Grey-tailed Tattler and, more rarely, Wandering Tattler. There are several sandbanks exposed at low tide where you may see shorebirds such as Beach Stone-curlew, Lesser and Greater Sand Plover, Common Greenshank, Bar-tailed Godwit, Red-necked Stint, Whimbrel and Eastern Curlew. The Cooktown Botanic Gardens have a series of well-marked walking tracks that extend to the coast and mangroves at Alligator Creek, adjacent to Finch Bay. Cape York specialists Black-winged Monarch (summer) and Tropical Scrubwren occur here on occasion; look at the start of the walking track near the north end of Finch Bay. This trail is also good for Wompoo and Rose-crowned Fruit-Dove, Red-winged Parrot, Blue-winged Kookaburra, Cicadabird (the southern limit of the north-western subspecies), Little Shrike-thrush, Great Bowerbird, Oriental and Channel-billed Cuckoo (both summer), and Dusky, Yellow-spotted, White-throated and Blue-faced Honeyeater. Likely night birds here include Papuan Frogmouth and Large-tailed Nightjar. Mount Cook National Park is within easy walking distance of the town centre, with a foot track leading to the summit of Mount Cook (431 m). Along the track, look for Buff-breasted Paradise-Kingfisher (summer), Grey Goshawk, Noisy Pitta, Lovely Fairy-wren, Macleay's Honeyeater, Tropical Scrubwren, Fairy Gerygone, Yellow-breasted Boatbill, Barred Cuckoo-shrike, Cicadabird and Grey Whistler. On a boat-trip up the Endeavour River, look for Little Kingfisher, Great-billed Heron, Mangrove Robin, Shining Flycatcher, Large-billed and Fairy Gerygone; the best areas are beyond the boat ramp at Marton, in the east arm of the river. Guided tours travel up this part of the river; check at the Information Centre. Along the Keating's Lagoon Track (1.5 km return) and from Keating's Lagoon bird hide, located 5 km south of Cooktown on the Cooktown Developmental Rd, look for waterbirds such as Radjah Shelduck, Green Cotton Pygmy-Goose, Wandering Whistling-Duck, Magpie Goose, Comb-crested Jacana, Black-necked Stork, Brolga and Sarus (rare) Crane. If you have time, it is worth searching for Buff-breasted Button-quail in the drier woodland areas around Cooktown, particularly just before the wet season.

White-streaked Honeyeater, another Cape York speciality, also occurs at Shipton's Flat, to the south of Cooktown. To get there, turn left 26 km south of Cooktown along the Helensvale Bloomfield Rd. After 14.9 km, turn right into the Shipton's Flat Rd. Look in areas of tropical eucalyptus woodland at and beyond the 6 km point

It is estimated that more than half a million Pied Imperial-Pigeons migrate to and from New Guinea to breed in coastal Qld sites each year. From August to March, they are a common sight across large areas of north Qld.

McIvor River Crossing

The area around the McIvor River crossing, 70 km north of Cooktown, is rich in birdlife. Anomalously, there is a range of species, such as Black-winged Monarch (Oct–late Apr; some over-winter), Chestnut-breasted Cuckoo (all year), Tropical Scrubwren and White-streaked Honeyeater, all normally associated with forested areas further north on Cape York. Other birds found at the crossing include Wompoo, Rose-crowned and Superb Fruit-Dove, Red-winged Parrot, Papuan Frogmouth, Azure Kingfisher, Buff-breasted Paradise-Kingfisher (summer), Noisy Pitta, Lovely Fairy-wren, Spectacled Monarch, Shining and Lemon-bellied Flycatcher, Yellow-breasted Boatbill, Grey Whistler and White-browed Robin. This is also an excellent site and the southern limit for the bright backed, pale-cheeked and black-vented Cape York subspecies of the Red-browed Finch (ssp. *minor*) and White-browed Robin. To get to the McIvor River crossing from Cooktown, drive north along the Cooktown Development Rd (the road to Hope Vale). After 36 km, turn left into Battlecamp Rd and travel past Isabella Falls. After 8.2 km, turn right onto McIvor Rd; the McIvor River crossing is another 24 km.

along the road. Other birds in this area include Lovely Fairy-wren, White-browed Robin, Northern Fantail, Graceful, Brown-backed and White-throated Honeyeater, Black Butcherbird, Leaden Flycatcher, Spotted Catbird, Victoria's Riflebird and Mistletoebird. Bennett's Tree Kangaroo and Striped Possum also occur in Shipton's Flat area, while Godman's Rock Wallaby can be found within Black Mountain National Park.

Rinyirru (Lakefield) National Park

Key species: Star Finch, Crimson Finch, Sarus Crane, Eastern Grass Owl, waterbirds
Other species: Yellow-tinted Honeyeater, Masked Finch, Black-throated Finch

Some 340 km north-west of Cairns, Rinyirru (Lakefield) National Park is the largest national park (537 000 ha) on Cape York and the second-largest in Qld. Protecting a diverse landscape, the park contains coastal estuaries, mangroves, mud flats, wetlands, grasslands, woodlands and sandstone hills in the south. Access to Rinyirru National Park is from Cooktown and Laura in the south or Musgrave Station in the north-west. There are several camping-grounds throughout the park, including Kalpowar Crossing and Hanns Crossing Campgrounds. Lakefield Rd (unsealed) runs through the middle of the park, but is impassable through much of the wet season.

The many waterholes and lagoons at Rinyirru National Park provide habitat for large numbers of waterbirds, including Radjah Shelduck, Wandering and Spotted (rare) Whistling-Duck, Green Pygmy-Goose and Pacific Black Duck, Black-winged Stilt, Black-fronted and Red-kneed Dotterel, Comb-crested Jacana,

Vegetated wetlands on Cape York support a great diversity of waterbirds (and numerous crocodiles!). Rinyirru National Park is especially good, with many accessible lagoons that are worth a careful scan.

Nankeen Night-Heron, White-necked, White-faced and Pied Heron, and Cattle, Little, Intermediate and Great Egret. In the southern section of the park, recommended waterholes include Caulders Lake (2WD) and Horseshoe Lagoon (4WD). In the north, Low Lake (access via Lakefield Rd 50 km east of Musgrave Station) is an outstanding birding site. In late spring and summer, the lake is also a good place for shorebirds including Common Greenshank, Marsh Sandpiper and Latham's Snipe (in grassy areas bordering the lake). In the dry season, large numbers of finches drink on the water's edge, particularly in an area 200 m east of the carpark. Species include Star, Masked, Black-throated and Double-barred Finch. Woodland birds found around Low Lake include Pale-headed Rosella, Lemon-bellied and Leaden Flycatcher, Red-backed Fairy-wren and Varied Sittella. Honeyeaters here include Brown, Dusky, Banded, White-throated, Brown-backed, Yellow, Rufous-banded and Blue-faced. Night birds that frequent the national park include Papuan Frogmouth and Large-tailed Nightjar, while Agile Wallaby, Northern Nail-tailed Wallaby and Little Red Flying Fox are three of the more common mammal species present.

The Marina Plains are located in the far north of the park. This is an extensive area of grassy floodplain dotted with termite mounds and stands of Gebang Palm, bordered by mangrove forests along Princess Charlotte Bay. The plains can be accessed via Marina Plains Rd (dry season only), with the road ending at the headwaters of the Annie and North Kennedy Rivers. In the open grassy areas, scan for Brolga, Sarus Crane, Australian Bustard, Emu, Little Curlew (uncommon in summer) and Australian Pratincole. Just on dusk, watch these same grasslands for Eastern Grass Owl; the Marina Plains is thought to be a stronghold for this rare species. The grasslands are also good for Masked, Black-throated, Star and Crimson Finch, and both Zitting and Golden-headed Cisticola occur here. At the mudflats and river banks and the Annie and North Kennedy rivers look for Black Bittern, Striated and Great-billed Heron, Black-necked Stork, White-bellied Sea-Eagle, Eastern Osprey, Brahminy Kite, Yellow White-eye, Large-billed and Mangrove (rare) Gerygone, Collared Kingfisher, Mangrove Robin, Broad-billed and Shining Flycatcher. There is camping at Annie River and Five Mile Creek.

Musgrave Roadhouse

Key species: Golden-shouldered Parrot, Black-backed Butcherbird, Masked Finch, Red Goshawk
Other species: Banded Honeyeater, Bar-breasted Honeyeater, Red-browed Pardalote, Pied Currawong

Musgrave Roadhouse is located 139 km north of Laura on the Peninsula Development Rd. Accommodation at Musgrave Roadhouse includes cabins and camping with barbeque facilities, showers and toilets provided: www.musgraveroadhouse.com.au.

The major target species in this area is the beautiful, yet sadly Endangered, Golden-shouldered Parrot, found only in a small area in central Cape York Peninsula. Your best chance to observe this rare species is during the late dry season when they drink at permanent water sources. A good place to concentrate your search is at the small farm dam near Windmill Creek, 19.5 km south of Musgrave (115 km north of Laura). Another place to look for Golden-shouldered Parrot is at the dam on the east side of the Peninsula Development Rd 26.5 km south of Musgrave (115 km north of Laura). Golden-shouldered Parrot will often feed along the roadside woodlands between these two sites, so keep your ears open and eyes peeled. The other main target species for the area is Black-backed Butcherbird: one of the few Cape York bird endemics that occupies tropical woodland (rather than rainforest). Start looking for them in the woodlands to the north of Laura. They are distinguished from Pied Butcherbird, also common on Cape York, by their white throat and black nape band. The above dam sites and the woodlands along the roadsides can also be good for honeyeaters, with Dusky, Banded, Yellow, White-throated, Bar-breasted, Blue-faced and Yellow-tinted Honeyeater all recorded. Other birds to watch for include Pale-headed Rosella, Red-winged Parrot, White-bellied Cuckoo-shrike, Grey-crowned Babbler, Squatter Pigeon, Red-backed Fairy-wren, White-throated Gerygone, Red-browed and Striated Pardalote, Weebill, Double-barred, Black-throated and Masked Finch, while Pied Currawong is common in the streets of Coen.

From Musgrave Station, travel 17.7 km east along Lilyvale Rd to the point where it turns north. The tall forest along this road (a continuation of Lilyvale Rd), particularly in the first few kilometres after turning north, is a potential site for the rare Red Goshawk. Other birds along the Lilyvale Rd include Collared Sparrowhawk, Red-winged Parrot, Channel-billed Cuckoo, Lemon-bellied Flycatcher, White-throated Gerygone, Little Woodswallow, Black-backed Butcherbird, and Yellow-tinted, Banded, Rufous-throated and Rufous-banded Honeyeater. Some 23 km east of Musgrave Station (5 km past the Lilyvale Rd turn-off), the Lotus Bird Lodge wetland holds a good selection of waterbirds, including Radjah Shelduck, Wandering Whistling-Duck, Green Pygmy-Goose, Black-necked Stork, Magpie Goose, Comb-crested Jacana, Glossy Ibis, Royal Spoonbill, Little Pied Cormorant, Eastern Swamphen and Eurasian Coot. Agile and Northern Nailtail Wallaby also drink at the wetland.

Artemis Station

The area south and east of Musgrave Roadhouse is private property, part of the working cattle-property called Artemis Station. Before entering Artemis Station, you must contact the station owners Tom and Sue Shepherd. The turn-off to the property is 23.5 km south of Musgrave along Dixie Rd. The station entrance is 1 km down the road. In 2007, they were the winners of the Queensland Landcare Conservation Award for their management of Golden-shouldered Parrot habitat. With an excellent knowledge of their local area, they occasionally take visitors out to see the birds. The station has limited accommodation.

Kutini–Payamu (Iron Range) National Park

Key species: Red-cheeked Parrot, Eclectus Parrot, Palm Cockatoo, Yellow-billed Kingfisher, Chestnut-breasted Cuckoo, Red-bellied Pitta (summer), Fawn-breasted Bowerbird, Northern Scrub-robin, Trumpet Manucode, Magnificent Riflebird, Green-backed Honeyeater, White-streaked Honeyeater, Tawny-breasted Honeyeater, White-faced Robin, Frilled-necked Monarch, Black-winged Monarch (summer), Yellow-legged Flycatcher, Tropical Scrubwren
Other species: Spotted Whistling-Duck, Double-eyed Fig-Parrot, Sooty

Tern, Marbled Frogmouth, Buff-breasted Paradise-Kingfisher, Wompoo Fruit-Dove, Superb Fruit-Dove, Rose-crowned Fruit-Dove, White-eared Monarch

Kutini–Payamu (Iron Range) National Park (49 000 ha) is one of Australia's most significant birding sites, protecting the largest area of lowland rainforest in Australia; testament to this is the fact that the above list of key species at Kutini–Payamu is the largest for any site in this book. To get there, approximately 750 km north of Cairns (by road), turn east 20 km north of the Archer River Roadhouse, following this road for another 94 km to the park's entrance. The park is most accessible in the dry and cooler months from June to October, when you can drive in (4WD required). During the wet season, the road into the park often becomes impassable, and the only way to visit is to fly into Lockhart River. Check the RACQ site for up-to-date information on road conditions: www.racq.com.au. Flights into Lockhart River are with SkyTrans Airlines, from Cairns, and there is limited car hire in Lockhart River. Camping is permitted at the Rainforest, Gordon Creek, Cooks Hut and Chilli Beach campgrounds. The only facilities provided are toilets at Chilli Beach and Cooks Hut. There is also bungalow accommodation at Portland Road and Lockhart River Airport, a supermarket in Lockhart River and café in Portland Rd; the park office is 3 km down Lockhart River Rd. Alcohol restrictions apply at the Lockhart River community.

Palm Cockatoos are a charismatic and characteristic Cape York species. Being both large and vocal, they are generally easy to locate. Ecotonal habitats: woodlands in proximity to rainforest and especially stands of the screw palm Pandanus spp. are their preferred habitat.

For rainforest species, the main areas for birding are along Portland Rd, from the Rainforest campground in the west, past Cooks Hut campground, to Gordon Creek campground in the east. One of the more productive approaches to locate rainforest bird species is to birdwatch along the rainforest edges (i.e. along Portland Rd itself). Watch for Eclectus and Red-cheeked Parrot, Double-eyed (Marshall's) Fig-Parrot, Wompoo, Rose-crowned and Superb Fruit-Dove, Chestnut-breasted and Oriental Cuckoo (summer), Trumpet Manucode, Spangled Drongo, Magnificent Riflebird, Yellow-breasted Boatbill, Spectacled, Black-winged (summer), Frill-necked and White-eared Monarch, Tropical Scrubwren, Little Shrike-thrush, Rufous Fantail, Tawny-breasted and Green-backed Honeyeater, Grey Whistler and White-faced Robin. Another good area of edge habitat is along the eastern end of the Old Coen Track (between the creek bed and the information sign and gate); the track starts at the Rainforest camping area, just west of Cooks Hut (10 km return). While birding the rainforest edge can be productive, some species are best seen by entering the rainforest; one such area is immediately behind the toilet block at the Cooks Hut camping area. This can be a reliable site for Northern Scrub-robin, while also providing opportunities for Red-bellied (summer) and Noisy Pitta,

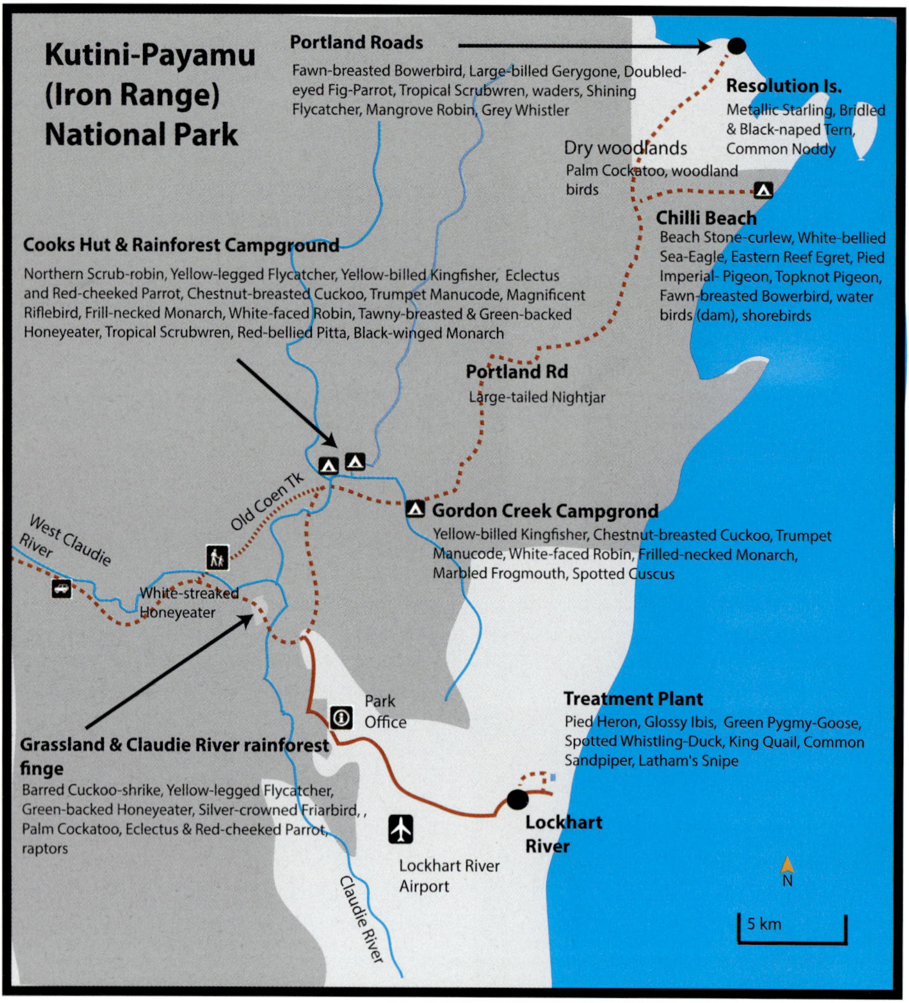

Yellow-legged Flycatcher, Buff-breasted Paradise-Kingfisher (Nov–Apr) and Yellow-billed Kingfisher.

Around the Gordon Creek campsite, look for Australian Brush-turkey (violet-wattled ssp. *purpureicollis*), Double-eyed Fig-Parrot (sometimes seen feeding in fig tree that overhang the west side of the campground), Chestnut-breasted Cuckoo, Trumpet Manucode, White-faced Robin, Frilled-necked Monarch and Tropical Scrubwren. A walk through the gallery forest along Gordon Creek is probably the best way to track down Yellow-billed Kingfisher. Other birds in this area of rainforest include Red-cheeked and Eclectus Parrot, Spotted Catbird, Magnificent Riflebird, Noisy and Red-bellied Pitta (summer), Yellow-breasted Boatbill, Green-backed Honeyeater and Yellow-legged Flycatcher. At night, the Gordon Creek camping area is a good site for Marbled Frogmouth, Large-billed Nightjar and Rufous Owl, as well as nocturnal rainforest

Claudia River, Iron Range. This is an excellent site for a range of birds because the forest is a little more open around the bridge. Among others, look here for Eclectus and Red-cheeked Parrot, Tropical Scrubwren, Green-backed Honeyeater, White-faced Robin, Yellow-legged Flycatcher, Frilled-necked Monarch and Trumpet Manucode.

mammals such as Striped Possum, Spotted and Common Cuscus and Cape York Melomys.

Immediately west of the Claudie River Bridge, there is good area of grasslands with a few scattered trees. Australian Swiftlet often hawks overhead here, while the grassy areas support Golden-headed Cisticola, Red-backed Fairy-wren and Red-browed Finch. Around the forest fringes look for Forest Kingfisher, Barred Cuckoo-shrike, Dollarbird, Rainbow Bee-eater, Yellow-legged Flycatcher, Fairy Gerygone, Brown-backed and Green-backed Honeyeater, Silver-crowned Friarbird and Mistletoebird. The open grasslands also offer a convenient area to view the larger rainforest species flying between areas of rainforest, such as Palm Cockatoo, Eclectus and Red-cheeked Parrot and Channel-billed Cuckoo (summer), and raptors such as Grey Goshawk (both white and grey morph are present in park), Pacific Baza, Little Eagle and, rarely, Red Goshawk. About 1.5 km west of the Claudie River Bridge, the road turns south-west: look here along the south side of the road for White-streaked Honeyeater, especially when the eucalypts are flowering.

It is worth spending some time birding around the Portland Roads township close to the coast. In the town's gardens, you can see Large-billed Gerygone, Double-eyed Fig-Parrot, Olive-backed Sunbird, Tropical Scrubwren, and Graceful, Tawny-breasted and Yellow-spotted Honeyeater. Some of the birds found in mangroves immediately north of the town include Broad-billed and Shining Flycatcher, Mangrove Robin, Grey Whistler, Large-billed Gerygone and Red-headed and Varied Honeyeater. Between the mangroves and the open woodland, Fawn-breasted Bowerbird regularly fly across the road ~200 m from the town. Other birds to be seen along the roadside include Lovely Fairy-wren, Red-browed Finch, White-throated Honeyeater and Rose-crowned and Superb Fruit-Dove. At low tide, search the mud flats directly in front of the town for Striated Heron, Eastern Reef Egret, Beach Stone-curlew, Whimbrel, Common Sandpiper and Grey-tailed Tattler, Collared and Little (uncommon) Kingfisher, and Crested and Lesser Crested Terns. Further out, check for seabirds such as Brown Booby, Great and Lesser Frigatebird, Common Noddy and Bridled Tern. Nocturnal species in the general area include Large-tailed Nightjar, Marbled and Papuan Frogmouth and Rufous Owl. Estuarine Crocodiles are an ever-present feature of the coastline and many inland water bodies on Cape York; with a bright torch, check for their brilliant orange eye-shine in the waters right in front of the town. Further north along the same road, in woodland ~5 km from town, look for Palm Cockatoo, Varied (Striated) Sittella, Great Bowerbird and Oriental Cuckoo (summer). At night, Large-tailed and White-throated Nightjar are regularly flushed up from the road when driving along Portland Rd, particularly between the Portland Roads township and the turn-off to Chilli Beach. Also at night, Bare-backed Fruit Bat and Spectacled Flying-Fox feed in the town gardens at Portland Rd.

The only member of the Nectariniidae family in Australia, the Olive-backed Sunbird is common around rainforest edges, mangroves and gardens. Males (like this one) show a strikingly iridescent blue-black throat and breast.

The Lockhart River treatment plant is located just north-east of the Lockhart River township. To get there from Lockhart River, take Piiramo Rd towards Quintell Beach and, after 500 m (from the edge of the town), turn left into Kattini St; this is a well-defined track and the plant is a further 500 m along here. Look for Pied Heron, Cattle Egret, Glossy Ibis, Australasian Grebe, Green Pygmy-Goose, Spotted Whistling-Duck (several records), King Quail (uncommon) and Australian Swiftlet. In summer, Common Sandpiper and Latham's Snipe may also be present. In the Lockhart River township, look for Red-browed Finch, Dollarbird, Rainbow Bee-eater and Australian Swiftlet.

From Chilli Beach, look out to sea for Bridled and Black-naped Tern and Common Noddy, especially circling over Ma'alpiku (Restoration) Island and Old Man Rock ~700 m off the coast. During the warmer months at dusk, large numbers of Metallic Starling flock off Chilli Beach and Restoration Island. Other birds that occur here include Lesser and Great Frigatebird, Australian Pied Oystercatcher, Beach Stone-curlew, White-bellied Sea-Eagle, Eastern Reef Egret, Pied Imperial-Pigeon and Topknot Pigeon. Shorebirds that frequent the area including Pacific Golden Plover, Lesser and Greater Sand Plover, Grey-tailed Tattler and Common Sandpiper. At night, Papuan Frogmouth and Large-tailed Nightjar may also be found. Just before you reach the Chilli Beach campsite, there is a small dam on the east side of the road 150 m from the intersection that is worth a look: Black Bittern, Little and Azure Kingfisher, Green Pygmy-Goose and Spotted Whistling-Duck (rare) have all been recorded here. Fawn-breasted Bowerbird occurs on the north-west side of the intersection of Cape Weymouth Rd and Chilli Beach Rd.

Weipa

Key species: Spotted Whistling-Duck, Lesser Frigatebird, Great Frigatebird, Red-headed Honeyeater, Great-billed Heron
Other species: Brolga, Sarus Crane, Black-necked Stork, Pied Heron, Radjah Shelduck

The habitat surrounding Weipa, the largest town on Cape York Peninsula, includes vine forest and woodlands, wetlands, mangroves and coastal dunes. Access to the town is via the Peninsula Development Rd, 816 km north of Cairns. Weipa is also serviced daily by air from Cairns. Facilities include a hotel, motel and caravan park, and there is bush camping at nearby Mapoon, and Wenlock and Pennefather Rivers.

Weipa is the only known Australia mainland roost for Lesser and Great Frigatebird: in the evening, huge numbers of birds fly over the town to roost in trees between Evans Landing and the Weipa port; this is a spectacle that no birdwatcher should miss. There are numerous wetlands and water ponds around the township, usually supporting good numbers of waterbirds, such as Brolga and Sarus Crane (uncommon), Black-necked Stork, Pied Heron, Radjah Shelduck, Wandering Whistling Duck, Grey Teal, Pink-eared Duck, Garganey (rare vagrant) and Green Pygmy-Goose. Weipa also has the distinction of being the only reliable place in Australia for Spotted Whistling Duck: a colonist from PNG first detected in Australia in the mid 1990s, they are now present in Weipa most of the time. Places to look include the small ponds on Land St near the corner of McLeod Drive and Central Ave, the Comalco ponds, the pond on Kerr Point Rd and the Awonga Point Treatment Plant on Andoom Rd, 2 km north of the Rocky Point.

Uningan Nature Reserve, on Andoom Rd 1 km north of Rocky Point, covers an area of over 2800 ha and includes habitats such as open woodland, palm forests and mangroves. The 9 km of walking tracks and footbridges constructed over the creeks in the area provide an opportunity to observe mangrove species such as Collared Kingfisher, Broad-billed and

Large areas of central Cape York, and especially around Weipa, support tall Darwin Stringybark *Eucalyptus tetrodonta* woodlands that grow on low-nutrient soils. Species that occupy this habitat include the rare Red Goshawk, as well as Palm Cockatoo, Rainbow Lorikeet, Blue-winged Kookaburra, Brown (Black) Treecreeper, Banded, Brown-backed, and White-throated Honeyeater and Black-backed Butcherbird.

Shining Flycatcher, Grey and Mangrove Golden (rare) Whistler, Mangrove and Large-billed Gerygone and Red-headed Honeyeater. Along the waterways, look for Nankeen Night-Heron, Black Bittern, Great-billed Heron, Yellow-billed, Azure and Little Kingfisher. Around Uningan, the woodlands, dominated by Darwin Stringybark, provide habitat for species such as Palm Cockatoo, Pale-headed Rosella, Red-backed Kingfisher, Laughing and Blue-winged Kookaburra, Lovely Fairy-wren, Yellow Oriole, Great Bowerbird, Black, Pied and Black-backed Butcherbird, Brown (Black) Treecreeper, and Banded, Tawny-breasted, Brown-backed, Graceful and White-throated Honeyeater. At night, Southern Boobook, Spotted Nightjar, Australian Owlet-Nightjar and Papuan and Tawny Frogmouth all forage within the reserve.

Lockerbie Scrub and the tip of Cape York

Key species: Pale White-eye, Palm Cockatoo, Trumpet Manucode, Magnificent Riflebird, Yellow-billed Kingfisher, White-streaked Honeyeater, Red-bellied Pitta, Black-winged Monarch, Frilled-necked Monarch, Northern Scrub-Robin, Fawn-breasted Bowerbird, Tropical Scrubwren
Other species: Mangrove Golden Whistler, Beach Stone-curlew

The tip of Cape York, some 950 km north of Cairns, offers some of the more interesting

Wenlock and Pennefather Rivers

The forest and vine-thickets along the Wenlock River at Batavia Outstation Landing (78 km north of Weipa on the Mapoon Rd) provides suitable habitat for Cape York specialists such as Yellow-billed Kingfisher, Frilled-necked Monarch (rare on the west side of Cape York), Magnificent Riflebird, Tropical Scrubwren, Yellow-breasted Boatbill and Trumpet Manucode. Search the riparian vegetation along the river for Little and Azure Kingfisher, and Great-billed Heron. In the surrounding woodlands, look for Fairy Gerygone, Tawny-breasted Honeyeater, White-bellied Cuckoo-shrike, Cicadabird, Brown (Black) Treecreeper, Spangled Drongo, Leaden and Lemon-bellied Flycatcher, Lovely Fairy-wren, Little (Gould's) Bronze-Cuckoo and Northern Fantail. In the evening, Large-tailed Nightjar and Papuan Frogmouth are sometimes seen, and the tree-dwelling Common Spotted Cuscus is found in the area. Riparian vegetation along the Pennefather River, west of the Batavia Outstation Landing, can be good for Great-billed Heron and Little Kingfisher. Beach Stone-curlew, Australian Pied Oystercatcher, Striated Heron, Eastern Reef Egret and Little Tern feed near the mouth of the river, where there is an attractive camping area.

and varied birding on Cape York. Aside from the Cape York specialties such as Palm Cockatoo, Trumpet Manucode, Magnificent Riflebird, Frilled-necked Monarch, Yellow-billed Kingfisher, White-streaked Honeyeater and Red-bellied Pitta (summer), the tip also provides an opportunity for Pale White-eye and Mangrove Golden Whistler on the offshore Torres Strait islands. Access in the wet season (Dec–Apr) is by air or sea only, while at other times 4WD is essential. There is a range of accommodation available, but services, fuel and supplies can be limited so travellers should be self-sufficient. Fishing charter-boats and water taxi services from Thursday Island may offer trips to nearby small islands in Torres Strait to search for Pale White-eye and other species.

The Lockerbie Scrub, located on Pajinka Rd, is an excellent area for many of the rainforest specialties; to get there from Bamaga, head north towards the Tip. Lockerbie (the ruins of the old Jardine Outstation) is 17 km from Bamaga opposite the turn-off to Punsand Bay. Several kilometres later, the track passes through the rainforest area known as the Lockerbie Scrub. This significant area of rainforest is 25 km long and 1–5 km wide and is the most northerly area of monsoonal rainforest in Australia. As a result, it acts as a corridor for migratory land birds moving between Australia and New Guinea via the Torres Strait. There are several good walks here, with the best being the Roma Flats walk. Interesting species recorded include Palm Cockatoo, Yellow-billed Kingfisher, Trumpet Manucode, Magnificent Riflebird, Noisy Pitta, Spectacled and Frilled-necked and Monarch, Shining and Yellow-legged Flycatcher, Yellow-breasted Boatbill, Northern Scrub-Robin, Lovely Fairy-wren, White-faced Robin, Tropical Scrubwren, Grey Goshawk, Australian Brush-turkey, Red-browed Finch and Fairy Gerygone. In summer, additional species to look for include Buff-breasted Paradise-Kingfisher, Red-bellied Pitta, Black-winged Monarch and Chestnut-breasted Cuckoo. Honeyeaters here include White-streaked, Tawny-breasted, Brown-backed, Graceful and Yellow-spotted Honeyeater, while pigeons and doves are also well represented, with regular records of Emerald Dove, Brown Cuckoo-Dove, and Wompoo, Superb and Rose-crowned Fruit-Dove. Red-necked Crake and King Quail occur in wet heathland areas within the scrub. At night, look and listen for Marbled and Papuan Frogmouth, Barking Owl and Large-tailed Nightjar. The area also supports some great mammals, including Sugar Glider, Common Spotted Cuscus (sometimes observed in roadside trees during the day) and Rufous Spiny Bandicoot.

White-streaked Honeyeater and Fawn-breasted Bowerbird occur south of Bamaga at Cowal Creek on Mutee Heads Rd (4.5 km west of its intersection with the Peninsula Rd). Both species are also found along the Jardine River, a further 7.8 km west of Cowal Creek on Pump Station Rd. In this same area, along the river banks, within the gallery forest and adjacent woodland, species include Black Bittern, Varied Triller, Spangled Drongo, Olive-backed Oriole, Black-backed

The Pale White-eye is a small island specialist restricted to islands in Torres Strait and the adjacent northern Great Barrier Reef. The most accessible site is Little Tuesday Island just north of Thursday Island.

Little Tuesday Island, southern Torres Strait. Access for day visits is via water taxi from Thursday or Horn Islands: contact the Torres Strait Island Regional Authority for permission. In addition to Pale White-eye, look for Beach Stone-curlew, Rose-crowned Fruit-Dove, Noisy Pitta, Dusky Honeyeater, Red-headed Honeyeater, Spectacled Monarch and Mangrove Golden Whistler.

Jardine River National Park

On your way to 'The Tip', it is worth detouring into Jardine River National Park (330 000 ha). In rainforest sections on the road to Captain Billy Landing, stop and look for Palm Cockatoo, Wompoo and Rose-crowned Fruit-Dove, Magnificent Riflebird, White-faced Robin, Tawny-breasted Honeyeater, Chestnut-breasted Cuckoo, Yellow-billed Kingfisher, Yellow-breasted Boatbill, Yellow-legged Flycatcher, Black Butcherbird, Little (Gould's) Bronze-Cuckoo, Trumpet Manucode, Azure Kingfisher, Papuan Frogmouth and Lovely Fairy-wren. Little Tern sometimes nest on the beach at Captain Billy Landing; also look here for Great and Lesser Frigatebird, especially in the late afternoon as they drift along the coastline overhead. In the western section of the park, White-streaked Honeyeater occurs at both Fruit Bat Falls and the Elliott Falls camping area, which is accessible from the Old Telegraph Track. Along the road to Ussher Point in the northern section of the park, look for Tropical Scrubwren, White-streaked and Tawny-breasted Honeyeater, Frill-necked Monarch, Yellow-billed Kingfisher and Lovely Fairy-wren.

Butcherbird, and Brown-backed, Tawny-breasted, Banded, Yellow, Yellow-spotted, White-throated and Graceful Honeyeater, and Silver-crowned Friarbird.

The other main target species for the tip of Cape York is Pale White-eye. They are found on many of the southern Torres Strait Islands (just off the coast of mainland Australia), particularly Mount Adolphus Is, Little Woody Is (just west of Possession Is) and Little Tuesday Is (possibly the most reliable site). All of these sites are accessible by water taxi from Horn Island (the trip takes about an hour), but most visitors to the tip of Cape York will need to first catch one of the ferry services to Thursday and Horn Island to connect with a pre-arranged water taxi or charter service. Aside from Pale White-eye on Little Tuesday Is, look for Mangrove Golden Whistler, Beach Stone-curlew, Pacific Golden Plover, Rose-crowned Fruit-Dove, Rufous Fantail (seasonal migrant), Spectacled Monarch, Noisy Pitta, and Dusky and Red-headed Honeyeater. On the way to the islands, look for seabirds such as Great and Lesser Frigatebird, Brown Booby, Lesser Crested, Crested, Common, Black-naped Tern, and Common and Black Noddy.

North Queensland

Regional overview

From Cooktown in the north to Townsville and Ayr in the south, the north Qld region boasts one of the largest avifauna lists of any region in Australia. This is in part because of the wide variety of habitats, such as riverine, wetland, woodland, lowland and upland rainforest, grassland and agricultural areas. Located on the coast, Cairns is the second largest city in the region, with a population of ~140 000. Although it is a regional town in its own right, facilities are also heavily focussed towards the tourist industry, meaning the area is very well serviced for visiting birdwatchers. Most of the 430 species recorded for the region occur on the eastern side of the Great Dividing Range; with markedly lower rainfall, the western slopes of the Great Dividing Range supports several different species. On a birdwatching trip around Cairns, the Atherton Tableland and the Daintree, you can expect to see in excess of 200 bird species. You can achieve most site visits with day trips from any of these centres.

In order to experience the birds of the region, included below are three itineraries. These can be undertaken individually, or combined to form a longer more extensive

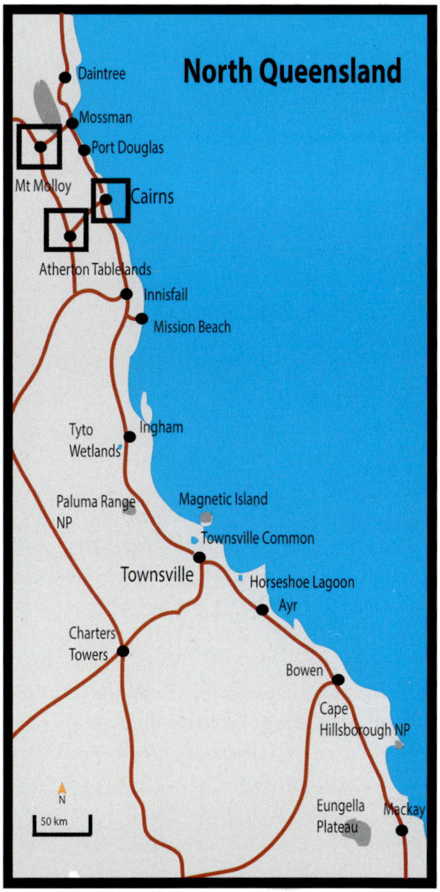

Queensland

trip. A recommended 3-day itinerary for Cairns involves:

- **Day 1.** Cairns Esplanade, Cairns Botanical Gardens, Centenary Lakes (Beach Stone-curlew, Broad-billed Sandpiper, Collared Kingfisher, Double-eyed Fig-Parrot, Varied Honeyeater, Mangrove Robin, Papuan Frogmouth, Little Kingfisher, Red-necked Crake, Lovely Fairy-wren, shorebirds).
- **Day 2.** Yorkeys Knob, Smithfield Treatment Plant, Cattana Wetlands, Cassowary House, Kuranda (Wandering Whistling-Duck, Plumed Whistling-Duck, Green Pygmy-Goose, Crimson Finch, Southern Cassowary, Pale-vented Bush-hen, Red-necked and White-browed Crake, Pied and White-eared Monarch).
- **Day 3.** Michaelmas Cay (Common Noddy, Black Noddy, Sooty, Bridled, Lesser Crested, Black-naped and Roseate Tern, Lesser and Great Frigatebird, seabirds and Wandering Tattler).

A recommended 5-day itinerary for areas to the north of Cairns involves:

- **Days 1–2.** Cairns to Kingfisher Park Birdwatchers Lodge, 90 km (Buff-breasted Paradise-Kingfisher (summer), Red-necked Crake, Noisy Pitta, Macleay's Honeyeater, Victoria's Riflebird, Papuan Frogmouth, Lesser Sooty Owl, Pale-yellow Robin).
- **Day 3.** (Staying in Kingfisher Park Birdwatchers Lodge) Visit Abattoir Swamp, Mount Molloy and Mount Carbine, 40 km (Squatter Pigeon, Australian Bustard, Northern Fantail, White-browed Robin, White-browed Crake, Lemon-bellied Flycatcher, the distinctive northern subspecies of Brown Treecreeper, Northern Fantail and Blue-faced Parrot-Finch (Jul–Oct)).
- **Day 4.** (Staying at Kingfisher Park Birdwatchers Lodge) Visit Newell Beach, Mossman River, Daintree, 60 km (Barn and Red-rumped Swallow: wet season only), Great-billed Heron, Beach Stone-curlew, Little Kingfisher, Black Bittern, Lovely Fairy-wren, Southern Cassowary, Rufous Owl, Pied Monarch).
- **Day 5.** (Staying at Kingfisher Park Birdwatchers Lodge) Visit Mount Lewis, 30 km (Golden Bowerbird, Tooth-billed Bowerbird, Chowchilla, Blue-faced Parrot-Finch (summer), Mountain Thornbill, Atherton Scrubwren, Bridled and White-cheeked Honeyeater, Fernwren, Grey-headed Robin, Bower's Shrike-thrush).

A 6-day itinerary for the Atherton Tableland and Mission Beach involves:

- **Day 1.** Cairns to Crater Lakes National Park and Yungaburra, 60 km (Atherton Scrubwren, Grey-headed Robin, Mountain Thornbill, Chowchilla, Pied Monarch, White-eared Monarch, Fernwren, Bower's Shrike-thrush, Yellow-breasted Boatbill).
- **Day 2.** Yungaburra to Curtain Fig Tree, Hasties Swamp National Park, 30+ km (Sarus Crane, Cotton Pygmy-Goose, Yellow-breasted Boatbill, Pied Monarch, Mountain Thornbill).
- **Day 3.** Mareeba Wetlands, Tinaroo Creek Rd, Big Mitchell Creek, Lake Mitchell, 50+ km (Buff-breasted Button-quail (slim chance), Green Pygmy-Goose, Black-necked Stork, Sarus Crane, Squatter Pigeon, White-browed Robin, Northern Fantail, the distinctive subspecies of Brown Treecreeper, Black-throated Finch).
- **Days 4–5.** Atherton Tableland to Mission Beach, 155 km (Southern Cassowary, Pale-vented Bush-hen, Yellow-breasted Boatbill).
- **Day 6.** Mission Beach to Cairns, 140 km.

North Qld has a tropical climate with distinct seasons: a 'winter' period of mild, pleasant temperatures and low rainfall (the dry season), and a 'summer' period of warm temperatures with increased rainfall (the wet season). In the hotter and wetter months (Oct–Apr), migratory species such as Buff-breasted Paradise-Kingfisher, Oriental Cuckoo, Channel-billed Cuckoo, Eastern Koel and Metallic Starling are present, as are many more migratory shorebirds. This is also the main breeding period for many local species, with some such as Noisy Pitta and Golden Bowerbird easier to observe at this time. In the cooler and drier, winter months (May–Sep), Brolga and Sarus Crane flock to the Tableland, feeding in the agricultural fields and roosting at a small number of key sites. At this time, winter-breeding species, such as White-eared Monarch can be easier to find and Victoria's Riflebird starts displaying. Some species from the high upland rainforests, such as Bridled Honeyeater, Grey-headed Robin, Fernwren and Bower's Shrike-thrush also move to lower altitudes at this time.

Most roads in the region are good, easily reached with 2WD, but a few areas (such as Mount Lewis and Shipton's Flat) require 4WD. Fuel is readily available along the main highways, with many service stations open 24 h a day. There are large numbers of hotels, motels and caravan parks scattered throughout north Qld. For information on accommodation see the Cairns eguide at www.cairnseguide.com or contact the Cairns Tourist Information Centre on the Esplanade, (07) 4047 9125.

Birding highlights

North Qld is a major destination for both national and international birdwatchers, with over half Australia's bird species occurring in the region. There are 13 bird species restricted to the Cairns and Wet Tropics region. Those that occupy high-altitude rainforests include Golden and Tooth-billed Bowerbird, Australian Fernwren, Atherton Scrubwren, Chowchilla, Mountain Thornbill, Bower's Shrike-thrush and Grey-headed Robin, while Lesser Sooty Owl, Bridled Honeyeater and Victoria's Riflebird occur more widely. Pied Monarch is largely absent from the lowlands and the highest altitudes while Macleay's Honeyeater occurs throughout.

The distribution of several other bird species is centred on north Qld, meaning that many of these are best seen here. These include Buff-breasted Paradise-Kingfisher (Oct–Mar), Sarus Crane, Great-billed Heron, Red-necked and White-browed Crake, Little Kingfisher, Buff-breasted Button-quail, Papuan Frogmouth, Australian Swiftlet, Lovely Fairy-wren, Yellow White-eye, Zitting Cisticola, Graceful, Brown-backed, Yellow-spotted and Yellow Honeyeater, Hornbill Friarbird, Yellow-breasted Boatbill, Blue-faced Parrot-Finch and White-browed Robin. Yellow White-eye, Zitting Cisticola, and Mangrove Golden Whistler (rare) are three species that occur around the northern Australian coastline and extend to near Ayr, south of Cairns. Distinct subspecies found in the area include Southern Boobook ('Red Boobook' ssp. *lurida*), Double-eyed Fig-Parrot ('Macleay's Fig-Parrot' ssp. *macleayana*), Crimson Rosella (ssp. *nigrescens*), Little Bronze-Cuckoo ('Gould's Bronze-Cuckoo' ssp. *russatus*), White-throated Treecreeper ('Little Treecreeper' ssp. *minor*), Yellow-throated Scrubwren (ssp. *cairnsi*), Yellow Thornbill (ssp. *flava*), Eastern Spinebill (ssp. *cairnsensis*), Noisy Friarbird (ssp. *corniculatus*), Pale-yellow Robin (ssp. *nana*), Brown Gerygone (ssp. *mouki*), Grey Fantail (ssp. *keasi*), Satin Bowerbird (ssp. *minor*), Spotted Catbird (ssp. *melonotis*), Bassian Thrush (ssp. *cuneata*), Varied Sittella ('Striated Sittella' ssp. *striata*), Brown Treecreeper ('Black Treecreeper' ssp. *melanotus*), Bassian Thrush (ssp. *cuneata*) and Grey Fantail ('Mountain Fantail' ssp. *keasti*). The list of rare and vagrant species recorded from north Qld is also extensive: it includes Dunlin, Common Redshank, Little Stint, Ruff, Little Ringed Plover, Grey Phalarope, Garganey, Red-footed Booby, Black-tailed, Franklin's, Slaty-backed and

Laughing Gull, White Tern, House Swift, Glossy Swiftlet, Collared Imperial-Pigeon, Grey Wagtail, White Wagtail, Oriental Reed Warbler, House Crow, Eye-browed Thrush and Isabelline Wheatear.

Cairns Esplanade

Key species: Beach Stone-curlew, Broad-billed Sandpiper, Collared Kingfisher, Double-eyed Fig-Parrot, Varied Honeyeater, Mangrove Robin

Other species: Whimbrel, Grey-tailed Tattler, Terek Sandpiper, Lesser and Greater Sand Plover, Grey Plover, other shorebirds, Little Tern, Shining Flycatcher

A focal point for both locals and visiting birdwatchers, the mud flats adjacent to the Cairns Esplanade are an internationally important habitat for resident and migratory shorebirds. Although several species are present year round, the best time for shorebirds is between August and April, with peaks between August and December, and again between March and April when species that winter further south are passing through. Best viewing is on an incoming tide, particularly in the evening when the sun is setting behind the observer. Tide charts are available from Cairns Tourist Information Centre: 51 The Esplanade, Cairns.

The boardwalk and viewing platforms along the Esplanade offer an excellent vantage point to watch large numbers of birds that are accustomed, and often indifferent to, the nearby observers. Some of the species that are present year round include Beach Stone-curlew, Australian Pied Oystercatcher, Eastern Reef Egret, Striated Heron, Eastern Curlew and Whimbrel. Spring–summer migrants include Pacific Golden and Grey Plover, Lesser and Greater Sand Plover, Black-tailed and Bar-tailed Godwit, Grey-tailed Tattler, Common Greenshank, Ruddy Turnstone, Great and Red Knot, Red-necked Stint and Marsh, Wood, Terek, Sharp-tailed and Curlew Sandpiper. Broad-billed Sandpiper and Asian Dowitcher turn up with sufficient frequency that it is worth checking the flocks of shorebirds to see if any are present. Uncommon shorebirds to look out for include Australian Pratincole, Little Curlew, Wandering Tattler, Sanderling, Oriental Plover, Ruff and Pectoral Sandpiper. A remarkable 15 species of terns have been recorded – you have a good chance of observing Little and Caspian Tern (both summer), Gull-billed Tern (Australian subspecies *macrotarsa* and summer-visiting Asian subspecies *affinis*) and Crested Tern, while less common species include Lesser Crested, Black-naped, Whiskered and White-winged Black Tern. Vagrant species recorded on the Esplanade (most recent sighting bracketed) include Common Redshank (April 1991), Dunlin (Jan 1983), Little Stint (Dec 1996), Long-toed Stint (Sept 2006), Laughing Gull (May 2010), Franklin's Gull (Nov 1993), Black-tailed Gull (April 2013), Barn Swallow (Nov 2006) and Eastern Yellow Wagtail (Nov 1987).

The foreshore trees and gardens bordering the length of the Esplanade provide further excellent birding opportunities, and again many of these species are relatively tame. Expect to find Varied and Yellow Honeyeater, Double-eyed Fig-Parrot, Rainbow and Scaly-breasted Lorikeet, Pied Imperial-Pigeon (Aug–May), Australian Swiftlet, Hornbill Friarbird, Spangled Drongo, White-breasted Woodswallow, Yellow Oriole, Australasian Figbird and Metallic Starling (Aug–May, but many present year round). Raptors that patrol the mud flats and perch on nearby pylons include Eastern Osprey, Brahminy Kite, White-bellied Sea-Eagle, Peregrine Falcon and Australian Hobby. It is also worth investigating the mangroves at the northern end of the Esplanade where you may find Collared Kingfisher, Mangrove Robin,

Metallic Starlings are a common migrant of rainforest edges, riparian habitats and gardens. Large numbers are present throughout the wet season. The period of absence from the Wet Tropics is quite short, as the last birds usually leave by early May and the first birds return by early August.

Shining Flycatcher, Black Butcherbird and Large-billed Gerygone.

Cairns Botanical Gardens and Centenary Lakes

Key species: Papuan Frogmouth, Little Kingfisher, Great-billed Heron (chance), Lovely Fairy-wren
Other species: Rufous Owl, Black Butcherbird, Pacific Baza, Bush Stone-curlew, honeyeaters, Cicadabird.

The Cairns Botanical Gardens (also known as the Flecker Botanic Gardens) are accessed via Greenslopes St from the south and Collins Ave from the north. In this area, which is rich in birdlife, there is a good selection of rainforest, botanic gardens, the two Centenary Lakes (one freshwater, the other salt) and the mangrove-lined Saltwater Creek, which flows through its centre. One excellent walk is the rainforest boardwalk (500 m), which passes through lowland swamp forest and melaleuca wetlands, connecting Flecker Botanic Gardens with the two lakes and creek. Adjacent to the gardens on the north side, the slopes of Mount Whitfield Conservation Park protects the closest rainforest to Cairns.

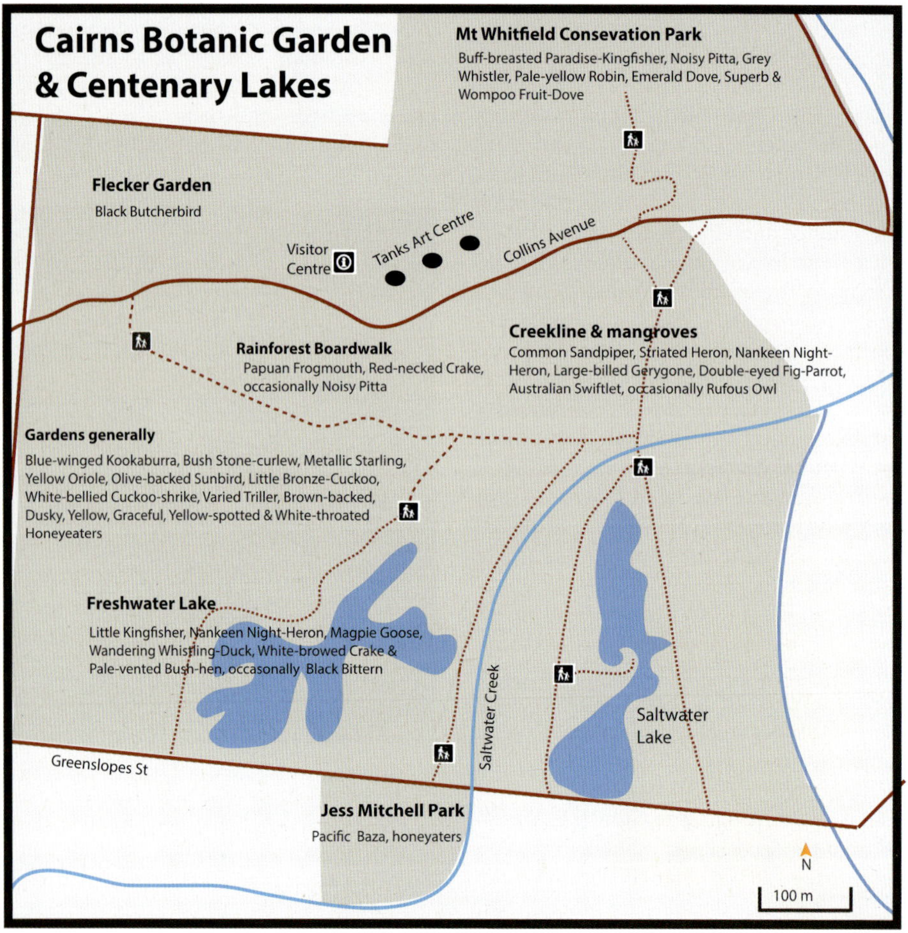

Bird lists and maps are available from the visitor centre on Collins St.

During the day, Papuan Frogmouth roost in trees along the Rainforest Boardwalk; they often roost near the third turn on the walk (when entering from the eastern end), so look in the trees ~30 feet up on the north side of the boardwalk. They also roost near

Jack Barnes Mangrove Boardwalk

The carpark for this excellent mangrove boardwalk is 1 km south of Cairns Airport on the east side of Airport Ave. There are two different mangrove walks. The Northern Boardwalk (700 m loop) passes through low mangroves to a canopy-tower that provides views across the treetops, and then to a viewing-platform on Middle Creek. The Southern Boardwalk (1.2 km return) passes through a tall and open mangrove forest to a canopy-tower near the mouth of Swampy Creek. Along both walks, look for mangrove specialists such as Shining Flycatcher, Mangrove Robin, Large-billed Gerygone, Little and Collared Kingfisher, Striated Heron, Black Butcherbird, Varied Honeyeater, Hornbill Friarbird, Lovely Fairy-wren (especially the low dense mangroves) and Olive-backed Sunbird. Saltwater Crocodile occasionally sun themselves on the banks of Middle Creek.

Mangrove habitats are generally 'under-visited' by birdwatchers because of difficult access. As well as several excellent mangrove boardwalks in the Daintree region, dedicated boat-based tours on both the Daintree and Mossman Rivers mean the Wet Tropics provides some excellent mangrove-birding opportunities.

the footbridge over Saltwater Creek. Other birds to look for along the boardwalk include Red-necked Crake, Little Kingfisher and, occasionally, Noisy Pitta, with the best times being soon after dawn or early in the evening.

Look along the creek for Common Sandpiper (summer), Striated Heron, Nankeen Night-Heron, Large-billed Gerygone, Little Kingfisher, Collared Kingfisher, Lovely Fairy-wren, Double-eyed Fig-Parrot, Australian Swiftlet and, occasionally, roosting Rufous Owl. There is also a chance of the much sought-after Great-billed Heron in this area. Freshwater Lake supports a range of interesting species, including Little Kingfisher (check low hanging twigs and branches in more shaded area of the lake), Black Bittern (uncommon), Nankeen Night-Heron, Magpie Goose, Wandering Whistling-Duck, Radjah Shelduck, Comb-crested Jacana and Australasian Darter, while White-browed Crake and Pale-vented Bush-hen sometimes feed around the vegetated margins of the lakes. A wide range of honeyeaters are found in the gardens, including Brown-backed, Dusky, Yellow, Graceful, Yellow-spotted and White-throated Honeyeater: most seek out flowering shrubs, such as *Grevillea* and *Melaleuca*. Other birds in the gardens include Blue-winged Kookaburra, Bush Stone-curlew, Metallic Starling, Yellow Oriole, Olive-backed Sunbird, Little Bronze-Cuckoo, White-bellied Cuckoo-shrike, Cicadabird and Varied Triller. Black Butcherbird regularly frequent the gardens on the north side of Collins Ave. Across Greenslopes Rd, it is worth investigating the gums in Jess Mitchell Park, where Pacific Baza occasionally roosts.

Within Mount Whitfield Conservation Park, there are two circuit walks: Red Arrow Circuit (1.5 km), and Blue Arrow Circuit (5.4 km). Both begin near the Tanks Art Centre on Collins Ave and both are good for birdwatching. Look along the trails for Buff-breasted Paradise-Kingfisher (a summer breeding site for this species), Noisy Pitta, Grey Whistler, Pale-yellow Robin, Lovely and Red-backed Fairy-wren, Yellow Oriole, Spangled Drongo, Brush Cuckoo, Emerald Dove and Superb and Wompoo Fruit-Dove.

Yorkeys Knob and Cattana Wetlands

Key species: Wandering Whistling-Duck, Plumed Whistling-Duck, Green Pygmy-Goose, Crimson Finch, White-browed Crake
Other species: Radjah Shelduck, Mangrove Robin, Collared Kingfisher, Shining Flycatcher, Restless Flycatcher

There are several excellent wetlands located north of Cairns including the Half Moon Bay Golf Course Wetland at Yorkeys Knob, Cattana Wetlands, Smithfield Treatment Plant and the Skyrail Wetland. Yorkeys Knob is 17 km north-west of Cairns. Take the Captain Cook Hwy north for 11.4 km before turning right into Yorkeys Knob Rd and after 5.2 km turn left into Wattle St. The Golf Course Lagoon is 500 m down this road. Cattana Wetland is off Dunne Rd in Smithfield, accessed by either Yorkeys Knob Rd or McGregor Rd.

Waterfowl at the Half Moon Bay Wetland include Wandering and Plumed Whistling-Duck, Green Pygmy-Goose and, occasionally, Radjah Shelduck and, very rarely, Cotton Pygmy-Goose. Look also for Australasian Darter, Nankeen Night-Heron, Black Bittern, Buff-banded Rail, White-browed Crake, Little Kingfisher and Pied Heron. In the areas around the lagoon, you may find Bush Stone-curlew, Little Bronze-Cuckoo, Oriental and Brush Cuckoo, Crimson Finch and Chestnut-breasted Mannikin, and summer rarities such as Eastern Yellow Wagtail and Barn Swallow. At the end of Wattle Street, a path leads between the golf course and the mangroves. In the mangroves here, you may find Mangrove Robin, Collared Kingfisher and Shining Flycatcher. Continue along the path and search the tidal flats at the mouth of Yorkeys Knob Creek. Beach Stone-curlew is sometimes present here; the mud flats are good for migratory shorebirds such as Eastern Curlew, Common, Marsh and Sharp-tailed Sandpiper, while Great-billed Heron occasionally roost in the larger mangroves bordering the creek.

Once a degraded sand-mine, the Cattana Wetlands (80 ha) have recently been developed into a wetland site, with bird hides, a viewing platform, boardwalks paths (420 m) and substantial rainforest plantings. At the wetland, you may find Pink-eared and Australian Wood Duck, Green Pygmy-Goose, Comb-crested Jacana, Red-kneed Dotterel, Glossy Ibis and Black-necked Stork–scan any muddy edges for a chance of Pale-vented Bush-hen, while White-browed Crake can at times be common among the waterlilies, especially from the southern shore. In the rainforest fringe, look for Shining Flycatcher, Yellow Honeyeater, Little Bronze, Brush and Channel-billed (summer) Cuckoo, Varied Triller, Lovely Fairy-wren and Crimson Finch. Overhead raptors such as Whistling Kite, Brahminy Kite and Pacific Baza are seen occasionally. Nearby, the Skyrail Wetlands (at the intersection of the Captain Cook Hwy and Cairns Western Aerial Rd) can be a good site for Green Pygmy-Goose and Glossy Ibis.

Kuranda and Cassowary House

Key species: Southern Cassowary, Red-necked Crake, Pied Monarch
Other species: Fork-tailed Swift, White-throated Needletail

Kuranda is 27 km north-west of Cairns on the Kennedy Hwy. Cassowary House is located along Black Mountain Rd 1 km north from the Kennedy Hwy. Run by Phil and Sue Gregory, it is a guest-house established for those with an interest in natural history, especially birds: www.cassowary-house.com.au, (07) 4093 7318.

Southern Cassowary often visits the grounds of Cassowary House in the morning and there is a good selection of other rainforest species here including Victoria's Riflebird, Spotted Catbird, Black Butcherbird, Macleay's, Yellow-spotted and Graceful Honeyeater, and, occasionally, Red-necked Crake. Along Black Mountain Rd near Cassowary House, look for

Barron Falls near Kuranda. Visiting in the Wet Season has its rewards (e.g. migrant Buff-breasted Paradise-Kingfishers are present), but expect it to rain! It is usually best to keep plans flexible in case access to some areas is restricted.

Wompoo and Superb Fruit-Dove, Double-eyed Fig Parrot, Barred Cuckoo-shrike, Lovely Fairy-wren, Yellow-breasted Boatbill, Chowchilla, Victoria's Riflebird and Spotted Catbird and White-eared, Spectacled, Black-faced and Pied Monarch. Nocturnal birding around the house and along Black Mountain Rd may produce Lesser Sooty and Rufous Owl, Australian Owlet-Nightjar and White-throated Nightjar. Mammals here include Musky Rat Kangaroo and White-tailed Rat, and Platypus are sometimes seen the creek. In Kuranda, the fig trees along the main street are a good place for Double-eyed Fig-Parrot, Scaly-breasted and Rainbow Lorikeet and Australasian Figbird.

The Barron Falls and Wrights Lookouts, located in the northern section of Barron Gorge National Park (3 km from Kuranda via Barron Falls Rd), offer extensive views across Barron Gorge. Australian Swiftlet are usually present here year round and, through the wet season, search the skies here for Fork-tailed Swift and White-throated Needletail, especially when the weather is unsettled around storm fronts. Taking a boat trip on the Barron River above Barron Falls is an excellent way to see Freshwater Crocodile. The dock is next to Kuranda train station.

Michaelmas Cay

Key species: Common Noddy, Black Noddy, Sooty Tern, Bridled Tern, Lesser Crested Tern, Black-naped Tern, Roseate Tern, Lesser Frigatebird, Great Frigatebird
Other species: White-tailed Tropicbird, Masked Booby, Red-footed Booby, Wandering Tattler

Michaelmas Cay (~1.5 ha) is one of the most important seabird nesting islands in the northern Great Barrier Reef. Located 43 km north-east of Cairns on the western tip of Michaelmas Reef, 38 species of bird have been recorded, with nine of those using the

Green Island

Green Island (12 ha) is 27 km offshore from Cairns and is one of the most accessible and popular coral cays on the Great Barrier Reef. The Boardwalk Walk (1.4 km return) on Green Island passes through Green Island National Park. This provides access to the tropical vine forest and leads to the south-east beach. Along the boardwalks, look for Buff-banded Rail, Pheasant Coucal, Emerald Dove and Rose-crowed Fruit-Dove. Large numbers of Pied Imperial-Pigeon roost on the island in summer. While returning along the beach, look for Eastern Reef Egret (usually white morph) and shorebirds such as Grey-tailed Tattler, Whimbrel and Ruddy Turnstone; Beach Stone-curlew may also occur. Offshore, and occasionally resting on beaches, look for seabirds such as Brown Booby, Common Noddy, and Black-naped, Bridled, Crested and Lesser Crested Tern. In spring, Eastern Osprey nest on the beacon near the main jetty. Green and Hawksbill Turtle and Dugong use the surrounding reef and seagrass meadows as a feeding ground. Commercial charter boats to Green Island depart from the Reef Fleet Terminal, located at the end of Spence Street in Cairns.

island for nesting purposes. An impressive 20 000+ birds use the island at the peak of the breeding season. In the Cairns area of the Great Barrier Reef, this is the only destination offering birdwatchers the opportunity to experience a seabird rookery with high species diversity and large populations. When visiting the cay, visitors are restricted to the beach to avoid disturbing the nesting birds. Seastar Cruises stop at Michaelmas Cay and allow you to land for several hours: www.seastarcruises.com.au, (07) 4041 6218.

The main breeding species on Michaelmas Cay are Common Noddy and Sooty and Crested Tern. Smaller numbers of Lesser Crested Tern also occur. Among the Common Noddy, you can occasionally find Black Noddy: a darker, smaller bird with a more striking white cap. Other species normally present in small numbers include Black-naped (often perched on pontoons), Roseate, Bridled and Caspian Tern, while Lesser and Great Frigatebird circle the island. Birds less commonly recorded on and around the cay include Short-tailed Shearwater, Wandering Tattler, White-tailed Tropicbird and Masked and Red-footed Booby. On the way out to the Michaelmas Cay, scan the sea for Brown Booby, Little and Gull-billed Tern, and Pomarine and Long-tailed Jaeger. Some magnificent coral reefs also surround Michaelmas Cay, so it is recommended that even very keen birdwatchers also devote some time to snorkelling!

Newell Beach and the Mossman River

Key species: Barn Swallow, Red-rumped Swallow, Little Kingfisher, Great-billed Heron, Beach Stone-curlew
Other species: Collared Kingfisher, Shining Flycatcher, Mangrove Robin

One hour north of Cairns, Newell Beach is a small town bordering the northern side of the Mossman River inlet. The beach, sandbar and mangroves in the area can be rich with birdlife. To get there from Mossman, travel 3.5 km north along the Mossman–Daintree Rd, before turning east along Newell Rd; Newell Beach is another 2 km. There is a caravan park with shaded camping sites and fully self-contained units. For longer stays, there are also beachfront houses for rent.

Newell Beach is one of the best sites in Australia to see swallows and swifts, particularly from late September to early March. Barn Swallows occur annually on the town's power lines, while the Red-rumped Swallow, generally considered a vagrant anywhere in Australia, has been recorded here more frequently than anywhere else in the country. One good area to look is near the Newell Water Tower on Andrew St. When they occur, Red-rumped Swallows aren't usually around for long, with the first records in late December, a spike in records in January and the occasional remaining bird in February and March. There are also several intriguing, yet unconfirmed, records of Pacific Swallow: a species not yet recorded with certainty in Australia. More common species include Welcome Swallow, Fairy, and Tree Martin and thus it is possible to have five (or theoretically six!) species of hirundine perched on the wires at once. Australian Swiftlet is also common at Newell Beach, and White-throated Needletail and Fork-tailed Swift can both occur in numbers during the wet season. To add to the excitement, House Swift, Glossy Swiftlet and Uniform Swiftlet have all been reported at least once in the area. If you do not find Barn and Red-rumped Swallow at Newell Beach, search for them on the fence lines on the approach into Newell Beach, and along Somerset Drive (located 5.2 km north of Mossman; the second road on the left after Newell Beach turn-off). They have also been seen at Cooya Beach, the township on the south side of the Mossman River and on the power lines at the intersection of Mossman-Daintree Rd and Cape Tribulation Rd.

On the beach near the mouth of the Mossman River, look for Greater and Lesser

Although Little Kingfishers are regularly seen around Cairns (especially in the Cairns Botanic Gardens), a birdwatching cruise on either the Mossman River or Daintree River greatly increases the chance of connecting with this species.

Sand Plover (summer), Bar-tailed Godwit, Grey-tailed Tattler, Common Sandpiper, Great Knot, Whimbrel, Eastern Curlew and Australian Pied Oystercatcher, and terns such as Little, Crested and Lesser Crested. Beach Stone-curlew are occasionally seen here, more-so on the south side (accessed by boat or via Cooya Beach); they also occasionally occur near the mouth of Saltwater Creek at the northern end of Newell Beach. (Another spot for Beach Stone-curlew is the beach just north of Yule Point, located on the Captain Cook Hwy 21.5 km south of Mossman, 3.5 km north of Oak Beach.)

Along the Mossman River, you can find Little Kingfisher and Great-billed Heron; the best method for locating them is to hire a boat and travel slowly west along the river (there is a boat ramp at the southern end of Newell Beach). This takes you through a range of habitats, including coastal estuary, mangroves, reedbeds, coastal rainforest and mangrove thickets. In the mangroves, look for mangrove specialists such as Collared Kingfisher, Shining Flycatcher, Mangrove Robin and Large-billed Gerygone. Along the river, several smaller tributaries branch north ~300 m from the boat ramp. Little Kingfisher are in this area, preferring narrow waterways with overhanging vegetation where they perch low to the water. Great-billed Heron are known to occur 1.5–2 km west of the boat ramp: look for this large heron in the area where rainforest appears to be intermixed with mangroves. Other birds found along the Mossman River include Eastern Osprey, White-bellied Sea-Eagle, Azure and Sacred Kingfisher, Striated Heron, Black Bittern, Hornbill Friarbird, Varied Triller and Black Butcherbird.

The Daintree

Key species: Little Kingfisher, Buff-breasted Paradise-Kingfisher (Nov-Apr), Great-billed Heron, Black Bittern, Lovely Fairy-wren, Southern Cassowary, Rufous Owl, Pied Monarch
Other species: Papuan Frogmouth, Yellow-breasted Boatbill, Noisy Pitta

The township of Daintree Village, 111 km north of Cairns via Mossman, is at the centre of the Daintree River catchment. The Daintree River and rainforest surrounding the township are home to sought-after and iconic species as Southern Cassowary, Little Kingfisher, Great-billed Heron, Black Bittern, Papuan Frogmouth, Lovely Fairy-wren and Noisy Pitta. There is a wide range of accommodation options in Daintree Village: www.daintreevillage.asn.au.

A good option for birdwatching along the Daintree River is by participating in a Daintree River Tour aboard a dedicated wildlife-watching boat. The dawn departures target birds along the river, accessing interesting sections of the river system (such as under the Barratt Creek Bridge). Some of the targeted species include Great-billed Heron, Little Kingfisher, Black Bittern and Papuan Frogmouth, and you should find Azure Kingfisher, Nankeen Night-Heron and Striated Heron, Black-necked Stork and Radjah Shelduck, and raptors such as Pacific Baza, Brahminy Kite, Eastern Osprey, and White-bellied Sea-Eagle. Advance bookings are recommended: daintreerivertours.com.au, 0417 651 929.

A walk along Stewart Creek Rd can be rewarding. Along here there is a chance of Pale-vented Bush-hen, Buff-breasted Paradise-Kingfisher (Nov–Apr, during which time they nest in termite mounds along the road), Papuan Frogmouth, Blue-winged Kookaburra, Macleay's Honeyeater, Fairy and Large-billed Gerygone, Black Butcherbird, Grey Whistler, Cicadabird, Lovely Fairy-wren, Victoria's Riflebird, Yellow-breasted Boatbill, Spotted Catbird, Golden-headed Cisticola, Double-eyed Fig-Parrot, Pied and Spectacled Monarch, Leaden and Shining Flycatcher. Buff-breasted Paradise-Kingfisher also nest in termite mounds near the Daintree Water Tower, and Lovely Fairy-wren occur along the track that leads up the hill behind the tower; search the scrub 100 m up the track. Other birds found around the Daintree Village include cuckoos such as Channel-billed Cuckoo, Eastern Koel, Little Bronze-Cuckoo, Brush Cuckoo, Pheasant Coucal and, occasionally, Oriental Cuckoo. Honeyeaters to look for include Yellow-spotted, Dusky, Graceful and Brown-backed, particularly when there is flowering *Grevillea* and *Melaleuca*. Watch out also for pigeons such as Bar-shouldered Dove, Pied Imperial-Pigeon (summer) and Wompoo Fruit-Dove. The Daintree Saltwater Barramundi Farm (located in Wonga 16 km south of Daintree) often has a good variety of waterbirds present including Radjah Shelduck, Wandering Whistling-Duck and Black-necked Stork. The farm is private property, but there good views can be obtained from the roadside.

The Daintree Ferry (operating from 6 a.m. to midnight daily) provides access to the Cape Tribulation area via Cape Tribulation Rd. Along this road there are three excellent boardwalks. The Jindalba Boardwalk, located 7.5 km from the Daintree Ferry, is a reliable location for Noisy Pitta. Along the 700 m circuit, you may also find Spotted Catbird, Wompoo Fruit-Dove, Pied Monarch and Buff-breasted Paradise-Kingfisher (Nov–Apr). The Marrdja Boardwalk (1 km loop), 26 km north of

The Daintree River mouth viewed from the Mount Alexandra lookout on Cape Tribulation Rd. A great stop to provide some context to the extent of the Daintree and Cape Tribulation rainforests, the best birding is usually to be had further north on well-maintained Jindalba, Marrdja and Dubuji Boardwalks.

the Daintree Ferry, passes through rainforest and mangroves then overlooks Noah Creek. It can be a good site for Southern Cassowary, and there is a chance of Little and Azure Kingfisher, Shining Flycatcher, Yellow-breasted Boatbill, Victoria's Riflebird and Double-eyed Fig-Parrot. Along the Dubuji Boardwalk (1.2 km loop), signposted just south of Cape Tribulation, search for Red-necked Crake, Black Butcherbird, Shining Flycatcher and roosting Rufous Owl. Southern Cassowary are regularly seen near the Cooper Creek crossing, 19 km from the Daintree Ferry, and Beach Stone-curlew often frequent the beach at the mouth of Coopers Creek, especially at low tide.

Kingfisher Park Birdwatchers Lodge

Key species: Buff-breasted Paradise-Kingfisher (Nov–Apr), Red-necked Crake, Noisy Pitta, Macleay's Honeyeater, Victoria's Riflebird, Papuan Frogmouth, Lesser Sooty Owl
Other species: Pale-yellow Robin, Lemon-bellied Flycatcher, Grey Whistler, Yellow-breasted Boatbill, Bush Stone-curlew, Pied Monarch

Approximately 90 km from Cairns, Kingfisher Park Birdwatchers Lodge (5 ha) is named after the Buff-breasted Paradise-Kingfisher that visit to breed within the grounds over the wet season. It is located in Julatten just off the Mossman–Mount Molloy Rd, 23 km from Mossman (11 km from Mount Molloy). A tremendous birding area, it is the also the perfect base for birding in the wider district. Birds that frequent the lodge grounds include Red-necked Crake, Noisy Pitta, Macleay's Honeyeater, Papuan Frogmouth, Lesser Sooty Owl and the aforementioned Buff-breasted Paradise-Kingfisher (Nov–Apr). The lodge consists of self-contained units and bunkhouses, camping and caravan sites set in a rainforest environment and is owned and operated by expert local birders: www.birdwatchers.com.au, (07) 4094 1263.

There are several tracks around the property that lead around the campsite and lodge and through rainforest into an open grassed orchard bordering Bushy Creek. The best way to birdwatch at the lodge is spend a few hours walking around these tracks, particularly early in the morning. Around the lodge and campsite, you can find Australian Brush-turkey, Orange-footed Scrubfowl, Pale-yellow Robin, Grey Whistler, Olive-backed Sunbird, Emerald Dove, Little Shrike-thrush, Barred Cuckoo-shrike, Spangled Drongo and Pheasant Coucal, and summer migrants such as Channel-billed Cuckoo, Eastern Koel, Dollarbird and Metallic Starling. Honeyeaters can be plentiful and include Macleay's, Yellow-spotted, Graceful, Scarlet and Dusky, as well as the occasional Bridled Honeyeater, most often found feeding in flowering *Grevillea* along Mount Kooyong Rd on the eastern boundary of the lodge.

A pair of Noisy Pitta has a territory next to the main driveway and, with some careful observations, the species can usually be found foraging on the rainforest floor or out onto the paths and lawned areas. The orchard and rainforest typically support a wide range of species. Buff-breasted Paradise-Kingfisher are present in the park between November and April and usually start nesting after the first wet season rains. Red-necked Crake feed around the small dam on the south side of the orchard, with the best time being just before nightfall, and Papuan Frogmouth usually roost in the larger trees on the east side of the orchard. Up-to-date information on these sought-after species can of course also be obtained from the lodge. Other rainforest birds include Brown Cuckoo-Dove, Wompoo and Superb Fruit-Dove, Double-eyed Fig-Parrot, Spotted Catbird, Striated Pardalote, Lovely Fairy-wren, Large-billed Scrubwren, Fairy and Large-billed Gerygone, Yellow-breasted Boatbill, Varied Triller, Cicadabird, Pied, Spectacled and Black-faced (summer)

Buff-breasted Paradise-Kingfishers are present in the Wet Tropics from November to April each year. Excellent sites for this striking species include the aptly named Kingfisher Park Birdwatchers Lodge, as well as Mount Whitfield Conservation Park (Cairns), the Daintree, the lower slopes of Mount Lewis and Paluma Range.

Monarch and Victoria's Riflebird. Birds of prey found in and around the park include Pacific Baza, Brahminy, Black and Whistling Kite, Nankeen Kestrel and Grey Goshawk. Frequent night birds include Lesser Sooty, Eastern Barn and Barking Owl, Australian Owlet-Nightjar and Bush Stone-curlew. Other wildlife found at the lodge includes Striped and Green Ringtail Possum, Fawn-footed Melomys, Platypus and Prehensile-tailed Tree Mouse, and butterflies such as Cairns Birdwing and Ulysses Butterfly.

Abattoir Swamp Environmental Park and surrounds

Key species: White-browed Crake, Northern Fantail, Blue-faced Parrot-Finch
Other species: Pale-vented Bush-hen, Latham's Snipe, Lovely Fairy-wren, Large-tailed Nightjar

At Abattoir Swamp, 6 km south of Kingfisher Park Birdwatchers Lodge (4.5 km from Mount Molloy) on the Mossman–Mount Molloy Rd, open woodlands surround a small car park, with a boardwalk through to a bird hide.

The woodland around the carpark can be productive, particularly when paperbark trees are flowering. Look for Northern Fantail, Scaly-breasted and Rainbow Lorikeet, Red-winged Parrot and Great Bowerbird, and honeyeaters such as Brown, Scarlet, Yellow-faced, Yellow, White-throated, Brown-backed, White-naped, Black-chinned ('Golden-backed' race *laetior*) and Blue-faced. From the bird hide, look for White-browed Crake feeding on the water's edge to the north-west of the hide, and you may also see Pale-vented Bush-hen (chance, but they tend to keep to cover), Buff-banded Rail, Baillon's and Spotless Crake, while the grassy fringes of the swamp hold Latham's Snipe (summer) and Brown Quail.

On the Mossman–Mount Molloy Rd, 2 km north of Abattoir Swamp, a small track leads east into long grass (it is located directly opposite Sides Rd). Blue-faced Parrot-Finch (often with Red-browed Finch) feed here on the seeding grasses from late July to early October, and White-cheeked Honeyeater can also be common.

Wessell Rd turns east from the Mossman–Mount Molloy Rd, 1 km south of Abattoir Swamp – follow this road for 700 m until it turns south. Northern Fantail and Large-tailed Nightjar (spring–summer) inhabit the drier areas of woodland, which are also good for Yellow-faced, White-throated and Brown-backed Honeyeater, and Red-backed and Lovely Fairy-wren.

Mount Lewis and Brooklyn Station

Key species: Golden Bowerbird, Blue-faced Parrot-Finch, Mountain Thornbill, Atherton Scrubwren, Bridled Honeyeater, Fernwren, Bower's Shrike-thrush
Other species: Tooth-billed Bowerbird, Victoria's Riflebird, Chowchilla, White-throated Treecreeper, Southern Boobook, White-cheeked Honeyeater

Many of the highest rainforest-clad mountains of the Wet Tropics World Heritage Area are inaccessible. The 28 km Mount

Lewis drive is a notable exception. Mount Lewis is an outstanding highland-rainforest area, and is a hotspot for animal diversity and endemism. The Australian Wildlife Conservancy (AWC) owns Brooklyn Station, including a section of Mount Lewis. The parking area and the walk up to the dam (mentioned below) are within this area. To get there from Cairns, travel north along the Captain Cook Hwy and turn left onto the Mossman–Mount Molloy Rd (also called Rex Hwy). After 17.5 km, turn right into Mount Lewis Rd, which climbs over 1200 m, winding its way through rainforest-clad ridges and spurs. The road is recommended for 4WD only and can be closed during wet weather; check conditions with Queensland Parks and Wildlife in Mossman, (07) 4098 2188.There is no camping in Mount Lewis National Park, but the area can be easily reached from the nearby Kingfisher Park Birdwatchers Lodge.

Before heading up the mountain, it is worth stopping near the World Heritage signs 3.3 km from Mossman–Mount Molloy Rd, a good area for lower altitude birdwatching – look here for Barred Cuckoo-shrike (feeding in fruiting fig trees), Emerald and Bar-shouldered Dove, Wompoo Fruit-Dove, Double-eyed Fig-Parrot,

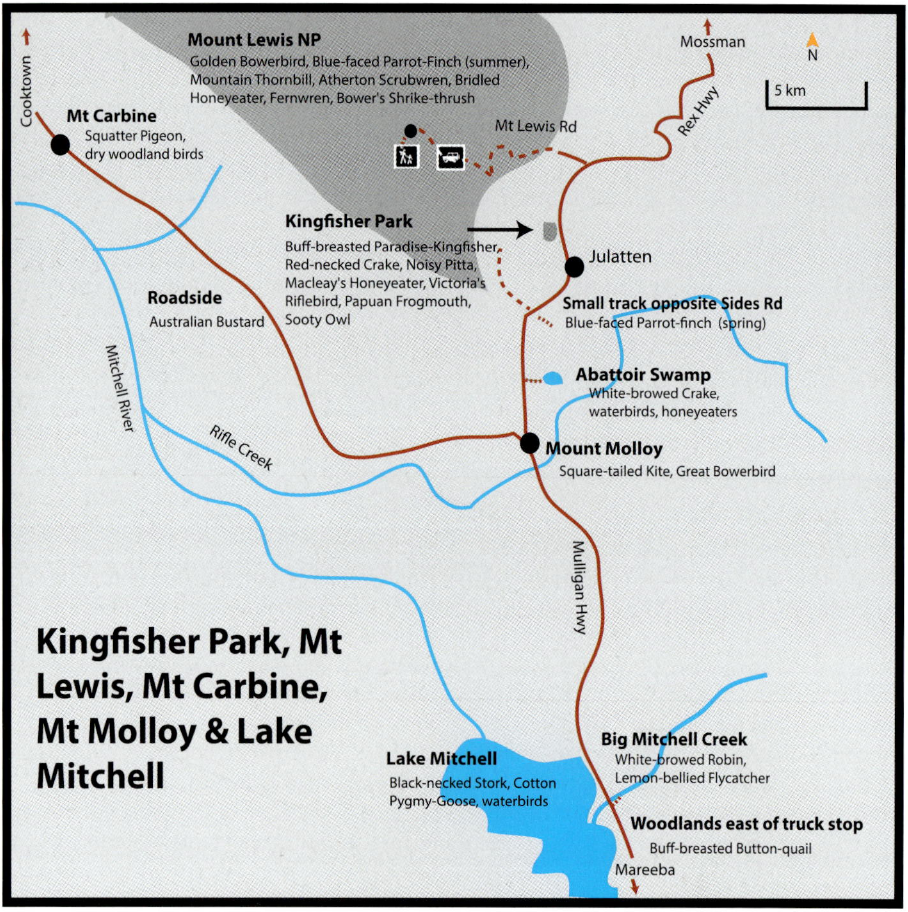

Buff-breasted Paradise-Kingfisher, Yellow-breasted Boatbill, Pied Monarch, Spangled Drongo and Grey Goshawk. Southern Cassowary is very occasionally observed on the Mount Lewis Rd between here the grassy clearing further up the mountain.

The main area for birding at Mount Lewis is a grassy clearing located on Mount Lewis Rd (11.6 km from the Mossman–Mount Molloy Rd) and the Dam Track, a steep walking path leads up from the clearing to a picturesque dam (4 km return). Small numbers of Blue-faced Parrot-Finch occur at the clearing during the summer months (usually arriving in mid-November and departing in April) and feed on the seeds of long grass in association with Red-browed Finch. Look also around the clearing for Bridled Honeyeater, Grey-headed Robin, Mountain Thornbill, Atherton Scrubwren and Varied Triller. Along the Dam Track, and near the dam at the top of the track, look for Golden and Tooth-billed Bowerbird, Spotted Catbird, Victoria's Riflebird, Chowchilla (listening for rustling noises in the undergrowth), Fernwren, Atherton and Yellow-throated Scrubwren, Grey (Mountain) and Rufous Fantail, Bassian Thrush, White-throated (Little) Treecreeper, Spectacled Monarch, Australian King-Parrot, Bower's Shrike-thrush and Lovely Fairy-wren. The impressive Boyd's Forest Dragon can be found along many of the tracks in the area and the diurnal Musky Rat Kangaroo is frequently seen here also. Spotlighting at night along Mount Lewis Rd may produce Lesser Sooty Owl and rainforest inhabiting 'Red Boobook' (ssp. *lurida* of the Southern Boobook) as well as Lemuroid Ring-tail Possum, Prehensile-tailed Tree Mouse and a variety of other small mammals.

Mount Molloy to Mount Carbine

Key species: Squatter Pigeon, Australian Bustard, Northern Fantail, White-browed Robin, Black-throated Finch

Other species: Red-tailed Black-Cockatoo, Red-winged Parrot, Pale-headed Rosella, Blue-winged Kookaburra

Mount Carbine and Mount Molloy are located 30 km apart on the Peninsula Development Rd (also known as the Mulligan Hwy). There is a range of accommodation in the area: mostly bed and breakfast accommodation and self-contained holiday units, with a number aimed specifically at birdwatchers, such as Kingfisher Park Birdwatching Lodge (see p. 96) and Bustard Downs (on East Mary Rd, (07) 4094 3110, www.bustarddowns.com.au).

Mount Molloy is a small town 100 km north of Cairns. Great Bowerbird inhabits the gardens of the township, and are usually easy to find. At the southern end of town, there are several sporting ovals, where Square-tailed Kite inhabits the woodlands on the eastern boundary; look here also for Red-winged Parrot, Pale-headed Rosella and Squatter Pigeon. The Rifle Creek Reserve (300 ha) and camping area is located on the Mulligan Hwy 1 km north of Mount Molloy. Bounded by gallery vine forest, riparian rainforest and tropical eucalypt woodlands, it is an excellent birding area, with well over 200 species of bird recorded, the most notable being Buff-breasted Button-quail (several records). Birds to look for include Emerald Dove, Blue-winged Kookaburra, Noisy Pitta, Spotted Catbird, Yellow-breasted Boatbill, Black Butcherbird, Barred Cuckoo-shrike and Pale-yellow Robin, and honeyeaters such as Graceful, Yellow-spotted, Yellow-faced, Yellow, Dusky, Scarlet, White-throated, Blue-faced and Macleay's. In the south-west corner of the camping area, a track leads down to the creek line, where it is worth searching for Pale-vented Bush-hen. An impressive 12 species of raptor have been recorded at Rifle Creek, including Square-tailed Kite, Pacific Baza, Spotted Harrier and Little Eagle. It is also worth birdwatching along Wetherby Rd, concentrating your efforts for ~4 km east of the reserve until you reach a lagoon.

Australian Bustard feed along Peninsula Development Rd near Maryfarms between Mount Carbine and Mount Molloy; try the grassy areas along East Mary Rd (the road to Bustard Downs) and West Mary Rd, 22 km north of Mount Molloy. Look here also for Squatter Pigeon, Red-tailed Black-Cockatoo, Red-winged Parrot, Pale-headed Rosella and Blue-winged Kookaburra.

Immediately north of the Mount Carbine Hotel on the Peninsula Development Rd, there is a track that leads south alongside a cricket oval to the walls of a small dam. Here you should find Squatter Pigeon, along with Pale-headed Rosella, Red-winged Parrot, Great Bowerbird, Apostlebird, Grey-crowned Babbler, Pied Butcherbird and Black-throated Finch. In December 2002, Australia's only reported Isabelline Wheatear was found on the cricket oval at Mount Carbine.

Big Mitchell Creek and Lake Mitchell

Key species: Buff-breasted Button-quail, White-browed Robin, Black-necked Stork, Cotton Pygmy-Goose (rare)
Other species: Black-throated Finch, Lemon-bellied Flycatcher, Pale-headed Rosella, Brown Treecreeper, waterbirds

Big Mitchell Creek is 16.5 km south of Mount Molloy (23 km from Mareeba). Park on the south side of the bridge and drop down into the dry creek bed, walking either east or west. White-browed Robin feed up and down the creek, and often respond well to whistling and pishing. The gallery forest along the Big Mitchell Creek supports Black-throated Finch (also in adjacent woodland with seeding grasses), Northern Fantail, Lemon-bellied Flycatcher, Pale-headed Rosella, Pied Butcherbird,

Drier woodlands dominate the landscape inland of the Wet Tropics. These habitats support species such as Australian Bustard, Squatter Pigeon, Red-tailed Black-Cockatoo, Red-winged Parrot, Pale-headed Rosella, Blue-winged Kookaburra and Black-throated Finch.

Red-tailed Black-Cockatoo, Rainbow Bee-eater (summer), Striated Pardalote, Fairy Gerygone, Little Shrike-thrush and Great Bowerbird. When the Forest Red Gum (*Eucalyptus tereticornis*) flowers along the creek, Banded Honeyeater and Little Lorikeet are attracted to the area. Interesting birds of prey recorded here include Square-tailed Kite and Grey Goshawk, while the very rare Red Goshawk has been recorded in the general vicinity on several occasions.

About 700 m north of Big Mitchell Creek, there is a causeway behind a gate on the west side of the road that leads out into Lake Mitchell (also known as Quaids Dam). This is an extensive wetland that attracts thousands of waterbirds. The causeway between the gate and out to 1 km from the road is a stock reserve and you can enter this area. The causeway can be a good place for Black-necked Stork, Black-winged Stilt, Comb-crested Jacana, Australasian Grebe, Glossy Ibis, Royal Spoonbill, Pied Heron, Australian Painted Snipe (rare, but several records), and waterfowl such as Wandering Whistling-Duck, Grey Teal, Hardhead, Radjah Shelduck (uncommon), Green and Cotton Pygmy-Goose and Magpie Goose. In spring and summer, if the water level is low, migratory shorebirds may congregate around the lake edges. Regular species include Marsh and Sharp-tailed Sandpiper,

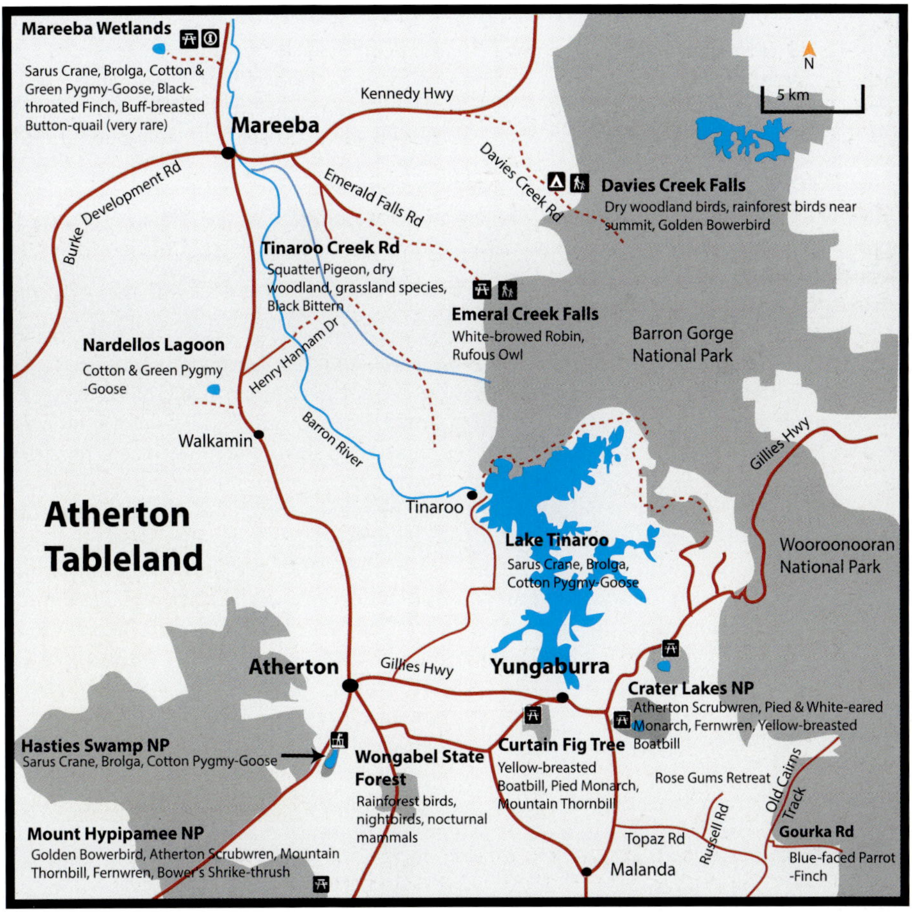

Queensland

Common Greenshank and Red-necked Stint, and, less commonly, Grey and Pacific Golden Plover, and in November 2003 a vagrant Grey Phalarope was recorded. Raptors here include Black and Whistling Kite, Eastern Osprey and White-bellied Sea-Eagle. The woodland around the lake supports Pale-headed Rosella, Brush Cuckoo, Grey-crowned Babbler, Rufous Whistler, Leaden Flycatcher, Blue-faced Honeyeater, Little and Noisy Friarbird, White-throated Gerygone, Black-throated Finch and Torresian Crow.

The very rare Buff-breasted Button-quail have been recorded alongside the Mulligan Hwy near Lake Mitchell – look in the woodlands at the base of the hillside east of a truck stop 18 km south of Mount Molloy (21 km north of Mareeba/1.5 km south of Big Mitchell Creek). Concentrate your search in the gravelly strip at the base of the slopes, where there is a mix of native grass tussocks and bare areas of gravel. Easily confused with the more common Painted Button-quail, when flushed, the Buff-breasted Button-quail appears as a large very plain buff-brown quail with distinctly darker (and contrasting) flight feathers. Buff-breasted Button-quail display a distinct preference for lightly grassed eucalyptus woodlands (especially near the base of these hills), whereas Painted Button-quail tend to prefer more thickly grassed woodlands with a denser cover of leaf-litter. Except in very rare eruption years, Buff-breasted Button-quail are enigmatic, with few or indeed no sightings in any year; when observed, these sightings are concentrated in the period November to January, especially after the first rains promote green grass growth, but before the ground cover becomes too dense.

Mareeba Wetlands

Key species: Sarus Crane, Brolga, Cotton Pygmy-Goose, Green Pygmy-Goose, Black-throated Finch, Buff-breasted Button-quail (slim chance)

Other species: Black-necked Stork, Glossy Ibis, White-browed Crake, Comb-crested Jacana, Wandering Whistling-Duck, Pacific Baza, Squatter Pigeon, Pale-headed Rosella

At the centre of the Mareeba Tropical Savannah and Wetland Reserve (2000 ha) is a series of 12 inter-connected lagoons, creeks and channels that attract significant numbers of waterbirds. For bird-watchers, the surrounding woodlands also hold a good selection of bush birds. From Mareeba, travel north towards Biboohra for 7 km before turning left into Pickford Rd to follow the signs the reserve. There is accommodation at the Jabiru Safari Lodge located within the reserve: www.jabiru-safarilodge.com.au, (07) 4093 2514. The reserve is open from 9 a.m. to 4:30 p.m. through much of the year, but is closed during the wet season (from December/January to March/April, depending on the conditions).

Waterbirds at Mareeba Wetlands include Sarus Crane and Brolga, Cotton and Green Pygmy-Goose, Wandering Whistling-Duck, Black-necked Stork, Glossy Ibis, White-browed Crake, Comb-crested Jacana and, in summer, shorebirds such as Marsh Sandpiper, Black-tailed Godwit, and Latham's Snipe. The very rare Buff-breasted Button-quail has been recorded within the reserve, but records are few and far between; search for them on the dry hillsides (see also the Big Mitchell Creek section, p. 100, for further tips including habitat preferences). Other woodland birds here include Emu, Painted Button-quail, Red-backed Fairy-wren, White-throated Gerygone, Squatter Pigeon, Pale-headed Rosella, Grey and Pied Butcherbird, Grey-crowned Babbler, Cicadabird, Brown (Black) Treecreeper, Lemon-bellied Flycatcher, Black-throated and Double-barred Finch and Chestnut-breasted Mannikin. This is also the location for a landmark Gouldian Finch re-introduction programme and free

The northern form of the Squatter Pigeon ssp. *peninsulae* remains fairly common in grassy woodland habitats. Key sites close to the Wet Tropics include Mount Molloy, Mount Carbine, Mareeba wetlands, Emerald Creek Falls and Tinaroo Creek Rd. Tame birds are also often present at Granite Gorge Nature Park.

flying birds can usually be found around the lodge. Pacific Baza often roost near the carpark, and Eastern Osprey nest on the power-poles on your way into the reserve. At night, spotlight along the roadside for nocturnal birds such as Eastern Barn Owl, Southern Boobook, Tawny Frogmouth, Australian Owlet-Nightjar, White-throated and Large-tailed Nightjar.

On the way into the Wetlands, it is worth stopping to look at a small wetland on Pickford Rd 1 km from the Peninsula Development Rd. Black-throated and Double-barred Finch, and Chestnut-breasted Mannikin are regularly seen drinking here, and Black Bittern has been recorded at the adjacent Two Mile Creek crossing.

Around Mareeba – Emerald Creek, Davies Creek, Tinaroo Creek Rd and Granite Gorge Nature Park

Key species: Rufous Owl, Golden Bowerbird, Squatter Pigeon, Black-throated Finch, Atherton Scrubwren

Other species: Masked Owl, Square-tailed Kite, White-browed Robin, Buff-breasted Button-quail (old records), Painted Button-quail

Emerald Creek and Davies Creek offer access to good areas of eucalyptus woodland and riparian forest and both areas are particularly good for dry-country birds. There is a nice area of rainforest near the top of the Kahlpahlim Rock. Emerald Creek is 14.2 km south of Mareeba on Emerald Falls Rd, whereas Davies Creek National Park is 25 km from Mareeba (28 km from Kuranda) along Davies Creek Rd. The gravel road into Davies Creek can be heavily corrugated and unsuitable for caravans.

Adjacent to Emerald Creek Picnic area are tracts of open woodland and riparian forest along a creek line. The creek line can be a particularly good site for Rufous Owl and White-browed Robin; walk up and down the creek, scanning the sheltered canopy because Rufous Owl are known to roost in trees near the picnic ground. Masked Owl also occur here. The Emerald Creek Falls walk (2 km return) can be rewarding. Along the trail there is a chance of Square-tailed Kite, Pied Currawong, Squatter Pigeon, Northern Fantail, Pale-headed Rosella, Weebill, White-browed Robin, Fairy Gerygone, Varied (Striated) Sittella, Rufous, Golden and Grey Whistler, Little Shrike-thrush, Lemon-bellied Flycatcher, and Scarlet, Yellow, White-naped, Yellow-faced and White-throated Honeyeater. Bridled Honeyeater also occurs in winter.

The Kahlpahlim Rock Walk (10 km return) in Davies Creek National Park can be undertaken as a return walk along the original Kahlpahlim Trail or as a circuit route returning via the Ridge Trail. The two trails converge on a ridge 1 km from Kahlpahlim Rock. Look for Pale-headed Rosella, Eastern Yellow Robin (around the carpark), White-cheeked Honeyeater, Grey-crowned Babbler, Painted Button-quail and, further up the trail, Spotted Catbird, Victoria's Riflebird, White-eared and Spectacled Monarch, Satin Flycatcher, Atherton Scrubwren, Scarlet and Bridled Honeyeater, Pale-yellow Robin and Chowchilla. Near the summit, there is a Golden Bowerbird bower alongside the track. At night, listen for Lesser Sooty Owl and White-throated Nightjar. The elusive Buff-breasted Button-quail has been recorded in the reserve, but the probability of a sighting is very low and woodlands near Big Mitchell Creek (p. 102) provide a better chance of an encounter. Mammals here include Northern Quoll and Northern Bettong. There is a camping area downstream from the Davies Creek Falls.

Tinaroo Creek Rd, accessed via the Kennedy Hwy 3 km south of Mareeba, is also a good place for dry woodland and grassland species. Along the roadside, Squatter Pigeon, Spotted Harrier, Black-throated Finch, Pale-headed Rosella, Yellow Honeyeater, Varied Sittella, Brown (Black) Treecreeper, Great Bowerbird, Double-barred Finch and, occasionally in winter, Sarus Crane. Driving the length of road to stop and search when you observe bird activity can be rewarding. A good site for Squatter Pigeon and Black-throated Finch is near the corner of Tinaroo Creek Rd and Henry Hannam Drive (8 km from the Kennedy Hwy). If you travel west along Henry Hannam Drive, you soon cross the Tinaroo Creek; a walk along the creek may produce Black Bittern, Pheasant Coucal, Blue-winged Kookaburra and Azure and Red-backed Kingfisher.

The privately run Granite Gorge Nature Park is located 14.5 km west of Mareeba. Best known as a place to see Mareeba Rock-wallaby, the park holds a nice selection of dry woodland species such as Squatter Pigeon, Pale-headed Rosella, Pheasant Coucal, Red-winged Parrot, Great Bowerbird, Grey-crowned Babbler, Little Friarbird, Lemon-bellied Flycatcher and, in 2011, a pair of Spinifex Pigeon were seen here, representing Australia's most easterly record for this species. To get there, follow Chewko Rd west out of Mareeba for 7 km to a well-signposted turn-off on the right.

Hasties Swamp National Park

Key species: Sarus Crane, Brolga, Cotton Pygmy-Goose
Other species: Magpie Goose, Plumed Whistling-Duck, Tawny Grassbird, Latham's Snipe

Hasties Swamp National Park (57 ha) protects a seasonal wetland that supports a diverse range of resident and migratory birds. More than 220 bird species have been recorded, with the best time for birdwatching during the late dry season (Aug–Nov) when other wetlands in the area dry out and Hasties Swamp becomes an important refuge for waterbirds. Birding highlights include Sarus Crane, large numbers of waterfowl and occasional rarities. Hasties Swamp is 4 km south of Atherton, off the Atherton–Herberton Rd; turn left into Hastie Rd and then right onto Koci Rd. The Nyleta Wetlands bird hide, which overlooks Hasties Swamp, is 1 km further down the road. Built on stilts, the two-storey elevated hide (with wheelchair access) offers uninterrupted views over the open swamp and shoreline areas. Sarus Crane and Brolga roost here during the late dry season (Oct–Nov). At times, large numbers of Magpie Goose, Plumed and Wandering Whistling-Duck, Green and, occasionally, Cotton Pygmy-Goose, and Pink-eared Duck occur. The swamp can also be good for herons and ibis, including Great, Intermediate, Little and Cattle Egret, White-necked Heron, and Glossy, White and Straw-necked Ibis. Other waterbirds include Black-necked Stork, Australasian Grebe, Australasian Darter, Comb-crested Jacana, Black-winged Stilt, and Red-kneed and Black-fronted Dotterel.

Search the reeds bordering the swamp for Tawny Grassbird, Australian Reed-Warbler, and Golden-headed Cisticola. Visiting summer shorebirds (arriving Aug–Sept) include Sharp-tailed, Marsh and Wood Sandpipers, while Latham's Snipe stop over at the swamp on southward migration from August to September, and again in March and April on northbound migration. Look on the fringe of the wetland for Buff-banded Rail, White-browed and Spotless Crake, and rarer species such as Pale-vented Bush-hen, Lewin's Rail and Australian Painted Snipe. A wide range of raptors occur in the area including White-bellied Sea-Eagle, Wedge-tailed Eagle, Pacific Baza, Square-tailed and Brahminy Kite, Swamp and Spotted Harrier and Grey Goshawk. The dry eucalypt forest surrounding the swamp contrasts with rainforest found elsewhere on the tableland. There is a track around the swamp's southern end and a nature walk winds through Blue Gum and Swamp Mahogany forest. Look for species such as Rufous Whistler, Spangled Drongo, Little Shrike-thrush, Crested Shrike-tit, Horsfield's and Little Bronze-Cuckoo and Black-eared Cuckoo (uncommon), and Yellow-faced, Lewin's, Scarlet, White-cheeked, White-naped and White-throated Honeyeater.

Mount Hypipamee National Park

Key species: Golden Bowerbird, Atherton Scrubwren, Mountain Thornbill, Fernwren, Bower's Shrike-thrush

Other species: Southern Cassowary, Pale-yellow and Grey-headed Robin, Bassian Thrush, Tooth-billed Bowerbird, Spotted Catbird, Victoria's Riflebird, Yellow-breasted Boatbill

Mount Hypipamee National Park (also known as The Crater) is 110 km south-west of Cairns (25 km south of Atherton). At an altitude of 1000 m, Mount Hypipamee protects significant areas of upland rainforest and is World Heritage listed. A variety of vegetation types occur in the park, which in turn support a diversity of birds. The park is also recognised as a hot spot for many species of possum.

Around the main carpark and picnic area, there is a chance of Golden Bowerbird, particularly in fruiting trees, as well as Southern Cassowary (uncommon, but sometimes regular), Australian Brush-turkey, Tooth-billed Bowerbird, Spotted Catbird, Victoria's Riflebird, Lewin's and Bridled Honeyeater and Grey-headed Robin. Look along the start of the Crater Track (800 m) for Fernwren, Emerald Dove, Chowchilla, Atherton and Yellow-throated Scrubwren, Mountain Thornbill, Brown Gerygone and Pale-yellow Robin. Two areas that are often productive along the walk are in the vicinity of the first bridge and around a small gully on the left after the second bridge. The track continues through rainforest to a viewing platform overlooking the crater, an alternative route back being the Dinner Falls Track (1.2 km return). Along both trails, other birds to look for include Brown Cuckoo-Dove, Superb and Wompoo Fruit-Dove (usually high in the canopy), Bassian Thrush, Eastern Whipbird, Yellow-breasted Boatbill, White-throated (Little) Treecreeper, Golden Whistler, Bower's Shrike-thrush and Grey and Rufous Fantail.

Another spot to look for Golden Bowerbird is along a 4WD track that leads from the Kennedy Hwy 3.8 km south of the Crater Rd turn-off. Here, travel 400 m up this track to the crest of the hill, and then walk down a further 25 m. Here, on the right-hand side, there is a viewing area, where Golden Bowerbird can sometimes be seen. Blue-faced Parrot-Finch has also been recorded along this track, particularly between the Kennedy Hwy and the hill's crest.

Spotlighting at night at Mount Hypipamee can be very rewarding, particularly along Crater Rd. Look for Southern Boobook (ssp. *lurida*), Rufous and Lesser Sooty Owl. Mount Hypipamee is also well known for its arboreal mammals, with Green, Herbert River and Lemuroid Ringtail Possum all being quite common.

Macleay's Honeyeater is endemic to the Wet Tropic region. The species is relatively common within its restricted distribution and occurs in rainforests and gardens at all but the highest altitudes (>1000 m).

There is also a chance of Striped Possum, Lumholtz's Tree-Kangaroo, Long-nosed Bandicoot, Long-tailed Pygmy Possum and the spectacular Northern Leaf-tailed Gecko. Musky Rat-kangaroo and Boyd's Forest Dragon are two other rainforest inhabitants that are most often encountered in the day.

Curtain Fig Tree and Yungaburra

Key species: Yellow-breasted Boatbill, Pied Monarch, Mountain Thornbill
Other species: Double-eyed Fig-Parrot, Wompoo Fruit-Dove, Spotted Catbird, Macleay's Honeyeater

The Curtain Fig Tree is on the outskirts of Yungaburra. From Yungaburra, head west along the Gillies Hwy for 2 km before turning left into Fig Tree Boulevard. An elevated boardwalk protects the enormous Curtain Fig Tree (50 m tall, and with a trunk circumference of 39 m), while allowing visitors uninterrupted views of the tree from all angles. It is a very popular tourist attraction, so it is worth getting there early to look for birds.

Walk slowly around the tree, or simply sit on one of the seats surrounding the tree, and look up. Around the tree there are regular sightings of Double-eyed Fig-Parrot, Wompoo Fruit-Dove, Spotted Catbird, White-throated

Wongabel State Forest

Located 8 km south of Atherton on the Kennedy Hwy, there are two excellent walking tracks in the reserve – the Forest Walk (750 m return) and the Heritage Walk (2.5 km return) – both flanked by Mabi Forest (also known as Notophyll Vine Forest) dominated by Hoop Pine. Species to look for along both of these walks include White-headed Pigeon, Brown Cuckoo-Dove, Wompoo Fruit-Dove, Spotted Catbird, Tooth-billed Bowerbird, White-throated (Little) Treecreeper, Atherton, Yellow-throated and Large-billed Scrubwren, Large-billed and Brown Gerygone, Mountain Thornbill, Chowchilla, Bridled and Macleay's Honeyeater, Yellow-breasted Boatbill, Bower's and Little Shrike-thrush, Rufous Fantail, Black-faced, Spectacled and Pied Monarch, Victoria's Riflebird, and Pale-yellow and Grey-headed Robin. In the drier woodland areas around the perimeter of the Wongabel State Forest, look for White-cheeked and White-naped Honeyeater, Jacky Winter and Brown (Black) Treecreeper. At night, Rufous Owl and Southern Boobook patrol the forests. The area can also be a good place for Lumholtz's Tree-kangaroo, Striped Possum and Red-legged Pademelon.

Dense upland rainforest surrounding Lake Eacham. A 3-km circuit track provides good opportunities for many of the Wet Tropics specialties including Atherton Scrubwren, Mountain Thornbill, Macleay's Honeyeater, Bridled Honeyeater, Grey-headed Robin, Chowchilla, Bower's Shrike-thrush, Pied Monarch, Victoria's Riflebird and Tooth-billed Bowerbird.

(Little) Treecreeper, Lewin's and Macleay's Honeyeater, Large-billed Scrubwren, Mountain Thornbill, Brown Gerygone, Eastern Whipbird, Little and Bower's Shrike-thrush, Black-faced, Spectacled and Pied Monarch, Yellow-breasted Boatbill (usually in the upper canopy), Victoria's Riflebird, Pale-yellow and Grey-headed Robin and Metallic Starling. At night, around the tree and along Fig Tree Boulevard, look for Rufous Owl and arboreal mammals such as Lumholtz's Tree-kangaroos and Green Ringtail Possum; the latter is also sometimes seen during the day, curled up in a ball and asleep on open limbs of rainforest trees.

Yungaburra lies in the heart of the Atherton Tableland, 68 km from Cairns. White-headed Pigeon and Australian King-Parrot are often common around the Yungaburra township: look for them in the larger trees along the main street. Other birds around the township include Pied Butcherbird, Pied Currawong, Macleay's, Blue-faced, Dusky, Lewin's, Brown and Scarlet Honeyeater, and Little and Hornbill Friarbird. North of Yungaburra, Lake Tinaroo can hold a good selection of waterbirds including Sarus Crane, Brolga (between late winter and the end of the dry season), Cotton Pygmy-Goose, Plumed Whistling-Duck and Glossy Ibis, while Brown Quail and Tawny Grassbird frequent the grassy areas around the lake. A good access point to the lake is the Education Centre at the end of Black Gully Rd.

Crater Lakes National Park

Key species: Atherton Scrubwren, Mountain Thornbill, Chowchilla, Pied Monarch, White-eared Monarch, Fernwren, Bower's Shrike-thrush,

Blue-faced Parrot-Finch

The elusive Blue-faced Parrot-Finch is confined to north Qld, where it is considered rare. In summer, they occur near Butchers Creek and Topaz, south-east of Yungaburra. Look along the roadside past Lamins Hill Lookout in Wooroonooran National Park. To get there from Yungaburra, travel east down the Gillies Hwy for 3.3 km and then travel 6.6 km along Barrine Rd to Topaz Rd. Continue 9.5 km until you come to Old Cairns Track. Here there is a also sign to Gourka Rd. Turn left and then (after ~500 m) right onto Gourka Rd. Usually mixing with Red-browed Finch, Blue-faced Parrot-Finch feed in the long grasses between the turnoff and the end of Gourka Rd (a distance of 7 km). It is also worth looking along the Old Cairns Track, which branches left off Gourka Rd. In summer, Blue-faced Parrot-Finch also occasionally occur in the grounds of the Rose Gums Wilderness Retreat east of Yungaburra, particularly around the Tree Houses: info@rosegums.com.au, (07) 4096 8360. The Retreat is accessed via Russell Rd, which runs off Topaz Rd (5 km from the Barrine Rd turn-off). Further north, Mount Lewis (p. 97) and several sites near Kingfisher Lodge (p. 96) provide additional opportunities for this elusive species.

Yellow-breasted Boatbill, Tooth-billed Bowerbird

Other species: Barred Cuckoo-shrike, Victoria's Riflebird, Spotted Catbird, Red-necked Crake

Crater Lakes National Park (980 ha) comprises two separate sections – Lake Barrine and Lake Eacham. Both sections protect crater lakes surrounded by upland rainforest. More than 180 bird species are on the park list. The turn-off to Lake Barrine is 60 km from Cairns along the Gilles Hwy, and the turn-off to Lake Eacham is 8 km further along the same hwy. There is cabin-style accommodation at Chambers Wildlife Rainforest Lodge (www.rainforest-australia.com).

Lake Eacham is surrounded by rainforest, with good birding around the day-use area and the walk to the ranger's office. When fig trees are fruiting in these areas, look for Barred Cuckoo-shrike and Double-eyed Fig-Parrot. The Lake Circuit Track (3 km loop) winds around the lake, giving you access to upland tropical rainforest. Endemics of the Wet Tropics to look for here include Atherton Scrubwren, Mountain Thornbill, Macleay's and Bridled Honeyeater, Grey-headed Robin, Chowchilla, Bower's Shrike-thrush, Pied Monarch, Victoria's Riflebird, Tooth-billed Bowerbird and White-throated (Little) Treecreeper. Other regular species include Wompoo and Superb Fruit-Dove, White-eared Monarch and Spotted Catbird. The small reddish form of the Southern Boobook (ssp. *lurida*) sometimes calls at night from within the day-use area. Other resident animals of interest include Musky Rat-kangaroo and Red-legged Pademelon, Saw-shelled Turtle, Boyd's Forest Dragon and Eastern Water Dragon.

The aquatic environment of Lake Barrine, another clear blue crater lake, is more suited to waterbirds than Lake Eacham. In the reeds along the shallow edges, look for Buff-banded Rail, Red-necked Crake and Azure Kingfisher, and species more typical of open water on the lake include Hardhead, Grey Teal, Pacific Black Duck, Wandering Whistling-Duck and Australasian and Great Crested Grebe.

Birding within 150 m of the forest bordering the top carpark is worthwhile; look here for Chowchilla, Fernwren, Yellow-breasted Boatbill, Large-billed and Yellow-throated Scrubwren and Spotted Catbird. Tooth-billed Bowerbird are found along the Twin Kauris Walk (160 m return), which leads from the lower carpark to the boardwalk at the base of two large Kauri Pines. The Lake Circuit Track (5 km loop) and the Rainforest Walk (600 m one way) are also very worthwhile. Flowering plants in gardens around the main buildings in the park attract Scarlet, Yellow-spotted, Bridled, Lewin's, Dusky and Macleay's Honeyeater. Spotlighting here provides opportunities for Red-legged Pademelon, Northern Brown and Long-nosed Bandicoot and, occasionally, Giant White-tailed Rat, while Musky Rat Kangaroo is frequently seen in the day.

In the grounds of Chambers Wildlife Rainforest Lodge, located next to Lake Eacham, you may see Emerald Dove, Victoria's Riflebird, Spotted Catbird, Tooth-billed Bowerbird, Grey-headed Robin and Chowchilla. A good spot for nocturnal mammals: Red-legged Pademelon, Sugar Glider, Long-nosed and Northern Brown Bandicoot, Giant White-tailed Rat and Striped Possum, Herbert River Ringtail and Coppery Brushtail Possum feed at night in the lodge grounds and the surrounding rainforest.

Mission Beach

Key species: Southern Cassowary, Pale-vented Bush-hen, Yellow-breasted Boatbill
Other species: Red-necked Crake, White-eared Monarch, Barred Cuckoo-shrike, Double-eyed Fig-Parrot

Mission Beach is 110 km south-east of Cairns and 40 km south of Innisfail. With a population of ~500 people, the township of Mission Beach is spread along a thin strip of land between the ocean and the hills and farmland behind. Although Mission Beach is home to many lowland rainforest species the most notable resident is the Southern Cassowary. There is a range of holiday accommodation at Mission Beach, and a visitor information centre: www.missionbeachtourism.com.

The lowland rainforest around Mission Beach supports the highest density of the endangered Southern Cassowary in Australia. Despite their size, they are a surprisingly cryptic species. You have a chance of seeing them along the two

The Southern Cassowary is an iconic species of the wet tropics. Although widespread within the region, it occurs at low densities. Individuals are most frequently observed in the Mission Beach and Daintree Rainforest areas. Cassowary House near Kuranda is also visited regularly by several birds.

main walking tracks in Djiru National Park (4123 ha), 8 km west of Mission Beach. The Lacey Creek Circuit (1.2 km), located on the El Arish-Mission Beach Rd, 7.5 m from Mission Beach Visitor Centre, does a small loop through a variety of lowland rainforest vegetation, crossing numerous arms of Lacey Creek. Much of the Musgravea Track (6 km) is along an old forestry track running alongside the Hull River and Jurs Creek. Another good spot to search is around the Licuala day-use area at the southern end of the Musgravea Track. To get there, turn north off the Tully–Mission Beach Rd onto the signposted unsealed 2WD road, 8 km south-west of Mission Beach. It is worth noting that all roads in the Mission Beach area travel through Southern Cassowary habitat, so there is a chance of observing them on any of the roads in the area.

In February 2011, category 5 Cyclone Yasi crossed the Qld coast near Mission Beach. With wind gusts of 290 km, much of the rainforest vegetation was shredded. As a result, Southern Cassowary regularly moves around the township and along Mission Beach itself in search of food. As many guest houses in Mission Beach are regularly visited by cassowaries, it can be worth enquiring about them when checking in to your accommodation. Other birds in Mission Beach include White-eared Monarch, Yellow-breasted Boatbill, Barred Cuckoo-shrike and Double-eyed Fig-Parrot. Look also in waterside thickets of vegetation for Red-necked Crake and Pale-vented Bush-hen. If you fail to find Southern Cassowary at Mission beach, try the patch of coastal rainforest at Etty Bay, near Innisfail as they sometimes forage among the tents at the caravan park.

Paluma Range National Park

Key species: Victoria's Riflebird, Northern Fantail, White-browed Robin, Golden Bowerbird, Chowchilla, Yellow-breasted Boatbill, Fernwren, Atherton Scrubwren, Mountain Thornbill, Bower's Shrike-thrush, Bridled Honeyeater
Other species: Noisy Pitta, Macleay's Honeyeater, Pale-yellow Robin, Tooth-billed Bowerbird, Russet-tailed Thrush

Rising 1000 m above the Big Crystal Creek floodplain, Paluma Range National Park (10 800 ha) straddles the summit and escarpment of the Paluma Range. The park forms the southernmost section of the Wet Tropics World Heritage Area, and is the southern range limit for many of the high-altitude rainforest bird species that occur in the region. Access to the park is via an old section of the Bruce Highway, 61 km north of Townsville or 40 km south of Ingham. There is a camping area at the base of the mountain along Big Crystal Creek; this can be busy during holiday periods, so advance bookings are advised. Here, there are species associated with drier eucalypt woodlands and gallery forest including Northern Fantail, White-browed Robin, Fairy Gerygone, Lemon-bellied Flycatcher and Grey Whistler. At night around the campsite, listen for both Large-tailed and White throated Nightjar. Similar species occur along Waterview Creek at the Jourama Falls camping area in the north of the park.

From the camping area, head up Mount Spec Rd and stop at the Little Crystal Creek picnic area. Here you may encounter Noisy Pitta, Little Shrike-thrush, Yellow-breasted Boatbill, Pale-yellow Robin, Azure Kingfisher and Scarlet and Yellow-spotted Honeyeater. In summer, Buff-breasted Paradise-Kingfisher occasionally nest in termite mounds near the picnic area. Further up the mountain, along the Witts Lookout Track (3 km return), starting 200 m beyond McClellands Lookout, look for Chowchilla, Fernwren, Mountain Thornbill, Yellow-throated Scrubwren and Russet-tailed Thrush.

The bird-feeders at Ivy Cottage, a café serving Devonshire Tea and other items in the Paluma, is well known for attracting

Victoria's Riflebird, and Macleay's, Lewin's, White-cheeked and Yellow-spotted Honeyeater. The H Track (1.1 km) starts at the small ring road near the old Paluma Police Station, finishing at the rear of Lennox Crescent. Along here, look for Australian Brush-turkey, Tooth-billed Bowerbird, Grey-headed Robin, Victoria's Riflebird, Spotted Catbird, Bridled Honeyeater, Lovely Fairy-wren, Satin Bowerbird, Barred Cuckoo-shrike, Bower's Shrike-thrush, Grey-headed Robin and Fernwren. Along the creek line in the same area, there is a chance of Red-necked Crake or Little Kingfisher. Similar species occur along the shorter Paluma Rainforest Walk (680 m), which starts opposite the Paluma Town Hall and finishes 200 m westward along Mount Spec Rd. Occasionally, Southern Cassowary move along the road in the township of Paluma. Night birds here include Lesser Sooty Owl, Papuan Frogmouth and Southern Boobook Owl (including ssp. *lurida*).

Golden Bowerbird occurs near the Birthday Creek Falls parking area, with a well-known bower located very near to the walking track. Along the Birthday Creek Falls Walk (1 km return), look also for Tooth-billed Bowerbird, Spotted Catbird, Bower's Shrike-thrush, Bridled Honeyeater, Atherton Scrubwren, Grey-headed Robin, Mountain Thornbill, Russet-tailed Thrush, Chowchilla, Yellow-breasted Boatbill, Spectacled and White-eared Monarch. Access this walk 4.6 km west of the Paluma along Mount Spec Rd; turn right onto the unsealed road that leads to Lake Paluma and travel another 6.8 km to a carpark and the start of the track.

Tyto Wetlands

Key species: Australian Little Bittern, Black Bittern, White-browed Crake, Australian Spotted Crake, Little Kingfisher, Eastern Grass Owl

Other species: Crimson Finch, Pale-vented Bush-hen, Red-backed Button-quail, Bush Stone-curlew

The Tyto Wetlands (named after genus that includes the Eastern Grass Owl) (90 ha) are located off the Bruce Hwy on Cooper St, 1 km south of Ingham. There is a wide range of accommodation in Ingham.

The main path is the Perimeter Walk (3.3 km loop), which takes you around the complete wetland and a series of viewing platforms and bird-hides. The main wetland is 750 m from the carpark. Here, a path leads to the left for 180 m, where Crimson Finch occurs. Black and Australian Little Bittern occur here; look for them perched in the sedges and bulrushes around the lake and occasionally observers are rewarded with views when they fly across open water from one vegetated area to another. From the hides, look for White-browed Crake and Comb-crested Jacana on floating vegetation including lily pads, while around the grassy/muddy fringes look for Australian Spotted Crake, Buff-banded Rail, Latham's Snipe (late spring and summer), Pale-vented Bush-hen (occasionally) and Glossy Ibis. Waterfowl here include Magpie Goose, Green and Cotton Pygmy-Goose, and Plumed and Wandering Whistling-Duck, while Little and Azure Kingfisher favour the quiet fringes of the wetland; look especially for Little Kingfisher in the small channels near the main bird hide. The grassy areas around the wetland hold Red-backed

Although Red-backed Button-quail occur in a wide variety of habitats, they typically frequent dense wet grasslands including those at Tyto Wetlands.

Button-quail, Tawny Grassbird, Australian Reed-Warbler, Golden-headed Cisticola and Bush Stone-curlew, and there is also a dawn-and-dusk viewing platform, designed to provide opportunities to observe Eastern Grass Owl; the best time to look for them is when they take flight just on or after dusk.

Townsville Town Common

Key species: Little Kingfisher, Black-necked Stork, Magpie Goose, Cotton Pygmy-Goose, Green Pygmy-Goose, Brolga
Other species: Pale-vented Bush-hen, White-browed Crake

The Townsville Town Common Conservation Park (3000 ha) is an expansive wetland system. Located 6 km north of Townsville, the park's entrance is reached from Cape Pallarenda Rd near the Rowes Bay Golf Club. Gates are open between 6:30 a.m. and 6:30 p.m. There is usually water throughout the Common from the break of the wet season until at least June, but levels then usually drop from July through August and into September. At that time, large numbers of waterfowl congregate in the remaining wetlands. The late dry season period extends from early September to December, when most of the lagoons have dried up. There is a wide range of accommodation in and around Townsville, as well as on nearby Magnetic Island (see the following account).

The Forest Walk (1.9 km return) starts just after the entrance to the Common. Look for open woodland birds such as Pheasant Coucal, Little Shrike-thrush, Satin Flycatcher, Spangled Drongo, and White-gaped, Yellow, Brown-backed, White-throated and Blue-faced Honeyeater. During the wet season, Brush, Oriental (uncommon) and Channel-billed Cuckoo may also occur. The Barramundi Pools, on the south side of the Forest Walk, are a good site for Little Kingfisher. Along the Freshwater Lagoon Rd, located 1 km from the entrance, there is a short walk that leads to the observation tower. This provides excellent views of the extensive Long Swamp. When it holds water, expect to find Magpie Goose, Australasian Grebe, Australasian Darter, Black-winged Stilt and Black-fronted Dotterel. From May onwards, Brolga begin to flock as other wetlands in the area dry out. Along Freshwater Lagoon Rd, search the grassy fringes for Brown Quail, Horsfield's Bushlark, Tawny Grassbird, Chestnut-breasted Mannikin and Crimson, Plum-headed and Black-throated Finch. Both Red-backed and Red-chested Button-quail are also reported from this area from time to time.

There are three bird hides (Jacana, Freshwater Lagoon and Bald Rock) located at Freshwater Lagoon. From any of these, look for waterbirds such as Magpie Goose, Green and Cotton Pygmy-Goose, Plumed and Wandering Whistling-Duck, Comb-crested Jacana and White-browed Crake. Around the wetland margins, search for Little Kingfisher, Blue-winged Kookaburra, Black-necked Stork, Black-winged Stilt, Australasian Darter, Latham's Snipe (mostly Sep–Oct), Nankeen Night-Heron, Black Bittern and Glossy Ibis. In the shallow fringes of the lagoon, you may see Buff-banded Rail and Pale-vented Bush-hen. Migratory shorebirds visit between September and April; during this period, you may find Pacific Golden Plover, Sharp-tailed and Pectoral (rare) Sandpiper, Bar-tailed and Black-tailed Godwit.

The Shelly Beach Circuit Track (9 km return) leaves from Bald Rock carpark and leads to mangroves that grow along Shelly Beach; search here for Striated Heron, Collared Kingfisher, Mangrove Robin, Large-billed and Mangrove Gerygone, Shining Flycatcher and Black Butcherbird. Great-billed Heron occur in the mangroves west of Shelly Beach along Bohle River.

The Many Peaks Range walking trails (such as the Mount Marlow Walk – 5 km one way) start at the Bald Rock carpark, taking you through thickets of vine forest and rainforest gullies: look for Rufous Fantail, Spectacled Monarch, Satin Flycatcher, Brown Cuckoo-Dove, Rose-crowned Fruit-Dove, Yellow-spotted Honeyeater and White-eared Monarch (winter); Allied Rock Wallaby also occurs here.

On your way to the Townsville Town Common, it is worth stopping at Soroptimist Park in Rowe Bay, located on Cape Pallarenda Rd, 3.5 km from the centre of Townsville. Search here for Varied Honeyeater and Mangrove Gerygone, and check along the beach for Beach Stone-curlew, especially near the creek-mouth. Another good site for Beach Stone-curlew is near the mouth of the Ross River (5 km south of Townsville). Scan the river's edge and the large sand spit on the south side of the river. The mangroves here are also excellent; they support Mangrove Honeyeater, Collared Kingfisher, Mangrove Gerygone, Striated Heron, Black Butcherbird and Mangrove Robin.

Magnetic Island

Key species: Bush Stone-curlew, Brown Booby, White-eared Monarch
Other species: Beach Stone-curlew, Little Kingfisher, Collared Kingfisher, Spectacled Monarch, Black-faced Monarch, Noisy Pitta

Reached by passenger and car ferry from Townsville, Magnetic Island is 8 km east of Townsville; from the ferry, look for Brown Booby roosting on the channel markers. Eastern Reef Egret and Striated Heron feed on the rocky shoreline near the Nelly Ferry Terminal. Common species on the

The Townsville Town Common is a vast area of coastal marshlands. Seasonal changes mean that expansive wet season wetlands shrink to waterbird refuges through the middle of the year before drying completely towards the end of dry season.

island include Scaly-breasted and Rainbow Lorikeet, Olive-backed Sunbird and Hornbill Friarbird. The 'special bird' on the island is the Bush Stone-curlew, which is both abundant and remarkably tame. They are readily found along the main esplanade in the township of Picnic Bay, while the Magnetic Island Golf Course is another good location for the species.

There is a good area of mangroves between Cockle Bay and Bolger Bay (immediately south of Picnic Bay), accessed from West Point. Birds that can be found here include Mangrove and Large-billed Gerygone, Shining Flycatcher, Sacred Kingfisher and, occasionally, Little and Collared Kingfisher, and Beach Stone-curlew (sometimes feeding around Bolger Bay). Eastern Osprey roost on the shipwreck *City of Adelaide* (located 100 m off the coast at Cockle Bay), and nest at the headland of Balding Bay on the north side of the island. White-eared, Spectacled and Black-faced Monarch and Noisy Pitta occur on the island, with a good place to look for them being the vine-thicket forest along Gustav Creek, located at the start of the Nelly Bay to Arcadia Walk (5 km

Horseshoe Lagoon

Horseshoe Lagoon is located 31 km west of Ayr (56 km south-east of Townsville) on the Bruce Hwy. Viewed from the Hwy, the lagoon is a fairly reliable place for Cotton and Green Pygmy Goose. White-winged Black, Whiskered, Caspian and Gull-billed Tern also occur here, and you should see Magpie Goose, Hardhead, Black-necked Stork, Australasian Darter, Brolga, Black-necked Stint and Comb-crested Jacana. Crimson Finch and Chestnut-breasted Mannikin are generally quite common in the grassy areas around the wetland. You can also view the wetland from Hodel Rd, located on the west side of the wetland, off the Bruce Hwy, 32 km west of Ayr (just east of Giru).

Just near Horseshoe Lagoon, there occurs an isolated southern population of Zitting Cisticola. This species can be reliably found along Hodel Rd, just east of Giru. Travel down Hodel Rd until a 'Wildlife Refuge Carrick Plains' is visible on the right. Frequenting the grasslands here are both Zitting and Golden-headed Cisticola. Zitting Cisticola also occur along Hodel Rd much of the way to the boat ramp at the roads end (which becomes Morris Creek Rd). Another area worth checking is beside Beach Rd, which heads north from the centre of Ayr. Search the south-side the road 15–16 km from Ayr, just before Alva Beach. Lastly, they also occur along the road to Jerona and at Cape Cleveland, in Bowling Green Bay National Park. A somewhat cryptic species, they are best observed during the breeding season (Dec–Apr), when the males give a zigzagging flight display combined with a regular *zitting* call (likened to repeated snips of a pair of scissors).

one way) at the end of Mandalay Avenue in Nelly Bay. Horseshoe Bay Lagoon Conservation Park is on the outskirts of Horseshoe Bay. An attractive lagoon, here you may find Little Kingfisher (uncommon), Blue-winged Kookaburra, Magpie Goose and Wandering Whistling-Duck. Mammals include Koala and Allied Rock-Wallaby. A significant population of Dugong, along with Green and Flatback Turtles, forage on the seagrass beds around the island.

Cape Hillsborough National Park

Key species: Yellow White-eye, Mangrove Golden Whistler, White-eared Monarch, Buff-breasted Paradise-Kingfisher
Other species: Collared Kingfisher, Broad-billed Flycatcher, Shining Flycatcher, Mangrove Gerygone, Rose-crowned Fruit-Dove

Cape Hillsborough National Park (1012 ha) is 50 km north-west of Mackay on the central Qld coast, ~1025 km north of Brisbane. Lowland rainforest and vine forest can be found along creeks, valleys and hillside gullies within the park. Several tropical birds reach the southern limit of their distribution in the area around Cape Hillsborough. To get there from Mackay, travel north on the Bruce Hwy for 20 km and turn right into Seaforth Rd. Travel a further 20 km and turn right into Cape Hillsborough Rd. It is then 10 km to the main park entrance. Basic camping is available at Smalley's Beach camping area, reached via a signposted gravel road off Cape Hillsborough Rd, and there is accommodation at Cape Hillsborough Nature Resort: www.capehillsborough resort.com.au.

The Diversity Boardwalk (1.2 km return) passes through paperbark woodland, mangrove communities and vine thicket. In the mangroves, look for Collared Kingfisher, Broad-billed and Shining Flycatcher, and Large-billed and Mangrove Gerygone. Although rare, both Yellow White-eye and Mangrove Golden Whistler have been recorded in the mangroves also. Other birds to look for along the boardwalk include Emerald Dove, Little Bronze Cuckoo, Forest Kingfisher, White-eared and Black-faced Monarch.

Along the Yuibera Plant Trail (1.2 km return) and longer Andrews Point Track (5.2 km return via the track or 2.8 km return via the beach) look for Rose-crowned and Wompoo Fruit-Dove, Pied Imperial-Pigeon, Red-tailed Black-Cockatoo, Pale-headed Rosella, Pheasant Coucal and Fairy Gerygone. Smalley's Beach (a muddy beach) can be good for Beach Stone-curlew, Striated Heron, Great and Eastern Reef Egret, Grey-tailed Tattler and Whimbrel; search the area 500 m from Central Point, near the campground. Other birds recorded at Cape Hillsborough include Orange-footed Scrubfowl, Buff-breasted Paradise-Kingfisher (at the southern edge of their range Nov–Apr), Bush Stone-curlew, Pied Imperial-Pigeon, Peaceful and

Despite being widespread in suitable habitat, the White-eared Monarch can be surprisingly elusive. Occupying a wide range of rainforest, gallery forest and vine thicket habitats, the species usually frequents the outer foliage of taller trees.

Yellow White-eye and Mangrove Golden Whistler

Aside from Cape Hillsborough National Park (see above), isolated populations of Yellow White-eye and Mangrove Golden Whistler have been recorded at several sites in central Qld. Both have been seen at the mouth of the Haughton River (search the mangroves just south of the township of Cungulla), while Yellow White-eye have been observed in mangroves along Plantation Creek (such as near the boat ramp at the end of Old Wharf Rd, 12 km east of Ayr). Another spot to look for Yellow White-eye is the mangroves at the mouth of Molongle Creek, just north of the boat ramp. Mangrove Golden Whistler has been recorded in the mangroves on Camp Island: www.campislandresort.com.

Bar-shouldered Dove, Brown Cuckoo-Dove, Little Bronze-Cuckoo, Blue-winged Kookaburra, Lewin's, Dusky, Yellow and White-throated Honeyeater, Little Shrike-thrush, Spangled Drongo, Black Butcherbird, Spectacled Monarch, Olive-backed Sunbird, Eastern Osprey, Grey Goshawk and Rufous Owl (rare).

Eungella Plateau

Key species: Eungella Honeyeater, Squatter Pigeon, Barred Cuckoo-shrike
Other species: Superb Fruit-Dove, Rose-crowned Fruit-Dove, Wompoo Fruit-Dove, Regent Bowerbird, Cicadabird. Also Platypus.

The Eungella Plateau is an isolated mountain range that lies close to the boundary for subtropical and tropical rainforest systems. As a consequence, it supports a variety of species from both vegetation types. Located 80 km west of Mackay, the key species in the area is the locally endemic Eungella Honeyeater.

The Eungella Honeyeater, first recognised as a separate species in 1983, is closely related to Bridled Honeyeater. Usually feeding high in the rainforest canopy, they can on occasions prove very difficult to get onto; early morning when they are most vocal is often best. There are a few good places for this species, all in rainforest on the western border of Eungella National Park. Look first along the 2WD Chelmans Rd, with the best area being right at the end of the road near some double gates. To get there, drive 15.6 km up Dalrymple Rd and then 2.9 km down Chelmans Rd. Note that Chelmans Rd can get slippery after rain. If you miss them here, they occur nearby close to the junction of Dalrymple Rd with Snake Rd. Another site for Eungella Honeyeater is along the first few kilometres of Diggin's Rd. This area is good for more general birding also. Here, you may also encounter Brown Cuckoo-Dove, Superb, Rose-crowned and Wompoo Fruit-Dove, Topknot Pigeon, Noisy Pitta, Lewin's, Dusky and Scarlet Honeyeater, Brown Gerygone, Cicadabird, Barred Cuckoo-shrike, Spectacled and Black-faced Monarch and Regent Bowerbird. The grassy eucalypt woodlands on the western side of Eungella Reservoir is a good site for Squatter Pigeon and other drier country species. While in the area, stop at the platform at Broken River for a good chance of Platypus; early in the morning and late in the afternoon are best.

Mackay

Key species: Beach Stone-curlew, Black Bittern, Pale-vented Bush Hen, White-browed Robin
Other species: Black-necked Stork, Radjah Shelduck, Red-whiskered Bulbul

Mackay is 390 km south of Townsville and 970 km north of Brisbane. There are several good birdwatching locations in or near Mackay including Mackay Regional Botanic Gardens, the mouth of the Pioneer River and Padaminka Nature Refuge. Being a large regional town, there is a wide range of hotel and B&B-style accommodation available.

On the south side of the Pioneer River, start at the eastern end of River Street and walk along the embankment towards the mouth of the river, looking along the way for Beach Stone-curlew, Black-necked Stork, Black Bittern, Striated Heron, Plumed Whistling-Duck, Radjah Shelduck, Brown Quail, and Caspian, Lesser Crested, Gull-billed and Whiskered Tern. There is 4WD access from Gordon Street East. Alternatively, from the north side of the Pioneer River, head south to the end of East Point Drive. The mangroves and estuary here can be excellent for migratory shorebirds including Grey and Pacific Golden Plover, Greater and Lesser Sand Plover, Terek and Curlew Sandpiper, Bar-tailed Godwit, Grey-tailed Tattler, Whimbrel, Eastern Curlew, Ruddy Turnstone and Common Greenshank. Flocks of Little Tern are usually also present in summer.

First opened in 2003, the Mackay Regional Botanic Gardens is one of Australia's newest botanic gardens. The main entrance is on Lagoon Street, accessed via the Bruce Hwy 2 km from the centre of Mackay. The area usually holds a good variety of waterbirds including Magpie Goose, Plumed and Wandering Whistling-Duck, Cotton Pygmy-Goose, Australasian Darter, Glossy Ibis, Comb-crested Jacana and, occasionally, Black-necked Stork. Honeyeaters here include Yellow, Brown, Bar-breasted and Blue-faced. There is also an isolated northern population of Red-whiskered Bulbul (introduced) in and around the gardens.

The Padaminka Nature Refuge (74 ha) is a private property just outside Mackay. It features a restored wetlands and remnant bushland. Some of the species recorded on the refuge include Black Bittern, Pale-vented Bush Hen, Pacific Baza, Grey Goshawk, White-browed Robin, White-eared Monarch, Lemon-bellied Flycatcher, Little Shrike-thrush, Tawny Grassbird, Large-tailed Nightjar and Powerful Owl. For further information: padaminka@bigpond.com, (07) 4959 3770.

Lady Elliot Island

Key species: Red-tailed Tropicbird, Bridled Tern, Sooty Tern, Black-naped Tern, Roseate Tern, Wandering Tattler, 'Capricorn Silvereye'
Other species: Wedge-tailed Shearwater, Great Frigatebird, Lesser Frigatebird, Beach Stone-curlew, shorebirds

Lady Elliot Island (42 ha) is the southernmost coral cay of the Great Barrier Reef. It is situated 85 km north-east of Bundaberg, and is serviced daily by flights from Bundaberg, Brisbane and the Gold Coast. Because of its small size, it is quite feasible to walk around the entire island in an hour or two. Lady Elliot Island Eco Resort offers a range of accommodation: www.ladyelliot.com.au.

Lady Elliot Island is an important seabird breeding site, with over 100 000 birds nesting during the spring–summer breeding season. Red-tailed Tropicbird has

Lady Elliot Island is the southern-most cay in the Great Barrier Reef and provides excellent opp\ortunities to observe a variety of tropical seabird species.

nested on Lady Elliot Island since 1983, with 23 known nest locations and up to 8 pairs nesting on the island at one time. Nesting occurs almost year round, with a small period of absence in April–May. Nesting terns include Black-naped, Bridled, Roseate (small breeding population on Leeward Beach), Crested and Sooty Tern, and Common and Black Noddy. There are also occasional records of Little, Lesser Crested and White Tern, and Grey Ternlet. While walking around the island, look for seabirds such as Wedge-tailed Shearwater (nesting), Great and Lesser Frigatebird, and Brown and Red-footed (rare) Booby. Summer shorebirds include Grey-tailed and Wandering Tattler (favouring the rocky south-east side), Grey and Pacific Golden Plover, Greater and Lesser Sand Plover, Great Knot, Sharp-tailed and Common Sandpiper, Red-necked Stint, Whimbrel, Bar-tailed Godwit, Ruddy Turnstone, while Double-banded Plover are regular visitors from February to September. Resident shorebirds include Beach Stone-curlew and Australian Pied and Sooty Oystercatcher. Rarities have included Little Curlew and Oriental Plover, with the airstrip being a good place to check.

The 'Capricorn Silvereye' (sometimes considered a full species), a large form of the Silvereye (ssp. *chlorocephalus*) and the only bird taxon endemic to the Great Barrier Reef, is a common resident on Lady Elliot Island. Other birds to look for include Eastern Reef Egret (both white and dark morphs), White-bellied Sea-Eagle, Buff-banded Rail, Sacred Kingfisher, Tawny Grassbird, Golden-headed Cisticola, Brown and Scarlet Honeyeater and, in summer, Horsfield's and Shining Bronze-Cuckoo and Channel-billed Cuckoo. The reef areas surround Lady Elliot Island are also renowned for the scuba diving and snorkelling opportunities. There is a resident population of Manta Ray, and every year,

Capricorn Yellow Chat and the Tortilla Plain

The Capricorn subspecies of Yellow Chat (ssp. *macgregori*) occurs on the Tortilla Plain, Curtis Island and the Fitzroy River delta in Capricornia, central Qld. The most accessible site is along Port Alma Rd near the saltworks, especially near the steal pits beside the road, just west of the saltworks. To get there, turn east off the Bruce Hwy onto Port Alma Rd, 32 km south of Rockhampton. The saltworks are another 10 km; there is just enough room to park on the verge near the steal pits, though take care as the road is narrow. The best time to look for Yellow Chat is from late December to March, particularly after seasonal rain. Another accessible site for the species is the salt flats and associated grasslands at Twelve Mile Creek, just south of Marmor. To get there, turn east of the Bruce Hwy 47 km south of Rockhampton and then right into Twelve Mile Rd immediately after you cross the railway lines. Continue for a further 2 km until you cross the creek line before finally turning left at this point down a 200 m long track. From here, look for Yellow Chat by walking ~2 km downstream. Zitting and Golden-headed Cisticola are also found in the area. Note that the wetlands here have a series of shallow braided channels with areas of mud and shallow water, so it can be a difficult walk, especially after rain.

between November and March, Green and Loggerhead Turtle nest on the island. Finally, Humpback Whale is common in the waters around the island from June to October.

Queensland Gulf Country and the Queensland outback

Regional overview

The Qld outback covers ~12% of the Australian land mass. Defined regionally as the north-west (Dinosaur, Fossil and Mining Country), central-west (Matilda Country) and south-west (Channel Country), habitats range from seasonally flooded wetlands adjacent to the gulf, through Spinifex-dominated woodlands in the central west, to Mitchell grass plains and lignum-choked watercourses in the south-west.

These areas are some of the most arid regions in the world; however, after significant rains, vegetation in the region transforms and birdlife similarly responds. Due to the enormous distances between sites, to do the area justice it may be necessary to make several trips (or one long trip) to the region.

For visiting the north-west section of outback Qld, a recommended 14-day itinerary involves:

- **Day 1.** Cairns to Cumberland Dam near Georgetown, 381 km (Squatter Pigeon, Black-throated, Masked, Plum-headed and Gouldian Finch, Pictorella Mannikin, Rufous-throated, Banded and Yellow-tinted Honeyeater)

- **Days 2–3.** Georgetown to Karumba, 372 km (Sarus Crane, White-breasted Whistler, Red-headed Honeyeater, Zitting Cisticola, Yellow White-eye, Star Finch, Arafura and Mangrove Grey Fantail)
- **Days 5–7.** Karumba to Mount Isa, 576 km (Kalkadoon Grasswren, Carpentarian Grasswren, Pictorella Mannikin, Spinifex Pigeon, Yellow Chat, waterbirds).
- **Days 8–10.** Mount Isa to Boodjamulla (Lawn Hill) National Park, 381 km (Purple-crowned Fairy-wren, Grey Falcon, Buff-sided Robin, Sandstone Shrike-thrush, Spinifex Pigeon).
- **Days 11–13.** Boodjamulla National Park to Cairns, 1087 km.

To explore the south-west section of outback Qld (including Innamincka, SA), a recommended 14-day itinerary involves:

- **Days 1–3.** Brisbane to Bowra, 800 km (Grey Falcon, Bourke's Parrot, Chestnut-breasted Quail-thrush, Hall's Babbler, Painted Honeyeater, Chirruping Wedgebill).
- **Days 4–5.** Bowra Station to Thargomindah, stopping at Eulo Bore and Lake Bindegolly on the way, 197 km (Bourke's Parrot, Chestnut-breasted Quail-thrush, Hall's Babbler, Slaty-backed Thornbill, waterbirds).
- **Days 6–8.** Thargomindah to Innamincka, SA, stopping at the Cooper Creek Crossing and Wilson River along the way, 372 km (Grey and Eyrean Grasswren, Gibberbird, Cinnamon Quail-thrush, Freckled Duck, Banded Whiteface).
- **Days 9–11.** Innamincka to Birdsville, 413 km (Eyrean and Grey Grasswren, Yellow Chat, Banded Whiteface, Gibberbird, Letter-winged Kite).
- **Days 12–14.** Birdsville back to Brisbane, 1500 km.

Both of the above itineraries are major undertakings, and require a 4WD and extensive planning.

The best time to visit the Qld outback is during autumn or spring, when both night and day temperatures are mild. In the north, they range from an average 24.7°C in winter to 37.2°C in summer. In the central-west and south-west, outback winter temperatures range from an overnight low of 6°C to a daytime high of 23°C, while summer temperatures are very hot with an average 35°C (sometimes reaching into the mid-40s). In the north-west section of outback Qld, there are well-defined wet

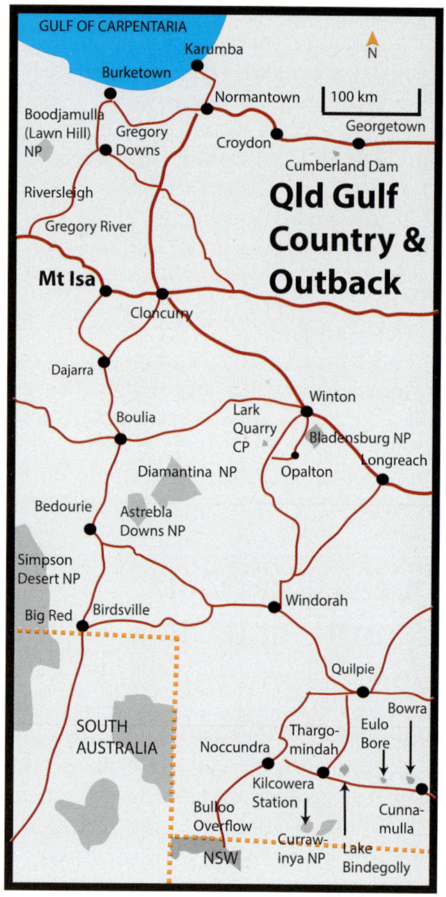

and dry seasons; the dry season is from May to November, with September the month of lowest rainfall, while the wet is from November to April, with March being the month with the most rain.

The roads in the Qld outback are generally good, with nearly all the major highways sealed; however, visiting most of the recommended birdwatching sites requires a 4WD – during the wet, rains can bring a dramatic rise in river and creek levels, and many of the roads in the north-west may become impassable. Visitors should be well equipped, self-sufficient and carry spare fuel, food and water. There is also limited communication, including limited mobile phone reception. To check local road conditions visit www.racq.com.au.

Birding highlights

There is one range-restricted endemic in the region: the Kalkadoon Grasswren. Other resident highlights include Sarus Crane, Inland Dotterel, Spinifex and Squatter Pigeon, Hall's Babbler, White-breasted Whistler, Buff-sided Robin, Spinifexbird, Grey, Eyrean, and Carpentarian Grasswren, Rufous-crowned Emu-wren, Purple-crowned and Splendid Fairy-wren, Yellow Chat, Red-headed, Rufous-throated, Banded and Yellow-tinted Honeyeater, Zitting Cisticola and Yellow White-eye. Regularly recorded birds for the area include eight finch species (Zebra, Painted, Star, Black-throated, Masked, Plum-headed (uncommon), Gouldian (rare) and Pictorella Mannikin (uncommon)) and an impressive 19 species of raptor, including rare species such as Letter-winged Kite, Grey and Black Falcon. The region may also provide the best chance of a Night Parrot: Australia's rarest bird. Summer migrants to outback Qld include Little Curlew, Oriental Plover and Oriental Pratincole. Interesting subspecies in the area include Striated Grasswren (ssp. *rowleyi*, a subspecies that may be elevated to full species status, known as Rusty Grasswren), Purple-crowned Fairy-wren (ssp. *macgillivrayi*), Australian Ringneck ('Cloncurry Ringneck' ssp. *macgillivrayi*), Squatter Pigeon (red-eyed ssp. *peninsulae*) and Chestnut-breasted Quail-thrush (ssp. *castaneothorax*).

Cumberland Dam (Georgetown)

Key species: Squatter Pigeon, Black-throated Finch, Masked Finch, Plum-headed Finch, Gouldian Finch (rare), Pictorella Mannikin, Rufous-throated, Banded Honeyeater, Yellow-tinted Honeyeater, Spinifex Pigeon
Other species: Chestnut-breasted Mannikin, Double-barred Finch, Zebra Finch, Brown Treecreeper, Varied Lorikeet

Cumberland Dam is an old mine site 21 km west of Georgetown (386 km west of Cairns) on the south side of the Gulf Developmental Rd; look for the old brick chimney adjacent to several dams (a large dam on the east side and a smaller dam to the west). Surrounded by open grassy woodlands scattered with termite mounds, Cumberland Dam is an excellent site for finches and honeyeaters, especially towards the end of the dry season when surface water becomes more limited. There is accommodation at the Goldfields Caravan Park in Georgetown and bush camping at Cumberland Dam itself.

Eight species of finch have been recorded at Cumberland Dam: Zebra, Double-barred, Black-throated, Masked, Plum-headed (uncommon and the northern limit for this species), Gouldian (rare) Finch, Chestnut-breasted and Pictorella Mannikin (uncommon). During the dry season, finches tend to drink on the east side of the small dam, especially early in the morning and usually first landing within any grassy areas near the fence line. On the south side of the small dam, there are several mid-sized shrubs lining the dam wall, providing shelter for honeyeaters before they too come to drink; look for Rufous-throated, Yellow-tinted, Banded, Grey-fronted and Blue-faced Honeyeater. Again, the best time for viewing is generally early in the morning. The larger of the two dams can be good for waterbirds such as Hardhead, Green and Cotton (uncommon) Pygmy-Goose, Australian Wood and Freckled (rare) Duck, Plumed Whistling-Duck (often present in large numbers on the north side), Nankeen Night-Heron, Black Bittern (uncommon), Comb-crested Jacana, Australasian Grebe, Painted Snipe (rare) and Black-fronted Dotterel.

In the wooded areas on the eastern side of the larger dam, look for Blue-winged Kookaburra, Red-backed Kingfisher, Red-browed Pardalote, Australian Bustard, Squatter (red-eyed) and Spinifex Pigeon, Paperbark Flycatcher, Rufous Songlark, Brown (Black) Treecreeper, Ground and White-bellied Cuckoo-shrike, Grey-crowned Babbler, Apostlebird, Peaceful and Diamond Dove and Pied Butcherbird. Occasionally in summer, Fork-tailed Swift forages overhead. Woodland parrots that are possible include Pale-headed Rosella, Red-winged Parrot and Varied Lorikeet (the south-east limit for this species). Just to the south of the dam wall, there is usually a Great Bowerbird's bower. At night, Southern Boobook and Spotted Nightjar are sometimes found around the dam.

Red-browed Pardalotes occur in a wide range of wooded habitats across arid and semi-arid Australia; nevertheless, they generally display a preference for smooth-barked eucalypts along watercourses.

Gulf Developmental Rd (Savannah Way)

The Gulf Developmental Rd is 440 km road that links the Cairns region to the Normanton region. While on the drive, look for signs of birdlife at the roadside. In the grassy woodlands between Mount Surprise and Georgetown, Red-backed Kingfisher, Ground Cuckoo-shrike, Little and Red-chested (uncommon) Button-quail, Red-browed and Striated Pardalote, White-throated Gerygone, Weebill and White-browed, Masked and Black-faced Woodswallow all occur. In grasslands between Georgetown and Normanton, keep an eye open for Emu, Australian Bustard, Sarus Crane and Brolga, and birds of prey such as Black-breasted Buzzard, Spotted Harrier, Wedge-tailed Eagle, Collared Sparrowhawk and Peregrine Falcon. The famous Gulflander railway links Croydon and Normanton and provides opportunity to observe the larger species from a moving train.

Karumba

Key species: Sarus Crane, White-breasted Whistler, Red-headed Honeyeater, Zitting Cisticola, Yellow White-eye, Star Finch, Arafura Fantail, Mangrove Grey Fantail
Other species: Mangrove Robin, Australian Pratincole, Little Curlew

Karumba is 72 km north of Normanton at the base of Cape York in the south-eastern corner of the Gulf of Carpentaria. The mangroves along the coastline hold a range of interesting species and, in the wet season, the grasslands surrounding Karumba transform into large ephemeral wetlands that can attract thousands of waterbirds. There are camping facilities at the Karumba Point Caravan Park and other accommodation options in town. Although the road into the town is sealed, it is low-lying, so can be flooded between December and March.

Between Normanton and Karumba, look along the roadsides for birds such Sarus Crane (first recorded in Australia near here in 1966), Brolga, Australian Bustard, Black-necked Stork, Glossy Ibis, Radjah Shelduck, Australian Pratincole (often in recently burnt grasslands), Oriental Plover (uncommon summer) and Little Curlew (summer).

Some of the most rewarding birding in Karumba is to be had in the mangroves and swampy grasslands beside the Norman River. Start by searching the mangroves north of the boat ramp behind Shellie's Beach, accessed via Norman St north of Karumba State School. Here you can expect to find Mangrove Gerygone, Rufous-throated, Yellow (western-most limit) and Red-headed Honeyeater, White-breasted and Mangrove Golden (uncommon) Whistler, Broad-billed and Paperbark Flycatcher, Mangrove Grey and Arafura Fantail, Mangrove Robin, Little Bronze-Cuckoo and Yellow White-eye. In the saline grassland flats between mangrove stands, look for Tawny Grassbird, Variegated Fairy-wren and Golden-headed and Zitting Cisticola. The latter is often confused with the more common Golden-headed Cisticola so a cautious approach to identification is necessary. Other areas worth searching for Zitting Cisticola include the grassland around the airport and similar habitats along Cemetery Road.

A good way of exploring the mud flats and mangrove-lined waterway of the Norman River is by boat. Ferryman Cruises (see www.ferryman.net.au) offers a service from the boat ramp in the centre of town, leaving daily at 9 a.m.; it is worth booking in advance. In addition to the mangrove species mentioned above, there is also a chance of sighting Striated and Great-billed Heron, Eastern Reef Egret, Australian Pied Oystercatcher, Black-necked Stork and Beach Stone-curlew (uncommon), and raptors such as Brahminy Kite, White-bellied Sea-Eagle and Eastern Osprey. The boat cruise also provides access to the sandbars at the mouth of the Norman River. This can be an important roosting site for migratory shorebirds such as Whimbrel, Eastern Curlew, Lesser and Greater Sand Plover, Grey Plover, Red-necked Stint, Sanderling, Red and Great Knot, Grey-tailed Tattler, and Common, Terek, Sharp-tailed and Broad-billed (uncommon) Sandpiper. The opportunity to observe Saltwater Crocodiles is another feature of the Ferryman Cruise. Another good area to observe shorebirds without the need for a boat is along the mud flats at Sunset Point, directly in front of Sunset Tavern.

An additional target species for Karumba is Star Finch, which is usually found near fresh water. Look for them at the small dam 50 m north of the intersection of the Karumba Developmental Rd and Col Kitching Drive (Karumba Point Rd), where they often occur in association with other finches such as Chestnut-breasted Mannikin, and Double-barred and Zebra Finch. Other sites include the grassland near the airport and behind Shellie's Beach, opposite Karumba Point Sunset Caravan Park, opposite the Karumba Health Centre (151 Walker Street), and near the pond opposite the rubbish tip. Within the township of Karumba, you may also find Red-tailed Black-Cockatoo, Red-collared Lorikeet, Red-winged Parrot, Red-backed Kingfisher, Channel-billed Cuckoo (summer), Pheasant Coucal, and Yellow-tinted and White-gaped Honeyeater. At night, listen for Barking Owl at the caravan park, where there are also large numbers of Agile Wallaby.

Spotted Nightjar is a migrant that is common in northern Australian through much of the dry season. The species is frequently encountered resting on outback roads at night.

Boodjamulla (Lawn Hill) National Park

Key species: Purple-crowned Fairy-wren, Buff-sided Robin, Pictorella Mannikin, Sandstone Shrike-thrush, Spinifex Pigeon, Grey Falcon
Other species: Australian Bustard, Black Bittern, Australian Pratincole, Ground Cuckoo-shrike, Grey-headed Honeyeater, Black-chinned Honeyeater

Boodjamulla (Lawn Hill) National Park (280 000 ha) abuts the Qld–NT border. By road, Boodjamulla is 340 km north-west of Mount Isa (via Riversleigh Rd) and 220 km south-west of Burketown (via Gregory Downs and Lawn Hill Rd).The entire road from Burketown and the last 115 km of road from Mount Isa are unsealed. In the dry season, these roads are often corrugated with areas of bulldust, and consequently they are generally unsuitable for conventional vehicles with caravans. The park camping area is well placed at the start of the walking trail system for Lawn Hill Gorge, and there is excellent private camping and cabin accommodation at nearby Adels Grove, situated on Lawn Hill Creek adjacent to the park.

On the drive into Lawn Hill, one passes through a vast expanse of undulating savannah grasslands and open woodlands. Along the roadsides, look for Australian Bustard, Brolga, Black-necked Stork, Australian and Oriental Pratincole (uncommon in summer), Budgerigar, Ground Cuckoo-shrike, Rufous Songlark and Crimson Chat. Birds of prey include Spotted Harrier and Black-breasted Buzzard. If entering from Gregory Downs, Pictorella Mannikin often drink on the creek line of Archie Creek, located 49 km from Wills Development Rd turn-off in Gregory Downs (37 km north-east of the campground at Lawn Hill). Other finch species along the same creek line include Long-tailed, Double-barred, Crimson and Zebra Finch.

Once in the park, a good place to start birding is along the spectacular deep blood-red gorge (carved out by Lawn Hill Creek) located between the Indarri Falls and the Upper Gorge Lookout. The gorge can be reached via the Indarri Falls Loop Walk (3.8 km return). Search in the areas of Pandanus and bamboo along the creek for Purple-crowned Fairy-wren. Other birds to look for in the riverine vegetation along the creek include White-bellied Sea-Eagle, Black Bittern (uncommon), Azure Kingfisher, Blue-winged Kookaburra, Channel-billed Cuckoo, Great Bowerbird, Buff-sided Robin, Olive-backed Oriole, Paperbark Flycatcher, Northern and Arafura Fantail and Crimson Finch. An enjoyable, and often productive, way to observe the birds along the river is to hire a canoe, and paddle to Indarri Falls (6 km return on flat water). Canoes can be hired from the east end of the campsite, and from nearby Adels Grove. Sandstone Shrike-thrush, Little Woodswallow and, occasionally, Spinifex Pigeon occur in the rocky escarpment at Boodjamulla, and may be seen at the Duwadarri Lookout located just west of the campground and along the circular Island Stack Walk (4 km return), which traverses the plateau in the heart of the gorge. Grey Falcon is apparently a rare resident at Boodjamulla. They are occasionally seen near the rangers' station – scan for them from any vantage point of the Duwadarri Lookout. Buff-sided Robin are common around the campground, while in the woodland around the campground you may encounter Northern Rosella (near the eastern extent of its range), Red-winged Parrot, Varied and Red-collared Lorikeet, and Red-browed Pardalote, and honeyeaters such as White-gaped, Yellow-tinted, Grey-fronted, Brown, Black-chinned (golden-backed), White-throated, Rufous-throated and, occasionally, Banded, Bar-breasted and Painted Honeyeater. A good selection of finches has been recorded near Lawn Hill Creek including Long-tailed, Crimson, Zebra and Double-barred and, less commonly, Pictorella and Chestnut-breasted Mannikin and Painted (uncommon) and Gouldian (rare) Finch. Other regular species at Boodjamulla include White-bellied Sea-Eagle, Silver-crowned and Little Friarbird, White-bellied Cuckoo-shrike, Black-faced, Masked and White-browed Woodswallow, and Variegated and Red-winged Fairy-wren. Night birds include Barking Owl, Southern Boobook and Spotted Nightjar (sometimes flushed at night along the dry sections of road into Boodjamulla). Notable fauna in the park include Purple-necked Rock-Wallaby, Rock Ringtail Possum, Freshwater Crocodile and both Olive and Black-headed Python.

A diverse shrub layer over a thick cover of spinifex near the Duwadarri Lookout in Boodjamulla (Lawn Hill) National Park. This area, and the adjacent sandstone rock stacks, can be good for Spinifex Pigeon, Grey-fronted Honeyeater, Rufous-banded Honeyeater, Sandstone Shrike-thrush and Little Woodswallow.

Mount Isa

Key species: Kalkadoon Grasswren, Carpentarian Grasswren, Painted Finch, Pictorella Mannikin, Spinifex Pigeon, Yellow Chat
Other species: Varied Lorikeet, Black-tailed Treecreeper, Grey-headed Honeyeater, Little Woodswallow, Red-browed Pardalote, Long-tailed Finch

Mount Isa (900 km west of Townsville) is a large mining town on the banks of the Leichhardt River in north-west Qld. Key sites near Mount Isa include the end of

Gregory River and Riversleigh

When visiting Boodjamulla National Park, it is worth also making a visit to the Gregory River, 55 km south on Riversleigh Rd. Along the river look for Purple-crowned Fairy-wren, Buff-sided Robin, Azure Kingfisher, Varied Lorikeet, Paperbark Flycatcher, Arafura Fantail, Long-tailed, Zebra and Double-barred Finch and Pictorella Mannikin. Spinifexbird and Spinifex Pigeon occur nearby at the world-heritage listed Riversleigh D fossil fields. Access is just past the Riversleigh Homestead. The Riversleigh deposits are one of the most important site for fossilised birds in the world, with the best known being 'Big Bird', a *Dromornithid* species that looks like a large Emu. This species dates back 25 million years, and can be viewed along the short Riversleigh Fossil Trail (800 m). The Miyumba Camping Ground is a small bush camp area adjacent to the Gregory River, 4 km south of Riversleigh D Site that provides good opportunities for an overnight stop.

Pamela St in town, Mica Creek, McNamara Rd and Lake Moondarra. The main target species for the region are Kalkadoon and Carpentarian Grasswren. The main access road into Mount Isa is the Barkly Hwy, while the airport has regular daily services from Brisbane, Cairns and Townsville and weekly services from Darwin. There is a range of accommodation in Mount Isa: www.mountisa.qld.gov.au/visitor-information.

Near Mount Isa, there are several good sites for Kalkadoon Grasswren. The closest is at the end of Pamela St, just 2 km from the town centre. From the end of the road, concentrate your search near the larger clumps of spinifex in the small valley between the two water tanks. Another well-known site for the grasswren is Mica Creek; to get there, from the intersection of Mica Creek Rd travel south down the Diamantina Developmental Rd (Hwy 83) for 6.5 km and park 100 m before the Mica Creek concrete causeway. From here, walk 150 m west, crossing a small gully to an area containing a wrecked car (in recent times there has also been a wrecked car adjacent to the road). Kalkadoon Grasswren occurs in the spinifex around the (second) wrecked car, along the track 50 m west of here, and in the hills along the gully to the north. Other birds at Mica Creek include Varied Lorikeet, Australian (Cloncurry) Ringneck, Cockatiel, Budgerigar, Diamond Dove, Spinifex Pigeon, Spinifexbird, Yellow-tinted, Grey-fronted and Grey-headed Honeyeater, Silver-crowned Friarbird, Grey-crowned Babbler, Little Woodswallow, Red-backed Kingfisher, Spotted Bowerbird, Red-browed Pardalote, Variegated Fairy-wren and Painted Finch. Raptors include Spotted Harrier, Wedge-tailed Eagle, Collared Sparrowhawk, Brown Goshawk, Brown Falcon and the very rare Grey Falcon. Kalkadoon Grasswren occurs along Sybella Creek, another 7 km further south down the Diamantina Developmental Rd, and, to the north of Mount Isa, they inhabit the hilltop around the Microwave Tower on the Barkly Hwy (34.5 km from the Mount Isa). There is a truck rest stop just after the tower that provides a convenient pull in.

McNamara's Rd has in recent times been the most reliable site for the rare Carpentarian Grasswren. To get there (from the Leichhardt River Bridge in Mount Isa), travel along the Barkly Hwy for 66.5 km, and turn right along McNamara's Rd (it is signposted 'CSC Lady Annie Operations'). Carpentarian Grasswren inhabits the Spinifex-covered plateaus along McNamara's Rd, particularly from 3.5 to 11 km from the Barkly Hwy. Sadly, the entire area was razed by wildfire in 2012 and it may be several years before this site is again suitable. As a guide, before the fires, the following sites were considered some of the best:

- 3.8 km from Barkly Hwy: stop in the left-side pullover, just before a small creek line and walk 400 m east of the road to search areas of Spinifex with open pebbly ground in between.
- 6.6 km from Barkly Hwy: stop at a breakaway on the right side of the road. Walk through an area with scattered termite mounds to search an area of Spinifex 400 m east of the road (near where some white quartzite protrudes through the Spinifex).
- 8 km from Barkly Hwy: this is the best-known site for Carpentarian Grasswren along McNamara's Rd and is marked with a small cairn (rock-pile) on the right side of the road (near a small track that leads right). Look in the larger patch of Spinifex in this area, particularly the well-vegetated gully and creek line 150 m east of the road. Look also in the patches of Spinifex adjacent to a creek just north-west of the cairn.
- 11 km from Barkly Hwy: search along the dry creek bed 400 m west of the road.

In general, glimpses of Carpentarian Grasswren are typically brief, because they tend to run and hop from one clump of Spinifex to the next. Other birds along McNamara's Rd include Spinifexbird, Spinifex Pigeon, Australian (Cloncurry) Ringneck, Black-tailed Treecreeper, Little Woodswallow, Little Button-quail, Crested Bellbird, Hooded Robin, Red-browed and Striated Pardalote, Inland Thornbill, Variegated Fairy-wren, and Grey-headed, Grey-fronted and Black-chinned (golden-backed) Honeyeaters.

Lake Moondarra is a large artificial lake on the Leichhardt River, 20 km north of Mount Isa. The main attraction here is the potential for larger numbers of waterbirds and shorebirds, but there are also a wide range of land birds that occupy the Spinifex-clad hills that surround the lake. Access to Lake Moondarra is from the Barkly Hwy 5 km north of Mount Isa. The lake is a further 15 km down this road. On your way to the lake, the first place to stop is Clearwater Lagoon, a natural water filtration system located on Lake Moondarra Rd 10 km from the Barkly Hwy. This lagoon is an excellent location for waterbirds. Here you may find Black-necked Stork, Wandering and Plumed Whistling-Duck, Green and Cotton Pygmy-Goose, Hardhead, Freckled Duck (rare), Hoary-headed, Australasian and Great Crested Grebe, and Little, Intermediate and Great Egret. Long-tailed, Zebra and Double-barred Finch and, occasionally, Pictorella Mannikin use the shallows of Clearwater Lagoon for drinking. Look for Painted Finch along the track to the Mount Isa Water Ski Club (it heads west just before you reach the first picnic area, 14 km from the Hwy); search for them near the base of the cliff faces (known as Blackrock) towards the end of the 1 km track. With luck, Purple-necked Rock-Wallaby may also been seen frequenting the rocky terrain here. Transport Bay, a pleasant area with several picnic grounds, is 15 km from

Dense and mature spinifex along McNamara's Road – the preferred habitat of the rare and declining Carpentarian Grasswren. Sadly, soon after this photograph was taken, this particular area was razed by fire.

the highway. Here you may encounter Little, Intermediate and Great Egret, Sacred Kingfisher, Spotted and Great Bowerbird (the range of the two species overlaps here), Rainbow Bee-eater, Paperbark Flycatcher, Black-tailed Treecreeper, Dollarbird, Black-eared Cuckoo, Red-browed Pardalote, and Black-chinned (golden-backed), Spiny-cheeked, Rufous-throated, Grey-fronted and Grey-headed Honeyeater. In the Spinifex-clad hills surrounding the lake, search for Spinifexbird, Spinifex Pigeon and Kalkadoon Grasswren (particularly the hilly area on the right side of Lake Moondarra Rd, 7 km from the Barkly Hwy). Warrina Park is situated below the Lake Moondarra dam wall at the end of Leichhardt River Rd. Large numbers of feral Indian Peafowl can be seen around this park, though whether they represent a truly feral population is debatable. The wide grassy bays on the western side of the lake often hold significant numbers of waterbirds including Black-necked Stork, Wandering and Plumed Whistling Duck, Green Pygmy-Goose and Glossy Ibis. In the build-up to the wet season, this area is especially good for migratory shorebirds such as Red-necked and Long-toed Stint, Common, Wood and Marsh Sandpiper, Black-tailed Godwit, Pacific Golden and Oriental Plover, Little Curlew (uncommon) and Oriental Pratincole. Resident Australian species include Red-kneed Dotterel and Australian Painted Snipe (several records). There is a chance of Yellow Chat (rare) and wet season vagrants have included Northern Pintail, Red-necked Phalarope

Lake Moondarra usually supports large numbers of waterbirds. Look here for Black-necked Stork, waterfowl, grebes, egrets and herons. The grassy surrounds provide habitat for a number of seed-eaters including the sought-after Pictorella Mannikin and occasional records of Flock Bronzewing.

Night Parrot

If you wish to search for the extremely elusive Night Parrot, then western Qld is as good a place as any. Said to inhabit a range of habitats, including spinifex grasslands and samphire flats in the arid and semi-arid areas, reports come from all mainland states, but there have been almost no confirmed observations for over 50 years. Three notable (and irrefutable) exceptions include a dead bird found in October 1990 on the Diamantina Developmental Rd, 36 km north of Boulia and a second dead bird found in Diamantina National Park (507 000 ha), 300 km south-west of Winton in December 2006 and a live bird photographed at an undisclosed location in the same general area in May 2013. In the 1990s, there were also several sight-records of Night Parrots along the Cloncurry Dajarra Rd. We'd suggest the most suitable habitat for Night Parrot along the Cloncurry Dajarra Rd is from 20 km south of the Barkly Hwy turnoff to 55 km south of the of the Barkly Hwy (with previous reports made at the 25 km and 37 km marks) but where the next confirmed sighting will be made is really anyone's guess!

and Eastern Yellow Wagtail. You can access the western side of the lake via a dirt track that leads east from the Barkly Hwy, 16.5 km from the Leichhardt River Bridge in Mount Isa; continue east along the track for 4.4 km. Interesting raptors recorded at Lake Moondarra include White-bellied Sea-Eagle, Little Eagle, Black-breasted Buzzard, Black Falcon, Australian Hobby and Red Goshawk (rare), while at night, Spotted Nightjar regularly hawk over the roads during the dry season.

Bladensburg National Park

Key species: Rufous-crowned Emu-wren, Spinifexbird, Spinifex Pigeon, Hall's Babbler, Chestnut-breasted Quail-thrush, Painted Finch, Striated (Rusty) Grasswren, Black Honeyeater
Other species: Australian Bustard, Horsfield's Bushlark, Pied Honeyeater, Rufous-throated Honeyeater, Grey-headed Honeyeater, Varied Sittella, Red-browed Pardalote

Bladensburg National Park (84 900 ha), 17 km south-west of Winton, is a large remote park with sandstone ranges, grassland plains and river flats. It is the north-eastern boundary range of Spinifex, and north-easterly populations of species such as Spinifexbird, Spinifex Pigeon, Rufous-crowned Emu-wren and Painted Finch. It also contains the most northerly population of Hall's Babbler. To get there, drive south from Winton on the Winton–Jundah Rd. After 7 km, turn left onto the 'Route of the River Gum', follow this for 17 km to Bough Shed Hole where there is a pleasant bush camping ground. Conventional vehicles can access Bladensburg, but during the wet season 4WD is recommended.

Much of the northern half of the park consists of reddish clay soils covered with Mitchell grasses and open woodlands. Look along the roadside for Emu, Australian Bustard, Brolga, Australian Pratincole, Cockatiel, Little Button-quail, Ground Cuckoo-shrike, Black-faced, White-browed and Masked Woodswallow, uncommon raptors such as Black-breasted Buzzard and Spotted Harrier and, at night, spotlight for Inland Dotterel and Spotted Nightjar. As you continue through the park, the landscape becomes more heavily vegetated, intermixed with areas of Spinifex and *Eremophila*. Here you may find Spinifexbird, Crimson Chat and Black, Grey-headed, Spiny-cheeked and, occasionally, Pied Honeyeater (especially when the *Eremophila* is flowering).

Skull Hole, located 10 km south of the Winton–Jundah Rd, is a water-filled rock hole ~20 m deep. Here there is a chance of Spinifex Pigeon, Budgerigar, Rainbow Bee-eater, Spotted Bowerbird, Varied (white-winged) Sittella, Grey-headed Honeyeater, and Painted and Zebra Finch. Hall's Babbler also occur along the creek line north of Skull Hole – search in areas of Gidgee woodland scattered along the creek ~2 km upstream from the waterhole. Just before the Skull Hole carpark, a dirt track leads south; 2.5 km down here is a dry creek bed. Hall's Babbler frequent the creek line, especially to the east. This same area can also be good for Chestnut-breasted Quail-thrush. A further 300 m along the track, there is an old fence line and open gate; turn right along the fence line to reach an area of Spinifex between the road and some escarpment. Rufous-crowned Emu-wren, Spinifexbird and Spinifex Pigeon all occur here, along with Hooded Robin, Red-browed Pardalote, Rainbow Bee-eater, Restless Flycatcher, Crested Bellbird and Painted Finch. Striated (Rusty) Grasswren (ssp. *rowleyi*), inhabit the southern section of the park along Opalton Rd, with a good site being 1 km south of Logan's Falls (near a cattle grid) in the Spinifex ~80 m west of the road.

Birdwatching around the Bough Shed Hole camping area can also be rewarding,

Spotted Harriers move over vast distances in search of suitable conditions. When conditions are good (higher rainfall, lush grass cover, an abundance of invertebrates and small vertebrates), the species is a regular sight as it patrols the grasslands on upswept wings.

with a range of species coming into drink along Surprise Creek, particularly near the northern section of the campground. Look for Australian (Cloncurry) Ringneck, Red-winged Parrot, Red-backed Kingfisher, Inland Thornbill, Striated and Red-browed Pardalote and Zebra Finch. Although scarce, there is also a possibility of Pictorella Mannikin (sometimes with Zebra Finch). Chestnut-breasted Quail-thrush have been recorded in the scree covered area immediately north-east of the northern section of the campground.

Opalton

Key species: Striated (Rusty) Grasswren, Rufous-crowned Emu-wren, Spinifexbird, Hall's Babbler
Other species: Black Honeyeater, Pied Honeyeater, Grey-headed Honeyeater, Painted Finch, Splendid Fairy-wren

Opalton is a tiny settlement in central Qld 100 km south of Winton. To get there from Winton, take the Winton–Jundah Rd. After 8 km, turn left onto the 'Route of the River Gum' (the road to Bladensburg National Park). Opalton is another 92 km. The road is mainly gravel, varying from good to very rough and corrugated. The low red-hills with Spinifex south of the town hold a range of desert species including an isolated population of Striated Grasswren (ssp. *rowleyi,* a subspecies that may soon be elevated to full species status, known as Rusty Grasswren). First recorded at Opalton region in the early 1970s, this population of Striated (Rusty) Grasswren is located several hundred kilometres from other populations. If you travel directly past the telephone box, you will come to the bush park where you can camp (washing water, as well as toilets and cold showers available). The Opalton Outpost Store has basic provisions, but fuel is not available.

The best sites to look for Striated (Rusty) Grasswren are in areas just south of Opalton. As you travel south through Opalton (from Winton), you pass a junction, with a turn-off to the store on your left. Continue past this junction and shortly after the road bears to the left (near the Opalton sign). As it straightens out again, there is a small flat-topped sandstone outcrop on the left-hand side of the road, just before it crosses a creek bed. Striated (Rusty) Grasswren inhabits the Spinifex between the creek and the outcrop, and Rufous-crowned Emu-wren, Spinifexbird, Spinifex Pigeon and Painted Finch also occur in this area. All species also occur south of the Opalton Outpost Store. From Opalton Rd, turn left after the phone booth, and continue past the store for 1 km. Look for the birds on the ridge (known as Caprock Ridge) on the west side of the road. Hall's Babbler inhabits in the Mulga woodland on the east side of the road just south of the airstrip. Look among the Mulga for Splendid Fairy-wren, Spotted Bowerbird, Inland Thornbill, Red-capped Robin, Crested Bellbird, and Black, Pied, Grey-headed, Brown and Spiny-cheeked Honeyeater. At night, listen for Spotted

Lark Quarry Conservation Park

If you do not find Rufous-crowned Emu-wren or Striated (Rusty) Grasswren at Bladensburg or Opalton, try for them at Lark Quarry Conservation Park (400 ha), 110 km south-west of Winton. Rufous-crowned Emu-wren occur in the Spinifex near the dinosaur trackways and in the Spinifex near the airfield, located halfway along the road to the trackways. The area around the airfield also holds Striated (Rusty) Grasswren (ssp. *rowleyi*). Halls Babbler occur in the Mulga around the airport, with a family group often feeding on the south side of the airport, behind the airfield's wind sock. Grey Falcon (rare) hunt in the area and, from time to time, they roost on the microwave tower located at the beginning of the road into Lark Quarry. Other birds recorded at Lark Quarry include Spinifex Pigeon, Little Woodswallow, Grey-headed Honeyeater, Spinifexbird and Painted Finch.

Nightjar and Australian Owlet-Nightjar. Striated (Rusty) Grasswren also occur west of Opalton Rd, just south and west of the airstrip.

Bowra

Key species: Grey Falcon, Bourke's Parrot, Chestnut-breasted Quail-thrush, Hall's Babbler, Painted Honeyeater, Chirruping Wedgebill
Other species: Spotted Nightjar, Major Mitchell's Cockatoo, Western Gerygone, Red-backed Kingfisher, Ground Cuckoo-shrike, Black Honeyeater, Pied Honeyeater, Plum-headed Finch, Red-browed Pardalote

Bowra (14 000 ha), 10 km north-west of Cunnamulla, was purchased by Australian Wildlife Conservancy (AWC) in 2010. Birds Queensland assists with the management of visiting birdwatchers. Mulga communities dominate the elevated ridges in the north of the property, while the southern section consists of alluvial plains and open woodlands of Gidgee and Coolabah. An impressive 212 bird species have been recorded on the property, and you have a reasonable chance of sought-after species such as the rare Grey Falcon, Hall's Babbler, Chestnut-breasted Quail-thrush, Bourke's Parrot, Chirruping Wedgebill and Painted Honeyeater. To get there from Cunnamulla, travel north along Humeburn Rd and turn west after ~5 km. There are camping facilities and accommodation (at the Cottage and Shearer's Quarters). To book, visit www.australianwildlife.org.

While driving the 6 km from the entrance gate to the homestead, look for resident species such as Major Mitchell's Cockatoo, Cockatiel, Little Corella, Red-winged Parrot, Blue Bonnet, Budgerigar, Spotted Bowerbird, Apostlebird, Chestnut-crowned Babbler, Hooded Robin, Crimson Chat, Restless Flycatcher, White-winged Triller, Rufous Songlark, Variegated Fairy-wren, and Zebra, Double-barred and Plum-headed Finch. Keep an eye open for Bourke's Parrot in the area and look for Grey Falcon (a

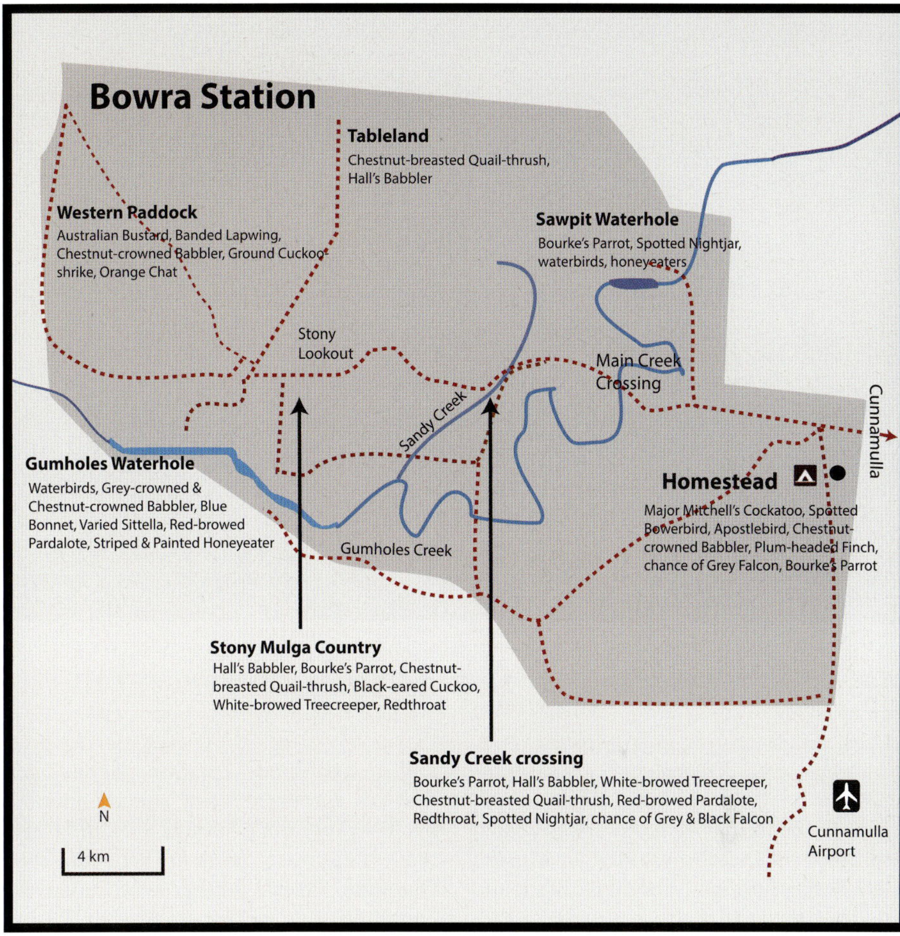

breeding resident), which occasionally soar over the homestead area.

Bourke's Parrot occur at the Sawpits Waterhole, located 4.8 km from the Homestead, where they drink at dawn, and dusk – during the day search the *Allocasuarina* woodlands just south of the waterhole. Waterbirds that occur here regularly include Plumed Whistling-Duck, Australian Wood Duck, Hardhead, Grey Teal, White-necked Heron and Black-fronted Dotterel. Mixed flocks of honeyeaters such as White-plumed, Singing, Blue-faced and Spiny-cheeked all regularly drink through the heat of the day. Just after dusk, Spotted Nightjar sometimes hawk over the dam. The Sandy Creek Crossing, located on Gumholes Rd 5 km from the homestead, provides some of the best birding on the property. During spring, in the narrow band of eucalypts along the creek, the call of Striated and Red-browed Pardalote can be all but continuous. Brown Treecreeper is also common here. In the Mulga adjacent to the creek, look out for Bourke's Parrot, Hall's Babbler,

Bowra is widely recognised as one of the best birding sites in south-western Qld. Some of the more interesting species that frequent the property include Grey Falcon, Bourke's Parrot, Painted Honeyeater, Hall's Babbler, Chestnut-breasted Quail-thrush and Chirruping Wedgebill.

White-browed Treecreeper, Chestnut-breasted Quail-thrush, Little Button-quail, Crested Bellbird, Striped Honeyeater, Redthroat, Splendid Fairy-wren, and Inland and Chestnut-rumped Thornbill. Black and Grey Falcon occasionally fly overhead and, after dark, Spotted Nightjar regularly rest on the roadside. On the north side of the creek, on a small ridgeline, there are several areas of *Eremophila* that, when flowering, attract nomadic nectarivores such as Black, Pied and White-fronted Honeyeater. Along a track 2.5 km beyond the creek, many of the property's woodland birds occur in a patch of Mulga (~500 m^2) known locally as the 'Stony Mulga Country'. Look here for Hall's Babbler, Bourke's and Mulga Parrot, Australian Ringneck, Chestnut-breasted Quail-thrush, Crested Bellbird, Black-eared, Horsfield's Bronze-Cuckoo, Pallid Cuckoo, White-browed and Brown Treecreeper, Red-capped and Hooded Robin and Western Gerygone, and smaller ground-dwelling passerines such as Splendid Fairy-wren, Redthroat, Southern Whiteface and Diamond Firetail. Further west along the Gumholes Rd, another track leads 2.4 km north to an area known as the Tableland. Search this area, particularly the lower slopes, for Chestnut-breasted Quail-thrush and Hall's Babbler. Continue along the main track to reach the Western Paddock circuit, which is a good area for grassland and open woodland species. Here you may find Emu, Australian Bustard, Banded Lapwing, Chestnut-crowned Babbler, White-browed Treecreeper, Chestnut-breasted Quail-thrush, Ground Cuckoo-shrike, Orange Chat, White-winged Fairy-wren, and Masked, White-browed, Little and Black-faced Woodswallow. Although rare here, Flock Bronzewing is a possibility (in 2008,

Lake Bindegolly

Lake Bindegolly National Park is situated 40 km east of Thargomindah. You can view the lake by either parking near the bridge on the Bulloo Developmental Rd (south-east side of the lake), or driving to the park entrance and walking to the lakeside bird viewing site (9 km return). Up to 60 species of waterbird have been recorded, including Australasian Shoveler, Grey Teal, Hardhead, Plumed Whistling-Duck, Freckled (rare), Blue-billed and Pink-eared Duck, Australasian, Hoary-headed and Great Crested Grebe, Australasian Darter, Glossy Ibis, Yellow-billed Spoonbill and Red-necked Avocet. Look along muddy, vegetated fringes for Baillon's, Spotless and Australian Spotted Crake, while in the samphire flats on the west side of the lake look for Orange Chat and White-winged Fairy-wren. In the adjacent scrub-covered sand hills, search for Blue Bonnet, Mulga Parrot, Major Mitchell's Cockatoo, Spotted Bowerbird, Chestnut-crowned Babbler, Black-faced Woodswallow and Singing, Pied, and Black Honeyeater. Both Grey and Black Falcon have also been seen.

Lake Bindegolly is a vast wetland that when full can support huge numbers of waterbirds. The surrounding shrublands can also be productive with sought-after species such as Black Falcon, Grey Falcon, Major Mitchell's Cockatoo, Pied Honeyeater and Black Honeyeater all occurring from time to time.

Queensland

several thousand occurred 25 km north of Bowra along the Charleville Rd).

Along the access road to Gumholes Waterhole, ~12 km from the homestead, look for Grey-crowned and Chestnut-crowned Babbler, Blue Bonnet, Varied (Black-capped) Sittella, Red-browed and Striated Pardalote, and Blue-faced, Striped and Painted Honeyeater. More than 50 species of waterbirds have been recorded at the waterhole, including such rarities as Australian Painted Snipe and more regular inland waterbirds such as Australian Spotted Crake, Plumed Whistling-Duck, Australasian Darter, Australasian Grebe, Black-winged Stilt, and Black-fronted and Red-kneed Dotterel. Finally, the birds of prey can be a highlight of any visit to Bowra, with a remarkable 19 species recorded. Aside from Grey Falcon, raptors regularly recorded include Peregrine and Black Falcon, Black-breasted Buzzard, Spotted Harrier, Square-tailed Kite and Collared Sparrowhawk. Nocturnal birds that frequent the property include Eastern Barn Owl, Southern Boobook, Tawny Frogmouth, Spotted Nightjar and Australian Owlet-Nightjar.

Eulo Bore

Key species: Bourke's Parrot, Chestnut-breasted Quail-thrush, Hall's Babbler, Slaty-backed Thornbill
Other species: Major Mitchell's Cockatoo, Red-backed Kingfisher, Black Honeyeater

Eulo Bore (also known as Nine Mile Bore) is 16 km east of Eulo on the Bulloo Developmental Rd (50.7 km west of Cunnamulla). The prominent landmark is a ~20 m diameter concrete tank fed by Artesian bore and a long cattle-watering trough. The dominant trees are Gidgee and Mulga, while there is also an abundance of larger dead trees that contain hollows. Around the bore, which sometimes overflows to create a small muddy wetland, you may find Australasian Grebe, White-necked Heron, Black-fronted Dotterel, Blue-faced and Singing Honeyeater, Variegated Fairy-wren, White-winged Triller, Diamond Dove, Common Bronzewing, Grey and Pied Butcherbird, Chestnut-rumped thornbill, Australian Raven, Southern Whiteface, and Red-capped and Hooded Robin. At dawn, and again at dusk, a variety of parrots drink here, notably Bourke's Parrot, but also Mulga and Red-winged Parrot, Australian (Mallee) Ringneck and Blue (Yellow-vented) Bonnet.

The Mulga woodlands surrounding within about a kilometre of the bore, notably to the north-east, west and south (the latter across the road) provide excellent birding opportunities. Here you may encounter Hall's, Grey-crowned and Chestnut-crowned Babbler, Chestnut-breasted Quail-thrush (especially on the gravel rises), White-browed Treecreeper (within Mulga), Slaty-backed Thornbill (rare), Spotted Bowerbird, Horsfield's Bronze- Cuckoo, Pallid Cuckoo, Varied Sittella, Brown, Black (seasonal), Grey-fronted (uncommon) and Spiny-cheeked Honeyeater, Noisy and Little Friarbird, Yellow-throated Miner, Crimson Chat and Splendid Fairy-wren. Although on the eastern limit of its range, there is also a chance of Chirruping Wedgebill. Night birds include Australian Owlet-Nightjar (sometimes flushed during the day from tree hollows), Spotted Nightjar and Southern Boobook, while birds of prey include Black-breasted Buzzard, Little Eagle, Whistling Kite, and Brown, Peregrine, Grey (rare) and Black Falcon. Similar areas 1.5 km north-west of the bore, and another 2 km north of the bore (bordering a usually dry creek line), can be equally productive. If you do not find Chestnut-breasted Quail-thrush near the bore, another good area to search is a distinctive ridge 3.5 km east of Eulo Bore. From the Bulloo Developmental Rd, you

Kilcowera Station

Situated in the Mulga rangelands, Kilcowera Station is some 90 km south of Thargomindah. Cardenyabba Lagoon, south of the homestead, can be a good place for Bourke's Parrot: they drink here in the morning and evening. Chestnut-breasted Quail-thrush and Slaty-backed Thornbill have been recorded in Mulga near the Youlangee Holding Yard, along with Red-backed Kingfisher, Black and Pied Honeyeater (especially where *Eremophila* is flowering), Crested Bellbird, and Masked, Black-faced and White-browed Woodswallows. From Kilcowera Station, you have access to the western edge of Lake Wyara and Currawinya National Park, 30 km east of the campground (see below).

can see a ridge to the south with telegraph lines running along it – Chestnut-breasted Quail-thrush are often observed between poles 255 to 257.

Currawinya National Park

Key species: Bourke's Parrot, Freckled Duck, Hall's Babbler, Chestnut-breasted Quail-thrush, Black Honeyeater, Pied Honeyeater, Slaty-backed Thornbill, waterbirds
Other species: Crested Bellbird, Black-eared Cuckoo, Australian Bustard, Australian Pratincole, Inland Dotterel, Little Button-quail

Currawinya National Park (154 870 ha) is on the NSW–Qld border next to the township of Hungerford, 170 km south-west of Cunnamulla. It is home to over 200 species of birds, including sought-after species such as Hall's Babbler, Chestnut-breasted Quail-thrush, Bourke's Parrot and Freckled Duck. Two large lakes, separated by only a few kilometres of sand dunes, are the centrepiece of the park. Lake Numalla is fresh-water

Emu-bush *Eremophila* spp. near Currawinya National Park. When in flower, this species provides a much sought-after nectar resource for Black Honeyeater and Pied Honeyeater. The groundcover of Mitchell Grass visible here is also suitable for Red-chested Button-quail (denser areas of cover) and Little Button-quail (more open areas).

and the slightly larger Lake Wyara is saline. Mulga covers the park's sand plains. The Park office is 4.5 km north of the Ten Mile Bore. Conventional access is possible, but 4WD is necessary to reach the lakes. There is bush camping at Ourimperee Waterhole behind the Woolshed or at sites along the Paroo River near Caiwarro.

The Ramsar-listed Currawinya Lakes are an important site for waterbirds, with as many as 250 000 waterbirds counted. No other wetlands in arid Australia consistently support such high numbers of waterbirds. Lake Wyara (salt water) supports large numbers of Grey Teal and Eurasian Coot, and is an important refuge for the rare Freckled Duck (of an estimated global population of 19 000, the mean estimate for the Currawinya Lakes is 2500). Other birds that occur in large numbers include Australasian Shoveler, Pink-eared Duck, Black-winged Stilt and Red-necked Avocet. The lake is also a significant breeding site for Australian Pelican, Black Swan Red-necked Avocet, Silver Gull and Caspian Tern; most breed along the western shoreline. In April 2008, for example, an estimated 30 000 breeding pairs of Australian Pelican were present. By contrast, Lake Numalla has less aquatic vegetation but more abundant fish and shrimp populations. As a result, predatory waterbirds such as Pied Cormorant, White-necked Heron, Australian White Ibis, Royal and Yellow-billed Spoonbill display a preference for Lake Numalla. Other interesting birds here include Brolga, Great Egret, Plumed Whistling-Duck, Glossy Ibis and Great Crested Grebe. Although Lake Wyara typically supports more waterbirds, Lake Numalla is an important waterbird refuge when Lake Wyara's water levels fall; indeed, when Lake Wyara was almost dry in 1985, Lake Numalla supported more than 100 000 waterbirds. The lakes can also host significant numbers of shorebirds and terns, including Black-tailed Godwit, Common Greenshank, Red-necked Stint, Marsh, Common, Sharp-tailed and Curlew Sandpiper, Australian Pratincole, and Caspian, Gull-billed, Whiskered and White-winged Black Tern. In the low samphire shrublands surrounding the lakes, look for Orange and White-fronted Chat, Black-faced Woodswallow, Southern Whiteface and White-winged Fairy-wren.

The Mulga woodland on the south-eastern side of the park supports Hall's Babbler, Bourke's Parrot, Chestnut-breasted Quail-thrush and Slaty-backed Thornbill (rare this far-east). Concentrate your search around 10 Mile Bore on Hungerford Rd. Generally, the Mulga around the southern end of the Old Thargomindah–Hungerford Rd, where it intersects with Hungerford is a good place for woodland birds. In addition to the 10 Mile Bore, look 6.3 km north of this intersection at the base of the Granites. The Hood Range is worth checking for Hall's Babbler, White-browed Treecreeper, Chestnut-breasted Quail-thrush and Pallid, Horsfield's Bronze- and Black-eared Cuckoo. The woodlands around the Old Caiwarro Homestead are worth investigating too, with several families of Hall's Babbler present near the homestead's airfield. In the Mulga, look also for Australian Ringneck, Mulga Parrot, Crested Bellbird, Little Button-quail, Chestnut-rumped Thornbill and Splendid Fairy-wren, and nomadic White-fronted, Black, Pied and Spiny-cheeked Honeyeaters. Around Ourimperee Waterhole, look for Emu, Spotted Bowerbird, Pied Butcherbird, Major Mitchell's Cockatoo, Blue Bonnet, Squatter Pigeon (very rare this far south), Chirruping Wedgebill (uncommon), Rainbow Bee-eater, Striated and Red-browed Pardalote, Brown Treecreeper, Restless Flycatcher, Zebra and Plum-headed (uncommon) Finch, Variegated Fairy-wren, Striped, Brown and Painted (uncommon) Honeyeater and Grey-crowned Babbler. Notably, all four Australian Babblers (Grey-crowned, Hall's, Chestnut-crowned and

Gibber is an Australian term for large open areas covered in small weathered pebbles. Many desert areas in the Lake Eyre Basin are characterised by this land form and it is here that the Gibberbird is at home.

White-browed) occur in the park, one of just a handful of places where this occurs. Raptors recorded at Currawinya include Black-breasted Buzzard, Wedge-tailed and Little Eagle, Whistling Kite and Black and Grey (rare) Falcon, while night birds include Spotted Nightjar, Australian Owlet-Nightjar, Barking Owl (rare) and Southern Boobook. Notable mammals include Red, Eastern Grey and Western Grey Kangaroo and Wallaroo, while in 2005 Bilby were re-introduced.

Birdsville Region

Key species: Eyrean Grasswren, Grey Grasswren, Yellow Chat, Banded Whiteface, Gibberbird, Letter-winged Kite
Other species: Australian Bustard, White-backed Swallow, Orange Chat, Crimson Chat, Black Falcon, Black-breasted Buzzard

Birdsville is a small, but iconic, outback town located in the Channel Country on the banks of the Diamantina River. The town is situated between the sands of the Simpson Desert and the gibber plains of Sturts Stony Desert, 1590 km west of Brisbane, and 720 km south of Mount Isa. Most notably, it is the northern starting point of the famous Birdsville Track. Birdsville has a legendary pub; during the annual horse-race meet, the population grows from 120 to 6000 for 2 days. Accommodation in Birdsville includes

Grey Grasswren in Queensland

Search the larger clumps of lignum between the Noccundra Hotel and the Wilson River for Grey Grasswren (ssp. *diamantina*), 80 km west of Thargomindah. They are also found in the lignum on the east side of the Cooper Creek bridge on Thargomindah–Innamincka Rd, 218 km west of Thargomindah – again concentrate your search in the larger clumps of lignum. The Thargomindah–Innamincka Rd is bitumen, easily accessible by conventional vehicle. Grey Grasswren are also found at Eyre Creek, in the overflow of Koolivoo Waterhole, Lake Machattie and Lake Mipia, 119 km north of Birdsville – look in the outer edges of dense lignum clumps. Grey Grasswren (ssp. *barbatus*, with darker streaking) occurs further south in lignum patches on the floodplains of the Bulloo Overflow on the NSW–Qld border. For details on access, see Sturt National Park, p. 236.

A huge red sand dune on the edge of the Simpson Desert. The dune vegetation provides habitat for sought-after species such as Eyrean Grasswren, while the adjacent open swale is suitable for Banded Whiteface.

the hotel-motel and caravan park. The climate is typical of the aridlands, with two notable extremes: day temperatures reaching into the mid 40°C in summer (Nov–Feb) and nights as low as 4°C in winter. During the autumn to spring period, milder day temperatures are more typical (15°C–35°C).

In summer, Yellow Chat feed near the Birdsville Bore overflow, a small wetland created by the bore adjacent to town. On the Birdsville Lagoon, look for waterbirds such Red-necked Avocet, Black-winged Stilt, Red-kneed Dotterel, Black-tailed Native-hen, Blue-billed and Pink-eared Duck, Grey Teal and Hardhead. Big Red is a huge sand dune ~37 km west of Birdsville on the Simpson Desert National Park Rd. It stands ~35 m above the plain and can be a reliable place for Eyrean Grasswren. A good area to look is within a 50 m by 300 m strip on the dune, ~100 m to the left of the road as you approach from Birdsville. Look among the larger clumps of cane grass, but they also occur in Spinifex; because they have a very soft high-pitched contact call, ideal conditions are when there is little or no wind. Banded Whiteface (usually around low shrubs) and Gibberbird (more open areas) inhabit the flat areas of gibber plain bordering Big Red and between the dunes to the west. Other birds to look for here include White-backed Swallow, Variegated and White-winged Fairy Wren, Orange and Crimson Chat, Yellow-throated Miner, Black-faced Woodswallow and Little Crow, and raptors such as Little Eagle, Black-breasted Buzzard and Black Falcon.

If you continue onto the Simpson Desert National Park (1 012 000 ha) from Birdsville, you will find the landscape dominated by parallel sand dunes that range in height from a few metres to 20 m. Big Red is the first and highest of some 1100 dunes that run north–south for hundreds of kilometres. Eyrean Grasswren is found on many of the dune crests, especially those that support cane grass. This drive is extremely challenging, strictly 4WD only and should really only be attempted in a convoy of well-prepared vehicles.

Gibberbird and the nomadic and eruptive Flock Bronzewing often occur just north of Birdsville, along the Eyre Developmental Rd between Birdsville and Bedourie, and further north along the Diamantina Developmental Rd between Bedourie and Boulia. If travelling these roads, look out for Australian Bustard, Australian Pratincole and Oriental Plover. Letter-winged Kite, an irruptive species that feeds mainly at night, frequents this area. Its preferred prey is the native Long-haired Rat. In 2009, after an extended period of rain, there was an eruption of rat numbers at Astrebla Downs National Park (17 000 ha), 100 km east of Bedourie. As a result, Letter-winged Kite was present in large numbers. Other birds found here, and within Diamantina National Park (507 000 ha), immediately to the north-east, include Inland Dotterel, Australian Bustard, Red-chested Button-quail, Plains-wanderer (rare), Flock Bronzewing, Yellow Chat (rare) and Gibberbird. See also the box text on Night Parrot (p. 129).

South-East Queensland

Regional overview

South-East Qld provides some of Australia's most spectacular birding. The region extends from Lamington National Park on the NSW–Qld border, Sundown National Park in the west, to the Great Sandy National Park in the north. South-East Qld supports approximately two-thirds of Qld's population, (~2.8 million people) and covers ~22 400 km^2. Brisbane is Australia's third largest city and the third-most popular destination for international tourists after Sydney and Melbourne.

Although large areas have been cleared for agriculture, there still reman some extensive rainforests within region's southern border ranges (an area known as the Scenic Rim), including those protected in Lamington National Park. The region includes one of the largest upland subtropical rainforest remnants in the world and supports Australia's most northerly cool temperate rainforests, which are dominated by Antarctic Beech. Aside from rainforest, habitats range from seashores, mudflats and mangroves, to drier eucalyptus forests and woodlands. The world's largest sand island, Fraser Island, and neighbouring areas of the mainland also support extensive areas of coastal heathland.

A suggested 11-day itinerary for South-East Qld is as follows:

- **Day 1.** Brisbane (Powerful Owl, Magpie Goose, Black-necked Stork, Lewin's Rail, Mangrove Honeyeater, Mangrove Gerygone, Collared Kingfisher, shorebirds).
- **Days 2–3.** From Brisbane, visit D'Aguilar National Park and Lake Samsonvale, 50 km (Sooty Owl, Marbled Frogmouth, Noisy Pitta, Australian Logrunner, Russet-tailed Thrush, Paradise Riflebird, Regent Bowerbird, King Quail, Red-backed Button-quail).

There are several sites in South-East Qld Cotton Pygmy-Goose occurs frequently: these include a large wetland in the southern Brisbane suburb of Forest Lake and Greenwood Lake at nearby Forestdale (neither are covered elsewhere in this book). Further afield, Lake Atkinson in the Lockyer Valley regularly hosts the species.

- **Days 4–6.** Brisbane to Lamington National Park, 110 km (Albert's Lyrebird, Rufous Scrub-bird, Paradise Riflebird, Australian Logrunner, Noisy Pitta, Regent Bowerbird, Russet-tailed and Bassian Thrush, Pale-vented Bush-hen, Marbled Frogmouth, Sooty Owl).
- **Day 7.** Lamington National Park to the Lockyer Valley, 165 km (Australian Painted Snipe, Cotton Pygmy-Goose, Black Falcon, Square-tailed Kite).
- **Days 8–9.** Lockyer Valley to the Conondale Ranges, 190 km (Marbled Frogmouth, Sooty Owl, Masked Owl).
- **Days 10–11.** Conondale Ranges to Inskip Point and Great Sandy National Park, 130 km (Black-breasted Button-quail, Beach Stone-curlew, Ground Parrot, Eastern Grass Owl, Southern Emu-wren).
- **Day 12.** Inskip Point to Brisbane, 250 km.

A shorter 6-day trip would be to visit Brisbane, D'Aguilar National Park, Lake Samsonvale and Lamington. Another worthwhile 3–4-day trip involves heading west to Girraween National Park (Superb

Lyrebird, Chestnut-rumped Heathwren, Spotted Quail-thrush, Turquoise Parrot, Plum-headed Finch, Diamond Firetail, Squatter Pigeon) and Sundown National Park (Painted Button-quail, Turquoise Parrot, Diamond Firetail, Spotted Bowerbird, Ground Cuckoo-shrike).

Brisbane has a subtropical climate, with warm to hot and humid summers and dry, mild winters. January is the warmest month, with a mean daily high of 30.3°C and mean daily low of 21.4°C, while July is the coldest month (21.9°C and 10.0°C). From November to March, thunderstorms are common over the region, with the more severe events often accompanied by torrential rain and high winds. South-East Qld is an excellent birdwatching destination year round, though summer migrants generally arrive between August and November and depart between March and May.

Brisbane and neighbouring coastal regions (e.g. the Gold Coast) are serviced by air, road and rail links with other Australian cities. There are several excellent birdwatching sites close to Brisbane, with all the sites reached easily by car. The city has an extensive network of bicycle and walking paths. Brisbane also boasts an extensive public transport network within the city, as well as connections to regional centres, interstate and to overseas destinations. A 2WD vehicle will get you to almost all the birding sites in the region. Daily scheduled flights to Brisbane and Gold Coast connect to all mainland Australian capital cities, and there are regular international air services.

Birding highlights

South-East Qld is a floristically diverse region that holds great numbers of bird species. Although there are no endemics in the area, the best sites for several Australian endemics are here: these include sites for Wandering Tattler, King Quail, Albert's Lyrebird, Black-breasted Button-quail, Regent Bowerbird and Rufous Scrub-bird. Other birding highlights include Pale-vented Bush-hen, Wompoo Fruit-Dove, Pale-headed Rosella, Ground Parrot, Pacific Baza, Red Goshawk, Sooty and Eastern Grass Owl, Marbled Frogmouth, Australian Logrunner, Paradise Riflebird, Russet-tailed Thrush, Southern Emu-wren, Eastern Bristlebird (very rare and declining), Green Catbird and Noisy Pitta. An interesting selection of subspecies are found in the region. These include the extremely rare (possibly extinct) Double-eyed Fig-Parrot ('Coxen's Fig-Parrot' ssp. *coxeni* recently considered by *Birdlife International* to be a separate species), Eastern Rosella ('Golden-mantled Rosella' ssp. *elecica*), Marbled Frogmouth ('Plumed Frogmouth' ssp. *plumiferus*), Pied Currawong (ssp. *graculina*), Australasian Figbird (green ssp. *vielloti*), Noisy Pitta (ssp. *versicolor*), White-throated Treecreeper (ssp. *metastasis*), Superb Lyrebird ('Edward's Lyrebird' ssp. *edwardi*), Fairy Gerygone (ssp. *flavida*), Striated Thornbill (ssp. *alberti*), Pale-yellow Robin (ssp. *capito*), Varied Sittella ('White-headed Sittella' ssp. *leucocephala*), Olive Whistler (ssp. *macphersoniana*) and Varied Triller (ssp. *leucomela*). Sadly the local form of the Star Finch (ssp. *ruficauda*) is probably extinct.

A wide range of Australian rarities have been recorded including Grey Ternlet, Franklin's and Laughing Gull, South Polar Skua, White-bellied Storm-Petrel, Broad-billed and Salvin's Prion, Stejneger's, Herald, Black, Westland and Kerguelen Petrel, South Island Pied Oystercatcher, Blue Rock Thrush and Grey Wagtail.

Greater Brisbane

Key species: Eastern Grass Owl, Powerful Owl, Magpie Goose, Black-necked Stork, shorebirds

Other species: Lewin's Rail, Mangrove Honeyeater, Mangrove Gerygone, Collared Kingfisher

Brisbane is Qld's capital city, and Australia's third largest city, with a metropolitan population of over 2 million. There are several excellent birding sites quite close to the centre of city. For visitor information, including accommodation options, see www.visitbrisbane.com.au

The Boondall Wetlands Reserve (1100 ha) protects Brisbane's largest wetland system. With a broad selection of habitats, including coastal forest, mangroves and estuaries, facilities here include an environment centre, elevated walkways and birdwatching platforms. Species to look for include Eastern Osprey, Brahminy Kite, Black-winged Stilt, Pale-headed Rosella, Little Lorikeet, Collared, Forest and Sacred Kingfisher, Mangrove Gerygone, Red-backed Fairy-wren, and Mangrove and Scarlet Honeyeater. The estuaries provide foraging areas for summer shorebirds such as Pacific Golden Plover, Eastern Curlew, Grey-tailed Tattler, Bar-tailed Godwit, Whimbrel and Great Knot. Immediately to the east of the Boondall Wetlands, the coastal wetlands and mangroves at Nudgee Beach support a good variety of birdlife. Here there are a series of boardwalks and bridges over Nundah Creek, and a bird hide within the mangroves that provides views over the mud flats. In summer, Pacific Golden, Grey and Lesser Sand Plover, Eastern Curlew, Whimbrel, Bar-tailed Godwit, Sharp-tailed and Curlew Sandpiper, Ruddy Turnstone and Great Knot forage on the exposed flats and roost nearby at high tide. Other birds to look for include Eastern Osprey, Striated and Nankeen Night-Heron, Collared Kingfisher, Rainbow Bee-eater, Mangrove and Brown Honeyeater and Mangrove Gerygone.

The Kedron Brook Wetlands (200 ha) are a series of saltwater to brackish lagoons near Brisbane Airport. Accessed off Toombul Rd, a walking track circles the wetlands and the grasslands are criss-crossed with mown tracks. Eastern Grass

The freshwater lagoons at Sherwood Forest Park support a range of waterbirds. Regular species include Dusky Moorhen, Purple Swamphen, Hardhead, Pacific Black Duck, Grey Teal, Darter, Australasian Grebe, Little Black Cormorant and Little Pied Cormorant, while less common species have included Australian Little Bittern, Pale-vented Bush Hen, Spotless Crake and Baillon's Crake.

Owl occurs at the wetlands, with the best time being evening twilight as they leave their roost to hunt. Other birds here include Brown Quail, Australasian Darter, Striated Heron, Spotless Crake, Buff-banded Rail, Lewin's Rail (occasional), Azure and Collared Kingfisher, Mangrove and Brown Honeyeater, Mangrove Gerygone, Tawny Grassbird and Chestnut-breasted Mannikin. Summer shorebirds that occur here include Latham's Snipe, and Common, Marsh, Sharp-tailed and Curlew Sandpiper.

The Tinchi Tamba Wetlands (380 ha) lie between Pine River and Bald Hills Creek, 25 km north of Brisbane. The wetlands can be a good site for Magpie Goose, Black-necked Stork, Striated Heron, Brolga, White-bellied Sea-Eagle and Mangrove Gerygone, and summer shorebirds such as Marsh and Sharp-tailed Sandpiper, Great Knot, Whimbrel, Eastern Curlew and Latham's Snipe. North of the Tinchi Tamba Wetlands, the rocky coastline on the peninsula at Redcliffe is often good for Wandering Tattler: try for them at Redcliffe Point (on the rocky outcrop just off the point), at Scarborough and Scotts Point. Other birds here include the occasional Grey-tailed Tattler, Eastern Curlew, Bar-tailed Godwit, Ruddy Turnstone and Common Tern (Nov–Dec). Sooty and Australian Pied Oystercatcher also occur and there has been a single record of a vagrant South Island Pied Oystercatcher (July 2001).

The Oxley Creek Common, located on Sherwood Rd, Rocklea, 10 km west of the Brisbane CBD, has a combination of swamps, grasslands, woodlands and plantation forest. Some of the more interesting species seen here include Brown Quail, Black-necked Stork, Latham's Snipe, Red-backed Fairy-wren, White-throated and Mangrove Gerygone, Tawny Grassbird, Double-barred and Plum-headed Finch (occasional), and Chestnut-breasted Mannikin.

Mount Coot-tha Forest (1500 ha) is just 15 min from the city and is Brisbane's largest open eucalypt forest. Access is from Sir Samuel Griffith Drive or Gap Creek Rd. In the west of Mount Coot-tha Forest, the dry and open eucalyptus forests around Gap Creek Reserve hold a good selection of birds. You have a chance of Square-tailed Kite, Powerful Owl, Australian Brush-turkey, Pheasant Coucal, Painted Button-quail, Spotted Quail-thrush, Cicadabird,

South-East Qld is a stronghold of the Eastern Grass Owl. Accessible sites include those at Kedron Brook Wetlands, Lake Samsonvale and along Cooloola Way in the Great Sandy National Park.

Little Shrike-thrush, Bell Miner and Speckled Warbler. The J C Slaughter Falls area is the best site close to Brisbane for Powerful Owl. At dusk, keep a lookout for the owl in the taller older trees around the upper car park. White-throated Nightjar hawks for insect (spring–summer) at dusk around the carpark; try also around a small dam 500 m along the Hovea Track.

The freshwater lagoons and reedbeds at Sherwood Forest Park is an excellent site for skulking birds. Australian Little Bittern, Pale-vented Bush Hen, and Spotless and Baillon's Crake have all been recorded. Access is via the car park off Jolimont Street.

South of the Brisbane River mouth, the Lytton Wader Roost consists of a large inter-tidal area bordered by mangroves. Park at the end of Wynnum North Rd, and then take the track to the bird hide. Depending on the tide height, there is usually a diverse mix of shorebirds, with high tide usually the best. Grey-tailed Tattler often perch in the dead mangroves on the eastern side of the roost, and there are mixed flocks of Black-tailed and Bar-tailed Godwit, Great and Red Knot, Eastern Curlew, Whimbrel, and Curlew, Terek, Marsh and Sharp-tailed Sandpiper. Rarer species that have been recorded here include Asian Dowitcher and Broad-billed Sandpiper. In the mangrove forest along the Wynnum Mangrove Boardwalk, look for Collared Kingfisher, Mangrove Gerygone and Mangrove Honeyeater. The Manly Boat Harbour breakwall is also a high-tide summer shorebird roost. Look here for mixed flocks of Lesser and Greater Sand Plover, Pacific Golden and Grey Plover, Sharp-tailed, Curlew and Terek Sandpiper, Black-tailed and Bar-tailed Godwit, Red-necked Stint, Grey-tailed Tattler, Common Greenshank, Ruddy Turnstone, Red and Great Knot, Whimbrel and Eastern Curlew. There have also been rare sightings of Asian Dowitcher. It is also a good site for terns, with records of Crested, Lesser Crested (winter), Gull-billed, Little, Whiskered and Caspian Tern.

D'Aguilar National Park (Mount Glorious)

Key species: Sooty Owl, Marbled Frogmouth, Noisy Pitta, Russet-tailed Thrush, Paradise Riflebird, Regent Bowerbird
Other species: Green Catbird, Rose-crowned and Wompoo Fruit-Dove, Australian Logrunner

Eagleby Wetlands

Eagleby Wetlands is a re-claimed area around a water treatment plant on the Albert River. It is located on the Gold Coast, 36 km south of Brisbane. Although small, this is a superb wetland where it is possible to locate 80+ bird species in a few hours. The wetlands feature a range of habitats including paperbark and brackish swamps, freshwater lakes, woodlands, cane fields and open grassland. A wide range of waterbirds include Magpie Goose, Plumed and Wandering Whistling-Duck, Australasian Grebe, Australasian Darter, Black-winged Stilt, Comb-crested Jacana, Striated Heron, Intermediate, Great, Little and Cattle Egret and, occasionally, Black-necked Stork. It is also good place for skulking crake and rails – look for Buff-banded Rail, Spotless, Australian Spotted and Baillon's Crake around any wetland fringes. Although rare, Pale-vented Bush-hen, Australian Painted Snipe and Pectoral Sandpiper have also been recorded. Raptors are well represented, with records of White-bellied Sea-Eagle, Eastern Osprey, Pacific Baza, Brahminy Kite, Brown Goshawk, Collared Sparrowhawk, Australian Hobby and Peregrine Falcon. The long list of other species includes Australian Brush-turkey, Brown Quail, Azure, Sacred, Forest and Collared Kingfisher, Rainbow Bee-eater (mainly summer), Mangrove Gerygone, Tawny Grassbird, Golden-headed Cisticola, Striated and Spotted Pardalote, Red-backed, Variegated and Superb Fairy-wren, Red-browed and Double-barred Finch, Chestnut-breasted Mannikin and Mangrove, Lewin's, Scarlet, Brown, White-throated, Striped and Blue-faced Honeyeater.

Many of the better parks for birdwatching around Brisbane have areas of 'unmanaged' habitat specifically set aside to promote opportunities for wildlife.

D'Aguilar National Park (36 400 ha) supports a variety of habitats including subtropical rainforests, dry and wet sclerophyll forests, eucalypt woodland, rocky outcrops and freshwater systems. D'Aguilar can be an excellent site for Sooty Owl and Marbled Frogmouth. The visitor centre located at Walkabout Creek on Mount Nebo Rd is 12 km from Brisbane's city centre. Despite its proximity to Brisbane, there are a range of camping options within the park: www.derm.qld.gov.au/parks/daguilar.

There are several good rainforest tracks starting at the Maiala picnic area, located on Mount Glorious Rd, just past the Mount Glorious township. A popular site for picnickers, it is worth visiting first thing in the morning to avoid the crowds. Regent and Satin Bowerbird sometimes perch in the higher branches of the remnant Hoop Pine plantation at the picnic ground. The Rainforest Circuit (2 km) winds its way through stunning subtropical rainforest. This track joins Cypress Grove (2.5 km return) and Greenes Falls (4.3 km return) walking tracks. Noisy Pitta occur from the start of the Rainforest Circuit, and you may also find Australian Logrunner, Russet-tailed Thrush, Paradise Riflebird, Rufous Fantail (summer), Golden Whistler, Rose Robin (winter), Green Catbird, Yellow-throated and Large-billed Scrubwren and Scarlet Honeyeater. The Greenes Falls Walk crosses Browns Rd. Check within the roadside vegetation along Browns Rd (all the way up to Mount Glorious Cottages) for Regent Bowerbird, Australian King-Parrot and fruit-eating pigeons such as Wompoo and Rose-crowned Fruit-Dove, Brown Cuckoo-Dove, Topknot and White-headed Pigeon. At night along Browns Rd, spotlight for Marbled Frogmouth and Sooty Owl, particularly near the intersection with Greenes Falls Walk. Another good place for Sooty Owl and Marbled Frogmouth is the Manorina parking area, 7 km south of Mount Glorious; listen for them along the Atrax Circuit (750 m return) and the longer Morelia Walking Track (6 km return). Here, during the day, look here for Pale-yellow Robin, Australian King-Parrot, and Red-browed and White-throated Treecreeper. Sooty Owl also occurs at Boombana picnic ground, located between the town of Mount Nebo and Jolly's Lookout. This site is also good for Regent Bowerbird, White-eared, Spectacled and Black-faced Monarch, Eastern Whipbird, Varied (White-headed) Sittella, Barred Cuckoo-shrike, Cicadabird, Yellow-throated Scrubwren and Pale-yellow Robin. The Rainforest Circuit (1.1 km) passes through an open eucalypt forest to subtropical rainforest. The longer Thylogale Walking Track (8 km return), connecting Jolly's Lookout with Boombana, passes through eucalypt and drier rainforests. Aside from excellent lunches and coffee served here, birding around the Mount Glorious Tea House can be

Queensland

rewarding, with species such as Australian King-Parrot, White-headed Pigeon and Pale-yellow Robin feeding within the gardens. Other birds recorded at D'Aguilar include White-throated Needletail, Oriental Cuckoo (rare summer visitor), Little and Rainbow Lorikeet, Little Shrike-thrush and Torresian Crow, and raptors such as Square-tailed Kite and Grey Goshawk.

Gold Creek flows south out of the southern section of D'Aguilar National Park. It is a particularly good creek line for Pale-vented Bush-hen: look for them along the creek crossing along Gold Creek Rd and near the Gold Creek Reservoir overflow. If you do not find them here, it is worth searching the creek crossing along Savage Rd.

Strangler figs *Ficus watkinsiana* are a feature of rainforests in south-eastern Qld. Along the rainforest walking trails in D'Aguilar National Park, look for Noisy Pitta, Yellow-throated and Large-billed Scrubwren, Scarlet Honeyeater, Australian Logrunner, Paradise Riflebird, Green Catbird and Russet-tailed Thrush.

Lake Samsonvale

Key species: King Quail, Red-backed Button-quail, Eastern Grass Owl, Australian Painted Snipe, Oriental Cuckoo, raptors
Other species: Latham's Snipe, Great Crested Grebe, Black Bittern, Australasian Bittern, Pale-vented Bush-hen

A good site for quail and raptors, the areas around Lake Samsonvale (also known as North Pine Dam), near the Samsonvale Cemetery, can offer superb bird watching opportunities. The lake is 35 km north-west of Brisbane and is accessed via Golds Scrub Lane off Mount Samson Rd. In addition to the Samsonvale Cemetery, there is a small picnic area above the shoreline at the end of the lane. SEQwater manages Lake Samsonvale and a permit is required to gain access to many of the grassy areas around the lake: www.seqwater.com.au.

From the cemetery, head down the mown path to Lake Samsonvale. Search the surrounding grassland for King Quail and Red-backed Button-quail (summer), particularly where the grass is chest-high. Numbers of quail are dependent on local conditions, the best being when the grassy areas are wet. If the lake is drying out, search around the edges of the remaining pools of water. Other grassland birds to look for include Eastern Grass Owl, Brown Quail, Red-chested Button-quail (rare), Tawny Grassbird, Golden-headed Cisticola, Chestnut-breasted Mannikin, and Red-browed and Double-barred Finch. An alternative access point to the grasslands

Point Lookout, North Stradbroke Island

Point Lookout overlooks the Pacific Ocean and is Qld's most easterly point. Its rocky outcrops are a perfect spot for land-based seabird watching. The birding can be good here in any conditions, but is usually best when strong easterly winds push birds onshore. Between autumn and early spring, scan the ocean for Indian Yellow-nosed, Black-browed, Buller's and Wandering Albatross, Southern and Northern Giant-Petrel, Great-winged and Providence Petrel, and Fairy Prion. In summer months, look for Wedge-tailed, Short-tailed, Fluttering, and Flesh-footed Shearwater, and rarer species such as Mottled and White-necked Petrel, and Little, Buller's and Streaked Shearwater. Other seabirds to look for include Lesser Frigatebird, Masked and Brown Booby, Long-tailed, Pomarine and Arctic Jaeger. A variety of terns occur at Point Lookout, sometimes roosting on the rocks below. These include Gull-billed, Caspian, Lesser Crested, White-fronted, Sooty, White and Common Tern, Grey Ternlet, and Common and Black Noddy. Also search the rocks below the lookout for Wandering Tattler and Sooty Oystercatcher. A nice selection of marine wildlife can be seen in the offshore waters here. Between May and November, Humpback Whale travel past the point on their annual migration between Antarctica and the warm inshore waters north of Fraser Island. There is also a chance of seeing Bottle-nosed and Pantropical Spotted Dolphin, Green and Loggerhead Turtle, and Manta Ray.

around Lake Samsonvale is the end of the Postman's Track, directly across the water from the Samsonvale Cemetery. The lake itself is home to many species of waterbird, with Great-crested Grebe usually common. Other interesting records include Australian Painted (rare) and Latham's (summer) Snipe, Black and Australasian Bittern, Comb-crested Jacana, Azure Kingfisher, Little Egret, Great Crested Grebe and Black-necked Stork, and skulking species such as Lewin's and Buff-banded Rail, Baillon's and Spotless Crake, and Pale-vented Bush-hen. Whiskered, Caspian and Gull-billed Tern are also regularly observed hawking over the water. Around the periphery of the Samsonvale Cemetery, woodland and wet forest birds include Rainbow Bee-eater (summer), Bar-shouldered Dove, Pale-headed Rosella, White-bellied Cuckoo-shrike, Little and Shining Bronze-Cuckoo, Brush Cuckoo, Pheasant Coucal, Painted Button-quail, Channel Billed Cuckoo, Barred Cuckoo-shrike, Olive-backed Oriole, Australian Owlet-Nightjar, Speckled Warbler, White-throated Gerygone and Variegated, Superb and Red-backed Fairy-wren. Lake Samsonvale is also a remarkably regular site for summer visiting Oriental Cuckoo in South-East Qld. Look in areas of tall *Acacia*, especially where there is evidence of abundant

Southport pelagic boat trip

The Southport pelagic boat trips head ~50 km offshore from the Gold Coast. Generally on the third Saturday of each calendar month, for details on forthcoming trips, visit www.sossa-international.org. Notably, Black-bellied Storm-Petrel is regular in winter and Tahiti Petrel regular in summer. Other sought-after species include Black, Black-winged, Cape, Gould's, Great-winged, Kermadec, Providence and White-necked Petrel, Northern Giant-Petrel, and Buller's (rare), Flesh-footed, Fluttering, Hutton's, Short-tailed, Sooty, Streaked and Wedge-tailed Shearwater. The trip can also be particularly good for terns, with records of Bridled, Common, Gull-billed, Little, Sooty and White Tern, Grey Ternlet, and Common and Black Noddy. Albatross, which are mostly recorded in winter, include Indian Yellow-nosed, Shy, Black-browed, Campbell, Buller's and Wandering. Other seabirds have included Broad-billed (rare June/July), Antarctic, Salvin's, Slender-billed and Fairy Prion, Wilson's, White-faced, and New Zealand (rare, with records in May 2011 and January 2013) Storm-Petrel, White-tailed and Red-tailed Tropicbird, Brown, Masked and Red-footed (rare) Booby, Arctic, Pomarine and Long-tailed Jaeger, and Brown and South Polar (rare) Skua.

caterpillars. One spot to look is the southern end of the Postman's Track, such as near where the track meet Kobble Creek, 4 km from the Mount Samson Rd turn-off. Finally, an impressive 19 species of raptor have been recorded at Lake Samsonvale including Eastern Osprey, Pacific Baza, White-bellied Sea-Eagle, Wedge-tailed and Little Eagle, Square-tailed and Brahminy Kite, Swamp and Spotted Harrier, Grey and Brown Goshawk, Collared Sparrowhawk, Australian Hobby, and Peregrine and Black Falcon, so it is worth keeping an eye on the sky while birding here.

Lamington National Park

Key species: Albert's Lyrebird, Rufous Scrub-bird, Paradise Riflebird, Australian Logrunner, Noisy Pitta, Regent Bowerbird, Russet-tailed Thrush, Marbled Frogmouth, Sooty Owl, Pale-vented Bush-hen

Other species: Bassian Thrush, Rose-crowned Fruit-Dove, Wompoo Fruit-Dove, Emerald Dove, Glossy Black-Cockatoo, Red-browed Treecreeper, Spotted Quail-thrush, Black-breasted Button-quail

Lamington National Park (20 590 ha) is situated within a vast forested region containing a portion of Australia's largest area of remaining subtropical rainforest, and the most northerly cool temperate Antarctic Beech forests. Much of the national park is 800 m or more above sea level and it is networked with over 160 km of marked walking tracks. Lamington is famous for its diversity of east coast rainforest species, including such sought-after species as Rufous Scrub-bird and Albert's Lyrebird.

The national park is divided into two sections: Green Mountains and Binna Burra. Both sections encompass major ecotourism lodges, with a range of accommodation and nearby national park camping facilities available. Limited bush camping is also possible between February and November in remote areas accessible only to bushwalkers. To get to Binna Burra from the Pacific Highway, take the Nerang turn-off to Beechmont, and then it is another 10 km from the roundabout. To get to Green Mountains from the Pacific Highway, take the Nerang turn-off to Canungra, and then take Lamington National Park Rd to O'Reilly's Plateau. For more information see www.derm.qld.gov.au. Before heading up to Lamington, it is worth searching for Pale-vented Bush-hen in the riparian vegetation along the Canungra Creek. A particularly good area to look is immediately west of the Coburg Rd concrete causeway, just south of Moriarty Park ~800 m from the Canungra town centre.

Within the Green Mountain–O'Reilly's precinct there are hand-tame Regent and Satin Bowerbirds, Australian King-Parrot Crimson Rosella and Lewin's Honeyeater

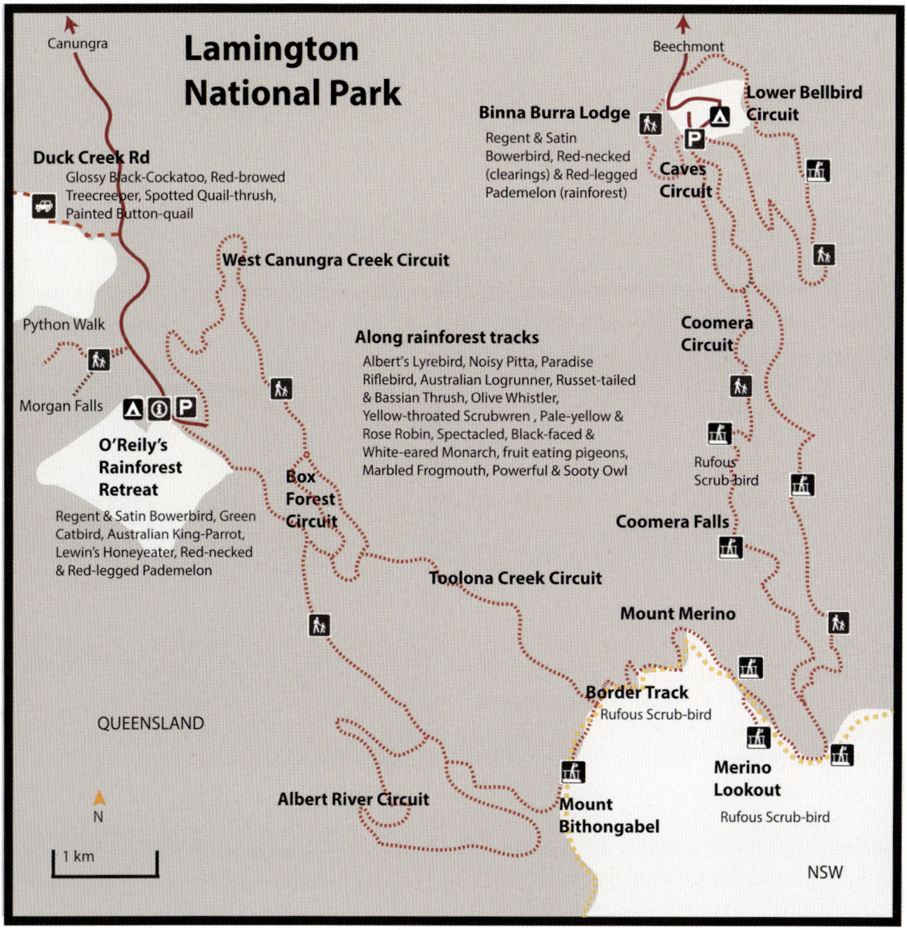

that frequent the feeding area near the store (and the camping sites) looking for handouts. Numbers of Australian Brush-turkey wander around the feet of visitors looking for further scraps. If the bowerbirds are not immediately obvious, check the trees at the entrance to O'Reilly's Guesthouse.

For general birding, well-known tracks that start from Green Mountain include the Box Forest Circuit (10.9 km), West Canungra Circuit (13.9 km), Toolona Creek Circuit (17.4 km), Albert River Circuit (20.6 km), Python Rock Walk (3.4 km) and Morans Falls Walk (4.6 km). Along these tracks, you may find rainforest specialists such Albert's Lyrebird, Noisy Pitta, Paradise Riflebird, Australian Logrunner, Russet-tailed and Bassian Thrush, Olive Whistler, Yellow-throated Scrubwren and Pale-yellow Robin. In spring, migrants that arrive including Rufous Fantail, Spectacled, Black-faced and White-eared Monarch, Shining Bronze-Cuckoo, Brush and Fan-tailed Cuckoo and fruit-eating pigeons such as Rose-crowned and Wompoo Fruit-Dove, Brown Cuckoo-Dove, Emerald Dove, Topknot and White-headed Pigeon. Although unlikely, remain alert to the possibility of the extremely rare (possibly extinct) Double-eyed (Coxen's) Fig-Parrot. Rufous Scrub-bird inhabit the rainforest along the Border Track (21.4 km one way), which connects Binna Burra and Green Mountains. Search in sections of Antarctic Beech forest, particular where there is grassy understorey, between Mount Bithongabel and Mount Merino and along the Mount Merino loop – the return hike from O'Reilly's is ~12 km). An extremely difficult bird to observe: to obtain a sighting, you will most probably need to sit quietly and very patiently.

At night along the parks tracks, there is a chance of finding Southern Boobook, Sooty and Powerful Owl and Marbled Frogmouth. A good spot to look and listen for Marbled Frogmouth and Sooty Owl is near Mick's Tower, located on the walk to Glow Worm Gully, a diversion at the end of the Wishing Tree Track (1.5 km return). The entrance to the track is located near the tennis courts at the rear of O'Reilly's guesthouse. The same two species occur along the main entrance road in the vicinity of the Duck Creek Road intersection. Red-necked Pademelon frequent the grassy areas at Green Mountain while Sugar Glider and Mountain Brushtail Possum often visit the feeders at O'Reilly's guest house after dark. Within the rainforest, Red-legged Pademelon, Fawn-footed Melomys and Long-nosed Bandicoot are all reasonably common.

Within the drier woodland areas of the Lamington, other species include Glossy Black-Cockatoo, Red-browed Treecreeper, Spotted Quail-thrush, Painted Button-quail, Variegated and Red-backed Fairy-wren, Buff-rumped Thornbill and White-naped Honeyeater. One productive area of drier woodland can be found along Duck Creek Rd (34 km): a 4WD road that is often impassable after heavy rain that leads to Beaudesert via Kerry. A good area to bird-watch is 2.5 km from the start of the track. Many of Lamington's rainforest species,

Regent Bowerbirds seek handouts at the Green Mountains day-use area, Lamington National Park. Bowerbirds are especially fond of sultanas, whereas the tame Crimson Rosellas and King Parrots prefer seed that is available from the store.

such as Paradise Riflebird and Albert's Lyrebird, occur along the first 500 m of Duck Creek Rd. Black-breasted Button-quail have also been recorded along this road, particularly in scrubby areas of rainforest with a Lantana understory, and there are historical records of Eastern Bristlebird, though that species is now thought to be locally extinct.

The Binna Burra section of the park provides many similar birding opportunities to those at Green Mountain, but has the advantage that it is often less busy than Green Mountain. Good walks in the rainforest here include the Coomera Circuit (17.5 km) and Lower Bellbird Circuit (12 km return). If you have limited time, a good short walk (500 m) from Binna Burra is to head down the Border Track, then cut back across to the Coomera Track for the return leg. Many of the rainforest specialists occur along here. From the Binna Burra side, Rufous Scrub-bird also occurs along the Coomera Walking Track – search the grassy heath on the way to the Coomera Falls Lookout (~1 km before the lookout) – the return hike from the Binna Burra is ~6 km. Black-breasted Button-quail have been recorded feeding along the roadside on the way up to Binna Burra, with a recent record from along Beechmont Rd, 9.5 km from Canungra.

The boulder-strewn West Canungra Creek accessed via walking trails from the Green Mountains section of Lamington National Park. Along this walk, look for Noisy Pitta, Albert's Lyrebird, Yellow-throated Scrubwren, Australian Logrunner, Pale-yellow Robin and Paradise Riflebird.

Lockyer Valley

Key species: Australian Painted Snipe, Cotton Pygmy-Goose, Black Bittern, Black Falcon, Square-tailed Kite, Red Goshawk (rare)
Other species: Glossy Ibis, Comb-crested Jacana, Black-necked Stork, Cockatiel, Ground Cuckoo-shrike, Plum-headed Finch

The Lockyer Valley is an area of fertile farmland lying to the west of Brisbane and east of Toowoomba. The largest town in the Lockyer Valley is Gatton. A diversity of resident and migratory birds frequent wetlands of the Lockyer Valley, including Lake Atkinson Dam, Lake Clarendon, Lake Dyer, Lake Galletly, Jahnke's Lagoon, Seven Mile Lagoon and Adare Road Lagoon. There is a range of accommodation available, including caravan parks at Lake Atkinson and Lake Dyer.

Lake Atkinson is 32 km north-east of Gatton, and usually holds an interesting selection of birds. Species that may be seen

Mount Tamborine National Park

An isolated population of Albert's Lyrebird occurs within the lush rainforest at Tamborine National Park (1160 ha), 80 km south of Brisbane within the Gold Coast hinterland; in the Witches Falls section of the Park (Qld's first national park, declared in 1908), try along the Witches Fall circuit (3.1 km return) and Witches Chase track (1.5 km); in the Palm Grove section, try along the Palm Grove circuit (2.6 km) and Jenyns circuit (4.5 km); in the Knoll section, search along the Sandy Creek circuit (2.6 km). The best time to look is in the cooler winter months, from about from mid-May, when the males are singing. Also, explore the trails in Tamborine National Park for Wompoo Fruit-Dove, Noisy Pitta, Russet-tailed Thrush, Green Catbird, Regent and Satin Bowerbird, Pale-yellow Robin, Australian Logrunner, Yellow-throated Scrubwren and Lewin's Honeyeater. At night, Mount Tamborine can also be good for Sooty Owl and Marbled Frogmouth. Although camping is not permitted in the national park, there is a wide range of accommodation on Tamborine Mountain: www.tamborinemtncc.org.au

here include Freckled Duck, Cotton Pygmy-goose, Glossy Ibis, Black Bittern, Whiskered, White-winged Black (uncommon summer visitor), Gull-billed and Caspian Tern. In summer, migratory shorebirds include Marsh, Curlew and Sharp-tailed Sandpiper and Pacific Golden Plover. Around the lake, look for Australian Pratincole and Ground Cuckoo-shrike in areas of short grassland while Square-tailed Kite is occasionally observed soaring overhead. Seven Mile Lagoon is connected to Lake Atkinson via a 1.2 km channel, and can hold large numbers of waterbirds such as Black-necked Stork, Latham's Snipe (summer), Glossy Ibis, Pink-eared, Musk and Blue-billed Duck, Australasian Shoveler and Comb-crested Jacana. Black Falcon is also seen here with some frequency. Rarer sightings have included Australian Painted Snipe, Little Curlew and Oriental Pratincole. Seven Mile Lagoon is situated on private property, but birds can be viewed from Jimba and Waters Rds, and Topaz Crescent. The Turf Farm near Seven Mile Lagoon, along Watson's Rd, occasionally hosts Banded Lapwing, while there are also several records of Oriental Plover here. Look nearby, along Boyce's Lane, for Ground Cuckoo-shrike and Channel-billed Cuckoo (summer).

Surrounded by cattle-grazing farmland and bordered by open forest, Lake Clarendon is 10 km north-east of Gatton. Waterbirds recorded here include Australasian Bittern, Comb-crested Jacana, Magpie Goose, Cotton Pygmy-Goose, Wandering and Plumed Whistling-Duck, Musk and Pink-eared Duck, Great Crested Grebe, Australasian Darter, Glossy Ibis, Black-winged and Banded Stilt (rare), and Red-necked Avocet. This lake is particularly good for birds of prey, with a chance of White-bellied Sea-Eagle, Little and Wedge-tailed Eagle, Swamp and Spotted Harrier,

Australian Hobby, Peregrine and Black Falcon; the very rare Red Goshawk has also been reported. Other good birds here include Brown Quail, Ground Cuckoo-shrike, Blue-winged Kookaburra, Grey-crowned Babbler, Cockatiel and Pale-headed Rosella. On the way to Lake Clarendon, the road passes Jahnke's Lagoon, a low-lying area of farmland on Lake Clarendon Rd 6 km from Gatton. When the lagoon holds water, look for Magpie Goose, White-necked Heron, Royal Spoonbill and Comb-crested Jacana, and summer shorebirds such as Marsh, Sharp-tailed and Pectoral (several records) Sandpiper.

Lake Dyer, 16 km south-east of Gatton, is a large dam on the outskirts of Laidley. The lake is considered a good site for Plum-headed Finch and Red-backed Button-quail (try near the boat ramp). Other interesting birds recorded here include Australian Painted Snipe (rare) and Latham's Snipe (summer), Brown Quail, Glossy Ibis, Magpie Goose, Plumed and Wandering Whistling-Duck, Pink-eared Duck, Red-backed Fairy-wren and Chestnut-breasted Mannikin.

Lake Galletly, in the grounds of the University of Queensland, 7 km east of Gatton, was created specifically as a habitat for waterbirds. Look for Magpie Goose, Plumed and Wandering Whistling-Duck, Blue-billed and Pink-eared Duck, Buff-banded and Lewin's (rare) Rail, Red-kneed and Black-fronted Dotterel, Little Grassbird and Golden-headed Cisticola. Another good birding area is around the Adare Road Lagoon, 2 km north of Gatton. The eucalyptus woodlands surrounding the lagoon support many woodland birds, including species that are uncommon in the Brisbane area. Some of the more interesting species that may be present include Cockatiel, Little Lorikeet, Little Bronze-Cuckoo, Rainbow Bee-eater, Black-chinned and Fuscous Honeyeater, Crested Shrike-tit, White-breasted Woodswallow, White-bellied Cuckoo-shrike, Restless Flycatcher, Jacky Winter and White-backed Swallow. On the lagoon itself, check for Cotton Pygmy-Goose, Plumed Whistling-Duck, Australasian Grebe and Australasian Darter. Within the township of Gatton, Freckled Duck has been recorded at Apex Lakes, sometimes roosting on the island, while other species include Cotton Pygmy-Goose, Wandering and Plumed Whistling-Duck, Glossy Ibis and Latham's Snipe (summer).

Ravensbourne National Park, 33 km west of Esk, (32 km north of Toowoomba) overlooks the Lockyer Valley. Small remnants of the rainforest and wet eucalypt forest are protected within the park. Best access is via National Park Rd, from the Esk–Hampton Rd, 32.4 km from Esk. Potential birding highlights here include Pacific Baza, White-headed Pigeon, Brown Cuckoo-Dove, Bar-shouldered Dove, Wompoo, Superb and Rose-crowned Fruit-Dove, Topknot Pigeon, Channel-billed Cuckoo, Noisy Pitta, Green Catbird, Regent and Satin Bowerbird, Varied Sittella, Cicadabird, Russet-tailed and Bassian Thrush, Black-faced and Spectacled Monarch, Paradise Riflebird, and Yellow-throated and Large-billed Scrubwren. For an interesting mix of rainforest and dry eucalypt woodland, it is worth visiting Redwood Park, immediately east of Toowoomba. The best birding is around the Redwood Park picnic area, located off the Warrego Hwy 5.6 km from Toowoomba where there is a chance of interesting species such as Grey Goshawk, Brown Cuckoo-Dove, Peaceful and Bar-shouldered Dove, Glossy Black-Cockatoo, Little Lorikeet, Shining Bronze-Cuckoo, Channel-billed Cuckoo, Dollarbird, Brown and White-throated Gerygone, Lewin's, Yellow-faced, White-naped, Brown and Scarlet Honeyeater, Eastern Yellow Robin, Eastern Whipbird, Varied Sittella, Black-faced Monarch, Rufous Fantail, Spangled Drongo, Cicadabird, Varied Triller, Olive-backed Oriole and

Sheep Station Creek Conservation Park

Sheep Station Creek Conservation Park (231 ha) is a relatively undisturbed open eucalyptus forest, located 7 km south-west of Caboolture and 40 km north of Brisbane. This is a good area for a range of dry woodland species that are generally uncommon in the region, including Painted Button-quail, Black-chinned and Fuscous Honeyeater, Dusky Woodswallow, Crested Shrike-tit and, occasionally in winter, Swift Parrot. Other birds recorded here include Square-tailed Kite, Grey Goshawk, Powerful Owl, Little Lorikeet, Glossy Black-Cockatoo, Little Lorikeet, Pale-headed Rosella, Australian King-Parrot, Brown Cuckoo-Dove, Pheasant Coucal, Channel-billed Cuckoo, Golden Whistler, Little Shrike-thrush, Restless and Leaden Flycatcher, and honeyeaters such as Scarlet, White-throated and Striped. Sheep Station Creek Conservation Park has six points of entry, with the most accessible being via Phelps Rd, off Caboolture River Rd 11 km from Caboolture.

Regent Bowerbird. This same reserve also provides habitat for Masked Owl.

Inskip Point

Key species: Black-breasted Button-quail, Beach Stone-curlew
Other species: Brown Quail, Fairy Gerygone, terns and shorebirds

Inskip Point Recreation Area is 250 km north of Brisbane on a peninsula to the north of Rainbow Beach. In addition to being a well-known site for Black-breasted Button-quail, Inskip Point is an excellent birding site more generally, despite the large numbers of campers passing through to catch the ferry across to Fraser Island. Accommodation is available at Rainbow Beach and Freshwater Lake, as well as a wide range of campgrounds at Inskip Point and Great Sandy National Park.

Inskip Point is a reliable site for the otherwise elusive Black-breasted Button-quail, which inhabit the more dense vegetation with a thick ground cover of leaf litter on the peninsula's coastal dunes. Park at the end of Inskip Point Rd near the roundabout – just before it becomes a dirt track. From here take the walking track that leads south-west – it's marked with a large sign that reads 'Pedestrian access only'. Black-breasted Button-quail are regularly seen along this track, although be prepared to put in a bit of time searching as each small covey moves around their territory daily. As with most button-quail, recent activity is often revealed by characteristic small circular scrapes in the leaf-litter called 'platelets'. If you do not find them here, take the northward track towards the nearby coast before turning right and following this for ~30 m to reach the 'beach fence'. Other birds within the vine scrub and the bordering mangroves include Noisy Pitta, Brown Quail, Australian Brush-turkey, Bar-shouldered Dove, Collared, Forest and Sacred Kingfisher, Spangled Drongo,

The coastal vine thicket towards the end of Inskip Point provides important habitat for the rare Black-breasted Button-quail. To assist with the search, look for the distinctive platelets (small circular areas from which leaf litter has been removed by scratching) that indicate the bird's recent presence.

Lewin's, White-cheeked, Mangrove, Scarlet and Blue-faced Honeyeater, Fairy and White-throated Gerygone, White-breasted Woodswallow, Varied Triller, Spectacled Monarch, Little Shrike-thrush, Large-billed Scrubwren and Red-winged and Variegated Fairy-wren. At night, listen for Large-tailed Nightjar and Bush Stone-curlew calling around the campgrounds at the east end of Inskip Point. Beach Stone-curlew usually frequent the very tip of the point, but if traffic is especially busy they move further

Many large sand inlets and sheltered bays along the Qld coast support a resident pair of Beach Stone-curlew. One such pair occurs at Inskip Point: if they aren't on the point, check the quieter south-facing beaches in the vicinity of the last carpark.

down the peninsula by a few hundred metres to quieter retreats on the beach. At low tide, shorebirds such as Lesser and Greater Sand Plover, Grey and Double-banded (winter) Plover, Grey-tailed Tattler, Eastern Curlew, Bar-tailed Godwit, Curlew Sandpiper, Whimbrel and, occasionally, Sanderling and Broad-billed Sandpiper roost on the sand bar. Terns to look for here include Common, Gull-billed, Caspian and Little, while raptors such as Eastern Osprey, Brahminy Kite, and White-bellied Sea-Eagle patrol the adjacent waterways.

Great Sandy National Park and Fraser Island

Key species: Ground Parrot, Eastern Grass Owl, King Quail, Southern Emu-wren
Other species: Brown Quail, Fairy Gerygone, Masked Owl, Marbled Frogmouth, Large-tailed Nightjar, shorebirds

The Great Sandy National Park (56 000 ha) is located between Noosa Heads (155 km from Brisbane) and Rainbow Beach (240 km

from Brisbane). The main area for birding is the Cooloola section of the park. It stretches from the north shore of the Noosa River, through vast areas of open wallum heathland, a few areas of coastal lowland rainforests to reach the mangroves of Tin Can Bay inlet.

In Great Sandy National Park, the wallum heath flats off Cooloola Way provides important habitat for Eastern Grass Owl and Eastern Ground Parrot (with the park representing the northern-most distribution for the latter species). To get there, turn off Rainbow Beach Rd onto the Cooloola Way 16 km from Rainbow Beach – a sign here, reading 'Great Sandy National Park, Cooloola Section' marks the way. Follow the Cooloola Way (gravel) for 3 km and then turn left near some overhead power lines onto a dirt track. Drive for 1.7 km and park next to where the power-lines turn right. Depending on conditions, this last 1.7 km may require 4WD or at least a high-clearance vehicle. Search here for Eastern Ground Parrot, which have a preference for low lying, ground-covering heath. Early morning or late afternoon visits are often best because the species calls at these times. The same habitat is also used by Southern Emu-wren (again, its northernmost distribution). As dusk falls, look for Eastern Grass Owl as they leave their roost sites to hunt. Lewin's Rail feed in areas of damp heath with their distinctive staccato call being a reliable indicator of their presence. In summer, the heathland can be good for aerial foraging White-throated Needletail, especially ahead of thunderstorm activity. Other interesting species found here include King and Brown Quail, Brush Bronzewing, Pheasant Coucal, White-cheeked, Scarlet and White-throated Honeyeater, Eastern Whipbird, Leaden Flycatcher, Tawny Grassbird and Golden-headed Cisticola.

The rainforest on the walk to Poona Lake (2.2 km return) from the Bymien Picnic Area supports a selection of lowland forest birds including Grey Goshawk, Rose-crowned and Wompoo Fruit-Dove, White-headed Pigeon, Emerald Dove, Green Catbird, Regent Bowerbird (winter), Noisy Pitta, Little Shrike-thrush, Brush Cuckoo (summer), Spectacled (summer) and White-eared Monarch, Rufous Fantail, Large-billed Scrubwren and Pale-yellow Robin. At night, spotlight along the track for Southern Boobook, White-throated and Large-tailed Nightjar (the park is approaching the southern-most limit for this species) and Marbled (Plumed) Frogmouth; the latter along the shorter Dundathu Walk (700 m).

The mud flats and sand flats on the Great Sandy Strait – a 70 km estuarine passage separating mainland Qld from Fraser Island – support an estimated 120 000 non-breeding shorebirds, including Bar-tailed Godwit, Eastern Curlew, Whimbrel, Grey-tailed Tattler, Lesser Sand Plover, Terek Sandpiper, Great Knot and Red-necked Stint, and rarer species such as Asian Dowitcher. Good access points to view the shorebirds can be found at Point Vernon near Hervey Bay, Boonooroo Point, 24 km

Wallum heath flats off Cooloola Way, Great Sandy National Park. This site provides ideal habitat for rare and hard-to-find species including King Quail, Lewins Rail, Eastern Ground Parrot, Eastern Grass Owl and Southern Emu-wren.

Wandering Tattler at Caloundra and Alexandra Headlands

A wader with a preference for rocky headlands, in summer, Wandering Tattler feed along the rock shelf at Wickham Point, between Shelly Beach and King's Beach in Caloundra. Commonly known as the Caloundra Headland, it's located ~90 km north of Brisbane. Other birds that feed on the rock shelf include Sooty Oystercatcher and Eastern Reef Heron, and look out to sea for Wedge-tailed Shearwater (summer). Another recommended site for Wandering Tattler is the Alexandra Headlands, located ~19 km north of Caloundra. Look for them feeding on the small rocky reefs in the surf-break immediately in front of where Alexandra Pde intersects with Buderim Ave.

Wandering Tattlers are a surprisingly localised migrant wader that occurs on select rocky headlands in eastern Qld and northern NSW. Caloundra and Alexandra Headlands in South-East Qld are two of the more accessible sites that are frequented by the species.

south-east of Maryborough, Mangrove Point clay pans (immediately north of the township of River Heads), Inskip Point and Tin Can Bay. All of these sites also provide opportunities for Black-necked Stork, Beach Stone-curlew, Mangrove Honeyeater, Mangrove Gerygone and Varied Triller.

On Fraser Island, the world's largest sand island (access by 4WD), Central Station is an excellent area of rainforest. Here there is an information centre, picnic area and camping facilities. A boardwalk passes through the rainforest along the banks of the Wanggoolba Creek and provides opportunities for sightings of Wompoo Fruit-Dove, Emerald Dove, Noisy Pitta, Russet-tailed Thrush, Large-billed Scrubwren, Brown Gerygone, Eastern Whipbird, Little Shrike-thrush and Lewin's, Dusky, Scarlet, White-throated and White-cheeked Honeyeater. Eastern Ground Parrot (rare) frequents the low wallum heath on the island, while Black-breasted Button-quail is patchily distributed, but can be common in coastal shrublands at both the northern and southern ends of the island. Night birds found on the island include Southern Boobook, Powerful Owl and Marbled Frogmouth. The island is also home to a large population of pure bred Dingo.

Conondale Ranges

Key species: Marbled Frogmouth, Sooty Owl, Masked Owl, Eastern Bristlebird, Black-breasted Button-quail

Other species: Pale-vented Bush-hen, Lewin's Rail, Russet-tailed Thrush, Bell Miner, Double-eyed Fig-Parrot (possibly locally extinct)

Conondale National Park (30 000 ha) and Imbil State Forest (17 000 ha) are part of the superb Conondale Ranges: a series of rainforest-covered mountains in the western sector of the Sunshine Coast hinterland. The ranges lie between the townships of Maleny, Kenilworth, Kilcoy and Jimna, 150 km north of Brisbane. There is camping at Charlie Moreland and Booloumba Creek camping areas, and at the Sunday Creek Environmental Centre.

Well known for its night birds, the Conondale Ranges are a regional stronghold for Marbled (Plumed) Frogmouth and Sooty and Masked Owl. Look for them in the forest edges along Booloumba Creek Rd, near the Booloumba Creek camping area, and along Sunday Creek Rd near the Charlie Moreland campground. Both roads are accessed from the Maleny–Kenilworth Rd. Sunday Creek Rd is gravel, but generally remains suitable for 2WD. Powerful Owl and Australian Owlet-Nightjar also occur locally. Pale-vented Bush-hen and Lewin's Rail are uncommon in the more dense vegetation and grassy thickets near water. Black-breasted Button-quail occur in the vine scrub dry rainforest and lantana along Little Yabba Creek: look near the Charlie Moreland campground. There is a good selection of walking tracks in the Conondale Ranges including the Fig Tree circuit (780 m return), Little Yabba circuit (1.5 km), Piccabeen circuit (3.5 km) and the Conondale Range Great Walk (56 km circuit walk – which can be broken up in shorter sections). Each walk accesses outstanding areas of rainforest and/or tall eucalypts forests, with opportunities for Emerald Dove, Wompoo and Rose-crowned Fruit-Dove, Green Catbird, Regent Bowerbird, Bell Miner, Dusky Honeyeater, Australian Logrunner, Barred Cuckoo-shrike, Oriental Cuckoo (uncommon in summer), Cicadabird, White-eared, Spectacled, Black-faced and White-eared (uncommon) Monarch, Leaden Flycatcher,

Cooroy Treatment Plant

Cooroy Treatment Plant is particularly good place to find uncommon skulking waterbirds such as Spotted, Spotless and Baillon's Crake, Lewin's and Buff-banded Rail, and Pale-vented Bush-hen. To get there, head north out of Cooroy along Mary River Rd; after ~1 km, turn left onto an unnamed track that leads along the east side of the Cooroy Pony Club. The Plant is another 250 m. Other birds to look for include Swamp Harrier, Latham's Snipe (summer), Wandering Whistling-Duck, Magpie Goose, Hardhead, Australasian Grebe, Intermediate Egret, Royal Spoonbill, Dusky Moorhen, Eastern Swamphen, Eurasian Coot, Azure and Forest Kingfisher, Tawny and Little Grassbird, Chestnut-breasted Mannikin and woodland species such as Variegated and Red-backed Fairy-wren, and Scarlet, Lewin's and Brown Honeyeater.

Bell Miner, Noisy Pitta, Paradise Riflebird, Pale-yellow Robin and Bassian and Russet-tailed Thrush. In areas of *Casuarina*, look for Glossy Black-Cockatoo. The ranges harbour the northernmost population of the Eastern Bristlebird; first discovered in the mid 1980s, sadly these populations are now perilously close to extinction.

If you do not find Marbled Frogmouth and Sooty Owl in the Conondale Ranges, they are also found in the Blackall Range, with a good area being the Mapleton Forest Reserve, especially in the tall forest along Mapleton Forest Rd ~2.5 km from Mapleton.

Toorbul Shorebird Roost

Key species: Shorebirds, Brolga, Black-necked Stork
Other species: Latham's Snipe, Collared Kingfisher, Mangrove Honeyeater, Mangrove Gerygone

Located on the western side of Pumicestone Passage, with Bribie Island to the east, the Toorbul Shorebird Roost regularly attracts a large selection of migratory shore-birds. The roost is easily visible from the Esplanade in Toorbul. The adjacent mud flats are also good for shorebirds, especially as the tide is rising pushing them towards the roost. To get there, turn off the Bruce Highway on to Pumicestone Rd 50 km north of Brisbane. Toorbul is located at the end of Pumicestone Rd, 13 km east of the Hwy.

The main roosting site (commonly known as Pelican Point) is 1.1 km south of the Toorbul Store (located at the inter-section of Freeman Rd and the Esplanade). Sandfly Bay, a small bay with mangroves at each end, is 0.7 km south of Pelican Point, while a smaller roosting site is located at the north end of the Esplanade. In spring and summer, look for Red-necked Stint, Sharp-tailed, Curlew and Terek Sandpiper, Black-tailed and Bar-tailed Godwit, Grey-tailed Tattler, Greenshank, Eastern Curlew, Whimbrel, Great and Red Knot, Lesser Sand Plover and, less commonly, Broad-billed and Pectoral Sandpiper and Grey and Pacific Golden Plover. Search the mangroves for Collared Kingfisher, Striated Heron, Mangrove Honeyeater and Mangrove Gerygone. It is also worth visiting Bishop's Marsh, located on Pumicestone Rd (the main road into Toorbul) 1.2 km from the Toorbul Store. Here you may find water-birds and shorebirds such as Marsh and Sharp-tailed Sandpiper, Latham's Snipe, Black-winged Stilt, Black-fronted Dotterel and Black-necked Stork; in the nearby paddocks, look for Brolga.

Durikai State Forest and Coolmunda Dam

Key species: Turquoise Parrot, Blue Bonnet, Squatter Pigeon, White-winged Fairy-wren, Regent Honeyeater (rare), Speckled Warbler
Other species: Yellow-tufted Honeyeater, Black-chinned Honeyeater, Fuscous Honeyeater, Plum-headed Finch, Diamond Firetail

The grassy woodlands at Durikai State Forest (3000 ha) are one of the closest points to Brisbane for dry-country species. It is located on the Cunningham Hwy, 40 km west of Warwick and 5 km east of Karara (200 km west of Brisbane). Coolmunda Dam is located on the Cunningham Hwy, 50 km west of Durikai State Forest, 13 km east of Inglewood.

Woodland species at Durikai include Squatter Pigeon, Turquoise Parrot, Little Lorikeet, Red-winged Parrot, Dusky Woodswallow, White-browed Babbler, Jacky Winter, Speckled Warbler, Buff-rumped and Yellow Thornbill, White-throated and Western Gerygone, Plum-headed and Double-barred Finch, and Diamond Firetail. The best places for birding are around waterholes that are scattered through the forest. A particularly good waterhole for honeyeaters is located 8.2 km west of the Thanes Creek Rd turn-off on the southern side of the Cunningham Hwy (immediately east of where the track crosses the Hwy). Here you may find Black-chinned, Yellow-tufted, Fuscous, White-eared and Regent (rare) Honeyeater, as well as Noisy and Little Friarbird. At Coolmunda Dam, look for grebes such as Hoary-headed, Australasian and Great Crested, and waterfowl such as Hardhead, Australasian Shoveler, and Musk, Pink-eared and Freckled (rare) Duck. Around the picnic areas, and along the first few kilometres of Mosquito Rd (accessed from the Cunningham Hwy directly opposite the dam), you may find Emu, Squatter Pigeon, Cockatiel, Blue Bonnet, Red-winged Parrot, White-winged Fairy-wren, Spiny-cheeked Honeyeater, Yellow-throated Miner, Apostlebird and Plum-headed Finch.

Girraween National Park

Key species: Superb Lyrebird, Chestnut-rumped Heathwren, Spotted Quail-thrush, Turquoise Parrot
Other species: Plum-headed Finch, Diamond Firetail, Squatter Pigeon

Situated in the Granite Belt along the Great Dividing Range, the habitat in Girraween National Park (11 800 ha) is dominated by eucalypt forests and heathlands. It is located 260 km south-west of Brisbane on the NSW–Qld border, at the northern end of the New England Tablelands. Bald Rock Creek and Castle Rock camping areas provide convenient places to use as a base within the park.

The Junction Track (3.7 km) follows the northern bank of Bald Rock Creek downstream to meet Ramsay Creek. This can be a good area for the elusive Chestnut-rumped Heathwren (Girraween is at the northern limit of their range), which favour the particularly heavily vegetated creek lines in the park. Along the Junction Track, there are a several spots to look for them: along the track ~20 m after the junction with the path to Granite Arch; the very end of the track; and the thickets bordering the swimming hole on Bald Rock Creek, below the picnic area. Along Junction Track, you may also encounter Spotted Quail-thrush and Turquoise Parrot, while the flowering shrubs along the creek may attract Yellow-tufted, White-eared, New Holland, and Yellow-faced Honeyeater. Other woodland species include Eastern Rosella, Dusky Woodswallow, Satin Bowerbird, Brown Treecreeper, White-throated Gerygone, Leaden Flycatcher, Painted Button-quail and Crested Shrike-tit. The area around Dr Roberts Waterhole can be a reliable site for 'Edward's Lyrebird', a long-tailed northern subspecies (*edwardi*) of the Superb Lyrebird; try in the vicinity of the Underground Creek junction. The Underground Creek track also provides opportunities for Spotted Quail-thrush, Scarlet Robin, Rufous Fantail, Cicadabird, Crested Shrike-tit, Buff-rumped Thornbill and, occasionally, Red-browed Treecreeper. Outside the park, excellent birding can be had along Old Wallangarra Rd, opposite Bald Mountain Winery. To get there from New England Hwy, travel 2 km east along Pyramid Rd, and then turn south along Old Wallangarra Rd. Look here for Turquoise and Swift Parrot (rare winter migrant), Plum-headed Finch, Diamond Firetail, Southern Whiteface, Scarlet and Yellow-tufted Honeyeater, White-browed Babbler (the eastern range limit for this species) and Hooded Robin. Other interesting birds recorded at Girraween include Peregrine Falcon, Glossy Black-Cockatoo and Southern Emu-wren. The park also represents the southern range limit of the Squatter Pigeon.

Sundown National Park

Key species: Painted Button-quail, Turquoise Parrot, Diamond Firetail, Crested Shrike-tit
Other species: Spotted Bowerbird, Ground Cuckoo-shrike

Sundown National Park (15 275 ha) is on the NSW–Qld border, 250 km south-west of Brisbane via Stanthorpe. This reserve, a rugged wilderness area dominated by open woodland, is particularly good for dry country birds, especially the reserve boundaries. There are camping areas at Broadwater and Nundubbermere Falls, both accessible to 2WD, while a 4WD track leads to camping areas at Red Rock Gorge, Reedy and Burrows Waterhole.

Some of the best birding is within the drier woodland areas near Broadwater campground (to the gate) and along the Western Circuit (4.5 km). Look for species such as Painted Button-quail, Turquoise and Red-winged Parrot, Eastern and Pale-headed Rosella (occasionally hybrids between the two occur in the park), Spotted Bowerbird, Speckled Warbler, White-throated Gerygone, Southern Whiteface, Grey-crowned and White-browed Babbler, White-browed and Dusky Woodswallow, Grey and Pied Butcherbird, Crested Shrike-tit, Red-capped and Hooded Robin, Plum-headed and Double-barred

Blue-faced Honeyeaters are a large and striking honeyeater that is common throughout much of eastern and northern Australia. At some sites the species is tame and begs for food scraps around picnic and camping areas.

Squatter Pigeon (Cement Mill Road)

When visiting Sundown and/or Girraween National Park, it is worth detouring to look for Squatter Pigeon at Cement Mills (accessed via the Cunningham Hwy and then Gore Rd, 63 km west of Warwick). This site is the southernmost limit for this species. Search for them is in the open woodland on either side of Cement Mills Rd, between 1.6 km and 3 km west of Cement Mills (7 km to 8.6 km west of Gore). Other birds to look here for include Plum-headed Finch, Diamond Firetail, White-backed Swallow, Hooded Robin, Restless Flycatcher and Speckled Warbler.

Finch, and Diamond Firetail. Ground Cuckoo-shrike and Square-tailed Kite are occasionally recorded. Over 17 species of honeyeater have been recorded, including Spiny-cheeked, Striped, Scarlet, Fuscous, Black-chinned and Blue-faced Honeyeater, and Little and Noisy Friarbird. Summer migrants include Rainbow Bee-eater and Dollarbird. The campground can also be good for night birds; listen for Tawny Frogmouth, Southern Boobook and both White-throated and Spotted Nightjar. Other areas to explore at Sundown include the Permanent Water Walk (2 km) and Ooline Creek Walk (4 km), where there is a chance of Azure Kingfisher.

New South Wales and the Australian Capital Territory

New South Wales (NSW) is Australia's most populous state, with some 7.3 million people. It has a land area of 801 600 km^2, with most of the population residing in the narrow coastal strip that extends from cool temperate areas on the far south coast to subtropical areas near the Queensland border. Geographically, New South Wales also surrounds the whole of the Federal Australian Capital Territory (ACT) and so this area is dealt with at the end of the section on Sydney, the central and southern coast. NSW has a wide diversity of landscapes, ranging from sandy deserts, rocky ranges and rolling downs in the west, undulating ranges and fragile, wooded grasslands on the western slopes of the Great Dividing Range, to lush temperate forests and rainforests, rugged mountain ranges and extensive coastal areas in the east. In addition to the ACT, regions covered in this chapter are Sydney and NSW South Coast, the Hunter Valley and NSW North Coast and the Western Slopes and Western Plains. The Laughing Kookaburra is the bird emblem of NSW.

Sydney, the Central and South Coasts

Regional overview

This region covers the Central Coast just to the north of Sydney, the Blue Mountains to the west of Sydney, down the spine of the Great Dividing Range, south through the Illawarra and Shoalhaven regions, then Eurobodalla and Sapphire Coast along the far south coast and inland to the Australian Capital Territory and the Snowy Mountains. The region supports diverse ecological habitats, from temperate and subtropical rainforests, woodland to wetland, coastal and highland heath, swamps and estuaries.

The NSW and ACT branches of Birdlife Australia represent the region and manage the Birdlife Australia Discovery Centre, located in the Newington Armoury Building 133, 1 Jamieson St, Sydney Olympic Park, (02) 9647 1033, (www.birdlife.org.au). Birding NSW aims to bring together all those with an interest in wild birds in Sydney, its surrounds and the state of NSW: www.birdingnsw.org.au. The Cumberland Bird Observers' Club conducts birdwatching activities throughout the greater Sydney region: www.cboc.org.au. The Illawarra Birders cover the area south of Sydney, from Stanwell Park in the north through to the areas of Wollongong, Shellharbour and Kiama: www.illawarra-birders.freei.me. The Blue Mountains Bird Observers are active in the Blue Mountains region: www.bmbo.org.au.

You can reach most of the Sydney sites mentioned using public transport. Further afield, areas include the Blue Mountains, Hawkesbury regions, the ACT and Mount Kosciusko, Booderee National Park, Barren Grounds and Nadgee and Mimosa Rocks on the far south coast.

You could easily spend 4–6 days exploring greater Sydney, with a suggested itinerary being:

- **Day 1.** Sydney – Bicentennial Park, Royal Botanic Gardens, Centennial Park (Tawny Grassbird, White-fronted and Gull-billed Tern, Southern Emu-wren, Red-whiskered Bulbul, Lewin's Rail, Buff-banded Rail, Spotless Crake, Baillon's Crake).
- **Day 2.** Maroubra and Long Reef Aquatic Reserve (seabirds).
- **Day 3.** Hawkesbury River Valley (Australasian Bittern, Baillon's Crake, Australian Spotted Crake, Spotless Crake, Regent Honeyeater).
- **Day 4.** Royal National Park (Rockwarbler, Pilotbird, Beautiful Firetail, Chestnut-rumped Heathwren, Sooty and Powerful Owl, Superb Lyrebird, Green Catbird, Southern Emu-wren, Tawny-crowned Honeyeater).
- **Days 5–6.** Blue Mountains (Rockwarbler, Pilotbird, Superb Lyrebird, Satin Bowerbird, Glossy Black-Cockatoo, Gang-gang Cockatoo, Beautiful Firetail, Satin Flycatcher, Crescent Honeyeater, Red-browed Treecreeper).

To explore the south coast of NSW, a suggested 14-day itinerary involves:

- **Days 1–2.** Sydney to Barren Grounds Nature Reserve and Budderoo National Park, 130 km (Eastern Ground Parrot, Eastern Bristlebird, Southern Emu-wren, Pilotbird, Crescent Honeyeater, Beautiful Firetail, Chestnut-rumped Heathwren).
- **Days 3–4.** Budderoo National Park to Morton National Park, 100 km (Eastern Ground Parrot, Beautiful Firetail, Southern Emu-wren, Eastern Bristlebird, Rockwarbler, Chestnut-rumped Heathwren).
- **Days 5–6.** Morton National Park to Nowra and Booderee National Park 70 km (Rockwarbler, Eastern Bristlebird, Eastern Ground Parrot, Glossy Black-Cockatoo, Beautiful Firetail, Square-tailed Kite, Green Catbird, Australian Logrunner).
- **Days 8–9.** Booderee National Park to Mimosa Rock National Park, 250 km (Hooded Plover, Powerful Owl, Little Tern).
- **Days 10–11.** Mimosa Rock National Park to Ben Boyd National Park, 60 km (Eastern Ground Parrot, Glossy Black-Cockatoo, seabirds)
- **Days 12–13.** Ben Boyd National Park to the ACT, 250 km (Gang-gang Cockatoo, Baillon's Crake, Spotless Crake, Australian Spotted Crake, Lewin's Rail, Superb Lyrebird, Pilotbird, Spotted Quail-thrush)
- **Day 14.** ACT to Sydney, 280 km.

In summer, the national park camping grounds can be very popular, so it is worth booking your campsite well in advance. For further information on camping see www.environment.nsw.gov.au. If you plan to visit Nadgee, you will need to contact the NSW National Parks and Wildlife office in Merimbula: FSCR@environment.nsw.gov.au, (02) 6495 5000.

Viewed from Echo Point, the Three Sisters is an iconic rock formation less than 2 h west of Sydney. Although not covered here, the Blue Mountains provide good additional birdwatching opportunities close to Sydney. For example, close to Echo Point, try the Scenic Railway and lower boardwalk for Wonga Pigeon, Superb Lyrebird, Brown Gerygone, Rufous Fantail (summer), Rose Robin, Satin Bowerbird and Bassian Thrush.

Sydney and the south coast of NSW are in a temperate zone, although the climate varies considerably, depending on the proximity to the coast or mountains. Although generally mild, temperature can be very high inland from the coast in the summer months and very cold in winter in the more alpine southern regions. The warmest month for the region is January, with the mean daily temperature range from 18 to 26°C (although it can exceed 40°C on days of extreme heat). The coldest month is July with a mean daily temperature range from 7 to 16°C. Despite high temperatures in summer, the coastal areas can be quite pleasant, because onshore breezes cool the area. The best period for birdwatching is in spring and early summer, when many migrants move into the area.

Excellent sealed highways provide easy access through Sydney and the south coast of NSW. The main road south along the coast from Sydney is the Princes Hwy, while the Pacific Hwy heads north along the coast. Both highways form sections of Hwy 1, which circumnavigates coastal Australia. Sydney is well serviced by metropolitan train services, while Cityrail runs commuter trains to and from Wollongong on a regular basis along the South Coast Line stations from Sydney to Nowra. Sydney, Canberra and the south coast are also well serviced by coach companies. A 2WD vehicle will get you to the majority of birding sites in the region. Sydney and Canberra have major international airports and there are also airports at Wollongong, Nowra, Moruya and Merimbula.

Birding highlights

Sydney and the central and south coast regions of NSW support one endemic species: Rockwarbler is restricted to the sandstone formations and nearby limestone formations on and around the Great Dividing Range, within a 250 km radius of Sydney. Metropolitan Sydney is also the best place to see Red-whiskered Bulbul (introduced) and the distribution of the Eastern Bristlebird is now mainly limited to the southern coastal and near coastal regions of NSW. Other notable species in the area include Australasian Bittern, Lewin's Rail, Hooded Plover, Topknot Pigeon, Eastern Ground and Swift Parrot, Glossy Black-Cockatoo, Gang-gang Cockatoo, Channel-billed Cuckoo, Eastern Koel, Powerful and Sooty Owl, White-throated Nightjar, Superb Lyrebird, Variegated Fairy-wren, Southern Emu-wren, Tawny-crowned and Scarlet Honeyeater, Bell Miner, Pilotbird, Chestnut-rumped Heathwren, Striated Fieldwren (south coast only), Red-browed Treecreeper, Australian Logrunner, Black-faced Monarch and Beautiful Firetail. For seabird watching, there is a good selection of species, most commonly Wedge-tailed, Short-tailed and Fluttering Shearwater, Indian Yellow-nosed, Campbell, Black-browed, Shy and Wandering Albatross, Northern and Southern Giant-Petrel, and Arctic, Pomarine and Long-tailed Jaeger.

Rarities recorded in the region include Canada Goose, Fiordland and Royal Penguin, South Polar Skua, Franklin's Gull, Black Tern, Antarctic and Westland Petrel, Broad-billed Prion, Little Stint, White-rumped and Buff-breasted Sandpiper, Asian Dowitcher, American Golden, Ringed and Semipalmated Plover, Dunlin, Eastern Yellow and Citrine Wagtail and House Crow.

As an international destination, Sydney provides some excellent birdwatching close to the CBD. Like these Rainbow Lorikeet, there is a chance to see many of our larger or more colourful species in the urban parklands around the city.

Greater Sydney

Key species: Tawny Grassbird, White-fronted Tern (winter), Southern Emu-wren, Red-whiskered Bulbul, Australasian Figbird, seabirds, shorebirds
Other species: Brown Quail, Striated Heron, Spotless Crake, Little Tern, Common Sandpiper, Lewin's Rail, Brown Honeyeater

With a population of over 4 million, Sydney is Australia's largest city. Despite this, there are several excellent birding locations very near to the CBD, including Maroubra, Long Reef Aquatic Reserve, Boat Harbour, Bicentennial Park and the Royal Botanic Gardens. As would be expected in such a large city, there are innumerable options for transport, meals and accommodation in Sydney: www.cityofsydney.nsw.gov.au.

In the Royal Botanic Gardens (30 ha), Buff-banded Rail feed in the garden beds around First Farm Garden. Tawny Frogmouth and a resident pair of Powerful Owl often roost in the larger trees in the gardens – ask the garden staff if they know the birds' current location. Other birds here include Australian White Ibis, Rainbow Lorikeet, Yellow-tailed Black-Cockatoo, Sulphur-crested Cockatoo, Channel-billed Cuckoo (summer), Australasian Figbird, Pied Currawong and Superb Fairy-wren.

Centennial Park (360 ha) has a good range of habitats, with large areas of planted woodlands, wetlands and some open grass-lands attracting a good selection of birds. Tawny Frogmouth often roost around the lily ponds and the paperbark swamp near the Duck Pond, and the reedy edges of these ponds can also provide opportunities for

Golden-headed Cisticola, Little Grassbird, Australian Reed-Warbler, Brown Quail and Latham's Snipe (summer). Waterfowl here includes Musk Duck, Hardhead and Pacific Black Duck. Other birds you may see include Yellow-tailed Black-Cockatoo, Sulphur-crested Cockatoo, Galah, Little Corella, Rainbow Lorikeet, Laughing Kookaburra, Pied Currawong, Australasian Figbird, Red-whiskered Bulbul, Fan-tailed and Channel-billed Cuckoo (both summer), White-plumed and New Holland Honeyeater, Yellow Thornbill and Superb Fairy-wren.

Mistral and Magic Points at Maroubra provide some of the best land-based seabird watching in Australia, with a list that rivals some of Australia's offshore pelagic trips. The best time for seabird watching is during stormy weather, especially strong easterlies when the birds are blown onshore. During winter months, you may see White-fronted Tern, Brown Skua, Fairy Prion, Great-winged Petrel and Shy, Black-browed, Indian Yellow-nosed, Wandering and Buller's Albatross (the last two are uncommon), Antarctic and Slender-billed Prion, Northern and Southern Giant Petrel,

With large wetlands, Centennial Park, provides excellent opportunities to get close to (and sometimes interact with) relatively tame waterbirds.

White-headed and Cape Petrel. During the warmer months there is a chance of Common Tern, Arctic, Pomarine and Long-tailed Jaeger, and Short-tailed, Wedge-tailed, Flesh-footed, Sooty, Fluttering and Hutton's Shearwater. There is a long list of rarities recorded at Maroubra, including Red-tailed and White-tailed Tropicbird, Common Noddy, Sooty, White and Arctic Tern, Grey Ternlet, Streaked and Little Shearwater and Gould's, White-necked, Providence, Soft-plumaged and Antarctic Petrel. Just south of the South Maroubra Surf Club, there is a collection of ponds, reedbeds and a stormwater outlet, where you may find Little and Tawny Grassbird, Brown Quail and, occasionally, Spotless Crake and Buff-banded and Lewin's Rail. The heath at Malabar Headland supports Yellow-tailed Black-Cockatoo, Southern Emu-wren, Little Wattlebird and Red-browed Finch. There are very occasional records of Eastern Ground Parrot, and in October 2003 an Orange-bellied Parrot was recorded (Australia's most northerly record – the bird was banded at Melaleuca in Tasmania as a fledgling in January of the same year).

The rock platform at Long Reef Aquatic Reserve (60 ha), 22 km north of Sydney (via Collaroy), is well known as a roosting site for shorebirds and seabirds. In summer, Red-necked Stint, Pacific Golden Plover, Ruddy Turnstone and Grey-tailed and Wandering (rare) Tattler roost here, while in winter you may find Double-banded Plover. Sooty Oystercatcher is resident here. The rock platform is also an excellent site for terns, such as Caspian, Crested and White-fronted and, rarely, Sooty and Little Tern, Grey Ternlet and Common and Black Noddy. The area can also be good for seabird watching, with records of Little Penguin, Indian Yellow-nosed, Black-browed, Shy and Wandering Albatross, Southern Giant-Petrel, and Short-tailed, Wedge-tailed and

Sooty Shearwater. The reserve is also the most southerly known east-coast breeding site for Eastern Osprey. Just south of the reserve, Red-whiskered Bulbul is found at Dee Why Lagoon, Surf Reserve and around the carpark at North Curl Curl Surf Club. In 2008, a House Crow (a presumed ship-assisted vagrant) was seen at Dee Why.

Boat Harbour is located on the south-west side of the Kurnell Peninsula. Here, species such as Striated Heron, Eastern Reef Egret, Great Knot, Pacific Golden and Grey Plover, Kelp Gull and Little Tern roost and feed on the beaches and inter-tidal rock platforms at Boat Harbour. A vagrant American Golden Plover has also been recorded.

Towra Point Nature Reserve (603 ha), located in Botany Bay, is the largest estuarine wetland in the Sydney region. Summer migrants to the site include Whimbrel, Eastern Curlew, Grey Plover, Grey-tailed Tattler, Bar-tailed Godwit and Ruddy Turnstone. It is also a breeding site for Little Tern. Although public access is restricted, there is a viewing platform overlooking Quibray Bay that can be accessed from Captain Cook Drive.

Sydney Olympic Park and Bicentennial Park

Key species: Lewin's Rail, Buff-banded Rail, Spotless Crake, Baillon's Crake
Other species: Australasian Grebe, Royal Spoonbill, Glossy Ibis, Great Egret, Nankeen Night-Heron, Red-necked Avocet, Black-winged Stilt, shorebirds (summer)

Sydney pelagic boat trips

The Sydney pelagic boat trips depart once a month from the Rose Bay Public Ferry wharf along New South Head Rd, Rose Bay. Bookings are essential: www.sydneypelagics.info, 0411 311 236. Over 80 species of seabird have now been recorded on these trips. Common species include Indian Yellow-nosed, Shy, Campbell, Black-browed and Wandering Albatross, Wedge-tailed, Flesh-footed, Fluttering, Hutton's, Short-tailed and Sooty Shearwater, Cape, White-headed, Providence and Great-winged Petrel, Southern and Northern Giant-Petrel and White-faced and Wilson's Storm-Petrel. Some of the rarities that have been recorded include Buller's, Grey-headed, Light-mantled Sooty and Royal Albatross, Buller's, Great, Streaked and Little Shearwater, and Black, Black-winged, Cook's, Gould's, Kermadec, Soft-plumaged, Tahiti, Westland, White-chinned and White-necked Petrel. Other seabirds recorded on the trips include White-tailed and Red-tailed Tropicbird, Brown and South Polar Skua, Arctic, Pomarine and Long-tailed Jaeger, and terns such as Bridled, Little, Sooty, White and White-fronted Tern, and Common and Black Noddy. Many of these species occur on a seasonal basis: www.sossa-international.org has an excellent month by month guide for all species recorded during nearby Wollongong pelagic excursions.

Sydney Olympic Park (430 ha) is located 16 km west of the Sydney CBD. Redeveloped for the 2000 Olympics, there are extensive areas of parkland (incorporating Newington Nature Reserve and Bicentennial Park). To get there, take the M4 motorway off which Sydney Olympic Park is clearly signposted. Birdlife Australia runs the Discovery Centre, located in the Newington Armory Building 133, Jamieson Street: www.birdsaustralia.com.au/visit-us/discovery-centre.

At Sydney Olympic Park, a good selection of waterbirds has been recorded at the Triangle Pond (between Bennelong Rd and the Waterbird Refuge). Scan the northern end of the pond, where there is a large clump of papyrus. Lewin's Rail, along with Buff-banded Rail, Spotless and Baillon's Crake and Australian Painted Snipe (very rare) have been recorded here; the best conditions are when the water level is low and there is a strip of mud in front of the vegetation. Other birds you may see here include Australasian Grebe, Royal Spoonbill, Glossy Ibis, Great Egret, Nankeen Night-Heron, Red-necked Avocet and Black-winged Stilt, and summer shorebirds such as Pacific Golden Plover, Red-necked Stint, Sharp-tailed and Curlew Sandpiper and Bar-tailed Godwit. It is worth being alert to the possibility of rarer species because there have been records of Ruff, Pectoral Sandpiper and Australian Little Bittern. Sydney Olympic Park also supports a small, but regionally important, population of White-fronted Chat – they inhabit the Coastal Saltmarsh of the Newington Nature Reserve. Bush birds that can be observed within the park include the Superb Fairy-wren, Silvereye, Yellow Thornbill, Crested Shrike-tit, White-browed Scrubwren, Golden Whistler, Red-whiskered Bulbul, Brown, New Holland and Scarlet (occasionally) Honeyeater, Nutmeg Mannikin, Red-browed and Double-barred Finch. Established in 2009, EagleCAM (see www.birdlife.org.au/visit-us/discovery-centre/eagle-cam) is a live video stream of a pair of White-bellied Sea-Eagles that breed in the bushland behind the BirdLife Australia Discovery Centre.

The wetlands and saltmarsh at Bicentennial Park (100 ha), adjacent to Sydney Olympic Park support several shorebird species. The best areas are the wetlands at the northern end of the park and the mud flats on Homebush Bay. Look here for Black-winged Stilt, Red-necked Avocet and Red-kneed and Black-fronted Dotterel, and summer migrants such as Sharp-tailed Sandpiper, Bar-tailed Godwit, Eastern Curlew and Pacific Golden Plover. In the mangroves along the boardwalk, there is a chance of Striated Heron, Common Sandpiper and Brown Honeyeater.

Dharug National Park

Key species: Australian Brush-turkey, Glossy Black-Cockatoo, Lewin's Rail, Masked Owl, Spotted Quail-thrush, Green Catbird, Rockwarbler

Other species: Gang-gang Cockatoo, Pilotbird, Satin Bowerbird, Brown Cuckoo-Dove

One of the best birding sites north of Sydney, Dharug National Park (14 834 ha) has a diverse range of habitats, including sandstone escarpment, dense eucalypt forest, dry ridge-top woodlands, swamp, heath and pockets of rainforest. The park is a 1-h drive from Gosford via Mangrove Mountain and Central Mangrove, or a 2-h trip from Sydney via Wisemans Ferry. The best birding is around the Mill Creek picnic and camping area. To get to Mill Creek, turn right from the car ferry and take Wisemans Ferry Rd for 5.5 km before turning left for the picnic and camping area.

Look for Glossy Black-Cockatoo, Rockwarbler and Spotted Quail-thrush along the Circuit Track (11 km, with some steep climbs). Rockwarbler occupies the sandstone outcrops, Spotted

Ku-ring-gai Chase National Park

Ku-ring-gai Chase National Park (14 882 ha) is located where the Hawkesbury River meets the sea, 30 km north of Sydney's CBD. Here, heathland predominates around sandstone ridges, dense forests occur on the slopes and mangroves can be found on the tidal mud flats. Look for Chestnut-rumped Heathwren in the scrubland on the Chiltern Track (well known for its wildflowers), 1 km from Chiltern Rd entrance. It is also excellent for honeyeaters: Scarlet, Fuscous, Lewin's, White-cheeked and Yellow-tufted Honeyeater, Little Wattlebird and Noisy Friarbird have all been recorded. Along this track, you may also see Brush Bronzewing, Yellow-tailed Black-Cockatoo, Pheasant Coucal, Grey Goshawk, Varied Sittella and Golden Whistler. Running off West Head Rd, the Waratah Track and Challenger Track pass through open woodland and heathland. In the late winter and spring, when the heath is in flower, look for Tawny-crowned Honeyeater as well as those species listed above. Other birds along here include Painted Button-quail, Chestnut-rumped Heathwren and Brush Bronzewing. Chestnut-rumped Heathwren also occur along the Mount Ku-ring-gai and Basin Tracks. Australian Brush-turkey hang around the Resolute Picnic Area and Swift Parrots sometime feed here when the eucalypts are flowering in late autumn and early spring. Other good birding sites in the park include Salvation Creek and West Head: both are fairly reliable areas for Glossy and Yellow-tailed Black-Cockatoo.

Quail-thrush feeds on the ground along elevated woodland ridges, while Glossy Black-Cockatoo forage primarily within stands of Casuarina. Along the Grass Tree Circuit Walk (2 km), look for Superb Lyrebird, Gang-gang Cockatoo, Pilotbird, Satin Bowerbird, Green Catbird, Brown Cuckoo-Dove, Wonga Pigeon, Fan-tailed and Brush Cuckoo, Cicadabird, Leaden Flycatcher, White-throated Treecreeper, Eastern Yellow Robin, Dollarbird (summer) and Australian Brush-turkey. When the *Angophora* (a large tree closely related to *Eucalyptus*) is flowering, there can be an abundance of nectarivores, including White-naped, Yellow-faced, Lewin's, Scarlet and White-eared Honeyeaters. Due to a large number of nocturnal tree-dwelling marsupials (such as Brush-tailed and Ring-tailed Possum, and Yellow-bellied, Greater, Sugar and Feathertail Glider), the park is also an excellent site for owls, with five resident species; in the Mill Creek area, there is a reasonable chance of both Masked and Powerful Owl. Other night birds found here include White-throated Nightjar, Australian Owlet-Nightjar and Tawny Frogmouth. Black Bittern feed along the creek. On the Wiseman Ferry Rd, just before the turn-off to the Mill Creek, there is a small bridge over Mill Creek (5.3 km from the Wiseman Ferry). The grassy area along north-east of the creek line occasionally harbours the elusive Lewin's Rail.

NSW Central Coast

Key species: Bush Stone-curlew, Regent Honeyeater, Swift Parrot, Noisy Pitta, Regent Bowerbird
Other species: Freckled Duck, Plumed Whistling-Duck, Wandering Whistling-Duck, Spotless Crake, shorebirds

The Central Coast region of NSW is an hour's drive north of Sydney, accessed via the Pacific Highway (also known as the F3 Freeway): look for the two major exits ('Gosford, Woy Woy, Terrigal' and 'Wyong, The Entrance, Yarramalong'). There is a wide range of accommodation options in the Central Coast: www.visitcentralcoast.com.au.

The extensive estuarine lakes around the southern section of Brisbane Water support large numbers of waterbirds and shorebirds, including Australian Pelican (largest colony in NSW at ~500 pairs), Australian Pied Oystercatcher and migratory shorebirds such as Eastern Curlew, Bar-tailed Godwit, Whimbrel and Grey-tailed Tattler. The Brisbane Water area also holds a relict population of Bush Stone-curlew: listen for them at night at St Huberts Island, Empire Bay, Umina Beach, Saratoga Wetland Reserve, Lintern St Wharf in Davistown and Bensville. When flowering, the Swamp Mahogany forests around Brisbane Water provide an important winter feeding habitat for the Regent Honeyeater and Swift Parrot.

The Tuggerah Sewage Treatment Works are located 1 km from the Tuggerah Railway Station along Wyong Rd (with the best access from Ibis Rd). They are surrounded by settling ponds, attracting a wide variety of waterbirds, including Musk Duck, Australasian Shoveler, Hoary-headed and Australasian Grebe, Australasian Darter, Buff-banded Rail, Spotless Crake and Black-winged Stilt. Waterfowl that occur rarely include Freckled, Blue-billed and Pink-eared Duck. There is also a rewarding area of bushland (within the perimeter fence), with patches of Swamp Mahogany. Birds to look for include Collared Sparrowhawk, Grey Goshawk, Tawny Grassbird, Lewin's and Scarlet Honeyeater, Red-whiskered Bulbul, White-breasted Woodswallow, Yellow Thornbill, Brown Gerygone and Chestnut-breasted Mannikin. These same trees have also hosted the Critically Endangered Regent Honeyeater on several occasions.

The Tuggerah Lakes Reserve (Pioneer Dairy Wetlands) is part of an Important Bird Area that attracts a multitude of waterbirds and shorebirds. The reserve is located on the Old Pacific Hwy, immediately south of Wyong. Birds to look for here include Black-necked Stork (several recent records), Plumed and Wandering Whistling-Duck, Musk and Pink-eared Duck, Australasian Shoveler, Australasian and Great Crested (uncommon) Grebe, Australasian Darter, Great and Intermediate Egret, Yellow-billed and Royal Spoonbill, Sharp-tailed Sandpiper, Latham's and Australian Painted Snipe (rare). The site can be good for quail, with records of all three NSW species: Brown, Stubble and King Quail (several records). Regent Honeyeater and Swift Parrot have also been recorded here in autumn and winter; search the south side of the reserve, accessed via South Tacoma Rd. The nearby McPherson Rd Swamp is located on the other side of the Old Pacific Hwy and can be particularly good for Buff-banded and Lewin's Rail, Australian Spotted, Spotless and Baillon's Crake, Black-tailed Native-hen and Tawny and Little Grassbird.

The Ourimbah RTA Reserve is located on the lower section of the Ourimbah Creek and consists mostly of rainforest with a few grassy areas. It can be accessed via Ourimbah Creek Rd, adjacent to the Pacific Hwy (F3), just west of the Old Pacific Hwy intersection. Birds to look for here include Brown Cuckoo-Dove, Topknot Pigeon, Bar-shouldered Dove, Gang-gang Cockatoo, Noisy Pitta, Green Catbird, Satin and Regent Bowerbird, Yellow-throated, Large-billed and White-browed Scrubwren,

Lewin's Honeyeater, Brown Gerygone, Black-faced Monarch, Leaden Flycatcher, Olive-backed Oriole, Eastern Yellow Robin, Red-whiskered Bulbul and Bassian Thrush.

Soldiers Point overlooks a rock platform that acts as a shorebird roosting site. Look here for Eastern Reef Egret, Pacific Golden and Double-banded Plover, Sooty Oystercatcher, Grey-tailed and Wandering Tattler, Ruddy Turnstone, Red-necked Stint, Great Knot, Curlew Sandpiper and Common, Little and Crested Tern. Offshore you may see Wedge-tailed and Short-tailed Shearwater and Australasian Gannet. Soldiers Point is located at the end of Soldiers Point Rd, accessed from the Central Coast Hwy in Norah Head.

Hawkesbury River Valley

Key species: Australasian Bittern, Baillon's Crake, Australian Spotted Crake, Spotless Crake
Other species: Australasian Shoveler, Pink-eared Duck, Speckled Warbler

The Hawkesbury River Valley is 50 km north-west of Sydney's CBD. Although day visits can be made from Sydney, there is a wide range of accommodation and camping in the Hawkesbury regions: www.hawkesburytourism.com.au or the Hawkesbury Visitor Information Centre: (02) 4578 0233. The valley has a good selection of freshwater lagoons that attract a wide variety of waterbirds, especially in summer and during periods of drought, when inland species move to coastal wetlands.

Pitt Town Lagoon can be a very good place for Australasian Bittern: look in the reedbeds to the right of the observation ramp. It is also good for summer shorebirds such as Sharp-tailed, Wood, Marsh and Pectoral Sandpiper, Ruff (rare) and Latham's Snipe. Other birds seen here include Australasian Shoveler, Pink-eared Duck, Glossy Ibis and Baillon's, Australian Spotted and Spotless Crake. The lagoon, 7 km north-east of Windsor, is accessed via Lagoon Rd just south of Pitt Town.

Bushell's Lagoon attracts large numbers of waterbirds, including Australasian Shoveler, Pink-eared Duck, Hardhead, Australasian, Hoary-headed and Great Crested Grebe, White-necked Heron, Great, Intermediate, Little and Cattle Egret, Glossy Ibis, Royal Spoonbill and Australian Little Bittern (the latter is rare). The lagoon, 6 km north of Windsor, is accessed via Putty and Freeman's Reach Rds.

Pugh's Lagoon can be a good place to look for Azure Kingfisher and Australasian Grebe. Other birds recorded here include Nankeen Night-heron, Baillon's Crake, Banded Lapwing and Whiskered Tern (summer). The lagoon is located on Old Kurrajong Rd, south of the Hawkesbury River.

Baker's Lagoon can be an excellent site for shorebirds in summer with typical species, including Red-necked Stint, Marsh, Curlew and Sharp-tailed Sandpiper. Other birds that occur here include Baillon's Crake, Brown and Stubble Quail, Cattle and Great Egret, Little Grassbird and Pied Butcherbird. The adjacent grassland areas produce sightings of Spotted Harrier, Banded Lapwing, Chestnut-breasted and Nutmeg Mannikin and Zebra Finch. The lagoon, 6 km west of Windsor, is reached via Cornwallis Rd.

Longneck Lagoon and Scheyville National Park (950 ha) has records of Australasian and Black Bittern, as well as Australasian Darter, Latham's Snipe (summer), White-bellied Sea-Eagle and Azure Kingfisher. The lagoon is located in the north-east section of Scheyville National Park (960 ha) and best accessed via Dural and Whitmore Rds. Bushland within the Scheyville National Park holds a range of birds that are often uncommon elsewhere in the Sydney region. These include Speckled Warbler, White-bellied Cuckoo-shrike, Swift and Turquoise Parrot (both rare), Peaceful and Bar-shouldered Dove, Grey Goshawk, Painted Button-quail,

Londonderry

When the Mugga Ironbark is flowering (mainly winter) in the streets of Londonderry, 10 km south of Windsor, look for Swift Parrot and Regent Honeyeater, along with the more usual Rainbow, Musk, Scaly-breasted and Little Lorikeet – a good spot to search is in the trees in the vacant block opposite the shopping centre. Just south of Londonderry, Regent Honeyeater and Swift Parrot have also been recorded at Castlereagh Nature Reserve (492 ha). Other birds here include Musk and Little Lorikeet, Variegated Fairy-wren, Speckled Warbler, Crested Shrike-tit, Golden whistler, Red-capped and Rose Robin, Buff-rumped and Yellow Thornbill, and Yellow-faced, White-eared, Fuscous, Scarlet, White-naped and Brown-headed Honeyeater. In summer, Olive-backed Oriole and White-throated Gerygone also occur here. At Mitchell Park, a remnant area of bushland on Cattai Creek, you may see Nankeen Night-Heron, Black Bittern and Pacific Baza (the latter two rare). Powerful Owl also roost along the creek and can occasionally be found perched up in denser cover. Access Mitchell Park via Cattai Rd then Mitchell Rd (5.6 km east of Pitt Town). The Turf Farms, on Freeman's Reach Rd north of Windsor, is a fairly reliable place for Banded Lapwing, as well as Brown, Stubble and King (rare) Quail are also occasionally reported here.

Dollarbird (summer), Crested Shrike-tit, Fuscous Honeyeater, Buff-rumped Thornbill, Rose Robin and Double-barred Finch. Red-whiskered Bulbul also occurs.

Royal National Park

Key species: Rockwarbler, Pilotbird, Beautiful Firetail, Chestnut-rumped Heathwren,
Other species: Sooty Owl, Powerful Owl, Superb Lyrebird, Green Catbird, Southern Emu-wren, Tawny-crowned Honeyeater

First established in 1879, Royal National Park (15 091 ha) is the world's second oldest national park (after Yellowstone in the USA). The park offers excellent birding in a variety of habitats, including temperate rainforest, woodlands, heath, mud flats, mangroves and beaches. Car-based and caravan camping sites are available at Bonnie Vale. Bush camping is permitted at limited sites, though bookings are essential: (02) 9542 0683. There is an entry fee of $11 per vehicle per day.

Audley, located on Farnell Avenue, 3 km from the Princes Hwy, is an excellent starting point for any visit to Royal National Park. From here, head to the northern section of the Lady Carrington Drive Walk (10 km one way). Rockwarbler occur within the woodlands here: search near rocky outcrops, such as the large sandstone overhang known as Gibraltar Rock. Also expect to find Sacred Kingfisher, Wonga Pigeon, Australian King-Parrot, Satin Bowerbird, Eastern Whipbird, Crested Shrike-tit and Varied Sittella. Forest raptors that frequent the area include Grey Goshawk, Collared Sparrowhawk and Pacific Baza (uncommon). There is a good area of temperate rainforest along the short walk to the Wattle Forest Picnic

Early morning light on the Bundeena Cliffs, Royal National Park. The cliff tops provide habitat for Rockwarbler, while the adjacent heathlands harbour Chestnut-rumped Heathwren, Southern Emu-wren, Tawny-crowned Honeyeater and Beautiful Firetail.

Area, which starts in Audley. Here, there is a chance of Pilotbird, Green Catbird, Satin Bowerbird, Black-faced Monarch (summer), Bassian Thrush, Rose Robin (winter), Lewin's Honeyeater, Variegated Fairy-wren and Yellow-throated and Large-billed Scrubwren. Noisy Pitta, rare south of Sydney, has been recorded along this walk on several occasions. Near Audley Weir and at Kangaroo Flat (just west of the weir) you may find Azure Kingfisher, Australasian Darter, Australasian Grebe, Buff-banded Rail and Australian Spotted Crake. There are occasionally records of Black and Australian Little Bittern. The temperate rainforest in the southern end of the Lady Carrington Drive Walk (the first 2 km of the walk) provides habitat for Superb Lyrebird, Topknot Pigeon, Brown Cuckoo-Dove, Green Catbird, Bell Miner, Lewin's

The temperate rainforest at the southern end of the Lady Carrington Drive Walk, Royal National Park is one of a number of sites near to Sydney where Sooty Owl has been recorded. Other night birds recorded here include Powerful Owl, Southern Boobook and Australian Owlet-Nightjar.

Honeyeater, Yellow-throated Scrubwren, Brown Gerygone, Rufous Fantail, Black-faced Monarch and Cicadabird. At night, look and listen for Sooty and Powerful Owl, Southern Boobook and Australian Owlet-Nightjar. Similar species occur along the nearby Forest Path Walk (4.5 km return).

Chestnut-rumped Heathwren, Southern Emu-wren, Tawny-crowned Honeyeater and Beautiful Firetail inhabit the coastal heath along the Mount Bass Fire Trail. The trail runs off Bundeena Drive, 2.6 km north its intersection with Sir Bertram Stevens Drive. Look here for Brown Quail, Yellow-tailed Black-Cockatoo, Pheasant Coucal, Variegated Fairy-wren, Little Wattlebird and New Holland, White-cheeked, Yellow-faced and White-eared Honeyeater. Like most heathland areas, a good time for birding is the early morning when there are higher levels of nectar within flowers on the heath. In summer, White-throated Needletail, and less frequently Fork-tailed Swift, may hawk for insects over the heathland, especially ahead of storm fronts. Similar species occur along the Curra Moors Trail, accessed via Sir Bertram Stevens Drive, 1 km north of Garie Beach Rd. Shorebirds and other wading species, including Eastern Curlew, Bar-tailed Godwit, Little and Great Egret, Striated Heron, Nankeen Night-Heron, Australian Pied Oystercatcher and, occasionally, Hooded Plover and Beach Stone-curlew, feed on the sandbars and mangroves at Maianbar, Bundeena and the Bonnie Vale campground.

Wollongong pelagic boat trip

Over 100 seabird species have been recorded on the Wollongong pelagic boat trips, representing over one-third of the world's seabirds. During the trip, the boat travels to the edge of the continental shelf ~20 nautical miles east of Wollongong. Pelagic waters are an unpredictable environment, yielding albatrosses and killer whales one day then little the next, so it can be hard to predict what may be seen. Nevertheless, the guides will work hard to find the best spots based on the conditions of the day and identify the birds for the beginners on board. That being said, more common species include Indian Yellow-nosed, Shy, Campbell, Black-browed and Wandering Albatross, Cape, Great-winged and Providence Petrel, Flesh-footed, Fluttering, Hutton's, Sooty, and Wedge-tailed Shearwater, Southern and Northern Giant-Petrel, and Wilson's and White-faced Storm-Petrel. Over the years, outstanding rarities have included New Zealand Storm-Petrel, Herald Petrel, Light-mantled Sooty Albatross and Sabine's Gull. Full-day trips depart Wollongong Harbour on the fourth Saturday of every month; advance bookings are essential. If you have 'target' species, then maximise your chances of seeing them by consulting the Southern Oceans Seabird Study Association Inc.'s month-by-month birding guide: www.sossa-international.org.

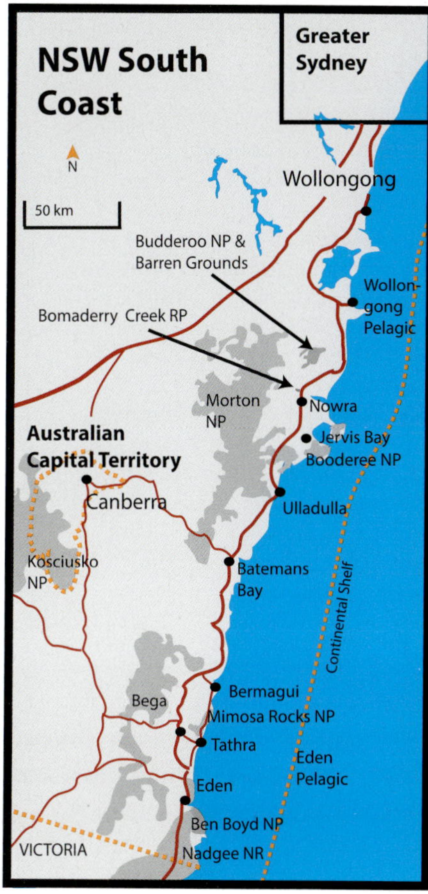

Barren Grounds Nature Reserve

Key species: Eastern Ground Parrot, Eastern Bristlebird, Southern Emu-wren, Pilotbird, Crescent Honeyeater, Beautiful Firetail, Chestnut-rumped Heathwren
Other species: White-throated Nightjar, Square-tailed Kite, Barking Owl, Brush Bronzewing

Barren Grounds Nature Reserve (1532 ha) is a superb birding site. Located high on the Illawarra Escarpment, and classified as a hanging swamp plateau, the habitat is dominated by heathland. Facilities include a picnic shelter, toilet and barbecues. The entrance to Barren Grounds is 19 km west of Kiama (16 km east of Robertson) off the Jamberoo Mountain Rd.

Immediately after entering the reserve, there is a house on your right. Stop and check the forest on your left for Pilotbird (listening for their penetrating call), Beautiful Firetail and, occasionally, Lewin's Rail, especially early in the morning. Eastern Bristlebird,

Barren Grounds is best known for its extensive heathlands that support important populations of Eastern Ground Parrot and Eastern Bristlebird. The small pockets of rainforest sheltered by the sandstone escarpment provide habitat for Brown Cuckoo-Dove, Superb Lyrebird, Pilotbird, Black-faced Monarch, Eastern Whipbird and Satin Bowerbird.

Budderoo National Park

If you don't see Eastern Bristlebird and Eastern Ground Parrot at Barren Grounds, continue 3 km along the Jamberoo Mountain Rd to the Budderoo Track (12 km) in Budderoo National Park (7120 ha). This is a fire trail that traverses heath, mixed eucalypt and banksia woodland. A particularly good spot to look is ~300 m from the road, where there is a fence and gate. The heath here is a little lower than at Barren Ground and consequently birds can be easier to see. Other birds to look for along the trail include Bassian Thrush, Southern Emu-wren, White-eared, Crescent and Tawny-crowned Honeyeater, and Beautiful Firetail and, in woodland areas further along the track, Gang-gang Cockatoo and Red-browed Treecreeper. The Minnamurra Falls Rainforest Centre is an excellent spot to see Superb Lyrebird (often foraging around the carpark) and along the Loop Walk (1.6 km) and Falls Walk (2.6 km) you may see Brown Cuckoo-Dove (summer), Topknot Pigeon (summer), Satin Bowerbird, Yellow-throated and Large-billed Scrubwren, Rose Robin, Bassian Thrush, Brown Gerygone and Lewin's Honeyeater.

Eastern Ground Parrot, Southern Emu-wren and Beautiful Firetail inhabit the heath along the western branch of the Griffith Trail Loop (8 km), particularly the section between the carpark and Natural Stone Bridge. At the beginning of the walk (near the old bird observatory), there is a fire trail that leads west down to a creek line (1 km return). The low-lying heath on this trail can be good for Eastern Ground Parrot: listen at dawn and dusk for their distinctive resonating call. A good spot for Eastern Bristlebird is where the Cooks Nose Lookout Walk branches west from the Griffith Trail, 1.5 km from the carpark. The heathland near the beginning of the eastern branch of the Griffith Trail is another reliable spot for Eastern Bristlebird, while Pilotbird occur in the woodland near the Illawarra Lookout. In spring, when the wildflowers are blooming, the Herbarium Walk (1.5 loop) is particularly good for honeyeaters such as Crescent, Lewin's, New Holland, Yellow-faced, White-cheeked White-eared, Fuscous and Scarlet Honeyeaters (the last two occasionally). Other regular species at Barren Grounds include Brown Quail, Superb Lyrebird, Brush Bronzewing, Brown Cuckoo-Dove, Gang-gang Cockatoo, Fan-tailed Cuckoo, Satin Bowerbird, Chestnut-rumped Heathwren (along the southern section of Griffith Trail, around the edges of forest clearings), Eastern Whipbird, Black-faced Monarch, Leaden and Satin Flycatcher and Eastern Yellow Robin.

Morton National Park

Key species: Eastern Ground Parrot, Beautiful Firetail, Southern Emu-wren, Eastern Bristlebird, Rockwarbler, Chestnut-rumped Heathwren

Other species: Brown Cuckoo-Dove, Superb Lyrebird, Pilotbird, Brush Cuckoo, Black-faced Monarch, Striated Fieldwren

Morton National Park (162 000 ha) supports a diversity of habitats, including rainforest and dry eucalypt forest as well as wet sedge and heathland on the plateaus. There is a campground at Gambells Rest. The Fitzroy Falls Visitor Centre is located 38 km north-west on Nowra on the Moss Vale Rd: (02) 4887 7270.

Within Morton National Park, the Little Forest Plateau is one of the largest heathland areas in NSW. To get there, take the Pointer Gap Rd towards Porters Creek Dam. After 10 km, turn left and drive a further 2.5 km to the sign for Little Forest Walking Tracks; the Rusden Head Track (10 km return) and Florance Head Track (6 km return). Along both walks, look for heathland specialties such as Eastern Ground Parrot, Southern Emu-wren, Variegated Fairy-wren, Striated Fieldwren, Chestnut-rumped Heathwren and Beautiful Firetail. Although not recorded for some years, it is worth looking for Eastern Bristlebird in areas where the heathland abuts woodland habitats. In the forested areas along both walks, look for Brown Cuckoo-Dove, Wonga Pigeon, Green Catbird, Superb Lyrebird, Pilotbird, Rockwarbler (southern limit, occurring near sandstone outcrops), Brush Cuckoo, Leaden Flycatcher, Black-faced Monarch, Crescent Honeyeater and Rufous Fantail.

The Erith Coal Mine Walk (2 km return) is located in the Bundanoon section of Morton National Park. Along this steep walk, which passes through woodland, heath and rainforest, look for Brown Cuckoo-Dove, Superb Lyrebird, Pilotbird, Grey Goshawk, Rufous Fantail, Rockwarbler (on the rock face near the waterfalls), Black-faced Monarch and Yellow-throated Scrubwren. Red-browed Treecreeper and Spotted Quail-thrush occur around Gambells Rest, particularly in woodlands around the ridgelines. Night birds in the Bundanoon section include Sooty and Powerful Owl: try listening for them from Sunrise Point and the Grand Canyon Lookout (both located on the Fairy Bower Falls Scenic Circuit).

There are several good walks starting at Fitzroy Falls. Birds to look for along the East Rim Track (6.7 km return) include Pilotbird, Superb Lyrebird, Gang-gang Cockatoo, Topknot Pigeon, Green Catbird, Crescent Honeyeater, Yellow-throated Scrubwren, Varied Sittella, Crested Shrike-tit, Rose Robin and Satin Bowerbird. Similar birds occur along the West Rim Track (4 km return). Along this walk, a steep staircase descends to a cool ferny gully known as the Grotto. Here, Rockwarbler can often be found along the rocky stream just before the Grotto. Along the Redhill Fire Trail (7 km return), which starts near the Twin Falls Lookout, located on the West Rim Track, Southern Emu-wren and Beautiful Firetail occur in the heathland areas. Finally, back at the Fitzroy Falls Visitor Centre, Bassian Thrush occasionally occur in and around the carpark.

Booderee National Park and Jervis Bay

Key species: Eastern Bristlebird, Eastern Ground Parrot, Glossy Black-Cockatoo, Beautiful Firetail, Square-tailed Kite
Other species: Southern Emu-wren, Tawny-crowned Honeyeater, Brush Bronzewing

The coastal heathlands, forests and marine environments at Booderee National Park (6400 ha) support a rich variety of birdlife, including Eastern Bristlebird and Eastern Ground Parrot. Booderee National Park has an excellent visitor centre that is open daily and a variety of camping options: www.environment.gov.au/parks/booderee.

Begin your search for Eastern Bristlebird near the historic ruins of the Cape St George Lighthouse, located at the end of Stony Creek Rd. Here, they are occasionally seen running across the road to the carpark, especially early in the morning:

also look and listen for them in the heath around the carpark and along the short Lighthouse Walk (100 m). A shy and cryptic bird, they spend most of their time hidden from view among low dense heath. The males, when alarmed, may move to a lookout perch a metre or so above the ground to give their distinctive loud and penetrating call and, if a female is in the territory, she may also respond. Other birds here include Southern Emu-wren, Variegated Fairy-wren and Tawny-crowned Honeyeater. In summer, scan the skies for White-throated Needletail. The lighthouse provides an excellent vantage point for seabird watching, with a chance of Australasian Gannet, Short-tailed, Wedge-tailed and Fluttering Shearwater, Northern Giant-Petrel, Arctic and Pomarine Jaeger and Shy, Indian Yellow-nosed and Black-browed Albatross.

Hyams Beach is a small town on the shores of Jervis Bay. Eastern Bristlebird and Eastern Ground Parrot inhabit the heathland behind the township. There is good access to the heath along Kallaroo Rd and near the intersection of Erowal Bay and Jervis Bay Rds, particularly on the walk east from the intersection to the Water Tanks (1.5 km). Watch also for Eastern Whipbird, Southern Emu-wren, Beautiful Firetail, Lewin's, White-cheeked, New Holland and Yellow-faced Honeyeater, Eastern Spinebill and Little Wattlebird. Glossy Black-Cockatoo sometimes feed in the forest fringe at the top of the walk near the Water Tanks. Eastern Bristlebird also occurs in the heathland bordering the start of Cave Beach Rd (the road to the Botanic Gardens), while Eastern Ground Parrot, Brush Bronzewing and Southern Emu-wren frequent heathland around the start of Wreck Bay Rd and along the road to the Jervis Bay Territory Airport, both located off Jervis Bay Rd, 3 km south of the visitor centre. It is also worth looking for Eastern Ground Parrot in the heath on the north-east side of the Booderee Botanic Gardens. The Booderee Botanic Gardens, a former annex of the Australian National Botanic Gardens, contains a significant collection of plants. This diversity means that bush birds are often plentiful here; possible species include Square-tailed Kite, Gang-gang Cockatoo, Bassian Thrush, Leaden Flycatcher, Varied Sittella, Scarlet and Lewin's Honeyeater and Variegated Fairy-wren.

Nowra

Key species: Rockwarbler, Glossy Black-Cockatoo, Green Catbird, Australian Logrunner
Other species: Yellow-throated Scrubwren, Large-billed Scrubwren, Masked Owl, Square-tailed Kite

Bomaderry Creek Regional Park (82 ha), 3 km north of Nowra, contains a nice mix of open woodland and remnant rainforest, as well as a sandstone gorge. Off the Princes Hwy, 300 m down Narang Rd, the entrance gate can be found on the left. Once in the park, there is the choice between the short Bomaderry Creek trail (1.4 km loop) and a more extended walk (5.5 km loop). The short walk takes you through dry woodlands, before descending into a cool ravine and lush rainforest. The sandstone gorge along the way provides perfect habitat for Rockwarbler, with Bomaderry Creek holding one of the most southerly populations of the species. A good spot to look for them is around the cliff-top lookout on the western side of the walk, 250 m from the weir. The species is usually found on the sandstone rock ledges and among the larger boulders, with their distinctive high-pitched call being a good indicator of their presence. The lookout also provides a view across the gorge and the surrounding forest. Channel-billed Cuckoo, Cicadabird and Olive-backed Oriole often frequent the trees here during late spring and summer, while Glossy Black-Cockatoo, Gang-gang Cockatoo, Satin Bowerbird and, occasionally, Square-tailed Kite, may be present year round. Mixed-species foraging flocks, including Varied Sittella, White-throated Gerygone, Striated Thornbill, White-throated Treecreeper and Leaden Flycatcher, and Yellow-tufted and Lewin's Honeyeater often frequent the bush behind the look-out. The longer walk continues from here through a rainforest gully, where you may see Wonga and White-headed Pigeon, Black-faced Monarch (summer), Brown Gerygone, Bassian Thrush, Rufous Fantail (summer), Scarlet Honeyeater and all three local Scrubwrens – Yellow-throated, Large-billed and White-browed. Search along the creek for Azure Kingfisher, which frequently perch over water near to the bank. Bomaderry's healthy population of Yellow-bellied Glider also attracts some of the larger owls, including Masked and Powerful Owl.

If you miss seeing Rockwarbler at Bomaderry Creek, try the sandstone overhangs along Ben's Walk (4.5 km return) in Nowra, which provide further good habitat for this species. Starting on the southern side of the Shoalhaven Bridge, walk south-west along the river via Paringa Park to Hanging Rock, where the Rockwarbler occurs. Here, Ben's Walk crosses the Nowra Creek, with Southern Emu-wren and

Gang-gang Cockatoos often spend hours at a time quietly foraging in the canopy on green gum (eucalypt) nuts: a mass of chewed 'fruit' on the ground below or the occasionally 'creaky door' call may reveal their presence.

Comerong Island Nature Reserve

This outstanding nature reserve encompasses part of the Shoalhaven Estuary and provides habitat for a wide selection of shorebirds. To get there, you can take the car ferry across Berry's Canal from the end of Terara Rd, 11 km east of Nowra. Alternatively, you can walk to the reserve across the sand spit at Shoalhaven Heads (park at the end of River Rd). In summer, you can expect to find birds such as Eastern Curlew, Whimbrel, Great and Red Knot, Bar-tailed Godwit, Grey-tailed Tattler, Terek Sandpiper and Grey and Pacific Golden Plover – the last two congregate on the beaches east of the main carpark. Winter-visiting Double banded Plover can be found in the same area. There is also a chance of rarer species, with past records including Oriental Plover, Pectoral Sandpiper and Broad-billed Sandpiper.

Variegated Fairy-wren inhabiting the grasslands along the walk.

Another southerly population, this time of the Australian Logrunner, occurs along Grotto Walk (2 km loop) in the Nowra area. Access is via Yurunga Drive on the northern side of the Shoalhaven River. The track begins at the lookout at the east end of Yurunga Drive and then leads down the escarpment and west along the river. Search for Australian Logrunner in the small stands of rainforest, especially near gullies. After 1 km, the walk climbs through rainforest habitat, where Green Catbird occur. Along the walk, there is also a chance of Glossy Black-Cockatoo, Australian King-Parrot, Azure Kingfisher and, in summer, Brown Cuckoo-Dove, Black-faced Monarch, Olive-backed Oriole, Cicadabird and Rufous Fantail. Once at the top, turn left and walk to another lookout (500 m) and back along the cliff top to the carpark, looking for Rockwarbler along the way. There is a wide variety of accommodation available in Nowra, including camping at the Nowra Wildlife Park.

Mimosa Rocks National Park

Key species: Hooded Plover, Powerful Owl, Little Tern
Other species: Azure Kingfisher, Channel-billed Cuckoo, Eastern Koel, Scarlet Honeyeater, White-headed Pigeon, Variegated Fairy-wren

Mimosa Rocks National Park (7300 ha) is located on the far south coast of NSW, half way between Tathra and Bermagui. Spotted Gum, Banksia woodlands and Mugga Ironbark dominate much of the habitat, with an understorey that includes Burrawang Cycad, while rocky coves and headlands, sandy beaches and lagoons are a feature of the coastline. Excellent bush camping is available at Aragunnu, Picnic Point and Gillards (the latter is suitable for caravans). There is also walk-in camping at Middle Beach.

Hooded and Double-banded (winter) Plover and Australian Pied Oystercatcher feed along the shoreline at Middle Beach, especially near Middle Lagoon, while Azure Kingfisher can be found along the water's edge further into the lagoon. In late spring and summer, look in the forest along the road into Middle Beach for Gang-gang Cockatoo, Topknot Pigeon, Channel-billed Cuckoo, Eastern Koel, Leaden Flycatcher, Cicadabird and Scarlet Honeyeater, while resident

Spotted Gum woodlands in Mimosa Rocks National Park. These coastal woodlands typically flower in winter, when they can attract large numbers of nectivorous species, including the Endangered Swift Parrot.

Eden pelagic boat trip

Seabirds recorded on Eden pelagic boat trips include a nice selection of common and rarer seabirds. These include Indian Yellow-nosed, Campbell, Black-browed, Shy, Buller's, Light-mantled Sooty, Southern Royal and Wandering Albatross, Hutton, Flesh-footed, Fluttering, Great (rare), Sooty, Short-tailed and Wedge-tailed Shearwater and Black, Black-winged, Cape, Cook's, Gould's, Great-winged, Kermadec, Mottled, Providence, White-headed and White-chinned Petrel. Other seabirds you may see include Little Penguin, Antarctic, Fairy and Slender-billed Prion, Black-bellied (rare), Wilson's and White-faced Storm-Petrel, Brown Skua, and Pomarine and Arctic Jaeger. Full-day pelagic boat trips out of Eden leave on the fifth Sunday of every month when they occur. Contact the Far South Coast Birdwatchers: www.thebegavalley.org.au/birdwatchers.html, birds@asitis.net.au.

species include Bell Miner and Variegated and Superb Fairy-wren. In winter Musk, Rainbow and Little Lorikeet and, occasionally, Swift Parrot feed in the flowering Spotted Gum and Coastal Banksia in the Park. Powerful Owl frequent the littoral rainforest on the road into Middle Beach, sometimes roosting in the gullies just before the campground and both Masked and Sooty Owl have been recorded in the taller eucalypt forests west of Middle Beach. At the spectacular Bithri Inlet, you can usually find Little and Caspian Tern, Australian Pied Oystercatcher and Eastern Curlew, while White-bellied Sea-Eagle and Eastern Osprey (uncommon) hunt for fish over the lagoon. Superb Lyrebird is common around the Aragunnu Campground, while Variegated Fairy-wren, near the southern limit of its distribution on the NSW coast, can usually be found along the boardwalk to the Mimosa Rocks.

Wapengo Lagoon supports similar species to those found at Bithri Inlet. The woodlands around the Picnic Point campground support Wonga Pigeon, Glossy Black-Cockatoo, Australian King-Parrot, Lewin's and Crescent Honeyeater and, in summer, Shining Bronze-Cuckoo, Channel-billed Cuckoo, White-throated Needletail, Dollarbird, Cicadabird, Black-faced Monarch, Leaden and Restless Flycatcher and Brown Gerygone. Also in summer, Little and, less commonly, Fairy Tern nest at Mogareeka Inlet at the mouth of the Bega River. Other birds that occur here include Caspian Tern, Great Egret and Striated Heron. Finally, White-headed Pigeon often roost on the power lines along the Tathra–Bermagui Rd during summer.

Ben Boyd National Park and Nadgee Nature Reserve

Key species: Eastern Ground Parrot, Eastern Bristlebird, Glossy Black-Cockatoo, Sooty Owl, Masked Owl
Other species: Striated Fieldwren, Brush Bronzewing, Tawny-crowned Honeyeater, Scarlet Honeyeater, Chestnut-rumped Heathwren, Southern Emu-wren

Ben Boyd National Park (10 486 ha) includes spectacular rocky headlands, heathland, open forest and coastal littoral rainforest. To get there, from the Princes Hwy take Edrom Rd, 8 km south of Eden. Nadgee Nature Reserve (17 116 ha) protects one of the largest undeveloped catchments in eastern Australia and preserves the only coastal Wilderness Area in NSW. At Ben Boyd National Park, there is camping at the Bittangabee and Saltwater Creek campgrounds; bookings are essential during Christmas and Easter school holidays. There is cottage accommodation at the Green Cape Lightstation (up to 12 people). Contact the National Parks and Wildlife Office in Merimbula: (02) 6495 5000. There is also a wide variety of motel and caravan park accommodation in Eden, Pambula and Merimbula: www.sapphirecoast.com.au.

Green Cape Lighthouse (in the far east of Ben Boyd National Park) is a superb seabird watching location. Among the many species reported from here are Brown Skua, Northern Giant-Petrel, Short-tailed (spring), Wedge-tailed, Fluttering and Sooty Shearwater, Fairy Prion, Common Diving-Petrel, Black-faced Cormorant, Australasian Gannet, and Crested and White-fronted Tern. It is a particularly good location for albatross, including Indian Yellow-nosed, Campbell, Black-browed and Shy and, more rarely, Southern Royal, Wandering and Buller's Albatross. From the lighthouse, look for Little Penguin around the rocks below; in November 2006, a Fiordland Penguin was observed. Migrating Humpback and Southern Right Whale move along the coast near Green Cape, with the best time for sighting these being between August and December. Birds such as Eastern Ground Parrot, Striated Fieldwren, Chestnut-rumped Heathwren, Tawny-crowned Honeyeater, Southern Emu-wren, Brush Bronzewing and Beautiful Firetail occur in the extensive area of heath around the lighthouse. Eastern Bristlebird has also been recorded here, though there have been few recent records. While driving through Ben Boyd National Park, keep your eyes open for Glossy Black-Cockatoo, particularly where there are patches of

Maxwell's Flora Reserve

It is worth stopping at Maxwell's Flora Reserve (370 ha), located just inside the NSW border 40 km south of Eden along Maxwell's Rd (access from the Princess Hwy). The Rainforest Walk (1.2 km return) descends through a superb temperate rainforest with a lush understory of tree ferns. Birds here include Brush Bronzewing, Satin Bowerbird, Pilotbird, Red-browed and White-throated Treecreeper, Lewin's and Crescent Honeyeater, Large-billed Scrubwren, Brown Gerygone, Eastern Whipbird, Rose Robin and Bassian Thrush. In summer, look also for Topknot Pigeon, Black-faced Monarch, Cicadabird, Rufous Fantail, White-throated Gerygone and Scarlet Honeyeater. Along the roadside on your way in to the reserve, look and listen for Glossy Black-Cockatoo that frequent areas of seeding Casuarina woodland. Maxwell's Rd is also the access road to Howe Flat (see Croajingolong National Park p. 270).

Casuarina woodland. Hooded Plover feed along the beach adjacent to the Saltwater Creek campground, where you should also be able to find Australian Pied and Sooty Oystercatcher and White-bellied Sea-Eagle. Finally, Superb Lyrebird are common around the Bittangabee Bay campground.

The main section of Nadgee Nature Reserve is 35 km south of Eden. The ranger station is reached via Old Bridge Rd and the Newtons Rd. From there, access to the reserve is by foot only. Of the 216 bird species recorded at Nadgee, the highlights include Eastern Bristlebird and Eastern Ground Parrot. Both inhabit the coastal heathland and bordering forest between the Little Creek estuary and Cape Howe. Nadgee is also a stronghold for Striated Fieldwren, as well as Brush Bronzewing, Tawny-crowned Honeyeater, Chestnut-rumped Heathwren and Southern Emu-wren. Most of the reserve's beaches support a pair of Hooded Plover. A small population of Little Tern breed at Nadgee Lake. Masked, Powerful and Sooty Owl have been recorded in the patches of old growth forest distributed throughout the reserve. To really make the most of the birding opportunities in this wilderness area overnight hikes are recommended.

Kosciusko National Park

Key species: Gang-gang Cockatoo, Pilotbird, Flame Robin, Scarlet Robin, Olive Whistler, Brush Cuckoo, Red-browed Treecreeper
Other species: Emu, Rose Robin, Spotted Quail-thrush

Mount Kosciuszko (2228 m) is Australia's highest mountain. The ski resorts of Thredbo, Selwyn Snowfields, Perisher and Charlotte Pass lay within the park. There is a major campground at Sawpit Creek, with facilities for caravans and cabins available for rent. There is also camping on the south side of the park, at Tom Groggin and Geehi Flats. During late spring and summer, the alpine and subalpine areas at Kosciusko National Park (690 000 ha) attract a range of migrant bird species that visit from lower elevations to breed within the park.

Around the Thredbo Village, you may find Gang-gang Cockatoo, Pied Currawong, Pilotbird, Olive Whistler, Satin Flycatcher, Flame Robin, and Yellow-faced and White-eared Honeyeater. A good area to see these species is along the Thredbo River Walk (4 km). The track crosses the Thredbo River several times on bridges as it follows the river downstream to the Thredbo golf course. Other good walks that support similar species include the Dead Horse Gap Track (5 km one way) and the Bridle Trail Loop (4 km). Additional birds to look for along these walks include Brush Cuckoo, Rose Robin, Satin Bowerbird, Red-browed and White-throated Treecreeper and Red-browed Finch. During summer, Pilotbird and Olive Whistler also occur in the larger thickets on the ski field. In the southern section of the Kosciusko National Park, around Geehi Flats and along the Old Geehi Walk (6 km return), look for Emu, Gang-gang Cockatoo, Australian King-Parrot,

Flame Robin are strong altitudinal migrants, moving to the high country to breed in spring and summer, and dispersing throughout the lowlands of south-eastern Australia during autumn and winter.

Wedge-tailed Eagle, Brush and Fan-tailed Cuckoo, Flame Robin, Leaden Flycatcher, Noisy Friarbird, Satin Bowerbird and Olive-backed Oriole. In the tall forests and gullies near Geehi and Tom Groggin, you may see Superb Lyrebird, Grey Goshawk, Pilotbird, Olive Whistler, Red-browed Treecreeper, Brush Cuckoo, White-throated Gerygone, Rufous Fantail, Eastern Whipbird, Satin Flycatcher, Golden Whistler, Spotted Quail-thrush, Rose Robin, Crescent and Lewin's Honeyeater and Bassian Thrush.

Australian Capital Territory

Regional overview

The Australian Capital Territory (ACT) is the capital of the Commonwealth of Australia and is the smallest self-governing internal territory (2358 km^2). Canberra, the national capital, is located 280 km south-west of Sydney and 660 km north-east of Melbourne. Apart from the city of Canberra, the ACT also contains large areas of national park, much of it mountainous and forested. There is a wide selection of accommodation in Canberra and the international airport has direct domestic services to most state capitals. The Canberra Ornithologist Group represents the ACT: www.canberrabirds.org.au. The Gang-gang Cockatoo is the bird emblem of the ACT.

In order to observe most of the ACT bird species, 2 to 3 days is recommended. The ACT is small enough that visitors could stay in Canberra and make day excursions to nearby birdwatching sites.

Birding highlights

Approximately 290 species of bird have been recorded in the ACT. Some of the sought-after species include Emu, Gang-gang Cockatoo, Superb Lyrebird, Pilotbird, Spotted Quail-thrush, Olive Whistler, Bassian Thrush, Rose and Pink Robin, Leaden Flycatcher, Crescent Honeyeater, Speckled Warbler, Double-banded Finch, and wetland species such as Baillon's Crake, Spotless Crake and Australian Spotted Crake. Some rare and uncommon species recorded in the ACT include Plumed Whistling-Duck, Freckled Duck, Lewin's Rail, Australian Painted Snipe, Long-toed Stint, Australian Little Bittern, Barking and Powerful Owl, Glossy Black-Cockatoo, Swift Parrot, White-headed Pigeon, Eastern Koel, Channel-billed Cuckoo and Regent, Painted, Scarlet and Black Honeyeater.

Canberra and surrounds

Key species: Gang-gang Cockatoo, Baillon's Crake, Spotless Crake, Australian Spotted Crake, Lewin's Rail, Superb Lyrebird, Pilotbird, Spotted Quail-thrush
Other species: Blue-billed Duck, Bassian Thrush, shorebirds

Lake Burley Griffin is located in the middle of Canberra. At its eastern end, the pools, channels and streams that make up Jerrabomberra Wetlands Nature Reserve (221 ha) provide excellent habitat for a good selection of waterbirds. The main centre of activity is Kelly Swamp, adjacent to the carpark on Dairy Rd. From the Bittern and Cygnus bird hides, look along any exposed mudflats for Black-fronted and Red-Kneed Dotterel, Baillon's, Spotless and Australian Spotted Crake and, less commonly, Lewin's Rail and Australian Painted Snipe (rare). In summer, look for shorebirds such as Sharp-tailed, Marsh and Wood Sandpiper and Latham's Snipe and rarer species such as Long-toed Stint and Pectoral Sandpiper. The entrance to the Fyshwick Sewage Treatment Works is located at the southern end of Dairy Flat Rd. Scan the ponds for Pink-eared, Blue-billed and Freckled (rare) Duck, Australasian

Shoveler and Chestnut Teal. In the reeds and grasslands surrounding the wetland, you may find Australian Reed-Warbler (summer), Golden-headed Cisticola, Little Grassbird, European Goldfinch, Double-barred and Red-browed Finch and Yellow-rumped Thornbill. The Australian National Botanic Gardens, situated on the slopes of Black Mountain, maintain a significant collection of native plants from all parts of Australia. Here, there is also a good selection of naturally occurring bush birds, including Gang-gang Cockatoo, White-winged Chough, Golden and Rufous Whistler, Olive-backed Oriole (summer), Eastern Yellow Robin and, occasionally, Rose and Pink Robin. Honeyeaters here include Crescent, Yellow-faced, White-eared and New Holland Honeyeater and Noisy Friarbird. Search the dense gully areas within the gardens for Bassian Thrush (mainly winter).

Tidbinbilla Nature Reserve (5400 ha), nestled between the Tidbinbilla and Gibraltar Ranges, supports a variety of habitats, including wetlands, grasslands, wet forests and dry woodlands. In the Mountain Creek area, look along the Lyrebird (2 km) and Cascade (1.8 km) Trails for Superb Lyrebird, Australian King-Parrot, Gang-gang Cockatoo, Red-browed and White-throated Treecreeper, Satin Bowerbird, Crescent, Yellow-faced, White-naped and Brown-headed Honeyeater, Pilotbird, Eastern Whipbird, Satin Flycatcher, Cicadabird, Crested Shrike-tit, Rose Robin, Olive and Golden Whistler and Bassian Thrush. In the reclaimed farmland and grassy woodlands at Turkey Hill (2.5 km from the Tidbinbilla Information Centre) on the west side of the Tidbinbilla Ring Rd, you may find Emu, Southern Whiteface, Rufous Songlark (summer), Yellow-rumped Thornbill and Speckled Warbler. It is also worth completing the Sanctuary Loop (2.1 km), where you can see recovery programs for Northern Corroboree Frog and Brush-tailed Rock Wallaby and captive populations of Magpie Goose, Brolga and Platypus. A good spot to find Spotted Quail-thrush in Tidbinbilla is along the Camel Back Trail (12.5 km), which follows a management trail through towering eucalypts before emerging into the subalpine Snow Gum woodland. Another good park for Spotted Quail-thrush is at Namadgi National Park, which borders Tidbinbilla southern boundary. They are regularly seen along Mount Franklin Rd (on the ACT and NSW border). The best approach to finding the species here is to drive slowly, looking for them foraging

Tree Fern clad gullies like this provide ideal habitat for wet forest species such as Wonga Pigeon, Superb Lyrebird, Pilotbird, Red-browed Treecreeper, Leaden Flycatcher and Olive Whistler.

in leaf-litter on the roadside, particularly just past the information centre at Bull's Head. Spotted Quail-thrush are also found along the Square Rock Walking Track (6 km return) and in the south the park, along Orroral Loop (1 km), Orroral Valley Heritage (7 km return) and Booroomba Rocks (4 km return) Walks. In addition to Spotted Quail-thrush, at Namadgi National Park, you may see Superb Lyrebird, Wonga Pigeon, Pilotbird, Red-browed Treecreeper, Leaden Flycatcher and Olive Whistler.

The Hunter Valley and NSW North Coast

Regional overview

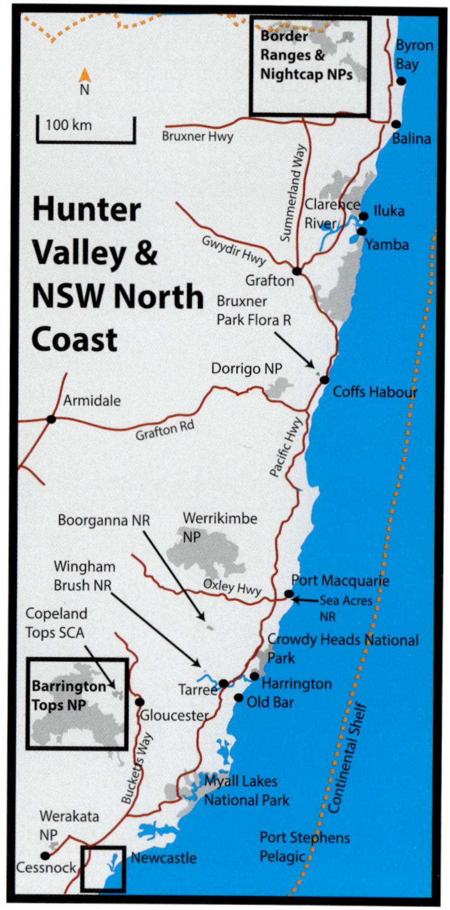

The Hunter Valley and NSW North Coast extend from Newcastle north along the eastern seaboard to the Qld border and cover the areas from the coastline to the top of the Great Dividing Range. The habitats here include subtropical, warm temperate and coastal rainforests, wet and dry sclerophyll forests, coastal woodlands, heathland and freshwater swamps, estuarine mudflats, wetlands and long sandy beaches. Through much of the region, the temperate rainforest at higher elevations along the spine of the Great Dividing Range, creates an ecological refuge for a variety of rainforest birds. This has led to superb birding opportunities within key national parks such as Barrington Tops, Werrikimbe, Dorrigo, Nightcap and the Border Ranges. At the southern end of the region, the Hunter River Estuary contains significant areas of habitat suitable for migratory and resident shorebirds, and includes such well known sites as Ash Island and Stockton Sandpit. North of Newcastle, the Myall Lakes are one of the state's largest coastal lake systems and contain protected areas of coastal littoral rainforest, while the Clarence River estuaries and salt marshes are of national importance. On the far north coast of NSW, in the vicinity of Ballina, there are several excellent shorebird sites. For birdwatchers the region has excellent representation in the Northern NSW branch of Birdlife Australia (www.birdlife.org.au) and the Hunter Bird Observers Club (www.hboc.org.au).

A suggested 10-day itinerary in the Hunter Region involves:

- **Days 1–2.** Newcastle to Myall Lakes, 110 km (Forest Raven, Regent Bowerbird, Gould's Petrel, seabirds).
- **Days 3–4.** Myall Lakes to Harrington and Crowdy Bay National Park, 120 km

(Beach Stone-curlew, Eastern Grass Owl, Regent Bowerbird, shorebirds).

- **Days 5–6.** Harrington to Barrington Tops National Park, 150 km (Rufous Scrub-bird, Olive Whistler, Red-browed Treecreeper, Russet-tailed Thrush, Pale-yellow Robin).
- **Days 7–8.** Barrington Tops National Park to Hunter River woodlands, 120 km (Swift Parrot, Turquoise Parrot, Regent Honeyeater, Black-chinned Honeyeater, Diamond Firetail).
- **Days 9–10.** Hunter River woodland to Newcastle, visiting Ash Island, Stockton Sandspit and Hunter Wetlands Centre, 50 km (Tawny Grassbird, Mangrove Gerygone, White-fronted Chat, Australasian Bittern, Eastern Grass Owl and, in summer, shorebirds and Eastern Yellow Wagtail).

To explore the north coast of NSW beyond Newcastle, a suggested 12-day itinerary involves:

- **Days 1–2.** Newcastle to Werrikimbe National Park, 300 km (Rufous Scrub-bird, Powerful, Sooty and Masked Owl, Barred Cuckoo-shrike).
- **Days 3–4.** Werrikimbe National Park to Dorrigo National Park, 280 km (Russet-tailed Thrush, Grey Goshawk, Australian Logrunner, Noisy Pitta, Paradise Riflebird, Regent Bowerbird, Sooty and Powerful Owl).
- **Days 5–6.** Dorrigo National Park to Clarence River, 215 km (Beach Stone-curlew, Mangrove Honeyeater, Black-necked Stork, shorebirds).
- **Days 7–8.** Clarence River to Nightcap National Park 160 km (Albert's Lyrebird, Marbled Frogmouth, Russet-tailed Thrush, Paradise Riflebird).
- **Days 9–10.** Nightcap National Park to Ballina, 110 km (Beach Stone-curlew, terns, shorebirds).
- **Days 11–12.** Ballina to Newcastle, 600 km.

Note that such a visit could also be done in reverse, approaching the north coast of NSW from Brisbane or the Gold Coast.

It is worth noting that during the summer holidays the national park camping grounds in northern NSW can be very popular, so advance bookings are advised. For further information on your camping options see www.environment.nsw.gov.au/nationalparks. Aside from camping in the parks, finding accommodation in the Hunter Valley and North Coast regions is straightforward, with numerous holiday towns lining the coast, including Newcastle, Forster, Port Macquarie, Coffs Harbour, Grafton, Ballina, Byron Bay and Tweed Heads.

A good time to visit the region is during spring and summer, when the weather is mild to warm and there is little rain. The north coast of NSW has a subtropical climate on the coast with hot summers, through to a sub-humid climate on the slopes and uplands. In summer, the maximum average monthly temperature ranges from 20.3°C at night to 30.9°C during the day. In the winter, temperatures are mild, averaging 18°C during the day.

Excellent sealed highways provide easy access. The main road along the coast is the Pacific Hwy (Hwy No. 1), which circumnavigates coastal Australia. A 2WD vehicle will get you to the majority of birding sites in the region, but some mountainous areas (such as Gloucester Tops) require crossing fords over rivers, which can easily become impassable after heavy rain, even to 4WDs. It is best to check local conditions before setting off to these places. There are several regional airports in the north coast of New South Wales. Ballina, Coffs Harbour, Grafton, Newcastle, Port Macquarie and Taree. All have scheduled services on commercial airlines.

Birding highlights

The variety of habitats of the north coast of NSW harbour a rich bird diversity, with several species restricted to rainforest habitats in the region. The Hunter Region is especially good for birding with habitats ranging from the aforementioned World Heritage Gondwanaland Rainforests to dry woodlands, significant wetlands, coastal heaths and the Pacific Ocean coastline. To date, there have been around 440 species recorded, or over half of all species recorded in Australia. The region is significant for the Australian Logrunner, Paradise Riflebird, Albert's Lyrebird, Rufous Scrub-bird (both the northern ssp. *rufescens* and southern ssp. *ferrieri*), Forest Raven ('Relict Raven' ssp. *boreus*) and northern subspecies of Eastern Bristlebird (ssp. *monoides*). There has been no confirmed report of the Critically Endangered Double-eyed Fig-Parrot (Coxen's Fig-Parrot ssp. *coxeni*) since the late 1990s and, sadly, despite its official status, it may already be extinct. The region is also at the junction of two races of Striated Pardalote – 'Eastern Striated Pardalote' (ssp. *ornatus*) and 'Black-headed Pardalote' (ssp. *melanocephalus*). The only Australian breeding population of Gould's Petrel (ssp. *leucoptera*) occurs on three small islands near the entrance to Port Stephens. Other notable species in the region include Black-necked Stork, Beach Stone-curlew (southern limit of distribution), Glossy Black-Cockatoo, Swift Parrot, Wompoo, Superb and Rose-crowned Fruit-Dove, Eastern Grass Owl, Marbled Frogmouth ('Plumed Frogmouth' ssp. *plumiferus*), Noisy Pitta, Green Catbird, Regent Bowerbird, Mangrove Honeyeater (southern limit of distribution), Regent Honeyeater, Eastern Yellow Wagtail, Red-browed Treecreeper, Olive Whistler (ssp. *macphersoniana*), and Bassian and Russet-tailed Thrush. Vagrants to the region have included Northern Pintail, Northern Shoveler, Lesser Yellowlegs, Kentish Plover, Oriental Plover, Buff-breasted Sandpiper, Sabine's Gull, South Island Pied Oystercatcher, Black-backed (White) Wagtail, Rose-coloured Starling and Yellow White-eye.

Newcastle and the Hunter River Estuary, including Ash Island

Key species: Black-tailed Godwit, Great Knot, Eastern Yellow Wagtail (chance in summer), White-fronted Chat, Australasian Bittern, Eastern Grass Owl
Other species: Tawny Grassbird, Mangrove Gerygone, Black-necked Stork, Magpie Goose, Wandering Whistling-Duck, migratory shorebirds

There are several excellent birding sites around Newcastle, including Ash Island, Stockton Sandspit and the Hunter Wetlands Centre. NSW's second largest city, Newcastle is located 160 km north of Sydney. When visiting Newcastle and the Hunter region, it is worth looking at some of the excellent birding information available from the Hunter Bird Observers Club: www.hboc.org.au (see the 'Publications' link for various birding routes brochures). Recognised as being of international importance, the Hunter Estuary (incorporating Ash Island, Hexham Swamp and the Kooragang Wetlands), has extensive areas of saltmarsh, mangroves, shallow and tidal ponds. To get to Ash Island, turn right off Maitland Rd onto the Ash Island Bridge (well signposted), 14 km north of Newcastle's CDB. There are information notices at several locations on the island and toilets and picnic facilities at the Old Schoolmaster's House and the Old Radar Station Estuarine Interpretive Centre.

Perhaps surprisingly, Ash Island is best known for the regular records of Eastern Yellow Wagtail: a summer migrant from the Northern Hemisphere with very few sightings in eastern Australia. Since 1998, between February and March, a few

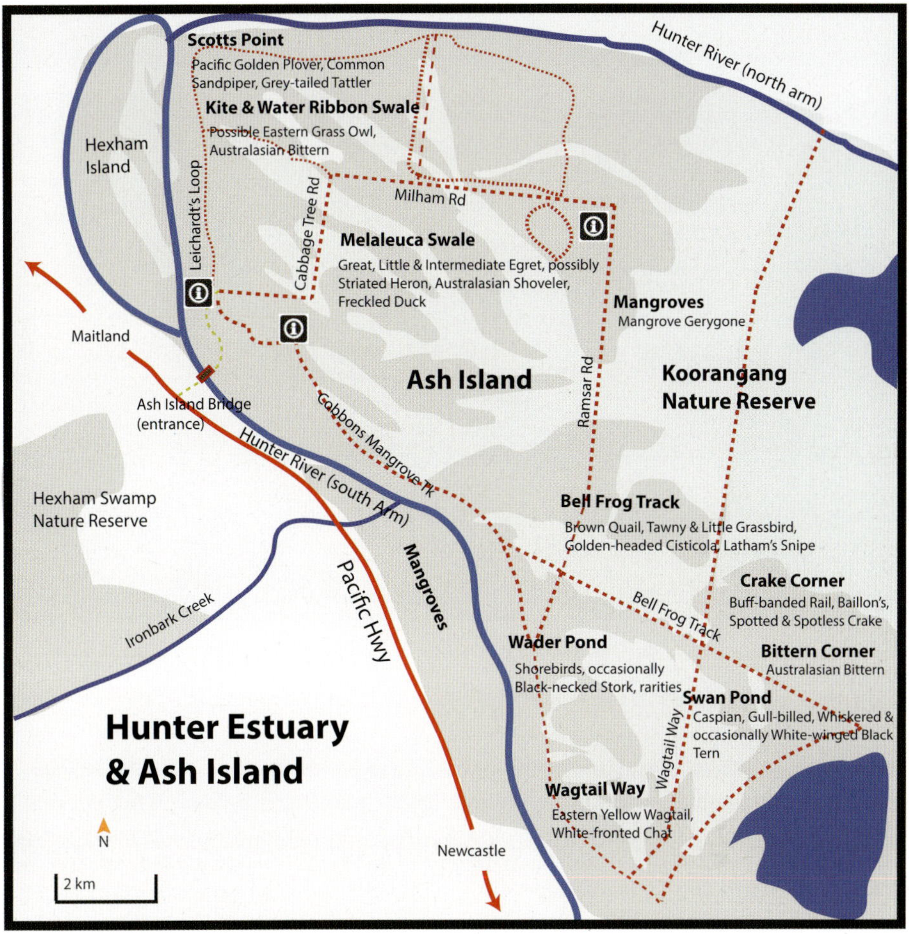

Eastern Yellow Wagtail have been present each year along 'Wagtail Way' and the railway line bordering Deep Pond. Green-headed Wagtails have also been recorded. Wagtail Way is also an excellent site to look for White-fronted Chat (in decline in coastal NSW) and for migratory shorebirds, particularly at the Wader and Swan Ponds. Resident (and locally nomadic) birds found here include Black-winged Stilt, Red-necked Avocet, Red-kneed and Black-fronted Dotterel and, occasionally, Black-necked Stork. The more common summer migrants, such as Common Greenshank, Sharp-tailed, Marsh, Curlew Sandpiper, Red-necked Stint, Pacific Golden and Grey (uncommon) Plover, Black-tailed and Bar-tailed Godwit, and Eastern Curlew generally arrive in September and leave in April (although some individuals over-winter, particularly Bar-tailed Godwit and Eastern Curlew). Local and national rarities have included Australian Pratincole, Australian Painted Snipe, Ruff, Little Curlew, Lesser Yellowlegs, Long-toed Stint, Broad-billed, Pectoral and Buff-breasted Sandpiper, and Northern Shoveler. Caspian, Gull-billed, Whiskered and, occasionally,

Wetlands on Ash Island, and a view down the locally famous 'Wagtail Way'. In addition to a great diversity of wetland birds, in recent years, between February and March, a handful of Eastern Yellow Wagtail have been located here. This is the only site in NSW where the species has been detected with any frequency.

White-winged Black Tern hawk for insects over Swan Pond, while Australasian Bittern sometimes use the reed-lined freshwater swamp at the north-east corner of Swan Pond. A good location to search for skulking Buff-banded Rail, Baillon's, Australian Spotted and Spotless Crake is at the appropriately named Crake Corner, located at the intersection of Wagtail Way and Bell Frogs Track. Check along the grassy areas on Bell Frogs Track for Brown Quail, Tawny and Little Grassbird, Golden-headed Cisticola and Latham's Snipe (summer). Scotts Point, in the far north-west corner of Ash Island, is an occasional high-tide roost for Pacific Golden Plover, Common Sandpiper and Grey-tailed Tattler. Although both rare, a potential site for Eastern Grass Owl and Australasian Bittern is just east of Scotts Point, at the Kite and Water Ribbon Swales. Search the mangrove in the area for Mangrove Gerygone. Other good birding spots on Ash Island include the Melaleuca Swale and Milham Pond, where you may find Great, Little and Intermediate Egret and possibly Striated Heron, Australasian Shoveler and Freckled Duck (rare). A group of four Australian Painted Snipe were found here in October 2011.

Stockton Sandspit is a reliable place to see the 'marine' migratory shorebirds, including Lesser Sand and Pacific Golden Plover, Terek and Curlew Sandpiper, Red and Great Knot, Bar-tailed and Black-tailed Godwit, Grey-tailed Tattler, Whimbrel and Eastern Curlew. Best accessed via a track that leads under the Stockton Bridge, it begins on Fullerton St, immediately south of the bridge (turn right to head under the bridge just after the roundabout). The best time to visit is between 2 and 3 h after high tide, when the beach in front of the Sandspit is used as a staging point for most of the shorebirds before flying north to feed

Port Stephens pelagic trips

Pelagic trips off Port Stephens are organised on an occasional basis and depart from the Public Wharf at Nelson Bay. Despite the occasional nature of these trips, the species list is impressive: Black, Cape, Gould's, Great-winged, Kermadec, Providence, Tahiti, White-headed and White-necked Petrel are all on the list as are Buller's, Flesh-footed, Great (vagrant), Hutton's, Short-tailed, Streaked and Wedge-tailed Shearwater. Other birds recorded include Indian Yellow-nosed, Shy, Wandering, Buller's, Campbell and Black-browed Albatross, Wilson's, White-faced, White-bellied and Black-bellied Storm-Petrel, Brown and Red-footed Booby, Brown Skua and Arctic, Long-tailed and Pomarine Jaeger. For details see www.sossa-international.org.

on the extensive mudflats of Fullerton Cove as the tide drops. The Sandspit is also a site for roosting terns such as Crested, Caspian, Gull-billed and Little, while Common and, occasionally, Whiskered and White-winged Black Tern hawk over the river and roost on the channel marker buoys. Mangrove Gerygone feed in the mangroves bordering the eastern side of the Spit. Check the oyster leases at high tide for roosting Grey-tailed Tattler and Terek Sandpiper. South of Stockton Sandspit (1.3 km along Fullerton St), there is a rusty shipwreck in the Hunter River. This acts as an intermittent high tide roosting site for Pacifica Golden Plover, Grey-tailed Tattler and Common Sandpiper, while Pacific Golden Plover, Striated Heron and Grey-tailed Tattler roost on the foreshore boulders 800 m further south of here.

Due the variety of habitats at the Hunter Wetlands Centre (45 ha of wetlands, paperbark woodland and grasslands, intermixed with small patches of rainforest) over 200 species have been recorded. To get there, turn off Maitland Rd (Old Pacific Hwy) on to Wallsend Rd 10 km from Newcastle's CBD. Follow Wallsend Rd until you come to a roundabout and the Hunter Wetlands Centre is on the right. Some of the more interesting species include Freckled Duck (there is a captive breeding program, but wild birds are occasionally recorded), Magpie Goose (a re-establishment program), Wandering Whistling-Duck, Black-necked Stork, Australasian Bittern, Grey and Brown Goshawk, Peregrine Falcon, Latham's Snipe, Scarlet, Brown and White-cheeked Honeyeater, Spangled Drongo (mostly winter), Tawny and Little Grassbird and Chestnut-breasted Mannikin. The wetlands centre is also recognised as being of international importance, listed under the Ramsar Convention.

Lower and Upper Hunter Valley woodlands

Key species: Swift Parrot, Turquoise Parrot, Regent Honeyeater, Black-chinned Honeyeater, Diamond Firetail
Other species: Little Lorikeet, Grey-crowned Babbler, White-backed Swallow, Brown Treecreeper, Speckled Warbler, Red-capped Robin, Crested Shrike-tit, Southern Whiteface

The Hunter Valley is a broad valley that holds some of the NSW's most significant subcoastal dry woodlands. Essentially there is a 'gap' in the Great Dividing Range in the far west of the valley, which may explain some of the more 'inland influences' observed in Hunter Valley's biota, even close to the coast. For example, it is the only coastal catchment that has populations of Weeping Myall and River Red Gum.

The woodlands around Cessnock, less than an hour's drive from Newcastle, are highly significant because they represent some of the last remaining large stands of bushland on the valley floor. In these remnants, species such as Black-chinned Honeyeater, Brown Treecreeper, Speckled Warbler, Little Lorikeet and Crested Shrike-tit persist. Honeyeaters such as Yellow-tufted, Fuscous and White-naped are also abundant, and the area is of significance for the Critically Endangered Regent Honeyeater and Endangered Swift Parrot. Regent Honeyeater have bred around Cessnock in most years of the past decade and it is one of the few places where concentrations of > 20 birds have been recorded in recent times. Grey-crowned Babbler has adapted to live in disturbed areas (e.g. golf courses) and is ubiquitous in the area. The Lower Hunter also sees occasional irruptions of more inland species such as White-browed and Masked Woodswallow, Rufous Songlark and White-winged Triller (all have bred here). A small population of Turquoise Parrot also persists. The best places to look for woodland birds are the various Crown lands around the township of Kurri Kurri, as well as the various sectors of Werakata National Park/State Conservation Area,

Dry woodlands near Goulburn River National Park, Upper Hunter Valley. Among other species, this area provides suitable habitat for Chestnut-rumped Heathwren and Rockwarbler (note the sandstone outcrop visible in the top right of the image).

especially around the villages of Pelton and Kitchener. Cessnock is in the heart of the Hunter Valley winery region and accommodation options are plenty.

The Upper Hunter woodlands (generally those areas west of about Jerry's Plains) are also very productive, though a little further west at ~2 to 3 h drive from Newcastle. As well as those species found in the Lower Hunter, there are populations of other sought-after woodland birds here, such as Turquoise Parrot, Diamond Firetail, Hooded Robin, Red-capped Robin, White-backed Swallow, Southern Whiteface and, seasonally, Painted Honeyeater. The area is also well known for regular records of Regent Honeyeater. At times, irruptive inland species such as Black and Pied Honeyeater are also found here. Spiny-cheeked Honeyeater is resident as far east as Bulga and there is a breeding population of Singing Honeyeater near Ulan: perhaps the only population in a coastal catchment in NSW. The best birding areas are around roadside forest remnants near towns such as Ulan, Denman, Wollar and Sandy Hollow (the latter is particularly productive). However, the best woodland birding is usually along Ringwood Rd, which bisects Goulburn River National Park between the Golden Highway and Wollar Rd, with habitat ecotones being especially good. Look for Emu and Brown Songlark in the cleared areas north of the national park. Check the sandstone sections in the southern part of this road for Chestnut-rumped Heathwren and Rockwarbler. Mogo Rd (Goulburn River National Park) is also worth investigating: there are past claims of Malleefowl in this area (camping is possible at the end of this road at Big River Camp).

Barrington Tops National Park

Key species: Rufous Scrub-bird, Olive Whistler, Red-browed Treecreeper, Russet-tailed Thrush, Pale-yellow Robin

Other species: Masked Owl, Sooty Owl, Glossy Black-Cockatoo, Satin Flycatcher, Bassian Thrush

Home to one of the largest temperate rainforests in Australia, Barrington Tops National Park (38 705 ha) contains one of four remaining populations of the vulnerable Rufous Scrub-bird. The main plateau reaches 1500 m above sea level, with steep slopes on all sides. One of the best places to concentrate your birdwatching effort is in the Gloucester Tops section of the park, with stops in both the lower altitude subtropical rainforests and higher up the range near the Gloucester Tops Picnic Area. There is a wide range of camping options, including the picturesque Gloucester River campground, suitable for camping and caravans, with toilets and barbecues. There is also a wide range of accommodation in Gloucester: www.gloucester.org.au. To get to Gloucester Tops, from Gloucester, drive 9 km south down Bucketts Way to Gloucester Tops Rd before following this for ~30 km to the Gloucester River campground at Sharpe's Creek.

Some of the best birding is around the Gloucester River campground and nearby along the Sharpes Creek Track, which can either be done as a 3 km return walk or a 4.5 km loop. The shorter trail takes you up Sharpe's Creek, while the longer loop continues up the hillside and then returns via the Gloucester Tops Rd. Look along the forest edge for species associated with lower altitude subtropical rainforests, such as Superb Lyrebird, Australian Brush-turkey, Wonga Pigeon, Yellow-throated and Large-billed Scrubwren, Satin Bowerbird, Green Catbird, Black-faced Monarch, Bell Miner, Satin and Leaden Flycatcher, Russet-tailed Thrush and Rose and Eastern Yellow Robin. At night, look and listen for Masked and Sooty Owl, both recorded around the hillside saddle on the Sharpes Creek Track. Glossy Black-Cockatoo feed along

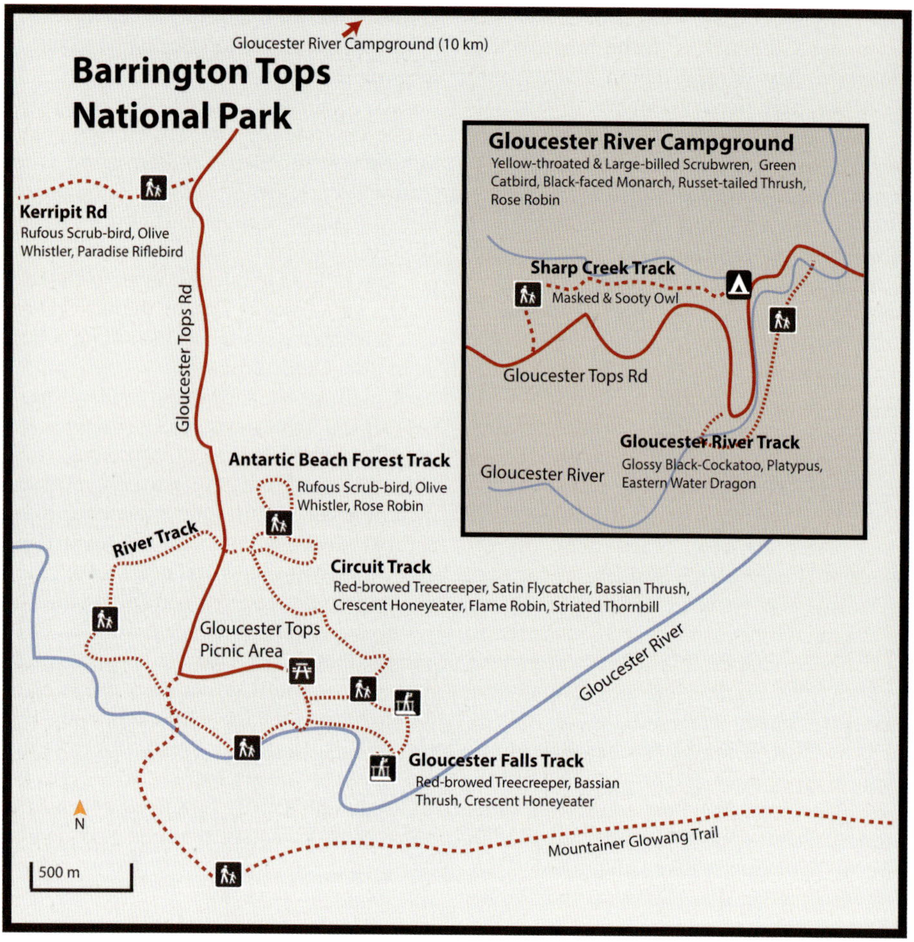

Gloucester River accessed via Gloucester River Track (2.5 km). Also watch the river for Platypus and along the river banks for Eastern Water Dragon. From the Gloucester River campground, take Gloucester Tops Rd to the top of the range. On the way, check rainforest areas for Paradise Riflebird. Stop when at the intersection with Kerripit Rd and search for Rufous Scrub-bird in the stands of Antarctic Beech along the first kilometre of Kerripit Rd. It is an extremely secretive species, which is 'easiest' to see (and hear) from late August to November; at other times of the year, they seldom call and can be hard to find. Returning to Gloucester Tops Rd, continue searching for Rufous Scrub-bird along the next few kilometres, all the way to the top of Gloucester Tops. Rufous Scrub-bird is found along the roadside just before the top and along the Antarctic Beech Forest Track, a 2.5 km loop walk that branches off Gloucester Tops Rd just before you reach the Gloucester Tops picnic area. Some of the other birds found at higher altitude, particularly in the warmer spring and summer months, include Olive

Satin Bowerbirds are frequent visitors to the picnic areas and campsites within Barrington Tops National Park.

Whistler, Red-browed Treecreeper, Satin Flycatcher, Bassian Thrush, Crescent Honeyeater, Flame Robin and Striated Thornbill. Look for these along the Circuit Track and the walk to the Gloucester Falls.

Jerusalem Creek, located in the southern-most section of the Barrington Tops (reached via Wangat Rd, which branches off the Chichester Dam Rd north of Dungog), is another good area for lowland rainforest species. Look along the Jerusalem Creek Trail (2 km return) for Superb Lyrebird, Russet-tailed Thrush, Pale-yellow Robin, Yellow-throated, Large-billed and White-browed Scrubwren and canopy birds such as Paradise Riflebird, Satin Bowerbird, Sacred Kingfisher, Black-faced Monarch, Leaden Flycatcher, Crested Shrike-tit and Rose Robin.

Myall Lakes National Park

Key species: Forest Raven, Regent Bowerbird, Gould's Petrel, seabirds
Other species: Green Catbird, Brown Cuckoo-Dove, Emerald Dove

Myall Lakes National Park (47 493 ha) is one of the state's largest coastal lake systems, with over 40 km of beaches. The area also supports some excellent areas of coastal rainforest. Myall Lakes is located off the Pacific Highway, 8 km north of Bulahdelah (100 km north of Newcastle).

From the Mungo Brush campground, investigate the Rainforest Walk (1.5 km loop). Winding its way through an isolated patch of littoral rainforest, a good selection of birds has been recorded here. These include Brown Cuckoo-Dove, Emerald Dove, Topknot Pigeon, Pheasant Coucal, Azure Kingfisher (look along the banks of the bay), Regent and Satin Bowerbird, Green Catbird, Blue-faced, Lewin's and Yellow-faced Honeyeater, Australasian Figbird, Large-billed Scrubwren, Eastern Whipbird, Black-faced Monarch, Rufous Fantail and Varied Sittella and, in summer, Channel-billed Cuckoo, Cicadabird and Olive-backed Oriole. The southern-most population of the Forest (Relict) Raven occurs here, especially near the edge of the rainforest. In 2005, a pair of Radjah Shelduck took up residency at the Mungo Brush campground, though in 2009, one of the ducks disappeared and the remaining duck then moved regularly between the campground, the Bombah Point Ferry Terminal and Bulahdelah. At White Tree Bay (just east of the Mungo Brush), you may see Leaden Flycatcher, Scarlet Honeyeater, Noisy Friarbird, Rufous Whistler, Varied Triller, Brush Cuckoo and Yellow Thornbill. To access Mungo Brush, follow the Mungo Brush Rd from Hawks Nest.

The township of Seal Rocks, located in the north-east section of Myall Lakes, is surrounded by coastal rainforest

Copeland Tops State Conservation Area

The diverse flora within Copeland Tops State Conservation Area (2201 ha) provides habitat for a good selection of rainforest birds. It is also the largest accessible area of 'dry rainforest' in the Gloucester district. Some of the best birding is around the old Mountain Maid Mine site, along the Hidden Treasure (3.3 km) and Basin Walks (7 km loop). Interesting species that you may encounter include Noisy Pitta, Green Catbird, Regent and Satin Bowerbird, Paradise Riflebird, Australian Logrunner, Cicadabird, Spectacled and Black-faced Monarch, Pale-yellow Robin and Bassian and Russet-tailed Thrush. Spring and summer months, and especially where there are tall fruiting trees, are particularly good for White-headed and Topknot Pigeon, Emerald Dove, Wompoo Fruit-Dove and Brown Cuckoo-Dove. The presence of some outstanding old growth forest within Copeland means the area is also excellent for owls. Powerful, Sooty and Masked Owl are all present, as are other night birds such as Australian Owlet-Nightjar and Tawny Frogmouth (a high proportion are of the rufous morph). Other birds recorded here include Superb Lyrebird, Australian Brush-turkey, Rufous Fantail, Scarlet and Lewin's Honeyeater, Large-billed Scrubwren and Rose Robin, and forest raptors such as Pacific Baza and Grey Goshawk are also sometimes seen. Copeland Tops is located 18 km west of Gloucester: to get there, follow Thunderbolts Way to Barrington, proceed over the single lane bridge at Barrington River and head west on the Barrington Tops Forest Rd to the historic town of Copeland. From here, turn left into Copeland Rd and follow this to the carpark.

and Banksia woodlands. Look around Treachery Camp (1 km south of Seal Rocks) for Regent Bowerbird, Green Catbird, Forest Raven, Variegated Fairy-wren and Eastern Whipbird. The ocean views from the Seal Rocks Lighthouse are excellent for seabird watching: an extensive list includes White-bellied Sea-Eagle, Eastern Osprey, Australasian Gannet, Black-browed Albatross, Flesh-footed, Short-tailed, Wedge-tailed and Fluttering Shearwater, Pomarine, Arctic and Long-tailed Jaeger, Common Noddy and White Tern. There is also a small possibility of Gould's Petrel – one of Australia's rarest seabirds – they mostly nest on Cabbage Tree Island, located near the entrance to Port Stephens (more recently, they have also bred on nearby Boondelbah and Little Broughton Islands). Look for Eastern Reef Egret and Grey-tailed Tattler feeding on the rocks below the headland near the lighthouse. It is also worth walking the Treachery Headland Track (1 km one way), which starts near the Treachery Camp carpark, leads through coastal rainforest and passes a giant fig tree with spreading buttress roots, before reaching the grassy headland. Here, you may see White-cheeked Honeyeater and, in summer, White-throated Needletail and Fork-tailed Swift are possibilities, especially in unsettled weather.

Werrikimbe National Park

Key species: Rufous Scrub-bird, Powerful Owl, Sooty Owl, Masked Owl, Barred Cuckoo-shrike
Other species: Glossy Black-Cockatoo, Wompoo Fruit-Dove, Olive Whistler

The World Heritage-listed Werrikimbe National Park (33 300 ha) protects a superb area of temperate and subtropical rainforest. There are campgrounds at Brushy Mountain, Mooraback and Plateau Beech. To access the area, take Forbes River Rd, which heads north off the Oxley Highway 49 km west of Wauchope.

Werrikimbe is an excellent place to see Rufous Scrub-bird. Good spots to look are along the two walks that leave the Brushy Mountain campground: the Brushy Mountain Loop Walk (2.9 km) and the longer Scrub Bird Walk (9 km return). Rufous Scrub-bird has a preference for dense, metre-high ground cover and has a remarkable ability to run rat-like beneath dense undergrowth, making it a very difficult bird to observe. A good technique used to obtain a sighting is to listen for the male's powerful territorial song and then sit patiently and quietly close to where the bird is calling until one scampers into view. Around the Brushy Mountain campground, look for Wompoo Fruit-Dove, Glossy Black-Cockatoo (primarily in stands containing Black Sheoke), Green Catbird, Regent Bowerbird, Australian Logrunner, Paradise Riflebird, Red-browed Treecreeper, Olive

The extensive forested ranges in the Werrikimbe National Park region support a high diversity of birds and other fauna, including sought-after species such as Wompoo Fruit-Dove, Glossy Black-Cockatoo, Sooty Owl, Masked Owl and Rufous Scrub-bird. Tiger Quolls are also a possibility.

Whistler, Barred Cuckoo-shrike and robins such as Flame, Pale-yellow and Rose. The Park has a good range of owls, with Southern Boobook and Powerful, Sooty and Masked Owl all recorded. The campground can also be a good place to see Tiger Quoll.

Harrington, Crowdy Bay National Park and Old Bar

Key species: Beach Stone-curlew, Eastern Grass Owl, Regent Bowerbird, shorebirds
Other species: Mangrove Gerygone, Forest Raven, Glossy Black-Cockatoo, Square-tailed Kite

The township of Harrington is located on the southern edge of Crowdy Bay National Park (10 001 ha), at the mouth of the Manning River, 335 km north-east of Sydney. There is a range of accommodation in the town, including caravan parks and motels. The national park is located immediately north of Harrington and has several dramatic headlands with extensive wetlands, rainforest pockets and coastal heathland (the latter featuring wildflowers such as Christmas Bells). There are several good camping grounds, including Diamond Head, Indian Head, Kylies Beach and Crowdy Gap Cultural Camp.

On the Harrington lagoon and sand flats (located directly in front of the town), look for Bar-tailed Godwit, Eastern Curlew, Pacific Golden Plover and less commonly Sanderling, Terek Sandpiper, Whimbrel and Grey Plover. A pair of Black-necked Stork sometimes feed at the southern end of the lagoon. Coastal raptors include Eastern Osprey and White-bellied Sea-Eagle and this area is approaching the southernmost limit for Brahminy Kite. Stop in the carpark where Beach St becomes Crowdy St and walk south along the Harrington Breakwater to Manning Point, a known roosting site for Beach Stone-curlew: the birds here are the southernmost known breeding pair in Australia. Other birds to look for around the breakwater include Sooty and Australian Pied Oystercatcher, Striated Heron, Eastern Reef Egret, Common Sandpiper, Double-banded Plover (winter) and roosting terns such as Crested, Common, Little, Gull-billed and Caspian. In the mangroves and adjacent casuarinas near the southern end of the breakwater, you may see Mangrove Gerygone, Brown and Striped Honeyeater and Yellow Thornbill.

On the east side of the Big 4 Harrington Holiday Caravan Park on Crowdy St, there is a small area of littoral rainforest that can provide some rewarding birdwatching. Species of note here include Brown Cuckoo-Dove, Emerald Dove, Regent Bowerbird, Scarlet Honeyeater, Cicadabird, Varied Triller, Spangled Drongo, Forest (Relict) Raven, White-bellied Cuckoo-shrike, Black-faced Monarch, Crested Shrike-tit, Australasian Figbird, Golden Whistler, Rufous Fantail and Red-backed Fairy-wren. Less common, but noteworthy, species include Noisy Pitta (winter), Rose-crowned Fruit-Dove, Barred Cuckoo-shrike and Spectacled Monarch. The rainforest species here are typical of the similar rainforest areas at Manning Point (south of Harrington) and near the Kylies Beach campground at Diamond Head (north of Harrington). In the forest at Diamond Head, there is also a chance of Glossy Black-Cockatoo and Square-tailed Kite and a slim chance of Swift Parrot, which occasionally visit in winter.

Around Crowdy Head (7 km east of Harrington), look for Striped and White-cheeked Honeyeater, Brown Quail, Buff-banded Rail, Spangled Drongo, Pheasant Coucal, White-throated Needletail (summer) and White-breasted Woodswallow. Also try for Sooty Oystercatcher and Striated Heron on the rocky headlands. Eastern Grass Owl (rare) reside in the wallum heathlands just north of Harrington. A walk along the fire

Sea Acres National Park

Sea Acres National Park (76 ha) is an outstanding remnant of coastal rainforest located 5 km south of Port Macquarie, on Pacific Drive, adjacent to Shelly Beach. Best access is along the boardwalk (1.3 km, elevated and wheelchair-accessible via the Sea Acres Rainforest Centre), where a long list of coastal rainforest species have been recorded. These include Australian Brush-turkey, White-bellied Sea-Eagle, Brahminy Kite (frequently around the headlands), Wompoo and Rose-crowed Fruit-Dove, Topknot Pigeon, Emerald Dove, Green Catbird, Regent and Satin Bowerbird, White-throated Treecreeper, Eastern Whipbird, Lewin's Honeyeater, Yellow-throated and Large-billed Scrubwren, Crested Shrike-tit, Golden Whistler, Spangled Drongo, Rufous Fantail, Black-faced and Spectacled Monarch, Pale-yellow and Eastern Yellow Robin, and Barred Cuckoo-shrike.

trails in this area provides opportunities for other species, including Brown Quail, Bar-shouldered Dove, Southern Emu-wren, Red-backed Fairy-wren, Scarlet and White-cheeked Honeyeater, White-throated Gerygone and, at night, White-throated Nightjar.

The extensive sandbars north of the township of Old Bar (located 40 km by road south of Harrington) are well known for resident and migratory shorebirds. To get there from the Pacific Hwy take the Old Bar Rd at Glenthorne for 11 km. Once in Old Bar take Mudbishops Point Rd north out of town (passing the airstrip) to access the sand spits via several foot tracks to the beach (the second carpark and track after the airstrip is best). Look on the sand bars for Beach Stone-curlew, Lesser Sand, Double-banded (winter) and Pacific Golden Plover, Eastern Curlew, Bar-tailed Godwit, Red Knot and Sanderling. Australia's second record of Kentish Plover was here. Look also for Common, Caspian, Crested and Little Tern, the last nesting in summer. There are also several records of Common Noddy and Sooty Tern.

Boorganna Nature Reserve and Wingham Brush Nature Reserve

Key species: Russet-tailed Thrush, Noisy Pitta, Regent Bowerbird, Green Catbird, Australian Logrunner, Spectacled Monarch
Other species: Wompoo Fruit-Dove, Emerald Dove, Brown Cuckoo-Dove, White-headed Pigeon, Rose Robin, Australian Brush-turkey

Conserving an important remnant rainforest, Boorganna Nature Reserve (396 ha) is the second oldest nature reserve in NSW (founded 1904). It is located 40 km north of Taree, 6.5 km west of the town of Comboyne. Wingham Brush Nature Reserve (10 ha) is a small area of subtropical lowland rainforest on the banks of the Manning River. The reserve is located 9 km west of Comboyne, 60 km south-west of Port Macquarie and 40 km north of Taree.

To reach the reserve, take the Innes View Rd, which branches off from the Comboyne to Wingham Rd just outside Comboyne.

At Boorganna Nature Reserve on the superb (but steep) walk down to Rawson Falls Lookout (1.7 km), there are five forest types: subtropical, warm temperate and dry rainforest, and wet and dry eucalypt forest. With such high variation in a relatively small area, the bird life is varied and abundant. Along here, look for Noisy Pitta, Regent and Satin Bowerbird, Green Catbird, Yellow-throated and Large-billed Scrubwren, Australian Logrunner, Spectacled Monarch, Bassian Thrush and Australasian Figbird. Rainforest pigeons here include Brown Cuckoo-Dove, White-headed and Topknot Pigeon. Rose-crowned and Wompoo Fruit-Dove also occasionally frequent the reserve when there are fruiting trees.

Topknot Pigeons are large striking fruit-eating pigeon of rainforests and scrubs. Flocks often commute between feeding and roosting sites, meaning the species is more often observed in flight than many other rainforest pigeons.

An important winter-foraging area for altitudinal migrants, Wingham Brush Nature Reserve is a remnant subtropical floodplain rainforest. Along the Graeme Allen Walk and Flying Fox Circuit (1.5 km), look for such birds as Australian Brush-turkey, Regent Bowerbird, Green Catbird, Wompoo Fruit-Dove, Emerald Dove, Brown Cuckoo-Dove, White-headed and Topknot Pigeon, and Rose Robin. It is an excellent site for Russet-tailed Thrush. Wingham Brush Nature Reserve also hosts a large colony of Grey-headed Flying-fox.

Dorrigo National Park

Key species: Russet-tailed Thrush, Grey Goshawk, Australian Logrunner, Noisy Pitta, Paradise Riflebird, Regent Bowerbird, Sooty Owl, Powerful Owl
Other species: Pacific Baza, Emerald Dove, Green Catbird, Pale-yellow Robin, Black-faced Monarch, Spectacled Monarch

Dorrigo National Park (11 902 ha), located 60 km south-west of Coffs Harbour, is situated on the subtropical slopes and spectacular eastern escarpment of the Great Dividing Range. Access is via Dome Rd off Waterfall Way, 2 km east of Dorrigo, where there is camping and accommodation. There are no campgrounds in the park, but permits to bush camp in remote areas of the park can be obtained from the park office.

The main walking track in the park is Wonga Walk (5.4 km circuit). Much of the walk provides excellent opportunities to search the understorey of the rainforest, while the occasional steep sections and canopy breaks offer views of both the mid-storey and canopy that are frequented by different species of bird. Look for ground-foraging birds such as Superb Lyrebird, Australian Brush-turkey, Pacific Baza, Russet-tailed and Bassian Thrush, Australian Logrunner, Noisy Pitta, Wonga Pigeon and Yellow-throated Scrubwren. Species that occupy

the mid-storey include Paradise Riflebird, Regent Bowerbird, Green Catbird, Black-faced and Spectacled Monarch, Eastern Yellow and Pale-yellow Robin, Rufous Fantail and Lewin's Honeyeater. The 'Walk with the birds boardwalk' (an elevated

A view across the rainforest canopy of Dorrigo National Park accessed from Dome Rd and Waterfall Way. The short loop walk below this lookout provide excellent rainforest birdwatching. Among others, look for Australian Brush-turkey, Wonga Pigeon, Noisy Pitta, Australian Logrunner, Regent Bowerbird, Green Catbird, Paradise Riflebird and Russet-tailed Thrush.

Bruxner Park Flora Reserve

This reserve (407 ha) contains an excellent area of dense rainforest and eucalypt forest not far from Coffs Harbour. Take the Pacific Hwy 4.4 km north of Coffs Harbour before turning west onto Bruxner Park Rd and travelling for another 4 km. It lies in a forested corridor that links the coast with the hinterland. The reserve is especially good for seasonal altitudinal migrants, particularly in winter. The best paths for birdwatching are along the Rainforest Walk (1.6 km loop) and Creek Trail (2 km return). Here, there is chance of Glossy Black-Cockatoo, Grey Goshawk, Collared Sparrowhawk, Brown Cuckoo-Dove, Wompoo and Rose-crowned Fruit-Dove, Green Catbird, Noisy Pitta, Pale-yellow Robin, Australian Logrunner, Spectacled Monarch, Little Shrike-thrush, Barred Cuckoo-shrike (uncommon) and Lewin's and Scarlet Honeyeater. There are picnic and barbecue facilities as Sealy Lookout, Halfway Creek and Swans Rd Crossing, where the walking trails begin.

boardwalk reached from the Glade Picnic Area, 1 km from the Rainforest Centre) provides further opportunities to observe canopy species. In spring and summer, the Wonga Walk is particularly good for fruit-eating pigeons, including Wompoo Fruit-Dove, Brown Cuckoo-Dove, and Topknot and White-headed Pigeon. Along the walk, look also for reptiles such as Land Mullet and the spectacular Southern Angle-headed Dragon. The Skywalk (70 m up and providing views across the forest) provides opportunities to observe Regent and Satin Bowerbird, Green Catbird, Australian King-Parrot and the fruit-eating pigeons. Red-necked Pademelon share the picnic areas with visitors near the Rainforest Centre.

Along the Rosewood Creek Circuit (5.5 km), which starts at Never Never Picnic Area and passes through warm-temperate rainforest, look for Grey Goshawk, Superb Lyrebird, Australia Brush-turkey, Eastern Whipbird, Brush Cuckoo and Varied Sittella.

Dorrigo National Park also supports large populations of nocturnal mammals, such as Short-eared Brushtail Possum and Yellow-bellied, Greater and Sugar Glider and, as a result, it is an excellent site for owls, particularly around the Never Never Picnic Area: listen for Sooty and Powerful Owls in this area. Other good walks in the park include the Blackbutt Track (6.4 km one way) and Casuarina Falls circuit (4.8 km return).

Clarence River Estuary

Key species: Beach Stone-curlew, Mangrove Honeyeater, Black-necked Stork, shorebirds
Other species: Eastern Grass Owl, Noisy Pitta, Regent Bowerbird

The Clarence River estuary, 650 km north of Sydney (270 km south of Brisbane), is noted for its rich diversity of migratory and resident shorebirds. The habitat along the lower reaches of the river consists of sand flats, mangroves, heathland and littoral rainforest. There is a range of accommodation options at Yamba and Iluka and camping at Woody Head in Bundjalung National Park.

Some of the best birdwatching is on Hickey Island (now not really an island), located on the southern side of the Clarence River, immediately north of Yamba. Park at the east end of Harbour St (a good spot for Pheasant Coucal) and walk in along a dirt track to the end of the 'island'. At low tide, the flats adjacent to mangroves support Beach Stone-curlew, Terek Sandpiper, Whimbrel, Great Knot, Grey-tailed Tattler, Eastern Curlew, Bar-tailed Godwit, Great Knot and Pacific Golden Plover. Mangrove Honeyeater also occurs here: one of the most southerly populations of the species. You may also find Eastern Osprey, Brahminy Kite, White-bellied Sea-Eagle, Mangrove Gerygone and Red-backed Fairy-wren. Peninsular Park is another good spot for shorebirds and mangrove species. Walk down the dirt road at the end of Shores Drive, where a track on the right-hand side leads to the Clarence River through littoral rainforest and mangroves. Many of the shorebirds mentioned above occur here. Low tide provides the best viewing conditions. Also here are Mangrove Gerygone, Yellow Thornbill, Mangrove, Striped, Blue-faced and Brown Honeyeater and Little Friarbird. The sand bars at Oyster Channel, located at the end of Carrs Drive (drive 2.5 km south of Yamba Rd), in summer usually support Terek Sandpiper, Grey-tailed Tattler, Whimbrel, Eastern Curlew and Bar-tailed Godwit. The bushland here holds Sacred Kingfisher, Grey-crowned Babbler, Little Shrike-thrush and Spangled Drongo. Search along the channel's edge for Azure Kingfisher, Striated Heron, Buff-banded Rail and Brahminy Kite. On the northern side of the Clarence River, Woody Head in Bunjalung National Park (4 km north of Iluka) is a summer roost site for Lesser and Greater Sand Plover, Pacific Golden Plover, Ruddy Turnstone and Grey-tailed Tattler. Wandering Tattler has also been reported, but is generally uncommon. The littoral rainforest at Iluka Nature Reserve can be good for Noisy Pitta, Regent Bowerbird and Emerald Dove in winter, while in summer, Rose-crowned Fruit-Dove, Spectacled and White-eared Monarch occur.

South of Yamba, Beach Stone-curlew sometimes feed at the mouth of the Sandon River, accessed via Brooms Head Rd and followed by a right turn onto Sandon Rd (just before the Brooms Heads Caravan Park). From here, it is 5 km to the mouth of the river. Along the drive, look for Emu (an endangered coastal population occurs) and Spotted Harrier. Further south of Sandon, Beach Stone-curlew also inhabit the mud flats at Wooli and Red Rock. Search the

The Australasian Darter is relatively common on larger wetland systems in eastern and northern Australia: wetland sites along the Clarence River valley provide ideal habitat.

heathland at Yuraygir National Park (just north of Wooli) for Eastern Ground Parrot, with the best area being along Diggers Camp Rd (the road to Wilson's Headland). Other birds you may see include Emu, Brolga, Glossy Black-Cockatoo, Southern Emu-wren, Tawny-crowned and White-cheeked Honeyeater, Brown Quail and, more rarely, King Quail. Although a rare species, the Eastern Grass Owl, also occurs in these heaths, with the best time to look for them being on dusk when they first take flight to forage for the evening. The freshwater wetlands alongside the Clarence River between Lawrence and Ulmarra and on Woodford Island (located 30 km east of Yamba), are particularly good for waterbirds such as Black-necked Stork, Magpie Geese, Wandering and Plumed Whistling-Duck, Great, Intermediate, Little and Cattle Egret, Glossy Ibis, Royal Spoonbill and Australasian Darter.

Ballina

Key species: Beach Stone-curlew, terns, shorebirds,
Other species: Eastern Osprey, Brahminy Kite, Mangrove Gerygone, South Island Pied Oystercatcher (vagrant)

The township of East Ballina and the village of South Ballina are located at the mouth of the Richmond River, 25 km south of Cape Byron (Australia's most easterly point). The major airport for the region, servicing Byron Bay and Lismore, is in Ballina. Ballina and Lennox Head offer a wide range of accommodation, including motels, guesthouses, caravan parks and camping grounds.

Angels Flat Rock, a rocky outcrop on the beach north of East Ballina, is a well-known roosting-site for terns and migratory shorebirds. Species that routinely roost here include Little, Common, Gull-billed and Crested Tern, while there are occasional sightings of Arctic, White-fronted, White-winged Black, Sooty and Bridled (rare, several records) Tern and Common and Black Noddy. Between September and April, look for shorebirds such as Wandering Tattler (uncommon, but regular) and Grey-tailed Tattler, Greater and Lesser Sand Plover, Pacific Golden Plover, Terek, Common and Curlew Sandpiper, Bar-tailed Godwit, Red-necked Stint, Ruddy Turnstone, Great and Red Knot, Whimbrel and Sanderling. Other birds you may find here include Eastern Reef Egret, Brahminy Kite and White-bellied Sea-Eagle. The nearby Chikiba Lake roost is another good area to view migratory shorebirds, particularly on spring tides through the warmer months, when many of the estuarine roosts may be inundated. East of the village of South Ballina, the tidal zones in the Richmond River Nature Reserve (254 ha), particularly Mobbs Bay, are another excellent site for resident and summer shorebirds: the best time is around low tide. Scan the southern riverbank and the mangroves for Beach Stone-curlew, Grey-tailed Tattler, Whimbrel, Bar-tailed Godwit, Striated Heron, Eastern Curlew, Common Greenshank, Terek Sandpiper and Black-winged Stilt. The roadside vegetation provides habitat for Lewin's, White-cheeked, Scarlet and Striped Honeyeater, Mangrove Gerygone and Eastern Whipbird. Raptors such as Eastern Osprey (nesting on an artificial nest pole), Brahminy Kite and White-bellied Sea-Eagle are regularly sighted. The 20 km stretch of sandy beach south of the Richmond River mouth (with access at Patches Beach east of Wardell) is notable for occasional visits from vagrant South Island Pied Oystercatcher. Usually associating with Australian Pied Oystercatcher, they can be differentiated by their shorter legs, a longer finer-bill and a greater extent of white on the wing. (Colour-marked Australian Pied Oystercatcher and Beach Stone-curlew should be reported to abbbs@environment.gov.au.)

Nightcap National Park

Key species: Albert's Lyrebird, Marbled Frogmouth, Double-eyed Fig-Parrot (possibly extinct), Russet-tailed Thrush, Paradise Riflebird
Other species: Sooty Owl, Wompoo Fruit-Dove, Rose-crowned Fruit-Dove, Superb Fruit-Dove, Australian Logrunner, Pale-yellow Robin

Nightcap National Park (8145 ha) is located 35 km north of Lismore. Home to a diversity of birds, it contains the largest known population of Albert's Lyrebird and there are historic records of the extremely rare (possibly extinct) Double-eyed (Coxen's) Fig-Parrot. There is a camping area at Rummery Park, with picnic areas nearby. The park is accessible from coastal towns such as Byron Bay, 50 km to the east.

Search for Albert's Lyrebird in the rainforest near the Terania Creek Picnic Area, particularly along the walk to Protesters Falls (1.4 km return). The best time to hear them is in late winter and early spring. By late spring (and into summer) this trail is good for viewing Wompoo,

Superb and Rose-crowned Fruit-Dove, Noisy Pitta, Green Catbird, Regent and Satin Bowerbird, Spangled Drongo and Black-faced Monarch. Other species present include Brown Cuckoo-Dove, Emerald Dove, Australian Logrunner, Pale-yellow and Eastern Yellow Robin, Spectacled Monarch, Little Shrike-thrush, Eastern Whipbird and Pied Currawong. At night, around the edge of the picnic area, listen for the distinctive call of Marbled (Plumed) Frogmouth. There is also a chance of Sooty Owl. To get to Terania Creek from Channon, take the Terania Creek Rd and follow the signs. The drive is 15 km along what is usually a good gravel road.

Some of the best examples of remnant subtropical rainforest in NSW are in the former Big Scrub Flora Reserve (now part of Nightcap National Park). This site covers an area of 196 ha, of which 148 ha are rainforest. The best birding is along the 1.5 km Big Scrub Loop and along the roadside for ~1 km of the Gibbergunyah Range Rd near the start of the walk. In summer, fruiting trees attract pigeons and doves such as Wompoo, Rose-crowned and Superb Fruit-Dove, Brown Cuckoo-Dove, Topknot, White-headed and Wonga Pigeon and Emerald Dove. Other species found here include Channel-billed Cuckoo (summer), Green Catbird, Regent Bowerbird, Paradise Riflebird, Australian Logrunner, Spectacled Monarch, Yellow-throated Scrubwren, Lewin's Honeyeater and Pale-yellow Robin. Although now possibly extinct, the Coxen's form of the Double-eyed Fig-Parrot has also been recorded here in the past. The Scrub Turkey Walk (6 km return, also known as Northern Boundary Trail or Big Scrub Rainforest Track) from Gibberagunyah Rd to Rocky Creek Dam is also worthwhile exploring. Along the elevated boardwalk beside Rocky Creek on the Platypus

The lush forests of Nightcap National Park support a dense cover of leaf litter that is perfect for ground-foraging birds such as Noisy Pitta, Albert's Lyrebird, Australian Logrunner and Russet Tailed Thrush.

Walk (1.5 km), just west of the Rocky Creek Dam Picnic Area, you may find Brown Cuckoo-Dove, Little Shrike-thrush, Australian Logrunner, White-eared and Spectacled Monarch, Scarlet Honeyeater and Australasian Figbird. On and around the dam, look for Comb-crested Jacana, Azure Kingfisher, Australasian Darter and Nankeen Night-Heron.

Along the short walk beside Boggy Creek to Minyon Falls, look for Australian Brush-turkey, Brown Cuckoo-Dove, Topknot Pigeon, Grey Goshawk, Paradise Riflebird, Green Catbird, Spectacled Monarch, Australian Logrunner, Rufous Fantail, Large-billed Scrubwren, Pale-yellow Robin, Russet-tailed Thrush and Leaden Flycatcher. At night, Marbled (Plumed) Frogmouth have been recorded here. Minyon Falls is a good area for Peregrine Falcon and possibly other soaring raptors. North of Boggy Creek, the forest around the Rummery Park campground is another good birding area that holds many of the above-listed species.

Border Ranges National Park

Key species: Rufous Scrub-bird, Albert's Lyrebird, Eastern Bristlebird (very rare), Russet-tailed Thrush, Sooty Owl, Double-eyed Fig-Parrot (possibly extinct)
Other species: Green Catbird, Paradise Riflebird, Spectacled Monarch, Cicadabird, Pale-yellow Robin, Australian Logrunner, Bell Miner

Border Ranges National Park (31 683 ha) (adjoining Queensland's Lamington National Park), has one of the highest concentrations of marsupial species in Australia. Reached via the Wiangaree turn-off on the Summerland Way, or the Barkers Vale turn-off on the Murwillumbah–Kyogle Rd, the Park is traversed by the 64 km Tweed Range Scenic Drive. There is camping at the Sheepstation Creek and Forest Tops camping areas, which provide water, shelter sheds, fireplaces and toilets. The best time to visit the park is in spring and summer, when birds return from wintering in lower altitude forests and the east-coast migrants arrive from northern Australia.

Rufous Scrub-bird occurs along the Brindle Creek Walk (5 km), between the Brindle Creek Picnic Area and Antarctic Beech Picnic Area. The best place to look is in where the ground cover thickens up along the walk, with a good area to search ~200 m from the start of the walk. This same track also provides opportunities for Albert's Lyrebird, Wompoo Fruit-Dove, Noisy Pitta, Regent Bowerbird, Bassian Thrush, Australian Logrunner, Paradise Riflebird, Olive Whistler and Bell Miner. Remain alert for the extremely rare (possibly extinct) Double-eyed (Coxen's) Fig-Parrot. Another place to look for Rufous Scrub-bird is around the Forest Tops camping area, particularly along a track that starts directly behind the campsite. Follow this track for ~200 m, and listen patiently for their loud high-pitched call. There is a good assortment of rainforest birds around the excellent Sheepstation

Noisy Pitta are an altitudinal migrant within NSW, mostly frequenting upland rainforest habitats in the warmer months and coastal and near-coastal scrubs and vine thickets in the cooler months.

Creek camping area, including Australian Brush-turkey, Green Catbird, Paradise Riflebird, Spectacled Monarch, Cicadabird, Pale-yellow Robin and Australian Logrunner. At night, look and listen for Sooty Owl. Also keep an eye out for Koala, lazing in the eucalypts by day around the campsite. A small and endangered population of Eastern Bristlebird is also found in the Border Ranges, mainly in the central part of the park; look along the Border Loop Walk, accessed via Lions Rd. Please report any sighting to Eastern Bristlebird Northern Population Recovery Team via the NSW Office of Environment and Heritage (www.environment.nsw.gov.au).

Western slopes and central tablelands

Regional overview

The western slopes and central tablelands region covers the inland slopes of the Great Dividing Range and includes some of Australia's best known birding sites, such as the Capertee Valley, Munghorn Gap Nature Reserve, Barraba, Pilliga Forest, Warrumbungle National Park and the Macquarie Marshes Nature Reserve. The occurrence of nectar-feeding birds is linked to the flowering of suitable plants and trees, so, when box–ironbark woodlands are in flower, the western slope can be transformed with an abundance of birds, all busily feeding. The favoured haunts for ground-feeding birds are open, grassy woodlands, one of the dominant habitats in the region.

To explore the western slopes and central tableland of NSW, a suggested 10-day itinerary from Sydney (including key target species) involves:

- **Days 1–2.** Sydney to Capertee Valley, 184 km (Turquoise Parrot, Regent Honeyeater, Painted Button-quail, Spotted Quail-thrush, Plum-headed Finch, Black-chinned Honeyeater, Diamond Firetail, Little Lorikeet, Speckled Warbler).
- **Day 3.** Capertee Valley to Munghorn Gap Nature Reserve, 117 km (Turquoise Parrot, Regent Honeyeater, Rockwarbler, Spotted Quail-thrush).
- **Days 4–7.** Munghorn Gap Nature Reserve to Warrumbungles National Park and Pilliga Forest, 223 km+ (Turquoise Parrot, Glossy Black-Cockatoo, Pale-headed Rosella, Blue Bonnet, Cockatiel, Barking Owl, Regent and Painted Honeyeater, Spotted Quail-thrush, Black-eared Cuckoo, Regent Honeyeater, Painted Button-quail).
- **Days 8–9.** Pilliga Forest to Barraba district, 217 km (Regent Honeyeater, Turquoise Parrot, Plum-headed Finch).
- **Days 10+.** Barraba area to Sydney, 480 km.

To see waterbirds, a 2-day extension to the Macquarie Marshes is also possible (best accessed while visiting Warrumbungles National Park).

With regards to birdwatching organisations, the northern and southern NSW and ACT branches of Birdlife Australia provide some coverage for the region: www.birdlife.org.au.

The western slopes and central highlands can provide pleasant weather at any time of year, though during summer months temperature can on occasion exceed 40°C. For this reason, autumn through winter to late spring are better seasons in which to visit. Much of the box–ironbark forest can flower in winter and early spring, providing excellent birding opportunities in these areas, as many species respond to this event for both the profuse nectar and the abundance of invertebrates that this event supports. In winter, overnight frosts can be severe, but days can still be mild. Several areas in the region are situated in rain shadows, meaning they are drier than surrounding areas.

Most of the highways in the region are sealed and, where not, are still usually suitable for 2WD. Most of the parks and reserves discussed in this section have one or more camping areas, with toilets, barbeque, fireplaces and picnic table facilities; some even have showers. Regional airports are scattered over the region with hubs for scheduled services at Narrabri, Coonabarabran, Bathurst, Tamworth, Parkes, Dubbo and Orange.

Birding highlights

The western slopes and central tablelands region is home to a remarkable diversity of birds. In particular, it is a stronghold for declining woodland birds, providing a refuge for many species that are now difficult to find elsewhere in Australia. Geographically, it is a transitional area between the sparsely vegetated western plains and the wetter and taller forests of the Great Dividing Range. This means it is inhabited by birds from eastern and western NSW, leading to a high diversity of species in a relatively small area. The box–ironbark woodlands and wooded grasslands are especially important for a wide range of declining species. Although there are no regionally endemic species, the western slopes and central tablelands are home to some of Australia's most sought-after bird species. These include the Critically Endangered Regent Honeyeater and Endangered Swift Parrot, and Vulnerable species such as Australian Painted Snipe. Other species that are characteristic of the region include Sacred and Red-backed Kingfisher, Turquoise Parrot, Gang-gang Cockatoo, Glossy Black-Cockatoo, Barking, Powerful and Masked Owl, Superb Lyrebird, Spotted Quail-thrush, Rockwarbler, Speckled Warbler, Black-chinned and Painted Honeyeater, Diamond Firetail and Plum-headed Finch. More widespread species that frequent the region include Budgerigar, Diamond Dove, Black-eared Cuckoo, Southern Whiteface, Hooded and Red-capped Robin, White-browed and Grey-crowned Babbler, Gilbert's Whistler, Yellow-tufted, Fuscous and Spiny-cheeked Honeyeater, White-backed Swallow and Brown Songlark. In summer, White-throated Nightjar, Rainbow Bee-eater, White-winged Triller, Rufous Songlark and Dusky and White-browed Woodswallow occur.

Barraba District

Key species: Regent Honeyeater, Turquoise Parrot, Plum-headed Finch
Other species: Western Gerygone, Hooded Robin, Ground Cuckoo-shrike

Barraba, 90 km north of Tamworth, is a renowned district for birdwatching. Over 190 species of birds occur in the district, including Turquoise Parrot and an important spring-summer breeding population of Regent Honeyeater; even the road signs on the way into town state 'Welcome to Barraba – a habitat of the Regent Honeyeater'. There is bush camping at Borah Reserve and a caravan park and motel in Barraba. The excellent *Bird Routes of Barraba* (the first bird route brochure to be produced in Australia) is available from the Barraba Information Centre: (02) 6782 1255.

The main birding locations in the district are along the old droving easements known as travelling stock routes, with many of them signposted for easy reference. Two of these are located on Borah Creek Rd, accessed from the Fossickers Way, 20 km south of Barraba. Tarpoly Reserve, which protects remnant white box woodland, is immediately after the turn-off. Here, there are regular sightings of Turquoise Parrot and Plum-headed Finch, as well as Striped Honeyeater, Western Gerygone

Macquarie Marshes

The Macquarie Marshes Nature Reserve (20 000 ha) is one of the largest semi-permanent wetlands in south-eastern Australia. With two distinct wetland systems (north and south), the marshes are located ~100 km north of Warren and 30 km west of Quambone. They are a major waterbird breeding area and an important refuge for a large number of bird species and the area is listed under the Ramsar Convention as a wetland of international importance. Because it is principally managed for its important wetland features, the reserve does not cater well for day-visitors or campers. Nevertheless, when conditions are suitable guided activities around the reserve are occasionally provided. Contact the Coonabarabran National Park office: (02) 6825 4364. When in full flood, the Macquarie Marshes can support up to 60 000 waterbirds, including large numbers of breeding Glossy, Straw-necked and Australian White Ibis, Intermediate, Great and Little Egret, Nankeen Night-Heron, White-necked Heron and Royal Spoonbill. The marshes also support populations of Brolga, Australian Little and Australasian Bittern, Freckled and Blue-billed Duck, Magpie Goose, Plumed and Wandering Whistling-Duck, Latham's (summer) and Australian Painted Snipe, and Baillon's and Australian Spotted Crake. There are also records of the regionally rare Black-necked Stork and Pied Heron.

Interesting woodland birds recorded here include Superb Parrot (rare), Cockatiel, Australian Ringneck, Blue Bonnet, Ground Cuckoo-shrike, Spotted Bowerbird, Painted and Striped Honeyeater, White-winged Fairy-wren and Diamond Firetail. A good place to base yourself when visiting the area is Willie Retreat, which adjoins the western perimeter of Macquarie Marshes and provides camping and bunkhouse accommodation: willie.retreat@bigpond.com, (02) 6824 4361.

Plumed Whistling-Duck are just one of the many waterbirds that can occur in large numbers when the Macquarie Marshes hold water.

and Hooded Robin. In spring and summer, Channel-billed and Pallid Cuckoo, Rufous Songlark and Dollarbird may also occur. After birding here, cross the grid at the southern end of Tarpoly Reserve and continue for 6.5 km to Borah Reserve. At the southern end of the 1 km long reserve, take the left-hand track to the creek. With permanent water, the adjacent grassy woodlands here can be superb for birding. Regent Honeyeater is regularly recorded here, along with other honeyeaters such as Fuscous, Black-chinned, White-plumed, Brown-headed, Striped, Spiny-cheeked, White-naped, Brown-headed and Yellow-faced. There is usually an assortment of seed-eaters, including Red-browed, Double-barred and Plum-headed Finch and Diamond Firetail, and the River Red Gum along the creek provide hollows for nesting Turquoise and Red-winged Parrot. In the creek-side vegetation look for Common Bronzewing, Peaceful Dove, White-bellied and Ground Cuckoo-shrike (occasional), Speckled Warbler, Buff-rumped, Yellow and Yellow-rumped Thornbill, Western Gerygone, Crested Shrike-tit, Restless Flycatcher, White-backed Swallow, Jacky Winter and Eastern Yellow Robin. In spring and summer, the number of species increases, with the addition of east-coast migrants such as Dollarbird, Rainbow Bee-eater, Olive-backed Oriole, Cicadabird, Rufous Songlark and Mistletoebird.

Another potential site for Regent Honeyeater is the box–ironbark woodland in the Nangahrah and Coonoor Travelling Stock Reserves. These are located on Bundarra–Barraba Rd, which begins 21 km east of Barraba at the Nangahrah Creek crossing. There are several strategically placed picnic shelters at good birding sites over the next 14 km: search areas of flowering Mugga Ironbark, especially those with Mistletoe. The forest here is also a good for Little and Musk Lorikeet, Brown Treecreeper, Hooded Robin and White-throated Gerygone. North of Barraba, Regent Honeyeater occasionally occur in the box–ironbark at Mille Creek Travelling Stock Reserve, along with Plum-headed Finch, Diamond Firetail, Dusky Woodswallow, Black-chinned Honeyeater and Rufous Songlark. This reserve begins on Fossickers Way, 3.3 km from Barraba.

Pilliga Forest

Key species: Turquoise Parrot, Glossy Black-Cockatoo, Pale-headed Rosella, Blue Bonnet, Cockatiel, Barking Owl, Painted Honeyeater, Spotted Quail-thrush, Black-eared Cuckoo
Other species: Black Falcon, Spotted Harrier, Crested Bellbird, Apostlebird, Spotted Bowerbird, Speckled Warbler

The Pilliga Forest (commonly known as the Pilliga Scrub) is the largest continuous remnant of semi-arid woodland in NSW (covering an area of over 500 000 ha). It is sometimes referred to as 'the place where east meets west', because it is situated between the wetter eastern coastal zone and the dryer plains of inland NSW. All the roads in the Pilliga Forest are dirt: during dry periods they are easily accessible by 2WD, but be cautious in the wet. A useful guide for birders is the *Bird Routes of Baradine and Pilliga*, available from both the Baradine and Coonabarabran Visitor Centres. When visiting the Pilliga Forest, a detailed local map of the area is strongly recommended (available at the Information Centre) because there are 2700 km of unsealed tracks and, although most are marked, it is easy to take a wrong turn. Discovery Centre at Baradine: (02) 6843 4011, Coonabarabran Visitors Centre: (02) 6849 2144.

A good starting point is Trapyard Dam in the Merriwindi State Forest (13 000 ha). Here, there is a usually a varied selection of honeyeaters such as Spiny-cheeked, Brown-headed, Black-chinned, Blue-faced

Open eucalypt woodland, such as this in the Pilliga Forest, provides habitat for a number of significant species, including Turquoise Parrot, Glossy Black-Cockatoo, Barking Owl, Speckled Warbler, Painted Honeyeater, Spotted Quail-thrush, Apostlebird and Spotted Bowerbird.

and Striped and, in spring, look for Painted Honeyeater feeding in Pilliga Box that surrounds the dam. This area can also be good for the smaller passerines such as White-throated and Western Gerygone, Speckled Warbler, Weebill and Buff-rumped, Yellow, Yellow-rumped, Inland and Chestnut-rumped Thornbill. Other birds to look for include Bar-shouldered Dove, Red-capped Robin and White-backed Swallow and spring/summer migrants such as White-winged Triller and Rainbow Bee-eater. To get to Trapyard Dam from the Gwabegar–Baradine Rd, turn left onto a dirt road 15 km north of Baradine: the dam is another 4 km further on.

Carmel Lagoon usually attracts a good selection of waterbirds, including Plumed Whistling-Duck, Pink-eared Duck, Australasian Shoveler, Australasian Grebe, Yellow-billed Spoonbill, White-necked Heron, Whiskered Tern, Black-tailed Native-hen, Bush Stone-curlew and, more rarely, Australian Painted Snipe. To get there from Baradine, drive west along Coonamble Rd for 17 km then turn north along Carmel Lane. The lagoon is a further 4.6 km on your right. If you continue along Carmel Lane (via a series of left turns, crossing the Teridgerie Creek, then looping back to Coonamble Rd), search the road-side vegetation for woodland species such as Blue Bonnet, Cockatiel, Apostlebird, Grey-crowned Babbler, Spotted Bowerbird, Rufous Songlark, Southern Whiteface and, occasionally, Ground Cuckoo-shrike. Also remain alert for Black Falcon and Spotted Harrier, which hunt over the grassy plains.

Glossy Black-Cockatoo and Turquoise Parrot regularly drink at Salt Caves Dam, located within Timallallie National Park (39 395 ha) especially later in the afternoon. Look also around the dam and the Salt Caves Picnic Area, for Australian Ringneck, Australian King-Parrot, Pale-headed Rosella (the southernmost population of this species where it occasionally hybridises with Eastern Rosella), Spotted Quail-thrush, Crested Bellbird, Black-eared Cuckoo, and White-browed and Grey-crowned Babbler. At night, Barking Owl can sometimes be vocal in this area. The last known report of Malleefowl at Pilliga (now probably locally extinct) was near Salt Caves. Koala are reasonably common. To get to the Salt Caves Dam, from Baradine, take Gwabegar Baradine Rd to Cumbil Rd, after 20 km turn into Pilliga Forest Way and travel another 19 km before turning right into County Line Rd, which leads to Salt Caves Dam.

Turquoise Parrots are widespread, though generally uncommon, in drier woodlands on the inland slopes of the Great Dividing Range. Good sites in NSW include Pilliga Forest, Warrumbungle National Park, Capertee Valley and Munghorn Gap Nature Reserve.

An impressive list of birds has been recorded in the grounds of Pilliga Pottery, including Turquoise Parrot, Glossy Black-Cockatoo, Grey-crowned Babbler, Hooded Robin, Speckled Warbler, Brown Treecreeper, Black-chinned Honeyeater and Diamond Firetail. To get there from Coonabarabran, take the Newell Highway north towards Narrabri. After 24 km, turn left into Borambitty Rd; Pilliga Pottery is well signposted. Finally, it can be worth birding in the streets of Baradine itself, because there is a chance of Spotted Bowerbird, Apostlebird, Blue Bonnet, Little Friarbird, Blue-faced Honeyeater, Diamond Firetail and Plum-headed Finch; Square-tailed Kite nest near town and occasionally soar overhead.

Warrumbungle National Park

Key species: Regent Honeyeater, Turquoise Parrot, Painted Button-quail
Other species: Peregrine Falcon, Speckled Warbler, Yellow-tufted Honeyeater, Black-chinned Honeyeater

The network of walking tracks in the Warrumbungle National Park (21 534 ha) provides excellent opportunities for bird-watching. Regent Honeyeater is a visitor to the area, with numbers fluctuating from year to year, depending on the abundance of flowering of gums. The Warrumbungles are 36 km drive west of Coonabarabran. There is a wide range of camping and picnic facilities available, see www.environment.nsw.gov.au. In January 2013, large areas of the park were burnt by wildfire and some of these areas may take some time to recover.

One of the better areas for Regent Honeyeater is within flowering Mugga Ironbark and White and Yellow Box around Camp Blackman, where you may also find Turquoise Parrot, Yellow-tufted, Yellow-faced and Blue-faced Honeyeater, and Noisy Friarbird. Look around the disused Woolshed on Tooraweenah Rd for Emu, Turquoise Parrot, Painted Button-quail, Pied Butcherbird, Spotted Bowerbird, White-browed Babbler, Speckled Warbler and Inland Thornbill. Around Camp Pincham (a walk-in camp from the Pincham carpark), Turquoise Parrot, Brown Treecreeper, Restless Flycatcher, White-bellied Cuckoo-shrike and Dusky Woodswallow all occur. The Wambelong Nature Track (1.1 km) begins from the eastern side of Canyon Picnic Area. It follows Wambelong Creek through a small sheltered gorge. In drier times, a few small waterholes usually remain along here, providing much-used drinking sites for bird life. Along this track, you are likely to encounter Red-winged Parrot, Varied Sittella, Pied Butcherbird, Blue-faced Honeyeater and Little and Noisy Friarbird. Keep your eyes on the treetops for Koala, often resting high in the gums beside the creek. Along the Burbie Canyon Circuit (3.8 km), look for Striped, Spiny-cheeked, Blue-faced, Black-chinned and Yellow-tufted Honeyeater. Peregrine Falcon also frequently nest on the rock face passed on the walk. Finally, Red-backed Kingfisher is regularly seen on the western approach to the park, especially along the Tooraweenah Rd.

The view from Pearson's Lookout across the Capertee Valley from Castlereagh Highway ~5 km south of the village of Capertee. These woodlands provide one of the most significant refuges for the Endangered Regent Honeyeater.

Capertee Valley

Key species: Turquoise Parrot, Regent Honeyeater, Painted Button-quail, Spotted Quail-thrush, Plum-headed Finch, Black-chinned Honeyeater, Diamond Firetail, Little Lorikeet, Speckled Warbler
Other species: Gang-gang Cockatoo, Glossy Black-Cockatoo, Rockwarbler, Black-eared Cuckoo, Painted Honeyeater, White-browed Babbler, Striped Honeyeater, Southern Whiteface

The Capertee Valley, 185 km from Sydney, is one of Australia's best-known birding locations. Most of the valley's special birds are not in national parks or reserves but on private properties and along the wooded roadsides and riparian vegetation, with the best area for birding along Glen Davis Rd, between Capertee and Glen Davis, and north along Glen Alice Rd. As much of the birding in the Capertee is done from the roadside, it is important that birders park well off these narrow roads (not in the middle!) to allow safe passage of other vehicles. There is accommodation at the Glen Davis campground and bush camping at Wollemi National Park. Several farms offer excellent cottage-style accommodation, with some stays specifically tailored for birdwatchers: www.bmbirding.com.au. There is a detailed Capertee Valley bird sites leaflet available at local stores and info outlets.

A good place to start any visit is at the Coco Creek Bridge on Glen Davis Rd 16 km east of Capertee. Regent Honeyeater has been recorded here and you can expect an assortment of woodland birds such as Gang-gang Cockatoo (seasonal), Glossy Black-Cockatoo, Musk and Little Lorikeet, Turquoise Parrot, Scarlet, Yellow-tufted, Painted and Fuscous Honeyeater, Noisy Friarbird, Dusky Woodswallow, Hooded and Eastern Yellow Robin. In summer, migrants such as Black-eared Cuckoo, Dollarbird and Rufous Songlark may also occur. This creek line is also often good for finches such as Plum-headed, Double-barred and Red-browed Finch and Diamond Firetail.

Genowlan Rd, accessed via Glen Alice Rd, 5 km north-west of Glen Alice, is another good spot to look for honeyeaters, particularly near the Genowlan Bridge

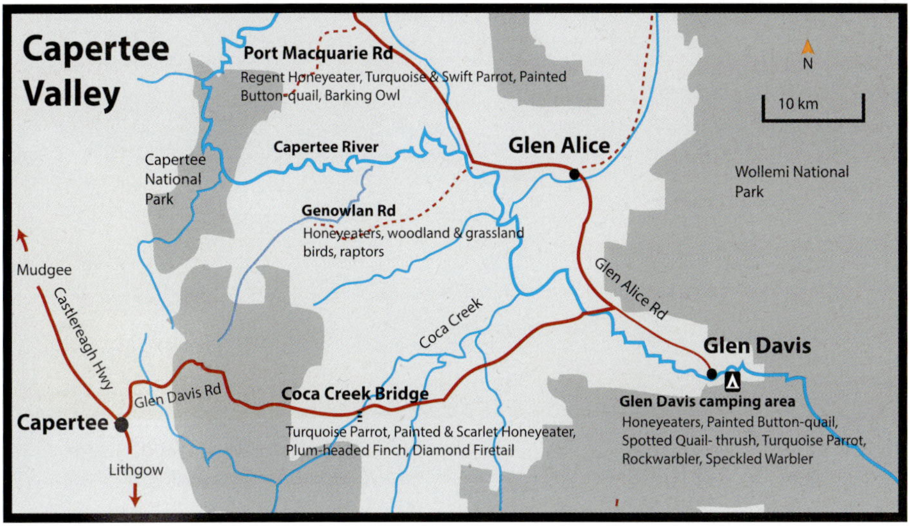

over the Capertee River. Here, you may see Regent, Scarlet, Striped, Black-chinned and Painted (summer) Honeyeater, as well as White-bellied Cuckoo-shrike (including frequent records of the dark form), Restless Flycatcher, Grey-crowned Babbler, Jacky Winter and Olive-backed Oriole (summer). In the clear grassy stretches of farmland in this area, look for Australasian Pipit, Rufous Songlark, White-backed Swallow, Horsfield's Bushlark, Southern Whiteface and Plum-headed and Zebra Finch. Raptors that are encountered in this area include Wedge-tailed and Little Eagle, Black-shouldered Kite, Brown Falcon, Brown Goshawk and, less commonly, Square-tailed Kite and Spotted Harrier.

Some of the best birding in the Capertee Valley can be had around the village of Glen Davis, particularly the bush surrounding the campground. An impressive 26 species of honeyeater have been recorded here, including Yellow-faced (winter), Yellow-tufted, Scarlet, Striped, Fuscous, White-plumed, Regent, White-naped (winter), Black-chinned, Brown-headed and Painted (summer). Search for Painted Button-quail and Spotted Quail-thrush in the wooded gravelly slope behind the campground. Along the start of the Pipeline Track (21 km, steep and strenuous!), linking Newnes over the mountain to Glen Davis, look for Turquoise Parrot, Rockwarbler, Superb Lyrebird, Speckled Warbler, Crested Shrike-tit, Eastern Whipbird, Bell Miner and Diamond Firetail. Other birds found around Glen Davis include Peaceful Dove, Little Lorikeet, Australian King-Parrot, Swift Parrot (occasionally), White-browed Babbler and Double-barred and Zebra Finch, and summer-migrants such as Sacred

Open River Oak (*Casuarina cunninghamiana*) woodland on the Capertee Valley. A favoured habitat for a range of species, including Regent, Scarlet, Striped and Black-chinned Honeyeater, as well as Restless Flycatcher, Grey-crowned Babbler, Jacky Winter, White-bellied Cuckoo-shrike and Olive-backed Oriole.

Kingfisher, Olive-backed Oriole, Rainbow Bee-eater, Fan-tailed and Channel-billed Cuckoo and Eastern Koel. Glen Davis is also a good area for night birds, including Eastern Barn Owl, Tawny Frogmouth and Australian Owlet-Nightjar and, less commonly, Powerful, Masked and Barking Owl and White-throated Nightjar: look for the latter in grassy slopes with gravel patches.

It is also worth visiting the newly created Capertee National Park (2800 ha). The park is located at the end of Port Macquarie Rd, reached via Glen Alice Rd near Bogee (26 km south of Kandos, 14 km north of Glen Alice). Formerly a grazing property known as Port Macquarie, the park is an important breeding area for Regent Honeyeater: look for them in the roadside trees near the park entrance. Other species that occur here include Painted Button-quail, Barking Owl, Little Lorikeet, Turquoise and Swift Parrot, Hooded Robin, Brown Treecreeper, Diamond Firetail, and Painted, Striped, Black-chinned and Yellow-tufted Honeyeater.

Munghorn Gap Nature Reserve

Key species: Turquoise Parrot, Regent Honeyeater, Rockwarbler, Spotted Quail-thrush
Other species: Black-chinned Honeyeater, Fuscous Honeyeater, Speckled Warbler, Hooded Robin

Munghorn Gap Nature Reserve (6800 ha) is located on the central western slopes of the Great Dividing Range. Noted for its high bird diversity, the reserve supports a range of eastern forest species, as well as those associated with the dryer western slopes of the Great Dividing Range. Some 164 species of bird have been recorded in the reserve, including the Critically Endangered Regent Honeyeater. Entry is via sealed roads off the Mudgee–Wollar Rd, 30 km north-east of Mudgee. Facilities include toilets and barbeques.

Regent Honeyeater have occurred around the Moolarben Day Use Area, located on Wollar Rd, 34 km north of Mudgee. Here, a short sign-posted path heads to a perennial spring that serves as focal point for the woodland birds. Other birds recorded here include Yellow-tufted, Black-chinned and Fuscous Honeyeater, Noisy Friarbird, and Musk and Little Lorikeet. At Honeyeater Flat, look for Rockwarbler (the north-western distribution edge for this species) near the water supply area: they can be found across the creek bed and up the adjacent ridges. Other birds that frequent the spring area include Speckled Warbler, Hooded Robin, White-winged Chough and, in summer, Dollarbird, White-winged Triller and Rainbow Bee-eater. The Castle Rocks Walking Track (8 km return) leads to rocky outcrops known as the Pagoda Formations. A good selection of birds may be seen along here, including Emu, Superb Lyrebird, Turquoise Parrot, Peaceful Dove, Varied Sittella, Brown Treecreeper, Yellow-faced (winter), White-eared, Spiny-cheeked, Scarlet and Striped Honeyeater, Little Friarbird, Spotted Quail-thrush (on the ridgelines), Cicadabird (summer), Varied Triller, Speckled Warbler, Rockwarbler, Buff-rumped, Yellow and Striated Thornbill, White-throated Gerygone and Diamond Firetail.

Western plains

Regional overview

The vast landscape of the western plains ranges from sandy deserts and rocky ranges, to rolling downs. The western plains cover a vast area bordered by Qld, SA and Vic on three sides. Sites covered in this section include Cocoparra, Murray Valley, Mutawintji, Sturt and Kinchega

National Parks, Round Hill and Nombinnie Nature Reserve and Binya State Forest, and the areas surrounding Lake Cargelligo, Leeton and Deniliquin.

To visit the western plains of NSW from Sydney, a suggested 14-day itinerary (including key target species) involves:

- **Days 1–3.** Sydney to Kinchega National Park, 1130 km (Major Mitchell's Cockatoo, Blue Bonnet, Chirruping Wedgebill, Chestnut-crowned Babbler, waterbirds).
- **Days 4–6.** Kinchega National Park to Mutawintji National Park, 250 km (Grey Falcon, Chirruping Wedgebill, Chestnut-crowned Babbler, Little Woodswallow).
- **Days 7–11.** Mutawintji National Park to Sturt National Park – with a side trip to Bollards Lagoon, 285 km+ (Eyrean and Grey Grasswren, Ground Cuckoo-shrike, Cinnamon Quail-thrush, Flock Bronzewing, Gibberbird, Grey Falcon).
- **Days 12–14.** Sturt National Park to Sydney, 1180 km.

To visit the south-west of NSW from Sydney, a 12-day itinerary involves:

- **Days 1–3.** Sydney to Deniliquin, 730 km (Plains-wanderer, Superb Parrot, Inland Dotterel (chance), Red-chested Button-quail, feral Ostrich (introduced)).
- **Day 4.** Deniliquin to Leeton, 230 km (Australasian Bittern, Superb Parrot, Glossy Ibis, Baillon's Crake, Australian Spotted and Spotless Crake).
- **Days 5–6.** Leeton to Cocoparra National Park, 55 km (Glossy Black-Cockatoo, Turquoise Parrot, Shy Heathwren, White-browed Treecreeper, Painted Honeyeater).
- **Days 7–10.** Cocoparra National Park to Lake Cargelligo, Round Hill and Nombinnie NR, 200 km (Red-lored and Gilbert's Whistler, Malleefowl, Black Honeyeater, Shy Heathwren, Orange Chat, Chestnut Quail-thrush, Southern Scrub-robin, Black-eared Cuckoo, Spotted Bowerbird, White-winged Fairy-wren).
- **Days 11–12.** Lake Cargelligo to Sydney, 550 km.

The western plains can be very hot in summer, with Sturt National Park often the hottest area in the state: however, the region does not suffer the summer humidity of northern Australia. In winter, temperatures can get very cold, dropping below zero at night. The western plains region is best visited in autumn or spring, when the daytime temperatures are pleasant and the nights are not too cold.

Most of the highways in western plains regions are sealed and, where not, are still usually suitable for 2WD. If you plan to visit the more remote areas of the state, such as Sturt National Park, 4WD is recommended. Most of the parks and reserves discussed in this section have one or more camping areas, with toilets, barbeque, fireplaces and picnic table facilities; some even have showers. Regional airports are scattered over the western sector of NSW with hubs for scheduled services at Wagga Wagga, Lightning Ridge, Broken Hill, Bourke and Tocumwal. With regards to birdwatching organisations, the northern and southern NSW branches of Birdlife Australia provide some coverage for the region: www.birdlife.org.au.

Birding highlights

The western plains are a truly vast area of cultivated lands (especially for wheat and sheep). Travelling west, farms (or sheep stations) become larger and cultivated lands are replaced by rangelands: here stock (mostly sheep) graze and browse native vegetation on expansive properties. Although the rangelands are often degraded, they nonetheless continue to support extensive areas of native vegetation. In the north, many of these rangelands support open or sparse arid woodlands, while in the south these rangelands give way to mallee habitats. Some of the birding highlights on the Western Plains include Emu, Malleefowl, Freckled and Blue-billed Duck, Australian Little and Australasian Bittern, Plains-wanderer, Inland Dotterel, Red-backed Kingfisher, Little and Red-chested Button-quail, Major Mitchell's Cockatoo, Red-tailed Black-Cockatoo (principally along the Darling River), Superb Parrot, Blue Bonnet, Flock Bronzewing, Red-lored and Gilbert's Whistler, Spotted Bowerbird, Ground Cuckoo-shrike, Grey Grasswren, Crested Bellbird, Hall's and Chestnut-crowned Babbler, Chirruping Wedgebill, Cinnamon and Chestnut-breasted Quail-thrush, Orange Chat, Gibberbird, and Painted, Black, Pied, Striped and Grey-fronted Honeyeater. Birds of prey include sought-after species such as Black-breasted Buzzard, Square-tailed Kite, Spotted Harrier, and Black and Grey Falcon.

Notable subspecies found across both regions include Blue Bonnet (red-vented ssp. *haematorrhous* and yellow-vented ssp. *haematogaster*), Crimson Rosella ('Yellow Rosella' ssp. *flaveolus)*, Chestnut-breasted Quail-thrush (eastern ssp. *castaneothorax*), Cinnamon Quail-thrush (ssp. *cinnamomeum*), Grey-fronted Honeyeater (ssp. *graingeri*), Shy Heathwren (eastern ssp. *macrorhyncha*), Grey Grasswren (ssp. *barbatus)*, Variegated Fairy-wren (ssp. *assimilis*) and Tawny Frogmouth (ssp. *brachypterus)*. Vagrants to the regions include Grey-headed Lapwing and White Wagtail. There is also a small feral population of Ostrich near Deniliquin.

Round Hill and Nombinnie Nature Reserves

Key species: Red-lored Whistler, Malleefowl, Black Honeyeater, Shy Heathwren, Orange Chat
Other species: Chestnut Quail-thrush, Southern Scrub-robin, Crested Bellbird, Black-eared Cuckoo, Gilbert's Whistler, Spotted Bowerbird, White-winged Fairy-wren

Round Hill Nature Reserve (13 629 ha) and the adjacent Nombinnie Nature Reserve (70 000 ha) are located 50 km north of Lake Cargelligo. A hotspot for mallee birds, the reserves protect isolated populations of Red-lored Whistler and Malleefowl. Although there are no formal camping facilities in either reserve, there is a bush camping area near the Whoey Tank. There are motels and a caravan park in Lake Cargelligo and hotel accommodation at nearby Mount Hope. It is worth noting that Nature Reserves within NSW (as opposed to National Parks) are often not open to the public, so if you plan to visit, contact the Cobar office of the National Parks and Wildlife office beforehand: (02) 6836 2692. Most of the roads at Round Hill are dirt and, under normal conditions, accessible via 2WD, but after rain 4WD may be required.

One of the most productive areas to locate mallee specialties such as Red-lored Whistler, Chestnut Quail-thrush, Shy Heathwren and Southern Scrub-robin is the mallee woodland bordering a disused, now over-grown, wheat paddock in Nombinnie Nature Reserve. To get there, turn west off the Lake Cargelligo–Mount Hope Rd on to Cactus Track (1.6 km south of the Euabalong–Mount Hope Rd T-section); a Nombinnie Nature Reserve sign marks its entrance. Travel west 1 km to the point where the track turns south; this is the north-west corner of the disused wheat-paddock (regenerating, but still relatively young and open). To find Red-lored Whistler, search along the small walking track that runs west (150 m) from here: listen carefully for their distinctive *'seesaw'* call. Gilbert's Whistler also occurs here, so a cautious approach to

Chat Alley

On route to Round Hill from Lake Cargelligo, stop and look for Orange Chat and White-winged Fairy-wren at a site known colloquially as 'Chat Alley': a shallow roadside surface-drain surrounded by saltbush. It is located on Wallenthery Rd, a linkage road to Lake Cargelligo–Mount Hope Rd, 3.7 km north of Lake Cargelligo–Euabalong Rd. Around Chat Alley, you may also find Emu, Banded Lapwing, Brown Songlark, Ground Cuckoo-shrike, Zebra Finch and open country raptors such as Black Falcon and Spotted Harrier.

Woodland dominated by *Myoporum* and *Allocasuarina* spp. in Round Hill Nature Reserve. This open woodland supports many species, including Black-eared and Pallid Cuckoo, Western Gerygone, Speckled Warbler, Chestnut-rumped Thornbill, Southern Whiteface, Striped Honeyeater, Spiny-cheeked Honeyeater, Grey-crowned Babbler, Apostlebird and Spotted Bowerbird.

identification is required. Other birds found in this area include Crested Bellbird, Black-eared Cuckoo, Red-capped Robin, Spiny-cheeked, Yellow-plumed and Grey-fronted Honeyeater, Inland and Chestnut-rumped Thornbill, Western Gerygone, White-browed and Masked Woodswallow and Splendid Fairy-wren.

Another excellent area for birding is in the northern section of Round Hill Nature Reserve around Whoey Tank, located on Whoey Tank Track, accessed via Euabalong–Mount Hope Rd, 3.7 km east of its intersection with Lake Cargelligo–Mount Hope Rd. The area around the tank is particularly good for open woodland species such as Peaceful, Bar-shouldered and Diamond Dove, Black-eared and Pallid Cuckoo, Spotted Bowerbird, Grey-crowned Babbler, Splendid and Variegated Fairy-wren, Western Gerygone, Speckled Warbler, Chestnut-rumped and Inland Thornbill, Southern Whiteface, Apostlebird, White-backed Swallow, and Black, White-fronted, Striped and Spiny-cheeked Honeyeater. At night, Spotted Nightjar sometimes hawk overhead for insects. Mallee parrots can also be common here: look for Major Mitchell's Cockatoo, Australian (Mallee) Ringneck, Mulga Parrot, Blue Bonnet and, occasionally, Cockatiel and Budgerigar. It is also worth birdwatching west along the Whoey Tank Track, (after 2 km it joins up with the Lake Cargelligo–Mount Hope Rd), where Red-lored Whistler, Shy Heathwren, Chestnut Quail-thrush and Southern Scrub-robin have all been recorded. Another area for Red-lored Whistler is along Lake Cargelligo–Mount Hope Rd, especially the section 5 to 7 km south of the Euabalong–Mount Hope Rd T-section. Malleefowl are sometimes encountered on the road here, especially early in the morning. Yet another area for Red-lored Whistler is around the intersection of the Marooba Fire Trail and the Nombinnie Track. To get there from Mt Hope, drive 12 km south down the Kidman Way. Turn west on to the Marooba Fire Trail for another 12 km to reach the Nombinnie Track heading north: search the mallee around this intersection and north along the Nombinnie Track for a further 3 km for the whistler.

Lake Cargelligo

Key species: Australian Spotted Crake, Baillon's Crake, Spotless Crake, Australian Painted Snipe, Freckled Duck, Blue-billed Duck, shorebirds
Other species: Great Crested Grebe, Buff-banded Rail, Glossy Ibis, Black-tailed Native-hen, Orange Chat

The township of Lake Cargelligo is located on the Lachlan River floodplain in the central west of NSW, 557 km from Sydney. There is a variety of accommodation options in the area, including hotels, a caravan park (Lakeview Caravan Park) on the edge of the lake and the Lake Cargelligo Weir camping area (accessed via Condobolin–Lake Cargelligo Rd, 18.5 km from the township).

A superb wetland, the Lake Cargelligo Treatment Works attracts large numbers of waterbirds. To get there, travel 1.2 km south-west down Condobolin–Lake Cargelligo Rd past the Lake Cargelligo showground, then turn south down Showground Rd for a further 1 km, to reach an access gate to the south-west side of the ponds. Search the edge of the reeds nearest the gate, where there is a chance of Australian Spotted, Baillon's and Spotless Crake and Buff-banded Rail. Other interesting birds recorded here include Australian Painted Snipe, Australasian and Australian Little Bittern, Freckled and Blue-billed Duck, Brown Quail, Australasian Darter, Glossy Ibis, Black-tailed Native-hen, Red-kneed Dotterel, Whiskered Tern, White-winged Fairy-wren, White-breasted Woodswallow and White-fronted and Orange Chat. In summer, there is also a chance of migratory shorebirds such as Wood and Marsh Sandpiper and, occasionally, Red-necked Stint, Sharp-tailed and Curlew Sandpiper. After periods of rain, two small ephemeral wetlands on the north side of Condobolin–Lake Cargelligo Rd (located just west of the Showground Rd turn-off), can provide opportunities for species such as Spotted, Baillon's and Australian Spotted Crake, Australian Painted Snipe (rare) and Red-kneed Dotterel.

Lake Cargelligo itself provides a valuable drought refuge for several wetland species, including Freckled and Musk Duck, Australasian, Hoary-headed and Great Crested Grebe, Australasian Bittern Australasian Darter and Black Winged

Owing to both their rarity and their preference for ephemeral wetlands Australian Painted Snipe are one of the more difficult-to-find shorebirds in Australia. Wetlands in the Murray–Darling Basin are the stronghold for the species, with Fivebough Wetlands, Lake Cargelligo and Macquarie Marshes all providing a chance of sightings if shallow, marshy conditions persist.

Stilt. At Lake Cargelligo Weir and along the Lachlan River, you may encounter Australasian and Great Crested Grebe, Australasian Darter, White-necked Heron, White-bellied Sea-Eagle, Cockatiel, Australian Ringneck, Sacred Kingfisher, Blue-faced and Striped Honeyeater, Yellow-throated Miner, Little Friarbird, Grey-crowned Babbler and Diamond Firetail.

Fivebough Wetlands and Leeton

Key species: Australasian Bittern, Brolga, Superb Parrot, Glossy Ibis, Baillon's Crake, Australian Spotted Crake, Spotless Crake
Other species: Brown Quail, Australian Little Bittern, Magpie Goose

The Ramsar-listed Fivebough Wetlands (400 ha) are 2 km north-east of Leeton on Petersham Rd. The 5 km of walking trails include viewing mounds and two excellent bird hides. The Fivebough and Tuckerbil Wetlands Trust hosts the annual Australian Birdfair: www.australianbirdfair.org.au. Leeton has a range of accommodation options, including hotels, a caravan park on Yanco Ave and camping grounds at the Leeton Showgrounds and Gogeldrie Weir.

Upon arrival at the main wetland carpark, take the All Weather Walking Trail (3 km) to the Bittern bird hide. Along the way, look for Brown Quail feeding in the grassy fringes of the track. This bird hide overlooks a portion of wetland specifically managed for the relatively rare Australasian Bittern: in season, their presence may be revealed by their repeated booming call, heard mainly at night. From the bird hide, you may also see Glossy Ibis, Magpie Goose, Australian Little Bittern (rare), Brolga, Red-kneed and Black-fronted Dotterel, Black-winged Stilt, Red-necked Avocet and Black-tailed Native-hen. Scan the exposed muddy edges for Baillon's, Australian Spotted and Spotless Crake. Summer shorebirds that visit the wetland from time to time include Latham's Snipe, Black-tailed Godwit, Common Greenshank, Wood, Marsh and Sharp-tailed Sandpiper, and Red-necked Stint. On rare occasions, Little Curlew, Pectoral Sandpiper, Long-toed Stint and Oriental Pratincole have been recorded. Also, in the summer months, look for Whiskered and White-winged Black Tern hawking for insects over the ponds. Two other summer migrants, White-throated Needletail and Fork-tailed Swift, occur occasionally. The shire council's sewage ponds are overlooked by a bird hide: a good spot for waterfowl such as Musk, Blue-billed, Freckled (rare) and Pink-eared Duck, Plumed Whistling-Duck and Australasian Shoveler. The Hooey Rd Lookout is located on Fivebough Swamp's eastern side and provides a vantage point to scan the wetland. The best time for viewing is during the final hours of daylight, when large numbers of Glossy Ibis may fly in to roost for the night. Numbers vary from year to year: the highest counts have been 20 000 (Nov 1995) and 15 000 (Nov 2005). It is worth investigating Tuckerbil Wetlands too, (also Ramsar-listed) accessed via Cantrill Rd, 12 km north-west of Leeton. The other major birding attraction in the Leeton district is the Superb Parrot. The species is perhaps best seen around the township itself. During spring and summer, look for them in roadside trees along Irrigation Way, 10.5 km west of the Leeton Information Centre. Crimson (Yellow) Rosella is also common around the roadside near Leeton, while Cockatiel, Ground Cuckoo-shrike and Red-backed Kingfisher all occur from time to time.

Cocoparra National Park and Binya State Forest

Key species: Glossy Black-Cockatoo, Turquoise Parrot, Shy Heathwren, White-browed Treecreeper, Painted Honeyeater
Other species: Crested Bellbird, Black-eared Cuckoo

Cocoparra National Park (8357 ha) and the adjacent Binya State Forest (4170 ha) are located 20 km east of Griffith on the Burley Griffin Way. There is excellent bush camping at Cocoparra at Woolshed Flat.

Cocoparra and Binya hold some of the most westerly populations of Turquoise Parrot. Along the Barry Scenic Drive, Moss Dam can be found on the left and Wattle Dam on the right. These two dams are well known as excellent birding sites, especially for Turquoise Parrot. Other birds that frequent the areas around the dams include Bar-shouldered Dove (western-most population), Australian Ringneck, Mulga Parrot, Rainbow Bee-eater (summer), Variegated Fairy-wren, Spiny-cheeked and Striped Honeyeater, Chestnut-rumped and Yellow Thornbill, Western Gerygone, Varied Sittella and Apostlebird. Turquoise Parrot may also be seen around Falcon Falls. A small population of the Endangered western population of Glossy Black-Cockatoo persists in Cocoparra, where they frequent Drooping Sheoke: look for them along the Mount Brogden Track (4 km), which starts from Binya Forest Drive off the Whitton Stock Route. Shy Heathwren, Black-eared Cuckoo and Crested Bellbird have been recorded along the Store Creek Track (4 km): along here you may also find Speckled Warbler, Striped Honeyeater, Chestnut-rumped, Yellow and Inland Thornbill, Western Gerygone, Masked and White-browed Woodswallow, Splendid Fairy-wren, Double-barred Finch and Diamond Firetail. White-browed Treecreeper and Painted Honeyeater (spring/summer) both feed among the roadside cypress pines near the intersection of Burley Griffith Way and Stock Route Rd, and north along Whitton Stock Route Rd. There are several other sites that offer good birding in the area, including Jack's Creek and Woolshed Flats on the western side of the national park and Spring Hill Picnic Area in the south-eastern corner of the national park. At these sites, the birding prospects include Emu, Major Mitchell's Cockatoo, Superb and Mulga Parrot, Australian Ringneck, Budgerigar, Spotted Harrier, Red-backed Kingfisher, Crimson Chat, White-backed Swallow and Diamond Firetail. At night, Spotted Nightjar can sometimes be heard calling around Woodshed Flats.

Deniliquin

Key species: Plains-wanderer, Superb Parrot, Inland Dotterel, Red-chested Button-quail, Ostrich (introduced)
Other species: Australasian Bittern, Australian Little Bittern, Little Button-quail, Barking Owl

Deniliquin is situated 725 km south-west of Sydney and 285 km north of Melbourne at the southern end of the Hay Plains. The town offers several accommodation options, including hotels and six caravan parks. The Deniliquin Visitor Information Centre is a useful resource: www.denitourism.com.au.

Deniliquin is one of the best places in Australia to observe the cryptic and elusive Plains-wanderer – a quail-like aberrant wader that frequents sparse native grasslands. The species has declined following extensive conversion of native grasslands for cultivation. The only real option to see them in the area is by participating in a tour run by local bird guide Phillip Maher: www.philipmaher.com. Phil also targets some of the district's more elusive species such as Red-chested Button-quail, Inland Dotterel and Australasian and Australian Little Bittern on these same tours. Upper Tocumwal Rd, running off the Riverina Hwy, 12 km west of Deniliquin, can be a reliable place to see Superb Parrot. They are usually present here year round: try in the stands of Yellow Box near Tuppal, 16 km from the highway. Nearer Deniliquin, Superb Parrot have been recorded along Aratula Rd, which heads north of the

Native grasslands in the Deniliquin and Oolambeyan National Park regions of NSW support some sought-after species, including Plains-wanderer, Red-chested Button-Quail, Inland Dotterel and Australian Pratincole, along with more common species such as Little Button-quail, Brown Songlark and Horsfield's Bushlark.

Riverina Hwy 7 km east of the township and they have been seen (and have bred) at the Island Sanctuary (7 ha), located near the town centre, beside Edward River. A good selection of other birds can be seen here, including Little and Long-billed Corella, Crimson (Yellow) Rosella, Purple-crowned Lorikeet, Azure Kingfisher, Intermediate Egret, Nankeen Night-Heron, Varied Sittella, Crested Shrike-tit and, in summer, Sacred Kingfisher, Dollarbird and Olive-backed Oriole, and Masked and White-browed Woodswallow. Waterbirds such as Baillon's Crake, Red-kneed and Black-fronted Dotterel, Black-tailed Native-hen and Banded Lapwing occur in and around the seasonal watercourses along Wirraway Drive, particularly near the cemetery, rice mill and sales yard. Look for Australasian Shoveler, Musk and Pink-eared Duck, as well as Australasian and Great Crested Grebe, in the settlement ponds of the sewerage works at the end of Macauley St.

Recently, a small feral population of Ostrich has been located on Moama–Barham Rd, between 57 and 59 km west of Deniliquin. If you cannot find them along the main road, try searching down Lashbrook Rd, which strikes north 59 km from Deniliquin: look especially in the paddocks on the right, some 500 m along this road. Although the Ostrich have bred at this site, the jury is still out as to whether the population can truly be considered 'self sustaining'.

Oolambeyan National Park

Key species: Plains-wanderer, Red-chested Button-quail
Other species: Superb Parrot, Bush Stone-curlew, Orange Chat, Banded Lapwing, Black-eared Cuckoo, Black Honeyeater

Once a grazing property that ran Merino Sheep, Oolambeyan National Park (21 851 ha) was purchased by the NSW Government in 2001 to ensure the protection of a significant area of native lowland grassland. Other habitats within the park include Black Box woodland, sand hills and lignum swamps. The park is 30 km south of Carrathool and the Sturt Highway, off the Conargo–Carrathool Rd through private property: this road is unsealed and impassable after rain. For more information about access contact the National Parks and Wildlife office in Griffith: (02) 6966 8100. There is group accommodation (for up to 32 people) at the homestead within the park.

The native grassland within Oolambeyan National Park provides vital habitat for the Endangered Plains-wanderer. The species is best observed at night while spotlighting. Red-chested and Little Button-quail, Stubble Quail, Spotted Harrier, Banded Lapwing, Australian Pratincole, Horsfield's Bushlark and Brown Songlark share these grasslands. In shrubby areas at the edge of the grasslands, look for Southern Whiteface and Yellow-rumped Thornbill, while White-backed Swallow is often seen overhead. The open woodland sections of the park and the sand hills dominated by native pine can be good for parrots such as Cockatiel, Budgerigar, Blue Bonnet, Australian Ringneck and Superb Parrot, as well as Painted, Striped and Black Honeyeater. The salt flats and lignum swamps in the park, such as those bordering Bromiumbong Swamp, support White-winged Fairy-wren and Orange Chat. Other birds recorded within the park include Black-faced Woodswallow, Black-eared Cuckoo (summer), Bush Stone-curlew, White-winged Triller, Rufous Songlark, Grey-crowned Babbler, Apostlebird, Rainbow Bee-eater (summer), Rufous Whistler, Crimson Chat, Chestnut-rumped Thornbill and Western Gerygone.

Red-chested Button-Quail are perhaps the most elusive of the 'small' Australian Button-quail. The species prefers more dense grasslands, especially those containing spear grass (*Stipa* spp.) or Mitchell Grass (*Astrebla* spp.). Within suitable habitat, they also appear to frequent the more lush (or indeed wetter) sites.

Kinchega (Menindee Lakes) National Park

Key species: Major Mitchell's Cockatoo, Blue Bonnet, Chirruping Wedgebill, Chestnut-crowned Babbler, waterbirds
Other species: Orange Chat, Bourke's Parrot, Red-backed Kingfisher, Striped Honeyeater, Red-tailed Black-Cockatoo

Kinchega National Park (62 000 ha) is located 111 km south-east of Broken Hill along the (sealed) road to Menindee. The unsealed roads in the park are closed when wet. The park largely encompasses Lake Menindee and Lake Cawndilla, while

Murray Valley National Park

Created in 2010, Murray Valley National Park (41 601 ha) combines 16 former state forests. There are several camping areas in the park, including several on the banks of the Murray and Edwards Rivers. The roads in the park are at times a confusing network of interconnecting tracks and shortcuts, so it is best to obtain a good map of the area in advance: try the Mathoura Tourist Information Centre. In summer, Superb Parrot, Gilbert's Whistler and Diamond Firetail occur in the Gulpa Island section of the park, especially those areas dominated by Bulloak. The woodlands here also support Emu, White-browed Babbler, Noisy Friarbird, Western Gerygone, Brown Treecreeper, Hooded Robin, Rainbow Bee-eater (summer), White-breasted, Masked and White-browed Woodswallow and Southern Whiteface. Anywhere along the Murray River, you may find Azure and Sacred Kingfisher, Great and Intermediate Egret, Australasian Darter, Swamp Harrier, Whistling Kite, Nankeen Night-Heron and Crimson (Yellow) Rosella.

the eastern edge of the reserve is formed by the Darling River. There is camping at the Darling River, Emu Lake and Lake Cawndilla campgrounds.

Along the Darling River, where it is bordered by massive River Red Gum, you will find Australian Ringneck, Crimson (Yellow) Rosella, Red-tailed Black-Cockatoo, Little Friarbird, Striped and White-plumed Honeyeater, White-breasted Woodswallow and large numbers of Whistling and Black Kite. Waterbirds include Australasian Darter, Great, Little Black, Pied and Little Pied Cormorant, Australian Pelican, White-necked Heron, and Great and Little Egret. Lakes Menindee and Cawndilla were at capacity in 2012, with water received from the 2011 Queensland floods. This was the first time in almost a decade that these lakes held water. In such conditions, the lakes attract large numbers of waterbirds, including Pink-eared, Freckled and Blue-billed Duck, Australasian Shoveler, Great Crested Grebe, Australasian Darter, Nankeen Night-Heron, Magpie Goose, Brolga and Black-tailed Native-hen. Shorebirds that frequent the lakes when conditions are right (low water levels with exposed mud banks) include Red-kneed and Black-fronted Dotterel and, occasionally, Sharp-tailed Sandpiper (summer). Above the shoreline of the Menindee Lakes, you may find Emu, Major Mitchell's Cockatoo, Little Corella, Cockatiel, Diamond and Peaceful Dove, White-fronted and Orange Chat, Spiny-cheeked Honeyeater and Apostlebird. Blackbush and Bluebush vegetation dominate the plains that surround the Menindee Lakes. Here, you may find Blue Bonnet, Chirruping Wedgebill, Chestnut-crowned Babbler, Black-faced Woodswallow, White-winged Fairy-wren, Singing Honeyeater, Southern Whiteface and Zebra Finch. In areas of mallee woodland, often associated with red sand dunes, look for Mulga Parrot and, more rarely, Bourke's Parrot,

Red-backed Kingfisher, Variegated Fairy-wren, Crimson Chat, Chestnut-rumped Thornbill, White-browed and Masked Woodswallow, Red-capped Robin and Rufous Songlark.

Mutawintji National Park

Key species: Grey Falcon, Chirruping Wedgebill, Hall's and Chestnut-crowned Babbler, Little Woodswallow
Other species: Bourke's Parrot, Major Mitchell's Cockatoo, White-backed Swallow

Sometimes described as a mini Flinders Ranges (South Australia), Mutawintji National Park (68 912 ha) is characterised by a series of rocky red gorges along River Red Gum-lined creeks, punctuated by rock pools and waterholes. It is located 130 km north-east of Broken Hill, and is accessed via the Broken Hill–White Cliffs Rd. There is an excellent campground at Homestead Creek. This is a remote area that can be very hot in summer, and the usual precautions for outback travel should be taken.

Upon entering the park, the road crosses saltbush and Mulga plains, where the roadside vegetation supports Emu, Chirruping Wedgebill, Chestnut-crowned Babbler, White-backed Swallow, White-browed, Masked and Black-faced Woodswallow, Southern Whiteface, White-winged Fairy-wren, Chestnut-rumped Thornbill and White-fronted Honeyeater. Yellow-throated Miner, White-winged Chough and Apostlebird frequent the Homestead Creek campground. Here, parrots occur in good numbers: look for Major Mitchell's Cockatoo, Little Corella, Galah, Cockatiel, Australian Ringneck, Blue Bonnet, Mulga Parrot and, more rarely, Bourke's Parrot. Along the River Red Gum-lined creeks, you may find Red-backed Kingfisher, White-breasted Woodswallow and White-plumed, Striped, Spiny-cheeked and, occasionally,

An iconic Australian species. Although rare in many coastal regions of NSW, Emu are common, sometimes abundant, in the rangelands of inland NSW.

Brown Honeyeater. Search the gorges for Little Woodswallow and Peregrine Falcon. Mutawintji also contains the most southern population of Hall's Babbler, with several family groups found in the northern section of the park. The gorges in the northern section of the park are also home to the only known NSW colony of the Endangered Yellow-footed Rock Wallaby. Birds of prey to look for include Little and Wedge-tailed Eagle, Spotted Harrier and the rare Grey Falcon.

Sturt National Park

Key species: Grey Falcon, Black-breasted Buzzard, Grey Grasswren, Ground Cuckoo-shrike, Cinnamon Quail-thrush, Flock Bronzewing, Gibberbird
Other species: White-backed Swallow, Orange Chat, Chirruping Wedgebill, Chestnut-crowned Babbler, Red-browed Pardalote, Black Honeyeater, Pied Honeyeater

Sturt National Park (340 000 ha) is located in the arid far north-west corner of NSW, 1050 km north-west of Sydney. The park protects a vast arid landscape of Mulga bushland, intermixed with gibber plain and ephemeral creeks lined with River Red Gum and Coolabah. There are several campgrounds in the park, including Dead Horse Gully near Tibooburra, Fort Grey in the far west of the park, Mount Wood near the Mount Wood Homestead and Olive Downs, set among Mulga trees on top of the 'Jump Up' Range. There is also accommodation at Mount Wood Homestead. Most roads in the park are gravel, with some sandy stretches. These can be driven with 2WD, although after heavy rain roads may even be closed to 4WD. This is a remote area, that can be very hot in summer and the usual precautions for outback travel should be taken.

The Gorge Loop Rd (53 km) passes through the eastern section of the park. Starting in the south, look around the Mount Wood campground and along Thomson Creek for Emu, Red-backed Kingfisher, Red-browed Pardalote, Crimson Chat, Black-faced Woodswallow and Apostlebird. There is also a possibility of rarer species such as Bourke's Parrot, Painted Finch and Grey Falcon. Beyond the campground, the road passes through a section of gibber: look here for Cinnamon Quail-thrush, Gibberbird, Orange Chat, Chirruping Wedgebill and Chestnut-crowned Babbler. Further on, the landscape along the road changes, traversing rolling hills and the wooded drainage line of Twelve Mile Creek, where is it worth looking for Red-backed Kingfisher, Blue Bonnet, Black-faced Woodswallow and Black-eared Cuckoo. The Gorge Loop Rd then connects with the Jump Up Loop (44 km). After 2 km, you come to the South Myers Tank, which contains water pumped from a bore. From the bird hide on the north-western corner, look for waterbirds such as Pink-eared and Freckled (rare) Duck, Plumed Whistling-Duck, Hardhead, Australian Spotted Crake, Black-tailed Native-Hen, Black-fronted and Red-kneed Dotterel, Australian Pratincole and Whiskered Tern (summer). During the course of the day, large numbers of Galah, Little Corella, Cockatiel, Zebra Finch and, in some years, Flock Bronzewing come to drink. Around the Olive Downs campground, you may see Chestnut-crowned Babbler, Black and Pied Honeyeater (seasonal influxes depending on local conditions), Chestnut-rumped Thornbill, Southern Whiteface and, at night, Spotted Nightjar. Further along the Jump Up Loop, look for Australian Bustard, Inland Dotterel

Leopardwood *Flindersia maculosa* and chenopod shrubland near Sturt National Park provide suitable habitat for Rainbow Bee-eater, White-winged Fairy-wren, Singing and Spiny-cheeked Honeyeater, Chirruping Wedgebill, Black-faced Woodswallow and White-backed Swallow.

(uncommon and probably best found by spotlighting at night), Spotted Harrier, Cinnamon Quail-thrush and Orange Chat. A morning walk around the Dead Horse Gully campground may produce Diamond and Peaceful Dove, Cockatiel, Blue Bonnet, Budgerigar, Red-backed Kingfisher, Rainbow Bee-eater, Chirruping Wedgebill, White-backed Swallow, White-winged Fairy-wren, Singing and Spiny-cheeked Honeyeater, and Red-browed Pardalote.

The western section of the park is accessed via Fort Grey Rd (the road to Cameron Corner). Along the road, look for Australian Pratincole, Black-faced Woodswallow, White-backed Swallow and Crimson Chat. Sturt National Park provides more sightings of Grey Falcon in NSW than just about anywhere in the country, so always remain alert to the possibility of encountering one of Australia's rarest raptors. At Fort Grey campground, on the edge of the ephemeral Lake Pinaroo, you may see Australian Ringneck, Bourke's Parrot, Budgerigar, Black-tailed Native-hen, Banded Lapwing, Ground Cuckoo-shrike, Red-browed Pardalote, Australian Owlet-Nightjar, Black-eared Cuckoo, White-breasted and Black-faced Woodswallow, Yellow-throated Miner, Orange Chat, Crested Bellbird and Chestnut-rumped Thornbill. Keep an eye open for rarer raptors such as Grey Falcon, Black-breasted Buzzard and Spotted Harrier. When full, Lake Pinaroo (a wetland of international importance) supports large numbers of waterbirds and shorebirds, including Brolga and Freckled and Blue-billed Duck.

Finally, Grey Grasswren occur in the larger patches of lignum on the floodplains of the Bulloo Overflow near the NSW–Qld border. One place to see them is on Narriearra Station. It is not within Sturt National Park, so for access contact the station owners to let them know you are coming and to discuss where to find Grey Grasswren: (08) 8091 3547.

Victoria

Geographically the smallest mainland state in Australia, Victoria is also the second most populous. To the north, Victoria is bounded by the Murray River and NSW, while South Australia adjoins the western boundary and Bass Strait and Tasmania lie to the south. Melbourne is the state's largest city and the state capital. Victoria rests at the southern end the Great Dividing Range, which stretches from far northern Queensland, along the east coast of Australia and terminates to the west of Ballarat in Victoria. The state's climate ranges from the wet, cool-temperate environments of Gippsland in the south-east, through the Victorian alpine areas that rise to almost 2000 m (snow covered in winter), to the treed deserts of the mallee country in the far north-west of the state. Melbourne Airport is the major domestic and international gateway to the state, while Avalon Airport, situated closer to Geelong, is the second busiest airport in the state. Victoria differs from other mainland states because it is situated further south and has major mountain ranges that have an east–west orientation. Its southerly position, and the influence of the adjacent ocean, has a moderating effect on climate, especially on, and to the south of, the Great Dividing Range. In winter, the weather here is typically cool to cold and wet, with snow mostly restricted to areas above 1200 m. In contrast, summer is warm to hot and dry. To the west and north of the Great Divide, the land flattens out to the dry inland plains. It is in the Mallee in the north-west of the state where the highest temperatures occur during summer, and where the annual rainfall drops below 300 mm. The endangered 'Helmeted Honeyeater', race *cassidix* of the Yellow-tufted Honeyeater is the Victorian bird emblem.

Northern Victoria

Regional overview

Northern Victoria supports a wide variety of habitats, including semi-arid mallee woodlands and dune systems, box–ironbark woodlands, native grasslands, impressive lake systems and the Murray River. Consequently, the region has an excellent diversity of bird species, including a suite of habitat specialists such the mallee specialists, a diversity of arid land birds at the southern edge of their range and a wide selection of species that occupy woodlands and grasslands in a Mediterranean climate. The region encompasses Hattah–Kulkyne, Murray-Sunset and Wyperfeld National Parks in the north-west, the Little Desert in the south-west, Terrick Terrick and Greater Bendigo National Parks in central Victoria and Chiltern–Mount Pilot National Park and the Warby Ranges in the north-east.

A suggested 10-day itinerary for the mallee regions of northern Victoria involves:

- **Days 1–2.** Melbourne to Little Desert National Park, 340 km (Malleefowl, Blue-winged Parrot, Elegant Parrot,

The coastal view looking west from Point Addis. Not only is this one of the best seabird watching points near Melbourne (~1h 20 min drive), the heathland around the carpark and lookout supports Rufous Bristlebird.

Slender-billed Thornbill, Rufous Fieldwren, Purple-gaped Honeyeater, Shy Heathwren).

- **Days 3–4.** Little Desert National Park to Wyperfeld National Park, 100 km (Emu, Malleefowl, Major Mitchell's Cockatoo, Regent Parrot, Redthroat, Southern Scrub-robin, Shy Heathwren, Gilbert's Whistler, Splendid Fairy-wren).
- **Days 6–7.** Wyperfeld National Park to Hattah–Kulkyne National Park, 150 km (Mallee Emu-wren, Striated Grasswren, Chestnut Quail-thrush, Malleefowl, Major Mitchell's Cockatoo, Regent Parrot).
- **Days 8–9.** Hattah–Kulkyne National Park to Murray-Sunset National Park, 50 km (Malleefowl, Red-lored Whistler, Striated Grasswren, Mallee Emu-wren, Black-eared Miner).
- **Day 10.** Murray-Sunset National Park to Melbourne via Lake Tyrell, 450 km (Rufous Fieldwren, Orange Chat, White-winged Fairy-wren).

With 5 days birding in the woodlands of north-east Victoria, a suggested itinerary involves:

- **Days 1–2.** Melbourne to Chiltern–Mount Pilot National Park, 290 km (Regent Honeyeater, Painted Honeyeater (spring–summer) and Black-chinned Honeyeater, Turquoise Parrot, Swift Parrot (winter), Diamond Firetail and other woodland birds).
- **Days 2–3.** Chiltern–Mount Pilot National Park to Warby Ranges National Park, 50 km (Swift Parrot (winter), Regent Honeyeater, Painted Honeyeater (spring–summer), Turquoise Parrot, Speckled Warbler).
- **Day 4.** Warby Ranges National Park to Heathcote–Graytown National

Park via Lake Mokoan, 150 km (Swift Parrot (winter), Painted Button-quail, Chestnut-rumped Heathwren, Speckled Warbler, Spotted Quail-thrush, Scarlet Robin, waterbirds).

- **Day 5.** Heathcote–Graytown National Park to Melbourne, 120 km.

With 4 days in northern central Victoria, a suggested itinerary involves:

- **Day 1.** Melbourne to Greater Bendigo National Park, 150 km (Spotted Nightjar, Shy and Chestnut-rumped Heathwren, Crested Bellbird, Purple-gaped Honeyeater).
- **Day 2.** Greater Bendigo National Park to Terrick Terrick National Park, 80 km (Plains-wanderer, Little Button-quail, Black Falcon, Australian Ringneck).
- **Day 3.** Terrick Terrick National Park to the Kerang Lakes and Goschen Bushland Reserve, 130 km (Australian Painted Snipe, Freckled Duck, Banded Stilt, Cockatiel, Budgerigar, Black and Pied Honeyeater, Crimson Chat, Little

Button-quail, Orange Chat, White-winged Fairy-wren).

- **Day 4.** Goschen Bushland Reserve to Melbourne, 350 km.

Northern Victoria extends from the inland slopes and foothills of the Great Dividing Range north and west across flat plains and semi-arid desert country. Almost the entire region falls within the vast catchment of the Murray–Darling Basin. The habitats to be found here range from temperate woodlands in the foothills of the Great Dividing Range to mallee woodland systems with low rainfall in the north-west. The region is characterised by a temperate climate with four distinct seasons. Winters are generally cool to mild, while summers can be very warm to hot. In summer, the mean daily maximum temperature is around 30°C with low humidity; however, this is subject to substantial daily variation and, on average, there are 17 days per year that exceed 35°C. Mean maximum winter temperatures are around 14°C. Frosts are commonplace in autumn and winter, with ~20 days per year featuring overnight minimums below freezing. Rain can occur all year round, but Autumn is typically the driest season and most rain falls in the winter months; July being the wettest month and March the driest. The Mallee and upper Wimmera are Victoria's warmest regions because they are most influenced by northerly winds blowing off the arid interior. The area is also one of the driest in the state with the annual median rainfall dropping below 300 mm. Victoria's highest maximum temperature of 48.8°C was recorded in Hopetoun on 7 February 2009 and temperatures above 40°C can occur at any time through the summer months.

The Victorian road network is extensive and well maintained, with several of the major roads linking cities being of freeway standards. Most other interconnecting roads are also sealed or, at least, of reasonable graded quality. V/Line operates bus and rail services to major regional centres (www.vline.com.au). Most of the larger national parks in the state have some 4WD tracks, but, owing to their reduced access, only the best birdwatching sites that require 4WD are covered here (e.g. Murray-Sunset National Park and several of the tracks in Hattah–Kulkyne, Wyperfeld and Little Desert National Parks). For information on accommodation and camping options visit www.parkweb.vic.gov.au and www.visitvictoria.com.

Birding highlights

Northern Victoria is especially well known for its spectacular parrots, including Major Mitchell's and Sulphur-crested Cockatoos, Galah, Little and Long-billed Corellas, Cockatiel, Budgerigar, Eastern and Crimson (Yellow) Rosellas, Blue Bonnet and Blue-winged, Elegant (rare), Turquoise, Mulga, Regent, Superb and Swift Parrot. Northern Victoria is also home to great diversity of interesting honeyeaters, including Regent, Painted, Yellow-plumed, Spiny-cheeked, Purple-gaped, Black-chinned, Yellow-tufted and Fuscous Honeyeater and Yellow-throated and Black-eared Miner. Species that display a distinct preference for mallee habitats include Malleefowl, Mallee Emu-wren, Striated Grasswren, Red-lored Whistler, Chestnut Quail-thrush, Shy Heathwren, Southern Scrub-robin and Splendid Fairy-wren. The heathland and salt-plain specialists of northern Victoria include the Rufous Fieldwren, Slender-billed Thornbill, Orange Chat and White-winged Fairy-wren. Less common raptors that are still typical of the region include Spotted Harrier, Wedge-tailed and Little Eagle, Peregrine and Black Falcon, while night birds include Barking and Eastern Barn Owl, Spotted and White-throated Nightjar, and Australian Owlet-Nightjar. Wetland species include Australian Painted

Snipe, Brolga, Glossy Ibis, White-necked Heron, Intermediate Egret, Black-tailed Native-hen, and Baillon's, Spotless and Australian Spotted Crake. The native grasslands of northern Victoria are some of the best examples of this threatened and declining habitat in the state. These grasslands support habitat specialists such as Australian Bustard (rare), Plains-wanderer, Inland Dotterel, Australian Pratincole, Banded Lapwing, Little Button-quail, Brown Songlark and Horsfield's Bushlark, although some of these species such as the Plains-wanderer and Inland Dotterel have very specific habitat needs and targeted searches at key locations when conditions are just right are necessary to ensure success (see below).

Other species typical of northern Victorian environments include Painted Button-quail, Rainbow Bee-eater, Grey-crowned, White-browed and Chestnut-crowned Babbler, White-bellied Cuckoo-shrike, Crimson Chat, Redthroat (in Victoria restricted to Wyperfeld NP), Crested Bellbird, Gilbert's Whistler, Pallid and Black-eared Cuckoo, Hooded and Red-capped Robin, Rufous Songlark, White-winged Triller, Speckled Warbler, Chestnut-rumped Heathwren, White-breasted, White-browed, Masked and Black-faced Woodswallow, Slender-billed, Inland and Chestnut-rumped Thornbill, Splendid and Variegated Fairy-wren, Western and White-throated Gerygone, Diamond Firetail and Zebra Finch. Some of the distinctive subspecies found in northern Victoria include Major Mitchell's Cockatoo (scarlet and yellow crested ssp. *leadbeateri*), Regent Parrot (eastern ssp. *monarchoides*), Crimson Rosella (Yellow ssp. *flaveolus*), Australian Ringneck (Mallee ssp. *barnardi*), Blue Bonnet (yellow-vented ssp. *haematogaster*), Rufous Fieldwren (ssp. *winiam*), Shy Heathwren (ssp. *cauta*), Spotted Pardalote (yellow rumped ssp. *xanthopygus*), Striated Pardalote (both ssp. *substriatus* and ssp. *striatus*), Slender-billed Thornbill (ssp. *hedleyi*), Chestnut Quail-thrush (ssp. *castanotus*), Varied Sittella (black-capped ssp. *pileata*) and Grey Currawong (black-winged ssp. *melanoptera*).

Spring–summer migrants include Sacred Kingfisher, Rufous Songlark, White-winged Triller, Rainbow Bee-eater, Olive-backed Oriole and Dollarbird. Migratory shorebirds can appear at any suitable wetland. Most species arrive from more northerly areas between August and November and depart between March and May. One winter visitor is the Swift Parrot, which migrates from Tasmania to arrive in coastal Victoria in late autumn before making its way to the box–ironbark forests of central and northern Victoria (and beyond as far north as southern Queensland).

Hattah–Kulkyne National Park

Key species: Mallee Emu-wren, Striated Grasswren, Malleefowl, Major Mitchell's Cockatoo, Regent Parrot
Other species: Freckled Duck, Apostlebird, Splendid Fairy-wren, Crested Bellbird, Chestnut Quail-thrush, Shy Heathwren, Masked Woodswallow

Hattah–Kulkyne National Park (48 000 ha) protects an extensive area of mallee vegetation in close proximity to semi-arid shrublands, native pine woodlands, Black Box forest and River Red Gum floodplains that incorporates an interconnected system of shallow lakes and ephemeral swamps. Some delightful camping opportunities in close proximity to the lakes (and along the nearby Murray River), a diversity of habitats and an abundance of wildlife (including many kangaroos and emus) ensures this is one of the most popular national parks in north-western Victoria. The park is located on the eastern side of the Calder Hwy and is best accessed via the small township of Hattah 35 km north of Ouyen and 65 km

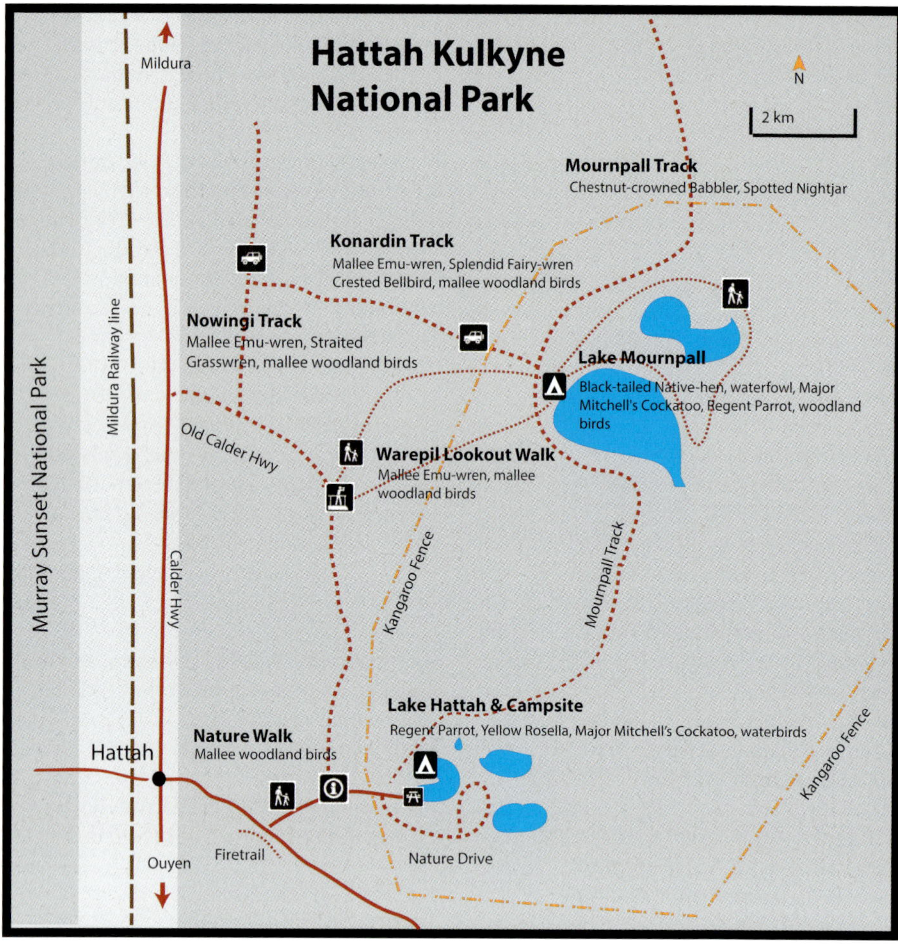

south of Mildura. The main entrance to the park is off the Hattah–Robinvale Rd, 4 km east of Hattah. The park supports a great variety of birdlife, including sought-after species such as Major Mitchell's Cockatoo, Regent Parrot, Malleefowl, Crested Bellbird and Apostlebird, and is one of the best places to see the Endangered Mallee Emu-wren and the elusive Striated Grasswren. There are two excellent campgrounds (Lake Hattah and Lake Mournpall) within the park and basic bush camping is available along the Murray River. There is a variety of accommodation options in both Mildura to the north and Ouyen to the south.

The best sections of mallee vegetation are in the area bordered by the Old Calder Hwy, Nowingi Track, Konardin Track and Mournpall Track. Here, with patience, Striated Grasswren and Mallee Emu-wren can be located. One of the better locations to search is towards the southern end of the Nowingi Track in areas that support large clumps of spinifex on either side of the track. Another good site to look for

A male Mallee Emu-wren perched atop a large clump of spinifex. This diminutive species can be difficult to locate because they often avoid observers by seeking shelter and remaining motionless within these large prickly grass tussocks.

these two spinifex inhabitants is along the Konardin Track ~500 m west from the kangaroo fence. A third area to try for the Mallee Emu-wren is in patches of spinifex along the Warepil Lookout Walk (0.6 km return) that sets off from the Lake Mournpall campground. The emu-wren has a reputation for being especially difficult to find; listen for the thin, very high pitched and almost insect-like contact call when conditions are still; at times the species also actively avoids observers by quietly retreating to spinifex clumps until the 'threat' has passed; do not ignore calls from areas that you have just passed through! Splendid Fairy-wren, which utters somewhat similar high-pitched calls, also occurs in the area but typically prefers dense shrub cover. Other species found in the mallee within the national park include Malleefowl (sometimes seen crossing vehicle tracks, particularly early in the morning), Crested Bellbird, Chestnut Quail-thrush, Shy Heathwren, Gilbert's Whistler, Inland and Chestnut-rumped Thornbill, and White-browed, Masked and Dusky Woodswallow. Nectar feeders are also well represented, with good opportunities to observe Yellow-plumed, Spiny-cheeked, White-fronted and Striped Honeyeater. At night, Southern Boobook, Tawny Frogmouth and Australian Owlet-Nightjar frequent both the Mournpool and Hattah Lakes campgrounds, while Spotted Nightjar (summer) hawks over mallee and adjacent areas, including the Mournpall Track north of the campground. An enjoyable walk that passes through a variety of habitats, including Black Box and River Red Gum woodlands, is the Mournpall Lake Walk (9.5 km loop). This track skirts the northern edge of Lake Mournpall before looping back through mallee dunes. Ground Cuckoo-shrike, a species that is very rarely recorded in Victoria, has been observed along this track. More common species include Hooded and Red-capped Robin, Jacky Winter, Southern Whiteface, Brown Treecreeper, Black-eared Cuckoo, Splendid and Variegated Fairy-wren, Blue-faced Honeyeater, Pied Butcherbird, and White-browed and Chestnut-crowned Babbler (the last preferring open areas with tall shrubs and native pine). Hattah–Kulkyne National Park is widely recognised as an excellent site to see the many wonderful inland parrots. These include Regent, Mulga, Blue-winged and Red-rumped Parrots, Blue Bonnet, Australian Ringneck, Crimson (Yellow ssp. *flaveolus*) Rosella, Cockatiel, Major Mitchell's and Sulphur-crested Cockatoos, Little Corella, Galah and, in 2012, there was a potential nighttime sighting of Night Parrot at Lake Mournpall. One the best places to see Regent Parrot and Crimson (Yellow) Rosella is at the Hattah campground itself; if time is short, check the mature Red Gums along the lake shore adjacent to the day use area.

The campground is also well known for hand-tame families of Apostlebird, Noisy Miner, Grey and Pied Butcherbird and White-winged Chough. In the grassland nearby, Emu and Western Grey Kangaroo can often be seen grazing. Red Kangaroo are common further north in the park and

Victoria

are observed during an early morning or late afternoon drive along the Mournpool Track to the north of the kangaroo fence. On the Hattah Lakes (recognised as internationally significant under the Ramsar Convention), there can be a wide variety of waterbirds. Look for Australasian Darter, Australasian, Hoary-headed and Great Crested Grebe, Pink-eared, Musk and Blue-billed Duck, Grey and Chestnut Teal, Eurasian Coot and Black-tailed Native-hen. Rarer species that have been recorded here include Freckled Duck, Australian Painted Snipe and Australasian Bittern. The lakes system also provides excellent opportunities for birds of prey: 18 species have been recorded, including Peregrine and Brown Falcon, Australian Hobby, Nankeen Kestrel, Brown Goshawk, Collared Sparrowhawk, Black and Whistling Kite, Wedge-tailed Eagle and White-bellied Sea-Eagle.

Murray-Sunset National Park

Key species: Malleefowl, Major Mitchell's Cockatoo, Red-lored Whistler, Striated Grasswren, Mallee Emu-wren, Black-eared Miner, Little Crow
Other species: Budgerigar (rainfall dependent), Regent Parrot, Elegant Parrot (chance), Scarlet-chested Parrot (rare visitor)

Dominated by mallee woodlands, Murray-Sunset National Park (633 000 ha) is one of the largest national parks in Victoria. In the east, it abuts Hattah–Kulkyne National Park at the Calder Hwy and extends west as a huge rectangle to the South Australian

A mature River Red Gum *Eucalyptus camaldulensis* on the shores of Lake Hattah. This particular location, adjacent to the day use area at Lake Hattah, can be an excellent site for parrots and cockatoos; look for Galah, Sulphur-crested Cockatoo, Little Corella, Major Mitchell's Cockatoo, Regent Parrot, Mallee Ringneck, Yellow Rosella and Red-rumped Parrot.

border. In the north-west, the park extends through native pine, semi-arid shrublands and some areas of abandoned grazing land to connect with the Black Box and River Red Gum floodplains of the Murray River. Not surprisingly, given the vast area involved and the diversity of habitats, more than 200 species have been recorded in the park. This list includes many of mallee specialists such as Malleefowl, Mallee Emu-wren, Striated Grasswren, Black-eared Miner and Red-lored Whistler, as well as a host of other species that are scarce or rarely encountered in Victoria. There are campsites at Pink Lakes, Mopoke Hut, Mount Crozier, Rocket Lake and the Shearers Quarters. All of the tracks within the park are best suited to 4WD, although conventional vehicles can usually safely access the Pink Lakes Rd from Linga in the south and Settlement Track to the Shearers Quarters in the north.

Red-lored Whistler is a mallee woodland specialist restricted to suitable habitats in north western Victoria and adjacent areas in SA and NSW (an isolated population also occurs at Round Hill and Nombinie Nature Reserves in NSW). In the Murray-Sunset NP (and at Gluepot, SA) the species displays a preference for mallee woodlands with a dense cover of spinifex and some low to medium shrubs.

Large areas of the Murray-Sunset offer excellent opportunities for many sought-after mallee specialists. One very productive strategy to locate species such as the Striated Grasswren, Mallee Emu-wren and Red-lored Whistler is to drive along tracks through suitable habitat, stopping every 500 m or so before conducting an area search on foot for 5–10 min before moving on to the next location. As a guide, look for areas of long unburnt mallee (the presence of small tree hollows is a good indicator) that support a good cover of spinifex. In the central Murray-Sunset, an excellent starting point for birdwatching is along Pheenys Track immediately east of its intersection with Underbool Track. This area is also one of the last refuges for the endangered Black-eared Miner in Victoria (though Gluepot Reserve, SA is a much more reliable location – see p. 372). Grey-fronted Honeyeater (a Victorian rarity) has been recorded near South Bore Track, but note that the similar Yellow-plumed Honeyeater is typically abundant in the park. Little Crow, another scarce bird in Victoria, occurs at low densities throughout the park. Driving slowly along tracks increases the likelihood of encountering Malleefowl, because sightings usually involve individuals crossing tracks. Between spring and autumn, large numbers of Budgerigar and Masked and White-browed Woodswallow may descend on the park to breed. In good years, they form noisy feeding flocks, with the woodswallows hawking for insects and feeding on flowering mallee, while the budgies concentrate on seeding spinifex. Any such aggregations are well worth a closer inspection because numerous honeyeaters, White-winged Triller and even the smaller grass parrots such as Blue-winged, Elegant and, very rarely, Scarlet-chested Parrots may join these flocks.

Pink Lakes have some great camping sites, which provide a good opportunity to see cockatoos and parrots such as Major Mitchell's Cockatoo, Cockatiel, Australian Ringneck, Mulga and Regent Parrot, and

Blue Bonnet. Other species here include Emu, Red-capped Robin, White-browed Babbler and Pied Butcherbird. Several short walks, Kline Nature Walk (4.8 km) and a circuit taking in Lakes Becking or Hardy (2 km), provide excellent opportunities for birdwatching. A good location for Striated Grasswren and Mallee Emu-wren is in the larger areas of spinifex along Pioneer Drive just north of the intersection with the Salt Bush Flat Track. Another good place to see the mallee parrots is at Wymlet Tank at the eastern end of the Honeymoon Hut Track. Here, you may find Australian Ringneck, Mulga Parrot, Blue Bonnet and Major Mitchell's Cockatoo as they come in to drink in the afternoon. Red-lored Whistler and Black-eared Miner occur in the mallee on the Honeymoon Hut Track, ~5 km west of Wymlet Tank. To reach Wymlet Tank, travel west via Trinita Rd (just south of the Hattah township), or via the Meridian Track from Walpeup.

Neds Corner (30 000 ha) is a Trust for Nature property that borders Lindsay Island and the northern section of Murray-Sunset National Park. An interesting assortment of birds can be seen in the low-lying saltbush plains that radiate out from the Black Box woodlands closer to the Murray River. This area is the only reliable site in Victoria for Inland Dotterel, which are occasionally seen at night standing on the gravel road into Neds Corner Station. Other birds here include Australian Pratincole, Banded Lapwing, Blue Bonnet, White-backed Swallow, White-browed Treecreeper, Chestnut-crowned Babbler, White-winged Fairy-wren and Orange Chat. Along the Murray River, visitors may encounter Caspian Tern, Great Egret, Australasian Darter and Red-kneed and Black-fronted Dotterel, while the River Red Gum and Black Box woodlands on the banks of the Murray River provide habitat for Blue-faced, Striped, Spiny-cheeked and White-plumed Honeyeater, Brown Treecreeper and the Yellow Rosella (ssp. *flaveolus* of the Crimson Rosella).

Small Mallee Reserves

Key species: White-browed Treecreeper, Malleefowl, Chestnut Quail-thrush, Black Honeyeater

Other species: Gilbert's Whistler, Splendid Fairy-wren, Southern Scrub-robin, Crested Bellbird, Purple-gaped Honeyeater, Tawny-crowed Honeyeater, White-fronted Honeyeater

In north-west Victoria, there is a good selection of relatively small mallee reserves, each supporting its own interesting selection of bird species. These reserves include Yarrara, Timberoo and Bronzewing Flora and Fauna Reserves and Gama Bushland Reserve.

Yarrara Flora and Fauna Reserve (110 ha), located 70 km west of Red Cliffs on the Red Cliffs–Meringur Rd (also known as the Millewa Rd), is an important refuge within Victoria for White-browed Treecreeper. They occur in sections of the reserve that support mature stands of *Casuarina*, with a good spot to look being along Yarrara South Rd, south of where it intersects (near the wheat silo) with Red Cliffs–Meringur Rd. Other birds that may be seen in this reserve include Little Button-quail, Cockatiel, Australian Ringneck, Mulga Parrot, Blue Bonnet, Australian Owlet-Nightjar, Gilbert's Whistler, Red-capped and Hooded Robin, Striped Honeyeater, Chestnut-rumped and Inland Thornbill, Splendid and Variegated Fairy-wren and, in spring and summer, Rainbow Bee-eater and Black-eared Cuckoo.

Timberoo Flora and Fauna Reserve (100 ha) is another important site for White-browed Treecreeper within Victoria. To get there from Walpeup (on the Mallee Hwy 40 km west of Ouyen), head south-east along the Hopetoun–Walpeup Rd for

13.3 km before reaching the Walpeup Lake Rd. Turn north here and travel 2.7 km to the Walpeup Lake camping area. One of the more reliable areas for the White-browed Treecreeper is in the Pine–Buloke woodlands on the north side of the lake. A variety of other species, including many of those listed for Yarrara FFR also occur.

Bronzewing Flora and Fauna Reserve (12 000 ha), just south of Timberoo, supports important populations of Malleefowl, Chestnut Quail-thrush, Southern Scrub-robin, Crested Bellbird, Gilbert's Whistler and Shy Heathwren. Other species include Purple-gaped, Yellow-plumed and White-fronted Honeyeater. The endangered Black-eared Miner also persists here and several captive colonies have in the past been released in an effort to preserve the species; take care not to confuse them with the superficially similar Yellow-throated Miner, which is also present in the reserve and at times associates with the Black-eared Miner. The best access to the reserve is from Merret Rd, 7 km west of the Sunraysia Hwy, where you will find bush tracks (4WD and sandy in parts) leading in a westerly direction into the reserve. After 1 km, the track reaches a large open area – a good place to bush camp. West of here (200 m), a large power line cutting runs north–south through the middle of the reserve. Some of the best birdwatching is often had through the centre of the reserve along this power-line easement. For a short distance, 2WD access to the reserve is possible via the roadside stop and track that leads west of the Sunraysia Hwy 7.5 km south of its intersection with the Calder Hwy.

Gama Bushland Reserve (2 ha and not signposted) is a tiny roadside reserve, located at the corner of the Gama-Sea

Woorinen Sands Mallee in Murray-Sunset; a vegetation type occupied by a number of habitat specialists, including Malleefowl, Mallee Emu-wren, Striated Grasswren, Red-lored Whistler and Chestnut Quail-thrush. The few Black-eared Miners that persist in Victoria are also found in this habitat.

Lake Rd and the Sunraysia Hwy. Here, there are several stands of Long-leaf Emu-bush, which, when flowering, attract nomadic nectar feeders such as Black (a rarity in Victoria), White-fronted and Pied Honeyeater (rare). Other birds that occur at Gama include White-backed Swallow, Blue Bonnet, Hooded Robin, Rainbow Bee-eater and Variegated Fairy-wren.

Wyperfeld National Park

Key species: Malleefowl, Major Mitchell's Cockatoo, Regent Parrot, Redthroat, White-browed Treecreeper
Other species: Southern Scrub-robin, Shy Heathwren, Gilbert's Whistler, Splendid Fairy-wren

Wyperfeld National Park (356 800 ha) is a substantial reserve in north-western Victoria that serves to protect vast areas of mallee vegetation, pine plains, Red Gum and Black Box woodlands and heathlands. At the centre of the park is a series of lake beds connected by Outlet Creek that fill only when the Wimmera River floods into Lake Hindmarsh to the south; these days, the wetlands are generally dry. Despite the usual lack of water, Wyperfeld is a haven for birdlife and an excellent camping destination for those looking for a taste of the outback. For birdwatchers, more than 20 species of parrot (including Major Mitchell's Cockatoo and Regent Parrot) occur within the reserve. It is the only reliable site in Victoria for Redthroat and one of the best sites in the state for both Malleefowl and White-browed Treecreeper. The northern section of the park is accessed via Patchewollock, while the southern section is accessed through either Hopetoun or Rainbow via Yaapeet. There are two excellent camping areas in the park: Wonga Campground 10 km from the southern entrance and Casuarina Campground 6 km from the northern entrance.

Major Mitchell's Cockatoo occur throughout Wyperfeld National Park, but are most frequently encountered in native pine *Callitris* spp. woodlands and at the interface between the reserve and adjacent farmland where they can occasionally be found feeding on the seeds of the paddy melon *Cucumis*. spp.

Before entering the park (either from the north or south), keep your eyes open for Spotted Harrier and White-backed Swallow as they fly over farm paddocks adjacent to the reserve. Blue Bonnet, Rufous Songlark (summer) and Yellow-throated Miner occur in roadside trees, while Brown Songlark (summer) perch on roadside fences adjacent to grain crops.

In northern Wyperfeld, small numbers of White-browed Treecreeper occur in the native pines at the beginning of the Meridian Track, particularly on the hillsides just south-west of the information sign. Note that the superficially similar Brown Treecreeper is very common here, so identification requires some care. The Casuarina Campground is an excellent site for many of the inland parrots – at dawn and dusk, you can expect to see Regent and Mulga Parrot, Blue Bonnet, Australian Ringneck, Major Mitchell's Cockatoo and Little Corella. Striped and Spiny-cheeked Honeyeater feed around the campground and Splendid Fairy-wren occupy the shrubby cover on the hill just behind the camp. You should also see Pied Butcherbird, Pallid Cuckoo (summer) and Red-capped Robin around the campground, while at night Southern Boobook, Eastern Barn Owl, Australian Owlet-Nightjar and Spotted Nightjar have all been recorded. Along the Meridian Track, look for spring–summer migrant species such as Crimson Chat, Masked and White-browed Woodswallow, White-winged Triller and Rufous and Brown Songlark. Two small grass parrots, the Blue-winged and Elegant (uncommon) Parrot have also been recorded here. The impressive Snowdrift Sandhill is another area that often produces sightings of Major Mitchell's Cockatoo, especially if paddy melons (a small weedy melon and favoured food item of the cockatoo) are present.

In southern Wyperfeld, just after entering the park, search the grassy areas around the Park Rangers' office for mixed flocks of small passerines that may include Southern Whiteface, Yellow-rumped and Inland Thornbill. In spring and summer, noisy flocks of White-browed and Masked Woodswallows sometimes occur. Typical species to be found around the Wonga Campground include Emu, Sulphur-crested Cockatoo, Regent and Red-rumped Parrot, Jacky Winter, Brown Treecreeper, White-winged Chough, Spotted (yellow-rumped) Pardalote, Sacred Kingfisher, Rufous Whistler, and Hooded and Red-capped Robin. Major Mitchell's Cockatoo also occur from time to time. Night birds are also a feature of the Wonga campground, with Southern Boobook, Eastern Barn Owl, Tawny Frogmouth and Australian Owlet-Nightjar all at least heard frequently. Interestingly, this area is also an historic site for the extremely rare Night Parrot (recorded 1911). The Discovery Walk (4 km loop – with a 1 km return extension to Lake Brambruk), which starts at the campground, is the most reliable site in Victoria to find the very localised Redthroat. Look (and listen for their almost canary-like call) in the heath near the start of the right branch of the walk. The species is also found between the 'Devil's Pool' and Lake Brambruk (in a small valley immediately before Lake Brambruk). The 'Devil's Pool' usually holds water, even during the driest periods of the year and, as a consequence, is a site favoured by birds for drinking; there is a reasonable chance of Regent and Mulga Parrot, Major Mitchell's and Sulphur-crested Cockatoo, Little Corella and Galah. At night, Spotted Nightjar has been known to hawk for insects overhead and drink 'on the wing' from these pools. On both sides of the Mount Mattingley Lookout, in the denser mallee shrublands, look for Southern Scrub-robin, Shy Heathwren, Splendid Fairy-wren and Inland, Chestnut-rumped Thornbill. When the heath is flowering, this area can also be teeming with honeyeaters such as Yellow-plumed, White-eared, Singing, New Holland, Brown-headed, Tawny-crowned and White-fronted; occasionally, the elusive Black Honeyeater (spring–summer) is also detected.

An excellent area of mallee woodland is along the start of the Dattuck Track (located off the Nature Drive) just past the Eastern Lookout. This area has provided many sightings of Malleefowl, particularly in patches of spinifex along the right-hand side of the track. Other birds to look for include Gilbert's Whistler (search between the track and the Eastern Lookout), Shy Heathwren, Crested Bellbird, Chestnut Quail-thrush, Splendid and Variegated Fairy-wren, Little Button-quail (in the grassy areas bordering the track), Australian Ringneck, Mulga, Regent and Blue-winged (occasionally) Parrot, and White-browed and Masked

Woodswallow (both spring–summer). During spring, the vegetation towards the end of the Lignum Track (located on the Nature Drive just before you get to the Eastern Lookout) often hosts mixed-species flocks of White-winged Triller, Rufous Songlark, Hooded and Red-capped Robin, Varied Sittella, and White-browed, Masked and Dusky Woodswallow. Brown Songlark (summer) can also be common in the surrounding grassy floodplain. Continuing along the Nature Drive (2 km past the Dattuck Track), the Lowan Track supports similar species to those listed above, but look also for Shy Heathwren, Chestnut Quail-thrush and Purple-gaped Honeyeater. Some 200 m down this track, there is a sign-posted Malleefowl mound. Another excellent walk in southern Wyperfeld is the Desert Walk (7 km loop), which passes through mallee, tea-tree and heathland. Redthroat and Slender-billed Thornbill (both very localised in Victoria) have been recorded in the heath along the walk and you may also encounter Gilbert's, Golden and Rufous Whistler, Shy Heathwren, Inland Thornbill, and White-fronted, Tawny-crowed, White-eared, Spiny-cheeked and Brown-headed Honeyeater. A good selection of raptors have been recorded at Wyperfeld, including Wedge-tailed and Little Eagle, Black-shouldered and Whistling Kite, Spotted Harrier, Brown Goshawk, Collared Sparrowhawk, Nankeen Kestrel, Australian Hobby and Brown, Peregrine and Black Falcon. The extremely rare Grey Falcon has also been reported on several occasions. Finally, a relict Victorian population of Australian Bustard persists, occupying recently burnt mallee vegetation and low heathland and adjacent farmland around the southern boundary of Wyperfeld. These birds are particularly elusive, despite their relatively

Extensive Loamy Sands Mallee in Wyperfeld National Park. This habitat type supports more shrubs and less spinifex than Woorinen Sands Mallee (p. 247). Malleefowl, Shy Heathwren, Splendid Fairy-wren, White-fronted and Yellow-plumed Honeyeater, and Southern Scrub-Robin are just some of the species that frequent this habitat.

Lake Tyrrell Wildlife Reserve

Lake Tyrrell Wildlife Reserve (21 000 ha) protects Victoria's largest salt lake and the surrounding vegetation. It is best reached via Baileys Rd (located off the Calder Hwy just north of the Sea Lake township), which joins up with Lake Tyrrell Rd: a dirt road that circles the lake. The saltbush and samphire shrublands around the lake edge support several fascinating habitat specialists, including Rufous Fieldwren, Orange Chat, White-winged Fairy-wren, White-backed Swallow, Blue Bonnet and Blue-winged Parrot and, more rarely, Elegant Parrot. Other birds here include Singing and White-fronted Honeyeater, Black-faced Woodswallow, Australasian Pipit and Brown Songlark. From Baileys Rd, head east along Lake Tyrrell Rd (looking for Orange Chat along the way) before crossing a small drainage line. The lightly wooded area, located just before you reach the Robinvale-Sea Lake Rd, can be a productive area for Blue Bonnet, Cockatiel, Budgerigar, White-browed Babbler, Variegated Fairy-wren and Brown Songlark.

large size; the best areas to search appear to be west of Lake Albacutya Regional Park in the vicinity of Telopea Downs.

Little Desert National Park

Key species: Malleefowl, Blue-winged Parrot, Elegant Parrot, Slender-billed Thornbill, Rufous Fieldwren, Purple-gaped Honeyeater
Other species: Shy Heathwren, Southern Scrub-robin, White-fronted Honeyeater, Tawny-crowned Honeyeater, White-backed Swallow

Little Desert National Park (132 000 ha) protects floristically diverse heathlands of banksia, sheoke, and tea-tree, with patches of broombush on the sandy ridges. Owing to the abundance of wildflowers, spring is a great time to visit the park when plants are in bloom and avian activity can be at a peak. Located 375 km from Melbourne, there are several roads leading south into the park off the Western Hwy. There are camping grounds south of Kiata, at Horseshoe Bend and at Ackle Bend south of Dimboola. There is also excellent accommodation and camping at the Little Desert Nature Lodge, 16 km south of Nhill, on the Nhill–Harrow Rd.

One of the most productive areas for birding in Little Desert is the *Banksia* heathland bordering the Nhill–Harrow Rd, especially where the road intersects with the Dahlenburgs Mill Track. This area is adjacent to farmland, 3.4 km south of the entrance to the Little Desert Nature Lodge. Search the knee-high heathland for Slender-billed Thornbill: a range-restricted species in Victoria. Take care with any identification because the superficially similar Buff-rumped Thornbill also occurs here, though generally occupies taller heathland. Another heathland (and saltbush/samphire) speciality, the Rufous Fieldwren, also occurs here in similar habitats to that of the Slender-billed Thornbill. Look also for Blue-winged and Elegant Parrot (the latter in spring and summer; perhaps the best site in Victoria),

Dusky, White-browed and Masked Woodswallow (all spring–summer), Purple-gaped, Tawny-crowned and White-fronted Honeyeater, Variegated Fairy-wren and White-backed Swallow (summer). Brown Songlark and Spotted Harrier often occur in the nearby paddocks. Similar species occur along the Salt Lake Track (sandy and requiring 4WD), including Slender-billed Thornbill and Rufous Fieldwren in the knee-high heath. Other birds around Salt Lake include Emu, Tawny-crowned Honeyeater and Australasian Pipit, while in areas of Stringybark woodland look for Jacky Winter, Purple-crowned and Musk Lorikeet, and White-fronted Honeyeater. If you still haven't managed to locate a Rufous Fieldwren, they also occur along the first 500 m of the Phillips Track (off the Nhill–Harrow Rd 8.8 km south of the Little Desert Nature Lodge) and near the intersection of the Nhill–Harrow Rd and McDonald Hwy (really just sandy track) 14.7 km south of the Little Desert Nature Lodge).

Mallee heathlands in Little Desert National Park are occasionally burnt by wildfire. Less than 12 months after fire, regenerating grasstrees and seedlings dominate. Such recently burnt habitat is occupied by migrant species such as Crimson Chat and White-winged Triller.

One of the most accessible areas to see a variety of birds in Little Desert is in the grounds of the Little Desert Nature Lodge. Here, you may find Diamond Firetail, Peaceful Dove, Southern Scrub-robin, Shy Heathwren, Variegated Fairy-wren, Red-capped and Scarlet Robin, White-fronted, Yellow-plumed and Tawny-crowned Honeyeater, Restless Flycatcher, Chestnut-rumped Thornbill and Purple-crowned Lorikeet. Nearby, the Stringybark Walk (1 km) and Stringybark Loop (2 km) are particularly good areas for nectar feeders such as Purple-gaped, White-fronted, Tawny-crowned, New Holland, White-naped and White-eared Honeyeater. Both walks commence from the Nhill–Harrow Rd, 1.5 km south of the Little Desert Nature Lodge. Finally, the Keith Hateley Nature Walk (1 km) at the Sanctuary Picnic Area, 10 km south of Kiata, provides access to extensive areas of broombush, which is a favoured habitat of the Purple-gaped Honeyeater. Similar patches of broombush occur just to the west of the nature walk, along the entrance road to the Kiata Campground and in the vicinity of the water bore 2 km south of the Kiata Campsite (accessed via the Salt Lake Track).

Goschen and Tresco West Bushland Reserves

Key species: Black Honeyeater, Pied Honeyeater (chance), Crimson Chat, Little Button-quail, Cockatiel, Budgerigar

Glenlee Flora and Fauna Reserve

Glenlee Flora and Fauna Reserve (590 ha) is on the Nhill–Jeparit Rd, 20 km north of Nhill. The Reserve protects grassy woodland dominated by buloke. Search along the roadside of the reserve for woodland species, with a good spot on the Nhill–Jeparit Rd 1.2 km west of the Gerang–Glenlee Rd. Here you may find Southern Whiteface, Diamond Firetail, Zebra Finch, Spotted (yellow-rumped) Pardalote, Red-capped and Hooded Robin and Variegated Wren, and a remarkable seven species of Thornbill: Brown, Inland, Chestnut-rumped, Striated, Buff-rumped, Yellow and Yellow-rumped. Another good area is around the bird hide and water tray in the middle of the eastern block of the reserve. To get there (from Nhill) take the first dirt track into the reserve – it is on your right-hand side immediately after the fence line. After 1.5 km, bear left at the forks and continue several hundred metres further on to the hide. Around the bird hide, a wide variety of nectar feeders have been recorded, including Purple-crowned Lorikeet and Yellow-plumed, White-fronted, Black (uncommon nomad), Striped, Black-chinned and Spiny-cheeked Honeyeater. Look also for Australian Ringneck, Little Button-quail, Dusky, White-browed and Masked Woodswallow, White-winged Triller, Rufous Songlark, Restless Flycatcher, Crimson Chat (summer), Varied (Black-capped) Sittella, Gilbert's Whistler, Brown Treecreeper and Grey (black-winged) Currawong. In the evening, listen for Bush Stone-curlew: a bird that has largely disappeared from Victoria yet hangs on in this reserve.

Other species: Blue Bonnet, Variegated Fairy-wren, Chestnut-rumped Thornbill, White-fronted Honeyeater

Despite its small size, Goschen Bushland Reserve (25 ha) is widely recognised as a 'hotspot' for inland birds that are often scarce or rare in Victoria. It is most productive in spring and early summer, especially after good seasonal rains. To get there from the Lake Boga township, take the Ultima–Lake Boga Rd until you reach the Donald–Swan Hill Rd intersection. After passing through the intersection, there is an unsealed vehicle track on the right, which leads into the reserve. The track does a loop through the reserve (past an old hall and disused tennis court) to re-join the road 250 m further west beyond the telecommunication tower. It is possible to bush camp at Goschen, but there is also a caravan park at nearby Lake Boga and a range of accommodation options in Swan Hill. The flora of Goschen is a mixture of mallee eucalypts and native grassland, interspersed with Long-leaf Emu-bush: a key food resource for many of the nomadic inland honeyeaters such as Black, White-fronted and, less commonly, Pied Honeyeater. A good area in which to look for these species is along the track between the tennis court and the tower. This area can also be good for Crimson Chat, White-browed, Masked and Dusky Woodswallow, Chestnut-rumped

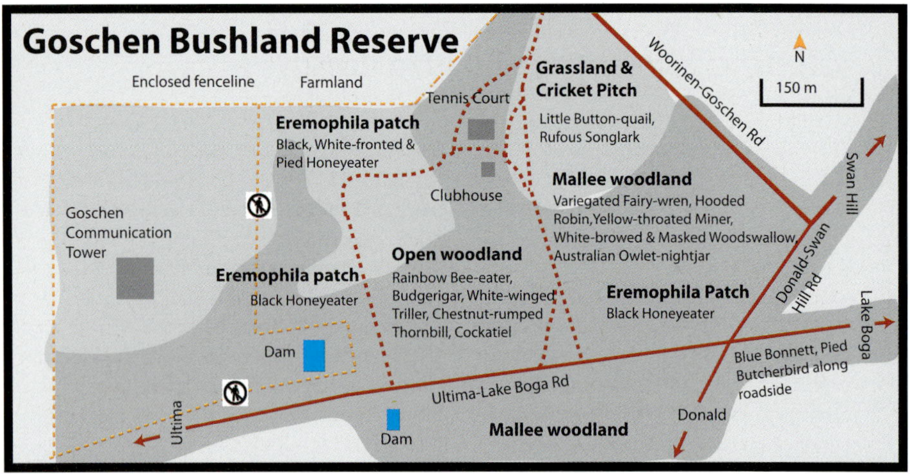

Thornbill, Hooded Robin and White-winged Triller. Little Button-quail are often flushed from the grassy areas east of the tennis court, particularly near an over-grown cricket pitch and here you can sometimes find Budgerigar, Cockatiel, Zebra Finch and Peaceful Dove. In the eastern-half of the reserve, look for Variegated Fairy-wren, Spiny-cheeked Honeyeater, Yellow-throated Miner, Red-capped Robin,

Open grassy woodland between the tennis court and tower, Goschen Bushland Reserve. Following good rains, the grass cover in the reserve can be extensive through spring and summer. At this time, look for Little Button-quail, Cockatiel, Budgerigar, Crimson Chat, Pied Butcherbird and Zebra Finch in this area.

White-browed Babbler, Varied Sittella and Brown Treecreeper, while Australian Owlet-Nightjar is sometimes flushed from tree hollows in this area during the day. On the drive to Goschen (from Lake Boga), and on many of the surrounding roads with remnant roadside vegetation, expect to see Blue Bonnet, Cockatiel, Pied Butcherbird and Yellow-throated Miner.

Tresco West Bushland Reserve (25 ha), known locally as the 'Daisy Patch', is located 3 km south-west of Lake Boga. To get there from Lake Boga, drive west along Lalbert Rd for 2 km to the Lakeside Golf Club; the sign for the reserve is another 200 m further along Lalbert Rd. Somewhat similar to Goschen, the habitat here includes several stands of Long-leaf Emu-bush. Check here for Black and Pied (uncommon) Honeyeater, while Little Button-quail occur in areas of native grassland. You may also see Blue Bonnet, Cockatiel, Budgerigar, Pied Butcherbird, Crimson Chat (summer chance), Variegated Fairy-wren, Zebra Finch and, occasionally, Red-backed Kingfisher. When there is water, the saline Golf Course Lake attracts Red-necked Avocet, Banded and Black-winged Stilt and flocks of hawking Whiskered Tern (spring–summer). An excellent area for woodland birds is along the south side of the lake, particularly the south-east corner. To get there, from the golf course, continue along Lalbert Rd for 1.2 km to McCubbin Rd. Here, turn left and travel for 1.3 km along this dirt track, before turning left again at a stone marker then continuing for a further 100 m. This general area can be good for Black Honeyeater, particularly if the stands of Long-leaf Emu-Bush are in flower. Other birds may include Pied Butcherbird, Crimson Chat (summer), Rainbow Bee-eater (summer), Singing, White-fronted and Spiny-cheeked Honeyeater, White-browed Babbler, Blue Bonnet, Cockatiel, Budgerigar and, occasionally, Red-backed Kingfisher and Zebra Finch.

Round Lake is situated 1 km west of the township of Lake Boga, on the Ultima–Lake Boga Rd. Here you may see Whiskered Tern (summer), Black-winged Stilt, Hardhead, Australasian Shoveler, Blue-billed, Pink-eared and Freckled (rare) Duck, Great Crested, and Hoary-headed and Australasian Grebe, as well as Great, Intermediate and Little Egret. When the water level is low, Australian Spotted and Spotless Crake may feed on the muddy edges close to cover. Approximately 1 km north along Long Lake Rd, there is yet another a good patch of Long-leaf Emu-bush (opposite a large green shed), which, when flowering (usually late spring and into summer), may attract Black Honeyeater. White-winged Triller and Rainbow Bee-eater have also been recorded in this area.

Kerang Lakes

Key species: Australian Painted Snipe (chance), Freckled Duck, Black Falcon, Wood Sandpiper (summer), Banded Stilt, Orange Chat, White-winged Fairy-wren, migratory waders (summer)
Other species: White-breasted Woodswallow, Gull-billed Tern, Great Crested Grebe, Australian Spotted Crake, Spotless Crake, Baillon's Crake, Glossy Ibis

In the 40 km stretch between Kerang and Lake Boga, there are some 57 lakes spread along either side of the Murray Valley Hwy. This wetland system supports a diversity of aquatic habitats from small ephemeral swamps to expansive and relatively deep waterbodies. Owing to the regions complex hydrology, waterbodies also range from freshwater marshes to hypersaline lakes, with some extremes in close proximity to one another. This concentration of diverse aquatic habitats means that the area can at times support vast numbers of waterbirds.

In the saltbush surrounding Lake Tutchewop (920 ha), you may see White-winged Fairy-wren and Orange Chat (summer). An excellent site to find these two species is along a track that skirts the western shore of the lake (between it and an irrigation channel). The track's southern entrance leaves the Murray Valley Hwy (just north of the Lake Steggals Rd intersection), while the northern entrance begins on Benjeroop–Tresco Rd (starting immediately after an irrigation channel 3 km east of the highway). The low saline vegetation and grassy areas around the lake shore also support Blue-winged Parrot, Blue Bonnet, Cockatiel, Black-faced Woodswallow, Brown Songlark, Fairy Martin, White-fronted Chat and Zebra Finch, as well as birds of prey such as White-bellied and Wedge-tailed Sea-Eagle, and Spotted and Swamp Harrier. Lake Tutchewop is also a significant inland site for migratory and resident shorebirds; species recorded here include Common Greenshank, Red-necked Stint, Marsh and Sharp-tailed and Sandpiper, Double-banded Plover (winter), Black-winged and Banded Stilt, Red-necked Avocet and Banded Lapwing. Waterfowl using the lake include Pink-eared and Freckled (rare) Duck, Grey Teal and Australasian Shoveler, while Caspian, Whiskered and, less frequently, Gull-billed and White-winged Black Tern may also occur.

Waterbirds can be plentiful on the permanent waterbody of Lake Boga (906 ha), with a chance of seeing Great Crested Grebe, Whiskered, Caspian and Gull-billed Tern, Australasian Shoveler, and Blue-billed, Pink-eared and Musk Duck. Around the shores, Blue-faced Honeyeater feed in the lakeside trees, while in spring and summer, White-breasted Woodswallow congregate on the powerlines.

Middle Lake (190 ha) Ibis Rookery, 6 km north of Kerang, is recognised as one of the largest permanent breeding colonies of ibis in Australia. Australian White and Straw-necked Ibis occupy the rookery in spring, when more than 20 000 birds can flock to the area to nest. A tower hide has been constructed to allow easy viewing. Other birds here include Glossy Ibis, Nankeen Night-Heron, Buff-banded Rail, Black-tailed Native-hen, Azure and Sacred Kingfisher, Swamp Harrier, Black and Whistling Kite, White-breasted Woodswallow, Blue-faced Honeyeater and Pied Butcherbird. In spring and summer, the reed beds can be alive with the song of Australian Reed-Warbler.

Foster Swamp (270 ha), adjacent to the Kerang Waste Treatment Plant at the end of Park St, Kerang, is also worth investigating, especially in summer when species such as Greenshank, Marsh, Wood, Curlew and Sharp-tailed Sandpiper and Pacific Golden Plover may frequent the site. Other birds that have been recorded here include Whiskered and Gull-billed Tern, Australasian Shoveler, Pink-eared and Freckled Duck, Black-tailed Native-hen, Red-kneed and Black-fronted Dotterel, Australian Spotted and Baillon's Crake, Black-winged and Banded Stilt, Red-necked Avocet and White-winged Fairy-wren.

The woodlands around the Loddon Weir, located on Weir Rd just north of Kerang are also worth investigating. Here you may see Gilbert's Whistler, Western Gerygone, Painted Honeyeater, Red-capped Robin, Grey-crowned Babbler, White-bellied Cuckoo-shrike, White-breasted Woodswallow and Chestnut-rumped Thornbill. In the same area, Back Swamp (21 ha), which consists of a series of wetlands between Kerang and the Loddon River, usually holds a good selection of birds. Interesting species recorded here include Glossy Ibis, Buff-banded Rail, Nankeen Night-Heron, Grey-crowned Babbler, White-winged and Variegated Fairy-wrens Little Friarbird and Blue-faced Honeyeater.

The Australian Painted Snipe is an elusive species that displays a distinct preference for ephemeral wetlands with vegetative cover such as lignum. Given their habitat requirements, the species appears to be highly mobile, and predicting where the next sighting will occur can be difficult. Nevertheless, Hird Swamp (400 ha) and Johnson Swamp (485 ha) Wildlife Reserves are two of the better sites for this species when conditions are suitable. Both wetlands are located off the Kerang–Leitchville Road, with Johnsons Swamp to the north being accessed via Day Lane, 15.4 km east of the Loddon Valley Hwy and Hird Swamp to the south being accessed via a dirt track adjacent to an irrigation channel 17 km east of the Loddon Valley Hwy. Australian Painted Snipe prefer shallow areas near to cover, and thus the swamps are at their best when large areas are flooded by ankle-deep water. All eleven species of waterfowl to be expected in the region have also been recorded here, including the rare Freckled Duck. During late spring and summer, Hird and Johnson Swamps can be excellent sites for waders such as Common Greenshank, Red-necked, Wood, Mash, Sharp-tailed and Curlew Sandpiper, Latham's Snipe and Pacific Golden Plover. Rarer summer visitors have included Ruff and Long-toed Stint, while Double-banded Plover is an occasional winter visitor. Hird Swamp is also an excellent site for Australasian and Australian Little Bittern; even if they can't be seen, both species can often be heard calling from the extensive reed beds in spring and summer. Other waterbirds include Glossy Ibis, Intermediate Egret, Buff-banded Rail, Spotless, Australian Spotted and Baillon's Crake, Black-tailed Native-hen, Red-kneed Dotterel, Black-winged Stilt, Red-necked Avocet, and Whiskered and White-winged Black Tern. Orange Chat and White-winged Fairy-wren frequent the surrounding vegetation. Snakes are fairly common at these wetlands and thus, if you intend to wade in the shallows in search of less common species, it is recommended that you don gumboots or waders.

Terrick Terrick National Park

Key species: Plains-wanderer, Little Button-quail, Red-chested Button-quail (rare), Inland Dotterel (rare), Black Falcon, Australian Ringneck

Other species: Stubble Quail, Painted Honeyeater, Grey-crowned Babbler, Chestnut-crowned Babbler

Terrick Terrick National Park (5890 ha) is a patchwork of native vegetation that persists in an area of wheat and sheep production. The largest block of the park protects open grassy woodlands surrounding Mount Terrick Terrick, while further to the east and north, reserved paddocks protect some of the largest and most-intact tracts of native grasslands to remain in Victoria. The park is located 230 km north of Melbourne and 7 km north of Mitiamo. There is a camping area near the Mount Terrick Terrick picnic area with toilets and picnic tables, but it is necessary to bring in your own water and cooking equipment.

Some of the park's best birding is around the open woodlands (dominated by White Cypress-Pine, Yellow and Grey Box), around the picnic area and the base of Mount Terrick Terrick. Here, you may see Australian Ringneck, Gilbert's Whistler, Painted Honeyeater (spring–summer), Brown Treecreeper, White-browed Babbler, White-winged Chough, Dusky Woodswallow, Red-capped Robin, Rufous Songlark (spring–summer) and Tree Martin (spring–summer). Smaller passerines include Yellow and Chestnut-rumped Thornbill, Western Gerygone, Mistletoebird, Southern Whiteface and Diamond Firetail. In the woodlands on the north side of Riegels Rock, additional species include Peaceful Dove, White-winged Chough, Restless Flycatcher, Black-chinned Honeyeater, Red-capped and Hooded Robin, Jacky Winter and Diamond Firetail, and summer migrants such as Black-eared Cuckoo, White-browed and Masked Woodswallow, and White-winged Triller. The most southerly population of the Chestnut-crowned Babbler is sometimes encountered northwest of Riegels Rock. One of the highlights of the Terrick Terrick National Park is the large number of raptors, with more

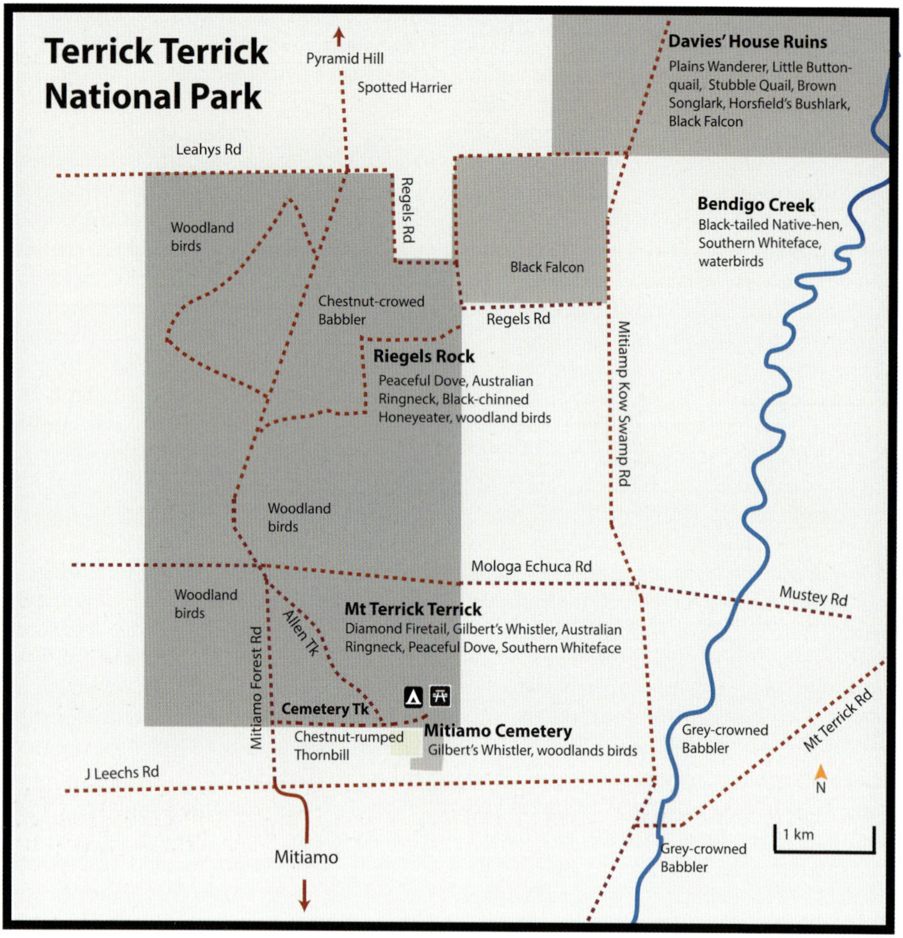

than 12 species regularly reported. Species that are frequently encountered within the woodland areas (including Reigels Rock) include Peregrine Falcon, Brown Goshawk, Collared Sparrowhawk and Wedge-tailed and Little Eagle.

A major drawcard for birdwatchers at Terrick Terrick National Park is the regular occurrence of Plains-wanderer in the grasslands sections of the park, especially near Davies Ruins on the Mitiamo–Kow Swamp Rd. Quail-like in appearance, Plains-wanderer are incredibly cryptic and rarely flush in daylight, as do quail. Rather, to have any chance of finding this species, a nighttime visit is all but essential. No vehicular access to the grassland sections of the park is permitted, because the fragile soil is vulnerable to compaction and erosion, so it is necessary to walk the grasslands with the aid of a torch or spotlight. Even within the grasslands, Plains-wanderers have quite specific habitat requirements, seeking areas of short grasses, with a diversity of herbs and small patches of naturally bare earth. Coupled with the fact that it is very easy to get lost at night in a featureless landscape, joining the Friends of Terrick Terrick National Park on one of their regular surveys is probably the best option to maximise the

Some of the most significant native grassland in Victoria, Terrick Terrick National Park. The low vegetative cover and bare open patches of dried clay are key habitat requirements of the enigmatic Plains-wanderer – if the vegetation becomes too dense, the birds will move elsewhere. This habitat also supports Black Falcon, Little Button-quail, Banded Lapwing, Brown Songlark and Horsfield's Bushlark.

chances of success (see www.home.vicnet.net.au/~friends). Other birds found on the grasslands and often encountered during spotlight walks include Little Button-quail, Stubble Quail and Eastern Barn Owl. The Red-chested Button-quail, which is generally rare in Victoria, also occurs here on occasion, though this species prefers wetter areas of taller spear grass (*Stipa* spp.) so is unlikely to be encountered in the same areas as the Plains-wanderer. Fat-tailed Dunnart, Curl Snake and Hooded Scaly-foot (a legless lizard) are all rare grassland specialists that may also be encountered. During the day, Brown Songlark (summer) and Horsfield's Bushlark occur, while Black Falcon may hunt overhead. Further to the north and east of the grasslands, around the Davies Homestead, are two additional grassland sections of the national park in an area known as the Patho Plains. Both sections are located on the Tomara Rd, with the best area near the intersection with Clee Rd. In spring and summer, especially when the grasslands are relatively sparse, these areas occasionally

An adult female Plains-wanderer. The species can only be reliably found by spotlighting, so joining the Friends of Terrick Terrick National Park on an organised evening gives the best chance of a sighting.

support three grassland-frequenting shorebirds: Australian Pratincole, Banded Lapwing and Inland Dotterel.

Along Bendigo Creek, there are several family groups of Grey-crowned Babbler, with a good area to search being along the creek north of the Echuca–Mitiamo Rd bridge, up to and around the old homestead (1 km north of the bridge), though the species is also occasionally found to the south of the bridge. When waters levels are high in the creek, the area south of Jungaburra Rd forms an extensive ephemeral wetland chain and can be superb for birding. Along the creek, you may encounter Brolga, White-necked Heron, Black-tailed Native-hen, Yellow-billed Spoonbill, Black-fronted Dotterel, Yellow-rumped Thornbill, Southern Whiteface and Zebra Finch, and summer migrants such as Sacred Kingfisher, Restless Flycatcher and Rufous Songlark.

Greater Bendigo National Park

Key species: Spotted Nightjar, Shy Heathwren, Chestnut-rumped Heathwren, Crested Bellbird, Purple-gaped Honeyeater
Other species: Brush Bronzewing, Variegated Fairy-wren, Inland Thornbill

Greater Bendigo National Park (17 000 ha) consists of mallee shrublands and box–ironbark woodland. The main sections of the park – Kamarooka and the Whipstick – are located immediately to the north of Bendigo (Victoria's fourth largest city). Honeyeaters are particularly diverse in the park, with an impressive 24 species having been recorded. In good seasons, honeyeaters are attracted to the sequential flowering of Kamarooka Mallee (Jan–Feb), Grey Box (Mar–May), Blue Mallee (Apr–Jun) and Yellow Gum (Apr–Nov) providing almost year-round access to nectar resources. Most roads are unsealed, but, unless wet, are suitable for 2WD. There is basic bush camping at Notley, Loeser and Shadbolt Picnic Areas, Rush Dam and Mulga Dam. Nearby, Bendigo has a wide range of accommodation options.

One excellent area for birdwatching is around the Distillery Dams, located in the Kamarooka section of the park. The site can be accessed just off Campbell's Rd, 600 m from the Bendigo–Tennyson Rd (13.5 km north of Huntly). Nectar feeders can be plentiful, especially when the Blue Mallee is in flower. Here, you may observe Purple-gaped, Yellow-plumed, White-plumed, White-naped (May–Aug), Brown-headed, Black-chinned, Tawny-crowned, White-eared, Yellow-tufted, Yellow-faced (May–Aug), Fuscous, White-fronted and, more rarely, Black Honeyeater. One of the few records of Brown Honeyeater in Victoria was also obtained here. Other birds to look for include Painted Button-quail, Brush and Common Bronzewing, Spotted Nightjar, Shy Heathwren, Red-capped and Hooded Robin, Crested Bellbird, Gilbert's, Golden and Rufous Whistler, Spotted (yellow-rumped) Pardalote and Variegated Fairy-wren. Tawny-crowned and Purple-gaped Honeyeater, which can sometimes be difficult to locate around the dams, also occur in the low mallee bordering Burnside Rd to the east of the Bendigo–Tennyson

A Varied Sittella of the nominate subspecies in the Greater Bendigo National Park. Because Sittellas regularly join mixed feeding flocks of small woodland birds, listening for their distinctive call can be a good way to locate birding hotspots.

Rd. Further north, along Camp Rd, east of the Bendigo–Tennyson Rd, the birding can also be rewarding. Stop at the Camp Rd Dam, 1.5 km from the intersection and check for Peaceful Dove, Eastern Yellow, Red-capped and Hooded Robin, Brown Treecreeper, Diamond Firetail, Jacky Winter, Crested Shrike-tit, Restless Flycatcher, and Chestnut-rumped, Buff-rumped, Yellow-rumped, Inland and Yellow Thornbill. Summer migrants in this area include Rainbow Bee-eater, White-browed, Masked and Dusky Woodswallow. At night, Spotted Nightjar may be seen hawking over the Camp Rd Dam and the nearby Mulga Dam (off Camp Rd, 800 m west of the Bendigo–Tennyson Rd).

The Whipstick section of the park is situated north of the township of Eaglehawk. Access is best gained by turning left (travelling west) off the Midland Hwy at Epsom (near Bendigo Pottery) and then right into Neilborough Rd to Shadbolt Picnic Area. The eastern end of Skylark Rd (north of Wards Rd) can also be a productive area for box–ironbark forest birding. Crested Bellbird, White-browed Babbler, Painted Button-quail and Black-chinned Honeyeater can regularly be observed in both of these areas and, when eucalyptus are flowering, they can be good for nectivores such as Musk, Purple-crowned and Little Lorikeet and, occasionally, Swift Parrot (winter). The low dense mallee along Neilborough–Eaglehawk Rd, just north of the Shadbolt Picnic Area provides habitat for the Shy Heathwren, Variegated Fairy-wren and Gilbert's Whistler. In the taller and more open areas of mallee, look for Purple-gaped, Tawny-crowned and Yellow-plumed Honeyeater and Crested Bellbird. Spotted Nightjar have a preference for stony rises and granite outcrops, particularly the areas harvested for eucalyptus oil (intermixed with knee-high

Red Ironbark *Eucalyptus sideroxylon* woodland around a fire dam in the Greater Bendigo National Park. Fire dams can be a magnet for many woodland birds, especially in warmer weather when large numbers of honeyeaters and other species may congregate to drink and bathe.

mallee); dusk is often the best time to locate this species because this appears to be when they call most frequently.

Although Shy Heathwren occurs in the northerly and westerly mallee sections of the park, the closely related Chestnut-rumped Heathwren occurs in the southern box–ironbark woodland sections of the park. A good area to look is around the intersection of Diamond Hill Rd and Kangaroo Gully Rd; search in the vicinity of a small track that heads south just east of the junction. Other birds here may include the nectivorous parrots mentioned above, Collared Sparrowhawk, Painted Button-quail, Crested Bellbird, White-browed Babbler, Dusky Woodswallow (summer), and Golden and Rufous Whistler. Chestnut-rumped Heathwren also inhabit the heath thickets at Salomon Gully Flora Reserve, Quarry Hill, just 2 km south of the centre of Bendigo. Access to the north end of the reserve can be gained via Boud Street, with a good site for the species being the heath around a small seating area 150 m into the walk.

Chiltern–Mount Pilot National Park

Key species: Turquoise Parrot, Swift Parrot (winter), Regent Honeyeater (very rare), Painted Honeyeater (spring–summer)
Other species: Painted Button-quail, White-throated Nightjar (spring–summer), Barking Owl, Square-tailed Kite (spring–summer), Black-chinned Honeyeater

Justifiably one of Victoria's most popular birdwatching locations, Chiltern–Mount Pilot National Park (4320 ha) is located 275 km north of Melbourne via the Hume Hwy. The reserve system protects some of the last remaining tracts of what was once an extensive box–ironbark forest that spanned much of central Victoria. As a birdwatching site, it is one of the most reliable locations for Painted Honeyeater, Swift Parrot and Turquoise Parrot in Victoria. It is also the last stronghold of the Regent Honeyeater in Victoria and the only site where birders now have a reasonable prospect of finding this rapidly declining species. There are many tracks that connect to the main access roads, most of which are suitable for 2WD. Although there are no camping facilities in the park, there are caravan parks and motel accommodation in the nearby historic towns of Chiltern and Beechworth.

Honeyeater Picnic Ground and Cyanide Dam, located in the eastern section of the national park off Cyanide Rd, are excellent areas to birdwatch. When the ironbark forest is flowering (winter and spring, but in some years flowering fails) you may find nectar-feeding species such as Swift Parrot (winter) and Little Lorikeet, and Fuscous, Yellow-tufted, Black-chinned and White-nape Honeyeater. Regent Honeyeater are occasionally found around the dam, most often in spring. Other birds to look for around the picnic area include Turquoise Parrot, and Scarlet and Red-capped Robin. In summer, Sacred Kingfisher, Olive-backed Oriole and White-throated Nightjar (hunting insects over the dam) also occur. The White Box Walking Track (8.5 km return) can be a very enjoyable walk, with opportunities for Painted Button-quail, Turquoise Parrot, Scarlet Robin and White-throated Treecreeper, especially along the ridge and in the valley south of the track (1 km into the walk). The track winds through healthy stands of White Box (the track's namesake). Keep your eyes open for Yellow-footed Antechinus, a common but shy marsupial, which often forages during daylight hours along logs and fallen branches. Summer provides opportunities to observe Square-tailed Kite, which is occasionally recorded flying low over the canopy hunting prey. At night, listen for Barking Owl: rare in Victoria, a pair or two favour the southern section of the park.

In spring, Painted Honeyeater occasionally feed in the mistletoe around Bartley's

Block: a partly cleared grassy woodland located on Howlong Rd, 2 km north of the township of Chiltern. Other honeyeaters that occur here include Yellow-tufted, Fuscous, Brown Headed, Black-chinned and, more rarely, Regent. Bartley's Block can be an excellent site for Turquoise Parrot and small passerines, such as Western and White-throated Gerygone (both summer), Speckled Warbler and Yellow, Yellow-rumped and Buff-rumped Thornbill. This is especially so in the vicinity of the wattles located at the top (south) end of the block. This area can also be good for robins, especially in winter when Scarlet, Flame, Red-capped, Rose and Eastern Yellow Robin may be encountered. Other birds here include, White-winged Triller (spring–summer), Brown and White-throated Treecreeper, and White-bellied Cuckoo-shrike. Elsewhere, two of the better areas to search for Regent Honeyeater are near the Magenta Mine and along Green Hill Rd and Klotz Track, because this area has been the focus of a re-introduction program in recent years. Be sure to note down any colour leg band combinations you see on the birds and report any sightings of this Critically Endangered bird to the national recovery team as soon as possible (1800 621 056). The two Chiltern Valley Dams can hold a diverse array of waterbirds, such and Musk, Pink-eared and the rare Freckled Duck, Australasian Shoveler, Hardhead, Australasian Grebe, Nankeen Night-Heron, and Intermediate and White-neck Heron. Check the muddy edges of the dams for Black-fronted Dotterel. Summer migrants that frequent the dam environs include Latham's Snipe, White-breasted Woodswallow, Rainbow Bee-eater, White-winged Triller, Rufous Songlark and, occasionally, uncommon species such as Red-backed Kingfisher and White-backed Swallow. Just outside the northern boundary of the park, Grey-crowned Babbler occur: look in shrubs for their large roost nests (a tell-tale sign of their presence) ~1 km west along Fishers Rd from the Chiltern–Howlong Rd.

Swift Parrots are winter migrants from Tasmania to the nectar-rich woodlands of south-eastern Australian mainland. Like the flowering events that they seek, their distribution is often patchy, but Chiltern-Mount Pilot National Park remains one of the more reliable winter sites in Victoria.

In the Mount Pilot section of the park, located on the Chiltern–Beechworth Rd, you may find Crimson Rosella (occasionally the yellow race occurs in mixed flocks), Common Bronzewing, Peaceful Dove, Scarlet Robin, Striated, Buff-rumped and Yellow Thornbill, Yellow-faced and Fuscous Honeyeater, Restless Flycatcher and Southern Whiteface. Spring–summer migrants that visit the area include Rainbow Bee-eater, Sacred Kingfisher, Leaden Flycatcher and White-throated Gerygone. On the higher ridges search for Spotted Quail-thrush, particularly in stony areas with native grass tussocks. From the lookout at the top of Mount Pilot, scan the

Victoria

sky for raptors such as Peregrine Falcon, Square-tailed Kite (mainly summer) and both Wedge-tailed and Little Eagle.

Warby Range State Park

Key species: Swift Parrot (winter), Regent Honeyeater (very rare), Painted Honeyeater (spring–summer), Turquoise Parrot, Speckled Warbler
Other species: White-bellied Cuckoo-shrike, Western Gerygone, Black-chinned Honeyeater, Barking Owl, Grey-crowned Babbler

The Warby Range State Park (7600 ha), located to the south-west of Wangaratta, protects a variety of woodland habitats that are particularly rich in birds. One excellent location in the southern Warby Ranges is the area around Wenham's Campground. Here, a large clearing surrounded by woodlands attracts a variety of birds. A morning or late afternoon walk around the edge of the clearing may produce sightings of Turquoise Parrot, Eastern and Crimson Rosella, Brown Treecreeper, Buff-rumped, Yellow-rumped, Brown, Striated and Yellow Thornbill, Speckled Warbler, and Western and White-throated Gerygone (the last two in spring–summer). Another excellent area for birdwatching is around Forest Camp, located in the more northerly Killawarra section of the park. Look here for Swift Parrot (winter), Australian King-Parrot (an uncommon visitor from foothill forests of the Great Dividing Range), Little and Musk Lorikeet, White-bellied Cuckoo-shrike, Rose and Red-capped Robin, Mistletoebird, Little and Noisy Friarbird, and Black-chinned, Yellow-tufted and Fuscous Honeyeater. This was one of the last strongholds for the now very rare Regent Honeyeater and occasional individuals are still sighted. At night listen for Southern Boobook, Barking Owl (rare), Australian Owlet-Nightjar and White-throated Nightjar (summer). The mistletoe-covered ironbark forest just west

Winton Wetlands (formerly Lake Mokoan)

Lake Mokoan (7800 ha), 7 km north of Benalla, was an artificial lake, created in 1971 when water from the Broken River and Hollands Creek was diverted into Winton Swamp. As an artificial water body, dozens of dead standing trees emerging from its depths characterised the lake. In 2010, the dam wall was decommissioned and efforts are now being made to return the are to a more natural wetland state. The resultant wetland system, known as Winton Wetlands, can attract significant numbers of waterfowl, including Plumed Whistling-Duck (generally rare), Musk, Australian Wood, Pink-eared, Pacific Black and Freckled (rare) Duck, Australasian Shoveler, Chestnut and Grey Teal, Hardhead, Australian Shelduck and Black Swan. It is also one of the better sites in Victoria to encounter the summer visiting Fork-tailed Swift. Other birds you may see here include Brown Quail, Latham's Snipe (summer), White-winged Black (summer), Gull-billed and Whiskered Tern, Glossy Ibis, Peregrine and Black Falcon, White-bellied Sea-Eagle and Diamond Firetail.

of Forest Camp can be a reliable place to see Painted Honeyeater (spring–summer) and has produced sightings of Regent Honeyeater. North-west of the Forest Camp a track leads to an old oval and area of regenerating forest. Search here for Western Gerygone, Gilbert's Whistler, Turquoise and Swift Parrot (winter), Red-capped and Hooded Robin, Diamond Firetail, Speckled Warbler and White-winged Triller (spring). Summer migrants to the Warby Ranges that occur throughout the park include Sacred Kingfisher, Olive-backed Oriole, White-winged Triller, Rainbow Bee-eater and Square-tailed Kite (scarce).

Further north, along the Ovens River (accessed off the Wangaratta–Yarrawonga Rd), search the riverside vegetation for Azure Kingfisher, White-bellied Cuckoo-shrike, Rose Robin and Golden Whistler and, in summer, Dollarbird, White-throated Gerygone and Leaden Flycatcher.

Heathcote–Graytown National Park

Key species: Swift Parrot (winter), Chestnut-rumped Heathwren, Speckled Warbler, Spotted Quail-thrush
Other species: Yellow-tufted Honeyeater, Black-chinned Honeyeater, Noisy Friarbird, Little Friarbird, Diamond Firetail, Painted Button-quail

Heathcote–Graytown National Park (12 833 ha) is located in north-central Victoria, 110 km north of Melbourne via the Hume Hwy. The southern section is accessed via the Heathcote–Nagambie Rd 10–20 km west of the Hume Hwy at Nagambie. Lying within the Rushworth Box–Ironbark Important Bird Area, the park at times supports significant numbers of Swift Parrot. There are basic camping facilities at the Dargile Picnic Area.

Some of the most mature woodland trees in the Heathcote–Graytown National Park area can be found along the access roads into the reserve. When in flower, these trees can support large numbers of honeyeaters, lorikeets and in winter Swift Parrots.

Dunach Nature Conservation Reserve

Dominated by box–ironbark forest and stringybark woodland, Dunach Nature Conservation Reserve (890 ha) lies adjacent to the Ballarat–Maryborough Rd between the townships of Clunes and Talbot. In spring and summer, it is an excellent site for honeyeaters, particularly the sought-after Painted Honeyeater. A good place for this species is in the forest that borders farmland in the south-west of the reserve. To get to the reserve, turn-left onto a dirt track 7 km from Clunes (a park sign marks the way). Take the southern branches of the track for 250 m until you reach an area of farmland on the south side of the track. Here, in summer, look for Painted Honeyeater in the larger eucalypts, particularly those infested with mistletoe. Other species that may be encountered include Yellow-tufted, White-eared, Fuscous, Black-chinned, Brown-headed, New Holland, Yellow-faced and Black Honeyeater, Red Wattlebird and Noisy Miner. The reserve is also an excellent site for robins, with opportunities for Red-capped, Hooded, Scarlet, Flame (winter) and Eastern Yellow Robin, as well as Jacky Winter. Other woodland species here include Musk, Little and Purple-crowned Lorikeet, Crested Shrike-tit, Speckled Warbler, Buff-rumped Thornbill, Dusky, Masked and White-browed Woodswallow, Restless Flycatcher and Mistletoebird.

In late winter and autumn, a good place to look for Swift Parrot is on the far north-east side of the park near Bailieston. To get there, from the Heathcote–Nagambie Rd (11 km west of Nagambie), turn north onto Cherry Tree Rd and follow this (through the small town of Bailieston) for 8.7 km to Bailieston Rd (gravel) before turning left and follow this road for 700 m until you reach a small creek line. Look here for Swift Parrot feeding on the ridgeline north-west of the creek line, particularly when the box and ironbark is flowering. Other birds here include Yellow-tufted, Brown-headed and Black-chinned Honeyeater, Noisy and Little Friarbird, Crested Bellbird, Dusky Woodswallow, White-bellied Cuckoo-shrike, Jacky Winter, Flame, Red-capped and Hooded Robin and Diamond Firetail. Heathcote–Graytown National Park is also a good place to see Chestnut-rumped Heathwren, particularly along the road to the lookout on Mount Ida. Search along the hillside just east of the road anywhere after the road turns sharply south (1 km from the summit). Other birds here include Speckled Warbler, Buff-rumped and Striated Thornbill, Restless Flycatcher and Scarlet Robin.

Southern Victoria

Regional overview

Southern Victoria is a wonderful area for birding, with a wide range of interesting

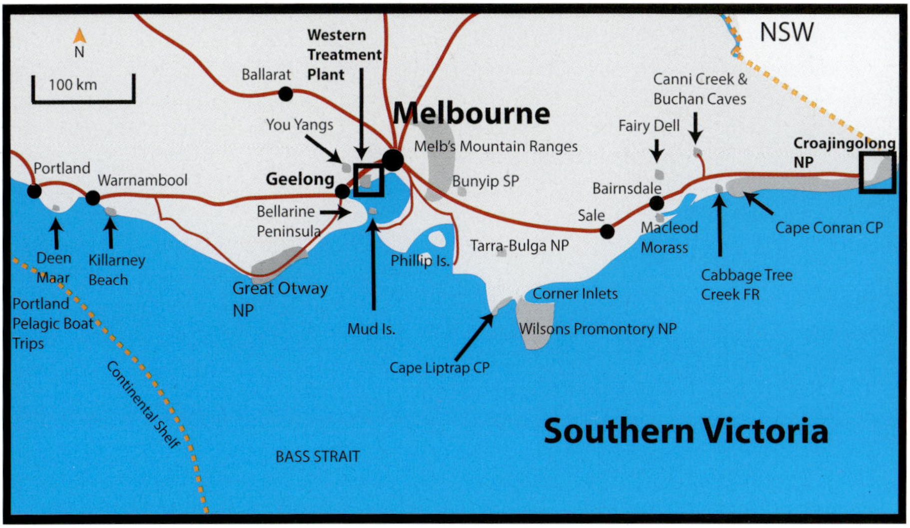

birding sites. It encompasses the entire coastline of Victoria, stretching from Croajingolong National Park in the far east, to Lower Glenelg National Park in the far west of the state. Habitats along the way include cool temperate rainforest,

Tall wet forests dominated by Mountain Ash *Eucalyptus regnans* are a feature of the Victorian central highlands. This habitat supports a wide range of species, including Sooty Owl, Yellow-tailed Black-Cockatoo, Superb Lyrebird, Red-browed Treecreeper and Eastern Whipbird.

temperate eucalyptus forest, heathlands, the coastline and open ocean, wetlands and mountain ranges. The sites discussed in this section are typically accessed from Melbourne, with those that are more distant requiring at least one overnight stay.

A suggested 6-day itinerary for visitors that wish to use Melbourne and Geelong as a central hub is:

- **Day 1.** Melbourne to Dandenong Ranges, Yellingbo and Toolangi 190 km return, (Superb Lyrebird, Sooty Owl, Rose Robin, Pink Robin, Olive Whistler, Pilotbird, Red-browed Treecreeper, Yellow-tufted Honeyeater).
- **Day 2.** Melbourne to the Western Treatment Plant and You Yangs Regional Park 50–60 km. You Yangs Regional Park to Geelong 27 km (Orange-bellied Parrot (extremely rare), Blue-winged Parrot, White-winged Black Tern, Cape Barren Goose, Brolga, waterbirds, waders, vagrants).
- **Day 3.** Geelong and Bellarine Peninsula 75 km return (Australasian Bittern, Latham's Snipe, Banded Stilt (seasonal), Eastern Curlew, Hooded Plover, Black-faced Cormorant, Freckled Duck).
- **Days 4–5.** Geelong to the Great Ocean Rd and Otway Ranges National Park 134 km (Rufous Bristlebird, Southern Emu-wren, Hooded Plover, Grey Goshawk, Olive Whistler, Crescent Honeyeater, Forest Raven, Bassian Thrush).
- **Day 6.** Otway Ranges National Park to Melbourne 180 km.

A suggested 9-day itinerary for East Gippsland involves:

- **Days 1–2.** Melbourne to Fairy Dell, Canni Creek and Buchan Caves reserves 356 km (Powerful Owl, Turquoise Parrot, Cicadabird, Hooded Robin, Red-capped Robin, Lewin's Honeyeater, Scarlet Honeyeater).
- **Days 3–6.** Buchan to Croajingolong National Park 200 km (Glossy Black-Cockatoo, Eastern Bristlebird, Pilotbird, Brown Gerygone, Scarlet Honeyeater, Rose Robin, Black-faced Monarch).
- **Days 7–8.** Croajingolong National Park to Cape Conran 128 km (Topknot Pigeon, Wonga Pigeon, Eastern Ground Parrot, White-throated Needletail, Masked Owl, Sooty Owl, Red-browed Treecreeper, Scarlet Honeyeater).
- **Day 9.** Cape Conran to Melbourne 404 km.

Southern Victoria is here delineated as all coastal areas up to and including the southern slopes of the Great Dividing Range. It extends for 950 km from Mallacoota in the far-east (near the NSW border) to Nelson in the far-west (near the SA border). At both the eastern and western extremes, the coastal environments are dominated by wilderness coastlines backed by large areas of temperate forests and eucalyptus woodland. Port Phillip Bay, Western Port Bay and the cities of Melbourne and Geelong are situated mid-way along the coastline. Despite the development, these areas still provide some truly stunning birding opportunities.

Southern Victoria is characterised by a temperate climate with four distinct seasons: autumn (cool nights to warm days; usually very pleasant), winter (cool and mild), spring (cool to warm, but usually pleasant) and summer (often hot with low humidity). The average daytime temperatures range from around 15°C during winter to 30°C during summer. Rain can occur all year round, but most of it falls in the winter months, with July being the wettest month and March the driest.

Excellent sealed highways provide easy access across the region. The Princes Hwy links Melbourne to all coastal areas within the state. Most of the main roads in southern Victoria are sealed. A 2WD vehicle will get you to the majority of birding sites in the region. V/Line operates bus and rail services to major regional centres. Information about V/Line is available at www.vline.com.au. For information on accommodation and camping options in southern Victoria, visit www.parkweb.vic.gov.au and www.visitvictoria.com. Daily scheduled flights to Tullamarine (Melbourne's main airport) operate from every mainland Australian capital city and many international cities. Avalon (closer to Geelong than Melbourne) provides additional scheduled services with several major airlines.

The Victorian and Melbourne branches of Birdlife Australia (www.birdlife.org.au) and Geelong Field Naturalist Club (home.vicnet.net.au/~gfnc) are very active birdwatching groups, with numerous monthly birdwatching activities for both members and visitors alike.

Birding highlights

Southern Victoria hosts a wide range of bird species, many of which are highly sought after by birdwatchers. Widespread, but nonetheless notable species, in the region include Emu, Brown Quail, White-bellied Sea-Eagle, Grey Goshawk, Peregrine Falcon, Gang-gang Cockatoo, Australian King, Swift, Orange-bellied and Blue-winged Parrot, Powerful and Masked Owl, Southern Emu-wren, Striated Fieldwren, Tawny-crowned and Crescent Honeyeater, Pink, Rose, Flame and Scarlet Robin, Spotted Quail-thrush, Olive Whistler, Bassian Thrush, Forest Raven, Satin Bowerbird and Beautiful Firetail.

Notable species that are mostly or entirely restricted to areas east of Melbourne include Wonga, Topknot and White-headed Pigeon, Brush Bronzewing, Glossy Black-Cockatoo, Eastern Ground and Turquoise Parrot, Brush and Channel-billed Cuckoo, Eastern Koel, Sooty Owl, White-throated Nightjar, Superb Lyrebird, Red-browed Treecreeper, Eastern Bristlebird, Pilotbird, Scarlet Honeyeater and Black-faced Monarch; many of the aforementioned species may be encountered during a summer birdwatching trip to Mallacoota and Croajingolong National Park. Notable species that occur west of Melbourne include Red-tailed Black-Cockatoo (Lower Glenelg National Park and forest reserves around Edenhope) and Rufous Bristlebird (coastal heathlands from Torquay to the South Australian border).

Shorebirds are also particularly well represented in the region, with the Western Treatment Plant in close proximity to Melbourne being a stand-out location, along with Swan Bay and various wetlands on the Bellarine Peninsula. Species include resident shorebirds such as Australian Pied and Sooty Oystercatcher, Red-necked Avocet, Banded and Black-winged Stilt, Red-capped and Hooded Plover, and Red-kneed and Black-fronted Dotterel. A suite of migratory shorebirds also occur, including Double-banded and Pacific Golden Plover, Red-necked Stint, Sharp-tailed and Curlew Sandpiper, Sanderling, Bar-tailed and Black-tailed Godwit, Red

Cicadabird, in this case an adult male, occupy forested habitats in eastern Australia and extend into Victoria to just east of Melbourne. Other species that share a similar distribution in the state include Large-billed Scrubwren, Brown Gerygone and Scarlet Honeyeater.

and Great Knot and Latham's Snipe. On the Victorian coastline, you can see Little Penguin, Shy, Black-browed and Indian Yellow-nosed Albatross, Arctic Jaeger, Brown Skua, Short-tailed and Fluttering Shearwater, Southern and Northern Giant-Petrel, Little, Fairy, Common, White-fronted, Whiskered and White-winged Black Tern, Pacific and Kelp Gull and Black-faced Cormorant. Wetland species such as Lewin's Rail, Baillon's and Spotted Crake, Australasian and Australian Little Bittern, Brolga, Freckled, Pink-eared and Blue-billed Duck, Australasian Shoveler, Cape Barren Goose and Azure Kingfisher are also possible. Introduced species that are more easily found in southern Victoria than elsewhere in Australia include Song Thrush, Eurasian Tree Sparrow, Common Greenfinch and Skylark (though the last two are also common in Tasmania).

Although there are no species restricted to the southern Victorian region, there are several range-restricted subspecies. These include the Red-tailed Black-Cockatoo (Endangered ssp. *graptogyne*), Southern Emu-wren (ssp. *malachurus* and *polionotum*), Rufous Bristlebird (ssp. *broadbenti* and *caryochorus*), Striated Fieldwren (ssp. *albiloris* and *bourneorum*), Brown-headed Honeyeater (dark ssp. *wombeyi*), Pink Robin (ssp. *inexpectata* and, in summer, Tasmanian ssp. *rodinogaster*), Olive Whistler (ssp. *hesperus* and *bathychroa*), Australian Magpie (White-backed ssp. *hypoleuca*), Pied Currawong (ssp. *ashbyi*) and Silvereye (summer migrant ssp. *lateralis* from Tasmania). The endangered Helmeted Honeyeater, subspecies *cassidix* of the Yellow-tufted Honeyeater, was once considered a full species.

Croajingolong National Park

Key species: Eastern Ground Parrot, Glossy Black-Cockatoo, Topknot Pigeon, Eastern Bristlebird, Pilotbird, Black-faced Monarch, Rose Robin, Scarlet Honeyeater

Other species: Eastern Reef Egret, Black Bittern, Striated Heron, Leaden Flycatcher, Cicadabird, Brush Cuckoo, White-throated Nightjar, White-headed Pigeon, Channel-billed Cuckoo, Eastern Koel

Croajingolong National Park (87 500 ha) is situated in far-eastern Victoria and spans almost 100 km of wild coastline from Sydenham inlet (Bemm River) to the NSW border. It is a spectacular national park that provides many opportunities for recreational activities, ranging from some of the most tranquil flat water canoeing in inlets closed to motorised water craft, to nature-watching in near-pristine coastal forests. Accommodation options range from camping in the national park and local caravan parks, to motel and hotel rooms in the nearby towns of Orbost and Mallacoota. For further details contact Parks Victoria www.parkweb.vic.gov.au or 13 1969. Although spectacular coastlines and great birdwatching opportunities exist at Wingan Inlet, Thurra River, Tamboon Inlet and Sydenham Inlet, because these are all remote locations (and some require 4WD), here we focus on accessible areas in the eastern section of the park around the coastal town of Mallacoota.

In the temperate rainforest along the Double Creek Nature Walk (1 km return), 7 km west of Mallacoota, look for Azure Kingfisher (creek line), Satin Bowerbird, Superb Lyrebird, Large-billed Scrubwren, Brown Gerygone, Crescent Honeyeater, Red-browed Treecreeper (ridge-top) and summer east-coast migrants such Brush Cuckoo, Cicadabird, Black-faced Monarch, Scarlet Honeyeater and Rufous Fantail. In summer months, Scarlet Honeyeater can also be relatively common at the Shady Creek Bushland Reserve (30 ha), on the western edge of Mallacoota, particularly near the Lakeside Drive Bridge. Within the gully, search for summer visiting Black-faced Monarch, Cicadabird, Rufous Fantail and Satin Flycatcher, and resident species such as Eastern Whipbird, Satin Bowerbird, Superb Lyrebird, Wonga Pigeon, Large-billed Scrubwren, Brown Gerygone and Lewin's Honeyeater. Night birds here include Australian Owlet-Nightjar and Southern Boobook, while Powerful Owl sometime roost in dense cover near the bridge.

Shorebirds feed on the exposed sand banks below the Stephenson's Point Lookout in Mallacoota. Summer migrants include Red-necked Stint, Curlew and Sharp-tailed Sandpiper, Eastern Curlew, Whimbrel, Bar-tailed Godwit, Great and Red Knot and Grey-tailed Tattler, while resident species include Australian Pied and Sooty Oystercatcher. Beach Stone-curlew, a vagrant in Victoria, has been recorded here on several occasions. Other birds that may be encountered include Pacific Gull, Caspian, Crested, Fairy and Little Tern (the last nesting on the sand spit during summer), and rarer species at the southern extremity of their range such as Eastern Osprey and Eastern Reef Egret (rocks at Bastion Point). Summer waders also occur congregate at the Goodwin Sands, located in the northern side of Mallacoota Inlet.

Along the coast to the south-west of Mallacoota, the area around Shipwreck Creek contains some outstanding coastal heathland. With effort and patience, Eastern Ground Parrot may be flushed while walking through the heath: concentrate in areas that support knee-high vegetation. Here, you can also see Southern Emu-wren, White-eared and Tawny-crowed Honeyeater, Chestnut-rumped Heathwren and White-throated Needletail (summer). In the coastal forest bordering the heath, look for Brush Bronzewing, Brown Quail, Jacky Winter, Beautiful Firetail, Restless Flycatcher and, occasionally, Grey Goshawk. In summer Sacred Kingfisher, Leaden Flycatcher and White-throated Gerygone also occur. At night, White-throated Nightjar is sometimes encountered along the road into the Shipwreck Creek. In the warmer months, the campsite at Shipwreck Creek is also a good place to see Lace Monitor: one of Australia's largest lizards. Seal Creek is located several kilometres west of Shipwreck Creek – a search along the beach here may produce Hooded Plover, with a pair usually feeding along the beach line near the creeks inlet. Other good areas of coastal heathland can be found on the Heathland Walk (1 km return) off Betka Rd (on the way to Shipwreck Creek) and in areas adjacent to the Mallacoota airstrip. In both areas, you may encounter Striated Fieldwren, Tawny-crowned Honeyeater, Brown Quail, Southern Emu-wren and, occasionally, Eastern Ground Parrot. (If the Eastern Ground Parrot is a sought-after target, and observers have no luck in the Mallacoota area, then it many also be worth detouring to Mueller River (in the western section of the park) to search for this species in the heath along Bald Hills Track.) In summer, Davies Creek (on Betka Rd, 1 km from Mallacoota) can be a good site for Cicadabird, Scarlet Honeyeater and Varied Sittella (orange-winged form).

Gipsy Point in the upper reaches of Mallacoota Inlet is one of the best sites in Victoria to see Topknot Pigeon (a common bird further north into NSW, but rare in Victoria). One excellent area to birdwatch is near the jetty for the Gipsy Point Lodge and along the Point Track (3 km return). This pleasant walk leads out on a peninsula of the Genoa River. Along here, you may find the aforementioned Topknot Pigeon (summer), Wonga Pigeon, Bell Miner, Azure Kingfisher and Satin Bowerbird. Rare east-coast visitors and stragglers to Victoria that have been recorded here include Brown Cuckoo-Dove, Channel-billed Cuckoo, Striated Heron and Black Bittern – the last two are best seen by hiring a boat from the Gypsy Point Lodge and using this to explore the riverside vegetation upstream of the boat ramp.

Croajingolong National Park extends along more than 100 km of remote ocean beaches and rocky headlands. The ocean beaches provide habitat for Hooded Plover and Pied Oystercatcher, while the coastal heathlands support Southern Emu-wren, Striated Fieldwren, Tawny-crowned Honeyeater, and occasionally Eastern Ground Parrot.

The woodland along the road up to Mount Genoa (accessed 2 km west of the township of Genoa on the Princes Hwy) contains just the right mix of *Casuarina* woodland and old growth eucalyptus forest to support the rare Glossy Black-Cockatoo. There are invariably a few scattered pairs and family parties along this road, but they can be difficult to locate given they spend extended periods of time feeding quietly. One approach to find them is to drive slowly up to the peak, stopping at each large stand of *Casuarina*, checking for chewed seed pods on the ground and looking and listening for their distinctive call. It is also worth visiting Genoa Falls, where you may encounter Rose Robin, Olive Whistler and Australian King-Parrot, among others. Eastern Water Dragon – a spectacular lizard with a prehistoric appearance – can also be observed in the warmer months basking on the rocks at the falls and along the creek line.

Howe Flat is located on the northern side of the Mallacoota Inlet and is well known among local birdwatchers as the only remaining Victorian site that supports the Endangered Eastern Bristlebird. The species inhabits the heath around the short boardwalk and foot track that leads south to the coast from the Howe Flat vehicle track. To get to Howe Flat Track, the best access (from the Princes Hwy in NSW) is via Maxwells Rd, then Mines Rd, Buckland Rd and the Lakeview Track (4WD only). Access can also be gained by boat across Mallacoota Inlet. In summer, there are usually good numbers of Scarlet Honeyeater and White-throated Gerygone in this area and the Turquoise Parrot is being reported with increasing frequency. Finally, White-throated Needletail occurs over much of Victoria during the summer months, but is most common over extensive wooded forests and coastal heaths in the far east; being highly mobile, there is no single site, and rather best areas are from about Orbost through to Mallacoota and the state border.

Cape Conran Coastal Park and Cabbage Tree Creek Flora Reserve

Key species: Topknot Pigeon, Eastern Ground Parrot, Masked Owl, Sooty Owl, Scarlet Honeyeater
Other species: Southern Emu-wren, White-bellied Sea-Eagle

Cape Conran Coastal Park (11 700 ha) is a popular reserve well known for its spectacular ocean beaches, extensive heathlands and tall coastal forests. In addition to the many birdwatching opportunities, the area has been the subject of an intensive feral predator control program (mostly targeting foxes), the success of which has led to small mammals such as Long-nosed and Southern Brown Bandicoot and Sugar Glider being a regular feature of the camping areas. Cape Conran is located 18 km from Marlo in east Gippsland and 396 km east of Melbourne. Cape Conran has camping and cabin-style accommodation (see parkweb.vic.gov.au) and there are several caravan parks in Marlo.

Cape Conran Coastal Park has a series of graded walks through coastal woodland and around the rocky coastline where you may see Sooty Oystercatcher, Ruddy Turnstone (summer) and Eastern Reef Egret (chance). The diminutive Hooded Plover regularly feeds on the beach at Cape Conran, while both Little and Fairy Tern feed along the coastal beaches and congregate on sandy spits adjacent to Sydenham Inlet. The Eastern Ground Parrot and Southern Emu-wren inhabit the coastal heathland near the beginning of the Estuary View Trail (1.5 km) immediately to the east of the Yeerung River Bridge, while Scarlet Honeyeater (summer), Rose Robin and Brush Bronzewing often frequent the dense stands of coastal eucalypts further along the trail. The delightful Beautiful Firetail is also occasionally observed near the Yeerung River Bridge. Turquoise Parrots were once considered very rare in East Gippsland, but in recent years appear to have made a comeback in the Cape Conran Coastal Park. One area where this species is regularly seen is near the junction of the Old Coast and Gravel Rds and along the Old Coast Rd (between Cape Conran and Bemm River). The Old Coast Rd can also be a productive area for night birds especially Masked Owl: the most accessible area is along the Old Coast Rd near where it intersects with the Cabbage Tree–Cape Conran Rd. Other night birds that occur in the Cape Conran Coastal Park include Sooty and Powerful Owl, Southern Boobook and White-throated Nightjar.

The nearby Cabbage Tree Creek Flora Reserve (1700 ha) is accessed from the Princes Hwy 22 km east of Orbost. The Reserve contains temperate rainforest and a unique stand of Cabbage Tree Palms (Victoria's only native palm). The forest surrounding the picnic area and short nature walk provides opportunities for a range of wetter forest species, including Superb Lyrebird, Pilotbird, Satin Bowerbird, Rose Robin, Bassian Thrush, Crescent and Scarlet (summer) Honeyeater, Brown Gerygone and Large-billed Scrubwren. During summer, increasing numbers of east-coast migrants that are otherwise rare in Victoria have also been reported from this small reserve. Such rarities include Channel-billed Cuckoo, Eastern Koel, Topknot and White-headed Pigeon and Spangled Drongo. In the drier sections of the reserve immediately to the north and

Victoria

south of the picnic ground, there have also been sightings of the Square-tailed Kite, Grey Goshawk, Turquoise Parrot and Spotted Quail-thrush.

Further to the west, the Snowy River estuary at Mots Beach and Frenchs Narrows in Marlo Coastal Reserve (1000 ha, accessed via the Marlo–Conran Rd), support healthy populations of Striated Fieldwren and White-fronted Chat. Look along the shoreline for White-bellied Sea-Eagle, Little and Caspian Tern, and Eastern Curlew. Dense cover adjacent to exposed mud flats may provide opportunities to see the elusive Lewin's Rail. Nearby, the coastal heathland adjacent to the Marlo aerodrome (sign posted off the Marlo–Conran Rd) provides habitat for Southern Emu-wren and Eastern Ground Parrot, while the adjacent woodland supports, among others, Beautiful Firetail.

Macleod Morass

In Bairnsdale the Macleod Morass (520 ha) is a deep freshwater marsh where you have a chance of observing Musk, Pink-eared and Blue-billed and Freckled (rare) Duck, Hardhead, Plumed Whistling-Duck (very rare), Royal Spoonbill, Black-winged Stilt and skulking species such as Baillon's, Australian Spotted and Spotless Crake, Lewin's and Buff-banded Rail and Australasian Bittern. Summer migrants that visit the wetland include Sharp-tailed, Wood (rare) and Marsh Sandpiper and Latham's Snipe. The best access to Macleod Morass is via the Macleod Morass boardwalk (500 m), which starts on Bosworth Rd, 1.3 km east of Forge Creek Rd.

Fairy Dell, Canni Creek and Buchan Caves reserves

Key species: Powerful Owl, Turquoise Parrot, Hooded Robin, Red-capped Robin, Scarlet Honeyeater, Black-faced Monarch, Spotted Quail-thrush, Cicadabird, Diamond Firetail
Other species: Leaden Flycatcher, White-throated Gerygone, Brown Gerygone, Rose Robin, Rufous Fantail

North-west of Bairnsdale, there are three interesting reserves: Fairy Dell Nature Conservation Reserve (80 ha), Canni Creek Reserve and Buchan Caves Reserves. All support a good selection of species that are approaching their westerly limit in Victoria. There are powered and unpowered campsites and self-contained units at Buchan Caves Reserves; bookings are required during holiday periods (www.parkweb.vic.gov.au). The nearby towns of Buchan, Bruthen and Bairnsdale provide additional accommodation options.

Fairy Dell Nature Conservation Reserve protects a small, but significant, pocket of warm temperate rainforest at the western extreme of this habitat's distribution. From Bairnsdale, take the Great Alpine Rd to Wiseleigh before turning left onto Deep Creek Rd. This is a narrow, winding dry-weather-only road that leads directly to the Fairy Dell. A loop walk takes you through the rainforest, dominated by lilly pillies, ferns and vines. The best time to visit is when the spring and summer migrants are present. During this period, there are regular sightings of Cicadabird, Black-faced Monarch, Leaden Flycatcher, Scarlet Honeyeater, White-throated and Brown

Gerygone, Rose Robin and Rufous Fantail. At night, Powerful Owl sometimes roost near the car park in dense cover.

Canni Creek Reserve (200 ha), which incorporates the Buchan Racecourse and Buchan Valley Golf Club, is located 21 km north of Nowa Nowa on the Bruthen–Buchan Rd. Turquoise Parrot and Diamond Firetail occur in the reserve, especially along the gravel racecourse access road and in the eucalypt woodland behind the stall and betting shed. Canni Creek Reserve is also the most southerly location for Hooded (resident around the golf course) and Red-capped Robin in eastern Victoria. Other birds that may be encountered include Spotted Quail-thrush, Chestnut-rumped Heathwren, Crescent, White-naped and Yellow-tufted Honeyeater, Crested Shrike-tit, Dusky Woodswallow, White-throated Gerygone, Southern Emu-wren and Restless Flycatcher.

Buchan Caves Reserve (285 ha), 5 km north of Canni Creek, is another site for Diamond Firetail. In this reserve, there is also a chance of Superb Lyrebird, Wonga Pigeon, Gang-gang Cockatoo, Australian King-Parrot, Satin Bowerbird, Eastern Whipbird, White-winged Chough, Speckled Warbler, Leaden and Restless Flycatcher, Golden Whistler, Brown Gerygone, Bell Miner, Scarlet, Flame, Rose and Eastern Yellow Robin, and Jacky Winter. Pleasant bush walks that provide good opportunities for birdwatching include the Spring Creek Walk (3 km return), Granite Pools Walk (4 km return) and the F J Wilson Walk (2 km return).

Cape Liptrap Coastal Parks

Key species: Eastern Ground Parrot, Southern Emu-wren, Olive Whistler, Beautiful Firetail, Striated Fieldwren
Other species: Blue-winged Parrot, Tawny-crowned Honeyeater, Crescent Honeyeater, Brush Bronzewing, Brown Quail

Cape Liptrap Coastal Park (4174 ha), 160 km east of Melbourne, stretches along the coast from Point Smythe near Inverloch

Tarra-Bulga National Park

Tarra-Bulga National Park (2015 ha) protects a remnant temperate rainforest in the Strzelecki Ranges. Among the wet forest birds are Superb Lyrebird, Olive Whistler, Pilotbird, Crescent Honeyeater, Eastern Whipbird, Crested Shrike-tit, Rufous Fantail, Eastern Yellow, Pink, Rose and Flame (summer) Robin, Large-billed Scrubwren and Bassian Thrush. At night, there is a chance of seeing Southern Boobook, Powerful and Sooty Owl. Excellent local walks include the Lyrebird Ridge Track (2.4 km return) and the Fern Gully Nature Walk (500 m loop) that leads to the Corrigans Suspension Bridge (1.2 km, one way). The high annual rainfall in the Strzelecki Ranges means that ferns are another feature of the area; more than 30 species occur along Fern Gully. The park is reached via the Princes Hwy at Traralgon by following Traralgon Creek Rd to Balook, or from Yarram via the Tarra Valley Rd or Bulga Park Rd.

to Waratah Bay near Walkerville. The park consists mainly of heathland, bordered by coastal tea-tree, Manna Gum and Messmate forests, with a shoreline of alternate sandy bays and rocky limestone outcrops. There is self-sufficient camping at Bear Gully.

Cape Liptrap Coastal Park contains extensive areas of heathlands that are home to a diverse range of birds, particularly along Walkerville Rd south of the Walkerville township. A good track for birding travels north from Walkerville Rd (1 km east from the intersection with Bear Gully Rd). Along the first kilometre, you may see Beautiful Firetail (often near the start of the track), Southern Emu-wren, Blue-winged Parrot, Tawny-crowned and Crescent Honeyeater, Brush Bronzewing, Striated Fieldwren, Brown Quail and Olive Whistler (in the gullies). There is a chance of flushing Eastern Ground Parrot, if you concentrate your search in knee-high heath. Beautiful Firetail, Southern Emu-wren and Blue-winged Parrot are also found in the heath around the Lookout carpark, on Walkerville Rd, 1.7 km west of the Fish Creek–Walkerville Rd (the main street in Walkerville). Mammals to look for here include Common Wombat, Black Wallaby and Echidna. Walkerville Rd, between the Walkerville township and the Walkerville Kiosk, is good for forest birds such as Satin Flycatcher (summer), Crescent Honeyeater, Olive and Golden Whistler, Eastern Whipbird, Crimson and Eastern Rosella and Brush and Fan-tailed Cuckoo. Bear Gully is also a reliable site for Olive Whistler, with the species often being common in the campground itself. Forest Raven also occurs in the campground, though note that both the Little and Australian Raven are common at Cape Liptrap. Along the coast, near the Bear Gully Campground, you may see Hooded Plover, Pacific Gull and Sooty Oystercatcher. The Eastern Reef Egret, which is rare in Victoria away from sites in far East Gippsland (p. 271), has been sighted here on occasion. Cape Liptrap provides good seabird watching from the headlands. Look close to shore for White-bellied Sea-Eagle, Australasian Gannet, Kelp and Pacific Gull, Black-faced Cormorant and Little Penguin, and further offshore for Shy Albatross, White-fronted Tern, and Short-tailed, Fluttering and Hutton's Shearwater. In winter and spring, Humpback and Southern Right Whale are also occasionally sighted.

Southern Emu-wren, in this case a male, frequent remnant heathlands scattered along the length of the Victorian coastline. Cape Liptrap is one such site, but the species can also be reasonably expected in Croajingolong, Wilsons Promontory and Port Campbell National Parks, as well as heathlands adjacent to both Anglesea and Portland.

Wilsons Promontory National Park and Shallow and Corner Inlets

Key species: Gang-gang Cockatoo, Eastern Ground Parrot, Crimson Rosella, Brush Bronzewing, Striated Fieldwren, Southern Emu-wren, shorebirds
Other species: Emu, Cape Barren Goose, White-bellied Sea Eagle, Pacific Gull, Blue-winged Parrot, Brown Quail

Wilsons Promontory National Park (50 512 ha), popularly known as 'the Prom', is a jewel within Victoria's reserve system. It is the most southerly point of mainland Australia and protects a large area of native vegetation on a wild promontory adjacent to Shallow and Corner Inlets. The national park includes remote ocean beaches and quiet inlets, a series of low mountain ranges, where towering granite boulders are a prominent feature, plus a variety of habitats ranging from expansive intertidal flats (within Shallow and Corner Inlets), extensive coastal heaths and rare pockets of cool temperate rainforest. Located ~200 km south-east of Melbourne, this is one of the state's most popular national parks. A fully serviced campground is maintained at Tidal River, but bookings through holiday periods and many long weekends are essential. Indeed, camping is so popular that a ballot system is used to allocate sites over the busy Christmas and Easter periods. Tidal River also has upmarket cabin accommodation (bookings essential) and a small general store (www.parkweb.vic.gov.au). Owing to the popularity of the park, there are growing number of accommodation options near the park entrance and back towards the small town of Yanakie.

Once in the park, along the main road some 10 km from the entrance, a short diversion along Five Mile Rd (the road to Millers Landing) provides access to low heathland where Eastern Ground Parrot, Southern Emu-wren and Beautiful Firetail have all occurred in the past. Starting at Five Mile carpark, the Millers Landing Link Track (2 km) winds its way through open heathy woodland and *Banksia* and Stringybark woodland to Millers Landing. Along here, look for Yellow-tailed Black-Cockatoo, Crimson Rosella, Little Wattlebird, and New Holland and Crescent Honeyeater. Located on the southern coast of Corner Inlet, the landing protects the southernmost stand of mangroves in the world. Although now very rare, there are some recent (last decade) sightings of the Eastern Ground Parrot along the fire breaks that cut through the heathlands adjacent to the Millers Landing walking track. Back on the main road to Tidal River, another short diversion can be made on the track to Cotters Lake. There, in the coastal woodlands, look for Brush Bronzewing, Crescent Honeyeater, Scarlet Robin and Bassian Thrush, while further along the track, in an areas of sedges and heathland, Striated Fieldwren and Southern Emu-wren occur. Another area worth a short stop is the regenerating grassy flats visible on either side of the main road just south of the Cotters Lake track. Here, large mobs of Eastern Grey Kangaroo, Wombat (often out in the day) and Emu are generally easy to find. Keep an eye out for less common mammals such as Echidna and the introduced Hog Deer. These open areas of rough grassland can also be good for Striated Fieldwren, Brown Quail and White-fronted Chat.

At the Tidal River campground, listen for the soft contact calls made by Yellow-tailed Black-Cockatoo and Gang-gang Cockatoo as they feed quietly among the Coastal Banksia and Coastal Wattle scattered throughout the camping sites. The trees around the general store and visitor's centre are a reliable area for Crimson Rosella. Radiating from the Tidal River campground and recreational area are a large number of excellent walking tracks, ranging from short nature walks to extended overnight hikes into the southern wilderness of the Prom. One excellent walk for birdwatching crosses Tidal River on the footbridge, strikes north across heathlands that support Striated Fieldwren and Southern Emu-wren and links with the Lilly Pilly Gully Nature Walk (5 km). Along this later track, you can expect species such as Eastern Yellow Robin, Crescent and New Holland Honeyeater, Superb Fairy-wren and Bassian Thrush. This is one of the better areas within the national park for Koala: look in the taller eucalypts about halfway along the walk. At night, Powerful Owl has also been recorded here. The South Norman/Biddy's Track (1.5 km) and Squeaky Beach via Tidal Overlook walks (3.5 km) pass through a mix of low coastal woodlands and open coastal heathlands, with occasional views to the rocky coast below. Species that may be encountered here include White-bellied Sea-Eagle, Blue-winged Parrot, Brush Bronzewing, White-fronted Chat, Striated Fieldwren and Forest Raven. There are usually a pair of the threatened Hooded Plover to be found on Norman Bay (especially the southern end) and the species occurs periodically on the other sandy ocean beaches at the Prom.

The offshore islands at Wilsons Promontory (some visible through binoculars or a spotting scope from Norman Bay) are significant breeding locations for Cape Barren Goose, Crested Tern, Sooty Oystercatcher, Silver and Pacific Gull (Victoria's only breeding site for Pacific Gulls), Short-tailed Shearwater, Fairy Prion and Common Diving-Petrel. Another excellent walk is the track to the top of Mt Oberon, where walkers are rewarded with a spectacular vista looking back over Norman Bay, Tidal River and beyond. Although much of the mountain was burnt by extensive fires in 2009, and some areas were subsequently damaged by extensive flooding in 2011, this walk still provides excellent birding and a chance to see species such as Brush Bronzewing, Crimson Rosella, Gang-gang Cockatoo, Brown and Striated Thornbill, New Holland, Crescent, White-naped and Yellow-faced Honeyeater, Rose Robin and Crested Shrike-tit. Beautiful Firetail sometimes congregate in areas with mature tea-tree and sword grass, where they feed on fallen seeds, while Pilotbird

The ocean beach adjacent to Shallow Inlet is one of just two regular sites for Sanderling flocks in Victoria. Check the wave-washed shoreline towards the end of the sandy spit (the other site is in western Victoria along Killarney Beach).

occur in the wettest gullies that are crossed by the Mt Oberon Track.

Shallow Inlet is on the shore of Waratah Bay, 20 km east of Walkerville. The Hourigan Camp Lane Walk (400 m) leads along a boardwalk, through a sheltered gully, to the sandy tidal flats of the inlet. At low tide, on a stroll along the beach you may see summer migrants such Pacific Golden Plover, Eastern Curlew and Whimbrel, while Sooty and Pied Oystercatcher are year-round residents. In the adjacent coastal sand dunes, Blue-winged Parrots occur. Corner Inlet – a Ramsar site also known as Nooramunga Marine and Coastal Park (30 170 ha) and Corner Inlet Marine and Coastal Park (28 500 ha) – supports intertidal mudflats, saltmarsh and the southern-most mangroves in the world. The area is of significance because it regularly supports up to 30 000 shorebirds of both migratory and resident species, representing some 20% of the Victorian shorebird population. This includes many species that are considered rare elsewhere in the state, including Grey Plover, Lesser Sand Plover, Eastern Curlew, Whimbrel, and Red and Great Knot. Other species that can occur in large numbers include Bar-tailed Godwit, Red-necked Stint, Curlew Sandpiper and Sanderling (the last especially on the ocean beach and sandy spit at Shallow Inlet). The inlets and the many small islands within Corner Inlet and around Wilson's Promontory are also a stronghold for Sooty and Australian Pied Oystercatcher. Waterbirds recorded

Phillip Island

Phillip Island is located ~140 km south-south-east from Melbourne by road. The Penguin Parade (www.penguins.org.au) is one of Australia's most popular wildlife attractions and tourist destinations. Hundreds of Little Penguins emerge from the sea each evening and march across the beach to their burrows in the nearby sand dunes. Cape Barren Goose mostly occur on small offshore islands in Bass Strait, but in spring and summer they venture to the lush pasture areas on Phillip Island in search of food – a pair is regularly seen grazing near the entrance to the Penguin Parade. Little Penguin also occur at the Nobbies, where, from the lookout, Sooty Oystercatcher, Black-faced Cormorant, Kelp Gull and Crested (breeding in summer) and White-fronted Tern (winter) can often be sighted below. Seabird watching from here may be rewarded with sightings of Short-tailed (Sept–Apr), Fluttering and Hutton's Shearwater, Shy and Black-browed and, occasionally, the largest of them all, Wandering Albatross. At the main carpark at Cape Woolamai, between late September and April, thousands of Short-tailed Shearwater return to their colony *en masse* at sunset. An estimated 1 million birds make Phillip Island their home, with most occupying breeding burrows around the cape. On the Cape Woolamai Loop Walk (8 km), you may also see Hooded Plover, Blue-winged Parrot and Peregrine Falcon.

Victoria

here include Cape Barren Goose, White-bellied Sea Eagle, Hooded Plover, and Crested, Caspian, Fairy and Little Tern. Three interesting parrots, Eastern Ground Parrot (uncommon), Blue-winged Parrot and Orange-bellied Parrot (very rare), have all been reported in coastal vegetation adjacent to Shallow and Corner Inlet.

Royal Park and Melbourne Zoo

Key species: Swift Parrot and other parrots, lorikeets and cockatoos, Song Thrush, Eurasian Tree Sparrow
Other species: Hardhead, Baillon's Crake, Australian Spotted Crake

Royal Park is the largest of Melbourne's inner city parks (181 ha) and is located just 4 km north of the CBD. For an urban park, it protects a surprising diversity of habitats, including areas of remnant vegetation and extensive replanting of indigenous vegetation. Specifically, the park includes areas of native grassland, two very bird-friendly wetlands and lightly timbered areas with eucalypts, casuarinas and acacias. Not only is it a popular park for outdoor recreation, but the native vegetation also attracts a surprising number of native birds. As an added bonus, Melbourne Zoo, with its impressive collection of captive birds and animals, is also here. Access is straightforward, with ample parking and direct links by both train and tram to Melbourne's city centre.

A good place to start is near the rear entrance to the Melbourne Zoo (near Royal Park train station), particularly around the golf course. In autumn and early winter, when the Spotted Gum along the tramline is flowering, this is an excellent area for nectar feeders. The more common species seen here include Rainbow and Musk Lorikeet, but less commonly, you may see Swift Parrot, as well as Little, Scaly-breasted and Purple-crowned Lorikeet. Other parrots and cockatoos include the common Red-rumped Parrot, Eastern Rosella, Sulphur-crested Cockatoo, Little and Long-billed Corella and, occasionally, Yellow-tailed Black-Cockatoo. The Trin Warren Tam-boore wetlands, on the western side of Royal Park, attract a good selection of waterbirds. Some of the species recorded here include Hardhead, Baillon's and Australian Spotted Crake, Purple Swamphen, Great Egret, White-faced Heron, Nankeen Night-Heron, Australasian Grebe, Common Sandpiper (rare), Latham's Snipe, Australian Reed-Warbler, Golden-headed Cisticola and Little Grassbird. Other birds recorded in the park include Peregrine Falcon, Collared Sparrowhawk, Flame Robin (winter), White-plumed Honeyeater,

Opportunities for children to interact with birds can be found at Melbourne Zoo and the Royal Botanic Gardens, while at Grant's Picnic Area in the Dandenong Ranges they can hand-feed parrots and cockatoos.

Spotted Pardalote, Crested Shrike-tit, Mistletoebird, Superb Fairy-wren, with occasional records of Little Eagle (mostly autumn), White-naped Honeyeater, Rufous and Brown Songlark and White-winged Triller (the last three mostly in summer).

Within the Melbourne Zoo grounds is the Great Flight Aviary: a huge aviary containing different established habitat zones, including woodland, wetland and rainforest. Some of the species held here include Southern Cassowary, Brolga, Black-necked Stork, Bush Stone-curlew, Royal Spoonbill, Eclectus Parrot and Red-tailed Black-Cockatoo. The main canteen area at the Zoo is a good site for wild Eurasian Tree Sparrow, which are patchily distributed through Melbourne. The Song Thrush is another introduced species that occurs in the zoo grounds (look in shrubby areas adjacent to lawns) that can be tricky to locate elsewhere in Melbourne. There is also a Bell Miner colony near the front entrance to the Zoo. (If you can't find either the Song Thrush or Eurasian Tree Sparrow here, they are also found (along with Bell Miner) at CERES Community Environment Park, a few kilometres to the north-east (on the corner of Roberts and Stewart Streets, Brunswick).

Yarra Bend Park

Just 5 km from Melbourne's CBD, Yarra Bend Park (260 ha) supports some excellent bushland patches in close proximity to Australia's second largest city. The reserve is an excellent site for parrots and cockatoos, particularly around the barbeque area east of the entrance to the Yarra Bend Golf Course and near Kane's Bridge. Here you may see Red-rumped Parrot, Rainbow, Musk and Little Lorikeet, Eastern and Crimson Rosella, Galah, Long-billed and Little Corella and Sulphur-crested Cockatoo. Both Gang-gang Cockatoo and Yellow-tailed Black-Cockatoo are occasional visitors, especially in winter. During the day, Nankeen Night-Heron roost near Kanes Bridge. A walk along the Macfarlane Burnett Walk (1 km loop) can be productive, watch for Laughing Kookaburra, Collared Sparrowhawk, Dusky Woodswallow (summer), Grey Butcherbird, Golden Whistler, Olive-backed Oriole (summer), Black-faced Cuckoo-shrike, Bell Miner, Spotted Pardalote and, at night, Tawny Frogmouth and occasionally Southern Boobook and Powerful Owl. Along the Yarra River, look for Sacred and Azure Kingfisher, Great Egret and Australasian Darter. Similar species are found near the Bellbird Picnic Area, where there is also a large colony of Grey-headed Flying-fox. Finally, in summer, especially ahead of thunderstorms, White-throated Needletail are occasionally encountered hawking for insects high over the sports fields. More rarely, Eastern Koel has been recorded using the large fig trees in the park (several summer records).

Victoria

Melbourne's mountain ranges (the Dandenongs, Toolangi and Yellingbo)

Key species: Superb Lyrebird, Sooty Owl, Rose Robin, Pink Robin, Olive Whistler, Pilotbird, Red-browed Treecreeper, Yellow-tufted (Helmeted) Honeyeater
Other species: Australian King-Parrot, Laughing Kookaburra, Eastern Whipbird, Brown Gerygone, Large-billed Scrubwren

Superb Lyrebird is a famous inhabitant of the Dandenong Ranges National Park (3000 ha), which is the closest location to Melbourne where it may be observed. The park is a 40 km drive from Melbourne's CBD via Burwood Hwy or Canterbury Rd. Sherbrooke Forest, on the southern section of the Dandenong Ranges, is one of the best spots to see Superb Lyrebird; look along the Sherbrooke Falls Walk (4 km) and Eastern Sherbrooke Forest Walk (7 km loop). In the Ferntree Gully section of the park, Superb Lyrebird occurs along the Lyrebird Track (1.5 km), particularly along the Ferntree Gully Creek and on the One Tree Hill Walk (3.7 km). The last two walking tracks can be accessed via public transport from Melbourne, by taking the train to Upper Ferntree Gully station then walking east adjacent to Burwood Hwy and Mountain Hwy for 800 m to reach the park entrance (the walking tracks commence at the top end of the carpark). Some of the lyrebirds are relatively used to people in these areas and a quiet approach may reward the observer with great views. Grants Picnic Area is a nearby site where hand-tame Crimson Rosella, Australian King-Parrot, Sulphur-crested Cockatoo and Laughing Kookaburra congregate for handouts from visitors. Although this is not everyone's ideal, it does provide great opportunities for children and those that are less mobile to interact with vibrant wildlife. Other common species in the Dandenong Ranges National Park include Golden Whistler, Eastern Whipbird, White-browed Scrubwren, White-throated Treecreeper and Crescent, Yellow-faced and White-naped Honeyeater. Less commonly, Lewin's Honeyeater, Brown Gerygone, Rose and Pink Robin, Olive Whistler, Pilotbird, Bassian Thrush, Large-billed Scrubwren and Red-browed Treecreeper also occur.

Tree fern gully within Mountain Ash forest. The gully provides habitat for Superb Lyrebird, Pilotbird, Eastern Whipbird and Pink Robin, while the tall trees in the background are suitable for Gang-gang Cockatoo and Red-browed Treecreeper.

Further afield, in the Toolangi State Forest (38 000 ha), (80 km by road from the Melbourne CBD and approached from the towns of either Toolangi or Healesville along

Myers Creek Rd), the Wirrawilla Rainforest Walk (800 m) and the adjacent unsigned Quarry Rd (walking track), provides access to some excellent cool temperate rainforest patches. In late spring and summer, this is a very pleasant walk that provides opportunities for wet forest species such as Superb Lyrebird, Rose and Pink Robin, Eastern Whipbird, Olive Whistler, Pilotbird, Bassian Thrush, Large-billed Scrubwren, Red-browed Treecreeper, and Lewin's and Crescent Honeyeater. Most of these species occur around the carpark and along the first 500 m of Quarry Rd. Wirrawilla is located 4 km down Sylvia Creek Rd, accessed from Myers Creek Rd 13 km from Healesville. This entire forest system is also a stronghold of the Sooty Owl. At night, listening and spotlighting for them in the taller wet forest along Myers Creek Rd may prove rewarding. The species is also found in the forest that surrounds the picnic ground at Badgers Weir (4.5 km east of Healesville along Don Rd), part of the Yarra Ranges National Park (76 003 ha). Further to the north, Masons Falls and Jehosaphat Gully in Kinglake National Park (22 360) (~40 km north-west of Healesville and ~50 km north north-east of Melbourne CBD) are additional excellent locations for Superb Lyrebird, Golden and Olive Whistler, Large-billed Scrubwren, Eastern Whipbird, Red-browed Treecreeper, Rufous Fantail (summer), and Flame and Rose Robin. In the drier eucalypt woodlands, especially on the upper slopes and ridgelines, Spotted Quail-thrush is a possibility.

Yellingbo Nature Conservation Reserve (9600 ha) provides a refuge for one of the last wild colonies of Helmeted Honeyeater (ssp. *cassidix* of the Yellow-tufted Honeyeater), one of Victoria's rarest birds and the state emblem. Other species found here include Bell Miner, Red-browed Finch, Beautiful Firetail, Powerful Owl and, in summer, Olive-backed Oriole. Public access is limited to the carpark at Parslows Bridge on Beenak Rd, so a better strategy for birding here is to join one of the guided tours that are occasionally organised by the Friends of the Helmeted Honeyeater group (www.helmetedhoneyeater.org.au).

A young male Superb Lyrebird 'working' the leaf litter. Masons Falls in Kinglake National Park and the Ferntree Gully section of Dandenong Ranges National Park are two sites where relatively tame lyrebirds are regularly observed.

Bunyip State Park and Glen Nayook Nature Reserve

Key species: White-throated Nightjar (summer), White-throated Needletail (summer), Sooty Owl, Southern Emu-wren, Black-faced Monarch (summer), Pilotbird, Olive Whistler, Beautiful Firetail
Other species: Brush Cuckoo (summer), Cicadabird (summer), Red-browed Treecreeper, Brown Gerygone, Superb Lyrebird, Wonga Pigeon, Bassian Thrush, Large-billed Scrubwren

Bunyip State Park (16 600 ha) is growing in popularity as a destination near to Melbourne that provides a diverse range of birdwatching opportunities in a wide range of habitats (wet and dry forests and heathlands). In the dense heathy woodlands along the Circuit Walk (3 km) in the Buttongrass Nature Reserve, you may encounter Southern Emu-wren, White-eared and Crescent Honeyeater, Striated Thornbill, Olive Whistler and Beautiful Firetail. To get there from the Gembrook–Tonimbuk Rd, turn off at Camp Rd (11 km east of Gembrook) and then proceed 200 m to a small carpark at Guide Track. Further to the north within Bunyip State Park, the helipad just off Back Creek Rd is a reliable site for White-throated Nightjar at dusk in the warmer months. White-throated Needletail are also often sighted from here, especially on summer evenings as they travel up the valley to roost. To get there from the Gembrook–Tonimbuk Rd, turn north onto Link Rd (6 km east of Gembrook) and after 7 km turn east along Back Creek Rd – the turn-in to the helipad is another 3.3 km. The area around the helipad is also a good site in summer for Cicadabird and Brush Cuckoo (summer). Good spots for Sooty Owl include the wetter forest areas along Link Rd and Ash Landing Rd (off Back Creek Rd, 1.3 km west of the helipad).

Glen Nayook Reserve is a remnant of a Mountain Ash forest on the east branch of the Tarago River. To get there, take the

You Yangs Regional Park

The You Yangs Regional Park (2000 ha) is 55 km south-west of Melbourne and 22 km north of Geelong. When the eucalypts are flowering (usually between April and September), search near the entrance for foraging Swift Parrot (uncommon), Musk, Little and Purple-crowned Lorikeet, and Yellow-faced, New Holland, White-naped, Black-chinned, Brown-headed and White-plumed Honeyeater. Several Tawny Frogmouth often roost in the larger trees near the visitor's centre, though they change location fairly frequently and some effort may be required to find them. In summer, one of the best birding areas in the park is on the far-west side along Hovells Creek, accessed via Sandy Creek Rd 2.6 km past the main entrance. The creek line attracts a range of species normally associated with drier woodland areas of northern Victoria. Here, look for Rainbow Bee-eater (spring–summer), Pallid Cuckoo (spring–summer), Dusky, White-browed and Masked Woodswallow (the last two mostly in summer), Restless Flycatcher, Jacky Winter, Brown Treecreeper, White-winged Triller, Spiny-cheeked and Black-chinned Honeyeater, Southern Whiteface, Diamond Firetail, Zebra Finch, occasionally, Red-backed Kingfisher (several summer records) and Black-eared Cuckoo (summer). Speckled Warbler occurs in the rocky areas of the park, while woodland areas support Painted Button-quail, Common Bronzewing, Crested Shrike-tit and Varied Sittella (orange-winged form). Resident raptors include Wedge-tailed and Little Eagle, Collared Sparrowhawk and Peregrine Falcon.

Princes Freeway from Melbourne and exit at Drouin to head towards Neerim Junction. From Neerim Junction, follow the signs to the Glen Nayook Reserve at the end of Paynters Rd. Here there is parking, toilets and picnic facilities. Along the Glen Nayook Rainforest Walk (1.5 km), which follows the Tarago River (to a gully where the river runs underground for 100 m), look for cool temperate rainforest species such as Brown Gerygone, Pilotbird, Superb Lyrebird, Wonga Pigeon, Bassian Thrush and Large-billed Scrubwren. During summer, it is the closest regular site to Melbourne for Black-faced Monarch. Other birds here include Collared Sparrowhawk, Australia King-Parrot, Gang-gang Cockatoo, Brush Cuckoo (summer), Red-browed Treecreeper, Lewin's, Crescent, White-naped and Scarlet (occasionally in summer) Honeyeater, Satin Flycatcher, Satin Bowerbird, Eastern Whipbird, and Rose and, less commonly, Pink Robin.

Western Treatment Plant

Key species: Orange-bellied Parrot (very rare), Blue-winged Parrot, White-winged Black Tern, a good diversity of raptors, huge numbers of waterbirds, waders and regular reports of vagrants
Other species: Cape Barren Goose, Brolga, Black Falcon, White-fronted Chat

The Western Treatment Plant (10 500 ha) is the largest sewage treatment facility in Australia and one of the best-known birding sites in the country. This is also one of the few well-known birding sites that birdwatchers have to themselves! Highlighting the importance of the site is the fact that at times, (summer) up to 20 000 shorebirds and 200 000 waterbirds may be present. All up, 270 bird species have been recorded in the plant and, indeed, in summer, observing more than 100 species in a day is achievable.

Owing to the extensive wetland systems and the frequent visitation and coverage by birdwatchers, the Western Treatment Plant has a reputation as one of the better sites for vagrant shorebirds and waterbirds in the country. Access to the plant is strictly by permit only (www.melbournewater.com.au) and a safety induction is mandatory before a permit is issued. Many local birdwatchers hold the necessary permit, and interstate and international visitors may find it easier to accompany a local birder for one-off visits. The Western Treatment Plant is located on the north-western shoreline of Port Phillip Bay, 50 km south-west of Melbourne. When approaching from Melbourne, turn left onto Point Wilson Rd from the Princes Freeway or, if coming from Geelong, take the Beach Rd turn-off. Permit holders are issued with a key, which provides access through several locked gates, as well as a detailed map (to which the below directions correspond).

In summer, at low tide, large numbers of shorebirds forage on the intertidal sand flats that border the plant to the south, such as the mouth of Little River, the 145W Outfall and (mainly on ebbing tides) the northern end of the lagoon at the Spit Nature Conservation Reserve (300 ha). The most common shorebirds include Red-necked Stint, Sharp-tailed and Curlew Sandpiper, Pacific Golden and Double-banded (winter) Plover, Australian Pied and Sooty (less common) Oystercatcher, Bar-tailed and Black-tailed Godwit, and Great and Red Knot. During high tide, large flocks of shorebirds move off the intertidal flats to roost on inland ponds within the plant. Some of the regular shorebird roost sites include the Paradise Rd Ponds, T-Section Lagoon, Walsh's Lagoon Pond 7, 35E Conservation Ponds and 270S Borrow Pit. Interesting species that occur here intermittently include Wood, Terek, Broad-billed and Pectoral Sandpiper, Ruff and Long-toed Stint. Other shorebirds that occur

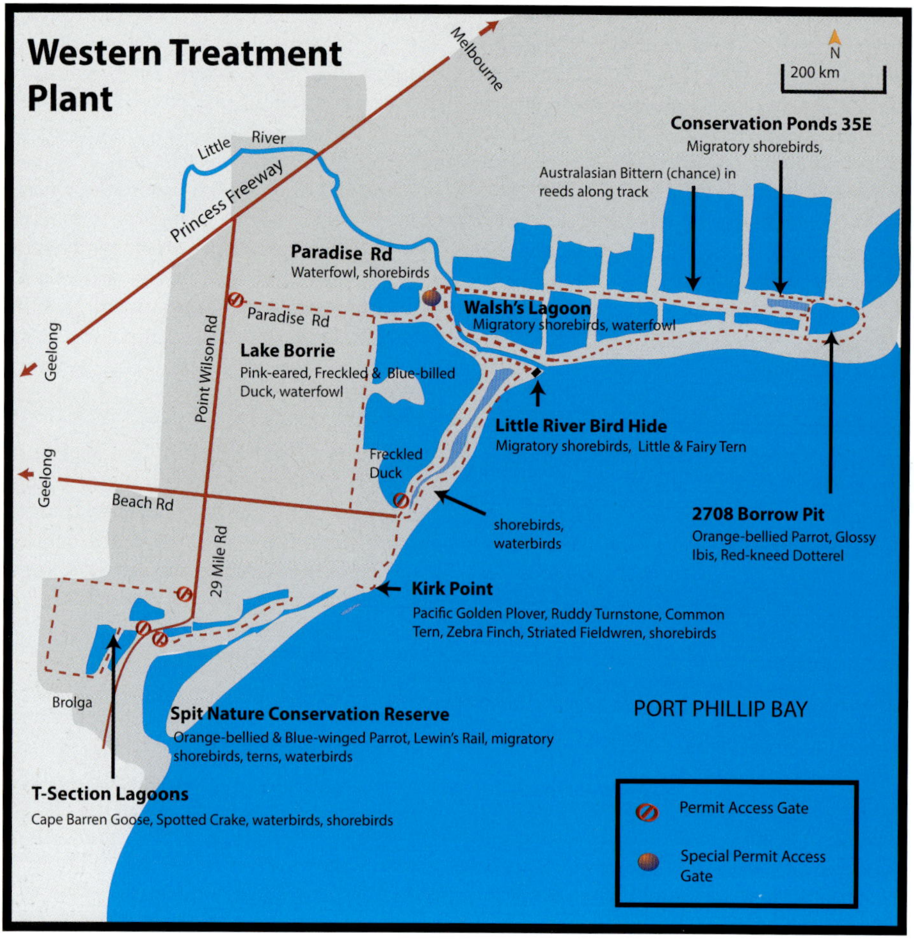

more regularly include Red-necked Avocet, Black-winged and Banded (nomadic) Stilt and Red-kneed and Black-fronted Dotterel. Pacific Golden Plover tend to prefer Kirk Point (at the end of Beach Road), feeding quietly among the rocks and seagrass wrack at low tide and roosting in nearby saltmarsh on high tide. On the point itself, check for roosting Crested, Caspian, Common, Little and Fairy Tern and also keep an eye out for Ruddy Turnstone, the odd Grey-tailed Tattler and Red Knot. Kirk Point is one of the few areas 'within' the Western Treatment Plant that is open to the general public and that can be reached without a key.

Back behind locked gates within the birdwatching permit area, Lake Borrie has a diverse range of waterfowl, including Pink-eared, Freckled (rare, but regular) and Blue-billed Duck, Chestnut and Grey Teal, Hardhead and Australasian Shoveler. Vagrant Northern Shoveler has been recorded here more than half a dozen times in recent decades. Whiskered and, less commonly, White-winged Black Tern (summer) feed over Lake Borrie and neighbouring ponds; if present, the latter species

Sunset over Lake Borrie – the largest of the 'ponds' at the Western Treatment Plant. The lake can host thousands of waterfowl, including many Australian Shoveler, and Pink-eared, Blue-billed and Musk Duck. Small numbers of Freckled Duck are also often present: check among the roosting waterfowl that are perched on fallen timber.

can be especially spectacular in autumn as they develop their striking breeding plumage. At low tide, scan the banks of the Little River and further to the south-west also scan the shores of the Spit Nature Conservation Reserve for Buff-banded and Lewin's Rail, and Australian Spotted, Spotless and Baillon's Crake. These species are often found skulking along muddy edges in close proximity to reedbeds and saltmarsh.

Other birds to look for at the Treatment Plant include Brolga (sometimes feeding in the saltmarsh at the end of Twenty Nine Mile Rd), Cape Barren Goose (especially T-section lagoons, but also grassy areas at the rear of Paradise Rd lagoons), Banded Lapwing (especially in paddocks with short grass along Avalon Rd) and Black-tailed Native-hen. Song birds that frequent the Treatment Plant include White-fronted Chat, Striated Fieldwren (along the track to Kirk Point), Zebra Finch, Skylark and Richards Pipit. A good diversity of raptors can be found throughout the plant, including Whistling, Black and Black-shouldered Kite, Swamp and Spotted (mostly summer) Harrier, Wedge-tailed Eagle, White-bellied Sea-Eagle, Australian Hobby, and Brown and Black Falcon.

Orange-bellied Parrot (winter) is (or at least was) a regular visitor to the Western Treatment Plant, although there have generally been fewer than 10 reports in recent years. This Critically Endangered species has declined rapidly over the past few years and the chances of observing them are increasingly slim. Potential areas include the saltmarsh along the Spit Nature Conservation Reserve, around the mouth of Little River and at the 270S Borrow Pit. With the species' demise in the wild now considered imminent, these once regular

Pink-eared Duck can on occasion occur in their tens of thousands at the Western Treatment Plant. Lake Borrie and neighbouring ponds usually host the largest concentrations, especially in summer.

feeding sites will soon be no more. The similar Blue-winged Parrot is also found at the Western Treatment Plant and some older birds may sport orange on their belly, so be careful with identification! Features that distinguish an Orange-bellied Parrot include the buzzing *'zzt-zzt-zzt'* call and their bright grass-green base colour. Be sure to report any sightings of Orange-bellied Parrot to Birdline Australia (www.eremaea.com) who will then pass the record onto the recovery team.

Geelong and Bellarine Peninsula

Key species: Australasian Bittern, Latham's Snipe, Banded Stilt, Whimbrel, Eastern Curlew, Hooded Plover, Black-faced Cormorant, Freckled Duck
Other species: Magpie Goose, Australian Spotted Crake, Spotless Crake, Baillon's Crake

The port city of Geelong is located on Corio Bay and the Barwon River, some 75 km south of Melbourne. The areas around

Mud Islands

Mud Islands (56 ha) are accessed by boat some 8 km east of Queenscliff in Port Phillip Bay. Dominated by Coast Saltbush on the seaward side and Austral Hollyhock on the dune ridges, the three islands are an important site for migratory shorebirds such as Bar-tailed Godwit, Red and Great Knot, Grey-tailed Tattler, Red-necked Stint, Common Greenshank, Ruddy Turnstone, Grey and Pacific Golden Plover and, occasionally, Greater and Lesser Sand Plover. Small numbers of the more common species remain through winter and, at the same time, Double-banded Plover (a winter migrant from NZ) arrive. In summer Fairy, Little and Common Tern feed in the shallows off the islands. Other species include Buff-banded Rail, Little Grassbird and Golden-headed Cisticola that all forage in the saltmarsh. There is a breeding colony of White-faced Storm-Petrel (~1000 pairs), which, along with Arctic Jaeger, can sometimes be seen on the way to the islands during a boat trip there. There are no scheduled ferry services to Mud Islands. Rather, groups such as the Friends of Mud Islands (www.vicnet.net.au/~fomi) organise day trips about once a month.

Geelong, including the adjacent Bellarine Peninsula, cover a diverse range of habitats, including coastal heathland, saltmarshes, bays, ocean, open pasture, rivers, marine estuaries, freshwater swamps and lakes, ephemeral salt lakes and saltworks. As the second largest city in Victoria, Geelong (and the Bellarine Peninsula) offers a wide range of accommodation options.

Belmont Common (64 ha) is an excellent site for a variety of waterbirds close to the city centre. A small bird hide provides excellent viewing opportunities across a small wetland along the entrance road to the Barwon Valley Golf Club off Barwon Heads Rd, just after it branches from the Princes Hwy. There are invariably a few Chestnut Teal, Pacific Black Duck, Eurasian Coot and Dusky Moorhen here. When the water levels are low, exposed areas of mud at the Common can provide excellent opportunities for hard-to-see waterbirds such as Buff-banded Rail, and Australian Spotted, Spotless and Baillon's Crake. In summer, Latham's Snipe can be numerous in the grass verges around the wetland. If the surrounding eucalypts are in flower, there is also an excellent chance that Purple-crowned Lorikeet will be present. Further out of town, along Drysdale (Port Arlington) Rd, the decommissioned Moolap and Point Henry Saltworks (500 ha) are visible to the north. At times, large numbers of Banded Stilt congregate here, along with Red-necked Avocet, Black-winged Stilt and Common Greenshank. Access to the saltworks is by permit only, but you can view the main ponds from the adjacent public roads.

Heading east onto the Bellarine Peninsula, Lake Connewarre and Reedy Lake make up a vast wetland complex that dominates the landscape. One of the best areas within this wetland system is Reedy Lake (180 ha), with various access points via Moolap Station Rd, Woods Rd, O'Halloran Rd or Fitzgerald Rd. Reedy Lake comprises a large freshwater lake densely covered in reeds. For birdwatching, this site is best when the water levels have receded somewhat to expose muddy shorelines. A host of waterbirds occur, including Australasian Bittern, Royal Spoonbill, Australian White and Straw-necked Ibis, Black Swan, Brolga, Australasian Shoveler, Hardhead, Red-kneed Dotterel, Magpie Goose and Swamp Harrier. On the southern side of this wetland is Hospital Swamp (enter via Lake Rd then Hospital Swamps Rd), which is another great place for waders and waterbirds. Brolga sometimes frequent the adjacent paddocks. In summer, the area can be teeming with migratory shorebirds if conditions are right (shallow waters and extensive mudflats). At these times, large flocks of Red-necked Stint and Sharp-tailed and Curlew Sandpiper may occur, with fewer numbers of Marsh Sandpipers, Common Greenshank and Black-tailed Godwit. Rarities reported here over the years have included Lesser Yellowlegs, Red-necked Phalarope, Ruff and Pectoral Sandpiper. Further downstream the wetland system, the Barwon River drains into Bass Strait at Barwon Heads. At low tide in spring and summer, the exposed sand flats and seagrass beds here in the estuary support Eastern Curlew, Bar-tailed Godwit, Red Knot, Common Sandpiper, and Caspian and Crested Tern. The best viewing areas are from the east side of the Barwon River, looking west, from Peers Crescent or the Ocean Grove Boat Ramp. For the keen birder, leave your car in Talbot Avenue, Barwon Heads and access the estuary from here on foot at low tide. Walking north-east along the exposed sand flats towards the saltmarsh and mangroves often yields good views of foraging Eastern Curlew. Make sure you begin the walk at the start of the low tide and be sure to return well before the tide turns to avoid being caught by the rising waters!

Lake Victoria (139 ha) lies to the east of the Barwon River Estuary and is reached from the west end of Emily St, Point Lonsdale. Take the sign-posted walking

Victoria

track, which follows the southern shoreline of Lake Victoria. Look out along the edges of the lake for Red-necked Avocet, Banded (seasonal) and Black-winged Stilt, as well as Hooded, Red-capped and Double-banded (winter) Plover. Little Egret and Fairy Tern also occasionally visit the lake. On the other side of the walking track is a brackish wetland that borders the golf course. In late summer, as water levels drop, it becomes a fantastic place to view and photograph Red-necked Stint, Sharp-tailed and Curlew Sandpiper, and Red-capped Plover. Check the local tide times and visit at high tide because stints and sandpipers that forage in nearby Swan Bay are forced off Swan Bay at this time and many of them move here to continue foraging. This is especially so in early Autumn when they are fattening up before their northward migration. Striated Fieldwren, Singing and Spiny-cheeked Honeyeater, White-fronted Chat, and Shining and Horsfield's Bronze-Cuckoo occur in the surrounding coastal vegetation. Flame Robin are often seen during winter.

From Point Lonsdale Lighthouse, look offshore for Black-faced Cormorant, Australasian Gannet, Short-tailed and Fluttering Shearwater and, especially during stormy weather, there is also a chance of Fairy Prion, Southern Giant-Petrel, and Shy and Black-browed Albatross. Along 13 Mile Beach near Black Rock, search the beach for Hooded, Red-capped and Double-banded (winter) Plover, Sooty Oystercatcher, Pacific Golden Plover and Ruddy Turnstone (summer). During summer, both Hooded and Red-capped Plover nest on this beach, so be mindful of nests or chicks and definitely keep dogs on leashes. Sanderling and Kelp Gull also occasionally occur here.

Edwards Point is a long saltmarsh-covered spit that projects south into Swan

The Bellarine Peninsula supports a great number of wetland habitats. Here, birdwatchers access a coastal saltmarsh site at Lake Connewarre.

Bay. The sandy tip is a favoured roosting site for migratory waders that feed in Swan bay during low tide. Many Little Pied Cormorant also roost here, as well as Double-banded Plover during winter. Enter from the eastern end of Beach Rd in St Leonards, allowing 2–3 h for the return walk. The saltmarsh along the walk supports Striated Fieldwren, Blue-winged Parrot and, very rarely in winter, Orange-bellied Parrot. Coastal birds that frequent the beach and the point include Caspian, Crested, Fairy and Little Tern, Pacific Golden, Grey and Double-banded (winter) Plover, Australian Pied Oystercatcher and, occasionally, Great Knot, and Greater Sand and Lesser Sand Plover. A flock of Eastern Curlew that use Swan Bay every summer often roost here as an alternative to the western shoreline of Swan Bay.

Finally, Lake Lorne in Drysdale, though small and somewhat urban, can hold a wide range of birdlife. Of note, it is a remarkably reliable site for Freckled Duck. This can be a difficult species to locate at the best of times, but it seems that whenever there are reports of this species in coastal Victoria some individuals find their way here. Other good species that are often recorded here include Blue-billed Duck, Black-fronted Dotterel, Australasian Darter and, in summer, Latham's Snipe (in the ankle-high grass around the edge of the drying wetland) and Shining Bronze-Cuckoo.

Great Ocean Road, Port Campbell National Park and Great Otway National Park

Key species: Rufous Bristlebird, Hooded Plover, Grey Goshawk, Olive Whistler, Rose Robin
Other species: Crescent Honeyeater, Forest Raven, Bassian Thrush, Tawny-crowned Honeyeater, Southern Emu-wren, Chestnut-rumped Heathwren

Regarded as one of the world's most scenic coastal drives, the spectacular Great Ocean Rd hugs the west coast of Victoria between Lorne and Apollo Bay. Continuing on from Apollo Bay, the road passes the Great Otway National Park (26 000 ha), which protects cool temperate rainforest, tall wet eucalypt forest and extensive coastal heathlands. Beyond here, the 'Shipwreck Coast' through Port Campbell National Park protects outstanding coastal features include the iconic Twelve Apostles, Loch Ard Gorge and London Bridge. There are good camping areas in the Great Otway National Park (www.parkweb.vic.gov.au) and numerous commercial accommodation options, ranging from caravan parks to upmarket guesthouses in the coastal towns along the Great Ocean Rd and the Shipwreck coast (www.visitvictoria.com).

Starting at the eastern end of the Great Ocean Rd, Rufous Bristlebird and Blue-winged Parrot (summer) feed in and around the carpark at Point Addis Lookout (between Torquay and Anglesea), while the lookout itself provides one of the better seabird watching points within a few hours drive of Melbourne. Here, expect Australasian Gannet, Crested and White-fronted Tern (winter) and Short-tailed (summer), Fluttering and, occasionally, Hutton's Shearwater. During and following onshore storms, there can be numbers of Indian Yellow-nosed, Black-browed and Shy Albatross. Less common species observed from here include Wandering and Sooty Albatross, and Cape, Great-winged and White-chinned Petrel. Peregrine Falcon breed on the ocean cliffs near the lookout and can occasionally be seen hurtling along the cliff edges. The Ironbark Basin Walk (4 km), located at the beginning of Point Addis Rd, takes you through a remnant Red Ironbark forest. Along here, look for Gang-gang and Yellow-tailed Black-Cockatoo, Painted Button-quail (look for the small cleared depressions, known as platelets and created while foraging, along the lower section of the walk), Bassian Thrush, Satin Flycatcher, Pink and Eastern Yellow Robin, Buff-rumped Thornbill (common around the carpark), and Crescent, White-naped, New Holland, Brown-headed, Yellow-faced and White-eared Honeyeater.

Further west, Anglesea Heath is famed for its wildflowers and hosts a rich variety of birds. One excellent area of heath is located adjacent to the corner of the Great Ocean Rd and O'Donohue Rd. Birds recorded here include Gang-gang Cockatoo, Blue-winged Parrot (spring–summer), Southern Emu-wren, Striated Fieldwren, Chestnut-rumped Heathwren (in the dense thicket bordering the heath), White-eared and Tawny-crowned Honeyeater, Scarlet Robin, Olive-backed Oriole (summer), Buff-rumped Thornbill and Forest Raven. Another good area of heath, which supports similar species, can be found along Harvey St, accessed via the northern end of O'Donohue Rd. Hooded Plover occur at several beaches along the Great Ocean Rd, with the sandy beach areas at Point Roadknight, Anglesea being one relatively reliable site.

The Sheoak Picnic Ground, at the end of Allendale Rd west of Lorne, is a superb area of temperate rainforest, where you may see Gang-gang Cockatoo, Australian King-Parrot, Satin Bowerbird, Bassian Thrush, Brush Cuckoo (summer), Olive Whistler, Brush Bronzewing, Crescent and Yellow-faced Honeyeater, Rufous Fantail, Rose Robin and Forest Raven. The best time for birdwatching here is early in the morning, when the sound of the dawn chorus is amplified within the valley. Similar species also occur along Sharps Track; it heads north just before the picnic ground. Night birds in the area include Southern Boobook,

Stringybark woodland (*Eucalyptus obliqua* and *E. baxteri*) on the more elevated areas of the Anglesea Heath. Here Chestnut-rumped Heathwren, Scarlet Robin and a diversity of honeyeaters can be expected. In the more open areas of heath, look especially for Southern Emu-wren and Striated Fieldwren.

Powerful and Masked (several records) Owl, and Australian Owlet-Nightjar. The forest around the picnic ground is also home to (often very vocal) Yellow-bellied Glider and Koala. Similar species are also found at the Grey River Picnic Area, located on the Grey River Rd, 6 km north of Kenneth River. Look for Koala in the trees along the way to the picnic ground.

West of Apollo Bay, Olive Whistler can be common in the campground at Blanket Bay, while Forest Raven, Pink Robin, Grey Goshawk (white morph), Bassian Thrush and Crescent Honeyeater occupy wet forest habitats. This area also has a large population of Koala. One of the last authenticated records of Eastern Ground Parrot (a dead bird found in 2003) in the Great Ocean Road region was recorded in the heath in the former Carlisle State Park (now part of the Great Otway National Park). Carlisle remains an excellent site for Blue-winged Parrot, Southern Emu-wren, Olive Whistler, Crescent Honeyeater, Golden Whistler, Scarlet Robin, Forest Raven and Beautiful Firetail. Some of the more productive areas of heath for bird-watching can be found along Carlisle–Gellibrand Rd and the Old Carlisle Track, 7 km west of Gellibrand. Located 3 km west of Lavers Hill on the Great Ocean Rd, the Melba Gully State Park is a small park that protect a rainforest gully. Well known for its colony of Glow Worms, during the day, look here for birds such as Gang-gang Cockatoo, Pink Robin, Bassian Thrush, Olive Whistler, Crescent Honeyeater and Forest Raven. At dawn and dusk, Platypus feed around the edge of Lake Elizabeth, located near the town-ship of Forrest. Osprey, rare in Victoria, has also been recorded at this lake, along with Grey (White) Goshawk, Gang-gang Cockatoo, Satin Bowerbird, Olive Whistler, Rose Robin and Bassian Thrush. At night, listen for Powerful Owl and Southern Boobook.

Cool temperate rainforest on a creek line in the Otway Ranges near Lorne. Species to be found here include Gang-gang Cockatoo, Olive Whistler, Rufous Fantail, Rose Robin, Satin Bowerbird and Bassian Thrush.

Further west, beyond the Great Otway National Park, the coastline is known as the Shipwreck Coast for its unforgiving nature that claimed numerous vessels in days gone by. The Port Campbell National Park includes spectacular coastal features, including the world-famous Twelve Apostles, as well as London Bridge, The Arch and Loch Ard Gorge. Surprisingly tame individuals of the usually elusive Rufous Bristlebird routinely feed in and around the carparks at all of these sites. The national park also

Portland and Port Fairy pelagic boat trips

Day-long pelagic trips regularly travel from the harbours of Portland and Port Fairy, visiting the continental shelf some 40–50 km off the coast in the Southern Ocean. These trips offer some of the world's best pelagic birdwatching. All 'Australian' albatross have been recorded, but those most often seen are Indian Yellow-nosed, Shy, Campbell, Black-browed, Buller's, Southern Royal, Antipodean and Wandering. Less common forms include Salvin's, Grey-headed (winter), Sooty (autumn), Light-mantled Sooty (winter/spring), and Northern Royal Albatross. The tube-nosed petrels can also be extremely well represented: in season, you may see Great-winged, White-headed, White-chinned, Cape, Southern and Northern Giant Petrel, while less commonly, Soft-plumaged, Cook's, Gould's and Blue Petrel, and Southern Fulmar are encountered. Common shearwaters can include Flesh-footed (autumn), Fluttering, Hutton's, Short-tailed (spring–autumn) and Sooty Shearwater. Other seabirds include Antarctic, Slender-billed and Fairy Prion, Wilson's, Grey-backed and White-faced Storm-Petrel, Common Diving-Petrel, Arctic, Pomarine and Long-tailed Jaeger, Brown Skua and, rarely, South Polar Skua. Terns include Arctic (rare), White-fronted and Common. Because these trips have been running since the late 1970s, there is now a long list of vagrant species that have been recorded. Some highlights include Atlantic, Barau's, Bulwer's, Grey, Kerguelen and Mottled Petrel, Buller's, Great, Little and Streaked Shearwater, and New Zealand Storm-Petrel. The Portland and Port Fairy Pelagics are generally organised as Birdlife Australia (www.birdlife.org.au) monthly activities, but, owing to their popularity, bookings are essential and at times it can be difficult to secure a place, especially at short notice.

The winter-visiting Cape Petrel is just one of over 60 species of species of seabird recorded on organised pelagic trips off Portland and Port Fairy.

supports expanses of coastal health around each of these destinations, with opportunities to see sought-after species such as Blue-winged Parrot (spring–summer), Southern Emu-wren, Striated Fieldwren and Tawny-crowned Honeyeater.

Finally, inland of the Shipwreck Coast, the Ralph Illidge Sanctuary (101 ha) – a Trust for Nature property located on the Cobden–Warrnambool Rd, 25 km west of Cobden – is an excellent site for Grey Goshawk (always white morph). Search the edge of the forest near the paddocks at the southern boundary of the property. Other birds here include Long-billed Corella (often abundant in nearby farmland), Gang-gang Cockatoo, Rufous Bristlebird, Bassian Thrush, Forest Raven, Rose Robin and, in summer, Rufous Fantail and Satin Flycatcher. Powerful Owl breed on the property and occasionally a roosting pair can be found in the acacias along the short creek walk adjacent to the picnic area.

Portland district

Key species: Red-tailed Black-Cockatoo, Lewin's Rail, Australasian Bittern, Rufous Bristlebird
Other species: Emu, Brown Skua, Cape Petrel, Southern Giant-Petrel, various albatross, White-fronted Tern, Arctic Tern, Azure Kingfisher

Portland, the site of Victoria's first permanent settlement in 1834, is located in the far south-west of the state, 360 km from Melbourne. The area boasts spectacular coastal scenery, ocean beaches, sheltered harbours, coastal wetlands and extensive coastal forests, especially along the Glenelg River. Extinct volcanos are also a feature of the district. There are numerous accommodation options in Portland and neighbouring towns, such as Port Fairy and Warrnambool to the east and Nelson to the west. There is also a wide selection of campgrounds within Lower Glenelg National Park and Discovery Bay Coastal Park (www.parkweb.vic.gov.au). In addition to excellent birding opportunities in the district, visitors can also enjoy fur seal colonies around Cape Bridgewater and an abundance of fauna in the coastal woodlands that includes Eastern Grey Kangaroo, Swamp and Red-necked Wallaby, Wombat and Koala. The forest along the Glenelg River also supports a large population of Yellow-bellied Glider: at night, listen for their distinctive shrieks and guttural rattles. In the winter months, Southern Right Whale occur in some of the sheltered bays (Warrnambool is the most reliable winter site for this species), while in summer Blue Whale congregate to feed on krill; at this time, a careful watch from Cape Bridgewater (or Cape Nelson, see below) may produce a sighting of the largest animal on Earth.

Portland Harbour, the only deep-water seaport between Adelaide and Melbourne, provides shelter for seabirds in rough weather, especially during winter. Look for Kelp and Pacific Gull, Crested and White-fronted (winter) Tern, Black-faced and Little Pied Cormorant, Australian Pied Oystercatcher and Little Penguin: all are regular in the harbor. In recent years, small numbers of Brown Skua have taken to resting on the Lee Breakwater during the winter months (the only mainland site where this occurs). Regular storm-driven species include Southern and Northern Giant-Petrel (both autumn to spring), and Black-browed and Shy Albatross, while rarer species have included Southern Fulmar, Cape Petrel and Arctic Tern. From June to October, Southern Right Whale occasionally enters Portland Harbour, sometimes coming within metres of the north side of the Lee Breakwater. The Portland Visitor Centre will have information on recent sightings.

Fawthrop Lagoon, located on Glenelg St (parking available 100 m past the

The only 'mainland' colony of Australian Gannets established on Point Danger in 1996 as a satellite colony of the large Lawrence Rocks colony (visible in the background). Access to the site can be gained off Madeira Packet Rd ~ 6 km from Portland. Occasionally Cape Gannet is also recorded.

Powerhouse Museum) in Portland is an excellent site for Lewin's Rail, as well as Spotless and Australian Spotted Crake. Search for these species from the Lagoon Walk (2.2 km), especially along the wooden walkway on the southern side of the wetland, in the vicinity of the railway line. Other birds recorded at Fawthrop Lagoon include Australasian Bittern (mainly winter), Royal Spoonbill, Swamp Harrier, Buff-banded Rail and Latham's Snipe (summer), and a good selection of water-fowl, including flocks of Chestnut Teal.

The spectacular coastal cliffs at Cape Nelson Lighthouse – part of Cape Nelson State Park (243 ha) – provide a superb vantage point for seabird watching. Accessible by either Cape Nelson Rd or the Scenic Rd along the coast, it is worth a careful scan at any time of year, but best results are typically obtained when there are strong onshore southerly and south-westerly winds blowing. The list of birds recorded from the point just beyond the lighthouse is impressive, with regular sightings of

At Cape Nelson, two excellent walks are the Sea Cliff Nature Walk (3 km) and the Lighthouse Loop Walk (6 km). Along either, there are good opportunities for Brush Bronzewing and other heathland and coastal shrubland birds.

Black-faced Cormorant, Crested and White-fronted Tern (winter), Arctic Jaeger (summer), Brown Skua (winter), Southern and Northern Giant-Petrel (both mostly winter), Cape Petrel (winter–spring), Fairy Prion, Short-tailed (spring–autumn) and Fluttering Shearwater, Black-browed, Shy and Indian Yellow-nosed Albatross, and White-bellied Sea-Eagle. Over the years, this well known sea-watching point has also produced sightings of less common species such as Southern Fulmar, Grey, Great-winged, White-headed and Blue Petrel, Common Diving-Petrel, White-faced Storm-Petrel, Sooty Shearwater and Wandering, Southern Royal, Buller's

Killarney Beach, Deen Maar and Logan's Beach

Killarney Beach is one of only a handful of sites in Victoria where summer-visiting Sanderling regularly congregate. The beach can be accessed at the end of Mahoneys Rd, 9 km east of Port Fairy, but also further to the east by taking the Basin Track (leading east of Mahoneys Rd) to one of several small carparks located behind the coastal dunes. Other birds that frequent this wave-washed beach include Hooded and Double-banded (winter) Plover, Sooty and Australian Pied Oystercatcher, Ruddy Turnstone, Red-necked Stint, Australia's three resident gull species, and Caspian and Crested Tern.

Deen Maar, on the western edge of Lake Yambuk, is a regular location for Magpie Goose. Part of the Framlingham Aboriginal Trust site, it is 18 km west of Port Fairy on the Princes Hwy. Other birds recorded at Deen Maar include Orange-bellied Parrot (now extremely rare), Eastern Grass Owl (several records, but very rare), Banded and Black-winged Stilt, Red-necked Avocet, Australasian Bittern (mostly winter), Lewin's Rail, Baillon's and Australian Spotted Crake, Latham's Snipe, Wood and Sharp-tailed Sandpiper (both summer), Red-kneed Dotterel, Striated Fieldwren and Southern Emu-wren. Access to Deen Maar is limited; contact the Eumeralla Backpackers at Yambuk, (03) 5568 4204, fram@standard.net.au. If access is not possible, the wetland can be viewed from the entrance, 3.8 km west of Yambuk on the Princess Hwy.

If visiting the area between June and October, be sure to also take time to visit Logan's Beach in Warrnambool, when calving Southern Right Whale are regularly just offshore from the viewing platform. From here, scan the ocean for seabirds, because there is always a chance of Kelp Gull, Black-faced Cormorant, and Shy, Black-browed and Indian Yellow-nosed Albatross. Along the beach, look also for Australian Pied Oystercatcher, Hooded Plover and Sanderling (summer). Logan's Beach is well sign-posted from the Princess Hwy.

and Grey-headed Albatross. Look also for mammals such as Australian Fur Seal, Bottlenosed and Common Dolphin, Southern Right (Jun–Oct) and Blue Whale (Dec–May). Along the cliffs, there are resident Peregrine Falcon that occasionally soar over the point. The heathlands and Soap Mallee woodlands surrounding the Cape Nelson Lighthouse provide some excellent opportunities for coastal bush birds, with good areas being around the lighthouse carpark, along the Sea Cliff Nature Walk (3 km) and the Lighthouse Loop Walk (6 km). Species include Rufous Bristlebird (sometimes in the carpark), Beautiful Firetail, Bassian Thrush, Brush Bronzewing, Olive Whistler, Pink Robin, Blue-winged Parrot (summer), Shining Bronze-Cuckoo (summer), Singing and New Holland Honeyeater, and Brown Thornbill.

The endangered south-eastern subspecies of the Red-tailed Black-Cockatoo (*graptogyne*) is restricted to Victoria's south-west and adjacent areas in South Australia. On occasions, the species occurs in the Lower Glenelg National Park (27 300 ha), 60 km west of Portland – good spots to look are along Croziers Rd, or Glenelg Drive (between Batterbys and Sapling Creek picnic areas). Other birds in Lower Glenelg National Park include Emu (common), Gang-gang Cockatoo, Forest Raven, Olive Whistler, Azure Kingfisher (along the Glenelg River) and Powerful Owl. Red-tailed Black-Cockatoo also occurs at Rennick State Forest, located on the Princes Hwy, adjacent to the South Australian border. Search along Old Caves Rd, on tracks on the west border of the forest, ~2 km south of the Princes Hwy. Further north, concentrations of Red-tailed Black-Cockatoo may be found seasonally in the Brown Stringybark forests in the Edenhope district (e.g. Dergholm State Park and Kadnook and Meereek State Forests).

Tasmania

Tasmania is an island state that has been isolated from the Australian mainland for ~10 000 years. With around 5400 km of coastline and 350 offshore islands, it covers an area of around 67 230 km^2. The island contains spectacular scenery and, impressively, almost half of the state remains cloaked in native vegetation. That much of this is protected by World Heritage status means Tasmania is recognised as one of Australia's premier ecotourism destinations. Tasmania's vegetation ranges from temperate rainforests, through wet and dry eucalypt forests to coastal heathlands and Buttongrass moorlands. Tasmania is Australia's most mountainous state– with 50 mountains reaching, or exceeding, an altitude of 1200 m (4000 feet). Tasmania's highest mountain is Mount Ossa, standing at 1617 m. Nineteen national parks protect large areas of the natural landscape, while an additional ~420 nature reserves are located within the state. For a comprehensive guide to Tasmania's national parks and reserves, visit Tasmania's Parks and Wildlife Service (www.parks.tas.gov.au). The Tasmanian Wilderness Area protects 1.38 million ha, or ~20 per cent of state, and is formally recognised through World Heritage listing.

In addition to its natural wonders, Tasmania also has a rich visible history, especially from the convict era, with sites such as the Port Arthur convict precinct on the must visit list. Being situated so far south and with nothing but clean ocean air striking the west coast, the island also has a reputation for high quality produce; a theme that extends to a large and still growing gourmet food market. The historic city of Hobart (founded in 1804) is Australia's second oldest capital city, after Sydney, and is located in the south-east of the state on the Derwent River. Daily scheduled flights to Hobart and Launceston (in the north-east) connect to most mainland Australian capital cities and there are smaller airports at Devonport, Wynyard (Burnie), Flinders Island and King Island. The *Spirit of Tasmania* ferry service operates between Melbourne and Devonport (www.spiritoftasmania.com.au). It is possible (and reasonably priced) to take your car across Bass Strait on the overnight ferry.

Tasmania has a mild temperate (oceanic) climate with four distinct seasons. Summer (Dec–Feb) is the hottest season – in Hobart the average maximum daily temperature is around 21°C and the minimum night-time temperatures around 11°C. In autumn (Mar–May), expect changeable weather conditions, ranging from warm to cold days, while birdwatching in winter (Jun–Aug) can be challenging because these months can be very cold and wet and access to many of the mountain areas is somewhat restricted due to snow. Even Mount Wellington, which sits adjacent to Hobart, is frequently snow-capped in winter. The weather in spring (Sept–Nov) is mild, although it is also the windiest time

The Green Rosella is one of Tasmania's 13 endemic bird species. Most, like the rosella, are relatively common and widespread and should be encountered if a variety of sites are included during a visit to the state.

of the year. Most birdwatchers visit during the late spring, summer and early autumn, when the temperatures are usually pleasant. However, if your focus is seeing the 13 resident endemic species, then these can be seen at any time of year.

Tasmania is a compact island, where excellent sealed and uncrowded highways provide relatively easy access to much of the state – indeed, the network of main highways means that most visitors that travel for a week or more end up completing a loop around the island. Hobart is ideally placed as a gateway to most of Tasmania's birding sites, with all sites easily reached by car. The Midland Hwy links Hobart to Launceston, the Bass Hwy travels across northern Tasmania, the Murchison, Ridgley and Zeehan Hwys travel to Queenstown down the north-west coast, while the Lyell Hwy links Queenstown back to Hobart. Tasmania's south is reached via the Huon and Channel Hwys. A 2WD vehicle will take you to the majority of birding sites in the region, but, if you intend to explore bush tracks, a 4WD is recommended (and at times essential). All major international and national rental car companies operate in Tasmania. For information on travelling to and around the state, including car hire, check the Discover Your Tasmania: www.discovertasmania.com. Travel information is also available from the Tasmanian Travel and Information Centre: www.hobarttravelcentre.com.au, (03) 6238 4222.

A comprehensive guide to Tasmania national parks, reserves and world heritage areas is available at www.parks.tas.gov.au. A valid park entry permit is required to visit Tasmania's national parks. Further information is available from the Parks and Wildlife Service: www.parks.tas.gov.au or (03) 6233 2621 or parkfees@parks.tas.gov.au.

With 5 days (or more) in and around Hobart, a suggested itinerary for birdwatchers involves:

- **Day 1.** Mount Wellington and Waterworks Reserve, ~50 km (Scrubtit, Tasmanian Scrubwren, Tasmanian Thornbill, Pink Robin, Black Currawong, Wedge-tailed Eagle, Swift Parrot (summer), Masked Owl).
- **Day 2.** Peter Murrell Reserves and Snug Tiers Nature Recreation Area, ~60 km (Forty-spotted Pardalote, Brown Quail, Yellow Wattlebird, Strong-billed Honeyeater, Tasmanian Thornbill, Dusky Robin, Scrubtit, Pink Robin, Olive Whistler, Bassian Thrush).
- **Day 3.** Goulds Lagoon Wildlife Sanctuary and Pitt Water Nature Reserve, 75 km (Latham's Snipe, Tasmanian Native-hen, Australian Spotted, Baillon's and Spotless Crake, Little Grassbird, Australian Reed-Warbler, Great Crested Grebe, Masked Owl, shorebirds).
- **Days 3–5.** Bruny Island ~120 km (Forty-spotted Pardalote, Swift Parrot, Eastern Ground Parrot, Tasmanian endemics, Little Penguin, Short-tailed Shearwater (Sep–Apr, Hooded Plover, Scrubtit).

A suggested 14-day itinerary covering the entire state of Tasmania, involves:

- **Days 1–5.** As above.

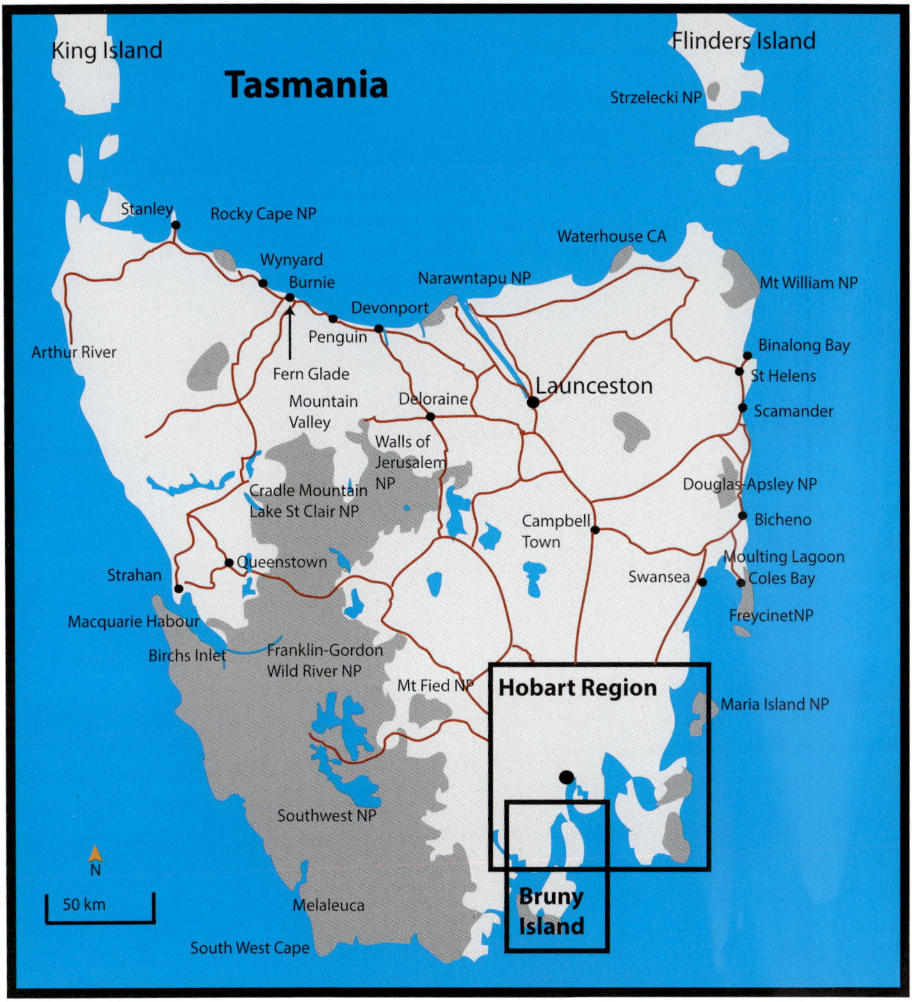

- **Day 6.** Melaleuca – Fly in, fly out (Orange-bellied Parrot (summer only), Eastern Ground Parrot, Southern Emu-wren and Striated Fieldwren).
- **Day 7.** Hobart to Mount Field National Park, 75 km (Black Currawong, Pink Robin, Scrubtit, Dusky Robin, Superb Lyrebird).
- **Day 8.** Cradle Mountain Lake St Clair National Park – south side, 130 km (Scrubtit, Black Currawong, Olive Whistler, Blue-winged Parrot and Morepork).
- **Days 9–10.** Strahan and Cradle Mountain Lake St Clair National Park – north side, 280 km (Eastern Ground Parrot, Southern Emu-wren, Striated Fieldwren, Scrubtit, Black Currawong, Olive Whistler, Pink Robin, Dusky Robin).
- **Days 11–12.** North West Coast including Arthur River and Burnie, 400 km

(Nankeen Kestrel, Little Penguin, Short-tailed Shearwater, seabirds, chance of Tasmanian Devil).

- **Days 12–13.** Burnie to Bicheno Region and Freycinet National Park, 324 km (Little Penguin, Hooded Plover, Australasian Bittern, Swift Parrot, Spotted Quail-thrush, Peregrine Falcon, Black Currawong, Beautiful Firetail, seabirds).
- **Day 14.** Freycinet National Park to Maria Island National Park, 120 km (Forty-spotted Pardalote, Swift Parrot, Cape Barren Goose, Hooded Plover, seabirds).

This itinerary will expose the visitor to a selection of the iconic national park areas within the state. Most visitors would also add a day visit to the historic precinct of Port Arthur on the Tasman Peninsula, (95 km south-east of Hobart).

Be aware that the roads in Tasmania can be steep and winding (including some highways), often becoming icy, especially during the colder months. If you are travelling from Melbourne via the *Spirit of Tasmania* ferry and arriving in Devonport, you may wish to begin your birding with the northern sites. If you have time, plan side trips to King Island for Orange-bellied Parrot (very rare), Common Pheasant, Wild Turkey, Indian Peafowl and California Quail (and gourmet food) or Flinders Island for Cape Barren Goose, Wild Turkey and Indian Peafowl (and great scenery).

Birding highlights

More than 260 bird species have been recorded in Tasmania, with over 180 recorded regularly. Birdlife Tasmania (www.birdlife.org.au) is Tasmania's premier birdwatching and ornithological conservation organisation. The endemic Yellow Wattlebird is the (unofficial) bird emblem of Tasmania and there are 12 further endemic Tasmanian species: Tasmanian Native Hen, Green Rosella, Morepork (formerly ssp. *leucopsis*, of Southern Boobook), Dusky Robin, Tasmanian Thornbill, Scrubtit, Tasmanian Scrubwren, Yellow-throated Honeyeater, Black-headed Honeyeater, Strong-billed Honeyeater, Black Currawong and Forty-spotted Pardalote. Most endemic species are quite widespread, but the Endangered Forty-spotted Pardalote is limited to the south-east of the state and Flinders Island. King Island Emu were extinct by 1802, while the endemic Tasmanian subspecies of Emu (ssp. *diemenensis*) became extinct in 1865. Mainland migrants whose breeding range is restricted to Tasmania include Orange-bellied Parrot (Critically Endangered) and the Swift Parrot (Endangered). Both species migrate to mainland Australia during the austral winter. Tasmania also has a wide range of endemic subspecies. These include Wedge-tailed Eagle (ssp. *fleayi,* Endangered in Tasmania, with an estimated population of 80 pairs), (Tasmanian) Masked Owl (ssp. *castanops*, sometimes considered a full species), Brown (Swamp) Quail (ssp. *ypsilophora*), Azure Kingfisher (ssp. *diemenensis*), Australian Owlet-Nightjar (ssp. *tasmanicus*), Eastern Rosella (ssp. *diemenensis*), Grey (Clinking) Currawong (ssp. *arguta*), Striated Fieldwren (ssp. *fuliginosus* in East Tasmania and ssp. *diemenensis* in West Tasmania), Spotted Quail-thrush (ssp. *dovei*), Superb Fairy-wren (ssp. *cyaneus*), Tawny-crowned Honeyeater (ssp. *chelidonia*) Grey Butcherbird (ssp. *cinereus*) and Australian Magpie (ssp. *hypoleuca*). Other sought-after species on the island and its surrounding waters include the entire global breeding population of Shy Albatross, as well as Southern Royal, Northern Royal, Antipodean and Wandering Albatross, Southern and Northern Giant-Petrel, Black-faced Cormorant, Kelp Gull, Australasian Bittern, Hooded Plover, Little Tern, Fairy Tern, Grey Goshawk (white morph), Olive

Whistler, Flame, Scarlet and Pink Robin, Spotted Quail Thrush, Forest Raven and Beautiful Firetail. Being a large island in the southern ocean, there have been numerous reports of vagrant penguins along the Tasmania coastline, including Southern Rockhopper, King, Snares, Royal and Gentoo. Vagrant individuals can occur at any time of year, but there is a peak in records in late summer when individuals haul out to moult.

A range of land-based bird species migrate (either wholly or partially) to the Australian mainland during the colder months (most leaving between March and May and returning August to November). In addition to the Swift Parrot and Orange-bellied Parrot, these species include Swamp Harrier, Blue-winged Parrot, Fan-tailed and Pallid Cuckoo, Shining and Horsfield's Bronze-Cuckoo, Dusky Woodswallow, Australian Reed-Warbler, Little Grassbird, Pink and Flame Robin, Striated Pardalote (Yellow-tipped ssp. *striatus*), Black-faced Cuckoo-shrike, Tree Martin, Welcome Swallow, Satin Flycatcher and Silvereye (ssp. *lateralis*). Northern hemisphere migrants include Short-tailed and Sooty Shearwater (both summer breeders in Tasmania), White-throated Needletail (summer visitor) and a selection of holarctic shorebirds. In winter, both White-fronted Tern and Double-banded Plover visit from New Zealand. Small numbers of White-fronted Tern also breed on islets off the southern Furneaux Group; the only place in Australia where they do so. Altitudinal migrants (moving down out of the mountain ranges during winter) include Black Currawong, Crescent Honeyeater, Olive Whistler, Flame Robin and, to a lesser extent, Pink Robin. By contrast, Green Rosella tolerates the higher

Tasmania supports some outstanding wilderness areas, such as this cool temperate rainforest in the upper reaches of the Franklin River. These lush, wetter habitats provide opportunities for many birds, including Tasmanian Thornbill, Tasmanian Scrubwren, Scrubtit and Pink Robin.

altitude all year round. Introduced species from the Australian mainland include Emu, Superb Lyrebird, Laughing Kookaburra, Galah, Long-billed and Little Corella and Rainbow Lorikeet. Four sought-after feral bird species exist on King Island – Indian Peafowl, Common Pheasant, Domestic Turkey and California Quail, while Indian Peafowl, Common Pheasant and Feral Turkey also occur on Flinders Island.

A wide selection of mammals can be seen on the island, notably Platypus, Short-beaked Echidna, Tasmanian Devil, Spotted-tailed and Eastern Quoll, Southern Bettong, Long-nosed Potoroo, Forester (Eastern grey) Kangaroo, Bennett's (Red-necked) Wallaby, Rufous-bellied Pademelon, Eastern-barred and Southern Brown Bandicoot, and Common Wombat. Where relevant, these species are mentioned in the text. Note that the Tasmanian Devil, once common across much of Tasmania, has suffered a significant decline in numbers due to the Devil Facial Tumour Disease. The disease was first detected in the north-east of the state near Mount William National Park and, by 2012, had spread across all but the very north-western tip of the state.

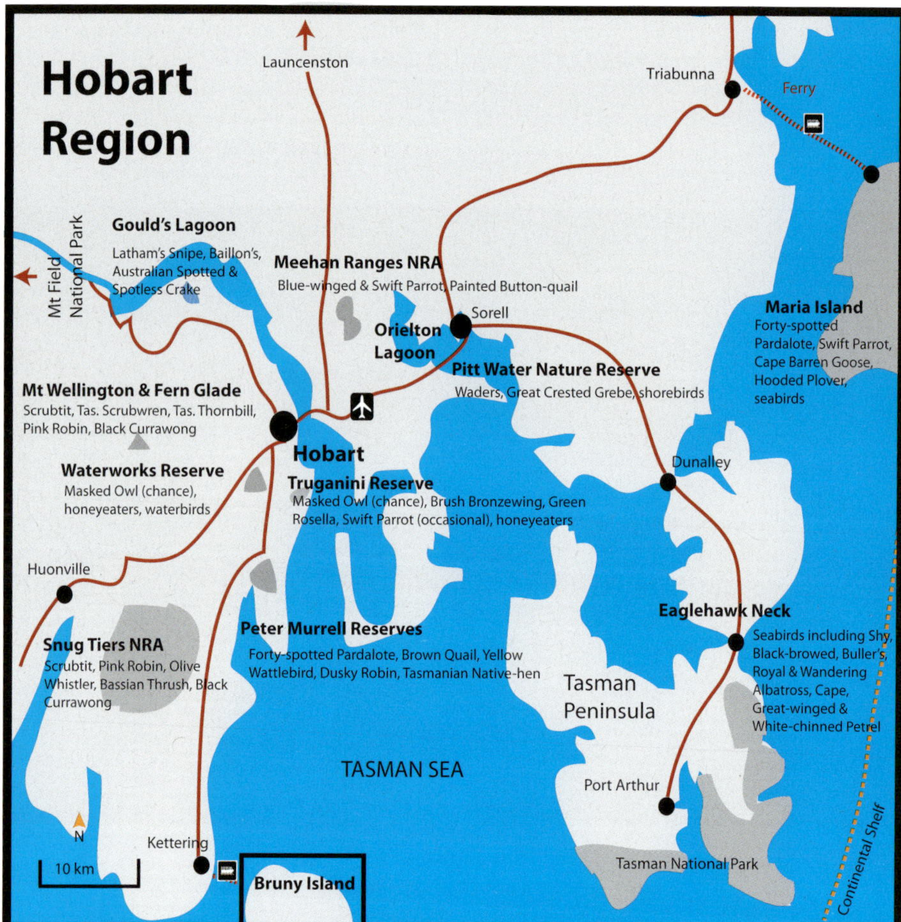

Tasmanian birding locations

Mount Wellington

Key species: Scrubtit, Tasmanian Scrubwren, Tasmanian Thornbill, Pink Robin, Black Currawong, Bassian Thrush,
Other species: Wedge-tailed Eagle, Grey Goshawk, Swift Parrot

A stunning adult male Pink Robin. This species is relatively common in wet forest habitats throughout Tasmania. Good sites around Hobart include Mount Wellington, Waterworks and Truganini Reserves and Snug Tiers Nature Recreation Area.

Mount Wellington (1270 m) dominates the skyline to the west of Hobart. In winter, the peak is frequently covered in snow. The 22 km road to the summit (Pinnacle Rd) winds up through dense eucalypt forest and patches of cool temperate rainforest, with semi-alpine to alpine vegetation at the top. Criss-crossed by excellent walking tracks, the lower slopes of Mount Wellington are thickly forested. To get there from Hobart, travel along Davey Street and Huon Rd for 8 km to the Mount Wellington turn-off and the start of the Fern Glade Walk. The Fern Tree General Store is just west of the turn-off. A range of shuttle buses travel to the top of Mount Wellington, while local buses (48 and 49) stop at Fern Tree.

The Fern Glade Track is one of the best walks for a selection of Tasmania's endemic species close to Hobart. While here, try to incorporate it with a walk to The Springs (3.8 km loop). The track passes through some of Mount Wellington's wetter forested gullies and the principal target species are Scrubtit, Tasmanian Scrubwren, Tasmanian Thornbill and Pink Robin. Of these, Scrubtit often proves the most difficult to find; look for them in the dense undergrowth and gullies along the walk – their diagnostic field marks include a whitish throat and white spot on the wing edges. Other birds to look for along the walk include Grey Goshawk, Brush Bronzewing, Green Rosella, Golden Whistler, Yellow-throated and Black-headed Honeyeater, Black Currawong, Yellow-tailed Black-Cockatoo, Forest Raven, Scarlet Robin and Bassian Thrush. In summer, many of the birds in the region also occupy areas higher up the mountain. On your way up to Mount Wellington's summit via Pinnacle Rd, stop anywhere you see birding activity, looking for Crescent and Yellow-throated Honeyeater, Yellow Wattlebird, Olive Whistler, Flame and Pink Robin, Bassian Thrush, Forest Raven and Black Currawong. Scan the skies for Wedge-tailed Eagle (rare) and White-throated Needletail (occasionally preceding thunderstorms in late summer and early autumn). Striated Fieldwren are sometimes seen around Big Bend, ~2 km before the summit. In summer, small flocks of Swift Parrot feed in the dry eucalyptus woodland to ~800 m altitude, especially in flowering Tasmanian Blue Gum.

Bruny Island

Key species: Forty-spotted Pardalote, Scrubtit, other Tasmanian endemics, Swift Parrot, Eastern Ground Parrot, Blue-winged Parrot, Hooded Plover, Tawny-crowned Honeyeater

Waterworks Reserve

First constructed in the 1860s, Waterworks Reserve in South Hobart, 4 km from the centre of Hobart, is a popular spot for picnicking and barbecues. Unbeknown to most visitors, it is also an excellent location for Masked Owl (ssp. *castanops*). The world's largest and most robust *Tyto* owl, the 'Tasmanian Masked Owl' is sometimes considered a full species. Look and listen for them along the Lower and Upper Reservoir Circuits, where Morepork, Australian Owlet-Nightjar and Tawny Frogmouth are also occasionally seen. To get there from the CBD, head south down Davey St for 2.3 km, turn-left into Lynton Ave and after 200 m, right again into Waterworks Rd – the reserve is another 1.5 km on your right. Daylight hours at Waterworks Reserve are a good time to see Black-headed, Yellow-throated, Crescent and Strong-billed Honeyeater, Eastern Spinebill, and Yellow and Little Wattlebird. On and around the reservoirs, look for Australasian and Hoary-headed Grebe, Tasmanian Native-hen, Grey and, occasionally, Black Currawong, Tasmanian Scrubwren, Tasmanian Thornbill, Dusky, Scarlet and Pink Robin, Beautiful Firetail and, in summer, Pallid and Fan-tailed Cuckoo, and Shining and Horsfield's Bronze-Cuckoo. The reserve is also one of the best spots for mammals near Hobart – at night, you might see Bennett's Wallaby, Common Brush-tail Possum, Eastern Barred Bandicoot, Southern Bettong, Rufous-bellied Pademelon and Eastern Quoll. The reserve is open to vehicles every day of the year, with the locked gates opening at 8.30 a.m. and closing at 4.15 p.m. Outside these hours, you need to walk in, gaining access through the turnstile adjacent to the main gate.

Other species: Little Penguin, Short-tailed Shearwater, Bassian Thrush

Separated from the main island of Tasmania by the D'Entrecasteaux Channel, Bruny Island is located off the south-east coast of state. The vegetation on the island consists mainly of dry eucalypt forest and cleared grazing land, with coastal heathlands at its southern end. Bruny Island (36 210 ha) is a superb birding location. It supports relatively large populations of Forty-spotted Pardalote and Swift Parrot and provides habitat for all 13 of Tasmania's endemic bird species. To get there, catch the Bruny Island car ferry from the township of Kettering, 30 km south of Hobart – the ferry runs almost hourly in daylight; see www.brunyislandferry.com.au for the schedule. Accommodation on the island includes guesthouses and backpacker hostels, a caravan park at Adventure Bay and cottage accommodation at INALA (www.inalabruny.com.au, (03) 6293 1217).

Camping areas are located at Neck Beach, Cloudy Bay and Jetty Beach. For more details, visit the Bruny Island information centre: www.brunyisland.org.au.

On the north side of the island, search for Forty-spotted Pardalote in woodland south of Dennes Point. Look specifically in the woodlands around the quarry and around the base of Waterview Hill, 3.5 km south of Dennes Point on the west side of Bruny Island Main Rd,. The pardalote has a clear preference for White (Manna) Gum, where it feeds among the foliage. Another excellent site for Forty-spotted Pardalote is in roadside woodlands along Missionary Rd, particularly near McCraken Creek, 1 km south of Barnes Bay (2 km north of the Bruny Island Main Rd turnoff). If you do not see Forty-spotted Pardalote at either of these sites, it is worth trying around INALA (202 ha): a Land for Wildlife farmstay property located north of Cloudy Lagoon on South Bruny Island. All Tasmania's 13 endemic species have been recorded on this property, while other highlights here include Wedge-tailed Eagle, Swamp Harrier, Grey Goshawk, Swift Parrot, Masked Owl (ssp. *castanops*)., Olive and Golden Whistler, and Beautiful Firetail. There are several patches of temperate rainforest here, where you may also encounter Pink Robin and Scrubtit.

'The Neck' is a narrow sand dune that links the two halves of the Island. This area is an important breeding site for Little Penguin and Short-tailed Shearwater (an estimated 240 000 pairs of the latter breed on the island). The best viewing area is near the Hummock steps (located on Bruny Island Main Rd, 24 km south of the ferry terminal), where there is a boardwalk and viewing platform. Peak viewing times are from September to February, when both penguin and shearwater arrive at last light. Along the coastline here, look also for White-bellied Sea-Eagle, Hooded Plover, Australian Pied and Sooty Oystercatcher, and Pacific and Kelp Gull. In the adjacent heathy woodlands, search for Brush Bronzewing, Green Rosella, Yellow-throated and Black-headed Honeyeater, Forest Raven and Bassian Thrush.

Situated on the southern tip of the Island is South Bruny National Park (5059 ha), which stretches along the coastline between Fluted Cape and the southern part of Great Taylors Bay. Impressively, all 13 Tasmanian endemics have been recorded within the park. To get there from the ferry, head towards South Bruny on Bruny Island Main Rd and, after ~29 km, you can either turn left to Adventure Bay to reach Fluted

Coastal woodlands surrounding Jetty Beach within South Bruny National Park. In addition to a delightful secluded camping area, these woodlands provide opportunities for many of the common Tasmanian endemics, including Green Rosella, Morepork, Yellow Wattlebird and Yellow-throated Honeyeater, as well as Crescent Honeyeater. Olive Whistler, Golden Whistler, Scarlet Robin and Bassian Thrush.

Cape, or right to Lunawanna and onto the Cape Bruny Lighthouse.

Birds can be plentiful around the Jetty Beach campground (accessed via Old Jetty Rd, reach from Lighthouse Rd 16.7 km south of Lunawanna), being regarded as one of the best places in Tasmania to see Dusky Robin, while you can also find Olive and Golden Whistler, Bassian Thrush, Strong-billed and Black-headed Honeyeater, Tree Martin, Satin Flycatcher and, at night, you may hear Morepork. Two good coastal walks – the Luggaboine Circuit (4 km) and its extension, the day-long Labillardiere Peninsula Circuit (15 km) – start at the Jetty beach campground. The Labillardiere Circuit does a clockwise loop of the peninsula. The coastal heathlands and dry sclerophyll forests along the walk contain abundant flowers (mostly spring–summer) that attract a good variety of birds. Eastern Ground Parrot occurs in the heathlands along the beginning of the walk and you may also see Brush Bronzewing, Green Rosella, Tasmanian Scrubwren, Tasmanian Thornbill, Olive and Golden Whistler, Scarlet and Dusky Robin, Bassian Thrush, Beautiful Firetail, and Tawny-crowned, Crescent, Strong-billed, Yellow-throated, New Holland and Black-headed Honeyeater and Yellow Wattlebird. Summer visitors include Swift Parrot (look for small flocks feeding in flowering Tasmanian Blue Gum, while listening for their distinctive '*kik kik kik*' contact call), White-throated Needletail and Satin Flycatcher. At the northernmost point on the walk, there are views of Partridge Island in the D'Entrecasteaux Channel, which protects one of the largest remaining colonies of the Forty-spotted Pardalote. The heathland directly behind the Jetty Beach campground, along the access road into Jetty Beach, holds many of the species mentioned above. In summer, look for the spectacular flowering Tasmanian Christmas Bell (*Blandfordia punicea*).

From the Cape Bruny Lighthouse, scan the oceans for seabirds, such as Pacific and Kelp Gull, Black-faced Cormorant, Shy, Black-browed and Indian Yellow-nosed Albatross and Short-tailed Shearwater. There is also always a chance of seeing less common species such as Cape and White-headed Petrel. Blue-winged Parrot, Tawny-crowned and Crescent Honeyeater and Brown Quail all occur in the short heathland around the lighthouse. Hooded Plover feed along the shores of Cloudy Bay and the Cloudy Bay Lagoon – search on foot for them along the first 3 km of the East Cloudy Head Track (12 km return), where it traverses sands of Cloudy Bay.

Birds to look for around the township of Adventure Bay include Tasmanian Native-hen, Hooded Plover, Green Rosella, Yellow-throated, Strong-billed and Black-headed Honeyeater, Yellow Wattlebird, Dusky Robin, Beautiful Firetail, Black Currawong and, in summer, Satin Flycatcher, Dusky Woodswallow, and Pallid and Fan-tailed Cuckoo. Summer is the time to look for Swift Parrot on Bruny Island, with the most reliable places to see them being the Adventure Bay foreshore and the Grassy Point Track, east of Adventure Bay. (They are also regularly seen on the north side of the island, particularly near the Robert Point Ferry Terminal and Church Bay, Killora.) Near Adventure Bay, there are several sections of some cool temperate rainforest, particularly along Coolangatta Rd near the top of Mount Mangana and along the short Mavista Nature Walk (1.5 km). The rainforest in both areas contains Scrubtit, Tasmanian Scrubwren,

Heathland interspersed with coastal woodlands on South Bruny Island support a wide range of birds. Some of the more notable species include Brush Bronzewing, Eastern Ground Parrot, Southern Emu-wren, Tawny-crowned Honeyeater, Olive Whistler, Dusky Robin and Beautiful Firetail.

Tasmania

Truganini Reserve

Truganini Reserve (43 ha) covers the south-east side of Mount Nelson, from the summit to Lower Sandy Bay and protects a wet gully of native bushland. There are two access points: via the Channel Hwy (Sandy Bay Rd) in Taroona, 7 km from Hobart's CDB (near bus stop 32), or via the Mount Nelson Signal Station lookout, located at the end of Nelson Rd, 8 km from Hobart's CBD. Facilities include toilets (at the Mount Nelson Signal Station), BBQs and a restaurant. A walking track follows the Cartwright Creek, running between the lookout at the Mount Nelson Signal Station and the Channel Hwy at Taroona, with the return walk taking about an hour and a half. Masked Owl (ssp. *castanops*) has been recorded along the tracks, 1 km up from the Channel Hwy. Look especially for roosting owls in the vicinity of the creek line (100 m either way); their presence often indicated by pellets and whitewash (droppings) on the ground. Morepork also occur in the reserve. The reserve is particularly good for Tasmania's endemic honeyeaters: Black-headed, Strong-billed and Yellow-throated Honeyeater, and Yellow Wattlebird (Australia's largest honeyeater) are all commonly seen. Other birds here include Brush Bronzewing, Musk Lorikeet, Eastern and Green Rosella, Tasmanian Scrubwren, Dusky Woodswallow, Olive and Golden Whistler, Forest Raven, Crescent Honeyeater, Little Wattlebird, Scarlet and Pink Robin and Beautiful Firetail. Summer additions include Pallid and Fan-tailed Cuckoo, Shining Bronze-Cuckoo, Swift Parrot (occasional) and Satin Flycatcher.

Green Rosella, Pink Robin, Bassian Thrush and Tasmanian Thornbill. At Adventure Bay, it is often possible to see Southern Right Whale from June to late October.

Peter Murrell Reserves

Key species: Forty-spotted Pardalote, Brown Quail, Yellow Wattlebird, Dusky Robin
Other species: Tasmanian Native-hen, Black-headed, Yellow-throated and Strong-billed Honeyeater

The Peter Murrell Reserves (277 ha) are located at Huntingfield, 3 km south of Kingston (15 km south of Hobart) on the Tinderbox Peninsula. They consist of three reserves: Peter Murrell State Reserve (133 ha), the Peter Murrell Conservation Area (135 ha) and a public reserve (9 ha). Access is from Huntingfield Ave in Huntingfield. To get there, turn left off the Channel Hwy into Huntingfield Ave, 2 km south of Kingston, travel 700 m and then turn left again down a dirt track (located immediately south of the Vodafone facility) – there is a Peter Murrell Reserves sign at the entrance. Follow the track down to the carpark near Penrhyn Pond.

The Peter Murrell Reserves were once a reliable location close to Hobart to find

A woodland path within Peter Murrell Reserve. The pale barked gums are White Gum *Eucalyptus viminalis*, a tree species favoured by the Endangered Forty-spotted Pardalote. Other species that frequent this habitat include Superb Fairy-wren, Brown Thornbill, Black-headed, Yellow-throated, Strong-billed and New Holland Honeyeater, Yellow Wattlebird, Scarlet Robin and Grey Currawong.

Snug Tiers Nature Recreation Area

Snug Tiers Nature Recreation Area (5600 ha) is located 8 km west of Snug (25 km south of Hobart). The Snug Falls Track (2 km) descends through bushland to a cool, fern-lined rainforest gully below Snug Falls. This is a nice walk to combine with a trip to Peter Murrell Reserves (located 10 km to the north). At the bottom of the walk, look for wet-forest species such as Scrubtit, Pink Robin, Olive Whistler and Bassian Thrush. You may also encounter Black Currawong, Green Rosella, Eastern Spinebill (winter), Crescent, Strong-billed and Yellow-throated Honeyeater, Scarlet Robin, Grey Shrike-thrush, Tasmanian and Brown Thornbill, and Grey Fantail. The buttongrass plains on the top of Snug Tiers are fairly reliable for Southern Emu-wren. To get there from Hobart, take the Channel Hwy south to Snug, then Snug Tiers Rd for ~1 km before a left turn onto Snug Falls Rd to the well-signposted carpark.

the endangered Forty-spotted Pardalote, though these days there may only be a few pairs left. Start searching for them in the large trees overhanging the carpark, but also try just upstream of here. Another spot to look is along Coffee Creek, particularly near the Penrhyn (White Gum) and Heron Ponds (also frequented by Tasmanian Native-hen). Concentrate your search for this small, leaf-sized bird in the foliage of the larger White (Manna) Gums, in which it forages. Both Spotted and Striated Pardalote are also common here, so take care with your observations, especially if juvenile Spotted Pardalote are present. The reserve contains a nice range of forest, heathland and Buttongrass moorland, perfect habitat for honeyeaters, so look for Black-headed, Yellow-throated, Strong-billed, Crescent and New Holland Honeyeater, Eastern Spinebill, and Yellow and Little Wattlebird. Other birds that may be seen include Brown (Swamp) Quail (ssp. *ypsilophora*), Green Rosella, Grey Butcherbird (ssp. *cinereus*), Grey (Clinking) Currawong, Superb Fairy-wren (ssp. *cyaneus*), Brown Thornbill, Scarlet and Dusky Robin (often foraging from exposed perches such as fence posts near the carpark) and, in summer, cuckoos such as Pallid and Fan-tailed, and Horsfield's and Shining Bronze-Cuckoo. At night, Tawny Frogmouth, Australian Owlet-nightjar and Masked Owl (ssp. *castanops*) have all been recorded. Eastern-barred Bandicoot and Long-nosed Potoroo are quite common in the Peter Murrell Reserves, especially adjacent to open grassy areas.

Goulds Lagoon Wildlife Sanctuary

Key species: Latham's Snipe, Australian Spotted, Baillon's and Spotless Crake, Little Grassbird, Australian Reed-Warbler
Other species: Tasmanian Native-hen, Australasian Shoveler, Hardhead

A noted birdwatching location, Goulds Lagoon Wildlife Sanctuary (8 ha) is a shallow lagoon, which is an important feeding and breeding ground for waterbirds. It is particularly important because many of the Derwent estuary wetlands have been destroyed and it is one of the last remnant wetlands of its type. It is located in Austins Ferry on Main Rd, 17 km north of Hobart, with the main parking area in Hestercombe Rd, on the west side of the lagoon.

Little Grassbird and Australian Reed-Warbler (summer) feed in the reedbeds around the lagoon, while Latham's Snipe (summer) frequents mud flats and adjacent grassy cover around the edges. When the water level is low, Australian Spotted, Baillon's and Spotless Crake that inhabit the reed beds may venture onto the exposed mudflats. You can also see Tasmanian Native-hen, Purple Swamphen, Eurasian Coot and Dusky Moorhen. Waterfowl to look for include Australasian Shoveler, Hardhead, Northern Mallard, Grey and Chestnut Teal, Pacific Black Duck and Black Swan. Other waterbirds that occur here include Hoary-headed Grebe, Australian Pelican, Great Egret, White-faced Heron, Masked Lapwing, Kelp Gull, and Little Black, Little Pied and Great Cormorant, which roost in the trees around the lagoon. Land birds recorded within the sanctuary include Swamp Harrier, Galah, Sulphur-crested Cockatoo, Musk and Rainbow Lorikeet, Swift Parrot (occasional in summer), Eastern Rosella, Laughing Kookaburra, Little and Yellow Wattlebird, Noisy Miner, Tree Martin (summer), Superb Fairy-wren, Grey Butcherbird, Spotted and Striated Pardalote, Common Greenfinch and European Goldfinch. Black-tailed Native Hen and Little Ringed Plover are two vagrant species to Tasmania that have also been recorded here. The lagoon also supports a healthy population of Eastern Banjo (Pobblebonk) Frog, which are

Close to Hobart, Goulds Lagoon is one of the most reliable sites for crakes and rails in Tasmania. Australian Spotted Crake is perhaps the most frequently reported of the group.

particularly active and vocal during their spring and summer breeding season.

Pitt Water Nature Reserve and Orielton Lagoon

Key species: Waders, Great Crested Grebe, Masked Owl, shorebirds
Other species: Cape Barren Goose, raptors

The shallow (average depth 1.3 m) Orielton Lagoon forms part of the Pitt Water Nature Reserve (790 ha). It is one of the most southerly major feeding grounds for waterbirds in Australia and is an important feeding ground for migratory and resident shorebirds. The lagoon lies within the Pitt Water–Orielton Lagoon Ramsar Site – one of 10 Ramsar wetlands in Tasmania – and is part of the South Arm Important Bird Area. Good places for viewing birds at the Orielton Lagoon include the Sorell Causeway (Tasman Hwy), the end of Forcett or Henry Streets in Sorell (Cemetery Point), Shark Point Rd in Penna (particularly where the Frogmore and Orielton rivulets cross the road) and the Waterview Bird Sanctuary, which can be accessed via Stores Lane in Sorell.

The shoreline and intertidal flats provide critical foraging habitat for birds such as Bar-tailed Godwit, Whimbrel, Eastern Curlew, Pacific Golden Plover (look at Cemetery Point), Common Greenshank, Red-necked Stint, Curlew and Sharp-tailed Sandpiper. More rarely, Double-banded Plover (winter), Marsh Sandpiper, Great and Red Knot, Grey Plover, Grey-tailed Tattler and Sanderling have also occurred. Some spectacular rarities have been recorded in the area, including Ruff, Little Stint, Pectoral and Buff-breasted Sandpiper, and Hudsonian Godwit. Resident shorebirds include Australian Pied and Sooty Oystercatcher, and Red capped Plover. These wetlands are also one of the best sites in Tasmania for Great Crested Grebe: look for them from the Sorell Causeway. You may also see Musk Duck, Hoary-headed Grebe, Great, Little and Cattle Egret, Caspian and Crested Tern and Kelp Gull (the last, first recorded in Tasmania in the 1950s, is now estimated to number ~8000 birds). Cape Barren Goose are occasionally seen feeding near the Pittwater Golf Course. Raptors to look for, soaring over the estuaries and adjacent land, include Swamp Harrier, White-bellied Sea-Eagle, Wedge-tailed Eagle and Brown Falcon. Masked Owl (ssp. *castanops*) has been recorded in the woodlands along the beginning of Pittwater Rd, which runs adjacent to the north-east side of the Hobart International Airport and turns south from the Tasman Hwy 500m before the western end of the Sorrell Causeway.

South of Orielton Lagoon, the extensive mud flats at Ralphs Bay (also part of the South Arm Important Bird Area), are another good area for migratory waders. Ralphs Bay is situated on the south-east side of the Derwent Estuary, accessed via South Arm Rd, 12 km from Hobart (just west of the township of Lauderdale). Look here for Australian Pied and Sooty Oystercatcher, Double-banded (winter) and Red-capped Plover, Red-necked Stint, Eastern Curlew, Red Knot and Bar-tailed Godwit. Australia's first reported Hudsonian Godwit was found here in 1992.

Maria Island National Park

Key species: Forty-spotted Pardalote, Swift Parrot, Cape Barren Goose, Hooded Plover, seabirds
Other species: Fairy Tern, Green Rosella, Yellow Wattlebird

Maria Island is a relatively large mountainous island (11 550 ha) located just off Tasmania's east coast. The entire island is a designated national park, including the tiny township of Darlington, which, save for a handful of national park rangers, has no permanent residents and consists mostly of historic stone buildings from the convict era. There are no roads, but the island does offer excellent bushwalking and bike-riding opportunities. The island is well known for its relatively tame mammals, with good opportunities to see Eastern Grey Kangaroo, Rufous-bellied Pademelon, Common Wombat and Brushtail Possum – it is also the site for one of the Tasmanian Devil 'insurance' populations; these are mostly captive animals not accessible to the public, but, in November 2012 a number were also released into the wild on the island. Given easy access via daily ferries from Triabunna (a 1.5 h drive north of Hobart with the crossing taking ~45 min), no traffic, relaxed accommodation and camping, the area is especially popular with families. Accommodation options on the island include staying in the historic penitentiary units (6–8 people per room) and camping. For bookings and information on the ferry timetable, call the East Coast Visitor Information Network on 03 6257 4772 and for camping options

Maria Island hosts an exceptionally tame population of Cape Barren Geese. As a consequence, they are usually one of the first species to be seen after alighting from the ferry. Although the stronghold is on Flinders and King Islands in Bass Strait, other SE Tasmanian sites for the species include Waterhouse Reserve and Pittwater Golf Course.

contact Maria Island National Park office (www.parks.tas.gov.au, (03) 6257 1420, maria.island@parks.tas.gov.au). There are no shops on the island, so you will need to bring your own food, but tank water is available. Bike riding is the popular form of transport, with bikes for hire on the island. Maria Island is also a superb birding location. One hundred and twenty-nine species of bird have been recorded on the island, all but one of Tasmania's 13 endemic species occur (the Scrubtit is the missing species) and it is a very reliable place to see the Endangered Forty-spotted Pardalote. The main birding sites are all within easy walking distance of the ferry landing at Darlington. Hooded Plover are occasionally seen immediately adjacent to the landing, while tame Cape Barren Goose (introduced to Maria Island in 1968) and Tasmanian Native-hen frequent the grassy areas around Darlington.

The Mount Maria Track (11 km return) begins on the left side of the road just after Counsel Creek, 1 km from Darlington. The beginning of this walk is an excellent area for Forty-spotted Pardalote, as well as other woodland species. The best area to look for the pardalote is the 1 km section of track between Counsel Creek and the first bridge. Tasmanian Blue Gum and White (Manna) Gum dominate this section of the walk, so look for the Forty-spotted Pardalote in the larger White Gums. Their presence is often indicated by the sounds of their soft '*where where*' call. This area is also a good place for the Swift Parrot, which are occasionally seen in small, noisy flocks during spring and summer. Other birds that occur here include Spotted and Striated Pardalote, Black-headed, Yellow-throated and Strong-billed Honeyeater, Yellow Wattlebird and, in summer, Shining Bronze-Cuckoo, and Fan-tailed and Pallid Cuckoo. After crossing the first bridge, the habitat changes, with a wetter understorey and trees such as Tasmanian Oak (Messmate *Eucalyptus obliqua*) and White Peppermint (*E. pulchella*) dominate. This is a good area for Green Rosella, Yellow and Little Wattlebird, and Grey Fantail. If you wish to continue to the summit for spectacular views, the total walking time is 6–7 h return.

Another walk where you may see Forty-spotted Pardalote is the Reservoir Circuit (4 km return from the ferry landing). It takes you through open woodlands into tall eucalypt forest, with the most reliable places to see Forty-spotted Pardalote in White (Manna) Gum (*E. viminalis*) near the reservoir at the south-east end of the circuit and between the reservoir and the Cement Works. Spotted and Striated Pardalote are common here, while other birds you may see include Tasmanian Scrubwren, Satin Flycatcher (summer), Flame Robin and Black Currawong.

Half way along the Fossil Cliffs Walk (2 km return from the ferry landing) you reach Cape Boulanger, where you can view Bird Rock. This is a roosting place for

Meehan Range Nature Recreation Area (Redgate section)

The Meehan Range is a series of hills that run parallel to the Derwent River just east of Hobart. An excellent location for dry sclerophyll-forest birds is the Meehan Range Nature Recreation Area (2700 ha), located in the Redgate section of the range. To get there, from the Tasman Hwy turn on to Cambridge Rd 9.5 km from Hobart and then immediately left on to Belbins Rd. After 300 m, park near the entrance gate. Two fire trails begin at this point. On the left is the Stringy Bark Gully Track (6.5 km return), rising gradually until, after 500 m, it comes to a junction. Take the right track and continue through Stringybark Gully and the Barilla Rivulet. The list of interesting woodland birds recorded in the Meehan Range is extensive, notably Blue-winged and Swift Parrot (summer), Green Rosella, Painted Button-quail, Wedge-tailed Eagle, Brush Bronzewing, Crescent, Strong-billed, Black-headed and Yellow-throated Honeyeater, Yellow Wattlebird, Satin Flycatcher (summer), Dusky Woodswallow (summer), Tasmanian Scrubwren, Scarlet, Flame and Dusky Robin, Forest Raven, Black and Grey Currawong, and Beautiful Firetail.

Tasmania

seabirds such as Australasian Gannet, Great, Black-faced and Little Pied Cormorant, Fairy Tern (occasional records), Caspian Tern and White-bellied Sea-Eagle. Ile du Nord supports 2000+ pairs of Little Penguin. Scan the seas for distant Short-tailed Shearwater, Black-browed and Shy Albatross, Fairy Prion and Common Diving-Petrel. Other birds to look for on Maria Island include Peregrine Falcon, Kelp Gull and White-fronted Chat. Look for Hooded Plover at Hopground Beach, accessed via the Painted Cliffs Walk (2.5 km return), where you may also see Ruddy Turnstone, Pied and Sooty Oystercatcher, Double-banded Plover (winter) and Red-capped Plover.

Mount Field National Park

Key species: Superb Lyrebird, Black Currawong, Scrubtit, Pink Robin, Dusky Robin,

Other species: Peregrine Falcon, Wedge-tailed Eagle, Morepork, Olive Whistler, Tasmanian Thornbill

Created in 1916, Mount Field National Park (16265 ha) was the first national park in Tasmania and just the second in Australia. Ranging in altitude from 158 m above sea level at the park entrance to 1434 m at the summit of Mount Field West, the park supports a great diversity of vegetation types. At the base of the mountain, visitors encounter tall Mountain Ash (*E. regnans*) forest that shelters large tree ferns. Further up the mountain, cool temperate rainforest bounds the Lake Dobson Rd, while towards the summit, alpine vegetation dominates. To get to Mount Field National Park (75 km west of Hobart), take the Brooker Hwy to New Norfolk. After New Norfolk, take the Glenora Rd and then the Gordon River Rd until you reach the entrance near

Lake Dobson, Mount Field National Park. Skirting the far shore of this pristine alpine lake, The Pandani Grove Nature Walk (best in summer) provides opportunities for Scrubtit, Tasmanian Thornbill, Tasmanian Scrubwren, Crescent Honeyeater, Yellow-throated Honeyeater and Black Currawong.

Westerway. There is a visitor centre at the entrance of the park that provides information on walks. Another shelter shed for day use is located at Lake Dobson and includes information on the longer walks and winter ski field. Camping and caravan facilities are available near the park entrance, including toilets, showers and BBQs. Because of the area's popularity, if you plan to stay in the camping area, it is worth booking well in advance.

With the exception of Forty-spotted Pardalote, all of Tasmania's endemic bird species have been recorded in the park. The best birdwatching is usually to be had along the Russell Falls Track and the longer linked walks known as the Russell Falls/Horseshoe Falls/Lady Barron Falls/Tall Trees circuit (2–4 h). These walks pass through stands of Stringybark *(E. obliqua)*, Mountain Ash, White Gum and sections of Antarctic Beech forest. Along here, there are opportunities to see Scrubtit, Tasmanian Thornbill, Tasmanian Scrubwren, Pink and Dusky Robin, Olive Whistler, Yellow-throated, Black-headed, Crescent and Strong-billed

Tasmanian pelagics

Occasional pelagic boat trips travel to the continental shelf off the east coast of Tasmania; the most regularly organised depart from Eaglehawk Neck, but there are also intermittent trips from St Helens and extended trips in the southern ocean. About 80 species have been recorded on these pelagic trips, with 20–25 species seen on a typical 1-day outing. Highlights include a wide selection of southern ocean seabirds, as well as the chance of seeing rare albatross, petrel and shearwater species. The most commonly seen albatross are Shy, Indian Yellow-nosed, Campbell, Black-browed, Buller's, Southern Royal, Northern Royal Antipodean and Wandering Albatross, while Salvins, Grey-headed, Sooty and Light-mantled Sooty Albatross are all seen occasionally. Petrel species that are recorded frequently include Cape (winter–spring), Great-winged, White-headed and White-chinned and, less commonly, Providence, Soft-plumaged, Mottled, White-necked (very rare), Cook's, Gould's, Grey and Westland (very rare). The frequently seen shearwaters are Short-tailed, Sooty, Hutton's and Fluttering, while rarities have included Great, Buller's and Little Shearwater. Five species of storm-petrel have been observed, with Wilson's, White-faced and Grey-backed Storm-Petrel being regular. There have also been rare records of Black-bellied Storm-Petrel and a single European Storm-Petrel (the only record for Australia). These trips are also noteworthy for records of Little Penguin, Southern and Northern Giant-Petrel, Fairy, Antarctic, Slender-billed, Salvin's (rare) and Broad-billed (rare) Prion, Common Diving-Petrel, Southern and South Polar Skua (vagrant), Arctic and Long-tailed Jaeger and White-fronted Tern. For details on the Eaglehawk Neck boat trips, visit (www.pelagics.com.au).

Tasmania

Honeyeater, Yellow Wattlebird, Grey (Clinking) and Black Currawong, and Forest Raven. As the name suggests, along the Lyrebird Nature Trail (15 min loop) you may see or hear Superb Lyrebird. Despite its iconic status on the east coast of the Australian mainland, the species is an exotic in Tasmania: it was first released in this very park in 1934. Native species that frequent the nature trail include Scrubtit, Tasmanian Scrubwren, Tasmanian Thornbill, Dusky Robin, Yellow-throated and Strong-billed Honeyeater, Black Currawong and, at night, look and listen for Morepork. The trail is located 7 km up the Lake Dobson Rd. Striated Fieldwren also occurs in the park, with the heathland just 1 km before Lake Dobson being a good site. In the alpine vegetation higher up the mountain, look for Black Currawong, Flame Robin, Forest Raven and Australasian Pipit. From the more exposed areas, scan the skies for birds of prey such as Peregrine Falcon and Wedge-tailed Eagle. Mount Field National Park is also an excellent place to see Tasmania's native mammals: Echidna, Platypus, Eastern Barred Bandicoot, Eastern Quoll, Rufous-bellied Pademelon, Common Brushtail Possum and Common Wombat occur here, and most are seen frequently around the campground near the park entrance. The last Thylacine (or Tasmanian Tiger) that was held at Hobart Zoo was trapped in the Florentine Valley area of the National Park in 1933.

Breeding only in Tasmanian waters, Shy Albatross are one of the most commonly encountered albatross species on pelagic trips off Eaglehawk Neck.

Melaleuca and the South-west Conservation Area

Key species: Orange-bellied Parrot, Eastern Ground Parrot, Southern Emu-wren,
Other species: Striated Fieldwren, Olive Whistler, Pink Robin, Morepork

Melaleuca is (at the time of writing) the last remaining breeding site for the Critically Endangered Orange-bellied Parrot. The area is situated in the South-west Conservation Area (~240 000 ha), forming a significant part of the Tasmanian Wilderness World Heritage Area. The habitat consists of Button-grass moorlands, tea-tree scrubland and woodlands. There are no roads in the area, with the only access to Melaleuca by sea via Port Davey or by air on chartered light aircraft flights (www.paravion.com.au, (03) 6248 5390), departing from Cambridge Airport near Hobart Airport; or on foot by walking one of two of Australia's most challenging coastal walks, the South Coast Track (85 km), or the Port Davey Track (70 km). There are no shops, so visitors need to be completely self-sufficient. Accommodation includes two basic, but large huts, for walkers and camping is also permitted. Because the huts cannot be booked in advance, any visitors should also pack a tent in case the huts are already at capacity.

With an estimated population of fewer than 30 wild birds, the Orange-bellied Parrot is one of the world's rarest species of bird. They are a migratory species, wintering in coastal Victoria and South Australia, arriving at Melaleuca in early October and departing after the breeding season, during

March and April. The species is undergoing a tragic decline and conservation managers have now accepted that extinction in the wild is both inevitable and imminent; it is now thought that the captive population holds the last hope for the species. At the time of writing, it is possible to watch the species (in season) from the temporary public hide and feeding table, located 1 km to the south of the airstrip. They feed here throughout the day, with the best times being early in the morning and late afternoon. Eastern Ground Parrot is reasonably common in the heathland around Melaleuca and is occasionally flushed from the walking tracks in the area. Other birds found in the Button-grass plains include Wedge-tailed Eagle, Blue-winged Parrot, Green Rosella, Southern Emu-wren, Tasmanian Scrubwren, Striated Fieldwren, Scrubtit, Tasmanian Thornbill, Olive Whistler, Dusky and Pink Robin, Yellow-throated, Crescent and Strong-billed Honeyeater, Beautiful Firetail and, at night, Morepork. Along the Melaleuca Inlet coastline, look for White-bellied Sea-Eagle, Hooded Plover and Australian Pied and Sooty Oystercatcher. For overnight visitors, Melaleuca is also a good place to see a range of Tasmanian mammals such as Eastern and Spot-tailed Quoll, Rufous-bellied Pademelon, Bennett's (Red-necked) Wallaby, Common Wombat and Common Ringtail Possum.

Perhaps Australia's most Critically Endangered bird, the vividly coloured Orange-bellied Parrot now breeds only at Melaleuca in south-west Tasmania. Visits must be in summer to observe the species here, because they migrate to the mainland during winter. Getting there requires chartering a light aircraft (group bookings are best) or undertaking a 70+ km walk!

Strahan and Birchs Inlet

Key species: Eastern Ground Parrot, Orange-bellied Parrot (very rare), Southern Emu-wren, Striated Fieldwren, Tawny-crowned Honeyeater, Azure Kingfisher
Other species: Morepork, Beautiful Firetail, Brush Bronzewing, Brown Quail, Double-banded Plover, Sanderling

Located on the shores of Macquarie Harbour, Strahan is a small historic township and former port on the west coast of Tasmania. The township is the gateway to the Lower Gordon River and the popular Gordon River cruises depart from here. The township is located 5 h from Hobart along the Lyell Highway, or 4 h from Devonport, though both roads are often narrow, steep and windy so expect it to take longer. Strahan Airport is located 3 km west of Strahan on Ocean Beach Rd and flights to Strahan are through Par-Avion (paravion.com.au). There is a wide range of accommodation options in Strahan, including caravan parks, backpackers, B&Bs and a hotel.

The heathland and Button-grass plains between the airport and the township along Ocean Beach Rd hold Eastern Ground Parrot, Southern Emu-wren, Striated Fieldwren, Tawny-crowned Honeyeater and Beautiful Firetail. It is also worth looking for these species in the heathland along Macquarie Heads Rd, which runs off Ocean Beach Rd immediately before you reach the airport. Other birds to look for here include Swamp Harrier, Brush Bronzewing, Olive Whistler, Brown (Swamp) Quail (ssp. *ypsilophora*), Crescent Honeyeater and, at night, listen for Morepork. A nice selection of shorebirds has been recorded at Ocean Beach (take Ocean Beach Rd), including Red-capped and Double-banded (winter) Plover, Sanderling, and Australian Pied and Sooty Oystercatcher.

Birchs Inlet is a narrow cove or coastal inlet on the south-west side of Macquarie Harbour. It lies ~38 km south of the town of Strahan and is only accessible via boat. Aside from Melaleuca, it is the only other study site for Orange-bellied Parrot in their breeding range (though sadly, as of 2012, it appears they no longer breed there). The inlet is part of the Melaleuca to Birchs Inlet Important Bird Area, which comprises a total area of 231 500 ha. The Button-grass plains also support Eastern Ground Parrot, which can be found in the heathland around the inlet. Other birds recorded here include Latham's Snipe (summer), Crescent Honeyeater, Blue-winged Parrot, Southern Emu-wren, Striated Fieldwren, Scrubtit and Beautiful Firetail. The Azure Kingfisher (an endemic Tasmanian subspecies *diemenensis* that is rarely sighted) may also occasionally be seen feeding around the shores of the inlet and along the banks of the Gordon River.

Northern coast

Key species: Little Penguin, Short-tailed Shearwater, other seabirds, Nankeen Kestrel, Peregrine Falcon, Pink Robin, Dusky Robin
Other species: Masked Owl, endemics, also Platypus

Heathland and Button-grass plains between the airport and the township of Strahan support an accessible population of Eastern Ground Parrot, along with Brown Quail, Southern Emu-wren, Striated Fieldwren, Tawny-crowned Honeyeater, Olive Whistler and Beautiful Firetail.

The Tasmanian north-west coast stretches from Port Sorell in the east, to Arthur River in the north-west. There is a wide range of accommodation along the coast (see www.tasmaniasnorthwest.com.au). The airport at Wynyard (Burnie) and Devonport service regular interstate and island flights.

The Nut State Reserve (59 ha) incorporates a sheer-sided rocky bluff, adjacent to the township of Stanley. Little Penguin occur on the northern foreshore of 'the Nut' (or Circular Head as it is also known), where a tour operates daily from October to March. This reserve also protects a significant colony of Short-tailed Shearwater, with an estimated 13 000 pairs. Birds arrive at their burrows just after dusk during spring and summer. Other seabirds to look for include Black-faced Cormorant, Australasian Gannet, Shy and Black-browed Albatross, Southern Giant-Petrel, Sooty Shearwater and White-bellied Sea-Eagle. Both Peregrine Falcon and Nankeen Kestrel breed on the Nut, with the kestrel being locally significant because there are thought to be only 10 or so breeding pairs in the state. Previously, the Orange-bellied Parrot (now very rare) has also been recorded here on migration.

There is an observation centre for viewing Little Penguin at Parsonage Point in Burnie, located at the western end of West Beach (linked to the town via a beach boardwalk). Other Little Penguin colonies can be found near the township of Penguin (17 km west of Burnie) and there is a viewing platform at the Lillico Beach Conservation Area (14 ha) 6 km west of Devonport.

Fern Glade Reserve (35 ha) is located 3.9 km from Burnie. To get there, turn on to Old Surrey Rd on the western side of the Emu River Bridge, 1 km east of Burnie. After 350 m, turn left on to Fernglade Rd – the reserve area is a further 1 km on and there is a picnic area, toilets and fireplaces. Begin by birdwatching around the picnic area and then walk up the trail beside the Emu River (500 m – with a longer, 1.5 km return extension). Along the way, look for birds such as Tasmanian Native-hen, Yellow-tailed Black-Cockatoo, Green and Eastern Rosella, Laughing Kookaburra, Yellow-throated, Strong-billed and Black-headed Honeyeater, Tasmanian Scrubwren, Scrubtit, Tasmanian Thornbill, Olive and Golden Whistler, Bassian Thrush, Forest Raven, Dusky, Pink and Scarlet Robin, and Beautiful Firetail. Fern Glade is also an excellent site for observing Platypus – although dawn and dusk are best, they can be active throughout the day in the shallows along Emu River. At night, there is a chance of hearing and seeing Masked Owl (ssp. *castanops*) and Australian Owlet-nightjar.

Narawntapu National Park (4349 ha), formerly known as Asbestos Range National Park, is 45 km east of Devonport. The freshwater lagoon in the park is perhaps the best place in Tasmania to see Blue-billed Duck. From the bird hide overlooking the lagoon, you may also find Musk Duck, Australasian Shoveler, and Australasian, Hoary-headed and Great Crested (uncommon) Grebe. The coastal heathland and dry sclerophyll woodlands around the lagoon support Wedge-tailed Eagle, Blue-winged Parrot, Green Rosella, Morepork, Striated Fieldwren, Dusky and Scarlet Robin, Yellow Throated Honeyeater, Yellow Wattlebird, Tasmanian Scrubwren and Olive and Golden Whistler. Look along Badgers Beach and Baker Beach for Hooded Plover, Australian Pied and Sooty Oystercatcher and White-bellied Sea-Eagle. Masked Owl has been recorded at Green Beach in the east of the park. Dubbed the 'Serengeti of Tasmania', Narawntapu National Park is well known as a place to see Tasmania's native mammals. A spotlight drive may produce sightings of Forester (Eastern Grey) Kangaroo, Bennett's (Red-necked) Wallaby, Rufous-bellied Pademelon, Common Wombat, Eastern and Spotted-tail Quoll. Platypus and Echidna also occur here. The

The Nut – a spectacular headland that shelters the small township of Stanley. A large colony of Short-tailed Shearwater occupy the seaward-facing slopes. Tours to see Little Penguin that also breed here operate through the warmer months.

national park was formerly one of the best places in the state to observe Tasmanian Devils, but, with the arrival of Devil Facial Tumour, the species is now extremely rare in the region.

Arthur River region

Key species: Azure Kingfisher, Tawny-crowned Honeyeater, Southern Emu-wren, also Tasmanian Devil and Spot-tailed Quoll

Other species: Grey Goshawk, Brown Quail, seabirds

Arthur River is a small coastal hamlet situated around the estuary of a large tannin-stained river of the same name. The lower reaches of the river provide the best opportunities to find the Endangered Tasmanian form of the Azure Kingfisher (ssp. *diemenensis*). There is a National Parks office, a small general store, some holiday rental accommodation, the Arthur River Cabin and Camping Park and several large and well-equipped national park camp-grounds (with toilets, showers and drinking water). To find the Azure Kingfisher, either join one of the regular Arthur River Cruises that travel a few kilometres upstream and land at several sites for short walks, or alternatively hire a small boat to do the same independently. The latter option can be good because it provides more flexibility to continue the search in key areas. Look especially in the small inlets and heavily vegetated narrow channels along the northern bank of the Arthur River in the first 2–3 km upstream of the vehicle bridge. A small series of waterfalls, and the often inundated walking track that leads to them from the river bank, is also an excellent area to search; most cruises stop here, but be sure to ask for specific directions if hiring your own boat because the markers (usually flagging tape) indicating the way can be difficult to find. Other birds that occur along the river include White-bellied Sea-Eagle, Grey Goshawk and variety of waterbirds. The lower reaches of the river can also be good for Platypus. Point Gardiner and the Edge of the World lookout are situated on the south side of the river mouth. Here, there are spectacular scenes along the remote western Tasmanian coast – scan the ocean for Shy, Black-browed and Indian Yellow-nosed Albatross, giant-petrels and the possibility of something less common. The beaches and rocky headlands here support Pied and Sooty Oystercatcher, Hooded Plover and Masked Lapwing.

The coastal heathlands adjacent to the sealed Arthur River Rd, from the Arthur-Pieman Conservation Area entrance south to Arthur River (a distance of 12 km), can be excellent for a range of sought-after species. Tawny-crowned Honeyeater, Southern Emu-wren, White-fronted Chat and Brown Quail are some of the species regularly encountered. In the coastal woodlands around Arthur River, Green Rosella, Yellow-tailed Black-Cockatoo, Morepork, Brown Thornbill, Tasmanian Scrubwren, Yellow-throated Honeyeater, Yellow Wattlebird, Black Currawong and Silvereye are all relatively common. The area is also one of the best sites for White-throated Needletail: in mid to late summer,

The spectacular windswept West Coast from the Edge of the World Lookout. Sooty Oystercatchers, Hooded Plover and Pacific Gull frequent the coastline, while various seabirds are often visible beyond the wide surf break.

Tasmania

scan the skies for this distinctive species. Finally, the Arthur River area is one of the last strongholds for the Tasmanian Devil. A slow drive along the Arthur River Rd at night may produce a sighting of this iconic animal. The area is a bit of a stronghold for mammals as there is also a good chance of the often elusive Spot-tailed Quoll. Other more common species include Echidna, Southern Brown Bandicoot, Bennett's (Red-necked) Wallaby, Rufous-bellied Pademelon and Common Wombat.

Cradle Mountain–Lake St Clair National Park

Key species: Scrubtit, Black Currawong, Olive Whistler, Blue-winged Parrot
Other species: Wedge-tailed Eagle, Morepork, Pink and Dusky Robin

Cradle Mountain–Lake St Clair National Park (161 400 ha) is part of the Tasmanian Wilderness World Heritage Area. The landscape in the park is dominated by temperate rainforest, Beech forest, alpine heathland and Button-grass moorlands. The national park protects a vast area of the central Tasmanian highlands and includes spectacular scenery such as the iconic Cradle Mountain. The park is essentially divided into two sections, as dictated by vehicle access, with Cradle Mountain located at the northern end of the park and Lake St Clair at the south. The 'Overland Track' is a world famous 65 km (6-day) walking track that links the northern and southern sections. Much of the park is within the alpine zone, and thus snow cover, especially on walking trails, may restrict access in winter. Because many bird species are altitudinal migrants that move to lowland areas in winter, it is best to avoid

Myrtle Beech *Nothofagus cunninghamii* forest is accessible at a number of sites in both sections of the Cradle Mountain–Lake St Clair National Park: look for Tasmanian Thornbill, Scrubtit, Tasmanian Scrubwren, Crescent Honeyeater, Olive Whistler, Pink Robin, Black Currawong and Bassian Thrush.

birdwatching in the colder months. The park is located 80 km from Devonport via Cradle Mountain Rd, or 150 km from Launceston, via the Bass Highway and the signposted route through Sheffield. At the Cradle Mountain end, accommodation includes the Waldheim Chalet and Waldheim Cabins ((03) 6492 1110, cradle@parks.tas.gov.au) and a private caravan park ((03) 6492 1395) just before the park entrance. At the Lake St Clair end, accommodation and camping are available at Cynthia Bay ((03) 6289 1137, frontdesk@lakestclairlodge.com.au). The park visitor centre ((03) 6289 1172) provides details on walks.

In the Cradle Mountain section, the Dove Lake Loop Track (6 km) sets out from the Dove Lake carpark and takes you through the stunning Ballroom Forest: an area of cool temperate rainforest nestled against the slopes of the mountain. Along here, you may see Green Rosella, Blue-winged Parrot, Olive Whistler, Scrubtit, Crescent, Yellow-throated and Strong-billed Honeyeater, Flame Robin, Yellow Wattlebird, Tree Martin and Black Currawong. The short Rainforest Walk (500 m) starts just behind the visitor centre and meanders through a patch of cool temperate rainforest. Along the walk, look for Black Currawong, Scrubtit, Tasmanian Scrubwren, Pink and Dusky Robin, Crescent Honeyeater, Yellow Wattlebird and Tasmanian Thornbill. Other short walks include Weindorfers Forest Walk (1 km) and the Enchanted Walk (1 km). Along the roadside into Cradle Mountain, keep an eye on the sky for Wedge-tailed Eagle. The park is also home to a wide range of Tasmanian mammals, including Tasmanian Devil (now rare), Spot-tailed and Eastern Quoll, Platypus, Echidna, Bennett's (Red-necked) Wallaby, Rufous-bellied Pademelon and Common Wombat. A good night drive for mammals is the route from the information centre to Dove Lake, stopping at Romney's Creek along the way. At Lake St Clair, there is good birding along the Watersmeet Track (3.5 km return) and it is worth continuing from Watersmeet to do the circular route that follows the Platypus Bay Track. This walk passes through several vegetation communities and provides opportunities for Morepork, Blue-winged Parrot, Green Rosella, Olive Whistler, Pink and Dusky Robin, Bassian Thrush, Scrubtit, Tasmanian Scrubwren, Tasmanian Thornbill, Crescent, Yellow-throated, Strong-billed and Black-headed Honeyeater, and Yellow Wattlebird.

Mountain Valley Wilderness Centre

An excellent place to see native marsupials is around the Mountain Valley Wilderness Centre (61 ha). It is a 'Land for Wildlife' reserve and an eco-style retreat, which is located in Loongana, 65 km south-west of Devonport. Tasmanian Devil, Platypus, Spot-tailed Quoll and Rufous-bellied Pademelon all occur on the property:(03) 6429 1394, info@mountainvalley.com.au, www.mountainvalley.com.au.

Waterhouse Conservation Area

Key species: Australasian Bittern, Cape Barren Goose, Hooded Plover, Latham's Snipe, Little Tern, Fairy Tern
Other species: Blue-winged Parrot, shorebirds

Located in Tasmania's north-east, and spanning the coast from Croppies Points

Tasmania

to Tomahawk Beach, Waterhouse Conservation Area (6953 ha) includes three main lagoons: Blackmans Lagoon, Big Waterhouse Lake and Little Waterhouse Lake. The vegetation in the reserve includes wetlands, heath and grasslands, coastal scrub and woodlands. It is accessed via Waterhouse Rd, 100 km north-east of Launceston. There are several campsites in the reserve and a caravan park with a shop in the township of Tomahawk.

Blackmans Lagoon can be a reliable site for Australasian Bittern. Listen for their booming call (given in the evening and early morning) in the wetland area behind the campsite. Access to the lagoon is via Blackmans Lagoon Rd (reached from Waterhouse Rd, 22 km east of Bridport). When entering the reserve by Homestead Rd, scan grassy areas for Cape Barren Goose. Australasian Bittern also occur in the expansive grassy areas around Big Waterhouse Lake (accessed via Homestead Rd just after you enter to the reserve), where you may also encounter Little Grassbird. Further along Homestead Rd, turn on to South Croppies Rd (the road to Little Waterhouse Lake). The heath and coastal shrublands along here support Tawny-crowned Honeyeater, Yellow and Little Wattlebird and, occasionally, Blue-winged Parrot. A careful scan around the perimeter of Little Waterhouse Lake may also produce sightings of Latham's Snipe (summer). The lakes in the reserve provide habitat for waterbirds such as Australian Pelican, Australian Shelduck, Musk Duck, Chestnut and Grey Teal, Hardhead (uncommon), Australasian and Hoary-headed Grebe, and Great Egret. Along the Waterhouse Conservation Area's shorelines, look for Little and Fairy Tern, Hooded, Pacific Golden and Red-capped Plover, Sanderling, Australian Pied and Sooty Oystercatcher, and White-fronted Chat. A good spot for these species is Herbies Landing, located off Homestead Rd, 5 km past the South Croppies Rd turn-off. Wedge-tailed Eagle is regular in the reserve and is often seen around Tower Hill. Other birds recorded include Brown (Swamp) Quail (ssp. *ypsilophora*), Banded Lapwing, Green Rosella, Swift Parrot (scarce), Dusky Woodswallow, Grey Butcherbird, and Black and Grey Currawong.

Bicheno district

Key species: Little Penguin, Hooded Plover, Australasian Bittern, Swift Parrot, Spotted Quail-thrush
Other species: Yellow Wattlebird, seabird colony (Governor Is)

Bicheno is located on the east coast of Tasmania off the Tasman Hwy, 175 km north of Hobart and 170 km south-east of Launceston. The beaches, offshore rocks, adjacent heath country and dry eucalypt forest provide a rich diversity of birding habitats. For accommodation details, contact the Bicheno Visitor Information Centre: (03) 6375 1500, bichenovic@bigpond.com.

Small-scale tours are provided at a number of Little Penguin colonies along the eastern and northern coastline of Tasmania. One such colony where evening tours are conducted is adjacent to the township of Bicheno: other options (not all covered in this book) include tours at Wynyard, Low Head, Burnie and The Neck on Bruny Island.

There is a Little Penguin colony near Diamond Island, 3 km north of the town centre. The best viewing time is just after dusk, when they come ashore to visit their burrows. A good way to see the penguin is to join one of the guided tours advertised through the Bicheno Visitor Information Centre. On a walk along the 3 km foreshore, you may see Hooded Plover and Australian Pied and Sooty Oystercatcher. Look out to sea for birds such as Australasian Gannet, Black-faced Cormorant, Kelp and Pacific Gull, and occasional White-fronted Tern and Short-tailed Shearwater. In spring, Swift Parrot feed in the flowering eucalypts around Bicheno. Governor Island Marine Reserve (50 ha), located on the east side of Bicheno, supports one of Tasmania's largest breeding colonies of Crested Tern (~2000 pairs), as well as nesting Silver Gull, Little Penguin, Sooty Oystercatcher and, occasionally, White-fronted Tern. It is also worth visiting the forests in the nearby Douglas-Apsley National Park. Apsley Waterhole, in the southern section can be accessed via Rosedale Rd off the Tasman Highway ~3km north of Bicheno. Here, there are opportunities for Musk Lorikeet, Green Rosella, Yellow-throated and Black-headed Honeyeater, Yellow Wattlebird, Beautiful Firetail and other forest birds. Also remain alert for Spotted Quail-thrush because they occur in forested areas with a grassy understorey within the park. At night, listen for Morepork.

Moulting Lagoon (4750 ha Ramsar site) is located 18 km south of Bicheno. Australasian Bittern have been recorded here, but you probably have a better chance of encountering Musk and Blue-billed Duck, Australasian Shoveler, Australian Shelduck, Hardhead, Great and Little Egret, Tasmanian Native-hen and Purple Swamphen. It is also a significant breeding and moulting site for Black Swan (as many as 10 000 birds have been recorded). In summer, the mud flats around Pelican Corner, located in the south-east of the lagoon, attract waders such as Eastern Curlew, Bar-tailed Godwit, Sharp-tailed Sandpiper and Common Greenshank. Around the wetlands, birds of prey such as White-bellied Sea-Eagle, Brown Falcon and Swamp Harrier also occur. A good viewing

A spectacular view across Moulting Lagoon accessed via Flacks Rd. Although not essential, a spotting scope can be useful here because the waterbirds are often spread over a vast area of suitable habitat.

area for the wetland is from Coles Bay Rd, 10 km from the Tasman Hwy turnoff. Here there is track on your right (a Greening Australia project sign marks the way), that leads down to a carpark and the water. Off Coles Bay Rd, there are also good entry points to the lagoon via Flacks Rd (16.5 km from the Tasman Hwy) and down River and Rocks Rd (18.7 km from the Tasman Hwy).

Freycinet National Park

Key species: Morepork, Spotted Quail-thrush, Hooded Plover, Black Currawong, Beautiful Firetail, seabirds
Other species: Morepork, Peregrine Falcon, also mammals including whales

First declared a national park in 1916, Freycinet National Park (16 803 ha) features spectacular red and pink granite formations and jagged granite ridge lines known as the Hazards. The park is located 125 km north-east of Hobart, near the townships of Swansea and Bicheno. A wide range of accommodation (hotel, motel, backpacker and cabin accommodation) is available at Coles Bay, while powered and unpowered campsites are available within the park:(03) 6256 7000, www.parks.tas.gov.au.

The view from Cape Tourville Lighthouse is breathtaking and also provides an excellent vantage point for seabird-watching. The list of species recorded here is extensive: Australasian Gannet, Black-faced and Great Cormorant, Kelp and Pacific Gull, Sooty Oystercatcher, Caspian Tern, White-bellied Sea-Eagle, Short-tailed (Sept–May), Hutton's and Fluttering Shearwater, Little Penguin and, less commonly, White-fronted Tern, Fairy Prion, Brown Skua, Cape Petrel, White-faced Storm-Petrel, Common Diving-Petrel, Southern Giant-Petrel, and Shy, Indian Yellow-nosed and Black-browed Albatross. Whales, including the Southern Right and Humpback and, less frequently, Long-finned Pilot Whale, are occasionally seen from here.

The Bluestone Bay Track (2 km return, 4WD) is accessed from Cape Tourville Rd, 4.8 km from the Coles Bay Rd turn-off. It winds its way through eucalypt woodlands to Bluestone Bay and Whitewater Wall. Along the track, keep your eyes open for Spotted Quail-thrush. It is often best located by its thin, high-pitched '*seep seep*' call. Other birds along the track include Green Rosella, Yellow-throated, Black-headed, Crescent and New Holland Honeyeater, Yellow and Little Wattlebird, Spotted Pardalote, Forest Raven, Golden Whistler, Scarlet and Dusky Robin and, in summer, Fan-tailed Cuckoo, Shining Bronze-Cuckoo, Black-faced Cuckoo-shrike, Dusky Woodswallow, Satin Flycatcher and, at night, listen for Morepork. In February 2009, a vagrant King Penguin was found at Bluestone Bay in an advanced state of moult. Look for Peregrine Falcon at Whitewater Wall.

Wineglass Bay is famous for its azure-coloured waters and white sandy beaches sheltered by wooded hills. Along the Wineglass Bay Track (2.5 km return) to the 'must visit' lookout, watch for Yellow-tailed Black-Cockatoo, Green and Eastern Rosella, Yellow-throated, Black-headed and Crescent Honeyeater, Tasmanian Thornbill, Grey Fantail and Black Currawong. In the gullies along the same walk, look for Beautiful Firetail, Tasmanian Scrubwren and Bassian Thrush. Similar bush birds are seen along Hazards Beach and Isthmus Track – a 6 km circuit that links up with the end of the Wineglass Bay Track. From the lookout, the Wineglass Bay Track descends slowly to the beach (2–3 h return). On a walk along the beach, you may see Hooded Plover and Australian Pied Oystercatcher. These species also occur on Friendly Beaches, located on the north-east side of the park. Mammals to look for at Freycinet National Park include

The view from Cape Tourville Lighthouse, Freycinet National Park. The coastal woodlands here provide excellent birdwatching: look for Yellow-tailed Black-Cockatoo, Green Rosella, Shining Bronze-Cuckoo (summer), Tasmanian Scrubwren, Yellow-throated, Black-headed and Crescent Honeyeater, Golden Whistler, Scarlet and Dusky Robin, Beautiful Firetail and Bassian Thrush.

Echidna, Eastern Quoll, Rufous-bellied Pademelon, Long-nosed Potoroo, Bennett's (Red-necked) Wallaby and Common Wombat.

King Island

Key species: Orange-bellied Parrot (very rare), Common Pheasant, Wild Turkey, Indian Peafowl, California Quail,
Other species: Hooded Plover, Scrubtit and other endemic King Is. subspecies, shorebirds

King Island (1098 km^2, population ~2000) is located 80 km north-west of Tasmania and 90 km south-east of Cape Otway in Victoria. The island (65 km long and 25 km wide) is mainly farmland, with small remnant areas of native forest. Accommodation on the island includes hotels, motels, self-contained apartments and cottages. The largest town is Currie, situated on the western side of the island, but there is also accommodation at Naracoopa on the east side (close to the main birding sites). The owners of Naracoopa Holiday Cottages are quite knowledgeable and helpful with the local birdlife, so this provides a good base for many visiting birdwatchers. Regular flights travel to King Island from Melbourne, Victoria and from Devonport and Burnie in Tasmania. For accommodation, flights and car hire contact King Island Tourism: www.kingisland.org.au, tourism@kingisland.net.au, (03) 6462 1355. Most of Tasmania's endemic species are found on King Island (with the exception of Forty-spotted Pardalote and Tasmanian Native-hen). There are six subspecies endemic to the island, including Green Rosella (ssp. *kingi*), Yellow Wattlebird (ssp. *kingi*), Dusky Robin (ssp. *kingi*), Black Currawong (ssp. *colei*) and Scrubtit (ssp. *greenianus*) – the King Island Brown Thornbill (ssp. *archibaldi*) was last recorded in 1971 and is thought to be extinct.

Perhaps surprisingly, for many Australian birdwatchers the most sought-after species on the island are wild populations of Wild Turkey, Common Pheasant and Indian Peafowl. The island is fox-free, so these species are quite common and are usually seen feeding along the roadside and out in the paddocks. A fourth exotic species, the

Tasmania

Californian Quail is another drawcard, but this species can be very elusive (see below). The Emu that frequent the south-west corner of the island are sadly farm escapees of the mainland form – the 'King Island' Emu (ssp. *ater*) became extinct in 1802.

At the end of their breeding season, the few remaining Orange-bellied Parrot migrate from south-west Tasmania to their wintering grounds in south-east Australia. On their way, a large proportion of the population stop and feed on the coastal glasswort saltmarsh on the southern shore of Sea Elephant Estuary, located in the south side of Lavinia Nature Reserve (6800 ha). To get there from Naracoopa, head north along Fraser Rd for 3 km, turn right onto Sea Elephant Rd, drive another 9 km until you reach the Sea Elephant River estuary, then walk across the saltmarsh to the beach; the best months to look for Orange-bellied Parrot are April and May. The Sea Elephant Estuary is also one of the better sites for the difficult to find California Quail; search along the fire trail that heads north just before you reach the end of Sea Elephant Rd. (California Quail have also been recorded at Tin Mine Rd.) Shorebirds found near the mouth of the estuary and along the beach include Hooded and Double-banded (winter) Plover, Sooty and Australian Pied Oystercatcher, and Black-faced Cormorant, and summer migrants such as Common Greenshank, Pacific Golden Plover, Sanderling, Bar-tailed Godwit, Eastern Curlew, Common Sandpiper (scarce) and Ruddy Turnstone. Other birds that you might expect to see around Sea Elephant Estuary include White-breasted Sea-Eagle, Fairy Tern, Musk Duck, Great Egret, Chestnut Teal, Blue-winged Parrot, White-fronted Chat and Golden-headed Cisticola (the only site for this species in Tasmania), and bush birds such as Black Currawong, Brush Bronzewing, Olive Whistler, Dusky Robin and Yellow Wattlebird. There is also a chance of seeing Nankeen Kestrel (rare in Tasmania). With an estimated population of less than 200 birds, the 'King Island', Scrubtit (ssp. *greenianus*) favours paper-bark swamps (in contrast to the 'mainland' Tasmanian Scrubtit, which favour dense gullies in rainforest and wet eucalypt forest). They are known only from a few sites, including Nooks Swamps (accessed via the 4WD Nooks Swamp Track) in Lavinia State Reserve and Collier's Swamp (on the far south of the King Island, accessed via Seal River Rd, just east of Big Lake). Short-tailed shearwater breeds in spring and summer within the reserve, with a colony at Cowper Point.

The near threatened Hooded Plover frequents sandy ocean beaches, especially where small areas of beach washed seaweed have accumulated. On King Island, search the shoreline of the Sea Elephant Bay.

Pegarah Forestry Reserve – a remnant of eucalypt forest that once covered most of the island's south – is not surprisingly one of the best areas for bush birds on King Island. Species to be found here include Yellow-tailed Black-Cockatoo, Green Rosella, Tasmanian Scrubwren, Tasmanian Thornbill, Yellow-throated, Strong-billed, Black-headed and Crescent Honeyeater, Yellow Wattlebird, Pink, Dusky and Flame Robin, Bassian Thrush, Fan-tailed Cuckoo (summer), Olive and Golden Whistler, and Forest Raven. Little Raven has also been

Flinders Island

Flinders Island (1333 km^2) is located 20 km north of Cape Portland on the north-east tip of Tasmania. The estimated population of Forty-spotted Pardalote on the Flinders Island is 70 birds and two reserves are set aside for their protection: Broughams Sugarloaf Conservation Area (280 ha) and Walkers Lookout (20 ha). In both reserves, the pardalote inhabits stands of White (Manna) Gum. Feral species such as Wild Turkey, Common Pheasant and Indian Peafowl are found in good numbers across the island. Chukar Partridge (while not on the official Australian bird list) occurs along the Coast Rd near Badgers Corner, west of the township of Lady Barron. Logan and Sellars Lagoon and Cameron Inlet are located in the south-east corner of Flinders Island and provide extensive habitat for waterbirds and shorebirds. Look here for Fairy Tern, Cape Barren Goose, Musk and Blue-billed (uncommon) Duck, Australian Pied and Sooty Oystercatcher, Double-banded (winter) Plover and Banded Stilt (in 2003 and again in 2012 when several thousand were recorded). Summer migrants to the area include Red-necked Stint, Common

Trousers Point beach and Strzelecki Peaks, Flinders Island. The woodlands on the distant foothills support seven of Tasmania's endemic bird species, while the sandy shoreline provides foraging areas for Australian Pied Oystercatcher, Hooded Plover and, in winter, Double-banded Plover.

Flinders Island (*Continued*)

Greenshank and Eastern Curlew. Besides Forty-spotted Pardalote, Flinders Island features eight other Tasmanian endemic bird species (Green Rosella, Morepork, Tasmanian Scrubwren, Tasmanian Thornbill, Yellow-throated and Strong-billed Honeyeater, Dusky Robin and Black Currawong). Look for them at Strzelecki National Park (4216 ha) and at the Darling Range Conservation Area (1400 ha), where you may also see Wedge-tailed Eagle, Swamp Harrier, Crescent Honeyeater, Flame and Scarlet Robin, Beautiful Firetail and, at night, there is a chance of seeing or hearing Morepork.

recorded here, with King Island supporting Tasmania's only known population of this species. Around the shores of Lake Flannagan (located off Cape Wickham Rd on the far north side of the island), look for Australian Shoveler, Chestnut Teal, Hoary-headed Grebe, Blue-billed Duck (rare), Banded Lapwing, Blue-winged Parrot and migratory shorebirds such as Double-banded (winter) Plover, Sharp-tailed and Curlew Sandpiper, and Red-necked Stint. The Yellow Rock Estuary, accessed via the north end of North Rd and South Yellow Rock Rd, is worth investigating. Here, you may see Cape Barren Goose, Australian Shoveler, Great Egret, Eurasian Coot and, at the river mouth, Hooded and Double-banded (winter) Plover, Sanderling, Australian Pied and Sooty Oystercatcher, and Fairy Tern. Orange-bellied Parrot has also been recorded in the past. For seabird-watching on King Island, it is worth visiting a few high points along the west coast, such as Seal Rocks, Fitzmaurice Point and the end of Charles St in Currie, where you may see Short-tailed Shearwater, Black-browed, Shy and Indian Yellow-nosed Albatross and, less commonly, Southern and Northern Giant-Petrel, Cape, Great-winged Petrel, Fairy Prion and Fluttering Shearwater. For great views of Little Penguin on King Island, visit the nightly penguin parade at Grassy Harbour.

South Australia

South Australia (abbreviated to SA) is located in the southern central portion of the continent. Covering some of the most arid areas within Australia, it has a total land area of 983 482 km^2. The state capital of SA is Adelaide, where 1.3 million of the state's 1.6 million people reside. The terrain consists largely of arid and semi-arid rangelands, with several mountain ranges, including the Mount Lofty Ranges and the Flinders Ranges. The south-west portion of the state consists of the sparsely vegetated (and inhabited!) Nullarbor Plain, fronted by the cliffs of the Great Australian Bight. South Australia borders every other Australian state and territory except Tasmania and the Australian Capital Territory. The states only endemic species, the Chestnut-breasted Whiteface, is restricted to the arid, gibber desert country north of Port Augusta and west of the Flinders Ranges and Lake Eyre. The Australian ('White-backed') Magpie is the bird emblem of SA.

Adelaide and the South-East

Regional overview

Although the south-east region covers only one-tenth of the state's land mass, it contains a great variety of productive birding sites. Stretching from Adelaide, down the Fleurieu Peninsula to Kangaroo Island, and across the Coorong to the border with Victoria, and north to the Mallee shrublands in the Riverland district, the region contains a diverse range of habitats. Most of the sites discussed in this section are easily accessible from Adelaide. Sometimes referred to as the '20-min city', the airport, Adelaide Hills and Dry Creek Saltfields are all less than half an hour away by car.

A suggested 8-day itinerary from this region involves travelling from Adelaide to the Fleurieu Peninsula and Kangaroo Island.

- **Day 1.** Using Adelaide as a base, visit the Adelaide Hills and the Mount Lofty Ranges, a round trip of ~50 km (Crimson (Adelaide) Rosella, Barbary Dove, Chestnut-rumped Heathwren, Bassian Thrush, Crescent Honeyeater).
- **Day 2.** Using Adelaide as a base, visit the Dry Creek Saltfields, a round trip of ~40 km (Slender-billed Thornbill, Banded Stilt, waders).
- **Day 4.** Adelaide to Fleurieu Peninsula, 100 km (Southern Emu-wren, Elegant and Rock Parrot).
- **Days 5–7.** Kangaroo Island, allowing for 300 km of driving on the island (Western Whipbird, Glossy Black-Cockatoo, Rock Parrot, Cape Barren Goose, Hooded Plover, Purple-gaped Honeyeater).
- **Day 8.** Return to Adelaide, 130 km.

An additional side trip to the Coorong National Park (Rufous Bristlebird, Cape Barren Goose, Malleefowl, Banded Stilt,

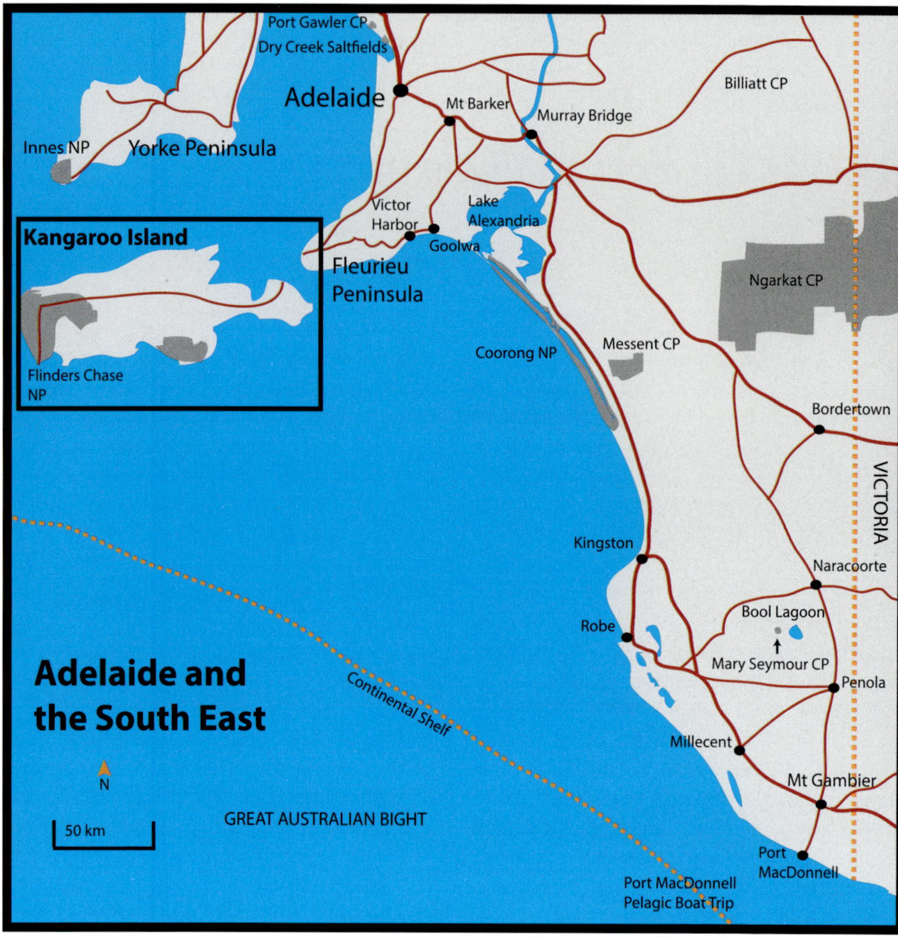

Lewin's Rail, Elegant Parrot, waders) and Bool Lagoon Game Reserve (Freckled Duck, Australasian Bittern, Australian Little Bittern, Magpie Goose) may also be appealing, though, to do this, allow for an additional 4+ days.

Two major birding organisations are well represented in the region: the South East SA regional group of Birdlife Australia (see www.birdlife.org.au) and Birds SA (see www.birdssa.asn.au).

The greater Adelaide, Kangaroo Island and south-east SA region has a Mediterranean climate, where most of the rain falls in the winter months. That being said, Adelaide is one of the driest capital cities in Australia. Mean monthly temperatures range from around 15–16°C in the winter to 25–29°C in summer, with occasional peaks over 40°C. At night, the temperature can vary from 0°C to 19°C throughout the year. Rainfall is unreliable, light and infrequent throughout summer. In contrast, winter has reliable rainfall. June is the wettest month of the year, averaging around 80 mm. The best season for birdwatching is usually spring (Sep–Nov), though autumn can also be good because

Eucalypt woodland in Cleland Conservation Park provides great woodland birding just a ~20 minute drive from the Adelaide CBD. Species you may encounter include Crimson (Adelaide) Rosella, Purple-crowned Lorikeet, White-throated Treecreeper, Crescent, Yellow-faced, White-naped and Brown-headed Honeyeater, Varied Sittella, Scarlet Robin and Grey Currawong.

temperatures are typically pleasant. Spring is usually the most productive, because there is an increase in bird activity, with the breeding season well underway, and many of the summer migrants (such as waders and cuckoos) are present. If you do plan to visit during summer (Dec–Feb), morning and late in the afternoon is the usually best time for birdwatching.

Excellent sealed highways provide easy access throughout large areas of the region. The Princes Hwy links Adelaide to Melbourne via the coast. Most of the main roads in the region are sealed and so a 2WD vehicle will get you to the majority of birding sites. Note, however, if you intend to explore some of the more remote locations, such as Ngarkat, a 4WD is recommended. Daily scheduled flights to Adelaide link to all Australian capital cities and there are regular international air services from many major international cities.

Birding highlights

There are no range-restricted endemic species to the region, though there are nonetheless several interesting species that may prove more easily located here than elsewhere. These include Cape Barren Goose, Australasian Bittern, Hooded Plover, Banded Stilt, Fairy Tern, Rock and Elegant Parrot, Rufous Bristlebird, Southern Emu-wren and Beautiful Firetail, and Mallee woodland specialists such as Malleefowl, Scarlet-chested Parrot, Mallee Emu-wren, Striated Grasswren, Western Whipbird, Black-eared Miner, Shy Heathwren, Chestnut-quail Thrush, Red-lored Whistler and Southern Scrub-robin. The south-east of South Australian has hosted some outstanding rarities over the years, including Northern Shoveler, Fiordland, Erect-crested, Royal, King and Northern and Southern Rockhopper Penguin, Antarctic and Grey Petrel, Broad-billed Prion, Great and Manx Shearwater, Little Curlew, American Golden Plover, Little Ringed and Ringed Plover, Baird's and Buff-breasted Sandpiper, Common Redshank, Lesser Yellowlegs, Hudsonian Godwit, Short-billed Dowitcher, Little Stint, Grey and Red-necked Phalarope, Franklin's and Sabine's Gull, Antarctic and Arctic Tern, South Polar Skua, Eastern Yellow, Citrine and Grey Wagtail, and Barn Swallow. Sadly, Barbary Dove has established a feral population in Adelaide and, in recent years, this population has been spreading across several suburbs.

Owing to the presence of several peninsulas, and the wetter and cooler Mount Lofty ranges that are isolated from other similar habitats further to the east, the Adelaide region supports a host of endemic or near-endemic subspecies. These include White-throated Treecreeper (ssp. *grisescens*), Southern Emu-wren (Mount Lofty Ranges Endangered ssp. *intermedius*), Crescent (ssp. *halmaturina*) and Yellow-faced (ssp. *samueli*), Honeyeater, Rufous Bristlebird (ssp. *broadbenti*), Rufous Fieldwren (ssp. *winiam*), Chestnut-rumped Heathwren (ssp. *parkeri*), Slender-billed Thornbill (ssp. *rosinae*), Spotted Quail-thrush (ssp. *anachoreta* – last recorded 1981 and probably extinct), Bassian Thrush (ssp. *halmaturina*), White-browed Scrubwren (ssp. *rosinae* and 'Spotted Scrubwren' *maculatus*), Grey Currawong ('Black-winged Currawong' ssp. *melanoptera*), Varied Sittella ('Black-capped Sittella' ssp. *pileata*), Western Whipbird ('Mallee Whipbird' ssp. *leucogaster*). There are also three forms of Crimson Rosella – 'Adelaide Rosella' ssp. *adelaideae*, 'Adelaide Rosella' ssp. *fleurieuensis,* and 'Yellow Rosella' ssp. *flaveolus* whose ranges meet in the Adelaide area. Further afield, owing to its isolation (~10 000 years) Kangaroo Island also hosts its own unique range of endemic subspecies. These include Glossy Black-Cockatoo (ssp. *halmaturinus*), Crimson Rosella (ssp. *melanoptera*), Red Wattlebird (ssp. *clelandi*), Purple-gaped (ssp. *cratitius*), Crescent (ssp. *halmaturinus*) and New Holland Honeyeater (ssp. *campbelli*), White-browed Scrubwren ('Spotted Scrubwren' ssp. *ashbyi*), Southern Emu-wren (ssp. *halmaturinus*) and Grey Currawong (ssp. *halmaturina*). There are also several introduced species to Kangaroo Island, including Australian Brush-turkey, Wild Turkey, Indian Peafowl, Emu (rare on the island), Cape Barren Goose, Crested Pigeon, Gang-gang Cockatoo and Laughing Kookaburra. The Kangaroo Island Emu (as opposed to the introduced Emu from the mainland) is now extinct: it was last recorded in 1827.

Greater Adelaide

Key species: Crimson (Adelaide) Rosella, Barbary Dove
Other species: Black-faced Cormorant, Peregrine Falcon, Long-billed Corella, Little Corella

Australia's fifth-largest city, the 'city of churches' as Adelaide is popularly known, is also a surprisingly green city, with the CBD bounded by a broad band of parkland. Within this precinct, one of the best birding areas is the Adelaide Botanical Gardens, located on North Terrace. It is particularly rewarding for parrots, with a good chance of Rainbow, Musk and, sometimes, Purple-crowned Lorikeet, Crimson (Adelaide) and Eastern Rosella, Red-rumped Parrot, and Long-billed and Little Corella. Look also for Striated Pardalote and Grey Fantail, and listen for the call of Laughing Kookaburra, especially early in the morning and at dusk. Raptors such as Peregrine Falcon (which breeds on tall buildings in the city centre) and Australian Hobby hunt over the gardens. A walk around Torrens Lake can be rewarding, especially the Weir Loop Walk. Around the lake you may see Australian Reed-Warbler (summer), Little Grassbird (summer), Australasian and Hoary-headed Grebe, Hardhead and Pink-eared Duck, Buff-banded Rail, Nankeen Night-Heron, Great Egret, Purple Swamphen and Tree Martin. The recently constructed Breakout Creek Wetlands lies along a stretch of the River Torrens between Henley Beach Rd and Tapley's Hill Rd (just next to Lockleys Oval) and provides good suburban birding. Here, you may see Black-winged Stilt, Australian Pelican, Purple Swamphen, Black-shouldered Kite, Collared Sparrowhawk and Nankeen Kestrel. The bustling beachfront suburb of Glenelg Beach can be worth a visit, because it is a good site for Crested and Caspian Tern, Black-faced, Great and Little Pied Cormorant and, occasionally, Eastern Osprey and White-bellied Sea-Eagle. Patawalonga Creek, a tidal creek immediately north of Glenelg, is also worthwhile, with possible species including all of the Glenelg species as well as Fairy Tern, White-fronted Chat, Singing and New Holland Honeyeater and Brown Songlark (summer), and raptors such as Brown Falcon, Nankeen Kestrel and Black-shouldered Kite. Barbary Dove, an uncommon introduced species to Australia (derived from the African Collared-Dove) occur in both the northern and southern suburbs of Adelaide. Reliable locations include O G Rd in Klemzig, Sampson Rd in Elizabeth Grove and Roscommon Rd in Salisbury Downs. At all sites, the species is often seen perched on powerlines, but, if planned control programs are successful, they may be more difficult to find in the future!

Red Wattlebirds are the largest honeyeater on the Australian mainland. A common species, they are frequently observed in well-vegetated urban areas.

Adelaide Hills and Mount Lofty Ranges

Key species: Chestnut-rumped Heathwren, Bassian Thrush, Crescent Honeyeater, Crimson (Adelaide) Rosella
Other species: Varied Sittella, Diamond Firetail, Hooded Robin, Painted Button-quail

The Mount Lofty Ranges are the range of mountains just to the east of Adelaide in South Australia, with the area nearest Adelaide also called the Adelaide Hills. The Mount Lofty Ranges are an 'island' of forest and woodland that receives relatively high rainfall (600–1200 mm), isolated from similar areas in eastern Australia by drier habitats. Only about 10% of the original 500 000 ha of native vegetation remains intact. Owing to habitat loss and fragmentation, several species have become extinct in the area including King Quail, Glossy Black-Cockatoo, Swift Parrot, Eastern Ground Parrot, Azure Kingfisher, Rufous Fieldwren, Regent Honeyeater and Barking Owl. The

Mount Lofty form of the Spotted Quail-thrush (ssp. *anachoreta*) was last recorded in 1981 near Spring Mount Conservation Park, and is now also likely to be extinct.

Conserving a vital area of bushland, Cleland Conservation Park (1000 ha) is one of the better birding spots in the Adelaide Hills. Just 12 km from the Adelaide city centre, to get there, take Glen Osmond Rd and the South Eastern Freeway. Exit the freeway at Crafers and drive north along Mount Lofty Summit Rd. The Mount Lofty Summit is a junction for many bushwalking trails. Birding along Waterfall Trail can be rewarding, especially for local endemic subspecies. Look for Crimson (Adelaide) Rosella, White-browed Scrubwren, White-throated Treecreeper, Varied (Black-capped) Sittella and Crescent Honeyeater, and, less commonly, Chestnut-rumped Heathwren, which are occasionally found in the heathy understorey below the summit. Because this park is a popular destination for locals and visitors, it is worth doing your birding early in the morning. Lorikeets and honeyeaters frequent the trees when flowering; look especially for Rainbow, Purple-crowned and Musk Lorikeet, and Yellow-faced, White-naped, New Holland and Brown-headed Honeyeater. Other birds at Cleland include Yellow-tailed Black-Cockatoo, Peregrine Falcon, Laughing Kookaburra, Fan-tailed and Pallid Cuckoo, Crested shrike-tit (rare), Brown and Striated Thornbill, Grey Currawong and Scarlet Robin.

Mount Lofty Botanical Gardens (97 ha), located on Summit Rd or Piccadilly Rd, in Crafers is the most reliable site to find Bassian Thrush. If you do not find Chestnut-rumped Heathwren at Cleland, they are more reliably found at Morialta Conservation Park (533 ha), 10 km north-east of Adelaide, and 3 km north of

Eucalypt woodland with a heathy understorey at Cleland Conservation Park. This habitat is ideal for the uncommon Mount Lofty form of the Chestnut-rumped Heathwren (ssp. *parkeri*).

Cleland. The park contains imposing ridges and deep gullies – tall wet forests are found in some gullies, while dry heath and Mallee occurs on the hilltops. Birds here include Sacred Kingfisher, Musk Lorikeet, Eastern and 'Adelaide' Rosella, Fan-tailed Cuckoo, White-throated Treecreeper, Crescent, Yellow-faced and White-naped Honeyeater, Scarlet Robin, Brown, Buff-rumped and Striated Thornbill, Varied Sittella, Golden Whistler and Mistletoebird (summer).

Dry open woodland species frequent the conservation parks north-east of Adelaide. At Kaiserstuhl Conservation Park (400 ha) you may see Emu, Crimson (Adelaide) Rosella, Diamond Firetail, Brown Treecreeper, Hooded Robin, Restless Flycatcher, Painted Button-quail, Rainbow Bee-eater, Jacky Winter, Crescent Honeyeater, White-browed Babbler, White-winged Chough and Buff-rumped Thornbill. Similar species are found at Para Wirra Recreation Park (1507 ha), where you may also encounter Brush Bronzewing, Peaceful Dove, Purple-crowned Lorikeet and Southern Whiteface.

Dry Creek Saltfields

Key species: Slender-billed Thornbill, Banded Stilt, waders, rarities
Other species: Elegant Parrot, Rock Parrot, Spotted Crake, Spotless Crake, Baillon's Crake

Each summer the inter-tidal mud flats and mangroves at the Dry Creek (formerly Penrice) Saltfields, located 20 km north of Adelaide, are used by large numbers of shorebirds as foraging and roosting areas. The lagoons, mangroves and saltbush at Dry Creek Saltfields extend for 10 km from St Kilda to Port Gawler. Located on the Gulf St Vincent, there are extensive areas of salt ponds, with the occasional island, as well as areas of saltmarsh, samphire and mangroves, especially on the western edge of the ponds. This wetland system is an important area for migratory waders, and there are regular reports of rare and vagrant species. Access is only through permission, and birdwatchers wishing to visit must first contact Birds SA through www.birdssa.asn.au. To get there from Adelaide, head north on Port Wakefield Rd, turning west into St. Kilda Rd, Thompson Rd or Plain Rd. Unfortunately, if you cannot gain access to the saltfields, the area is so expansive that there is usually little to be seen from these adjacent public roads.

Birding along the central (east to west) track is the easiest viewpoint to see waders. In summer, you should see Sharp-tailed and Curlew Sandpiper, Black-tailed and Bar-tailed Godwit, Common Greenshank and Red-necked Stint, Whimbrel, Marsh, Wood and Common Sandpiper, Grey-tailed Tattler, Grey and Pacific Golden Plover, and, less commonly, Great and Red Knot. Australian resident species include Red-capped Plover, Sooty and Australian Pied Oystercatcher, Red-necked Avocet, Banded and Black-winged Stilt (the last three sometimes in large numbers). Rarer species to look for include Ruff, Long-toed Stint, and Terek, Broad-billed and Pectoral Sandpiper, while the list of vagrants is impressive, including Buff-breasted Sandpiper, American Golden Plover, Hudsonian Godwit and Red-necked Phalarope. Terns are also well represented at Dry Creek, particularly in summer. Look for Whiskered (summer), White-winged Black (summer), Fairy, Caspian, Crested and Common Tern. There is also a single record of the rare Arctic Tern. Wetland birds can be plentiful, especially in late summer when they descend on the area as inland water bodies contract and dry up. You may see Australian Shelduck, Grey and Chestnut Teal, Musk and Freckled (rare) Duck, Great Crested Grebe, Great and Little Egret, and Australasian Darter. The area is also excellent for raptors, with visitors regularly observing Peregrine Falcon, Nankeen Kestrel, Australian Hobby, Black-shouldered Kite, White-bellied Sea-Eagle, Whistling and Black Kite, Swamp Harrier and, sometimes, Spotted Harrier. In areas of samphire heath and salt-marsh, look for White-winged Fairy-wren, Slender-billed Thornbill and White-fronted Chat – a good place for White-winged Fairy-wren is near the entrance on the eastern boundary of the saltfields. The western-most track runs parallel to some good areas of mangrove and saltmarsh, where it is worth looking for Grey-tailed Tattler, Nankeen Night-Heron and Black-tailed Native-hen, and skulking marsh birds such as Spotted, Spotless and Baillon's Crake. The saltmarsh adjacent to the mangroves is also a good spot for Slender-billed Thornbill. Small groups of both Elegant and Blue-winged Parrot sometimes perch on fence posts around the saltbush marshes. The public track to the mangroves of Port Gawler Conservation Park also hosts the occasional Rock Parrot and regular groups of Slender-billed Thornbill, especially near the gate that leads into the saltfields. Just to the south of Dry Creek, it is worth investigating Greenfields Wetland, located on the corner of Port Wakefield Rd and the Salisbury Hwy. Search here for Australasian Bittern (rare, winter), Buff-banded Rail, Baillon's, Spotted and Spotless Crake, Glossy Ibis, Red-kneed Dotterel, and Banded and Black-winged Stilt. During summer, the wetlands have held some rare shorebirds, including Long-toed Stint, Wood and Pectoral Sandpiper, and Australian Painted Snipe.

The Price Saltfields (130 km north of Adelaide) are also worth visiting. They contain many of the species found at Dry Creek, including waders such as Sharp-tailed, Pectoral, Marsh and Wood Sandpiper, Grey Plover, and Red and Great Knot, as well as Rock and Elegant Parrot, Slender-billed Thornbill and White-winged Fairy-wren. Rarities also turn up at Price, such as a Short-billed Dowitcher (Mar 2000) and Buff-breasted Sandpiper (Dec 2012). Again, for access contact Birds SA (www.birdssa.asn.au).

Banded Stilt: an Australian endemic with a remarkable breeding strategy largely reliant on the occasional flooding of inland salt lakes, including Lake Eyre. When not breeding, the species is typically concentrated at select near coastal sites, including the salt fields near Adelaide, the Coorong and Bird Lake near Port Augusta.

Fleurieu Peninsula

Key species: Southern Emu-wren, Elegant Parrot, Rock Parrot, Tawny-crowned Honeyeater
Other species: Beautiful Firetail, Chestnut-rumped Heathwren, Hooded Plover, Sooty Oystercatcher, Australian Spotted Crake, Spotless Crake, seabirds

The Fleurieu Peninsula is a 40-min drive south of Adelaide. Heading along the eastern side of the peninsula, the road passes through the townships of Goolwa, Port Elliot and Victor Harbour. At the end of the peninsula lies Cape Jervis and from here you can catch the ferry to Kangaroo Island. Bus and coach services depart Adelaide daily to most major towns within the region.

Port Gawler Conservation Park and Buckland Park Lake

Port Gawler Conservation Park (430 ha), 45 km north of Adelaide, is a reliable site for Slender-billed Thornbill. Located north of the Dry Creek Saltfields, access is via Port Gawler Rd, 36 km north of Adelaide. Search the samphire flats just before the mangroves, ~1 km before you get to the Gawler River. Here, you may also see Elegant and Rock (uncommon) Parrot, Blue Bonnet, White-winged Fairy-wren, White-fronted Chat, Little Grassbird, Brown Songlark (summer), White-browed Scrubwren and Singing Honeyeater. Similar birds frequent the samphire flats behind Thompson Beach and Webb Beach, which is best accessed via Prime Beach Rd, with the turn-off 58 km north of Adelaide, just south of the township of Dunlin. A walk along North Trail (5 km return – Thompson Beach to Bakers Creek) and South Trail (3.6 km return – Thompson Beach to Third Creek) will cover a good range of habitats. The area can be particularly good for summer migratory waders, including Lesser and Greater Sand Plover, Pacific Golden, Grey Plover, Common Greenshank, Ruddy Turnstone, Red-necked Stint, Great and Red Knot, Sharp-tailed Sandpiper and Bar-tailed Godwit. Double-banded Plover regularly occur here in winter. Nearby Buckland Park Lake has an impressive list of birds, with waterfowl sometimes present in huge numbers. Birds to look for include Hoary-headed and Little Grebe, Eurasian Coot, Little and Great Egret, Grey and Chestnut Teal, Hardhead, Australasian Shoveler, Pacific Black, Pink-eared, Musk, Blue-billed Duck. Rarer species recorded include Australian Painted Snipe, Long-toed Stint, Freckled Duck and vagrant Northern Shoveler, Little Ringed and Ringed Plover, and Baird's Sandpiper.

Newland Head Conservation Park (1037 ha), situated 15 km south-west of Victor Harbour (99 km south of Adelaide), is one of more than 20 nature reserves on the Peninsula. Hooded Plover and, occasionally, Sooty Oystercatcher use Waitpinga Beach between the carpark and the mouth of the creek estuary. There is a camping ground (basic facilities) off Dennis Rd just before you reach the Waitpinga Beach carpark at Newland Head. Look around the camping area for Beautiful Firetail and, more commonly, Red-browed Finch, White-browed Babbler and Golden Whistler. Elegant and Rock Parrot occur at Newland Heads, with a good place to search along the Heysen Trail to the Clifftop. Here, there is also a chance of seeing heathland birds such as Chestnut-rumped Heathwren, Beautiful Firetail and Tawny-crowned Honeyeater. The headlands provide an excellent vantage point for seabird-watching. In season, you may observe Black-browed, Shy and Indian Yellow-nosed

Albatross, Great Skua, Northern and Southern Giant-Petrel, Fluttering and Short-tailed Shearwater, Black-faced Cormorant, Australasian Gannet and Little Penguin. Watch out also for birds of prey, including White-bellied Sea-Eagle, Eastern Osprey and Peregrine Falcon.

The Endangered Fleurieu Peninsula form of the Southern Emu-wren (ssp. *intermedius*), along with Beautiful Firetail, are found at Deep Creek Conservation Park (4500 ha). The reserve is located 13 km east of Cape Jervis (104 km south of Adelaide). Concentrate your search along Tent Rock Rd and the Deep Creek Waterfall Hike, starting from Tent Rock Rd. There are several camping grounds at Deep Creek, including the Trig, Stringybark, and Cobbler Hill Campgrounds. A reliable place to see Southern Emu-wren is the Mount Compass Wetland, along the boardwalk opposite the school. Southern Emu-wren (ssp. *intermedius*), as well as Chestnut-rumped Heathwren, Beautiful Firetail and Elegant Parrot also occur at Cox Scrub Conservation Reserve, 17 km east of Mount Compass on the Adelaide–Goolwa Rd. In Victor Harbour, it is worth visiting the treatment ponds at the end of Crozier Rd, where Australian Spotted, Spotless and Baillon's Crake, Lewin's Rail and, in summer, Common Sandpiper and Latham's Snipe, have been recorded. There is a Little Penguin colony on the south side of Granite Island, with access via a bridge at the end of Flinders Parade.

The Aldinga Scrub Conservation Park (300 ha) contains a diverse range of plant communities, from sedgeland and closed heaths to Pink Gum woodlands and Mallee scrub. One hundred and sixty-six bird species have been recorded here, including Elegant Parrot, Crested Shrike-tit, Golden Whistler, White-browed Babbler and, in summer, Rufous Songlark, White-throated Gerygone, Dusky Woodswallow and Rainbow-bee-eater. The Aldinga Scrub is on the coastline of Gulf St Vincent between the townships of Aldinga Beach and Sellicks Beach, 46 km south of Adelaide.

Goolwa

Key species: Cape Barren Goose, Lewin's Rail, Elegant Parrot, Rock Parrot
Other species: Chestnut-rumped Heathwren, Sooty Oystercatcher, Tawny-crowned Honeyeater, Australian Spotted Crake, Spotless Crake, seabirds, waders

Goolwa is 18 km east of Victor Harbour on the western shore of the Murray Mouth. There are several interesting bird sites around the town, including the Signal Point Bird hide, the Goolwa Barrage and the mud flats near the Beacon 19 boat ramp. To get there from Hutchison St (Port Elliot Rd) in Goolwa, turn south into Oliver Street, cross over the train line and turn left at the T-junction into Barrage Rd. Follow Barrage Rd around the water's edge to the Signal Point bird hide on the left. The shoreline near the hide provides habitat for Purple Swamp-hen, Dusky Moorhen and Black-winged Stilt and, at times of low water, look for the secretive Australian Spotted and Spotless Crake. In summer, Latham's Snipe is a regular migrant to this area. The area 300 to 400 m east of the bird hide can be rewarding, with uncommon summer migrants such as Ruff, Wood Sandpiper and Lewin's Rail occasionally recorded here. Continue east to the Goolwa Barrage carpark. There is freshwater above the barrage and saltwater below. On the barrage, there are often roosting Australian Pelican and Black-faced, Pied, Little Pied, Great and Little Black Cormorant. Just before the barrage, a signed track leads 2.3 km east to the Beacon 19 boat ramp. During the summer months, migratory waders such as Sharp-tailed Sandpiper, Red-necked Stint, Common Greenshank, Bar-tailed and Black-tailed Godwit and

Pacific Golden Plover feed on the mud flats here at low tide. Resident waterbirds include Black-winged Stilt, Red-necked Avocet and Crested, Fairy, Caspian and Whiskered Tern and, more rarely, Gull-billed Tern occur in summer. Along the main river channel, you may see Great Crested Grebe, Musk Duck, Hardhead, Australasian Shoveler, and Grey and Chestnut Teal. An extended 4 km walk to the Murray Mouth and the ocean beach begins at Beacon 19. Singing Honeyeater and Little Grassbird are found in the coastal vegetation, while raptors such as Nankeen Kestrel, Swamp Harrier, and Whistling Kite often hunt overhead. Look for Australian Pied Oystercatcher, Sanderling and Red-necked Stint (the latter two in summer) along Goolwa Beach, particularly west of the carpark. Goolwa was also the location of Australia's first recorded Citrine Wagtail (May 1987). The exposed sand bars in the estuary at the mouth of the Murray River attract summer waders such as Sanderling, Grey-tailed Tattler, Eastern Curlew, Whimbrel, Black-tailed and Bar-tailed Godwit, as well as Fairy and Crested Tern.

The Goolwa Effluent Ponds, located on the north side of Kessel Rd, attract a variety of waterbirds include Pink-eared, Blue-billed and Freckled (rare) Duck, Australasian Shoveler, Hardhead, Black-tailed Native-hen and Red-kneed Dotterel. Look for Australian Buff-banded Rail, and Baillon's, Spotless and Australian Spotted Crake around the vegetated fringes of the ponds.

Hindmarsh Island is reached from Goolwa via Randall Rd. Follow the road signs to Mundoo Channel Drive and the Murray Mouth, looking for Cape Barren Goose and Banded Lapwing along the way. Opposite the Murray Mouth is Bird Island and Sugars Beach. In late spring and summer, scan here for migratory waders such as Bar-tailed Godwit, Eastern Curlew, Whimbrel and Red Knot. Australian Pied and Sooty Oystercatcher also occur, while there is a chance of Elegant and Rock Parrot in the coastal vegetation. To view Scab Channel, it is worth walking across the samphire flat at the end of Mundoo Channel Drive. The Murray Mouth carpark can be a reliable place to find Rock Parrot, particularly after the spring breeding season – walk around in the low sand dunes between the road and the water's edge.

Kangaroo Island

Key species: Western Whipbird, Glossy Black-Cockatoo, Rock Parrot, Cape Barren Goose, Hooded Plover, Purple-gaped Honeyeater

Other species: Australian Brush-turkey, Indian Peafowl, White-bellied Sea-Eagle, Eastern Osprey, Sooty Oystercatcher, Banded Stilt, Fairy Tern, Crimson Rosella, Southern Emu-wren, Crescent Honeyeater, Tawny-crowned Honeyeater, Shy Heathwren, Beautiful Firetail, Bassian Thrush, Bush Stone-curlew, Little Penguin

Kangaroo Island also hosts its own unique range of endemic subspecies. These include Glossy Black-Cockatoo (ssp. *halmaturinus*), Crimson Rosella (ssp. *melanoptera*), Red Wattlebird (ssp. *clelandi*), Purple-gaped (ssp. *cratitius*), Crescent (ssp. *halmaturinus*) and New Holland Honeyeater (ssp. *campbelli*), White-browed Scrubwren ('Spotted Scrubwren' ssp. *ashbyi*) and Grey Currawong (ssp. *halmaturina*). There are also several introduced species to Kangaroo Island, including Australian Brush-turkey, Wild Turkey, Indian Peafowl, Emu (rare on the island), Cape Barren Goose, Crested Pigeon, Gang-gang Cockatoo and Laughing Kookaburra.

With a total length of 155 km, Kangaroo Island is the third largest island in Australia (after Tasmania and Melville Island). 'KI', as it is affectionately known, is located 130 km south of Adelaide (16 km off the coast of

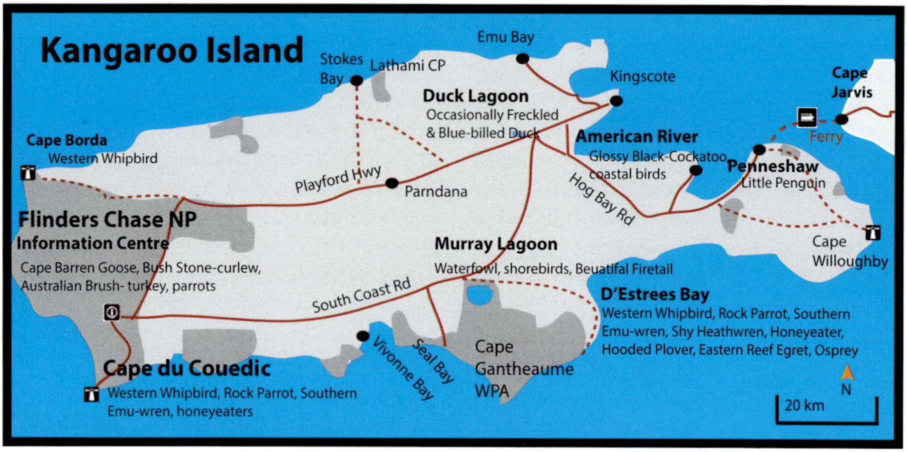

the township of Cape Jervis). The island is a delight for nature lovers, with over a third of the island protected by national and conservation parks. An impressive 267 bird species have been recorded on the island and, due to its geographical isolation (> 10 000 years), there is a high level of endemism at the subspecies level. Accommodation options on the island include hotels, motels, cabins and bush camping, see www.tourkangarooisland.com.au. Information about the island's national and conservation parks is available at www.environment.sa.gov.au. For information on ferry services to the island, see www.sealink.com.au.

At American River, Glossy Black-Cockatoo (ssp. *halmaturinus*) regularly feed on the seeds of Drooping Sheoke along the Fish Cannery Walk (3 km return). In Autumn, they prepare to chose a nesting hollow – nesting from April through June and into October. At American River, you may also encounter Laughing Kookaburra (introduced to the island in 1926), Golden Whistler, Crescent Honeyeater (ssp. *halmaturinus*) and Grey Currawong (ssp. *halmaturina*). Coastal birds occurring along the American River foreshore include Eastern Osprey, White-bellied Sea-Eagle, Cape Barren Goose (introduced to the island in the 1920s; look for them in grassy areas), Caspian Tern, Sooty and Australian Pied Oystercatcher, Eastern Reef Egret and Royal Spoonbill. In summer, migratory waders such as Eastern Curlew and Common Greenshank may also occur.

Flinders Chase National Park (33 000 ha) is home to the local form of the Western Whipbird (ssp. *lashmari*). Look for this renowned skulker at the eastern end of Weir Cove Rd near the Weir Cove carpark and eastern end of the Weir Cove Hike (3 km). To get there, turn east from the Cape Du Couedic Rd 500 m before you reach the Cape Du Couedic Lighthouse – the carpark is another 500 m. From here, you may also see Brush Bronzewing, Dusky Woodswallow (summer), Southern Emu-wren (ssp. *halmaturinus*, Australia's largest race), White-browed ('Spotted' ssp. *ashbyi*) Scrubwren, Red Wattlebird (ssp. *clelandi*), and Purple-gaped (ssp. *cratitius*), Tawny-crowned, White-eared and Crescent Honeyeater. The lookout at the Cape Du Couedic Lighthouse provides a good vantage point for seabird-watching, particularly in winter and spring; possibilities include White-bellied Sea-Eagle, Australasian Gannet, Brown Skua (winter),

Caspian Tern, Short-tailed Shearwater and Indian Yellow-nosed, Shy and Black-browed Albatross. In September 2006 and 2007, numbers of Antarctic Tern, a vagrant to the Australian mainland, roosted on the rock platform in front of the lookout. Around the Information Centre at the entrance to the park, look for Cape Barren Goose, Australian Brush-turkey (introduced in 1936), Bush Stone-curlew, Gang-gang Cockatoo (introduced 1947 and 1956), Bassian Thrush (rare), Crimson Rosella (ssp. *melanoptera*), and Purple-crowned and Rainbow Lorikeet. This same area is an excellent site for mammals: look for Koala, Western Grey Kangaroo, Tammar Wallaby and Southern Brown Bandicoot around the picnic area, while Australian Sea Lion and New Zealand Fur Seal frequent the coastline. The very short Discovery Walk takes you from the Visitor Centre to the Walking Trail Information Shelter overlooking Black Swamp. In May, the swamp becomes alive with Cape Barren Geese selecting nesting territories.

At D'Estrees Bay, look for Western Whipbird, Rock Parrot, Southern

Western Whipbird is one of the more elusive species on Kangaroo Island. Preferring to keep to dense cover, sometimes the observer can be rewarded by waiting and watching quietly for a bird to pass through a gap in the vegetation.

Emu-wren, Shy Heathwren, and Purple-gaped, Crescent, New Holland (ssp. *campbelli*) and Brown-headed Honeyeater in the Mallee bordering the bay (accessed via D'Estrees Bay Rd) in Cape Gantheaume Conservation Park (24 316 ha). At Wheatons Beach and Point Tinline, both located at the southern end of D'Estrees Bay Rd, scan the coastline for Eastern Osprey, Eastern Reef Egret, Ruddy Turnstone, Hooded Plover, Sooty and Australian Pied Oystercatcher, and Fairy Tern. In summer, migratory shorebirds such as Red-necked Stint and Ruddy Turnstone may also occur.

Murray Lagoon (2000 ha) is just north of Cape Gantheaume Conservation Park (signposted on the main road from Kingscote to Flinders Chase). The lagoon attracts large numbers of waterfowl, including Australian Shelduck, Australasian Shoveler and Blue-billed and Musk Duck, Banded and Black-winged Stilt, and summer waders such as Red-necked Stint, Sharp-tailed Sandpiper and Eastern Curlew. Rarer species have included Freckled Duck (sometimes in numbers) and Pectoral Sandpiper. To view the lagoon, walk to Bald Hill (1 km) or take the Timber Creek (1.5 km) and longer Curley Creek (11 km) tracks. Along these walks, look also for Beautiful Firetail (particularly in the grassy areas along Timber Creek) and the introduced Indian Peafowl.

Other spots worth visiting on Kangaroo Island include Lathami and Baudin Conservation Parks (the later holding a small population of Glossy Black-Cockatoo), Duck Lagoon, which occasionally holds Freckled and Blue-billed Duck (accessed via Duck Lagoon Rd near Cygnet River). For woodland birds, visit Lashmar Conservation Park and the Chapman River (where there is bush camping) on the far east side of the island near Antechamber Bay. Note that, if you do not see Western Whipbird at Weir Cove or D'Estrees Bay, they are also found at the Remarkable Rocks, Cape Borda Lighthouse

View to Remarkable Rocks, Flinders Chase National Park, Kangaroo Island. The heathland visible on the headland supports Southern Emu-wren and Western Whipbird, while the Mallee in the foreground provides habitat for Purple-gaped Honeyeater.

on the north side of Flinders Chase and Seal Bay Conservation Park (which supports one of Australia's largest Australian Sea Lion colonies). In Penneshaw, there is a Little Penguin colony next to the ferry terminal; guided tours are run each evening (www.kipenguincentre.com.au). Watch out for Rosenberg's Goanna on roadsides throughout the island. In June, Southern Right Whale make their annual visit to the protected waters around Kangaroo Island, and then head back south to Antarctic waters in October.

Coorong National Park

Key species: Rufous Bristlebird, Cape Barren Goose, Malleefowl, Banded Stilt, Lewin's Rail, Elegant Parrot, shorebirds

Other species: Southern Emu-wren, Beautiful Firetail, Fairy Tern, Sooty Oystercatcher, Striated Fieldwren

Coorong National Park (50 000 ha) protects a 100+ km long, shallow saline lagoon, which is separated from the Southern Ocean by the sand dunes of the Younghusband Peninsula. As a habitat for numerous species of migratory birds, and as a refuge for birds in times of drought, the Coorong is of great importance at both a national and international scale; one consequence of this has been the area's listing under the Ramsar Convention. There is a great diversity of birds, with over 200 species recorded, including many types of waders and waterfowl. Indeed, over summer an estimated 122 000 shorebirds, 45 000 ducks and 2000 Cape Barren Geese make use of the wetland system. There is camping at 42 Mile Crossing, Parnka and Long Point, and Godfreys Landing (the latter accessible by boat only).

Some of the best birding in the Coorong is around the township of Salt Creek. The marshy environment near the Salt Creek Crossing (the area around the Princes Hwy Bridge) can be a good place to observe the shy and secretive Australian Spotted Crake and Spotless Crake, Buff-banded Rail and Lewin's Rail. In the coastal shrublands near Salt Creek, search for Southern Emu-wren and Beautiful Firetail. Banded and Black-winged Stilt and Red-necked Avocet often feed just to the south of Salt Creek, especially towards the centre of the lagoon. The walk along Lakes Nature Trail (3 km loop), which starts from a carpark on the Loop Rd, is particularly worthwhile. The track provides views across the ephemeral Pipe Clay Lake and Salt Lake, both lined with samphire, saltbush and scattered mallee. Malleefowl sometimes feed along Loop Rd (Old Melbourne Rd) and near Tea Tree Crossing. Indeed, ~3.7 km from the service station on the Princes Hwy, there is a Malleefowl mound on the north side of Loop Rd. This road is also a good place to look for Rufous Bristlebird, particularly in areas of coastal heath bordering sand dunes. Other sought-after species found in the Mallee and woodland here include Emu, Collared Sparrowhawk, Australian Ringneck, Beautiful Firetail, Southern Scrub-robin, Shy Heathwren, Purple-gaped Honeyeater, Scarlet Robin, Brush and Common Bronzewing, Peaceful Dove, Masked, White-browed and Dusky Woodswallow, and, occasionally, Fork-tailed Swift (summer).

Parnka Point is the narrowest point between the northern and southern Coorong lagoon. It is a good place to look for both Elegant and Blue-winged Parrot. The Critically Endangered Orange-bellied Parrot has also been recorded in the past, though sadly chances of a winter sighting are now very slim (see p. 319). Other common species include Grey and Chestnut Teal, Crested, Fairy, Whiskered and Caspian Tern, and Southern Emu-wren, while the threatened Hooded Plover occurs on the beach near Parnka Point; to protect the last species, the beaches are closed to vehicles from the 24 October to Christmas Eve each year. In the same vicinity, another good birding area is the lagoons south of Hacks Point. Jack Point Pelican Observatory Walk (1.2 km return) overlooks a cluster of Islands where Australian Pelican, Crested and Caspian Tern, and Silver Gull have established large breeding colonies. Policeman's Point provides a good view into the lagoons, where you can see shorebirds such as Black-winged and Banded Stilt, Red-necked Avocet, Red-necked Stint, Curlew and Sharp-tailed Sandpiper, Common Greenshank, Marsh Sandpiper, Bar-tailed and Black-tailed Godwit, Red and Great Knot, Common Sandpiper, and Grey and Pacific Golden Plover. Vagrant, Red-necked Phalarope (several records) and Grey Phalarope (single record), have also been recorded here. Sanderling, which prefer waved-washed beaches, feed along much of the ocean beach-side of Younghusband Peninsula. Other birds that occur here include Fairy Tern, Whiskered Tern, Banded Lapwing (uncommon), Masked Lapwing, Red-capped Plover, and Sooty and Australian Pied Oystercatcher. The far north of the Coorong, near the Murray mouth, also provides good birdwatching. This area can be accessed via the unsealed, but well-graded, Long Point or Mark Point Rds. This is near the Tauwitchere Barrages, which connect Tauwitchere Island with Pelican Point and where Lake Alexandrina enters the Coorong. Interesting species that occur here include Gull-billed, Whiskered, Caspian, Fairy, Little (rare) and Crested Tern, as well as Australian Shelduck, Grey and Chestnut Teal, Pacific Black Duck, Australian Pelican, Little Black and Pied Cormorant, Great Egret,

The Coorong at Salt Creek. The shallows of the lagoon here can be good for Banded and Black-winged Stilt, and Red-necked Avocet, while the marshy areas near the Salt Creek crossing can be especially good for crakes and rails.

White-faced Heron and, occasionally, Little Egret. Summer shorebirds to look for include Common Greenshank, Red-necked Stint, Sharp-tailed Sandpiper, and, less commonly, Pacific Golden Plover and Sanderling.

The Messent Conservation Park (12 250 ha), 7 km north-east of Salt Creek, consists of mixed shrublands, sedgeland, grassland and Mallee woodland. Access to the park is gained via Salt Creek Rd. Most of the tracks through the park are 4WD, although a 2WD vehicle can reach the south-west corner along Salt Creek Rd, following the western edge of Martin Washpool Conservation Park. Some of the best birding can be had along a firebreak that starts near the west corner of the reserve and bisects the Park in an easterly direction. Interesting species that occur here include Brown Quail, Elegant Parrot, Shy Heathwren, Purple-gaped, Tawny-crowned, Spiny-cheeked and White-fronted Honeyeater, Rufous Bristlebird, Southern Scrub Robin, Striated Fieldwren, Southern Emu-wren, Variegated Fairy-wren, Tree Martin and Beautiful Firetail.

The Southern Scrub-Robin is widespread through Mallee habitats in southern Australia. Generally, they prefer areas of Mallee with a moderate to dense cover of low shrubs.

Port MacDonnell pelagic boat trip

The Port MacDonnell pelagic boat trips are organised by Birds SA: general@birdssa.asn.au, www.birdssa.asn.au. A great variety of southern ocean seabirds has been observed on these 1-day excursions to the continental shelf, including Indian Yellow-nosed, Shy, Salvin's, Campbell, Black-browed, Grey-headed, Buller's, Sooty, Southern Royal, Northern Royal, Antipodean and Wandering Albatross, Gould's, Great-winged, White-chinned, and White-headed Petrels and Flesh-footed, Fluttering, Great (very rare), Hutton's, Sooty and Short-tailed Shearwaters. Other seabirds recorded include Little Penguin, Southern Fulmar, Fairy, Antarctic and Slender-billed Prion, Black-bellied, Grey-backed, White-faced and Wilson's Storm-Petrel, and Arctic, Long-tailed and Pomarine Jaeger.

Several hundred seabirds, mostly Shy Albatross, but also Black-browed, Indian Yellow-nosed and Wandering Albatross, around a working fishing trawler during a South Australian pelagic trip.

Bool Lagoon Game Reserve

Key species: Freckled Duck, Magpie Goose, Australasian Bittern, Australian Little Bittern

Other species: Latham's Snipe, Brolga, Blue-billed Duck, Baillon's Crake, Spotless Crake, Australian Spotted Crake

Bool Lagoon Game Reserve (3023 ha) and the adjacent Hacks Lagoon Conservation Reserve (~180 ha) are located 28 km south of Naracoorte. The lagoons usually support a diverse range of waterbirds and regularly host more than 20 000 individuals. Numbers tend to peak in spring and again

Robe and Little Dip Conservation Park

The historic seaside town of Robe is on Guichen Bay 340 km south-east of Adelaide. Watching for seabirds from the coast near Robe can be rewarding, with good vantage points near a stone obelisk on Cape Dombey (built in 1852) and at the nearby Robe Lighthouse. Seabirds to look for include Australasian Gannet, Black-faced Cormorant, Northern and Southern Giant Petrel, Short-tailed and Fluttering Shearwater, White-bellied Sea-Eagle, Cape Petrel, and Black-browed, Shy, Indian Yellow-nosed and Wandering Albatross. On the various walks around the Lighthouse, you may see Rufous Bristlebird, Brush Bronzewing and Forest Raven. Just south of Robe, Little Dip Conservation Park (1977 ha) protects a large area of coastline and the foreshore of Lake Eliza. Hooded Plover occur on the beaches, while Rufous Bristlebird inhabits the vegetated foreshore dunes in the park. The Critically Endangered Orange-bellied Parrot has visited the park during winter, feeding on Two-horned Sea-Rocket. Along the Freshwater Lake Walk (1.8 km loop), look for Musk Dusk, Latham's Snipe and Brown Quail.

in early autumn as large numbers of migrating waders visit the lagoon. To get there from Naracoorte, head south down the Naracoorte–Penola Rd for 18 km, and then turn right at Bool Lagoon Rd; the main access to the lagoon is then 9 km. Water levels in drought years can be very low. To enquire about the current water levels contact the SA DEH Naracoorte District Office on (08) 8762 3412. The Hacks Lagoon Campground provides excellent facilities, including toilets and barbeques. The main walks at Bool Lagoon are the Tea-Tree Boardwalk (1 km return), the short Gahnia Walk (300 m return), the Pat-om Walk (1.4 km return), which skirts the deep waters on the eastern edge of the Hack Lagoon, and Gunawar Walk (1.5 km loop).

Bool Lagoon is a known breeding site for both the Australasian and Australian Little Bittern. The two species occupy the tall sedges and rushes around the Lagoon, although they will sometimes feed in more open swampy vegetation. Listen for their distinctive calls from the Pat-om and Gunawar Walks. Here also expect to see Little Grassbird, Golden-headed Cisticola and Australian Reed-Warbler (summer). The lagoons are also periodically an important site for the Australian Painted Snipe, particularly from late December to April, when up to 30 individuals have been recorded. In summer, Latham's Snipe feed in the grassy fringes of the lagoon. The lagoons are of high value as a drought refuge and summer habitat for ducks such as Hardhead, Freckled, Pacific Black, Musk, Blue-billed and Pink-eared Duck, Plumed Whistling-Duck (rare), and Grey and Chestnut Teal. Both Cape Barren and Magpie Goose have been re-introduced at Bool Lagoon and are usually easily found. Other waterbirds include Brolga, Yellow-billed and Royal Spoonbill, Eurasian Coot, Purple Swamphen, Australian White, Straw-necked and Glossy Ibis, Great and

Intermediate (uncommon) Egret, Nankeen Night-Heron, and Little Pied, Little Black and Great Cormorant.

Little Bool Lagoon, accessed via Moyhall Rd, is an important migratory wader habitat during late summer to autumn when mud flats in Bool and Hacks have dried out. At this time, when conditions are right, scan for Bar-tailed and Black-tailed Godwit, Common Greenshank, Curlew, Marsh, Sharp-tailed Sandpiper, Grey and Pacific Golden Plover and uncommon waders such as Pectoral Sandpiper, Ruff, Wood Sandpiper, Long-toed Stint and Australian Pratincole. Rarities recorded within the lagoon system have included Grey Phalarope, Oriental Pratincole and Common Redshank. At the fringes of the reedbeds, bordering mud flats, look for Baillon's, Spotless and Australian Spotted Crake. There can also be high concentrations of birds of prey, including Wedge-tailed Eagle, Swamp and (less commonly) Spotted Harrier, White-bellied Sea-Eagle (uncommon), Australian Hobby, and Brown, Peregrine and Black Falcon. Other birds to look for include Whiskered Tern, Forest Raven, Fork-tailed Swift (scarce in summer), Blue-winged Parrot, and Stubble and Brown Quail. The diminutive Southern Emu-wren frequents the sedge along the Gahnia Walk and between here and the Tea-tree Boardwalk. The view from Big Hill provides an opportunity to look over the wetlands, with Brolga regularly seen in Black Rush Swamp, immediately west of Big Hill.

The south-eastern Australian population of Brolga is relatively small, with most birds concentrated in western Victoria and adjacent areas of South Australia. Bool Lagoon is one site that regularly supports the species.

Ngarkat Conservation Park and Billiatt Conservation Park

Key species: Malleefowl, Mallee Emu-wren, Striated Grasswren, Red-lored Whistler, Black-eared Miner, Chestnut Quail-thrush, Slender-billed Thornbill
Other species: Rufous Fieldwren, Shy Heathwren, Southern Scrub-robin, Major Mitchell's Cockatoo

Ngarkat (270 000 ha) and the smaller Billiatt (59 200 ha) Conservation Parks provide important habitat for a range of threatened bird species. Between them, these two reserves represent some of the largest remaining areas of intact native vegetation in the southern settled areas of South Australia. There are nine campsites at Ngarkat – with Pertendi Hut and Pine Hut Soak having toilet facilities. If you have a 2WD vehicle, then you can drive the Pinnaroo to Bordertown Rd, but the other areas mentioned are all very sandy in places and 4WD is essential. Check www.environment.sa.gov.au before visiting, because a permit or trip intention log may be required for entry.

Several rare Mallee specialists are resident in Ngarkat Conservation Park,

Mary Seymour Conservation Park

Mary Seymour Conservation Park (360 ha) is located ~8 km south-west of Bool Lagoon on Bool Lagoon West Rd. Habitats in the park include *Eucalyptus* woodlands, *Melaleuca* scrub and ephemeral swamps (usually dry). The reserve is notable for the range of night birds that have been recorded. These include an isolated population of Barking Owl, as well as Eastern Barn Owl, Southern Boobook, Tawny Frogmouth, Australian Owlet-Nightjar and Spotted Nightjar (the last rare). Other birds here include Crested Shrike-tit, Beautiful and Diamond Firetail, Varied (Black-capped) Sittella, Brown Quail, Painted Button-quail, Grey Currawong, Forest Raven, Purple-crowned Lorikeet, Black-chinned and Tawny-crowed Honeyeater, Southern Emu-wren, and, in summer, Fan-tailed Cuckoo and Shining and Horsfield's Bronze-Cuckoo. Just north of Mary Seymour Conservation Reserve, Big Heath Conservation Park is also worth visiting.

Mature heathlands and small patches of Mallee vegetation, Ngarkat Conservation Park. A variety of threatened species occupy this habitat, though each displays subtle preferences for habitats of differing fire age classes and structure. Key species in Ngarkat include Malleefowl, Mallee Emu-wren (the last population in SA), Slender-billed Thornbill and Chestnut Quail-thrush.

including Mallee Emu-wren, Red-lored Whistler and Malleefowl. Concentrate your search in the south-west of the Park, especially around Jimmy's Well, near the north-west corner of the intersection of Bucks Camp and Mount Rescue Tracks. To get there from Bucks Camp, head north 7.6 km along the Bucks Camp Track to the intersection with the Mount Rescue Track. Then continue north along the Bucks Camp Track for another 1.5 km until you see a gate on the west side representing the beginning of a 1 km walk into Jimmy's Well (a track best hiked). Malleefowl, Rufous Fieldwren, Shy Heathwren and Western Whipbird (though sadly this latter species is probably now extinct in the reserve) have all been recorded in this area. Ngarkat is the last 'stronghold' of the Mallee Emu-wren in SA, though, sadly, with fewer than 50 known pairs, stronghold may be a misnomer –look in mature open heath along the southern half of the Baan Hill Track. To get there from Keith, head east along Emu Flat Rd for 8 km and then turn north onto Dark Island Rd and drive 19 km to the southern entrance of Ngarkat. At the entrance, the Dark Island Rd becomes the Bann Hill Track. Another track runs parallel for a few 100 m (east side) to Bann Hill Track immediately after the entrance and then bears right. Search along this track for Mallee Emu-wren. Slender-billed Thornbill and Rufous Fieldwren live in the low-lying heath near Pertendi Hut in Ngarkat. Look along the Scorpion Springs Walking Trails, near Pine Hut Soak in the north-east section of the Park, for Malleefowl, Chestnut Quail-thrush, Southern Scrub-robin and Red-lored Whistler.

In Ngarkat Conservation Park the Slender-billed Thornbill appears to reach peak densities about seven years after fire and declines thereafter as the heath becomes more mature.

Billiatt Conservation Park comprises sand plains and hills covered with Mallee vegetation. Running north to south through the Park, access is restricted to the Lameroo–Alawoona Rd (2WD). Despite this, there is excellent birding to be had along this road. The southern boundary of the park is 35 km north of Lameroo. A good spot for birding is on the west side of the road 42 km north of Lameroo (30 km south of Alawoona), where there is a large roadside pull-in (just 400 m before the road turns east). Look here for Golden and Gilbert's Whistler, Black-eared Cuckoo, and White-eared, Purple-gaped, Tawny-crowned, Spiny-cheeked, White-fronted and Yellow-plumed Honeyeater. Black-eared Miner has also been reported here (look on the higher sand dunes in the east side of the road), though take care with your identification because Yellow-throated Miners and their hybrid progeny are more common. It is also worth searching for Western (Mallee) Whipbird in this same area, though they have not been recorded in Billiatt for some years. Red-lored Whistler and Striated Grasswren prefer the habitat 2 to 3 km further north, between 43.5 and 46.5 km from Lameroo (around 25.5 to 28.5 km from Alawoona). Approximately 36 km north of Lameroo (1 km after you enter the park from the south), turn west onto a small bush track and pull off the road; this is a good area for birding, and it is well worth climbing to the nearby 'trig point' (133 m), which provides views over the park. Keep a look

Bordertown

The community parklands of Bordertown, and nearby at Mundulla, are home to several families of Bush Stone-curlew. Good locations to see them are the cemetery at the intersection of Brown Terrace and Crecoona Terrace. There is also a 'wild' pair in the enclosure of the wildlife park, situated just near the turn-off from the Dukes Hwy at the eastern entrance to the Bordertown. As a curiosity, the wildlife park also has a colony of white Western Grey Kangaroo, along with Emu and Red Kangaroo, Red-necked and Parma Wallaby and a variety of birdlife. On the outskirts of Mundulla, 10 km south of Bordertown, Bush Stone-curlew frequent Moot-Yang-Gunga Swamp. The swamp is accessible via the Bordertown–Mundulla Rd or from the Mundulla Oval. On the western outskirts of Bordertown, on Pigeon Flat Rd, a wetland borders the eastern section of the Bordertown treatment plant. Freckled Duck occasionally loafs here, along with more common Pink-eared and Blue-billed Duck, Australasian Shoveler and Hardhead.

out for Malleefowl, Southern Scrub-robin, Chestnut Quail-thrush, Shy Heathwren and Crested Bellbird. While driving through the park, stop and investigate any areas with bird activity; with this strategy and several stops you may find Regent Parrot, Brush Bronzewing, Major Mitchell's Cockatoo, Mulga Parrot, Purple-crowned Lorikeet, Grey Currawong, Hooded and Red-capped Robin, Restless Flycatcher, Varied (Black-capped) Sittella, and Variegated and Splendid Fairy-wren.

Arid lands: the Flinders Ranges, Eyre Peninsula and the South Australian outback

Regional overview

This section provides coverage for three large areas: the Flinders Ranges, South Australian outback, and Eyre Peninsula. In combination, these areas provide great opportunities for birding in arid environments. The Flinders Ranges commence some 250 km north of Adelaide at Mount Remarkable National Park, extending north through the Flinders Ranges National Park to the Gammon Ranges National Park, 630 km north of Adelaide. North of here, as far as the Northern Territory border, the South Australian outback is vast, beautiful and, from a birder's viewpoint, a rewarding place to visit. The Eyre Peninsula and Nullarbor Plain cover more than 2000 km of southern coastline, from Port Augusta to the Western Australian border. All three areas are extremely isolated and remote, containing over 80% of the state's land mass, but less than 1% of its population.

A suggested 14-day Flinders Ranges and outback itinerary would involve a substantial road trip that incorporates travelling to the Flinders Ranges then looping up the

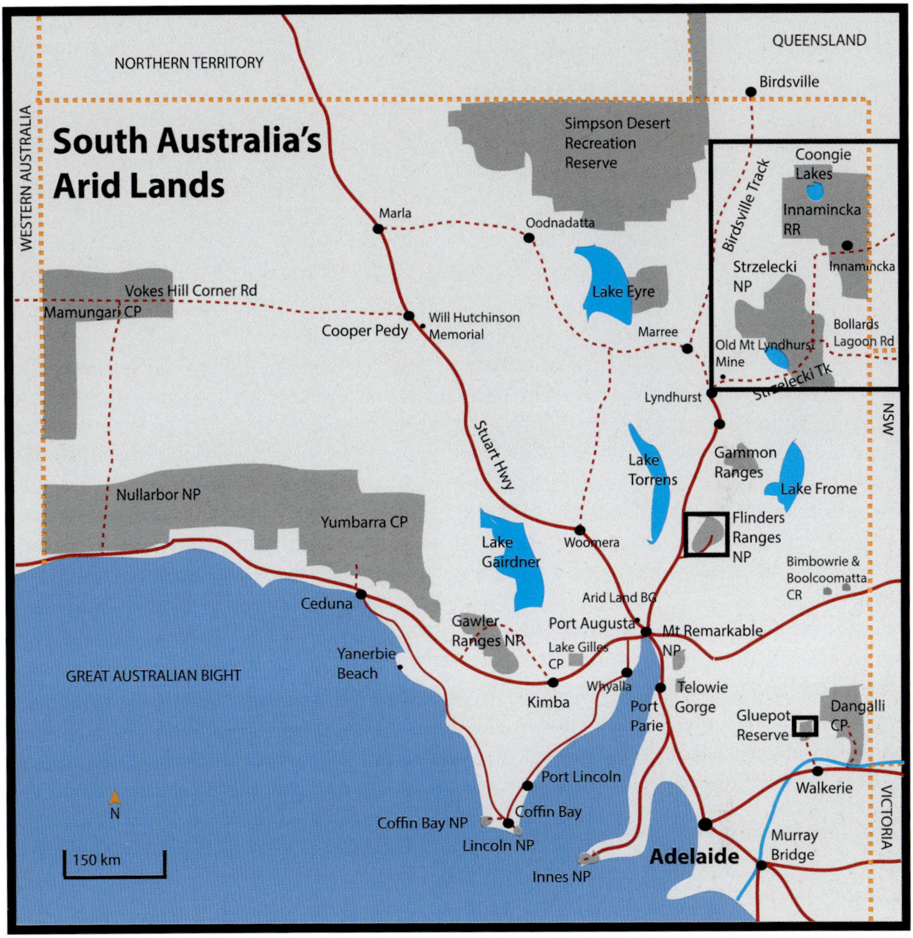

Strzelecki Track before returning via the Birdsville Track:

- **Days 1–3.** Adelaide to Flinders Ranges National Park, 400 km (Short-tailed Grasswren, Grey-fronted Honeyeater, Redthroat, Chirruping Wedgebill, Elegant Parrot) (2WD possible).
- **Days 4–5.** Flinders to Old Lyndhurst Mine, 200 km (Chestnut-breasted Whiteface, Thick-billed Grasswren, Rufous Fieldwren, Cinnamon Quail-thrush) (2WD possible).
- **Days 6–9.** Old Lyndhurst Mine to Innamincka, along the Strzelecki Track, 475 km (Eyrean Grasswren, Banded Whiteface, Letter-winged Kite, Inland Dotterel, Gibberbird, Cinnamon Quail-thrush) (4WD essential).
- **Days 10–12.** Innamincka to Birdsville, via Cordillo Downs, 420 km (Eyrean and Grey Grasswren, Yellow Chat, Banded Whiteface, Gibberbird, Letter-winged Kite) (4WD essential).
- **Days 12–14.** Birdsville to Marree along the Birdsville Track, 520 km

(Flock Bronzewing, Eyrean and Grey Grasswren, Gibberbird, Orange Chat, Yellow Chat, Letter-winged Kite, Black-breasted Buzzard) (4WD essential). From Marree you can then travel back to Adelaide.

There is also the option of detouring from the Strzelecki Track to Bollard Lagoon Rd, and then continuing onto Sturt National Park in NSW (p. 234). It is also recommended to do a 2–3 day trip to Gluepot Reserve, where there is a chance of seeing Black-eared Miner, Red-lored Whistler, Scarlet-chested Parrot, Malleefowl, Striated Grasswren, Chestnut-quail Thrush, White-browed Treecreeper and Major Mitchell's Cockatoo. Danggali Conservation Park is also well worth a visit, with similar species to those found at Gluepot – add an additional 2 days.

Low annual rainfall characterises all the Flinders Ranges, South Australia's vast outback and the Eyre Peninsula, with the mean annual rainfall between 150 and 250 mm. The area is subject to prolonged periods of drought, where rainfall is unpredictable; when rain does fall, it may cause flooding and road closures. Daytime temperatures range from warm to hot, with extremes in summer that exceed 40°C. From a birdwatching perspective, the best time to visit is during the autumn, winter and spring, when the days are generally mild to warm; be aware, though, that the nights can be very cold, especially in winter. During hotter periods, morning and evening are the best time for birdwatching, with a good strategy being to drive some of the longer stretches through the heat of the day.

There are three distinct areas in the Flinders Ranges: southern, central and northern. The southern and central Flinders Ranges are accessible by 2WD, with most roads sealed, while a 4WD is necessary to reach the northern Flinders Ranges. In the South Australian outback, you can take several routes, including the sealed Stuart Hwy, the unsealed Birdsville Track and the Strzelecki Track. Both the Birdsville and Strzelecki Tracks lead into Queensland, while the Stuart Hwy heads into the Northern Territory. On the Strzelecki Track, stops need to be made for fuel and supplies at both Lyndhurst and Innamincka. On the Birdsville Track, the only place to stop for fuel is the Mungerannie Roadhouse, approximately halfway between Marree and Birdsville.

When driving in arid areas, travellers must be well prepared, because there can be limited opportunities to get food, fuel and water. The risks of remote area travelling should not be underestimated – birdwatchers have died in the areas covered in this section because they were poorly prepared for remote area travel. Distances between towns are often large –plan your route carefully and be fully aware of the distances involved. Take particular care when venturing off the beaten track. Watch out for stock and native animals crossing the road, even on sealed roads, especially around sunset and sunrise. Birds of prey, such as Wedge-tailed Eagle, Black Kite and Black-breasted Buzzard, feed on carcasses along the roads. Avoid the Strzelecki and Birdsville Tracks during the hotter months because conditions are likely to be difficult – temperatures will be high and water in short supply. Respect the rights of traditional landholders and ensure you seek permission and/or permits where these are required. Take extreme care when birdwatching on foot away from your vehicle; the landscapes are vast, often with few landmarks and it can be easy to become disorientated. To access most conservation reserves in the region covered here, a 'Desert Park pass' is required; visit www.environment.sa.gov.au/parks/visitor/desertpass.html. Before any trip, road conditions should be checked at www.transport.sa.gov.au/quicklinks/northern_roads/northern.asp.

Like many areas in inland Australia, good seasons in the South Australian outback are characterised by flocks of Budgerigars. The most memorable experiences are usually to be had around waterholes, where flocks of all sizes may congregate to drink.

Birding highlights

Sought-after species that occur in the Flinders Ranges include the endemic Short-tailed Grasswren, as well as Thick-billed Grasswren (ssp. *modesta*), Redthroat, Grey-fronted Honeyeater and Rufous Fieldwren. Species of rocky crags and eucalypt-lined riverbeds include Little Woodswallow, Red-browed Pardalote and Painted Finch, which frequent the northern Flinders Ranges. Purple-crowned Lorikeet and Chestnut-rumped Heathwren (ssp. *pedleri*) occur in the southern Flinders Ranges. Elegant Parrot breeds in tree-lined watercourses, while Blue-winged Parrot is present as a non-breeding winter visitor.

True arid-land specialists that frequent the South Australian outback include Inland Dotterel, Letter-winged Kite, Grey Falcon, Black-breasted Buzzard, Spotted Harrier, Flock Bronzewing, Chestnut-breasted and Banded Whiteface, Thick-billed, Eyrean and Grey Grasswren, Gibberbird, Cinnamon Quail-thrush (ssp. *cinnamoneum*), Chirruping Wedgebill, Little Woodswallow, Plains-wanderer, Yellow Chat and Painted Finch. In exceptional years, when Lake Eyre and the Coongie Lakes fill with water, these wetland systems can attract huge numbers of waterbirds, including massive breeding colonies of Banded Stilt and Australian Pelican.

From an ornithological perspective, Eyre Peninsula, the Gawler Ranges and the Nullarbor Plain can be defined as where east meets west, because, here, several eastern species reach their western range limit while the reverse is true of several species often considered endemic to south-west

WA. This latter group include Blue-breasted Fairy-wren, Western Yellow Robin and Rufous Treecreeper, while Rock Parrot and Western Whipbird are also fairly widespread within suitable habitat. Isolated populations of several species on the Eyre Peninsula include Southern Emu-wren (ssp. *parimeda*), Western Grasswren (recently separated from Thick-billed Grasswren and elevated to full species), Western Gerygone and Yellow-tailed Black-Cockatoo. The Nullarbor Plain holds several sought-after specialities, including Inland Dotterel, Nullarbor Quail-thrush and Blue Bonnet ('Naretha Parrot' ssp. *narethae*), while Scarlet-chested and Princess Parrot are regularly recorded at Mamungari Conservation Park in the Great Victorian Desert.

Mount Remarkable National Park and Telowie Gorge Conservation Park

Key species: Painted Button-quail, Chestnut-rumped Heathwren
Other species: Crimson (Adelaide) Rosella, Diamond Firetail

Part of the southern Flinders Ranges, and offering spectacular scenery, Mount Remarkable National Park (16 000 ha) is within easy reach from Adelaide. There is an intriguing mix of flora and fauna, ranging from those adapted to the wetter southern regions of South Australia through to species more typical of the arid north. At last count, 124 bird species had been recorded in the park. During the summer months, high temperatures can make birdwatching difficult, whereas mild temperatures from April to October are more comfortable. There is camping at Mambray Creek, and accommodation is available at Alligator Lodge and Mambray Creek Cabins.

A good place to begin any visit to Mount Remarkable is Alligator Gorge, reached via Wilmington on Main North Rd. Painted Button-quail are regularly recorded along the Gorge Circuit Hike (2 km return) and the southern section of the Ring Route Track. A good strategy for finding them is to search for fresh platelets (circular saucer-like scrapes on the ground) in woodland with a closed canopy, some understorey and leaf-litter. Chestnut-rumped Heathwren (ssp. *pedleri*), an isolated subspecies limited to the southern section of the Flinders Ranges, is found in the heathy woodland above Alligator Gorge. Also, watch for Peregrine Falcon, Bassian Thrush (Endangered ssp. *halmaturina*), Yellow-rumped and Inland Thornbill, Diamond Firetail and Brown Treecreeper. The Daveys Gully Hike (2.4 km return) and longer Sugar Gum Lookout Hike (8 km return) start near the Mambray Creek campground. Watch for parrots such as Elegant Parrot, Australian Ringneck, Crimson (Adelaide) Rosella (ssp. *adelaidae* – here at its the northernmost range limit) and Purple-crowned Lorikeet. Other birds to look for include Redthroat, White-browed Babbler, Shining Bronze-Cuckoo, Black-chinned (uncommon in open woodland), Brown-headed, Singing and Spiny-cheeked Honeyeater, Variegated Fairy-wren, Brown and Rufous Songlark (both spring–summer), and Inland, Chestnut-rumped, Yellow and Yellow-rumped Thornbill. Night birds, including Eastern Barn Owl, Southern Boobook and Australian Owlet-Nightjar, are relatively widespread in the park.

Telowie Gorge Conservation Park (1967 ha) is located 25 km north of the coastal township of Port Pirie. One of its main drawcards for birdwatchers is Grey-fronted Honeyeater, with the Nukunu Walking Trail being an excellent area to search for the species. In the dense undergrowth along this walk, you can also find Southern Scrub-robin and Chestnut-rumped Heathwren, especially between 300 m and 600 m from the carpark. Other species that occur along the gorge include Elegant Parrot, Purple-crowned Lorikeet, Peaceful Dove, Diamond Firetail, the local subspecies of the

Bassian Thrush (rare), Red-capped Robin and Variegated Fairy-wren, and raptors such as Little Eagle, Spotted Harrier and Collared Sparrowhawk. There is a colony of the beautifully patterned Yellow-footed Rock Wallaby in Telowie Gorge; these are best seen in the early morning or late afternoon, when individuals move down to browse and graze.

Flinders Ranges National Park

Key species: Short-tailed Grasswren, Grey-fronted Honeyeater, Redthroat, Chirruping Wedgebill, Elegant Parrot

Other species: Wedge-tailed Eagle, Australian Ringneck, White-winged Fairy-wren, Inland Thornbill

Located 450 km north of Adelaide, Flinders Ranges National Park (91 330 ha) is an iconic Australian national park. The park comprises a wide range of habitat types, including dry woodlands, Mallee and River Red Gum-lined river and creek beds. Autumn brings warm, mild days and cool nights, and winter can be cool, while spring can be a carnival of colour with many flowering plants. Summer months can get very hot, often making birdwatching

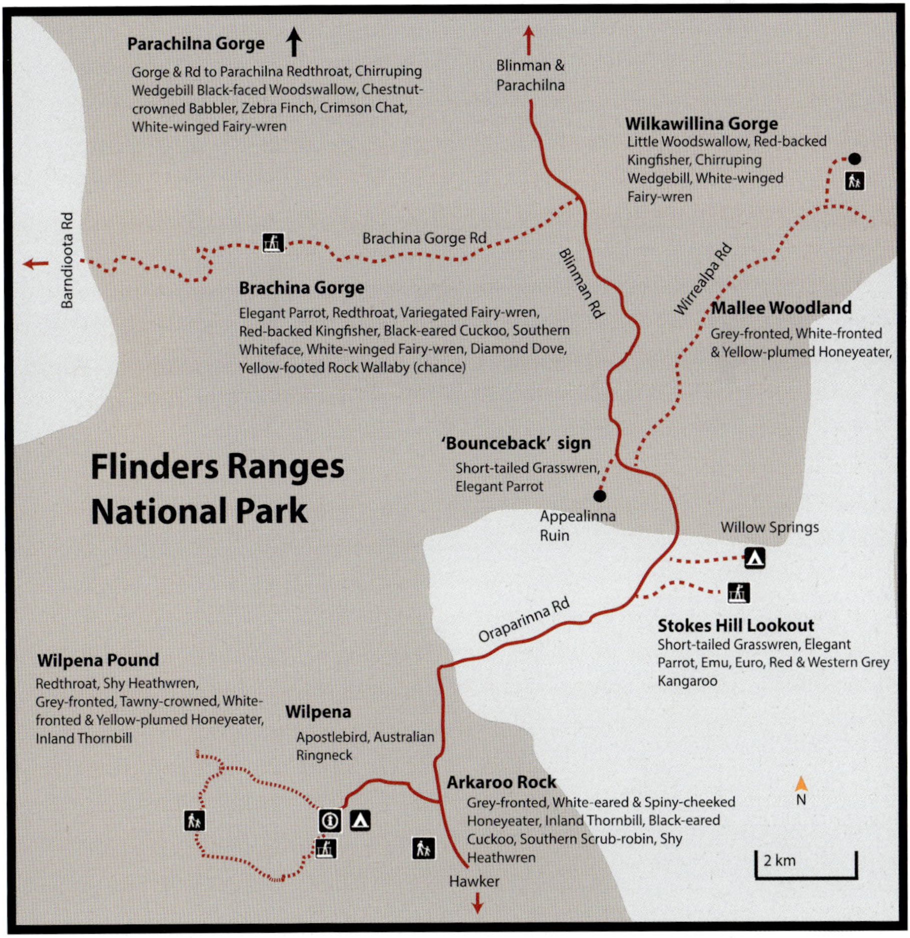

A typical scene in the Flinders Ranges. Huge old River Red Gums along a dry creek bed. These habitats often support numbers of parrots, including Adelaide Rosella, Australian Ringneck and Elegant Parrot.

difficult. Access to the centre of the park at Wilpena is gained via a sealed road from Hawker. Other areas in the park are also usually accessible to 2WD vehicles, unless there has been significant rain. There is a wide range of camping options in the park; contact the Wilpena Visitor Centre : (08) 8648 0048 or www.environment.sa.gov.au. Petrol and other supplies are available in Hawker, Wilpena, Blinman and Parachilna. There is also camping and self-contained accommodation at Willow Springs Station, 21 km north-east of Wilpena Pound. Willow Springs, adjacent to Stokes Hill, has both camping and self-contained accommodation and is a base to search for the Short-tailed Grasswren www. Skytrekwillowsprings.com.au.

Short-tailed Grasswren occur in the bare Spinifex-covered slopes at the Stokes Hill Lookout, 13 km north of Wilpena. The birds occur over most of Stokes Hill, but a search of the north side of the ridgeline south-east of the car park, down to the gully, dam and the main road is often the most productive. Like most grasswrens, the species can be skulking, so walk slowly and quietly, listening for its high-pitched melodious series of squeaks. If you do not see Short-tailed Grasswren here, another spot to search is along Appealinna Ruins Rd, 5.5 km north of the turn-off to Stokes Hill. Search the Spinifex from ~1 km onwards from the turn-off, especially the area in front of the 'Bounceback–Bush Bird Barometers' signage. Elegant Parrot and Emu also occurs at both Stokes Hill Lookout and along Appealinna Ruins Rd. Interesting mammals that occur at both sites (and indeed throughout much of the Flinders Ranges) include Euro, Western Grey and Red Kangaroo.

Wilpena Pound is a huge natural amphitheatre 14 km long and 8 km wide and bounded by two mountain ranges. Vegetation in the pound includes native pine, Mallee woodland, heathland and River Red Gums along creek lines. Birding around the information centre can be rewarding, with regular sightings of Apostlebird, Yellow-throated Miner, Striated Pardalote, Inland Thornbill and Australian Ringneck (the Flinders Ranges being a hybrid zone between the 'Port Lincoln Ringneck' and 'Mallee' Ringneck). There are several well-established walking tracks such as the Wangarra Lookout Hike, Hills Homestead Walk, Mount Ohlssen Bagge Hike and the longer St Mary Peak Hike, which all provide excellent opportunities for birdwatching within the Pound. Birds to look for include Redthroat, Inland, Chestnut-rumped and Yellow Thornbill, Red-capped Robin and White-browed Babbler. Shy Heathwren frequent the Mallee areas, particularly in the southern section near Saint Mary Peak, and a wide range of honeyeaters including Grey-fronted, Tawny-crowned, White-eared, White-fronted, Singing, White-plumed, Brown-headed, Spiny-cheeked and Yellow-plumed all occur. Other birds to look for include White-winged Triller (summer), Brown Treecreeper, Spotted and Striated Pardalote, and Grey Butcherbird. The walk to Arkaroo Rock (located 17 km south of Wilpena; 1 h return) can be good for Grey-fronted, White-eared, Spiny-cheeked and Brown-headed Honeyeater and Southern Scrub-robin, as well as Inland Thornbill, Black-eared Cuckoo (spring) and Shy Heathwren.

Brachina Gorge, an important refuge for the Yellow-footed Rock Wallaby, is also a good site for Elegant Parrot, Redthroat and Variegated Fairy-wren. While in the gorge, look for whitewash on cliff-faces, indicating the presence of roosting, and perhaps nesting, Peregrine Falcon. On the west side of the gorge entrance, at the edge of the national park, look for Red-backed Kingfisher, Black-fronted Dotterel, Black-eared Cuckoo, White-browed Babbler, Southern Whiteface, White-winged Fairy-wren, Diamond Dove, and Grey-fronted, Spiny-cheeked and White-plumed Honeyeater. Night birds that frequent the area include Southern Boobook, Tawny Frogmouth and Owlet Nightjar. Bunyeroo Gorge (7 km south of Brachina Gorge) and Parachilna Gorge (10 km east of Parachilna, 22 km west of Blinman) hold similar species to Brachina Gorge. Chirruping Wedgebill frequent the flatter areas west of Parachilna Gorge, particularly those areas of sparse Acacia shrublands. Other birds to look for in this habitat include Black-faced Woodswallow, Chestnut-crowned Babbler (uncommon), Zebra Finch, White-winged Fairy-wren and the nomadic Crimson Chat.

Wilkawillina Gorge is located 45 km north-east of Wilpena. To get there from the Stokes Hill turn-off, drive 4 km north along the Oraparinna Rd (also known as the Blinman Rd), and then turn east along the Wirrealpa Rd (the road to Arkaroola). After 18 km, turn north onto Wilkawillina Rd – the gorge is another 6 km. Wilkawillina Gorge is a River Red Gum lined creek-bed and gorge, surrounded by ridges and hills with a cover of *Callitris* pines. Along the river before entering the gorge, especially if there are pools of water present, you can find Red-backed Kingfisher, Elegant Parrot, Rainbow Bee-eater (summer), Peaceful and Diamond (uncommon) Dove, Little Corella, Rufous Songlark, Redthroat and Striated Pardalote. White-winged Fairy-wren, Chirruping Wedgebill, Southern Whiteface and, occasionally, Crimson Chat can be found in the moonscape area of scattered bluebush, wattle and saltbush north-west of the creek line just before you enter the gorge (as well as along Wilkawillina Rd, on the drive into the gorge). In the gorge itself, look for Little Woodswallow (rare, but occasionally seen near the gorge entrance), Wedge-tailed Eagle, White-browed Babbler, Grey Shrike-thrush, Redthroat, Tree Martin, and White-plumed, Spiny-cheeked and, occasionally, Grey-fronted Honeyeater. Raptors recorded here include Whistling Kite, Nankeen Kestrel and Peregrine Falcon. Sturt Desert-Pea sometimes flowers along the Gorge track, and you may see mammals such as Yellow-footed Rock Wallaby, Red Kangaroo and Euro. On your drive in to the gorge, it is worth stopping at the Mallee woodlands along Wirrealpa Rd (located between 8 and 12 km from the Wilpena Rd turn-off) to look for Grey-fronted, Yellow-plumed and White-fronted Honeyeater, Masked and White-browed Woodswallow, and other bush birds. Along Wilkawillina Rd, there are scattered areas of bluebush and saltbush, where you may see White-winged Fairy-wren, Chirruping Wedgebill and Southern Whiteface.

Vulkathunha-Gammon Ranges and Arkaroola Wilderness Sanctuary

Key species: Elegant Parrot, White-browed Treecreeper, Grey-fronted Honeyeater, Chirruping Wedgebill, Painted Finch
Other species: Inland Dotterel, Red-browed Pardalote, Redthroat, Red-backed Kingfisher

Vulkathunha-Gammon Ranges National Park (128 000 ha) is 630 km to the north of Adelaide and 70 km east of Leigh Creek. Opportunities for overnight stays include bush camping or self-contained accommodation. Contact the Wilpena Visitor Centre:

Flinders Ranges, SA. Mature spinifex on gravelly slopes and rises are the preferred habitat of the Short-tailed Grasswren.

(08) 8648 0048 or visit www.environment.sa.gov.au. Arkaroola is a privately owned wilderness sanctuary that adjoins the north-east boundary of Vulkathunha-Gammon Ranges National Park. At Arkaroola, there is camping and accommodation, fuel, basic supplies and water. All main access roads to the park are gravel and are suitable for 2WD vehicles (with care).

White-browed Treecreeper, Red-backed Kingfisher and Red-browed Pardalote frequent the Mount McKinlay Creek, a good spot for birdwatching being where the creek turns south 6 km from the Leigh Creek entrance to the park. Look for Chirruping Wedgebill, Australian Ringneck and Chestnut-rumped Thornbill at the beginning of the walk at Italowie Gap. Birds recorded along the Balcanoona Creek Hike and Weetootla Gorge (6 km one way, and part of the Weetootla Hike) include Painted Finch (scarce), Little Woodswallow (scarce), Redthroat, Elegant Parrot, Budgerigar, Peregrine Falcon, Black-eared Cuckoo, Crested Bellbird, and White-fronted and Grey-fronted Honeyeater. Begin the walk at the Weetootla Campground Trailhead (Checkpoint B1). Other birds commonly encountered in the Gammon Ranges include Inland Dotterel (mostly seen along the roadsides while driving), Inland and Chestnut-rumped Thornbill, Spiny-cheeked and Singing Honeyeater, White-winged, Splendid and Variegated Fairy-wren, and, at night, Boobook Owl and Australian Owlet-Nightjar.

Bimbowrie Conservation Reserve and Boolcoomatta Reserve

Thick-billed Grasswren (ssp. *modestus*) occur at Bimbowrie Conservation Park (72 000 ha): a significant extension to their previously known range. Look for them in low shrublands dominated by bluebush, 2 km and then again 10 km north-east of the Bimbowrie Station. Other interesting birds recorded in the reserve include Redthroat, Slender-billed Thornbill, Rufous Fieldwren, Cinnamon Quail-thrush, Chestnut-crowned Babbler, Chirruping Wedgebill, Pied Honeyeater and White-winged Fairy-wren (and, in the rocky hills, Yellow-footed Rock Wallaby). The turn-off to Bimbowrie Conservation Park is at Olary on the Barrier Hwy, 117 km west of Broken Hill, 395 km north-east of Adelaide. Boolcoomatta Reserve (63 000 ha) lies adjacent to the Bimbowrie Conservation Reserve. A former sheep station, it is managed by Bush Heritage Australia – for details see www.bushheritage.org.au. The flat grasslands on the reserve support a significant population of the Endangered Plains-wanderer. Other birds recorded include Chirruping Wedgebill, Cinnamon Quail-thrush, Inland Dotterel, Slender-billed Thornbill, Gibberbird and Grey Falcon (rare).The turn-off to Boolcoomatta Reserve is at Cockburn on the Barrier Hwy, 74 km west of Broken Hill, 438 km north-east of Adelaide.

Arkaroola Wilderness Sanctuary (61 000 ha) supports a population of regionally endemic Short-tailed Grasswren; search in the Spinifex-clad valley near Siller Lookout on the Ridgetop Trail, as well the Armchair Basin, Mount Gee, Mount Painter and the Streitzberg Ridge on the northern side of Mount Ward. Grey-fronted Honeyeater, Little Woodswallow, Red-browed Pardalote and Painted Finch (scarce) occur at Echo Camp Waterhole. Other birds to look for at Arkaroola include Elegant, Mulga and Red-rumped Parrot, Blue Bonnet, Australian Ringneck, Budgerigar, White-browed and Dusky Woodswallow, Rainbow Bee-eater (spring–summer), Diamond and Peaceful Dove, Striated Pardalote, Red-capped Robin, Inland and Chestnut-rumped Thornbill, Spiny-cheeked, Singing, White-fronted and White-plumed Honeyeater, Variegated Fairy Wren and Zebra Finch. The sanctuary also provides a refuge for the delightful Yellow-footed Rock-Wallaby.

The Chestnut-breasted Whiteface is the sole South Australian endemic restricted to sparse shrublands with a substrate of gibber. Although there are several sites along the Stuart Hwy near Coober Pedy, the Old Lyndhurst Mine Site on the Strzelecki Track is generally the most reliable site for the species.

Old Mount Lyndhurst Mine

Key species: Chestnut-breasted Whiteface, Thick-billed Grasswren, Rufous Fieldwren, Cinnamon Quail-thrush, Chirruping Wedgebill
Other species: Gibberbird, Redthroat, White-winged Fairy-wren, White-backed Swallow, Chestnut-crowned Babbler, Crimson Chat

Chestnut-breasted Whiteface inhabits the southern end of the Strzelecki Track, particularly near old Mount Lyndhurst Mine. Other interesting birds in the same area include Thick-billed Grasswren, Rufous Fieldwren, Cinnamon Quail-thrush and Chirruping Wedgebill. Before arriving, it is important to contact the owners of Mount Lyndhurst Station to gain permission for site entry and to pay a modest entry fee: (08) 8675 7796. This site attracts many Australian and international birdwatchers, so it is very important to minimise your impact; concern for the habitat means there is no camping at the site. There is hotel accommodation and camping at the Lyndhurst Hotel: www.lyndhursthotel.com.au or (08) 8675 7781.

The old mine site is one of the most reliable place to see the South Australian endemic Chestnut-breasted Whiteface. Approximately 27.5 km east of Lyndhurst, a track leads north (left) at a point marked by a small rock-cairn. Begin by checking both sides of the track near the intersection. Southern Whiteface also occurs in this area, and Thick-billed Grasswren inhabit the western side of the track. Travel ~500 m down this track to park near a small creek line. Chestnut-breasted Whiteface are often found in this general location, particularly in low hilly areas covered with gibber and shingle rocks. To the west, another good place to search for them is the large flat area (on the south side of the creek) in front of the old Mount Lyndhurst Mine. They have been found throughout the area, from around the back of the old mine through to the road. Other birds that frequent the creek line and shrubs around the mine site include Rufous Fieldwren, Redthroat, White-winged Fairy-wren, Cinnamon Quail-thrush, Chestnut-crowned Babbler and White-backed Swallow. As a general rule, looking for Chestnut-breasted Whiteface requires some perseverance, so be prepared to spend some time searching. The species often associates in mixed flocks with Zebra Finch and White-winged Fairy-wren, with much of their time spent foraging for seeds and insects on the ground; as a consequence, they can be difficult to detect. Thick-billed Grasswren live along the creek that runs parallel to the track from the Strzelecki Track to the old mine, and in adjacent smaller creek lines. If you do not see Chestnut-breasted Whiteface on the north side of the Strzelecki Track, they also occur in the rocky hills and adjacent flat areas on the other side of the Strzelecki Track.

Strzelecki Track

Key species: Eyrean Grasswren, Banded Whiteface, Letter-winged Kite, Inland Dotterel, Gibberbird
Other species: Ground Cuckoo-shrike, Black breasted Buzzard, Cinnamon

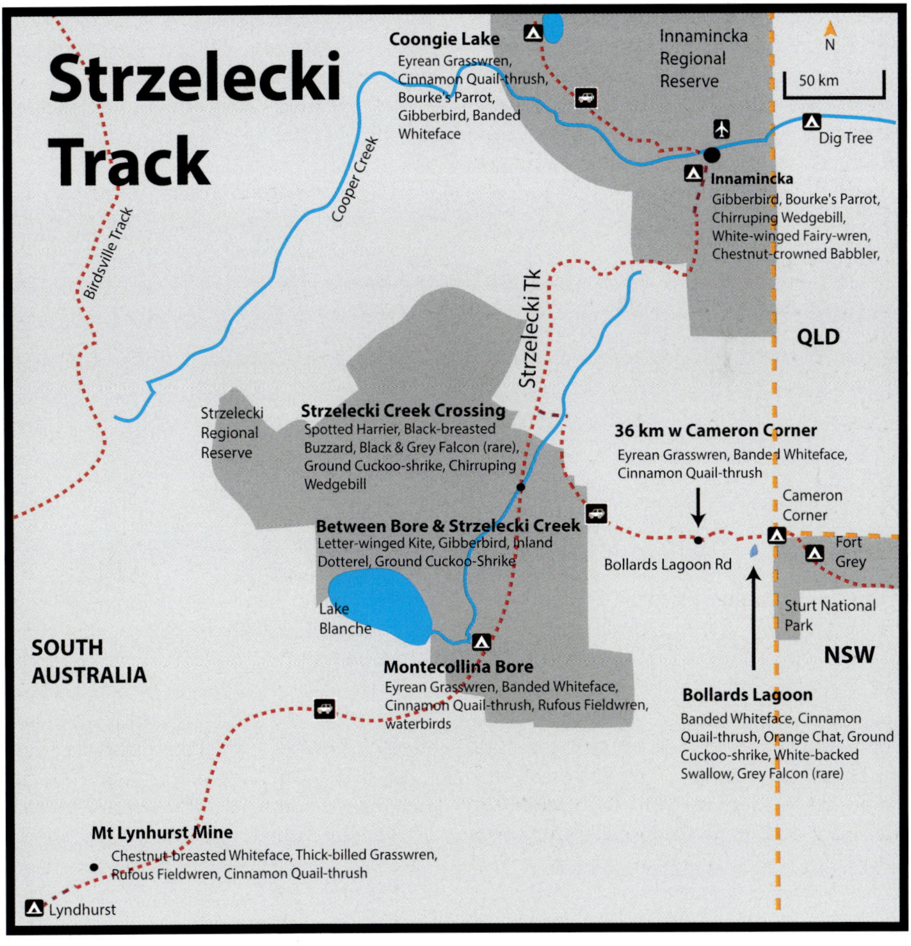

South Australia

Quail-thrush, Rufous Fieldwren, Chirruping Wedgebill

The Strzelecki Regional Reserve (1 200 000 ha) is located on the Strzelecki Track approximately halfway between Lyndhurst and Innamincka. There is bush camping here beside the Strzelecki Creek. It is possible to drive the track with a 2WD vehicle, although 4WD is highly recommended.

Montecollina Bore is located within the Strzelecki Regional Reserve, 220 km from Lyndhurst and 254 km from Innamincka. Eyrean Grasswren inhabit the Sandhill Canegrass that is scattered between the bore and the road, and on the sand hummocks 100–200 m west of the bore. Other birds found around the bore include Banded Whiteface, Cinnamon Quail-thrush, Rufous Fieldwren, White-winged fairy-wren, White-backed Swallow, Little Corella, Crimson and Orange Chat, and Zebra Finch. The overflow wetland at the bore attracts a variety of waterbirds such as Australian Spotted and Spotless Crake, Red-kneed Dotterel, Australian Shelduck, Nankeen Night-heron and Black-winged Stilt.

During certain years, Letter-winged Kite feed along the Strzelecki Track between Montecollina Bore and the Strzelecki Creek crossing (274 km from Lyndhurst and 157 km from Innamincka). Along the track, you will see isolated stands of the eucalypt 'Coolabah'; Letter-winged Kite roost and occasionally nest in these trees. A good spot to search is 20 km north of Montecollina Bore, where there is also a chance of finding Eyrean Grasswren, Banded Whiteface and Cinnamon Quail-thrush.

Among more common raptor species, Spotted Harrier, Black-breasted Buzzard,

Scattered eucalypts on the floodplains associated with the various creeks of the Lake Eyre Basin are favoured roosting sites of the elusive Letter-winged Kite.

The Inland Dotterel is a dry country 'shorebird' most at home on gibber plains, sparse native grasslands and herb fields. Their cryptic behaviour and plumage means they are easily overlooked in these environments.

Black and Grey Falcon (rare) roost and nest in the larger gums along the usually dry Strzelecki Creek. Other birds recorded along the creek include Red-backed Kingfisher, Blue Bonnet, Blue-winged Parrot (rare winter visitor), Ground Cuckoo-shrike, Chirruping Wedgebill, White-backed Swallow, Black-faced Woodswallow, Chestnut-crowned Babbler, Pied Honeyeater, Orange Chat, Red-browed and Striated Pardalote, and, after dark, Eastern Barn Owl and Australian Owlet-Nightjar. Another spot to search for Banded Whiteface is 9 km north of the Strzelecki Creek crossing, where there is a chance of finding Eyrean Grasswren and Pied Honeyeater. More generally, while driving the Strzelecki Track, watch for Gibberbird, Inland Dotterel and Banded Lapwing on and along the roadside.

Bollards Lagoon Rd

Key species: Eyrean Grasswren, Banded Whiteface, Cinnamon Quail-thrush, Grey Falcon
Other species: Chestnut-crowned Babbler, Ground Cuckoo-shrike, White-backed Swallow, Letter-winged Kite

Traversing the eastern section of the Strzelecki Desert, Bollards Lagoon Rd runs between Merty Merty Homestead and Bollards Lagoon Homestead in north-east South Australia. You can access the lagoon from the Strzelecki Track (320 km from Marree, 146 km from Innamincka) or through Sturt National Park via Cameron Corner in NSW. A 4WD is recommended. The closest camping ground is Fort Grey in Sturt National Park, NSW or at Cameron Corner (the border post where Qld, NSW and SA meet).

The favoured habitat of Eyrean Grasswren is the Sandhill Canegrass that grows as large sprawling tussocks on the top and flanks of the large red dunes. If you enter Bollards Lagoon Rd from Cameron Corner, Eyrean Grasswren occur on the larger dunes 36 km west of Cameron Corner (72 km east of Merty Merty Station). Concentrate your search in the canegrass at the top of the dunes either side of the road. The grasswrens are usually seen dashing or making short quick flights across the open sand between clumps of suitable cover; listen also for their distinctive far-carrying Goldfinch-like call. Check the flats between the dunes here for Cinnamon Quail-thrush and Banded Whiteface. Surrounded by Mulga, saltbush and River Red Gum, the usually dry Bollards Lagoon is 6 km west of Cameron Corner. Search around the lagoon for Banded Whiteface, Chestnut-crowned Babbler, Cinnamon Quail-thrush, Orange and Crimson Chat, Ground Cuckoo-shrike, Flock Bronzewing (rare), White-backed Swallow, Crested Bellbird and Grey Falcon (rare). Other birds recorded along Bollards Lagoon Rd include Letter-winged Kite (rare), Diamond Dove, Red-backed and Sacred Kingfisher, Little Button-quail, Horsfield's Bronze-Cuckoo, Red-browed Pardalote, Variegated and White-winged Fairy-wren, and Zebra Finch.

Innamincka and the Coongie Lakes

Key species: Eyrean Grasswren, Gibberbird, Cinnamon Quail-thrush, Freckled Duck, Banded Whiteface
Other species: Grey Falcon, Letter-winged Kite, Bourke's Parrot, shorebirds

Situated on the banks of the Cooper Creek, the township of Innamincka lies 459 km from Lyndhurst at the end of the Strzelecki Track, and 373 km from Thargomindah on the Adventure Way in Qld. There is a hotel and caravan park, cabins are available in town and bush camping is permitted along Cooper Creek. Coongie Lakes National Park (26 669 ha) is located entirely within the Innamincka Regional Reserve (1 354 193 ha). There is a pleasant camping area around the south-east corner of the lake. You will need a 4WD to reach Coongie Lake, accessed via Coongie Lake Track 100 km north of Innamincka. The famous Burke and Wills Dig Tree is located beside the Cooper Creek 70 km upstream of Innamincka.

Gibberbird inhabit the gibber plains and plateaus around Innamincka. One regular site is at the Innamincka Airport, located on Cordillo Rd 4 km north of town. Around the town, you may also encounter Chestnut-crowned Babbler, Chirruping Wedgebill, White-winged Fairy-wren and Bourke's Parrot. Another spot to search for Gibberbird is near the intersection of the Adventure Way and Dillons Hwy (sometimes known as the New Strzelecki Track) 3–4 km south of the township. The waterholes along the Cooper Creek, such as Queerbedie, Minkie, Malkonbar, Burke and Cullyamurra, all provide habitats for waterbirds such as Great Crested Grebe, Intermediate, Little and Great Egret, and Nankeen Night-Heron. Check the River Red Gums for Red-browed Pardalote and Red-backed Kingfisher, while raptors that

Mixed shrubs and Sandhill Canegrass atop sand dunes is the preferred habitat of the elusive Eyrean Grasswren. Although first collected in 1874, the species was then 'lost' until its rediscovery in dune systems east of Poeppel's Corner, SA, in 1976.

frequent the Cooper Creek include Black-breasted Buzzard, Peregrine Falcon, Brown Goshawk and Collared Sparrowhawk. Finally, there are several pairs of Barking Owl around the township; their calls are a familiar sound from dusk until dawn.

Fed by Cooper Creek, the Coongie Lakes are one of Australia's most pristine and unpolluted wetlands. Seldom dry, the wetlands attract large congregations of waterbirds and shorebirds, with over 70 000 recorded at one time. Freckled Duck, Pink-eared Duck and Australian Pelican regularly breed at the lakes. In spring and summer, the shallow south-west corner is a feeding area for shorebirds such as Common Greenshank, Marsh, Wood, Sharp-tailed and Curlew Sandpiper, Red-necked Avocet and Black-winged Stilt, and rarer species such as Ruff and Australian Painted Snipe. The Sandhill Canegrass-covered dunes next to Lake Drive are inhabited by Eyrean Grasswren, Cinnamon Quail-thrush, Bourke's Parrot, Chirruping Wedgebill, Chestnut-crowned Babbler, White-winged Fairy-wren and Black Honeyeater (nomadic), while in the flat expanses of gibber near the dunes search for Gibberbird and Banded Whiteface. In the River Red Gum along the north-west branch of the Cooper Creek, you can see Pied Butcherbird, Red-browed Pardalote, Spiny-cheeked, and Painted (rare) Honeyeater, and it is worth looking for Inland Dotterel along Coongie Rd, particularly in the evening and at night. Seventeen raptor species have been recorded in the vicinity of the Coongie Lakes, notably Grey Falcon (a rare breeding resident), Letter-winged Kite and Black-breasted Buzzard. Other birds found here include Barking Owl, Plains-wanderer (rare), Australian Pratincole, Flock Bronzewing (highly nomadic) and, when dry, it is worth searching for Yellow Chat in the grassy lake bed of Lake Toontoowaranie. The first ever specimen of the Critically Endangered Night Parrot was collected near Coongie Lakes by Charles Sturt in 1845, and there were two possible sightings in 1987.

Birdsville Track

Key species: Flock Bronzewing, Eyrean Grasswren, Grey Grasswren, Gibberbird, Orange Chat, Yellow Chat, Letter-winged Kite, Black-breasted Buzzard
Other species: Cinnamon Quail-thrush, Grey Falcon, Inland Dotterel, Banded Whiteface

An iconic 520 km overland track that links Marree and Birdsville, the Birdsville Track is seen as a right of passage for many interested in four-wheel driving. These days, especially after it has been graded, the 'track' does not necessarily pose any great challenge to the driver other than the usual precautions for remote-area travel, which should always be followed. Located between the Stuart Stony Desert to the east and the Simpson Desert to the west, the habitat along the track includes vegetated aridlands, cane grass-covered sandhills, eucalypt-lined watercourses (mostly dry) and rocky gibber plains. Although it is possible to drive the Birdsville with a 2WD vehicle, a 4WD is highly recommended, particularly along the northern section. It is best not to attempt the Birdsville Track during the hotter months from November to March (unless you are very well prepared) because the temperature can rise to 45°C. The only fuel and supplies depot on the Birdsville Track is the Mungerannie Roadhouse, where there is also accommodation and a good campground.

Eyrean Grasswren are found on the sand hills near the Mungerannie Hotel (203 km from Marree and 314 km from Birdsville), especially on the eastern side of the track just south of Derwent Creek. Along with Banded Whiteface and Grey Falcon (rare), Eyrean Grasswren have also been recorded along a track that leads west, 1 km north of Mungerannie; search the area 9.8 km from the Birdsville Track. The permanent Mungerannie Wetland extends along Derwent Creek. The main waterbody is ~600 m in length. At times, good numbers of aquatic birds occupy the waterhole, including Baillon's, Spotless and Spotted Crake, Buff-banded Rail, Nankeen Night-Heron Black-winged Stilt, Red-necked Avocet, Glossy Ibis, Yellow-billed and Royal Spoonbill and Purple Swamphen, and waterfowl such as Pink-eared and Musk Duck, and Grey Teal, Other species that drink here or frequent the vegetated edges of the wetland include Budgerigar, Australian Reed-Warbler, Little Grassbird and Orange Chat. If you miss the Eyrean Grasswren at Mungerannie, they also occur on Lake Surprise Dune (20 km south of the turn-off to Koonchera Dune). Along the Inner Birdsville Track, 12 km from the southern start of the Inner Birdsville Track, you enter Goyder's Lagoon. Grey Grasswren frequent the lignum and are sometimes visible from the track. Other species recorded in the more open areas here include Australian Bustard, Australian Pratincole, Gibberbird and Inland Dotterel.

Like many granivores found in the arid interior, Flock Bronzewing numbers fluctuate considerably in response to seasonal conditions. In especially good times, following prolonged periods of above average rainfall, the species can occur in aggregations that number in the thousands.

In the hotter months of the year, Paniburra Bore can be a good site for Yellow Chat and, sometimes, Flock Bronzewing. In late spring and summer, migratory shorebirds may occur at the bore, with a chance of seeing Sharp-tailed Sandpiper and, occasionally, rare species such as Oriental Plover and Ruff. Australian resident shorebirds that also occur here include Black-winged Stilt, Red-necked Avocet, Banded Lapwing and Red-kneed Dotterel. The turn-off to the Paniburra Bore is signposted on the Birdsville track. Flock Bronzewing, Eastern Grass Owl and Letter-winged Kite have been recorded at the Mirra Mitta Bore and wetland (279 km from Birdsville and 241 km from Marree). Anywhere along the Birdsville Track, keep your eyes open along the roadside for birds such as Australian Bustard, Inland Dotterel (semi-nocturnal and occasionally seen standing in the middle of the track at night), Flock Bronzewing (scarce) Grey Falcon (rare) and Letter-winged Kite, Cinnamon Quail-thrush, White-backed Swallow, Bluebonnet and Blue-winged Parrot, and Cockatiel.

Grey and Eyrean Grasswren occupy their preferred habitats in the vicinity of the Koonchera Dune and waterhole. The turn-off from the Birdsville Track is 118 km from Birdsville, 400 km from Marree. Take this well-graded track for another 20 km (although, beware, the last 500 m can be sandy). Flock Bronzewing and Letter-winged Kite are occasionally observed around the waterhole, and waterbirds recorded include Red-necked Avocet, Plumed Whistling-Duck, Pink-eared and Freckled Duck, Glossy Ibis, Royal and Yellow-billed Spoonbill, Black-tailed Native-hen, Black-fronted Dotterel, Black-winged Stilt and Caspian Tern. In an area 1 km east of the waterhole, look for Eyrean Grasswren in Sandhill Cane Grass on top of the Koonchera Dune – a distinctive

Lignum flats in the vicinity of Koonchera Waterhole are the preferred habitat of the Grey Grasswren. Other species to occur here include White-winged Fairy-wren, Variegated Fairy-wren and Orange Chat.

high-pitched double note alarm call may reveal their presence. Grey Grasswren inhabits the Lignum areas, ~1–2 km east of waterhole. Other species recorded in the area include Blue Bonnet, Blue-winged Parrot (approaching the northern limit for this species), Cockatiel, Australian Pratincole, White-backed Swallow, Cinnamon Quail-thrush, White-winged and Variegated Fairy-wren, Orange Chat, Gibberbird and Black Honeyeater (nomadic). Eastern Barn and Eastern Grass Owl (Australia's rarest and most elusive owl) have been reported in the expansive flats containing Lignum.

Chestnut-breasted Whiteface and the Stuart Hwy

The rare Chestnut-breasted Whiteface is known to occur at two sites along the Stuart Hwy in northern South Australia. The first is 12 km south of Coober Pedy on the gibber plain to the east of the Will Hutchinson Memorial; concentrate your search in gibber 500 m to 1 km from the memorial. The second site is on the west side of the Stuart Hwy 32 km north of Marla – here, they occasionally occur in mixed flocks with Banded and Southern Whiteface. Other birds to look for along the Stuart Hwy, particularly north of Marla, include Inland Dotterel, Black-breasted Buzzard and Bourke's Parrot.

Innes National Park

Key species: Western Whipbird, Malleefowl, Rock Parrot, Purple-gaped Honeyeater
Other species: Hooded Plover, Shy Heathwren, Fairy Tern, Eastern Osprey, White-bellied Sea-Eagle

On the southern tip of Yorke Peninsula, Innes National Park (9100 ha) is 300 km west of Adelaide. Coastal heath dominates the cliff tops to merge further inland with extensive Mallee woodlands, saltlakes, and grasslands. These habitats support birds such as Western (Mallee) Whipbird, Malleefowl, Rock Parrot and Shy Heathwren. Coastal birds at Innes include Fairy Tern, Eastern Osprey, White-bellied Sea-Eagle and Hooded Plover. Western Whipbird has played a significant role in the history of the park – its rediscovery here in 1962 led to the area being declared a national park.

Western Whipbird, although elusive, is widespread within the park, because its habitat remains relatively undisturbed. The best place to look for them is within the dense coastal vegetation along Browns Beach Rd between the turn to West Cape Rd and Dolphin Beach, particularly near the bend in the road between 2.5 and 3.5 km from the West Cape turn-off. They are also along the main road from (and including) Stenhouse Bay heading west. This is a notoriously skulking species, so the best technique is to scan the larger of dense and low scrubby vegetation very stealthily.

Other birds to look for include Southern Scrub Robin, Crested Bellbird, Brush Bronzewing, White-browed Babbler, Purple-gaped, Spiny-cheeked, and Tawny-crowned Honeyeater, and Variegated Fairy-wren. Western Whipbird also occupy the heathy area on the south side of the West Cape Lighthouse track. They are most often-heard singing in the 2 hours following sunrise and the hour before sunset, particularly during breeding season, so these are the times to listen for their distinctive duet call. In addition, here in the heath, you can see Rock Parrot; concentrate your search in short open coastal heath. While at the West Cape, look for coastal species such as those listed above and Australian Gannet, Pacific Gull, Sooty Oystercatcher, Short-tailed, Fluttering and Flesh-footed (uncommon) Shearwater, and Shy and Black-browed Albatross. Hooded Plover feed on the beaches at Pondalowie Bay, as do Pacific Gull. In rocky coastal areas, Ruddy Turnstone and Eastern Reef Egret can also be found. A good area for Malleefowl is the Mallee around the historic township of Inneston, where they can occasionally be found wandering around the ruins. The Thomson-Pfitzner Plaster Trail Hike (7.6 km return), which follows the old wooden railway line that runs from Inneston to Stenhouse Bay, also produces sightings of Malleefowl on a regular basis. Other birds along here include Shy Heathwren, Purple-gaped Honeyeater, Southern Scrub-robin, Grey Currawong and White-browed Scrubwren. Finally, it can also be worth looking for Malleefowl in the Mallee along Pondalowie Bay Rd between the Cape Spencer Rd and the Inneston turn-off. Western Grey Kangaroo and Emu are widespread throughout the park, Southern Right Whale pass along the coast during May to September and Bottlenose Dolphin frequent the shallows all year round.

Gluepot Reserve

Key species: Black-eared Miner, Red-lored Whistler, Scarlet-chested Parrot, Malleefowl, Striated Grasswren, Chestnut-quail Thrush, White-browed Treecreeper
Other species: Southern Scrub-robin, Shy Heathwren, Pied Honeyeater, Striped Honeyeater, Black Honeyeater, Yellow-plumed Honeyeater, Gilbert's Whistler

A very special birding location, and part of the Australia's largest block of intact Mallee, Gluepot Reserve (51 300 ha) was purchased

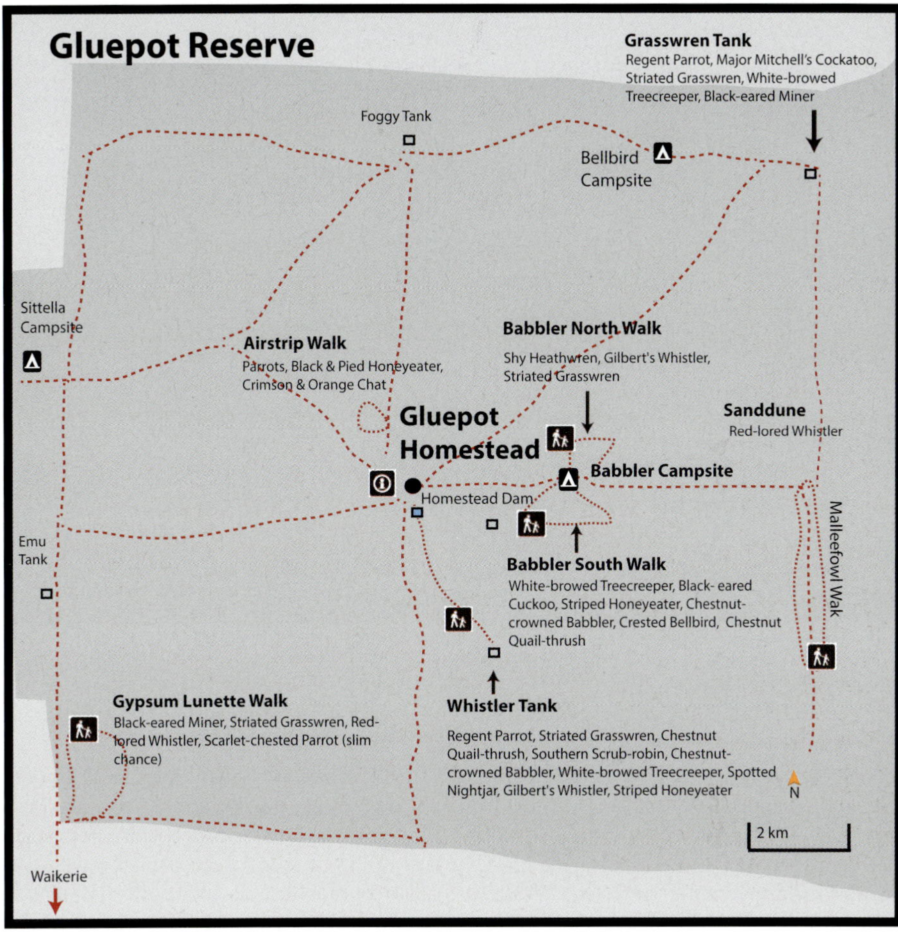

by Birdlife Australia in 1997. Located 64 km (1.5 h drive) north of Waikerie, it is part of the Riverland Biosphere Reserve on the SA–NSW border. The turn-off to Gluepot is on Morgan Rd, 30 km east of Morgan (84 km west of Renmark and 17 km north Waikerie). From Morgan Rd, travel 1.5 km to the first gate and then follow the well-signposted route. The last 50 km is via a well-maintained dirt track that is traversable by conventional 2WD vehicles and caravans in dry weather. One of the ornithological drawcards is that Gluepot Reserve and neighbouring properties support the last large population of the Black-eared Miner in Australia. Because of widespread habitat loss and genetic introgression as a result of interbreeding with the much more common and widespread Yellow-throated Miner, they are now Endangered. There is bush camping at Gluepot, with three well laid-out campsites (Babbler, Sittella and Bellbird). Visitors to Gluepot need to be self-sufficient, bringing all water, food and fuel. There is an entry fee of $5 per day per vehicle, campers $10 per night per vehicle. The excellent visitor information centre contains a noticeboard with the latest bird sightings.

Fourteen walking tracks provide access to a range of habitats within the reserve. The

Spring at Gluepot Reserve. This relatively open and long-unburnt Mallee habitat is a critical resource for the Endangered Black-eared Miner. Other species to occupy this Chenopod Mallee include Malleefowl, Southern Scrub-robin, Chestnut Quail-thrush and Gilbert's Whistler.

Babbler Campsite South Walk (2.9 km) passes through *Casuarina* and Mallee woodlands, providing opportunities to observe birds such as White-browed Treecreeper, Black-eared Cuckoo (spring–summer), Striped Honeyeater, Chestnut-crowned Babbler, White-browed Babbler, Crested Bellbird, Restless Flycatcher, Grey Butcherbird, and Hooded and Red-capped Robin. Regent (especially late summer through autumn) and Mulga Parrot often pass through the campsite itself. Nocturnal birds to listen for include Southern Boobook and Australian Owlet-Nightjar. On the Babbler Campsite North Walk (2.4 km), search for Shy Heathwren, Gilbert's Whistler and Striated Grasswren. The Grasswren favours areas of Mallee with a dense ground cover of Spinifex halfway along the walk – listen for their high-pitched and somewhat grating call.

The Whistler Tank Walk (6 km) commences from the carpark on Track 8 (1.5 km from the visitor centre). The walk passes through a range of habitats, including Mallee with an understorey of Spinifex, Black Oak, and *Acacia* and *Senna* shrublands. Keep a look out for Striated Grasswren (in areas of Spinifex), Chestnut Quail-thrush, Southern Scrub-robin, Chestnut-crowned Babbler, White-browed Treecreeper (Black Oak), Brown Treecreeper (mature Mallee) and Spotted Nightjar (sometimes flushed along the track). A hide (at the 2.5 km mark) overlooks a watering point that provides opportunities for Regent Parrot, Mulga Parrot, Mallee Ringneck, and Yellow-plumed and Striped Honeyeater. The Track 8 T-section is ~9 km east of the visitor centre (4 km east of the Babbler Campsite at the end of Track 8). A good spot for Red-lored Whistler is ~1 km north from this intersection (when heading towards the Grasswren Tank). Park your car near any of the larger dune ridgelines that run east–west (there is a dune ~700 m from the t-section), and walk west into the Mallee. Some of the best areas for the Red-lored Whistler are ~500 m of the track, in areas with a relatively open canopy and a complex understorey of shrubs and Spinifex – take extra care when walking off tracks because it is easy to become disorientated and lost in the Mallee. Around the Grasswren Tank, located in the north-east corner of the publicly accessible section of the reserve, you may find Major Mitchell's Cockatoo, Gilbert's Whistler, Shy Heathwren, Crested Bellbird, Yellow-plumed Honeyeater and, occasionally, Black-eared Miner.

The Gypsum Lunette Walk (5.5 km return), starting from the main Waikerie to Gluepot road, just south of the Gluepot entrance, passes through mature open Chenopod Mallee and, before entering, Mallee with Spinifex covered dunes. This walk can be good for Black-eared Miner, Striated Grasswren and Red-lored Whistler. Since the establishment of Gluepot, records of Scarlet-chested Parrot have been intermittent – in the first 5 years after the reserve were declared, there were just two confirmed sightings, but, in more recent years, small flocks have been observed in spring and summer. Several breeding records have also been confirmed. The species is one of the most sought-after birds in Australia, given there are no sites where it could be considered a certainty. In Gluepot, it can turn up anywhere, but recent records include along the Gypsum Lunette Walk, the Whistler Tank bird hide, the Homestead Dams and Track 8 along the western edge of the Birdseye Block.

The Chestnut Quail-thrush is relatively common at Gluepot Reserve – listen for their distinctive piping song at dawn and their thin high-pitched contact call throughout the day.

The old airstrip is worth checking for parrots such as Australian Ringneck, Mulga, Blue-winged and Elegant (rare) Parrot, Budgerigar and Cockatiel (the last two seasonal, often occurring in reasonable numbers after a wet winter–spring season). The old airstrip is also a good site for seasonal species such as Black and Pied Honeyeater, and Crimson and Orange Chat occur in reasonable numbers after a wet winter and spring. Other birds you may see in Gluepot include Emu, Major Mitchell's Cockatoo, Pallid Cuckoo, Red-backed Kingfisher, Chestnut-rumped and Inland Thornbill, Southern Whiteface, White-winged Triller, Dusky, Masked and White-browed Woodswallow, Hooded and Red-capped Robin, Jacky Winter, Splendid Fairy-wren, Apostlebird, Rainbow Bee-eater (summer) and honeyeaters such as Yellow-plumed, Grey-fronted (rare), White Fronted and Spiny-cheeked. Malleefowl are present at Gluepot, and most often seen crossing vehicle tracks. The reserve rangers may be able to give some advice regarding the location of active mounds.

Danggali Conservation Park

Key species: Scarlet-chested Parrot, Major Mitchell's Cockatoo, Red-lored Whistler, Black-eared Miner, Chestnut Quail-thrush
Other species: Emu, White-browed Treecreeper, Grey-fronted Honeyeater, White-fronted Honeyeater

The vast wilderness of Danggali Conservation Park (250 000 ha) is situated

South Australia

90 km north of Renmark in the Murray Mallee region of SA. There is accommodation at the Canopus Shearer's Quarters (pre-booking required), and three designated camping sites nearby. Contact the Danggali Park Office: (08) 8595 8010. Access to the park from Renmark is gained via the Old Wentworth Rd.

The Nanya's Pad Drive (90 km, 2WD), Tipperary Drive (100 km, 4WD) and the Target Mark Walking Trail (10 km return) all provide an opportunity to search for the rare Mallee specialists such as Black-eared Miner (very rare), Scarlet-chested Parrot (very rare), Malleefowl, Red-lored Whistler and Striated Grasswren. The best area for most of the above species is within the old-growth Mallee, particularly on the southern boundary of the park. Scarlet-chested Parrot typically breed in old-growth Mallee habitats but show a preference for foraging in areas of diverse shrubland that are recovering from fire (2–10 years since fire). Keep a look out for other species that frequent the Mallee including Red-backed Kingfisher, Regent and Mulga Parrot, Australian Ringneck, Major Mitchell's Cockatoo, Chestnut Quail-thrush, Apostlebird, Chestnut crowned Babbler, Black-eared Cuckoo, Southern Scrub-robin, Crested Bellbird, Gilbert's Whistler and White-browed Treecreeper, and Purple-gaped, Pied, Black, Yellow-plumed and Grey-fronted Honeyeater.

Mallee habitats in Dangalli Conservation Reserve are characterised by an understorey of chenopod (bluebush and saltbush) shrubs, with areas of Spinifex largely restricted to the southern half of the reserve. Scarlet-chested Parrots appear to show a preference for Mallee with a diverse understorey of shrubs, and there is also some association with areas that have been recently burnt.

Australian Arid Land Botanic Gardens and Port Augusta

Key species: Chirruping Wedgebill, Redthroat, Rufous Fieldwren
Other species: Elegant Parrot, Pied Honeyeater, Black Honeyeater, White-fronted Honeyeater, White-backed Swallow, White-winged Fairy-wren

Located 5 km north of Port Augusta on the Stuart Hwy, the Australian Arid Lands Botanic Gardens (250 ha) is an outstanding display of Australia's arid zone flora. The diverse native habitat here results in high levels of bird activity, with over 100 species now recorded in the gardens. Open from 7.30 a.m. until sunset 7 days per week, there is a series of well-defined walks that provide access throughout the site. Over 150 varieties of *Eremophila* shrubs have been planted along the Eremophila Loop (800 m). Emu-bush, as it is popularly known, produces a mass of flowers that provide a favoured food source for nomadic Pied, Black and White-fronted Honeyeater. Chirruping Wedgebill occur near the visitor's centre (and are relatively easy to observe), while Redthroat occur near the gazebo on the red sand dunes along the Regional Loop (1.4 km). Other birds to watch for include Red-backed Kingfisher, Southern Whiteface, Black-faced Woodswallow, Crimson Chat, White-backed Swallow, Elegant and, less commonly, Blue-winged Parrot. In the saltbush plains and woodlands along the Red Cliff Loop (4.5 km), watch for Rufous Fieldwren, Brown and Rufous Songlark, White-browed Babbler, Red-capped Robin, Pallid and Black-eared Cuckoo, Crimson and White-fronted Chat, and it is not unusual to see the three resident species of Fairy-wren: White-winged, Variegated and Splendid.

When passing through Port Augusta, stop and visit Bird Lake: a series of shallow saline wetlands adjacent to the Princes Hwy 3 km south of the town centre. Banded Stilt, often in large numbers, Black-winged Stilt, Red-necked Avocet are usually present, as are waterfowl such as Australasian Shoveler and Pink-eared Duck. Public access is possible from a side road next to the Bridgestone Tyre Centre.

Australia's only truly feral population of Ostrich occurs on and around Mount Arden Station, located on the Stuart Hwy 19 km north of Port Augusta. This is private property and visitors are, unfortunately, not encouraged. As an alternative, Ostriches occasionally occur near the northern section of Yorkeys Crossing, a road that navigates the top of the Spencer Gulf. Access is via the Stuart Hwy 6 km north of Port Augusta. Search the shrublands on the west side of the crossing 7 km from the turn-off.

Whyalla Conservation Park and Iron Knob Rd

Key species: Western Grasswren, Redthroat, Rufous Fieldwren, Slender-billed Thornbill

Other species: Crested Bellbird, Black-eared Cuckoo

Whyalla Conservation Park (960 ha) is located 10 km north of the township of Whyalla and ~70 km south-west of Port Augusta. The park is home to more than 80 bird species, including the Western Grasswren. The main entrance to the park is located on the Lincoln Hwy, and there is a northern entrance on Speedway Rd, accessed via Iron Knob Rd). Although the main tracks are suited to conventional vehicles (except after heavy rain), there is no access for caravans. Iron Knob Rd lies between Whyalla and Iron Knob.

Look for Western Grasswren around the base of Wild Dog Hill: a large sandstone outcrop in the park's north-west corner. With a preference for areas of Western Myall with an understorey of Low Bluebush, they are regularly recorded at the base and on the lower slopes of the hill. As with most grasswrens, this can be a cryptic and elusive species that requires some effort to locate. Other species that frequent Wild Dog Hill include Emu, Mulga Parrot, Blue Bonnet, Crested Bellbird, Variegated Fairy-wren, Chestnut-rumped Thornbill, Mistletoebird, Grey Butcherbird, Horsfield's Bronze-Cuckoo, Black-eared and Pallid Cuckoo. In the salt-plain areas of the park, look for White-winged and Splendid Fairy-wren, as well as less common species such as Rufous Fieldwren, Redthroat and Slender-billed Thornbill. There is a small dam on the central southern boundary, which may hold water for extended periods – at these times, it is worth checking for visiting waterbirds. Mammals in the park include Euro and Red and Western Grey Kangaroo. While in Whyalla, it is worth visiting the Whyalla Wetland. Interesting species reported here include Common Sandpiper (summer), Glossy Ibis, Banded and Black-winged Stilt, Red-kneed Dotterel, Black-tailed Native-hen and Australian Reed-Warbler (summer).

If you miss the Western Grasswren at Whyalla Conservation Park, another area to search for them is along Iron Knob Rd near the abandoned train station of Middleback (i.e. between rail posts 252 and 257), 16.5 km north of Whyalla. Other species that occur here include Crimson and White-fronted Chat, Rufous Fieldwren, Brown Songlark, Slender-billed Thornbill and Redthroat. A third location for the Western Grasswren is along Myall Creek, where it crosses either the Eyre Hwy or the Lincoln Hwy.

Lake Gilles Conservation Park

Key species: Western Yellow Robin, Blue-breasted Fairy-wren, Rufous Treecreeper
Other species: Malleefowl, Southern Scrub-robin, Crested Bellbird, Chestnut Quail-thrush, Spotted Nightjar

Lake Gilles Conservation Park (42 023 ha) is located on the Eyre Hwy between Kimba and Iron Knob. The park is a semi-arid Mallee landscape that incorporates salt-lakes, sandy rises and isolated rocky hills. It is a good place to see a range of Mallee specialist species, including three species (often regarded as south-west Western Australian endemics) at the eastern edge of their range: Blue-breasted Fairy-wren, Western Yellow Robin and Rufous Treecreeper. Although rare, there is also a chance of Scarlet-chested Parrot. There is bush camping at Lake Gilles and at Refuge Rocks, and hotel accommodation at Kimba.

The Eyre Hwy runs through the southern half of the park. It is worth stopping in any habitat that appears 'birdy', especially areas of old growth Mallee, Cypress Pine with bluebush and Spinifex in the understorey. Without leaving the highway, one good area for Blue-breasted Fairy-wren, Western Yellow Robin and Rufous Treecreeper is 6 km west of the eastern boundary of the park (17 km east of the western border, 53 km from Iron Knob).

Away from the highway, the most reliable site for these three specialities is along the Lake Gilles Rd (accessed 70 km west of Iron Knob, and 17 km east of Kimba) – 8.5 km north of the Eyre Hwy. Under normal conditions, this is a 2WD road. Other regular species here include Southern Scrub-robin, Crested Bellbird, Chestnut Quail-thrush, Gilbert's Whistler, and White-fronted, Yellow-plumed, White-eared, Brown-headed, Spiny-cheeked and Purple-gaped Honeyeater. Along Lake Gilles Rd, 8.5 km from the Eyre Hwy (1.5 km before you reach the lake), search for Blue-breasted Fairy-wren where the Mallee interfaces with bluebush habitats. Variegated Fairy-wren and Splendid (Turquoise) Fairy-wren are also found here, so some care with identification is required. Continuing north along this road, you come to the edge of Lake Gilles. Blue-breasted Fairy-wren also occurs around the campsite on the shores of Lake Gilles – while Variegated Fairy-wren tends to prefer the open Mallee areas. Rufous Treecreeper can be quite common too, especially in the fringe area between the forest and the lake shore. Malleefowl

Although threatened, Malleefowl remain quite widespread. Most encounters with this iconic species are had when individuals cross vehicle tracks in suitable habitat.

frequent the Mallee around Lake Gilles and are sometimes seen crossing the roads. Other species that are present include Grey Currawong, Inland Thornbill, Golden and Rufous Whistler, Hooded and Red-capped Robin, Mulga Parrot, Australian (Port Lincoln) Ringneck and Purple-crowned Lorikeet. Nocturnal species include Spotted Nightjar, Australian Owlet-Nightjar, Tawny Frogmouth and Southern Boobook.

Blue-breasted Fairy-wren and Western Yellow Robin also occur at nearby Refuge Rocks (known locally as Secret Rocks), 44 km south-east of Kimba on the Whyalla Rd. There is a picnic area next to the rocks, with nearby walking trails and a designated camping area. A further 5 km east of Refuge Rocks, a north–south 4WD track crosses Whyalla Rd. Blue-breasted Fairy-wren, Western Yellow Robin and Rufous Treecreeper are found along this track, with the first two regularly recorded 3 km north of the Whyalla Rd, while Rufous Treecreeper prefers areas along the southern track, again ~3 km from Whyalla Rd.

Lincoln National Park

Key species: Western Whipbird, Rock Parrot, Western Yellow Robin, Cape Barren Goose, Blue-breasted Fairy-wren
Other species: Bush Stone-curlew, Hooded Plover, Eastern Osprey, White-bellied Sea-Eagle, Western Gerygone, Southern Emu-wren, White-browed Scrubwren, Diamond Firetail

Over 123 bird species have been recorded in Lincoln National Park (31 531 ha), located 364 km south-west of Port Augusta on the south-east tip of the Eyre Peninsula. The main species to target in the park are Western (Mallee) Whipbird, Rock Parrot, Cape Barren Goose, Western Yellow Robin, and Blue-breasted Fairy-wren. There is camping and cottage accommodation on Cape Donington. Contact the local national parks office at 75 Liverpool St, Port Lincoln or (08) 868 83111. There are toilet facilities and rainwater tanks at most camping areas. A camping permit is required from the Tourist Information Centre in Port Lincoln.

A good place to look for Western Whipbird is in the scrub and tall heath near Pillie Lake along Donington Rd in Lincoln National Park. The species is often in full song between July and October, meaning it is easier to find during this period, when compared with other times of the year. This area also provides good opportunities for Blue-breasted Fairy-wren, Western Yellow Robin, Southern Scrub-robin, Purple-gaped and Tawny-crowned Honeyeater, and Diamond Firetail. Western Whipbird occur nearby along Wanna Rd (the road to Memory Cove); look on either side of the road up to 3 km from the Donington Rd turn-off, particularly near the tracks that lead east, and near the electricity station 2.2 km from the turn-off. Birding at Donington Point can be rewarding, with Emu, Rock Parrot, Southern Scrub-robin and Blue-breasted Fairy-wren all resident. Along the coast on the Wanna Track, look for Rock Parrot and Diamond Firetail (particularly near the final water bore on the track), while Eastern Osprey nest on the rocky stack at Wanna Head. Another spot to look for Western Whipbird is the dense heath near the Taylor's Landing carpark; again, their distinctive call is usually the best mean of locating the species. You will also have a good chance of seeing Blue-breasted Fairy-wren (on the edge of Mallee vegetation), Brush Bronzewing, Southern Scrub-robin, Western Yellow Robin and Grey Currawong. A good walking track leads south-west from Taylors Landing to the Pillie Lake carpark (part of the Investigator Trail), where much of the walk passes through suitable habitat for the Western Whipbird. Blue-breasted Fairy-wren, Purple-gaped Honeyeater and Rufous Fieldwren occur in the heath on the north side of the

Thickets of coastal heath and short Mallee eucalypts (shown here on Kangaroo Island) are also the preferred habitat of the Western Whipbird in Lincoln National Park.

road near the Woodcutter's Beach carpark, and coastal birds here include Fairy Tern and Black-faced Cormorant.

Western Whipbird also frequents the dense coastal Mallee at Memory Cove Wilderness Protection Area (9336 ha), which is accessible by 4WD only. A good place to look for them is along the walking trail from the Memory Cove Camping Area to the Cove (1 km return), particularly on the hill along this walk. Look for Rock Parrot as they fly over heath. The cove area provides great views out over the oceans, where you may sight White-bellied Sea-Eagle and Eastern Osprey, as well as Australian Sea Lion, Bottlenose Dolphin and Southern Right Whale (the latter between May and October). Birding around the campground can be rewarding, with a good chance of observing Blue-breasted Fairy-wren, Western Yellow Robin, Chestnut Quail-thrush, Southern Scrub-robin, Purple-gaped, Tawny-crowned, Brown-headed and New Holland Honeyeater, and White-browed (Spotted) Scrub-wren. The beach area supports Hooded Plover and Sooty Oystercatcher. The Endangered Eyre Peninsula form of the Southern Emu-wren (ssp. *parimeda*) inhabits the cutting-grass swamps and coastal heath habitats along the coast road.

Cape Barren Geese feed at Big Swamp, 20 km north-west of Port Lincoln on the Flinders Hwy. They also occur along the Lincoln Hwy, including near the turn-off to Louth Bay (21 km north of Port Lincoln), and near the creek mouth at the end of Hirschausen Rd (the turn-off is 16 km north of Port Lincoln). Other birds at Big Swamp may include Hoary-headed Grebe, Banded and Black-winged Stilt, Red-necked Avocet, Pink-eared and Musk Duck, Australasian Shoveler, Black-tailed Native-hen, Black-fronted and Red-kneed Dotterel, Glossy Ibis, White-faced and White-necked

Heron, Royal and Yellow Spoonbill, and, in summer, migratory waders such as Sharp-tailed Sandpiper and Common Greenshank. In Port Lincoln, Rock Parrot feed along the foreshore on the Parnkalla Walking Trail, and the coastline near the Kirton Point Caravan Park can be a good area for Blue-breasted Fairy-wren.

Coffin Bay National Park

Key species: Rock Parrot, Hooded Plover, Sanderling, Blue-breasted Fairy-wren, Western Yellow Robin
Other species: Eastern Osprey, White-bellied Sea-Eagle, Purple-gaped Honeyeater, Tawny-crowned Honeyeater

More than 150 species of bird have been recorded within Coffin Bay National Park (31 000 ha), including Rock Parrot, Sanderling (summer), Hooded Plover, Blue-breasted Fairy-wren, Western Yellow Robin, Eastern Osprey and White-bellied Sea-Eagle. The park is 50 km west of Port Lincoln and can be accessed via a sealed road to Yangie Bay. Beyond Yangie Bay, access is via a 4WD track that continues north-west along the Coffin Bay Peninsula. There is a good selection of camping areas within the park.

Rock Parrot frequents the Point Avoid Headland. A good place to look is the valley immediately north of the Avoid Point lookout; as they fly back and forth from the nearby islands, morning is often the best time to see them. On the Point Avoid Headland, look for Cape Barren Goose, Common and Brush Bronzewing, and Tawny-crowned and Purple-gaped Honeyeater, while Almonta and Gunyah Beaches immediately east of the Point support Hooded Plover, Australian Pied and Sooty Oystercatcher, Sanderling (summer), Red-necked Stint (summer) and Fairy Tern. Seabirds include Australasian Gannet and Pacific Gull, and pelagic species such as Shy and Black-browed Albatross and Short-tailed Shearwater. Birds of prey that frequent the area include White-bellied Sea-Eagle, Eastern Osprey and Peregrine Falcon. The Yangie Bay Hike (2 km loop) provides access to the shrublands that encircle the bay. Species that occur here include Western Yellow Robin, Blue-breasted Fairy-wren, Diamond Firetail, White-browed (Spotted) Scrubwren and Brush Bronzewing.

In summer, Yangie and Kellidie Bay may hold good numbers of migratory

Yanerbie Beach

From late spring to early autumn each year, migratory shorebirds roost at Yanerbie Beach, 18 km south of Streaky Bay. The waders' roost is on the rocks at the western end of the beach, while nearby cliffs allow a close approach for excellent views without disturbing the birds. Some of the interesting migrants recorded here include Grey and Pacific Golden Plover, Grey-tailed Tattler, Whimbrel, Ruddy Turnstone, Lesser and Greater Sand Plover, Great and Red Knot, Oriental Plover and Sanderling. Resident shorebirds include Hooded Plover, Australian Pied and Sooty Oystercatcher. Malleefowl (with several active mounds) occur in the Yanerbie Scrub north-east of Yanerbie Sands, a large area of sand dunes that extend behind the beach for 4.8 km inland.

shorebirds, including Grey Plover, Eastern Curlew, Sanderling, Sharp-tailed Sandpiper, Common Greenshank, Red-necked Stint and Ruddy Turnstone. Watch out for Australian Pied and Sooty Oystercatcher, Red-necked Avocet, Fairy Tern, Eastern Reef Egret, Buff-banded Rail, Swamp Harrier, Eastern Osprey and White-bellied Sea-Eagle. Eyre Peninsula Southern Emu-wren (Endangered ssp. *parimeda*) occur in the extreme south of the peninsula, with Kellidie Bay Conservation Park (adjacent to the township of Coffin Bay) being a good place to search for them; look especially in the heath along Airport Lane (the road to Wangary).

Yumbarra Conservation Park

Key species: Scarlet-chested Parrot, Malleefowl, Western Yellow Robin, Rufous Treecreeper, Chestnut Quail-thrush
Other species: Black-eared Cuckoo, Pied Honeyeater

Yumbarra Conservation Park (324 352 ha) is 30 km north-west of Ceduna. Forming a woodland corridor that links the Great Victoria Desert to the northern Eyre Peninsula, it is one of South Australia's most pristine Mallee reserves. The park derives its name from one of the rock holes that fill with water after rain and provide an important water source for birdlife. Over 100 bird species occur at Yumbarra, including the rare and much sought-after Scarlet-chested Parrot. The tracks in the park require 4WD because they occasionally traverse large sand dunes. There are no facilities within Yumbarra, so you need to be self-sufficient, including carrying all water. A permit issued by the SA DEH is required to access the park: www.environment.sa.gov.au. There is camping at Googs Lakes, and a caravan park and motel accommodation in Ceduna.

In late summer and early autumn, Scarlet-chested Parrot drink at Inila Rock Waters and Yumbarra Rockhole on the south side of Yumbarra Conservation Park. To reach the waterholes from the Eyre Hwy, 28 km west from Ceduna, turn north onto Foggos Rd towards Koonibba. Take this road for 17 km until you reach the Dingo Fence (the road is especially rough over the last km). Pass through the gate and turn left (west), driving along the northern side of the fence for ~2.3 km until you reach a track leading north, following this for 3.3 km until you see the sign for the Inila Rock Waters. When there is water here, Scarlet-chested Parrot may regularly come to drink. From here, take the track that heads east for 6.5 km to Yumbarra Rockhole. Note that there are two other management tracks here. One heads north, and another that heads north-east, signposted (after 100m) 'no access'. On occasions, Scarlet-chested Parrots also feed along the track between Inila Rock Waters and Yumbarra Rockhole. Concentrate your search between 3.5 km to 4.5 km east of Inila. They tend to be seasonal visitor (perhaps from the Great Victoria Desert region?), with August to April being the best time to look. There

The Brown Falcon is the most widely reported of any Australian bird species, being recorded in over 95% of all Australian Bird Atlas grid cells.

is also a good selection of other parrots within the reserve, including Major Mitchell's Cockatoo (scarlet-crested ssp. *mollis*), Mulga Parrot, Australian (Port Lincoln) Ringneck and Purple-crowned Lorikeet. Other species regularly sighted in the park include Malleefowl, Red-backed Kingfisher, Western Yellow Robin, Rufous Treecreeper, Black-eared Cuckoo, Chestnut Quail-thrush, Southern Scrub-robin, Shy Heathwren, Crested Bellbird, Gilbert's Whistler, and Pied and White-fronted Honeyeater.

Mount Ive Station (Gawler Ranges)

Key species: Western Grasswren, Short-tailed Grasswren
Other species: Major Mitchell's Cockatoo, Elegant Parrot, Scarlet-chested Parrot, Ground Cuckoo-shrike, Black-eared Cuckoo, Rufous Treecreeper, Grey-fronted Honeyeater

Mount Ive Station (90 000 ha) is a sheep station situated 200 km west of Port Augusta in the northern section of the Gawler Ranges. On the property, there is a camping and accommodation in shearers' quarters, as well as a kiosk and fuel supplies. The Hiltaba–Iron Knob Rd that passes Mount Ive Station provides 2WD access, and within the Gawler Ranges National Park, the Minnipa–Yardea Rd and the Old Paney Scenic Route are considered suitable for 2WD vehicles in dry conditions. Other tracks within the park require a 4WD.

Western Grasswren occurs on the Station, favouring lowland areas dominated by blackbush and bluebush shrublands. A good place to look for them is just to the east of the shearers' quarters; a short 5 minute walk from the homestead and accommodation area. Pass through the gate that leads past the dam to the old shearing shed and head north-east of here to search the flats and shallow gullies. The Western Grasswren occurs throughout this area, but especially where the three paddocks join. Whereas Western Grasswren prefer bluebush flats, the Short-tailed Grasswren (ssp. *pedleri*) favours the Spinifex-clad rocky hills. Good sites to look for the latter species include the summit areas of both Mount Ive and Mount Scott, near the homestead. Short-tailed Grasswrens are also found in the rocky hills around the southern section of Lake Gardner (Australia's fourth largest salt-lake), particularly near a rocky area known locally as Kev's Castle. Access to these areas needs to be arranged with Mount Ive Station (a gate key and mud map out to Lake Gairdner are supplied on request). Other birds that occur here include Major Mitchell's Cockatoo, Elegant and Scarlet-chested Parrot (rare), Ground Cuckoo-shrike, Black-eared Cuckoo, Rufous Treecreeper, Grey-fronted Honeyeater, Varied Sittella, Crimson Chat, Splendid Fairy-wren and Painted Finch (several records).

Mamungari Conservation Park and Vokes Hill Corner Rd

Key species: Scarlet-chested Parrot, Princess Parrot, Inland Dotterel, Cinnamon Quail-thrush, Malleefowl
Other species: Ground Cuckoo-shrike, Major Mitchell's Cockatoo, White-browed Treecreeper, Australian Bustard

Located in the southern part of the Great Victoria Desert and the northern Nullarbor Plain, the Mamungari Conservation Park (2 128 945 ha – formerly named the Unnamed Conservation Park) is a wilderness reserve whose flora and fauna have remained almost untouched. Ornithological highlights in the region include the Scarlet-chested Parrot, Princess Parrot and Malleefowl. Access is by 4WD only. To get to Mamungari from the Eyre Hwy, 42 km west of the Nullarbor Roadhouse, take the Cook Rd to Cook (105 km).This road passes through the Nullarbor Regional Reserve. The Cook to Vokes Hill Corner Rd (255 km) starts on the northern side of Cook and, with effort (including birding stops), you can travel its full length in a day. (Note that alternatively Mamungari Conservation Park is accessible via the Anne Beadell Hwy from Coober Pedy and Laverton in Western Australia). Due to the park's remoteness, you need to be well equipped and experienced in outback travel. Please contact the SA DEH office in Ceduna for information on permit requirements: www.environment.sa.gov.au.

Before heading north to Mamungari, search the Cook Airfield (located west along the Trans Access Rd) for Inland Dotterel and Orange Chat. Inland Dotterel are often nocturnal, so the best time to look is early evening or by spotlighting at night, especially in the open flat areas with a sparse cover of shrubs. To travel to Mamungari, head due north out of Cook (past the rubbish tip and quarry) along the Vokes Hill Corner. The first part of the road passes through treeless limestone plains scattered with bluebush. Inland Dotterel and Nullarbor Quail-thrush frequent the roadside, particularly the area 20 to 60 km north of Cook. In this same area, watch also for Orange Chat, White-winged Fairy-wren, Brown Songlark, Rufous Fieldwren and Spotted Harrier. Some 60 to 90 km north of Cook, watch for the distinctive 'Naretha' Bluebonnet (spp. *narethae*), sometimes considered a separate species limited to the Nullarbor Plains. Travelling further north, the terrain begins to change as you enter the Great Victorian Desert. Shrubs and bushes start to appear, and an increasing number of trees herald the first sand dunes. Watch for woodland birds such as Mulga Parrot, Jacky Winter, Hooded Robin,

Nullarbor National Park

Nullarbor National Park (2 847 041 ha), located near the head of the Great Australian Bight, is a good place to observe Nullarbor Quail-Thrush. Directly north of the Nullarbor Motel Hotel, a track heads north (passing the airstrip in the right). After 8 km, search the area just before a dam and an east turn in the track. Another place to look is 1.2 km east of Nullarbor Motel Hotel, where a north–south track, beside a fence delineating the boundary of the Nullarbor National Park, is cut by the Eyre Hwy. Nullabor Quail-thrush occur in the saltbush along this track. Search along the first few kilometres of the track to the south of the Hwy, and between 6 and 8 km along the track to the north of the highway. Other birds to watch out for include Inland Dotterel (especially in the evening and at night), Major Mitchell's Cockatoo, Orange Chat, Slender-billed Thornbill, Rufous Fieldwren and Stubble Quail. If you are there between May and October, it is worthwhile looking for Southern Right Whale, which gather along the coast of the Nullarbor National Park to give birth. There is a viewing platform 26 km from the Nullarbor Roadhouse at the 'Head of Bight' whale-watching area.

Rufous Treecreeper and Variegated Fairy-wren. A good spot to see Australian Bustard is around the Oak Valley Rd intersection. Scarlet-chested Parrot has been reported regularly from ~10 km north of this intersection and especially between 140 km to 170 km north of Cook.

Towards the northern section of the Cook to Vokes Hill Corner Rd, the road enters Mamungari Conservation Park. Here, the track crosses numerous sand hills and can become overgrown. Watch for Scarlet-chested Parrot along the roadside, especially in the last 10 km before you reach the Vokes Hill Corner (244 km from Cook). After 255 km, you come to the Anne Beadell Hwy, an unsealed dirt road that bisects the park east to west. If you have not already seen Scarlet-chested Parrot, your best strategy is to turn west and look along the roadside between Vokes Hill Corner and 140 km west of the corner, particularly in areas of Mallee vegetation. Areas to concentrate your search include 5 km west, between 50 to 60 km west, 100 to 110 km west, and 140 km west of Vokes Hill Corner. Princess Parrot has been reported in the past within the Marbled Gum woodlands that are intermixed with Black Oak and Desert Oak, between 90 km and 125 west of the corner. Other birds at Mamungari include Malleefowl, Ground Cuckoo-shrike, Major Mitchell's Cockatoo, Mulga Parrot, Rufous and White-browed Treecreeper (especially in Black Oak woodlands), Slaty-backed Thornbill, Crimson Chat and Gilbert's Whistler.

Western Australia

Western Australia (WA) is a huge state, covering the western third of the Australian continent. Indeed, if Western Australia were a separate country, it would be the 10th largest in the world. It shares land borders with the Northern Territory and South Australia, while the Southern Ocean and Great Australian Bight bound its southern shores, and the Indian Ocean bounds its western shores. The capital city is Perth. Despite its enormous area (~2 500 000 km^2), aside from the south-west corner, where most of the ~2.3 million people live, it is sparsely populated. Southern WA has a temperate climate, the vast inland areas are arid with freezing winter nights and daytime temperatures soaring into the 40s C in summer, while the north has a tropical climate characterised by wet (Dec–Apr) and dry (May–Nov) seasons.

Over 550 species of bird have been recorded in WA. BirdLife WA is the only major birdwatching organisation in the state and provides coverage for the entire area: www.birdlife.org.au, wa@birdlife.org.au, (08) 9383 7749. Visitors to the BirdLife WA office are welcome. Here, you can obtain local advice, peruse library books and collect regional guides: Peregrine House, 167 Perry Lakes Drive, Floreat 6014, (office hours are weekdays 9:30 a.m. to 12:30 p.m.). Regional guides are also accessible here: www.birdlife.org.au/locations/birdlife-western-australia/bird-guides-wa. You can view and report unusual sightings to BirdLifeWA (www.birdswa.org.au/sightings.htm) and Birdline WA (www.eremaea.com). For general tourist information, visit www.westernaustralia.com. The Black Swan is the bird emblem of WA.

Southern Western Australia

Regional overview

This section presents information about birdwatching locations around Perth and south-west WA. It also covers a range of locations in the greater southern WA region, in an area known as the 'Golden Outback'. Perth is situated on the shores of the Indian Ocean, and is one of the most remote capital cities in the world (second only to Honolulu), in terms of distance to its nearest neighbour.

Because of its geographic isolation, the temperate south-west corner of WA boasts a diversity of bird species (and subspecies) located nowhere else in the world. Habitats across the south-west are diverse, with the vegetation mainly comprising forests, woodlands, Mallee, shrublands and coastal heathland. In the forests and woodlands, the dominant eucalyptus species are Jarrah, Karri, Tingle, Tuart and Wandoo: all unique to WA. These support a diverse range of fauna and other flora species. Immediately

to the east of Perth, and the Swan Coastal Plain, the Darling Scarp (also referred to as the Darling Ranges), is a low escarpment that runs north–south. The uplands of the scarp are dominated by tall dry eucalypt forests. Beyond the Darling Scarp, the annual average rainfall declines dramatically. This is an agricultural zone, known as the wheatbelt, where only remnant patches of the native vegetation remain. The south-west region also has extensive rocky and sandy shores, coastal cliffs, wetlands, and streams and river systems. Over 1500 species of native flora occur in the region, including well-known and well-represented genera such as *Banksia*. The best time to see plants is between September and November, when the wildflowers are in full bloom. For information on WA's national parks and reserves, see www.dec.wa.gov.au.

South-west WA has a temperate (Mediterranean) climate that is characterised by four seasons – summer, autumn, winter and spring. Owing to temperate weather conditions, south-west WA can be worth visiting year round. About 900 mm of rain falls per year, with most falling between May and September. The south-west region's average maximum temperate is around 32°C in summer and 14°C in the winter. The city of Perth boasts more sunny days per year than any other Australian capital city – sunshine is usually the standard order in the region. Although high temperatures are common in summer (Dec–Feb), the hot weather is seldom oppressive for long, and the region does not suffer from the high humidity typical of northern WA. However, it is worth avoiding the hot interior of southern WA (and other desert areas which can be oppressive) in summer.

Excellent sealed highways provide easy access across the southern areas of WA. The Great Eastern Hwy (via the Coolgardie–Esperance Hwy) links up with the Eyre Hwy that continues across the Nullarbor Plain to South Australia and Adelaide. The Brand Hwy links up with the North West Coastal Hwy that continues north up the west coast of WA. Northern WA can also be accessed via the Great Northern Hwy.

There is a train and bus station right in the centre of Perth, as well as plenty of accommodation in the form of hotels and hostels. Free buses get you around the CBD, where you can access many of Perth's parklands and coastal regions. For Perth tourist information, contact the WA Visitor Centre: www.bestofwa.com.au, (08) 9483 1111. The Indian Pacific train travels twice each week from Perth to Sydney via Adelaide. Perth is well serviced by coach, with regular services to the Albany, Esperance and Kalgoorlie regions, up and down the west coast, via Geraldton and Broome, and onto Darwin. A 2WD vehicle will get you to the majority of birding sites in the region, but if you intend to explore remote locations, such as Neale Junction, a 4WD is essential.

Daily scheduled flights connect to Perth from every Australian capital city. For flight and passenger information: www.perthairport.com.au.

A 9-day stay in south-west WA is about the minimum time required to provide good opportunities to see all of the sought-after south-west specialities. A suggested itinerary involves:

- **Day 1.** Greater Perth, Victoria Dam (Baudin's Black-Cockatoo, Carnaby's Black-Cockatoo, Red-capped Parrot, Western Rosella, White-breasted Robin, Red-winged Fairy-wren, Red-eared Firetail, Western Thornbill, Western Spinebill, Western Wattlebird, waterfowl, shorebirds).
- **Day 2.** Perth to Dryandra Woodland, 200 km (Bush Thick-knee, Elegant Parrot, Western Yellow Robin, Rufous Treecreeper, Blue-breasted Fairy-wren, Western Thornbill, Crested [Western] Shrike-tit and possibly the iconic Numbat).
- **Days 3–4.** Dryandra to Fitzgerald River National Park, 300 km (Malleefowl, Western Whipbird, Western Bristlebird, Western Ground Parrot (very rare), Shy Heathwren, Southern Scrub-robin, Southern Emu-wren, Purple-gaped Honeyeater).
- **Days 5–6.** Fitzgerald River NP to Albany region, including Cheynes Beach, Two Peoples Bay, Albany, Stirling Ranges, 170 km (Noisy Scrub-bird, Western Bristlebird, Western Fieldwren, Western Whipbird, White-breasted Robin, Crested [Western] Shrike-tit, Red-eared Firetail, Western Ground Parrot).
- **Days 7–8** Albany to Cape Leeuwin (via Lake Muir) and Cape Naturaliste, 370 km (Western Corella, Australasian Bittern, Red-tailed Tropicbird [small chance], Rock Parrot, Bridled Tern, Fairy Tern, Eastern Reef Egret, White-breasted Robin, Red-eared Firetail).
- **Day 9.** Return to Perth, 260 km.

With an additional day, you could add a visit to Rottnest Island (Common

Pheasant, Indian Peafowl (perhaps just one left), Banded Stilt, and Rock Parrot). From Albany (if you have time, or if you are heading east across the Nullarbor Plain), plan side trips to Cape Arid National Park and Neale Junction (4WD required) in the far-west west (allow at least 7 days).

Birding highlights

The ecology and habitat of south-west WA is isolated from other more mesic areas in eastern Australia by the expansive desertlands of the Great Southern, Gibson and Great Sandy Deserts. Biogeographically, the south-west has become an island refuge as the climate has dried. This isolation has led to a great diversity of endemic flora and fauna; indeed, the combination of both high species richness and high endemism means south-west WA is a biodiversity hotspot. The list of unique bird species and subspecies is impressive, meaning south-west WA is a popular place for birdwatchers to visit. Extensive vegetation clearance for agriculture, especially in the wheatbelt, has destroyed habitats for native birds, so that even once abundant species of the wheatbelt have undergone precipitous declines in parts of their range (e.g. Yellow-plumed Honeyeater in the Kellerberrin district). As a result, south-west WA has a high concentration of Endangered species when compared with many areas of Australia.

Western Wattlebird is one of 16 bird species endemic to south-west WA. Although generally common, the species can be difficult to find, given that it is somewhat mobile – good sites around Perth include King's Park Botanic Gardens, Thomson's Lake (near the southern carpark) and Bungendore Park.

Sixteen bird species are currently recognised as being endemic to south-west WA. These are Carnaby's (Short-billed) and Baudin's (Long-billed) Black-Cockatoo, Western Corella, Red-capped Parrot, Western Rosella, Noisy Scrub-bird, Western Bristlebird, Red-winged Fairy-wren, Western Thornbill, Western Wattlebird, Western Spinebill, White-breasted Robin and Red-eared Firetail. Owing to taxonomic developments, the three recent additions are the Western Ground Parrot (formerly ssp. *flaviventris* of Eastern Ground Parrot), Western White-naped Honeyeater (formerly ssp. *Chloropsis* of White-naped Honeyeater) and Western Fieldwren (formerly ssp. *Montanellus* of Rufous Fieldwren); though the last species is more contentious than the others. Several species that have their distributions largely limited to south-west WA also occur in the Eyre Peninsula region of South Australia. These include Western Yellow Robin, Rufous Treecreeper and Blue-breasted Fairy-wren. Similarly, Western Whipbird also occurs in South Australia, and there are historic records in the Mallee region of Victoria. The recently split Western Quail-thrush (*marginatum* split from Chestnut-breasted Quail-thrush) and Nullarbor Quail-thrush (*alisteri* split from Cinnamon Quail-thrush) also occur in the region, while Crested (Western) Shrike-tit (ssp. *leucogaster*) is a clear candidate for a future elevation to full species status. Another endemic to WA, the Dusky Gerygone, occurs as far south as the Coral Coast, near Carnarvon, while the Western Grasswren (formerly ssp. *textilis* of Thick-billed Grasswren) is found in the same area (and also occurs in SA). Striated (Sandhill)

Grasswren ssp. *oweni,* which occurs in the far east of the state (also NT and SA), is another candidate for full species status.

Aside from those birds mentioned above, sought-after species in southern WA include Malleefowl, Square-tailed Kite, Banded Stilt, Hooded Plover, Red-tailed Tropicbird, Fairy Tern, Little Shearwater, Soft-plumaged Petrel, Major Mitchell's Cockatoo, Princess, Scarlet-chested, Rock and Elegant Parrot, Purple-crowned Lorikeet, Grey Honeyeater and Splendid Fairy-wren. As with the endemic species, the isolation of south-west WA has led to very high degree of endemism at the subspecies level. These include Little Shearwater (ssp. *tunneyi*), Cape Barren Goose ('Recherche Cape Barren Goose' ssp. *grisea,* south-east WA coastline), Purple (Western) Swamphen (ssp. *bellus*), Painted Button-quail (ssp. *scintillans,* Abrolhos Is.), Red-tailed Black-Cockatoo (Forest Red-tailed Black-Cockatoo ssp. *naso* in the south, ssp. *samueli* in the north), Western Corella (uncommon ssp. *pastinator* in the south, common ssp. *derbyi* in the north), Regent Parrot (ssp. *anthopeplus/westralensis*), Australian Ringneck (Twenty-eight Parrot ssp. *semitorquatus* south-west of Perth, Port Lincoln Parrot ssp. *zonarius* north and east of Perth), Blue Bonnet (Naretha Blue Bonnet ssp. *narethae*), Elegant Parrot (ssp. *carteri*), Splendid Fairy-wren (Banded Fairy-wren ssp. *splendens*), Southern Emu-wren (ssp. *westernensis*), Shy Heathwren (ssp. *whitlocki*), White-browed (Spotted) Scrubwren (ssp. *maculatus*), Western Gerygone (ssp. *fusca*), Yellow-throated (Dusky) Miner (ssp. *obscura*), New Holland Honeyeater (ssp. *longirostris*), White-cheeked Honeyeater (ssp. *gouldii*), Singing Honeyeater (ssp. *virescens*), White-browed Babbler (ssp. *ashbyi*), Western Whipbird (Mallee Whipbird ssp. *oberon* in Mallee woodland, ssp. *nigrogularis* in coastal heathland), Chestnut Quail-thrush (ssp. *fordianum*), Grey (Western) Shrike-thrush (ssp. *rufiventris*), Australian (Western) Magpie (ssp. *dorsalis*), Australian (Western) Raven (ssp. *perplexus*), Grey Fantail (ssp. *preissi*), Scarlet Robin (ssp. *campbelli*), Western Yellow Robin (ssp. *griseogularis*), Little Grassbird (ssp. *thomasi*), Welcome Swallow (ssp. *carteri*), Dusky Woodswallow (ssp. *perthi*) and Silvereye ('Western Silvereye' ssp. *gouldi*). The south-west forms of the Rufous Bristlebird (ssp. *litoralis,* last seen 1940s) and Lewin's Rail (ssp. *clelandi,* last recorded 1932) are both extinct.

Introduced species only (or more easily) seen in the south-west WA when compared with the remainder of Australia are Laughing Dove, Mute Swan and Common Pheasant. Species native to eastern Australia that have been introduced to south-west WA include Rainbow Lorikeet (common metropolitan area), Sulphur-crested Cockatoo (found at Lake McLarty and near Guildford), Long-billed Corella (common around Perth), Laughing Kookaburra (common across the south-west), and Red-browed Finch (best are the rose gardens at Melville Nursery (also a good spot for a snack or meal) at 40 Masonmill Rd near Victoria Dam, but also Victoria Dam in Korung National Park, Bickley and Ellis Brooks, and along Schipp Rd near Piesse Brook). Rarities recorded in southern WA include Northern Shoveler, Northern Pintail, Leach's Storm-Petrel, Antarctic and Barau's Petrel, King, Southern Rockhopper, Northern Rockhopper, Fiordland, Erect-crested and Macaroni Penguin, Cape Gannet, Grey Heron, Oriental Honey-Buzzard, Ringed and Little Ringed Plover, Little Stint, Buff-breasted Sandpiper, South Polar Skua, Antarctic Tern, Black-tailed, Laughing and Franklin's Gull, Black-capped Kingfisher and White Wagtail.

Greater Perth

Key species: Western Spinebill, Western Wattlebird, Carnaby's Black-Cockatoo
Other species: Freckled Duck, Tawny-crowned Honeyeater, waterfowl, shorebirds

Located on the banks of the Swan River, Perth has a surprising number of good birding sites within the metropolitan area or within an hour's drive of the CBD. The city is well serviced by air, and has road and rail links with Adelaide in SA. Most visiting birdwatchers enter WA by flying to Perth or alternatively visit as part of an extended road trip from the eastern states. The capital city provides a convenient gateway to the south-west WA birding sites.

Kings Park and Botanic Garden (401 ha) is located next to the Swan River within walking distance (1.5 km) from the Perth CBD. Protecting a significant bushland remnant, the park is an excellent place to observe White-cheeked, New Holland, Brown and Singing Honeyeater, Western Spinebill, and Western and Red Wattlebird, particularly in spring when the *Banksia* are flowering. Carnaby's Black-Cockatoo occur here in summer, autumn and winter, visiting the park from the wheatbelt east of Perth. They have developed a taste for the seeds of pine trees; listen for their distinctive '*wee-ya*' call given when flying. Other birds to look for include Little Corella, Rainbow Lorikeet, Australian Ringneck, Laughing Dove, Pallid (May–Nov) and Fantail (winter) Cuckoo, Shining Bronze-cuckoo (spring), Variegated Fairy-wren, Varied (Black-capped) Sittella, Western, Yellow-rumped and Inland Thornbill, Weebill and Western Gerygone.

Matilda Bay Reserve (21 ha) is located just south of Kings Park, between Hackett Drive and the river. The reserve includes Pelican Point, a sandy peninsula that has an exposed stretch of beach, low sand dunes and a lagoon. This is a high-tide roost for migratory species such as Red-necked Stint and Sharp-tailed and Curlew Sandpiper, and, less commonly, Great and Red Knot, Grey and Pacific Golden Plover, and Lesser Sand-Plover. There is also a record of Red-necked Phalarope. In spring, Fairy Tern occasionally nest at Pelican Point.

Herdsman Lake Regional Park (370 ha) is located 6 km from Perth's CBD in the suburb of Herdsman. The park contains a very large shallow lake, with large reed beds in its centre and a narrow strip of paperbark around its fringe. Access to the lake is via Powis St, 5 km from the city. There are two bird hides on the eastern shoreline and a pleasant boardwalk next to the wildlife centre. The Herdsman Lake Wildlife Centre is located at the corner of Flynn St and Selby St. Australian Little Bittern are occasionally recorded at Herdsman Lake, with most records occurring during late spring and early summer, particularly near the Wildlife Centre on the south side of the lake. The lake is a particularly good site for waterfowl, especially in autumn and early winter before the first good rains of the season. Species include Pink-eared, Freckled (rare, usually Jan–May), Blue-billed and Musk Duck, Hardhead and Australasian Shoveler. Waterbirds and shorebirds to look for include Australasian, Hoary-headed and Great Crested Grebe, Australasian Darter, Little and Great Egret, Nankeen Night-Heron, Glossy Ibis, Yellow-billed Spoonbill, Buff-banded Rail, Red-kneed and Black-fronted Dotterel, and Black-winged Stint. In summer, migratory shorebirds such as Common Greenshank, Long-toed and Red-necked Stint, and Sharp-tailed and Wood Sandpiper may also occur if the water level gets low enough. Quite a few rarities have been seen at the lake, including Wandering Whistling-Duck, Pectoral Sandpiper, Oriental Pratincole, White-winged Black Tern and Eastern Yellow Wagtail. In the woodlands bordering the lake, you may see Long-billed and Little Corella, Australian Ringneck, Sacred Kingfisher, Rainbow Bee-eater (summer), Yellow-rumped Thornbill, Western Gerygone and Tree Martin, while Australian Reed Warbler and Little Grassbird frequent the reedy fringes. Nearby, Lake Monger (70 ha) is another large urban lake just 5 km from Perth CBD. Access to the lake is via Vincent St in the south or Dodd St in the north, with a viewing platform on the northern side. Although the lake is more heavily used for recreational purposes than Herdsman Lake, it is still an excellent site for waterbirds such as Great-crested Grebe, Hardhead, Australasian Shoveler, and Blue-billed, Pink-eared and Musk Duck.

Australasian Shoveler can be remarkably common at some urban wetlands in Perth (in contrast to urban wetlands in the eastern states). Because these sites have many cyclists, joggers and walkers, some flocks have become remarkably tame: Herdsman Lake, and the nearby Lake Monger, are two such sites worth a visit.

Thomsons Lake Nature Reserve (538 ha) is an expansive wetland, with adjacent native woodlands, 25 km south of the Perth CBD. The main entrance to the reserve is off Russell Rd, 4 km from the Rockingham Rd turn-off. The lake is shallow and seasonal, with a large area of open water when full, but is now often dry for up to 5 months of the year (Dec–Jan to May–Jun). In excess of 20 000 waterbirds have been counted on occasion. Species most commonly seen include Australian Shelduck, Pacific Black Duck, Grey Teal and Australasian Shoveler. Shorebirds visiting the lake, mainly in summer, include Red-necked Avocet, Banded and Black-winged Stilt, Common Greenshank, Black-tailed Godwit, Red-necked and (less commonly) Long-toed Stint, and Curlew, Sharp-tailed, Wood and Marsh Sandpiper, while the generally uncommon Pectoral Sandpiper is almost annual here. Rarities recorded here have included Little Stint and Oriental Pratincole. When the right amount of mud is exposed, it can be an excellent site for Buff-banded Rail and Spotless, Australian Spotted and Baillon's Crake (though the right conditions may only last for a few weeks in early summer). Australasian Bittern and Australian Little Bittern have also been recorded. The woodland surrounding the lake hold Sacred Kingfisher, Red-capped Parrot, Australian

Penguin Island

Penguin Island (12.5 ha), part of the Shoalwater Islands Marine Park (6658 ha), is located 700 m off the coast of Rockingham, 42 km south of Perth. Ferry tours to the island operate from Mersey Point from mid-September to early June, leaving on the hour. There is no access to the island between July and mid-August to protect the breeding penguins from disturbance. See www.penguinisland.com.au. There are no facilities. The main target species on the island are Little Penguin, Bridled Tern and Fairy Tern, all of which are seen easily in summer as they breed within the park. Aside from these, a good selection of other species have been recorded, including Buff-banded Rail and Ruddy Turnstone, Caspian, Crested and Roseate (uncommon) Tern, Australasian Gannet, Arctic Jaeger, Brown Skua, Wedge-tailed and Flesh-footed Shearwater, and Osprey. Rarities that have been recorded include Red-tailed Tropicbird and Franklin's Gull.

Ringneck, Black-faced Woodswallow, Splendid Fairy-wren and Rainbow Bee-eater (summer), and the lake is considered one of the few breeding sites in Perth's metropolitan area for Swamp Harrier.

Rottnest Island

Key species: Rock Parrot, Wedge-tailed Shearwater, Banded Stilt, Eastern Reef Egret, Bridled Tern, Common Pheasant, Indian Peafowl (almost extirpated and possibly not 'tickable')
Other species: Osprey, Banded Lapwing, Sanderling, Red-capped Robin, Fairy Tern, Roseate Tern

Rottnest Island is located 18 km from Fremantle off the WA coast. With a total area of 1900 ha, the island comprises limestone and sand that is dominated by low scrubby vegetation, with several large permanent salt lakes. A popular tourist destination (over half a million visitors annually), most people visit for the day only. Daily ferries depart from Perth City, Fremantle and Hillarys Boat Harbour. The Rottnest Island Visitor Centre provides maps, bus timetables and accommodation information: www.rottnestisland.com, (08) 9372 9730. A good way to explore the island is to hire a bicycle from Rottnest Island Bike Hire or through the ferry companies before you leave – a full circuit of the island is a 22 km ride.

At Thomson Bay Settlement, Indian Peafowl, which were introduced between 1910 and 1915, have in recent years been systematically removed; at the time of writing, there are now just one or two males left that frequent the township area. Along the Thomson Bay coastline, you may see Rock Parrot, Australian Pied and Sooty Oystercatcher, and summer migrants such as Sanderling, Grey-tailed Tattler and Grey Plover. Several pairs of Osprey nest at Rottnest each year. The nest at Salmon Point (on the south side of the island) is estimated to be 70 years old.

The Rottnest Island Golf Course is perhaps the most reliable site to see Common Pheasant (introduced in 1928) – places to look include the low heath between the first and the second fairways, between the sixth tee and seventh

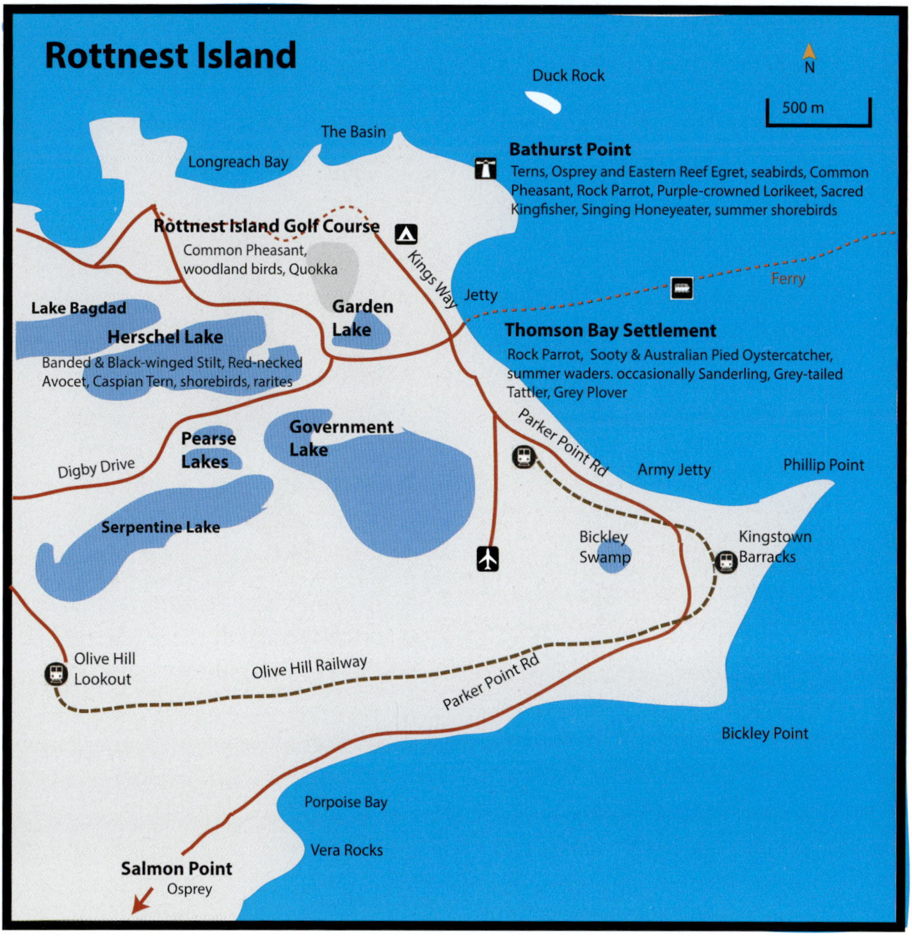

green, and near the clubhouse. Other birds here include Banded Lapwing, Rainbow Bee-eater (summer), White-fronted Chat, Golden Whistler, White-browed (Spotted) Scrubwren, Red-capped Robin and Western Gerygone (the last two species have vocalisations that differ significantly from mainland birds). An estimated 10 000 Quokka inhabit the island. Despite being largely nocturnal, they can usually be spotted seeking shade under the trees around the Golf Course, at Garden Lake, located just south of the Golf Course, or seeking handouts in the Settlement.

The Bathurst Point Lighthouse is located immediately north of the Thomson Bay settlement. From here, you can scan out to sea and along the shoreline for Bridled, Roseate (uncommon) and Fairy Tern, Osprey and Eastern Reef Egret. There is also a possibility of Indian Yellow-nosed Albatross, Southern Giant Petrel and Little Shearwater (all winter), and Wedge-tailed Shearwater (arrive Sept, depart Apr) and Huttons Shearwater (especially on passage in autumn and late winter-early spring). The heath and woodlands behind Bathurst Point, near the Bathurst tennis

Hillarys Boat Harbour pelagic trips

The Hillary's Boat Harbour pelagic trips are seabird-watching boat trips run by BirdLife WA. These 1-day trips depart from the Hillarys Boat Harbour, 20 km north of Perth, and pass Rottnest Island on the way to the Rottnest Trench. A particular highlight of most trips is the number of Soft-plumaged Petrels that are observed. Other regular species include Shy, Black-browed and Indian Yellow-nosed Albatross, Wedge-tailed, Hutton's, Little and Flesh-footed Shearwater, Southern and Northern Giant Petrel, White-faced Storm-Petrel, Cape and Great-winged Petrel. Rarities have included Arctic Tern and Kerguelen Petrel. Eastern Reef Egret is also possible on the breakwater. These trips run infrequently. To enquire about upcoming dates and berths contact Birdlife WA www.birdlife.org.au, wa@birdlife.org.au, (08) 9383 7749.

Generally considered a rarity in eastern Australia, Soft-plumaged Petrel are sighted regularly on pelagic trips off south west WA. Dedicated sea-watchers may also chance upon this species from sites such as Cave Point Lighthouse, near Albany and Cape Naturaliste.

courts and the camping area, provides habitat for Common Pheasant, Rock Parrot (chance), Sacred Kingfisher and Singing Honeyeater. Here, Singing Honeyeaters are ~20% heavier than mainland birds, and were once considered a separate species. Rottnest Island's waste treatment ponds are located just to the west of the tennis courts. Look here for Rock Parrot and Australian Shelduck. In summer, shorebirds such as Common Sandpiper, Common Greenshank and Ruddy Turnstone may also occur.

A series of salt lakes are located at the eastern end of the island, west of the Thomson Bay settlement. These contain brine shrimp, which in turn support a large numbers of resident and migrating bird species. Up to 20 000 Banded Stilt have been recorded, along with lesser numbers of Black-winged Stilt, Red-necked Avocet, Red-capped Plover and Caspian Tern. In summer, migratory shorebirds, including Red-necked Stint, Ruddy Turnstone, Sharp-tailed and Curlew Sandpiper, Grey-tailed Tattler and Grey Plover, congregate at the lakes. There is also a chance of rarities such as Red-necked Phalarope and Little Ringed Plover. A good spot to view the lakes is along Digby Drive, which runs between Government, Pearse and Herschel Lakes.

Wungong Dam

Key species: Red-capped Parrot, Western Rosella, Carnaby's Black-Cockatoo, White-breasted Robin, Red-winged Fairy-wren, Red-eared Firetail, Western Thornbill
Other species: Red-tailed Black-Cockatoo, Western Corella, Square-tailed Kite

Wungong Dam is located 5 km south-east of Armadale, and 40 km south-east of Perth. In the past, it was an excellent site to see a range of the south-west endemic species close to Perth, although in recent years the reliability of some species (e.g. Red-eared Firetail) has decreased somewhat. The habitat consists of Jarrah and Marri eucalypt forests, intermixed with areas of Wandoo woodlands. To get there from Perth, take the Albany or Tonkin Highway or Kwinana Freeway to Armadale, then follow Albany Hwy through to Bedfordale, a further 6 km. Turn right into Admiral Rd then proceed 3 km to Wungong Dam. The dam is open to visitors until 5 p.m. in winter and 6 p.m. in summer.

Begin by birdwatching around the lawn and picnic area near the main entrance to the dam (on the east side of Admiral Rd, though the woodlands on the west side of the road can also be worth a quick stroll). The lawn area especially can be good for parrots and cockatoos, such as Red-capped and Elegant Parrot, Australian Ringneck, Western Rosella, Red-tailed and Carnaby's Black-Cockatoo, Western Corella and Galah. Smaller passerines here include Western, Inland and Yellow-rumped Thornbill, Western Gerygone, White-browed (Spotted) Scrubwren, Splendid Fairy-wren, Western White-naped, New Holland and Brown Honeyeater, and Western Spinebill. Although they occur in a variety of habitats, Red-winged Fairy-wren, White-breasted Robin and Red-eared Firetail show a preference for the dense riparian vegetation along the Wungong Brook (River), located south of the lawn on the west side of Admiral Rd. Red-eared Firetail also feed in the thickets and grassed areas around the main dam. Other regular species at Wungong Dam include Laughing Kookaburra, Scarlet Robin, Golden and Rufous Whistler, and Shining Bronze-Cuckoo. Birds of prey such as Wedge-tailed Eagle, Brown Goshawk and Square-tailed Kite (between September and March, but uncommon) also occur.

Mandurah district

Key species: Banded Stilt, Red-necked Avocet, Bar-tailed Godwit, Great and Red Knot, Pacific Golden Plover, Common Tern, Fairy Tern
Other species: Wedge-tailed Shearwater, Blue-billed Duck, Buff-banded Rail, Little Egret, Little Grassbird

With a population of almost 90 000, Mandurah is the second-largest city in WA, and is located 72 km south of Perth. Approximately twice the size of Sydney Harbour, the waters of the Peel Inlet and Harvey Estuary lie at the centre of Mandurah. These expanses of open water provide habitat for up to 100 000 waterbirds. There is a wide range of accommodation options in Mandurah (see www.visitmandurah.com), but the area can be visited from Perth as a day trip.

The Creery Wetlands Nature Reserve (95 ha) is home to a myriad of bird life, and is a key habitat for migratory and resident shorebirds. Located on the north-east side of the Peel Inlet (south-east of the Mandurah Bypass bridge), there are several walking trails, boardwalks and viewing platforms around the wetland. The shoreline of Creery Island provides a roosting areas for migratory shorebirds such as Bar-tailed Godwit, Eastern Curlew, Great

Knot and Pacific Golden Plover (all mainly in summer), and resident species such as Banded Stilt, Red-necked Avocet, Buff-banded Rail, Australian Pied Oystercatcher, Great and Little Egret, and Yellow-billed Spoonbill.

Take a walk along the beach front at Coodanup Reserve (98 ha), which stretches from Creery Island east along Peel Parade to the mouth of the Serpentine River, and look in the samphire shrublands bordering the shoreline for White-fronted Chat and Little Grassbird. The shoreline itself is particularly good for migratory shorebirds, with an extensive list, including Grey Plover, Common Greenshank, Red Knot, Red-necked Stint, Bar-tailed Godwit, Sanderling, Grey-tailed Tattler, and Marsh, Curlew, Sharp-tailed and Broad-billed (rare) Sandpiper. A wetland located at the south end of Coodanup at the end of John St (and known as John St Reserve, though local birds call it the 'Nairns') can support Buff-banded Rail, Spotted Crake, Fairy Tern and many of the shorebirds mentioned above. The Austin Bay Nature Reserve (1400 ha) is also excellent (though more difficult to access), with the extensive mud flats providing habitat for similar species to those found at the Coodanup. Less common and rare species that have been reported here include Grey-tailed Tattler, Long-toed Stint, Pectoral Sandpiper and Ruff. Royal Spoonbill has also been recorded here: an uncommon species in south-west WA. To get there, turn west off the Forrest Hwy onto Greenlands Rd, 17 km from Mandurah. Travel along Greenlands Rd for 1 km, before taking the dirt road for a further 1 km to the reserve.

Look for terns and shorebirds around Mandurah harbour and near the Dolphin Pool (accessed via Dolphin Drive and Breakwater Parade, west of Mandurah central area). Species included Caspian, Crested and Fairy Tern, while there are occasional records of Common and rare records of Roseate and Arctic Tern. Bar-tailed Godwit, Eastern Curlew, Ruddy Turnstone, Grey Plover and Common Sandpiper (the last on the breakwater wall) are also possible.

One of the best areas for Whimbrel and Eastern Curlew is the Erskine area, encompassing Samphire Cove, Erskine Foreshore and Soldier's Cove. Erskine Foreshore is the most convenient for access and has the added bonus of a chance for Common Tern in summer. The nearby Erskine Lakes hosted a female plumaged Northern Pintail

Northam

With records dating back to the early 1900s, Northam Weir is the only location in Australia where you can see an arguably wild population of introduced Mute Swan (though even these have been managed by the introduction of new blood). The weir is situated on the Avon River in township of Northam, 97 km east of Perth. To get there from Perth, take the Great Eastern Hwy for 92 km, and then turn onto Mitchell Ave – Northam is another 4 km. Other waterbirds at the Weir include Black Swan, Hardhead, Australian Shelduck, Blue-billed Duck, Australasian Darter, Great Egret, Yellow-billed Spoonbill, Black-fronted Dotterel and Black-winged Stilt. Western Corella is also common here.

(Australian vagrant) for extended periods in both 2010 and 2011. A small suburban wetland, this site can be accessed via Mandurah Bypass, Sticks Boulevard and Abbotswood Park Way.

Robert Point is located in Halls Head, a coastal suburb of Mandurah, on the west-side of the Mandurah Estuary. On the ocean beach at Robert Point (north end of Halls Parade), look for Greater Sand Plover, Sanderling (both summer) and Common (scarce summer), Fairy, Crested and Caspian Tern. Out to sea, look for Wedge-tailed and, occasionally, Flesh-footed Shearwater.

Lake McLarty Nature Reserve

Key species: Long-toed Stint, Pectoral Sandpiper and other shorebirds, waterbirds

Other species: Australian Spotted Crake, Red-capped Parrot, Regent Parrot (uncommon), Stubble Quail (spring-summer), Emu

Situated 90 km south of Perth on the eastern side of the Harvey Estuary, Lake McLarty Nature Reserve (219 ha) can support large numbers of waterbirds and shorebirds. Part of the Peel-Yalgorup System Ramsar site, it is also recognised as an Important Bird Area. To get there, turn west off the Forrest Hwy, 26 km south of Mandurah and travel another 4 km down Mills Rd to find the lake on your right. Lake McLarty is a shallow lake that often dries during early to mid-summer. While drying, extensive mud flats are formed that are ideal for foraging shorebirds; this is especially so in the last month or so before the lake completely dries, so up-to-date local information on the

Korung National Park and Victoria Reservoir

Around Victoria Reservoir (sometimes referred to as Victoria Dam) in Korung National Park (6000 ha), located just 25 km from Perth's CBD, with some searching, you can see 11 of the 16 south-west WA endemics. These are Carnaby's and Baudin's Black-Cockatoo, Red-capped Parrot, Western Rosella, Red-winged Fairy-wren, Western Thornbill, Western Wattlebird, Western White-naped Honeyeater, Western Spinebill, White-breasted Robin and Red-eared Firetail. To get there from Welshpool Rd East, take the Tonkin Hwy (10.5 km south of the Perth Domestic Hwy) and head 8 km east to Canning Rd. Turn right and, after 450 m, turn right again on to Masonmill Rd. After 600 m, you will see the dam access road. Take this road to the last carpark, then follow the walking tracks down to the reservoir. The best birding is along the access road and towards the bottom of the valley near the reservoir, and around the lawn area at the base of the dam. There is also a chance of uncommon species, such as Square-tailed Kite, Red-tailed Black-Cockatoo (ssp. *naso*), Spotless Crake, Western Yellow Robin and Red-browed Finch (introduced to WA and scarce here).

lake's status can be useful. The list of shorebirds is long, with Greater Sand and Pacific Golden Plover, Common Greenshank, Red Knot, Sharp-tailed, Curlew, Wood and Marsh Sandpiper, Bar-tailed and Black-tailed Godwit and Red-necked Stint all regularly seen. Generally considered rarities, both Pectoral Sandpiper and Long-toed Stint are also recorded here almost every year, while, less frequently, Broad-billed Sandpiper and Ruff appear. True rarities have included Little Ringed Plover, Inland Dotterel, Little Stint, Asian Dowitcher, Latham's Snipe and Oriental Pratincole. The lakes can also support large numbers of waterfowl. Freckled (very rare), Pink-eared, Musk and Blue-billed Duck, Hardhead and Australasian Shoveler are all regular. Other waterbirds recorded include Australian Spotted Crake, Nankeen Night Heron (scarce), Black-winged and Banded Stilt, Red-necked Avocet, and Whiskered Tern, while less common species have included Royal Spoonbill and White-winged Black Tern (the last occurs almost annually and is most likely in summer). Banded Lapwing can be common in nearby farmland (e.g. along Greenlands Rd), but are usually rare on the lake itself.

Stands of paperbark fringe the north and east margins of the lake, while the western shore supports open eucalypt woodlands. Look here for Red-capped and Regent Parrot, Australian Ringneck, Shining Bronze-Cuckoo, Laughing Kookaburra, Sacred Kingfisher, Rainbow Bee-eater (summer), Splendid Fairy-wren, Western White-naped Honeyeater (uncommon, mostly winter) and Black-faced Woodswallow, and raptors such as Osprey, White-bellied Sea-Eagle, Little and Wedge-tailed Eagle, Swamp Harrier and Peregrine Falcon.

Bungendore Park Bushland Reserve

Key species: Rufous Treecreeper, Western Yellow Robin, Red-capped Parrot, Western Rosella, Western Thornbill

Other species: Square-tailed Kite, Western Spinebill, Western Wattlebird, Western White-naped Honeyeater

Bungendore Park Bushland Reserve (498 ha) is situated on the western edge of the Darling Scarp, with sweeping views to the Swan Coastal Plain. Owing to the large number of south-west WA endemic bird species, the reserve, located just off the Albany Hwy, is a convenient location to visit if undertaking the longer drive to Albany. The reserve is located 3 km south-east of Armadale and can be accessed from Admiral Rd (just past the Bedfordale Hall and the turn-off to Armadale Christian College). The vegetation in the park consists mainly of open Jarrah-Marri forest, with several smaller areas of Wandoo woodland and heathland and *Banksia* thickets surrounding granite outcrops.

The best birding spots in Bungendore Park are on Dryandra Drive and Casuarina Rd, and along the many excellent walking trails, most of which have bird names, such as Whistler Walk, Robin Ramble, Spinebill Stroll and Honeyeater Hike. A good area for Western Yellow Robin is the woodlands on the southern side of the Dryandra Drive, 100 m from the entrance (entering from Admiral Rd). Listen for their pleasant call (a series of slow piping whistles) ~50 m in from the road. Rufous Treecreeper is generally uncommon in the carpark, but, with a bit of effort, they can be found along Casuarina Rd. They tend to prefer Wandoo woodland and mixed Jarrah–Marri–Sheoke woodland, especially those areas that are long unburnt. Along any of the reserve tracks, you may see the WA endemics such as Red-capped Parrot, Western Rosella, Western Thornbill, Western White-naped Honeyeater, Western Wattlebird and Western Spinebill. Carnaby's and Baudin's Black-Cockatoo also occur here, with Carnaby's and Red-tailed Black-Cockatoo

Jarrah–Marri woodland with grasstrees at Bungendore Park. This forest type provides habitat for Red-capped Parrot, Western Rosella, Western Thornbill and Rufous Treecreeper. When the woodlands are in flower, look also for Western White-naped Honeyeater, Western Wattlebird and Western Spinebill.

breeding locally, while Baudin's Black-Cockatoo breed further south, but move north in the autumn and winter months to feed and roost. Elegant Parrot is occasionally recorded in ones and twos along the edges of watercourses. Other birds here include Varied (black-capped) Sittella, Scarlet Robin, Spotted and Striated Pardalote, Inland and Yellow-rumped Thornbill, Weebill, Western Gerygone, White-cheeked Honeyeater, Splendid Fairy-wren, Wedge-tailed Eagle (mainly southern edge of the park), Little Eagle, Square-tailed Kite (spring and summer), Brown Goshawk and Peregrine Falcon. Mammals at Bungendore Park include Western Grey Kangaroo, Western Brush Wallaby and Western Quoll (or Chuditch).

Western Rosella are relatively common through south-west WA and occur in most of the larger reserves that protect woodlands around the fringes of Perth.

Wearne Rd

Key species: Western Yellow Robin, Rufous Treecreeper, Blue-breasted Fairy-wren, Western White-naped Honeyeater

Wearne Rd is an excellent birding site, likened by some to 'Dryandra in miniature' (see p. 402). Wearne Road is on the left ~37 km south of the corner of Jarrahdale Road and Albany Highway (or ~62 km south-east of Armadale) and just after the sign 'North Bannister 2 km'. The road itself is poorly signed and strikes east from the highway: it is suitable for 2WD vehicles, although it is sometimes a little corrugated. The woodlands on the roadside are becoming well known as one of the better Wandoo woodlands sites easily accessible off Albany Hwy and are a convenient stop for travellers on route to and from Albany. Look in the roadside woodlands where there are good opportunities for Purple-crowned Lorikeet, Elegant Parrot, Rufous Treecreeper, Hooded and Western Yellow Robin, Painted button-quail, Restless Flycatcher, Blue-breasted Fairy-wren, and Western White-naped and Yellow-plumed Honeyeater. Crested (Western) Shrike-tit, although rare, has been recorded here.

Dryandra Woodlands

Key species: Western Yellow Robin, Rufous Treecreeper, Blue-breasted Fairy-wren, Western Thornbill, Crested (Western) Shrike-tit
Other species: Bush Stone-curlew, Elegant Parrot, Red-capped Parrot, Purple-crowned Lorikeet, Painted Button-quail, Hooded Robin, Restless Flycatcher

The Dryandra Woodland (28 000 ha) is one of the prime places in south-west WA for viewing many of the endemic bird species. Located 164 km south-east of Perth, to get there from the Albany Hwy, turn left at North Banister along Wandering Rd (100 km south of Perth) and continue for 67 km to reach the main entrance. The reserve is also signposted on the Great Southern Hwy at Cuballing and Narrogin. Accommodation is available at the Dryandra Village, located off Tomingley Rd, 3 km from the park entrance. At the Village, there are self-contained cottages and dormitories. For

Serpentine National Park

Serpentine National Park (4500 ha) is 50 km south-east of Perth. To get there, travel 26 km south of Armadale, and then take the South West Hwy into Falls Rd; a small entry fee is payable. Birdwatching around the Serpentine Dam Picnic Area provides an opportunity to observe species that frequent Jarrah forest, including Baudin's Black-Cockatoo, Western Rosella, Red-capped Parrot, Red-winged Fairy-wren, Western Spinebill and Western White-naped Honeyeater. Other species that are sometimes present include Red-tailed Black-Cockatoo, Australian Ringneck, Splendid Fairy-wren, Golden Whistler and Scarlet Robin. The short walk to the Serpentine Falls (200 m) is worthwhile, with Red-eared Firetail occasionally seen below the falls.

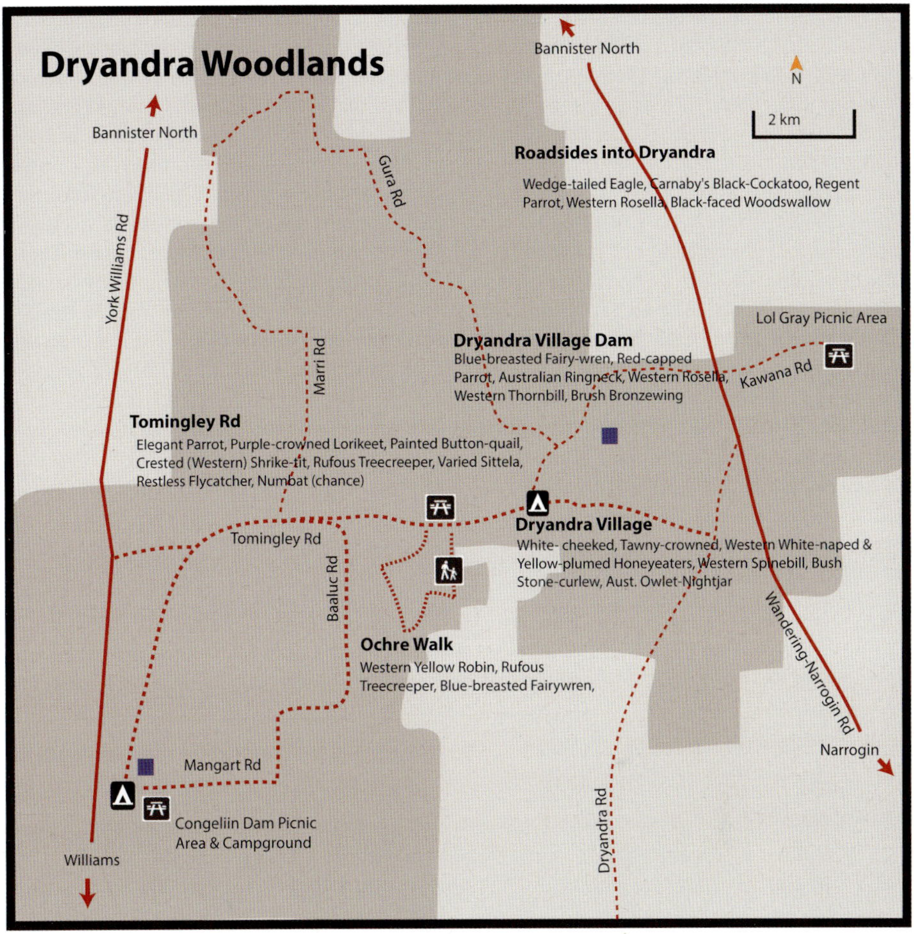

bookings (08) 9884 5231. There is also camping at the DPaW managed Congelin Campground. The woodlands consist of extensive stands of eucalypt, including Wandoo, Jarrah and Marri, and an understorey of native shrubs and spectacular wildflowers in spring. Dryandra Woodlands is a very good area for native mammals, including Numbat, Woylie, Bilby, Brush Wallaby, Tammar Wallaby and Western Grey Kangaroo. A re-introduction program for several Endangered mammal species is also underway and guided spotlighting tours of the open range pens that hold Burrowing Bettong, Rufous Hare Wallaby and Western Barred Bandicoot can be booked at the Village. On your way into Dryandra, look along the roadsides for Wedge-tailed Eagle, Carnaby's Black-Cockatoo, Western Rosella and Black-faced Woodswallow.

Dryandra Village is an excellent area for honeyeaters, especially in the morning, when they can be seen drinking from the caretaker's birdbath. Here, there is a chance of New Holland, Singing, White-cheeked, White-eared, Tawny-crowned, Western White-naped, Yellow-plumed, Brown and Brown-headed Honeyeater, as well as

Western Spinebill. At night, search around the village for Bush Stone-curlew, Southern Boobook, Eastern Barn Owl, Australian Owlet-nightjar and Tawny Frogmouth. There are also several recent records of Masked Owl in the reserve, which is otherwise rare in south-west WA.

Dryandra Village Dam is located ~1 km north-east of the village on Kawana Rd. The best time to visit is just after dawn, when large numbers of Red-capped Parrot, Australian Ringneck and mixed flocks of honeyeaters (Tawny-crowned, Yellow-plumed, White-eared and Brown) come into drink. The dam is also excellent place to find the Blue-breasted Fairy-wren. A secretive species, search the low shrubs in the south and east of the dam. Other birds to look for here include Brown Goshawk, Western Thornbill, Brush Bronzewing, Western Gerygone, White-browed (Spotted) Scrubwren and Grey (Western) Shrike-thrush.

Ochre Walk and Arboretum is located 3 km west of Dryandra Village on Tomingly Rd. The Dryandra Arboretum is a collection of native trees at the start of the Ochre Walk (5 km return). Look here for Rufous Treecreeper. Western Yellow Robin is found along the start of the walk, especially the woodlands ~150 m south of Tomingly Rd (near where you can see a small ridge line on the east side of the track).

One of the best strategies for birdwatching in Dryandra is to travel up and down

Wandoo woodland along Tomingley Rd, Dryandra. This is a preferred habitat of the elusive Crested (Western) Shrike-tit. Other species that occur include Painted Button-quail, Elegant Parrot, Purple-crowned Lorikeet, Western Thornbill, Rufous Treecreeper, White-browed Babbler, Restless Flycatcher, Scarlet Robin, Jacky Winter and Dusky Woodswallow.

Wagin Treatment Plant and Lakes

On the way down to Albany, it is worth visiting the lakes and treatment plant at Wagin. The treatment plant can be accessed 1 km from the centre of Wagin off the Great Southern Hwy, via Vine St, followed by a right turn into Cowcher Rd; the treatment plant is on your left. Look here for Black-tailed Native-hen, Australian Spotted Crake, Wood Sandpiper (summer), Chestnut Teal and Black-fronted Dotterel. The plant is fenced, but good views can be obtained from the road. When holding water, the lakes just south of Wagin (including Parkeyerring, Little Parkeyerring, Quarbing and Little Norring) may host Hooded Plover, Banded Stilt, Common Greenshank, Red-necked Stint, Sharp-tailed and Curlew Sandpiper, and Gull-billed and Whiskered Terns.

Tomingley Rd a few times. Due to the quality and openness of the Wandoo eucalypt trees along the road, you have a good chance of seeing a wide range of the woodland species. Stop anywhere there is bird activity and look for species such as Sacred Kingfisher, Elegant Parrot, Purple-crowned Lorikeet, Painted Button-quail, Crested (Western) Shrike-tit, Restless Flycatcher, White-winged Triller, Varied (black-capped) Sittella, White-browed Babbler, Rufous Treecreeper, Dusky Woodswallow, Scarlet and Red-capped Robin, Jacky Winter, Rainbow Bee-eater (summer), Yellow-plumed Honeyeater, Western and Inland Thornbill, Grey Butcherbird and Tree Martin. Raptors to look for include Collared Sparrowhawk and Square-tailed Kite (rare). Although rare, at Dryandra, there is a chance of Malleefowl. Tomingley Rd is also probably your best chance of seeing the rare Numbat (WA's faunal emblem). A diurnal species, keep your eye on the open wooded gullies, mainly on the south side of the road. Other recommended birding trails in Dryandra are the Wandoo Walk (1 km return), particularly around the Old Mill Dam and the Woylie Walk (5.5 km return).

Donnelly River

Key species: Baudin's and Carnaby's Black-Cockatoo, Rufous Treecreeper, White-breasted Robin, Western Rosella, Red-winged Fairy-wren, Western Spinebill, Western White-naped Honeyeater, Crested (Western) Shrike-tit

Near Bridgetown, the Donnelly River cottages are a great family-oriented birding site, with most of the typical Karri forest birds to be found around the village. Formerly an old mill village, semi-tame wild Emu and Western Grey Kangaroo roam the grounds, while Australian Ringneck Parrots ('Twenty-eights') come to verandahs to be hand fed. Species to be expected within the village grounds, and in the adjacent, forest include Purple-crowned Lorikeet, Baudin's and Carnaby's Black-Cockatoo, Rufous Treecreeper, White-breasted and Scarlet Robin, Western Rosella, Red-winged and Splendid Fairy-wren, Golden Whistler,

In many reserves, confiding White-breasted Robins are a feature of picnic areas and carparks. Good sites include Walpole-Nornalup National Park, Donnelly River, Porongurup National Park, Two Peoples Bay Nature Reserve and Cheynes Beach.

Grey Currawong, Western Spinebill and Western White-naped Honeyeater. The Crested (Western) Shrike-tit has also been recorded in the forest immediately behind the cottages. With such a diversity of species in close proximity to accommodation and other activities (a forest swimming lake, tennis, basketball and beach-volleyball courts and a flying fox), this can make for a very relaxing birding experience. There are also records of Masked Owl nearby along Rifle Range Rd (Bridgetown area is apparently a hotspot for Masked Owl in south-west WA), and Red-eared Firetail occur around many of the nearby creek crossings. Visit www.donnellyriver.com.au

Walpole-Nornalup National Park

Key species: Red-capped Parrot, Western Rosella, Rufous Treecreeper, White-breasted Robin, Red-eared Firetail, Black Bittern (rare)
Other species: Carnaby's and Baudin's Black-Cockatoo, Western White-naped Honeyeater, Crested (Western) Shrike-tit

Walpole-Nornalup National Park (20 000 ha) is best known for the towering Red Tingle and Karri trees. It is located 430 km south of Perth on the South Coast Hwy, and surrounds the townships of Walpole, Nornalup and Peaceful Bay.

The park features the Valley of the Giants Tree Top Walk (600 m) – a spectacular lightweight bridge that ascends 40 m above the forest floor, thus providing an excellent vantage point for viewing birds in the forest canopy. The walk then links up with the Ancient Empire boardwalk. Search the surrounding forest for Carnaby's and Baudin's Black-Cockatoo, Purple-crowned Lorikeet (particularly when the eucalypt trees are flowering), Australian Ringneck, Red-capped Parrot, Western Rosella, Fan-tailed Cuckoo, Laughing Kookaburra, Rufous Treecreeper, Western White-naped Honeyeater, Spotted and Striated Pardalote, Golden Whistler, Scarlet and White-breasted Robin, Red-winged and Splendid Fairy-wren, and Red-eared Firetail. Similar species occur along the circular Hilltop Rd, and near the Hilltop Lookout, both accessed 2.5 km east of Walpole on the South Coast Hwy. The road passes through an impressive Red Tingle forest. It is also worth birdwatching along the Giant Tingle Tree Loop Walk (800 m), located 2 km past the Hilltop Lookout.

The Knoll Scenic Drive (4 km) is an attractive circular scenic road that passes Nornalup and Walpole Inlets and Coalmine Beach. Described as where 'the forest meets the water', it is accessed 3 km east

The Tree Top Walk is a nature-based tourism attraction located in the heart of the Valley of the Giants. This provides an excellent vantage point from which to scan the surrounding forest: Square-tailed Kite is just one of the possibilities.

of Walpole along South Coast Hwy. One productive approach here is to park in the carpark at the end of Knoll Scenic Drive, and walk around the circular section, scanning the forest bordering the road. Look for Red-tailed and Baudin's Black-Cockatoos, Crested (Western) Shrike-tit, Varied Sittella, Rufous Treecreeper and Spotted Pardalote. In the Peppermint woodlands near the Coalmine Beach Holiday Park, located at the beginning of the Knoll Scenic Drive, you may see Western Rosella, Brush and Common Bronzewing, Red-winged Fairy-wren, Scarlet and White-breasted Robin, and Red-eared Firetail. The coastal heathland areas in the park, such as those along Bellanger Beach, support Southern Emu-wren. Great Crested Grebe occur on the larger waterways within the park. Although very rare in south-west WA, there is a possibility of Black Bittern: there have been several records in the area. Nearby, there is a recent one-off record of Western Ground Parrot at Conspicuous Beach.

Lake Muir Nature Reserve

Key species: Australasian Bittern, Australian Little Bittern, Australian Spotted Crake, Western Corella
Other species: Western Yellow Robin, White-breasted Robin

Lake Muir Nature Reserve lies near the Muirs Hwy, 55 km south-east of the town of Manjimup. It is a large (4600 ha) seasonal lake (often dry by autumn) surrounded by sedgelands, tea tree thickets and eucalypt woodland. There is a range of lagoons within the reserve, including Lake Muir (easily accessed via the Muirs Hwy), Tordit-Gurrup and Byenup Lagoons (east of Lake

Muir), and Red Lake and Cowerup Swamp (north of Lake Muir).

Lake Muir is a significant drought refuge for large numbers of waterbirds; it is regularly used by 20 000 waterbirds with up to 51 000 having been counted. You may see Banded Stilt, Eurasian Coot, Yellow-billed Spoonbill and White-faced Heron, and waterfowl such as Blue-billed, Freckled (rare) and Pacific Black Duck, and Grey and Chestnut Teal. Lake Muir is regularly used in autumn as a moulting site by thousands of Australian Shelduck, and is one of their most important moulting sites in southern WA. Counts of up to 10 Australasian Bittern have been made in the area and the species probably breeds in the surrounding sedge swamps such as Kulinilup, Tordit-Gurrup and Byenup. Australian Little Bittern and Australian Spotless Crake are among several other significant waterbirds that have been recorded breeding. Lake Muir is also used by small numbers of shorebirds, notably Red-necked Stint, when conditions are right.

The area around Lake Muir is a reliable place to see 'Muir's Corella', an isolated subspecies (*pastinator*) of the Western Corella. Formerly inhabiting open woodlands throughout south-west WA, following clearance for agriculture, the number of Muir's Corella was reduced at one stage to a perilous 100 individuals, though, fortunately numbers have since increased somewhat. Look for them along Thompson's Rd on the west side of the lake, and Hanekamp Rd on the east side. The woodlands around the lake also support an interesting range of other parrots and cockatoos. These include Carnaby's and Baudin's Black-Cockatoo,

Vasse-Wonnerup Estuary

Close to Busselton, the Vasse-Wonnerup Estuary (17 800 ha) is recognised as a wetland of international importance under the Ramsar Convention and is part of the Busselton Wetlands Important Bird Area. At times over 20 000 waterbirds make use of the estuary. A series of nature trails (to be known as the Busselton Wetlands Trails) are being developed to provide access points for birdwatchers – for information, contact the Shire of Busselton (www.busselton.wa.gov.au). Look along the edge of the estuary for Black-winged and Banded Stilt, Red-necked Avocet, Australian Darter, Glossy Ibis, Yellow-billed Spoonbill, Buff-banded Rail, Little Egret and Nankeen Night-Heron, and waterfowl such as Australasian Shoveler, and Blue-billed and Musk Duck. The Vasse-Wonnerup Estuary also has WA's largest Black Swan breeding colony. In summer, the estuary attracts migratory shorebirds such as Pacific Golden and Greater Sand Plover, Black-tailed Godwit and, occasionally, Wood and Terek Sandpiper. Several rarities have been recorded here, including Northern Pintail (Sept 2012) and Franklin's Gull (Feb 2009). Access can at times be difficult, given the vastness of the estuary. One good area is the mud flats visible from Coast Rd, just east of Layman Rd.

Red-tailed Black-Cockatoo, Purple-crowned Lorikeet, Western Rosella, and Port Lincoln and Elegant Parrot. Other woodland species recorded here include Brown Quail, Square-tailed Kite, Western Spinebill, New Holland and Western White-naped Honeyeater, Southern Emu-wren, White-winged Triller, Restless Flycatcher, and Western Yellow, Scarlet and White-breasted Robin.

If you do not see 'Muir's Corella' at Lake Muir, they are also found near the small township of Rocky Gully, located along the Muirs Highway, 35 km east of Lake Muir (65 km west of Mount Barker). Here, good places to look for them include the northern side of Muirs Highway, 1 km west of the township (near where the highway intersects with Mills Rd), and on the west side of the Rocky Gully Frankland Rd, accessed 500 m west of town.

Leeuwin-Naturaliste National Park

Key species: seabirds, Red-tailed Tropicbird (slim chance), Rock Parrot, Red-eared Firetail
Other species: Hooded Plover, Bridled Tern, Fairy Tern, Eastern Reef Egret, White-breasted Robin (also NZ Fur Seal)

Leeuwin-Naturaliste National Park (9800 ha) stretches 120 km between Cape Naturaliste in the north to Cape Leeuwin in the south. Cape Naturaliste is 36 km from Busselton and 261 km from Perth. The park features a rugged coastline of cliffs and rocky shorelines, and supports an area of Banksia, Jarrah and Marri woodland. There are three campgrounds with basic facilities at Contos, Point Road and Boranup. There is also a wide range of accommodation options in Dunsborough, Yallingup, Margaret River and Augusta. Most roads in the area are sealed and all gravel roads are suitable for 2WD.

Cape Naturaliste Lighthouse, located 13 km north-west of Dunsborough, stands on a 100 m bluff that overlooks Geographe Bay, providing panoramic views of the Indian Ocean. Walk out onto the cape and scan the ocean for seabirds such as Australasian Gannet, Southern and Northern Giant-Petrel, Indian Yellow-nosed and Shy Albatross, Flesh-footed, Wedge-tailed (spring–summer), Little and Hutton's Shearwater (especially Aug), Soft-plumaged and Great-winged Petrel, and Pomarine and Arctic Jaeger (both Oct–Apr). Around the lighthouse, look for Brush Bronzewing, Carnaby's and Baudin's Black-Cockatoo, Red-eared Firetail, Splendid Fairy-wren and White-browed (Spotted) Scrubwren. The Cape Naturaliste Track (3.2 km) travels through small limestone 'pinnacles' on route to a small whale watching lookout to the west (another more substantial platform is situated to the north of the lighthouse), with the best time for Humpback Whale from September to December and Southern Right Whale from June to October. The lower heath here can be excellent for Southern Emu-wren. Finally, take the dirt track to the right, just before the Lighthouse Ring Rd, down to the beach (suitable for 2WD with care). This provides better access to the beach where there are usually Hooded Plover immediately south of the Cape. Other interesting shorebirds such as Grey-tailed Tattler and Eastern Curlew occur occasionally.

Sugarloaf Rock Nature Reserve, just south of Cape Naturaliste, is a small rocky nature reserve that provides habitat for nesting seabirds. It is a spectacular feature in its own right and, in the past, Red-tailed Tropicbird (as many as 34 pairs) have bred here between spring and early autumn, representing the most southerly breeding location for the species in the world. Sadly, there have been few recent breeding records (the colony was abandoned in the mid to late 2000s), and now only single birds are reported from time to time. The turn-off for Sugarloaf Rock is off Cape Naturaliste Rd, 3 km before you get to the Cape. Aside from a chance of Red-tailed

Sugarloaf Rock in Leeuwin-Naturaliste National Park, until recently hosted a small colony of Red-tailed Tropicbird. There remains a slim chance of a tropicbird, but the viewpoint provides opportunities for other seabirds, including Indian Yellow-nosed Albatross, Flesh-footed Shearwater and Australian Gannet.

Tropicbird, many of the seabirds possible off Cape Naturaliste may also be seen over the waters around Sugarloaf Rock (though they tend to occur at greater distances from this observation point).

Meelup Regional Park (577 ha) is located on the Geographe Bay side of Cape Naturaliste. Along the coastal section of the park, look for dry country species such as Western Rosella, White-breasted and Scarlet Robin, Golden Whistler and Splendid Fairy-wren. The park is accessed via Meelup Beach Rd, with the turn-off located on Cape Naturaliste Rd, 9 km before the lighthouse.

Cape Leeuwin, located where the Southern Ocean meets the Indian Ocean, provides an excellent opportunity for seabird watching. Potential highlights include Bridled and Fairy Tern, Flesh-footed, Hutton's and Little Shearwater, White-chinned and White-headed Petrel, Southern and Northern Giant-Petrel, and Indian Yellow-nosed, Black-browed and Shy Albatross. Around the rocky bays at the Cape, you may see Sooty and Australian Pied Oystercatcher, and Eastern Reef Egret. The Lighthouse grounds are open to public, though an entrance fee is payable. Look for Rock Parrot feeding on the lawns of the lighthouse, and between the entrance building and the lighthouse, while Southern Emu-wren and Splendid Fairy-wren occur in the surrounding heathland. Rock Parrot can also be found outside the lighthouse fence (no fee payable), near the freshwater pool below the 'petrified waterwheel'.

The Boranup Karri Forest is a small isolate of the more extensive Karri forest a 100 km drive to the east. It lies between Caves Rd and the coast. One recommended walk is to the Boranup Lookout. Along here, observers have a reasonable chance of Western Rosella, White-breasted Robin,

Porongurup National Park

Located 20 km east of Mount Barker, the stands of Karri trees around the Tree-in-the-Rock day-use area in the Porongurup National Park are a very reliable area for Rufous Treecreeper. The park is also a good spot for parrots and cockatoos, including Purple-crowned Lorikeet, Australian Ringneck, Elegant and Red-capped Parrot, Western Rosella, and Red-tailed, Carnaby's and Baudin's Black-Cockatoo; indeed, the road up to Tree-in-the-rock day use area is one of the best for the last species in the region. Other birds you may find here include Western Yellow and White-breasted Robin, Red-winged Fairy-wren, Western and Inland Thornbill, White-browed Scrubwren, Western Gerygone and Striated Pardalote.

Western White-naped Honeyeater, Rufous Treecreeper and Red-winged Fairy-wren. Other species include Red-tailed, Baudin's and Carnaby's Black-Cockatoo. Crested (Western) Shrike-tit has also been reported within this small isolate from time to time; listen especially for the sounds of bark being stripped and lifted from eucalypts. Access to Boranup Forest is a short drive south of Margaret River via Caves Rd.

Two Peoples Bay Nature Reserve

Key species: Noisy Scrub-bird, Western Bristlebird, Western Whipbird
Other species: Carnaby's Black-Cockatoo, White-breasted Robin, Southern Emu-wren, Red-eared Firetail

Two Peoples Bay Nature Reserve (4750 ha) is 35 km east of Albany, and is reached via Two Peoples Bay Rd. The habitat at Two People Bay is largely heathland and shrubland. In 1961, the bay became internationally renowned as the site of the rediscovery of the then presumed extinct Noisy Scrub-bird. Western Bristlebird and Western Whipbird also occur here, along with a good selection of other heathland species. Two Peoples Bay is also home to Australia's most Critically Endangered mammal, the Gilbert's Potoroo, with just ~40 individuals left in the wild. Like the Noisy Scrub-bird, it was thought to be extinct until it was re-discovered on the slopes of the reserve in 1994. There is an excellent visitor centre located on Two People Bay Rd. Please note

The Noisy Scrub-bird is widely recognised as one of the most elusive passerines in Australia. The combination of a deafening song combined with a skulking (almost ghost-like) behaviour invariably leaves many birders frustrated – 'it was right in front of me yet I couldn't see it'.

that call playback to attract birds is not allowed in the reserve. To avoid damage to highly sensitive habitats, visitors must also keep to the walking tracks. There is no on-site accommodation or camping allowed, but the reserve is within easy reach from both Albany and Cheynes Beach.

The area between Little Beach and the visitor centre is a well known location for Noisy Scrub-bird, Western Whipbird and Western Bristlebird. All three species are most active at dawn and dusk. Concentrate you search around the carpark at Little Beach, particularly the shrublands above here. Western Bristlebird prefers the very low coastal heath, whereas the other two species tend to occupy taller, denser shrubs. The bristlebird is occasionally seen running across the road to the traffic island in the centre of the carpark. Another good spot to find the three skulking species, particularly Noisy Scrub-bird, is along a fire trial 400 m west of the Little Beach carpark. It heads south, and up into the heath. Concentrate your search near the junction in the track ~200 m from the road. Other birds to look for here, and in the heath around the carpark, include Rock Parrot, Southern Emu-wren, White-breasted Robin, Tawny-crowed Honeyeater, Western Wattlebird,

The flora-rich coastal heathlands at Two Peoples Bay and Cheynes Beach not only support a number of bird species that are endemic to the south-west: the tiny Honey Possum also occurs in these habitats and may occasionally be spotted during the day visiting the nectar-rich flowers of *Banksia* spp. and *Adenanthos* spp., among others.

Betty's Beach Reserve

A particularly good location for the Western Whipbird (ssp. *nigrogularis*), Betty's Beach Reserve (80 ha) is located on the edge of North Point facing Two Peoples Bay. To get there, turn south off the South Coast Hwy onto Homestead Rd, 35 km east of Albany. After 12 km, turn south again onto Betty Beach Rd and follow this for a further 4 km. The best place to look for Western Whipbird is on Boulder Hill some 750 m west of the beach. Here, walk north off Betty's Beach Rd up the Boulder Hill Track (4WD only). The whipbird tends to prefer the areas where scrub-thickets meet open heath. Betty's Beach Reserve is also a reliable place to find Rock Parrot and, less commonly, Noisy Scrub-bird and Western Bristlebird. Other species include Western Rosella, Southern Emu-wren, Dusky Woodswallow, Tawny-crowned and New Holland Honeyeater, Western Spinebill, White-browed (Spotted) Scrubwren, Tree Martin and Red-eared Firetail.

White-browed (Spotted) Scrubwren and Red-eared Firetail (occasionally seen near the toilet block). Along the coastline and out into Two Peoples Bay, you may see Flesh-footed Shearwater (summer), Australasian Gannet, Pacific Gull, Sooty Oystercatcher, Eastern Reef Egret and Brown Skua (winter).

Another location worth investigating is along the 4WD fire trail (5 km return, marked with a 'water' sign) that heads south to Sinker Reef off Two People Bay Rd, ~500 m west of the visitor centre carpark. Birdwatch in the heath along the edge of the track, looking and listening for Noisy Scrub-bird, Western Bristlebird, Red-capped Parrot, Carnaby's Black-Cockatoo, Brush Bronzewing, Rufous (Western) Fieldwren, Golden Whistler, Western Spinebill, White-cheeked and Tawny-crowned Honeyeater, Shining Bronze-Cuckoo and Fan-tailed Cuckoo. A fruitful area is around the carpark, 1 km from the Two People Bay Rd, and also investigate the fire trail (located a further 300 m down the track; it branches east for several kilometres before reaching the coastline. If you don't see Noisy Scrub-bird at the above locations, look near Lake Gardiner, located 1.6 km west of the visitor centre. Noisy Scrub-bird, Western Bristlebird and Western Whipbird also occur at Cheynes Beach (below).

Cheynes Beach

Key species: Noisy Scrub-bird, Western Bristlebird, Western Whipbird, Western Ground Parrot (very rare)
Other species: Red-eared Firetail, White-breasted Robin, Southern Emu-wren, Carnaby's Black-Cockatoo

The small coastal township of Cheynes Beach is 404 km south-east of Perth, and 65 km east of Albany. Cheynes Beach is a superb birdwatching location, where, with luck and patience, you can catch a glimpse of the notoriously skulking trio of Noisy Scrub-bird, Western Bristlebird and Western Whipbird (ssp. *nigrogularis*). All three species are found in Arpenteur Nature Reserve (88.9 ha), located immediately east of the Cheynes Beach Caravan Park. Bordered by the Southern Ocean, coastal heathland and shrublands are the dominant habitat around the town. Due to the risk of damage to highly sensitive habitat through both trampling, but also the spread

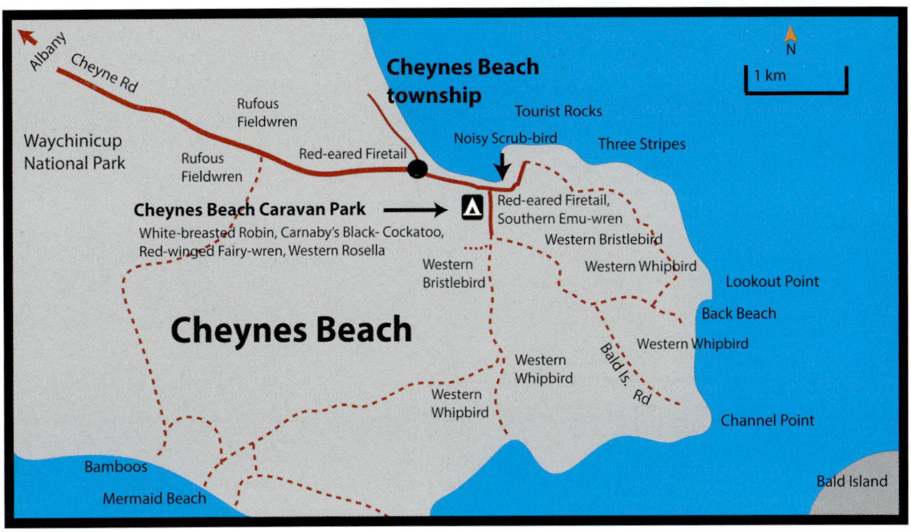

Western Australia

of 'dieback' (*Phytophthora cinnamomi*), please stay on the roads and walking tracks in the area. Facilities at Cheynes Beach include accommodation at the Cheynes Beach Caravan Park (ask the owners to view the dossier of recent bird sighting information and, if you make any interesting sightings, contribute to it yourself for future visitors), and there is a small bush camping area near the inlet of the Waychinicup River.

To obtain a good view of the Noisy Scrub-bird is one of the great challenges of Australian birdwatching. The species occupies dense, long-unburnt vegetation along Cheynes Beach Rd, particularly the section between the Bald Island Rd (the road to the caravan park) and the boat ramp. One of Australia's loudest songbirds, a good technique for locating them is to walk along the road just after dawn and listen for their call; easily recognised by its great volume! Once located, stop, wait and watch from a distance in the hope the bird moves into the open (including dashing across the road). Along the road, they often call near the first culvert (an enclosed drain under the road) and, further down, near a small track that heads north to the beach. A good spot to search for Western Bristlebird is the heath on the east side of Bald Island Rd, particularly were it branches, ~150 m from the entrance to the caravan park. (Bald Island Rd becomes a walking track just after the entrance to the caravan park.) Western Bristlebird have also been recorded along a short walking track that runs north, directly behind the caravan park. Western Whipbird has been seen in the heath along Bald Island Rd, mainly on the north side of the road slightly further south, on the track to Back Beach and Lookout Point.

Aside from the three skulking species, a host of other species can also be seen at Cheynes Beach. White-breasted Robin are

This gravel access road at the end of Cheynes Beach Rd is one of the more reliable sites for Noisy Scrub-bird. An excellent strategy to obtain a sighting is to simply position yourself near to where the photographer was standing for this photograph and watch and wait patiently for a territory holder to make the dash across the road.

common around the caravan park; look near the entrance and around the cabins in the north-east corner. They also occur along the tracks that lead to Cheynes Beach Rd. Red-eared Firetail may be seen along the beginning of Bald Island Rd, particularly between the road and caravan park (and they occasionally perch on the powerlines within the park). Carnaby's Black-Cockatoo roost and feed in the large pines in the centre of the caravan park. Southern Emu-wren (ssp. *westernensis*) occur in the heath near the intersection of Cheynes Beach Rd and Bald Island Rd, and around the rocks at the end of Cheynes Beach Rd. Purple-crowned Lorikeet can be quite common; they feed in flowering eucalypts on the east side of the caravan park. Other birds you may see include Brush Bronzewing, Wedge-tailed Eagle, Spotted Harrier, Black-shouldered Kite, Western Spinebill, White-checked, New Holland and Tawny-crowned Honeyeater, Fan-tailed Cuckoo, Red-winged Fairy-wren, White-browed (Spotted) Scrubwren, Inland Thornbill and Dusky Woodswallow. At night, listen for Tawny Frogmouth and Southern Boobook, while in spring and summer Spotted Nightjar is occasionally seen along Cheynes Rd. Some of the mammals found at Cheynes Beach include Quokka, Western Ringtail Possum, Honey Possum (occasionally feeding on flowering *Banksia* during daylight hours), Western Brush Wallaby, and Quenda (Southern Brown Bandicoot).

On the way into Cheynes Beach, you pass through Waychinicup National Park (4200 ha). Stop and search good-looking areas of heath in the park for Rufous (Western) Fieldwren, Southern Emu-wren and Red-winged Fairy-wren. A particularly good spot to search is around the turn-off of a track that heads south, 2 km west of the Cheynes Beach Caravan Park. The Endangered Western Ground Parrot has been reported here, though the last record appears to have been in 2004 and it is now thought to be locally extinct.

Albany Region

Key species: Red-winged Fairy-wren, Red-eared Firetail, White-breasted Robin, seabirds, shorebirds
Other species: Southern Emu-wren

Located on the northern edge of Princess Royal Harbour, Albany is a port city ~420 km south-east of Perth. Offshore islands near Albany provide breeding sites for seabirds such as Flesh-footed Shearwater, Great-winged Petrel and Little Penguin. The area usually also has a good selection of shorebirds and waterbirds. See the Albany Visitor Centre: www.amazingalbany.com.au, info@amazingalbany.com, (08) 9841 9290.

The windswept coastline and rugged headlands at the Gap provide a vantage point for superb seabird watching. The Gap is located in Torndirrup National Park (4000 ha) on the southern side of Frenchman Bay, 20 km south-west of Albany. To get there from Albany, take Frenchman Bay Rd for 16 km before a right turn into The Gap Rd. The Gap is another 2 km; here, there are several excellent vantage points for seabird watching, with the best time for viewing when there are a strong onshore winds and heavy seas. If it is too windy here, try the nearby Cave Point Lighthouse, which offers more shelter for observers. Some of the seabirds you may see include Australasian Gannet, Flesh-footed, Hutton's and Little Shearwater, Great-winged and Soft-plumaged Petrel, Southern Giant-Petrel, and Shy, Black-browed and Indian Yellow-nosed Albatross. Offshore islands here provide breeding sites for birds such as the Flesh-footed Shearwater, Great-winged Petrel and Little Penguin. On your way to the Gap, stop and check the shallow sand flats at Rushy Point on the south-west shore of Princess Royal Harbour. During the summer months, shorebirds recorded here include Ruddy Turnstone, Grey-tailed Tattler, Common Greenshank, Bar-tailed Godwit, Great

Knot, Red-necked Stint. More rarely, Long-toed Stint, Grey and Pacific Golden Plover, and Greater Sand Plover may also occur. Double-banded Plover is an uncommon winter visitor. Resident species here include Sooty Oystercatcher, Black-winged Stilt, and Fairy, Crested, and Caspian Tern. Rushy Point is located in Little Grove, which is accessed via Bay View Drive from Frenchman Bay Rd, 7 km south of Albany. The adjacent bushland is a good spot to look for Red-eared Firetail and there are occasional records of Crested (Western) Shrike-tit.

Lake Powell Nature Reserve (143 ha) is located off the Lower Denmark Rd ~500 m west of the Elleker–Grassmere Rd junction. Walk along the Grassmere Drain to a track that leads to a bird hide. Australasian Bittern has been recorded here. Look also for waterfowl such as Australian Shelduck, Australasian Shoveler, Grey Teal and Pink-eared Duck. The surrounding bushland holds a range of WA endemics such as Western Rosella, Red-eared Firetail, White-breasted Robin and Red-winged Fairy-wren, often seen along the access track to the bird hide. In Albany itself, the Ellen Cove Boardwalk (3 km), a wheelchair-accessible trail between the Middleton Beach carpark and the end of Princess Royal Drive, winds its way through coastal heath. Along the walk, look for Brush Bronzewing, Southern Emu-wren, Red-winged Fairy-wren, Red-eared Firetail and White-breasted Robin. Lake Seppings in Middleton Beach (best accessed from the carpark on Golf Links Drive) is also good for waterbirds, including Blue-billed Duck, with some interesting boardwalks around the lake edge. The walking track to the north of the Golf Links Drive carpark can be a good spot for Red-eared Firetail and White-breasted Robin, along with other bush birds. It is possible to walk across the golf course from the carpark (take care on the fairways!) to access Middleton Beach; Rock Parrots can often be found here feeding quietly in the low sand dune vegetation. Continuing north-east along Golf Links Drive (which becomes Emu Point Drive) for several km before taking Clark

Albany pelagic boat trips

Occasional seabird trips travel out of Albany to the continental shelf. These trips have turned up a variety of species, including Wilson's, White-faced and Black-bellied Storm-Petrel, Kerguelen, Great-winged, White-headed, Soft-plumaged and Grey Petrel, Flesh-footed, Sooty, Little and Hutton's Shearwater, Antarctic and Slender-billed Prion, and Wandering, Black-browed, Campbell, Shy, Indian Yellow-nosed, Grey-headed and Sooty Albatross. To find out if there are any scheduled trips, contact Birdlife WA www.birdlife.org.au. Alternatively, you may wish to join one of the whale watching trips organised out of Albany, though these tend not to leave the Sound and thus mostly provide opportunities for Indian Yellow-nosed Albatross and Australasian Gannets. For more information, visit the Albany Visitor Centre www.amazingalbany.com.au

St and Swarbrick St to Emu Point Boat Harbour. Here, scan behind and north of the harbour, because the tidal flats are often good for waders. Among other species, it is one of the better sites for Lesser Sand Plover in the south-west. While here, a short walk in the adjacent coastal vegetation may also provide sightings of Southern Emu-wren.

Stirling Range National Park

Key species: Western Whipbird, Southern Scrub-robin, Purple-gaped Honeyeater, Western Yellow Robin, Blue-breasted Fairy-wren, Western Fieldwren
Other species: Square-tailed Kite, Spotted Harrier, Western Thornbill, Red-eared Firetail, Red-winged Fairy-wren, White-breasted Robin, Crested (Western) Shrike-tit

The south-western form of the Crested Shrike-tit (ssp. *leucogaster*) is often mooted as a potential 'split', which would see it elevated to a full species. Generally scarce and elusive, good sites to search include Dryandra Woodlands (Tomingley Rd) and Stirling Range National Park (and especially in the grounds of the neighbouring Stirling Range Retreat).

Located 337 km south-east of Perth, Stirling Range National Park (116 000 ha) is 64 km long (east to west) and 18 km wide. It is renowned for its rich flora, with over 1500 different plant species recorded, including 16 orchid species (best seen from September to November). Accommodation is available at the Stirling Range Retreat, opposite the Bluff Knoll turn-off, which has a range of hostel dorms and self-contained two bedroom chalets, as well as powered and unpowered campsites: info@stirlingrange.com.au, (08) 9827 9229. There is also camping within the national park at Moingup Springs see www.dec.wa.gov.au.

Western Whipbird (ssp. *oberon*) occurs in the Mallee heathland along the road to Mount Trio, particularly the upper section. They are best seen and heard just after dawn: listen for their distinctive four-note creaking cartwheel call. Other birds that occur around Mount Trio include Square-tailed Kite, Western Fieldwren, Purple-gaped, Yellow-plumed and Tawny-crowned Honeyeater and Western Wattlebird. The walk to the top of Mount Trio (2 km return) is also worthwhile. In spring, Mountain Bellflower provide a splash of colour on the higher slopes. The walk up to Toolbrunup Peak (4 km return, difficult grade) is often regarded as the best hiking trail in the Stirling Ranges. Along this trail, you may encounter Baudin's Black-Cockatoo, Red-capped Parrot, Western Rosella, Red-winged and Blue-breasted Fairy-wren, Western Spinebill, Western Thornbill, White-breasted Robin and Red-eared Firetail.

A good diversity of birds occur along the Ongerup Creek Walk (4 km return), which is the first firebreak on your left as you turn into Bluff Knoll Rd. Look for Purple-crowned Lorikeet, Scarlet and Western Yellow Robin, Tawny-crowned Honeyeater, Splendid Fairy-wren, Western Thornbill and Red-eared Firetail. Little and Wedge-tailed Eagle, Spotted Harrier and Square-tailed Kite have also been recorded

Wave Rock and Dragon Rocks Nature Reserves

One of Western Australia's most iconic landforms, Wave Rock is located immediately east of Hyden in the Hyden Wildlife Park (160 ha). It is a good location for smaller passerines such as Redthroat, Striated and Spotted Pardalote, Weebill, and Chestnut-rumped, Inland and Yellow-rumped Thornbill. Other species that occur here include Emu, Mulga and Elegant Parrot, Australian Ringneck, Western Rosella, Purple-crowned Lorikeet, Rainbow Bee-eater (summer), Singing, Spiny-cheeked and White-eared Honeyeater, Pallid Cuckoo, White-browed Babbler and Black-faced Woodswallow. Malleefowl occur in the region, with a resident population at Dragon Rocks Nature Reserve (32 000 ha), 20 km south-east of Hyden. This reserve is within one of the largest areas of continuous Mallee remnants in the WA's wheatbelt. At Dragon Rock, you may also find Carnaby's Black Cockatoo, Western Rosella, Blue-breasted Fairy-wren, Purple-gaped Honeyeater and Western Yellow Robin. The reserve is also a release site for translocated Numbats.

along here. In the Mallee sections along the walk, look for Southern Scrub-robin and Purple-gaped Honeyeater, while there is a chance of Western Whipbird in the same habitat. Another good birding trail is the Kanga Walk (4.4 km return), which is the first firebreak on your right as you turn in to Bluff Knoll Rd. Along the walk, there are good sections of Mallee and Wandoo woodlands. Again, look for Southern Emu-wren, Southern Scrub-robin and Western Whipbird in the Mallee, and honeyeaters such as White-eared, Purple-gaped and Tawny-crowned. In the section of Wandoo woodland, you may see Carnaby's Black-Cockatoo, Elegant Parrot, Rufous Treecreeper, Western Thornbill and Blue-breasted Fairy-wren. Birding around the Stirling Range Retreat is excellent. Sharing a boundary with the park, it is located off Chester Pass Rd, just north of the turn-off to Bluff Knoll Rd. Around the retreat, look for Elegant and Regent Parrot, Blue-breasted Fairy-wren, Crested (Western) Shrike-tit (sometimes right next to the office), Southern Scrub-robin and Western Yellow Robin.

Fitzgerald River National Park

Key species: Malleefowl, Western Whipbird, Western Bristlebird, Western Ground Parrot (rare)
Other species: Carnaby's Black-Cockatoo, Regent Parrot, Elegant Parrot, Shy Heathwren, Western Fieldwren, Southern Scrub-robin, Southern Emu-wren, Purple-gaped Honeyeater

Fitzgerald River National Park (329 882 ha) is one of the largest and most botanically significant national parks in Australia. It is recognised internationally through the UNESCO Biosphere Program, and has been declared an Important Bird Area. Located 419 km south-east of Perth, camping within the park is available at Four Mile Beach, Hamersley Inlet, Quoin Head, Fitzgerald Inlet and St Mary Inlet. Adjacent to the park, there is accommodation at Fitzgerald River National Park B&B and the Quaalup Homestead Wilderness Retreat. Note that most of the roads within the park are not suitable for caravans, but they are permitted at Quaalup from the west. The park information centre can be contacted on (08) 9835 5043.

The South Coast Hwy has a good birdwatching stop around the Fitzgerald River Bridge, 34.5 km east of Jerramungup. A walk along river, and through the regenerating vegetation south-west of the bridge, may produce an encounter with the Western Whipbird: listen for their distinctive 'cart-wheel' call. Malleefowl also occur in the area, particularly the open grassy section 500 m south of the road. Other birds to look for include Southern Scrub-robin, Purple-gaped, White-eared and Brown-headed Honeyeater, Varied Sittella, Blue-breasted Fairy-wren, Spotted Pardalote and Purple-crowned Lorikeet. There are records of Western Ground Parrot in Fitzgerald River National Park, mostly from the heath south of the South Coast Hwy, 64 km east of Jerramungup, particularly on the eastern side of the Hamersley River (often dry), although much of this area has recently been burnt and the parrot may now be absent from here.

Western Whipbird occur around the park information centre located on Quiss Rd, 18.5 km east of Jerramungup. The same area you may also find Western Spinebill, Western Wattlebird, Purple-gaped Honeyeater, Southern Emu-wren

Heathland and short Mallee in Fitzgerald National Park. A remarkable number of birds frequent this habitat, including Malleefowl, Brush Bronzewing, Carnaby's Black-Cockatoo, Shy Heathwren, Blue-breasted Fairy-wren, Tawny-crowned Honeyeater, Purple-gaped Honeyeater, Southern Scrub-robin and Western Whipbird.

Cocanarup Timber Reserve

Cocanarup Timber Reserve (8853 ha) is located immediately north-east of Fitzgerald River National Park, 16 km west of Ravensthorpe. The Salmon Gum woodlands in the reserve support a good selection of parrots and cockatoos. It is an important south-coast nesting site for Carnaby's Black-Cockatoo, which often roost in trees along the banks of the Phillips River. You may also observe Western Rosella and Regent, Red-capped and Elegant Parrot. It is also good spot for night birds, with records of Southern Boobook, Eastern Barn Owl, Tawny Frogmouth, Australian Owlet-Nightjar and Spotted Nightjar. Cocanarup Timber Reserve is also a release site for translocated Numbats.

and Mistletoebird. Further south along Quiss Rd, an excellent birding site is around the Fitzgerald River National Park B&B, 10 km from the information centre. The mix of Mallee and native meadows around the B&B attracts a superb selection of birds, including Western Whipbird, Malleefowl (there are several active mounds here), Emu, Stubble Quail, Square-tailed Kite, Wedge-tailed Eagle, Carnaby's Black-Cockatoo, Regent and Elegant Parrot, Brush Bronzewing, Pallid Cuckoo, Crested Bellbird, Southern Scrub-robin, Brown Songlark, Shy Heathwren, Splendid and Blue-breasted Fairy-wren, and Tawny-crowned, Purple-gaped, Brown and New Holland Honeyeater. For Western Whipbird and Shy Heathwren, concentrate your search in the Mallee either side of Quiss Rd, 3 km south of the B&B, particularly around the park pay station.

The coastal inlets and bays in the national park support species such as Hooded Plover, Eastern Osprey, White-bellied Sea-Eagle and Great Crested Grebe. In summer, Sanderling, Grey Plover, Sharp-tailed Sandpiper and Red Knot may also occur here. Sites worth visiting include Bremer Bay, Wellstead Estuary, Gordon Inlet and Culham Inlet (located on the far-east side of the park). Beaufort Inlet, outside the park, is another good area for waterbirds and shorebirds. To get there from Jerramungup, travel 57 km down the South Coast Hwy to Boxwood Hill, turn left on to Borden Bremer Bay Rd and after 13 km turn south on to Millers Point Rd; the bay is a further 5 km.

Esperance region

Key species: Cape Barren Goose, Hooded Plover, Little Penguin, Flesh-footed Shearwater, Brown Quail, Rock Parrot, Red-eared Firetail
Other species: Freckled Duck, Short-tailed Shearwater, Peregrine Falcon, waterbirds

Esperance is a large town located on the Southern Ocean coastline, 720 km south-east of Perth. There are several excellent birdwatching areas around Esperance, including Woody Island Nature Reserve, Lake Warden and Monjingup Lake Nature Reserve. There is a wide range of accommodation in Esperance: see the information centre in Esperance www.visitesperance.com

Woody Island Nature Reserve (240 ha) is located within the Recherche Archipelago. On the 15 km boat trip across and back, search the ocean for seabirds: most commonly Australasian Gannet, Pacific Gull, Black-faced Cormorant and Arctic Jaeger (summer), and there is a chance of Little Penguin, Brown Skua (winter), Flesh-footed (they nest on Woody Island) and Short-tailed Shearwater, Great-winged Petrel, and Shy, Indian Yellow-nosed and Black-browed Albatross. Much of the island is densely wooded, with several walking tracks that lead variously to Skinny Dip Bay, the Summit and Muttonbird look-outs. While on the island, there is a chance of seeing Brown Quail, Peregrine Falcon, Sooty Oystercatcher, Brush Bronzewing, Rock Parrot, Sacred Kingfisher, Golden Whistler and Red-eared Firetail (near the water tank, as well as the bird bath in the Settlement). Woody Island has a small eco-stay facility, and there are campsites, barbeque facilities and a kiosk. See the Woody Island Information Centre: www.woodyisland.com.au, (08) 9071 5757.

Lake Warden (699 ha) regularly supports up 30 000 waterbirds and is a wetland of international importance under the Ramsar Convention. It is located 5 km north of Esperance, between the two main access roads to the town: the South Coast Hwy and Coolgardie-Esperance Hwy. The best birding is usually along Stearne Rd along the north shore of the lake, where you may see Hooded Plover, Banded Stilt, Red-necked Avocet, Fairy Tern, Hardhead, Musk and, more rarely, Freckled Duck. Scan the grassy areas around the lakes edge for feeding (Recherche) Cape Barren Goose. In the shrublands around the lake, you may also encounter Western Fieldwren, while the surrounding bushland

Peak Charles National Park

Rising 650 m above the surrounding plains, Peak Charles is a granite mount, surrounded by dry sand-plain heathland and salt-lake systems. The *Hakea*-dominated heathland attracts a myriad of nectivores, including Tawny-crowned, White-cheeked, New Holland, Yellow-plumed, White-eared, Brown-headed and Purple-gaped Honeyeater and Western Spinebill. Other birds here include Australian Bustard, Southern Scrub-robin, Golden Whistler, Rainbow Bee-eater (summer), Black-faced Woodswallow, Shy Heathwren, Inland Thornbill and Blue-breasted Fairy-wren, and raptors such as Wedge-tailed and Little Eagle, and Spotted Harrier. Like many expansive arid areas, a good technique to find birds is to drive slowly, stopping when you see or hear signs of activity. Peak Charles is located 202 km north-west of Esperance via the Coolgardie–Esperance Hwy, Lake King–Norseman Rd and Peak Charles Rd. Peak Charles campground has no facilities, so, if you intend to visit, make sure you are self-sufficient.

supports Western Rosella, Red-capped Parrot, Western Wattlebird, Western Spinebill and Red-eared Firetail. Hooded Plover and Banded Stilt are also regularly recorded on the adjacent Pink Lake, 3 km west of Esperance. Nearby, the entrance to the Esperance Golf Course (intersection of Dickerson Rd and Coolgardie-Esperance Hwy) is another good site for (Recherche) Cape Barren Goose; several Masked Lapwings (rare in southern WA) have also established here in recent years.

In the bushland around Monjingup Lake Nature Reserve, look for Sacred Kingfisher, Elegant and Red-capped Parrot, Brush Bronzewing, Western Wattlebird, Yellow-throated Miner (dusky ssp. *obscura*), Brown and New Holland Honeyeater, Western Spinebill, Red-eared Firetail, Common Bronzewing and White-winged Triller. To get there from Esperance, follow the South Coast Hwy to the sign to Telegraph Rd. Monjingup Lake Nature Reserve is on both sides of Telegraph Rd, with the first entrance on the left-hand side, and the second entrance further up on the right-hand side (of Telegraph Rd).

Cape Arid National Park

Key species: Western Ground Parrot, (Recherche) Cape Barren Goose, Chestnut Quail-thrush, Red-eared Firetail
Other species: Sanderling, Hooded Plover, Carnaby's Black-Cockatoo, Australian Bustard, Western Fieldwren

Cape Arid National Park is a large (279 832 ha) reserve of great significance to the ongoing conservation of birds in south-west WA. The reserve is located 730 km south-east of Perth and 120 km east of Esperance. The coastal areas of the park include sandy beaches, rocky headlands, *Banksia* woodlands and heathlands, while further inland, there are low woodlands and shrublands. Cape Arid is at the eastern limit of distribution for several south-west WA endemic species, including Carnaby's Black-Cockatoo, Western Ground Parrot, Red-eared Firetail, Western White-naped Honeyeater, Western Wattlebird and Western Spinebill. The most accessible campsite is at Thomas River, with 2WD access, barbeques, toilets and water tanks; other campsites, such as Mount Ragged, Jorndee Creek and Seal Creek, are only accessible to 4WD.

The park is home to one of the world's rarest and most Endangered birds, the Western Ground Parrot. Elevated to full species status in 2010 (when molecular studies revealed its genetic distinctness from the Eastern Ground Parrot), current estimates suggest that fewer than 140 birds remain in the wild, with the majority now surviving in Cape Arid National Park. The parrots' preferred habitat is sparely vegetated and low-lying heathland, unburnt for at least 15 years. Such areas of heath occur along Poison Creek Rd, north and east of Mount Arid. The best way to see them is to contact the Friends of the Western Ground Parrot (www.western-ground-parrot.org.au) to participate in a survey. The roadside heath along Poison Creek Rd is also home to Australian Bustard, Western Fieldwren, Tawny-crowned and White-cheeked Honeyeater, and Southern Emu-wren. The best time for birding in the heath is usually early in the morning, because overnight open flowers fill with nectar, providing a burst of feeding activity at daybreak. Early in the morning, and at dusk, is also the best time to listen for the resonating whistles of the Western Ground Parrot: the most reliable means of detecting their presence.

The *Banksia* woodlands around the Thomas River campground are home to Purple-crowned Lorikeet, Brush Bronzewing and Blue-breasted Fairy-wren. When flowering, these woodlands attract White-cheeked and New Holland Honeyeater, Western Wattlebird and

Western Spinebill. On the Thomas River estuary, particularly near the bar, look for Hooded Plover, Australian Pied and Sooty Oystercatcher and, in summer, shorebirds such as Sanderling, Red-necked Stint, Bar-tailed Godwit, and Grey Plover. In 2009, and again in 2010, Beach Stone-curlew was recorded here. The rare western subspecies (*grisea*) of the Cape Barren Goose (known as 'Recherche Cape Barren Goose') occasionally visit the bay from their only breeding site in the Recherche Archipelago – a group of 105 islands located off the south coast of Cape Arid National Park. The total population is estimated to be ~650 birds. Other birds at Thomas River include Carnaby's Black-Cockatoo, White-bellied Sea-Eagle, Caspian Tern, Pacific Gull, Black-faced Cormorant (a species that is rare in south-west WA west of Hopetoun) and Rock Parrot.

The Boolenup Walk (4 km return) starts west of the Rangers Office, off Tagon Rd. It winds its way to the brackish Lake Boolenup and passes through coastal heath, *Banksia* woodland and paperbark swamp. Along the walk look for Red-eared Firetail (favouring the paperbark swamp), Western White-naped Honeyeater, Western Wattlebird, and, occasionally, Square-tailed Kite (summer or autumn). Mount Ragged is surrounded by Mallee woodlands. Look here for Mulga Parrot, Purple-crowed Lorikeet, Chestnut Quail-thrush (ssp. *fordianum*), Shy Heathwren, Redthroat, Crested Bellbird, and Purple-gaped and Yellow-plumed Honeyeater, and keep an eye out for Malleefowl. The Mount Ragged Walk (3 km return) takes you to the top of Tower Peak.

Carnaby's Black-Cockatoo is the more widespread of the 'white-tailed' black-cockatoos in south-west WA and is usually the easiest of the two to find, given they occupy a wide range of habitats.

Kalgoorlie and Coolgardie region

Key species: Freckled Duck, Gilbert's Whistler, Redthroat, Shy Heathwren, White-fronted Honeyeater, Chestnut Quail-thrush (both ssp. *fordianum* and ssp. *clarum*)

Other species: White-browed Treecreeper, Tawny-crowned Honeyeater, Western Yellow Robin

Boorabbin and Goldfields Woodlands National Parks, and Victoria Rock Nature Reserve, are part of a 152 500 ha reserve network that straddles the Great Eastern Highway, 60 km west of Coolgardie. The extensive sand plains in these reserves support heathland plants such as *Grevillea*, Honey-myrtle and *Hakea*. These habitats are occupied by Gilbert's Whistler, Redthroat, Chestnut Quail-thrush, Shy Heathwren, White-fronted and Tawny-crowned Honeyeater, Inland and Chestnut-rumped Thornbill, and Blue-breasted Fairy-wren. Victoria Rock Nature Reserve is dominated by a large granite outcrop, and surrounded by Salmon Gum and Sheoke woodlands. Look here for Western Rosella, Shy Heathwren and Western Yellow Robin. Facilities include barbeques and, toilets, and bush camping is permitted.

Rowles Lagoon, and the smaller Muddy, Clear and Carnage Lakes, are considered an 'island refuge', providing a freshwater resource in an otherwise arid region.

Sandford Rocks Nature Reserve

Sandford Rocks Nature Reserve (806 ha) contains a complex mosaic of exposed granite outcrops and pools that are surrounded by shrublands and woodlands. It is situated 11 km north-east of Westonia on Boodarockin Rd. Species that frequent the dry woodlands include Emu, Painted Button-quail, Regent, Mulga and Elegant Parrot, Black-eared Cuckoo, Crested Bellbird, Southern Scrub-robin, Spiny-cheeked Honeyeater, Splendid and Blue-breasted Fairy-wren, Shy Heathwren, Redthroat and Chestnut-rumped Thornbill. Malleefowl has been recorded in the northern section of the park. Night birds include Southern Boobook, Tawny Frogmouth, Australian Owlet-Nightjar and Spotted Nightjar. Picnic facilities are available.

After major flooding, these lakes can host significant breeding events for water-birds. The lagoon is located 65 km north of Coolgardie, via Coolgardie North Rd through Kunanalling and Kintore. Notably, it is the only significant freshwater wetland in a region where most wetlands are saline. There is camping at Rowles Lagoon, with barbeques, picnic tables and toilets. When the lakes are full, Freckled Duck visit and breed, with as many as 90 individuals recorded. You may also observe Blue-billed, Musk and Pink-eared Duck, Black-fronted and Red-kneed Dotterel, Red-necked Avocet, Black-winged Stilt, Whiskered Tern, Banded Lapwing, Black-tailed Native Hen, Australian Spotted Crake, Great Crested Grebe and Glossy Ibis.

Goongarrie National Park is located at the junction of the eucalypt–Mulga line; to the south lie eucalypt woodlands and to the north lie Mulga shrublands. Goongarrie is accessed from Pianto Rd, which runs off the Meekatharra Rd, 30 km south of Menzies. Birds to look for at Goongarrie include Budgerigar, Cockatiel and Bourke's Parrot, White-browed Treecreeper, Red-backed Kingfisher, Pied Butcherbird, Ground Cuckoo-shrike, Black-eared Cuckoo and Crimson Chat, and, in the salt lake, shrub-lands in the north of the park, you may see White-winged Fairy-wren, Slender-billed Thornbill and Southern Whiteface. The Chestnut Quail-thrush (ssp. *clarum*) that occurs here is also known as Chestnut-backed Quail-thrush and may be a (slim) contender for elevation to full species status in the future.

Eyre Bird Observatory

Key species: Malleefowl, Major Mitchell's Cockatoo, Chestnut Quail-thrush, Sanderling, shorebirds
Other species: Western Yellow Robin, Blue-breasted Fairy-wren, Southern Scrub-robin, Shy Heathwren

Established in 1977, the Eyre Bird Observatory is Australia's oldest bird observatory. Over 240 bird species have been recorded within a 20 km radius of the homestead; this radius includes significant areas of habitat for seabirds, shorebirds and coastal Mallee species. The home-stead and observatory is located below the coastal escarpment, 500 km east of Norseman (280 km from the SA border). To get there, travel 16 km east of Cocklebiddy Roadhouse to the observatory turn-off (signposted). From here, it is 18 km to a lookout and then another 13 km (4WD) to the observatory. By prior arrangement, the caretakers can ferry overnight guests with 2WD from the lookout to the observatory. There is accommodation in the historic Old Telegraph Building. Info: www.birdlife.org.au/visit-us/observatories/eyre, eyrebird-obs@bigpond.com, (08) 9039 3450.

Within walking distance of the Observatory, you may see Malleefowl, Major Mitchell's Cockatoo, Purple-crowned Lorikeet, Brush Bronzewing, Southern Scrub-robin, Shy Heathwren, Western Yellow Robin, Chestnut Quail-thrush (ssp. *fordianum)*, Black-eared and Pallid Cuckoo, White-fronted Chat and Blue-breasted Fairy-wren, and a myriad of honeyeaters such as Purple-gaped, White-fronted, Yellow-plumed, White-cheeked, White-eared, Spiny-cheeked and Brown Honeyeater. Back nearer to the highway,

Naretha Blue Bonnet and Nullarbor Quail-thrush

The Nullarbor Quail-thrush and the isolated western race of the Blue Bonnet (Naretha Blue Bonnet ssp. *narethae*) occur along the Cocklebiddy–Rawlinna Rd, which heads north from the Cocklebiddy Roadhouse, 16 km west of the turn-off to Eyre Bird Observatory. A good spot to search for both species is 10 km north of Cocklebiddy; the Nullarbor Quail-thrush occupies open shrublands while the Blue Bonnet often perch in stands of Sheoke (*Allocasuarina*) in this area. Other birds that occur along the Cocklebiddy–Rawlinna Rd include Spotted Harrier, Australian Bustard, Inland Dotterel, Little Button-quail, Ground Cuckoo-shrike, Orange Chat, Rufous Fieldwren (ssp. *campestris*) and Slender-billed Thornbill. In 2011, huge flocks of Budgerigar were recorded here. Note that the Cocklebiddy–Rawlinna Rd passes through Arubiddy Station and the station owners should be contacted before entry: contact details are available at the roadhouse. Both Naretha Blue Bonnet and the Nullarbor Quail-thrush also occur on the Trans Access Rd (the road that follows the Trans-Australian railway), particularly near the remote localities of Rawlinna and Naretha (located 130 km north of Cocklebiddy). Search in the Mulga woodlands on the southern end of the Connie Sue Hwy, ~25 km north of Rawlinna. When in the area, also keep a lookout for Scarlet-chested Parrot, which are occasionally found along the southern section of the Connie Sue Hwy. If you have time to search, Naretha Blue Bonnet are also found within the *Acacia* shrubland that runs adjacent (~1 km to the north) of the Eyre Hwy between Mundrabilla and Eucla (just west of the SA border).

in more arid shrublands, Ground Cuckoo-shrike is observed with some frequency while nomadic Orange and Crimson Chat occur on occasion.

Weekly bird counts are carried out along the 13 km of Kanidal Beach, which is a good location for migratory and resident shorebirds, notably Sanderling, Ruddy Turnstone, Pacific Golden, and Double-banded (winter) Plover, and Common Sandpiper. The remarkably intensive effort has, over the years, also produced sightings of rarities such as Oriental Plover, Little Stint, Baird's and Buff-breasted Sandpiper, Red-necked Phalarope and Arctic Tern. Between August and October, Southern Right Whales calve and nurse their young along the coastline.

Neale Junction Nature Reserve

Key species: Princess Parrot, Scarlet-chested Parrot
Other species: Australian Bustard, Striated (Sandhill) Grasswren, Black-eared Cuckoo, Rufous-crowned Emu-wren, Pied Honeyeater, Black Honeyeater, Little Woodswallow

Neale Junction is an isolated and remote nature reserve located in the Great Victoria Desert, 1600 km from Perth and 430 km from Laverton. The reserve is an excellent site for inland parrots, and one of the more reliable sites to see two of Australia's most sought-after bird species: the Princess Parrot and Scarlet-chested Parrot. The reserve is located on the Anne Beadell Hwy (also known as the Serpentine Lakes Rd), where it intersects with the Connie Sue Hwy. To get there from Laverton, travel 190 km east along the White Cliffs–Yamarna Rd to the abandoned Yamarna Homestead. From there, travel another 240 km west along the Anne Beadell Hwy to Neale Junction. You will need to allow at least a day to drive from the Yamarna Homestead to Neale Junction. All of the usual remote outback travel precautions apply! Within the reserve, there is a campsite 500 m west of the junction and, although there is a rainwater tank, you'll need to be self-sufficient. Neale Junction is an extremely remote reserve and the Anne Beadell Hwy is a 'highway' in name only. For local advice on road conditions, you may wish to contact the Laverton Police (08) 9031 1000. To travel from Laverton to Yamarna, you will need to obtain a permit from the WA Aboriginal Lands Trust (08) 9235 8000, alt@dia.wa.gov.au. Permits may take up to 8 weeks, so start your preparation early.

Before you reach Neale Junction, the road passes through Yeo Lake Nature Reserve (160 km from Neale Junction). There is a campground here, where facilities include a rainwater tank, barbeque, tables and toilet. Bourke's Parrot can be quite common around Yeo Lake, while open, stony areas around the campsite can be good for Spotted Nightjar. You may also encounter Chiming Wedgebill, White-winged Fairy-wren, Masked, Little and Black-faced Woodswallow, Pied and Black Honeyeater, Southern Whiteface and Zebra Finch.

The habitat in Neale Junction Nature Reserve consists of Marbled Gum on Spinifex flats, with pockets of Mulga shrublands. Parallel east–west sand dunes dissect the reserve. Start looking for Princess and Scarlet-chested Parrot along the Anne Beadell Hwy from just beyond the western boundary of the reserve and, from there, to the junction itself; focus particular attention on areas dominated by Marbled Gum. Most recent observations have occurred between 5 and 15 km west of the junction. Both species have also been seen immediately north of the junction, along the Connie Sue Hwy, and further east along the Anne Beadell Hwy (such as shrublands 65 km east of the junction, and the shrublands along the highway near the Wanna Lakes and SA border). Both the Princess Parrot and Scarlet-chested Parrot are highly nomadic, so it is worth checking bird reports (see page 581) to see if there have

Arid shrublands within Yeo Lake Nature Reserve support White-winged Fairy-wren, Chiming Wedgebill, Masked, Little and Black-faced Woodswallow, Southern Whiteface and Zebra Finch. The flowering *Eremophila* shrub in the foreground is a favoured food plant of nomadic Pied and Black Honeyeater.

been any recent sightings. Other parrots and cockatoos at Neale Junction include Mulga and Bourke's Parrot, Budgerigar, Cockatiel, Australian Ringneck, Galah and Major Mitchell's Cockatoo. Striated (Sandhill) Grasswren (ssp. *oweni*, which may be elevated to full species status in the future) and Rufous-crowed Fairy-wren occur at Neale Junction Nature Reserve, mainly in the northern section where the Spinifex areas border the Connie Sue Hwy; if you have no luck here, try a few kilometres south of the Junction in areas of larger Spinifex adjacent to the Connie Sue Hwy. Other birds to look for include Australian Bustard, Red-backed Kingfisher, White-backed Swallow, Black-eared Cuckoo, Crested Bellbird, Redthroat, Ground Cuckoo-shrike, Variegated and Splendid Fairy-wren, Chestnut-rumped and Slaty-backed Thornbill, Grey-fronted and White-fronted Honeyeater. At night, listen for Spotted Nightjar and Southern Boobook. Thorny Devil, a spikey prehistoric-looking lizard, has been seen around the campsite and is commonly observed crossing tracks; they are slow moving and sit, rather than flee, in the face of danger so take care when driving.

North Western Australia

Regional overview

The northern half of WA covers a vast area and provides some outstanding birdwatching opportunities. It consists of four geographic regions, with a combined area of 1 535 853 km^2. The Mid-West

(472 336 km^2) is located on the west coast and extends for ~200 km to the north and south of Geraldton. The coastal portion of the region experiences a Mediterranean climate with hot, dry summers and cool, wet winters, while the inland portion, although hot and dry in summer, is usually cold and dry in winter. The Gascoyne region (138 000 km^2) is located further to the north and encompasses Carnarvon, the Shark Bay World Heritage Area and Ningaloo Reef. The Gascoyne has a moderate arid to semi-tropical climate, which is generally warm all year round. A feature of the Pilbara (502 000 km^2) is stunning landscapes of coastal flats, red-earthed plains and mountain ranges. The climate of the Pilbara largely spans hot and dry deserts, yet nearby to the north there are wet monsoonal areas. This area provides a diverse mix of desert species, as well as a good range of migrants as birds pass through from the south in the dry season and north in the wet season. Often described as the 'jewel in Australia's crown', the Kimberley region (423 517 km^2) extends from Broome on the Indian Ocean

to the coastal areas at the Northern Territory border. This area contains ancient mountain ranges bisected by magnificent sandstone and limestone gorges. The Kimberley has a tropical monsoon climate, receiving ~90% of its rainfall during the short wet season (Dec–Apr). Although the weather in the Kimberley during the wet summer months can be oppressive (and road access to the Kimberley proper being restricted), if you can handle the conditions, it can be a rewarding time for birdwatching, with many of those hard-to-find migrants returning to northern WA; as a consequence, the wet is also the best time for finding rarities.

WA's roads are generally good, with excellent sealed highways, such as the North West Coastal Hwy and the Great Northern Hwy, providing easy access up and down the state. However, the further you travel from major centres, the more difficult local road conditions can be. It is recommended that you use 4WD on unsealed roads through most of WA's north and west. The north is particularly prone to flash floods in the wet season, so, before travelling, it is worth checking the current road conditions. For daily updates visit www.mainroads.wa.gov.au. In planning your itinerary in northern WA, standard driving times can be very difficult to estimate because the quality of the unsealed tracks and roads vary considerably.

The Bureau of Meteorology, www.bom.gov.au, has detailed weather information concerning average monthly min and max temperatures for specific sections of WA, which is an excellent resource to plan birdwatching trips. This website is also a must visit during the wet season to monitor the status of possible cyclones in the North West. Conditions along the Gibb River Road, which takes you through the heart of the Kimberley, depend on the time of year, with travel only recommended from around May until the end of October. The Gibb River Road is routinely closed for long periods through the wet season. Daily scheduled flights to Broome, Kununurra, Exmouth, Monkey Mia and Karijini operate from Perth. Transwa is WA's regional public transport provider, with coach services to destinations along the west coast and in the north. Visit www.transwa.wa.gov.au.

A huge flock of Oriental Pratincole congregates to feed on grasshoppers among the termite mounds on Roebuck Plains, North Western Australia.

For general tourist information, including details about flights and coach services, visit www.westernaustralia.com.

Many birdwatchers who visit northern WA do so as part of an extended driving trip from Darwin, across to Broome, and then down to Perth (or the reverse of this route). Another common method to see the north is to fly to Broome, and using that town as a base, travel around the Kimberley (see itinerary below).

A suggested 9+-day itinerary from Perth to Kalbarri National Park, Monkey Mia, Carnarvon and return is:

- **Day 1.** Perth to Geraldton, 400 km.
- **Day 2.** Geraldton to Kalbarri National Park, 150 km (Malleefowl, Blue-breasted Fairy-wren, Western Yellow Robin).
- **Days 3–4.** Kalbarri National Park to Monkey Mia 350 km (Western Grasswren, White-winged Fairy-wren (Dirk Hartog black and white ssp. *leucopterus*), Chiming Wedgebill, Rufous Fieldwren).
- **Days 5–7.** Monkey Mia to Carnarvon, 330 km (Dusky Gerygone, Yellow White-eye, White-breasted Whistler, Mangrove Grey Fantail, shorebirds and terns).
- **Days 8–9.** Carnarvon back to Perth.

If you have time, plan a side trip east to the Cue region (Western Quail-thrush, Banded Whiteface, Grey Honeyeater, Western Bowerbird, Bourke's Parrot, Black and Pied Honeyeater, Inland Dotterel, Slaty-backed Thornbill, Chiming Wedgebill), adding an additional 3 to 4 days. Another side trip, which requires some organisation in advance, is a trip to the Houtman Abrolhos. Situated ~60 km off the coast of Geraldton and accessible via either charter float plane or overnight charter vessel, the islands support many seabirds, including Lesser and Common Noddy, Roseate Tern and Little Shearwater. An endemic form of the Painted Button-quail (ssp. *scintillans*) also occurs.

A suggested 16-day itinerary in and around the Kimberley region is:

- **Days 1–3.** Broome, (shorebirds, mangrove specialties, including White-breasted Whistler, Mangrove Golden Whistler, Dusky Gerygone and Red-headed Honeyeater as well as Yellow Chat and rarities). A side trip to Barred Creek could also be included (Beach Stone-curlew, Lesser Frigatebird, Lemon-bellied 'Kimberley' Flycatcher).
- **Days 4–7.** Broome via Derby and the Gibb River Road to Mitchell Falls National Park, ~700 km (Black Grasswren, Kimberley Honeyeater, White-quilled Rock-Pigeon, Partridge Pigeon, Sandstone Shrike-thrush, Pictorella Mannikin).
- **Days 8–9.** Mitchell Falls to Parry Lagoons Natures Reserve, ~400 km (Star Finch, Chestnut-breasted and Yellow-rumped Mannikin, Yellow Chat, Swinhoe's Snipe (wet season chance)).
- **Days 10–12.** Parry Lagoons to Wyndham, 40 km (Gouldian Finch, Mangrove Golden Whistler, White-breasted Whistler, Lemon-bellied Flycatcher, Chestnut-backed Button-quail, Arafura Fantail).
- **Days 13–14.** Wyndham to Kununurra, 100 km (Yellow-rumped Mannikin, Star Finch, Buff-sided Robin, Black Bittern, Australian Little Bittern (chance), White-browed Crake, Barn Swallow (wet season)).
- **Day 15–16.** Kununurra to Lake Argyle, 70 km (Yellow Chat, Purple-crowned Fairy-wren, Spinifex Pigeon, White-quilled Rock-Pigeon, Gouldian Finch).

The Karijini National Park is in the heart of the Pilbara. Because the region is invariably accessed via a road trip from either Perth or Broome, some of the above itineraries are relevant while in transit. Once in the Pilbara, to do the area justice (and owing to the vast distances largely on unsealed roads) a minimum 6-day stay is recommended (Grey Honeyeater, Striated (Pilbara) Grasswren , Western Bowerbird, Spinifex Pigeon).

For the adventurous (and only those well prepared for remote-area 4WD travel), head further inland to Jupiter Well near the NT border for a chance at Princess Parrot, Grey Falcon, Pied and Black Honeyeater. Allow at least 7 additional days and note that any such trip is a serious undertaking, given the remoteness of the region and the often (very) poor condition of 4WD tracks.

Birding highlights

Four range-restricted endemic bird species are found in central and northern WA: Black Grasswren, Kimberley Honeyeater, Dusky Gerygone and Western Quail-thrush. The Quail-thrush has only recently been elevated to full-species status and was formerly ssp. *marginatum* of Chestnut-breasted Quail-thrush. The Pilbara form of the Striated Grasswren (ssp. *whitei* is also candidate for full species status. With its range limited to the Pilbara region, it would become an additional species endemic to Western Australia. Northern WA has a wide selection of other sought-after species. In the Kimberley, look for Chestnut-backed Button-quail, Red Goshawk, Spinifex and Partridge Pigeon, White-quilled Rock-Pigeon, Varied Lorikeet, Barn Swallow (wet season), Yellow Chat, Buff-sided Robin, Mangrove Golden and White-breasted Whistler, Lemon-bellied Flycatcher, Arafura Fantail, Crested (Northern) Shrike-tit, Sandstone Shrike-thrush, Purple-crowned Fairy-wren, Gouldian and Star Finch, and Pictorella and Yellow-rumped Mannikin. In the Pilbara, you may see Oriental Plover (wet season), Grey Falcon, Spinifex Pigeon, Pheasant Coucal, Striated (Pilbara) Grasswren, Grey, Pied and Black Honeyeater, and Western Bowerbird. In the Gascoyne, look for shorebirds and terns, Western Grasswren, White-winged Fairy-wren, Rufous Fieldwren, Chiming Wedgebill, White-breasted Whistler, Mangrove Grey Fantail and Yellow White-eye. Princess Parrot, although extremely rare, is also possible. In the Mid West, search for Lesser Noddy (Houtman Abrolhos only), Malleefowl, Painted Button-quail, Inland Dotterel, Bourke's Parrot, Slaty-backed Thornbill, Banded Whiteface, Blue-breasted Fairy-wren, Grey, Black, Pied and White-fronted Honeyeater, Chiming Wedgebill, Western Quail-thrush, Western Yellow Robin and Western Bowerbird.

Some of the interesting waterbirds found in northern WA include Black Bittern, Australian Little Bittern, White-browed Crake, Magpie Goose, Green Pygmy-Goose, Wandering Whistling-Duck, Radjah Shelduck, Glossy Ibis, Black-necked Stork and Brolga. The area is also renowned for the huge numbers and great diversity of shorebirds. Because juvenile and immature individuals of many species 'overwinter' in this special area, many species can be found year round on the shores of Roebuck Bay. These include Broad-billed, Marsh and Curlew Sandpiper, Red-necked Stint, Common Greenshank, Common Redshank (uncommon to rare), Asian Dowitcher, Bar-tailed Godwit and Pacific Golden Plover. Other species are more strictly wet season visitors and include Swinhoe's Snipe, Wood and Sharp-tailed Sandpiper, Long-toed Stint, Oriental Plover, Oriental Pratincole and Little Curlew. Australian resident shorebirds, including Australian Pratincole, Black-winged Stilt, Black-fronted Dotterel and Masked Lapwing, are generally present year round.

Northern WA has a very long list of endemic subspecies. These include White-quilled Rock-Pigeon (darker ssp. *albipennis*, in the Kimberley), Partridge Pigeon (yellow-faced ssp. *blaauwi*, Kimberley), Spinifex

Pigeon (red-bellied ssp. *ferruginea*, Pilbara), Peaceful Dove (ssp. *clelandi*, Pilbara), Bar-shouldered Dove (ssp. *hedlandi)*, Galah (ssp. *roseicapillus*), Collared Kingfisher (ssp. *pilbara*, coastal Pilbara), Blue-winged Kookaburra (ssp. *cliftoni*, Pilbara), Rainbow Pitta (ssp. *johnstoneiana*, Kimberley), Black-tailed Treecreeper ('Allied Treecreeper' ssp. *wellsi*, Pilbara), Variegated Fairy-wren (ssp. *rogersi*, Kimberley), White-browed Scrubwren (ssp. *balstoni*, Pilbara), Rufous Fieldwren (ssp. *rubiginosis*, coastal Pilbara, and ssp. *wayensis*, arid WA), Dusky Gerygone (ssp. *christophori* Pilbara, and ssp. *tenebrosa* Kimberley), Green-backed Gerygone (ssp. *darwini*, Kimberley), Southern Whiteface ('Western Whiteface' ssp. *castaneiventris*), Black-faced Cuckoo-shrike (ssp. *subpallida*, Pilbara), Weebill (ssp. *ochragaster*, Pilbara), White-plumed Honeyeater (ssp. *carteri*), Varied Triller (ssp. *macrura*, Kimberley), White-breasted Whistler (ssp. *carnarvoni*, coastal Pilbara and ssp. *lanioides*, coastal Kimberley), Mangrove Golden Whistler (ssp. *melanura*, coastal Pilbara), Mangrove Robin (ssp. *cinereiceps*), Lemon-bellied Flycatcher ('Kimberley Flycatcher' ssp. *tormenti*, coastal Kimberley*)*, Australian Magpie (ssp. *longirostris*), Western Bowerbird (ssp. *carteri*, North West Cape in the Pilbara), Horsfield's Bushlark (ssp. *woodwardii*, Pilbara, and ssp. *halli*, Kimberley), and Yellow White-eye (ssp. *balstoni*). A few further subspecies are found in northern WA and western NT, including Spinifex Pigeon (ssp. *plumifera*), Purple-crowned Fairy-wren (ssp. *coronatus*), White-plumed Honeyeater (ssp. *calconi*), Grey Butcherbird ('Silver-backed Butcherbird' ssp. *argenteus* (along with Top End form), considered a possible candidate for full species status) and Long-tailed Finch (ssp. *acuticauta*). Finally, there are several subspecies that are limited to WA's offshore islands. These include Painted Button-quail (ssp. *scintillans*, Abrolhos Island), Lesser Noddy ('Australian Lesser Noddy' ssp. *melanops* Houtman Abrolhos Islands (the subspecies breeding on Ashmore Reef remains unresolved)), White-winged Fairy-wren ('Black and White Fairy-wrens' ssp. *leucopterus*, Dirk Hartog Island, and ssp. *edouardi*, Barrow Island), Variegated Fairy-wren (ssp. *bernieri*, Bernier Island), Southern Emu-wren (ssp. *hartogi*, Dirk Hartog Island), Rufous Fieldwren (ssp. *hartogi*, Dirk Hartog Island, and ssp. *dorrie*, Dorrie Island). The local forms of the Little Corella (ssp. *westralensis*) and Pheasant Coucal (ssp. *highami*) are also considered by some authorities to represent unique subspecies in the region.

A wide range of rarities, most originating from within the East Asian–Australasian flyway or further afield have been recorded

Finches are a feature of north Western Australia's avifauna. An early morning vigil at a dry season waterhole in the Kimberley region may reward the observer with sightings of up to six or more species of finch. Here, Long-tailed Finch and Zebra Finch drink at a roadside gravel scrape between Broome and Derby.

in the northern WA. These include Northern Pintail, Eurasian Wigeon, Great Shearwater, Abbott's Booby, Javan and Chinese Pond Heron, Grey Heron, Cinnamon Bittern, Red-legged Crake, Pin-tailed Snipe (annual in small numbers), Little Stint, Spotted Redshank, Nordmann's Greenshank, Eurasian Curlew, Pheasant-tailed Jacana, Little Ringed Plover, Semipalmated Plover, Black-tailed, Black-headed, Franklin's and Sabine's Gull, House Swift, Glossy Swiftlet, nest-swiftlets, Eurasian Hoopoe, Fairy, Hooded and Blue-winged Pitta, Dark-sided, Asian Brown, Narcissus and Blue-and-white Flycatcher, Arctic Warbler, Island Monarch, Red-throated and Pechora Pipits, Grey and White Wagtail, Red-rumped Swallow, Tiger shrike, Oriental Reed-Warbler, Black-billed Magpie and Rosy (Rose-coloured) Starling. Night Parrot has also been recorded in the Pilbara, with the most recent being of three birds reported at Minga Well, 200 km north-east of Carnarvon in April 2005.

Kalbarri National Park

Key species: Malleefowl, Blue-breasted Fairy-wren, Western Yellow Robin, Rufous Fieldwren

Other species: Black, Pied and White-fronted Honeyeater

Considered one of WA's most spectacular reserves, Kalbarri National Park (183 004 ha) is situated some 590 km north of Perth. It contains a wide variety of landscapes: extraordinary river gorges border the Murchison River, towering sea cliffs, heathlands that abound in wildflowers, *Acacia* shrublands and vast sand plains. Although there are no camping areas within the park, the park surrounds the township of Kalbarri, where facilities include a wide range of accommodation, shops, fuel outlets and banking.

Kalbarri National Park has a range of excellent walks. Along the Loop Walk (8 km), which starts at the Natures Window (37 km east of the Kalbarri township), you may be rewarded with sightings of Emu, Redthroat, Inland and Chestnut-rumped Thornbill, White-browed Babbler, Red-capped Robin, Black-faced Cuckoo-Shrike, Southern Scrub-robin, Black-eared Cuckoo, Little Woodswallow, White-backed Swallow and Splendid Fairy-wren. Kalbarri is famous for its diversity of wildflowers (over 1000 species) which, between

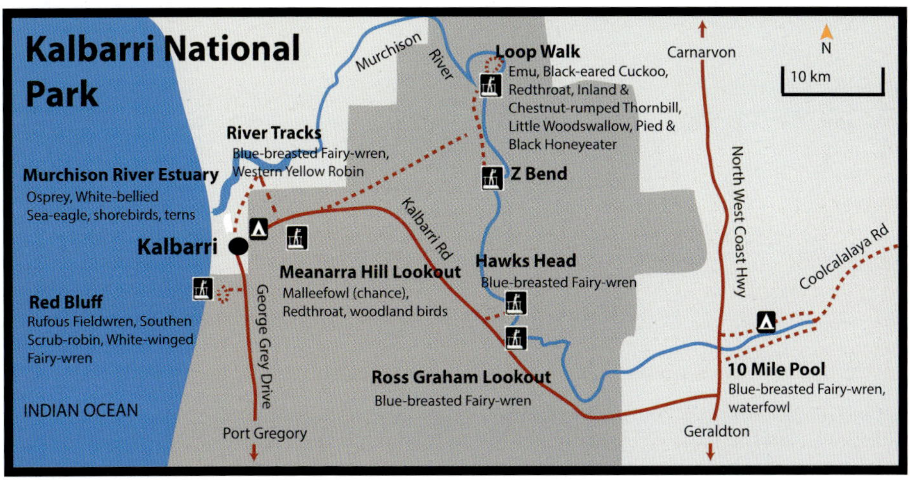

July and October, attract nectivores such as Black, Pied, White-fronted, Tawny-crowned, Spiny-cheeked, White-plumed, Grey-fronted and Brown Honeyeater. Keep your eyes open for Malleefowl: they are occasionally seen crossing the road, particularly around Meanarra Hill, 6 km east of the Kalbarri township. Scan the skyline anywhere in Kalbarri for raptors: Whistling Kite, Brown Goshawk, Collared Sparrowhawk, Little and Wedge-tailed Eagle, Nankeen Kestrel and Australian Hobby are probably the most common, while Osprey and White-bellied Sea-Eagle frequent the coastline.

Two of the most sought-after species at Kalbarri are Blue-breasted Fairy-wren and Western Yellow Robin. Here, at the very northern edge of their distribution, they occur in the shrublands that border the Murchison River immediately north of the Kalbarri township; take Grey St (which becomes River Rd), on the north-side of town, until it becomes a dirt track (Scenic River Drive), and from here onwards search the shrubs on either side of the track. Southern Scrub-robin is also common here. If you do not find the fairy-wren here, they also occur along Hawks Head and Ross Graham walks.

Another sought-after species at Kalbarri is Rufous Fieldwren. With a preference for areas of low coastal heath, search south of the Kalbarri township near Red Bluff (where you can also find Southern Scrub-robin, White-winged, Splendid and Variegated Fairy-wren and White-fronted Chat). You may also see them along a dawn or dusk walk along the Mushroom Rock Nature Trail (3 km loop) or the Bigurda Trail (8 km

Kalbarri George, Kalbarri National Park. Vantage points like this are especially good for scanning for raptors. Some of the more common species in the national park include Whistling Kite, Brown Goshawk, Collared Sparrowhawk, Little Eagle, Wedge-tailed Eagle, Nankeen Kestrel, Brown Falcon and Australian Hobby.

one way), which take you through coastal heath. (Note that there is some uncertainty regarding the identification of fieldwrens at Kalbarri, with some authorities considering individuals in this area as an intergrade between Western and Rufous Fieldwren. For certainty, look for Western Fieldwren in south-west WA and Rufous Fieldwren further north around Shark Bay.) Along these walks, look also for White-cheeked, Singing Honeyeater and Variegated Fairy-wren. For shorebirds, scanning the estuary at the Murchison River mouth can be rewarding, particularly in summer, with a chance of Grey Plover, Common and Marsh Sandpiper, Grey-tailed Tattler, Whimbrel, Ruddy Turnstone, Sharp-tailed Sandpiper and Sooty Oystercatcher. You may see terns hawking over the estuary, such as Caspian and Crested, while, less commonly, Fairy and Roseate Tern may roost on the nearby sand spits. Although uncommon, the park also represent the northern distribution of Carnaby's Black-Cockatoo. They are sometimes seen along the Murchison River. Another spot to look for Blue-breasted Fairy-wren, and a good birding spot generally, is Ten Mile Pool. It is located on the North West Coast Hwy at the Murchison River Crossing, 80 km to the west of Kalbarri. Camping is allowed along the river.

Houtman Abrolhos

Key species: Lesser Noddy, Common Noddy, Roseate Tern, seabirds, Painted Button-quail (ssp. *scintillans*)
Other species: Wedge-tailed Shearwater, Little Shearwater, White-faced Storm Petrel, Bridled Tern, Fairy Tern, Eastern Reef Egret

The Houtman Abrolhos islands are located in the Indian Ocean, 60–80 km off the coast of Geraldton. Clustered into three main groups (Wallabi, Easter and Pelsaert), the islands (192 in total) are scattered along a north–south axis that is some 100 km long. Pristine waters support an enormous number of seabirds, with perhaps three and a half million individuals being present. The islands are also among the most important breeding sites for seabirds in Australia, with around 35 breeding species recorded, including the largest (~49 000 pairs) of just three breeding colonies of the Lesser Noddy in Australia (the others are at Ashmore Reef and Adele Island and number just 50 and 10 pairs, respectively).

Capitalising on the bountiful waters around the Houtman Abrolhos are seabirds such as the Common and Lesser Noddy, Bridled, Sooty, Roseate, Fairy, Caspian and Crested Tern, Pacific Gull, Brown Skua (winter), Pomarine and Arctic Jaeger (summer), White-faced and Wilson's Storm-Petrel, and Wedge-tailed, Streaked (rare), Little, Flesh-footed and Hutton's Shearwater. The best way to see these birds is to charter a live-aboard vessel in Geraldton and head out to Pelsaert, Wooded and Morley Islands. On your way over to Houtman Abrolhos (a 60 km sea crossing), scan the open ocean for Shy and Indian Yellow-nosed Albatross and other seabirds, including many of the species listed above. Once at the islands, look for Osprey, White-bellied Sea-Eagle and Eastern Reef Egret, and shorebirds such as Bar-tailed Godwit, Ruddy Turnstone, Sanderling, Red Knot, Red-necked Stint, Grey-tailed Tattler, Greater Sand Plover, and Australian Pied and Sooty Oystercatcher.

The 'Abrolhos Painted Button-quail' (ssp. *scintillans*) is endemic to five islands in the Wallabi Group; they are generally easy to find on East Wallabi Island, where you may also see Spotless Crake, Tree Martin, White-browed Scrubwren and Silvereye, as well as Tammar Wallaby. Information about the vessel hire is available at www.abrolhosislandcharters.com.au. It is also worth contacting Birdlife Australia WA (www.birdlife.org.

au) whose members occasionally organise excursions to the islands. There is also an airport on East Wallabi Island, with chartered planes available from Geraldton Air Charter (www.geraldtonaircharter.com.au), though a boat would still then be required to access Pelsaert, Wooded or Morley Island where the Lesser Noddy breed.

Cue and Nallan Station

Key species: Western Quail-thrush, Banded Whiteface, Grey, Black and Pied Honeyeater, Western Bowerbird, Bourke's Parrot, Inland Dotterel
Other species: Slaty-backed Thornbill, Chiming Wedgebill

The small goldfield township of Cue (also known as the 'Queen of the Murchison') is on the Great Northern Hwy, 650 km north of Perth. Located in a fascinating region, its outback landscape is dominated by Mulga shrublands. After rain, the red sandy soils can be carpeted in native grasses and a spectacular display of flowers (mainly between August and October). For information on accommodation and local services try the Cue information centre on (08) 9963 1936, Dorsett Guest House, 6 Austin St, Cue, or www.cue.wa.gov.au.

Some of the region's best birding is on Nallan Station (~100 000 ha): a working cattle and sheep station accessed via the Great Northern Hwy, 11 km north of Cue. Here, you will find homestead-style accommodation and camping. For bookings: nallanstn@bigpond.com, (08) 9963 1054. In the Mulga around the property (such as near the airstrip), birding highlights include Bourke's and Mulga Parrot, Western Quail-thrush and Black, Pied and White-fronted Honeyeater. There is also possibility of Grey Honeyeater, especially in the denser stands of Mulga. Bourke's Parrot is most easily found in the half light of dawn and dusk, when they drink at the various wells on the property, notably at Judas Well and Jacksons Well (ask for directions at the homestead). Western Quail-thrush are perhaps best seen on the tracks on the east side of property and around Marshalls Paddock (located 7.5 km north-east of the homestead), where there is also some chance of encountering Banded Whiteface and Inland Dotterel. Other birds to look for include Crimson Chat, Varied Sittella (black-capped ssp. *pileata*), Chiming Wedgebill, White-browed Treecreeper, White-browed and Grey-crowned Babbler, and Spotted Nightjar, while Western Bowerbird occur around the homestead itself.

Nallan Lake is situated on the eastern side of the Great Northern Hwy, 20 km north of Cue. Here, you will find a pleasant picnic area and campsites. After periods of good rain, the lake attracts a large and diverse range of waterbirds such as Hoary-headed and Great-crested Grebe, Pink-eared and Musk Duck, Black-winged Stilt, Black-tailed Native-hen, and Red-kneed and Black-fronted Dotterel. In summer, scan for waders such as Red-necked Stint, Marsh, Sharp-tailed and Wood Sandpiper and Common Greenshank. Less common summer visitors include Long-toed Stint and Little Curlew. Another wetland worth investigating is Lake Austin, straddling the Great Northern Hwy, 22 km south of Cue. When the water conditions are right, you may see Red-necked Avocet, Banded and Black-winged Stilt, and Whiskered and Gull-billed Terns. The samphire heath, saltbush and Mulga around the lake hold Orange and Crimson Chat, Rufous Fieldwren, White-winged Fairy-wren, Slender-billed Thornbill, Black and Grey Honeyeater, and there is a chance of seeing Inland Dotterel.

Walga Rock is situated on Austin Downs Station (160 000 ha), 47 km south-west of Cue. To get there from the Great Northern Hwy at Cue, head north-west along the Beringarra Cue Rd (Robinson Rd) and, after 12.2 km, turn left towards Austin Downs

Station; it is another 36 km from here. Some 2.5 km from the Beringarra Cue Rd intersection, stop just before the entrance to Austin Downs Station. Here, search the flat areas with lower vegetation 200 m before the cattle grid for Banded Whiteface, Chiming Wedgebill and Inland Dotterel. Continue to Walga Rock, where you may encounter Little-Button-quail, Little Woodswallow, Western Bowerbird, Redthroat, Crimson Chat, White-backed Swallow, Ground Cuckoo-shrike, and Slaty-backed and Chestnut-rumped Thornbill. Other birds to look for in the area include Emu, Bush

Grey Honeyeater in Western Australia

Grey Honeyeater inhabits the Mulga woodlands across a vast area of inland NT, SA and WA. Because they often feed on the fruits of mistletoe, unlike many honeyeaters, they are not so strictly tied to nectar-bearing plants. The species is generally uncommon to rare, but, even when present, can also be quite cryptic; as a consequence, despite its non-descript plumage, it is one of the most 'sought-after' honeyeaters for birdwatchers. In WA, the Cue region (see p. 436), Karijini National Park and Tom Price (see p. 445) are some of the best areas to seek out this species. Some other recommended locations in WA for the Grey Honeyeater include:

- ***Paynes Find***. This township is located on the Great Northern Hwy, 430 km north-east of Perth. Immediately opposite the town, a dirt road (4WD recommended) heads south-east to Maranalgo Station. From Paynes Find, drive along this road for 32 km, passing the entrance to the Maranalgo Station and the north-east edge of Lake Moore, before stopping 150 m before a cattle grid. Grey Honeyeaters have been known to frequent the Mulga woodlands on either side of the road in this area.
- ***Geraldton–Mount Magnet Rd.*** The expansive Mulga woodlands along this road are suitable for Grey Honeyeater, particularly the 120 km section between Yalgoo and Mount Magnet. Recent sightings have been had 58.3 km east of Yalgoo (65.4 km west of Mount Magnet), and in the Mulga immediately west of the Yalgoo.
- ***Wanjarri Nature Reserve*** (53 000 ha). The Wanjarri Shearing Shed camping area is 16 km north-east of the Goldfields Hwy, which itself is 58 km north of Leinster (100 km south of Wiluna). Search for Grey Honeyeater, especially among mistletoe, around the Old Wanjarri homestead and shearing shed. An 'A' Class Nature Reserve, the area has other attractions for the birdwatcher, with over 120 species recorded. For access, first obtain permission from the Kalgoorlie office of WA DEC www.calm.wa.gov.au, (08) 9021 2677.

The vast Mulga woodlands of inland Australia are the favoured habitat of the Grey Honeyeater.

Stone-curlew, Red-backed Kingfisher, Grey-crowned and White-browed Babbler, Crested Bellbird, Hooded and Red-capped Robin, and raptors such as Wedge-tailed Eagle and Black-breasted Buzzard. There have also been occasional sightings of the rare Grey Falcon. In addition to some great birding, Walga Rock contains a large and impressive gallery of Aboriginal hand paintings.

The 'Granites', a 15-m high rocky escarpment, is located just east of the Great Northern Hwy, 6.5 km north of Mount Magnet, 65 km south of Cue. On the flats below the escarpment, look for Western Quail-thrush, Slaty-backed Thornbill (in relatively tall Mulga woodland), Chiming Wedgebill, Crested Bellbird and Redthroat, while Little Woodswallow often frequent the escarpment. The large expanses of Mulga woodlands along the Great Northern Hwy between Mount Magnet and Cue, and between Cue and Meekatharra, are suitable habitat for Grey Honeyeater, with a recent sighting ~40 km south of Cue (40 km north of Mount Magnet). A stop along here in any area that appears particularly 'birdy' may yield results.

Monkey Mia, Shark Bay and Dirk Hartog Island

Key species: Western Grasswren, White-winged Fairy-wren ('Dirk Hartog Black and White Fairy-wren' ssp. *leucopterus*), Chiming Wedgebill, Rufous Fieldwren
Other species: Osprey, Southern Scrub-robin, Crested Bellbird, shorebirds, also mammals

The Shark Bay World Heritage Area (2.2 million ha) is located 60 km south of Carnarvon and some 800 km north of Perth. It encompasses the Peron Peninsula, Dirk Hartog Island and the waters of Shark Bay. The northern half of the Peron Peninsula is reserved in Francois Peron National Park (52 500 ha), while the township of Monkey Mia is situated on the southern boundary of the park on the shores of Shark Bay. Accommodation in the area includes the Monkey Mia Resort (which offers units, backpacker accommodation and camping), Hamelin Pool Caravan Park, camping grounds in Francois Peron National Park, hotels and caravan parks in Denham and station-style accommodation at Tamala Station www.tamala.com.au. See also www.sharkbay.org.

Monkey Mia (24 km north-east of Denham, 151 km from the North West Coastal Hwy) is world famous for the pod of wild Bottlenose Dolphins that are habituated to humans. It is also an excellent place for the Western Grasswren (formerly *textilis* ssp. of Thick-billed Grasswren), which are usually easy to observe around the main carpark to the dolphin-viewing area. They also occur along the Wulyibidi Yaninyina Trail (2 km loop), which starts at the same carpark and winds its way through the surrounding sand dunes before passing through *Acacia* shrublands to the beach. This walk also includes a bird hide that overlooks an artificial waterhole and lookout. (Note the Western Grasswren extends to SA and also reliably found around Iron Knob in that state p. 377.) Aside from Western Grasswren, along the walk, look for Emu, Southern Scrub-robin, Crested Bellbird, Chiming Wedgebill, Black-eared Cuckoo, White-browed Babbler, White-winged and Variegated Fairy-wren and, occasionally, nomadic honeyeaters such as Pied, Black and White-fronted. On the foreshore, you may see Eastern Reef Egret, Pacific Gull, and Fairy, Lesser Crested, Crested, Caspian and, less commonly, Roseate and Gull-billed Terns. Between September and May, numbers of migratory shorebirds such as Grey Plover, Grey-tailed Tattler, Bar-tailed Godwit, Curlew Sandpiper, Eastern Curlew and Red-necked Stint feed along the foreshore; at other times of year, there are usually a few 'overwintering' individuals present. There are visitor fees for entry to Monkey Mia Reserve. Visit www.monkeymia.com.au, (08) 9948 1366.

Western Grasswren also occur around Denham; search for them around the south-west corner of Little Lagoon, located immediately west of the intersection of Monkey Mia Rd and Stella Rowley Drive (3 km north of Denham). In the areas of saltmarsh around the lake, it is worth looking for Chiming Wedgebill and White-winged Fairy-wren. Along the mangrove-lined creek that connects Little Lagoon with Shark Bay, also look for Mangrove Grey Fantail and Yellow White-eye: this is the southern limit of distribution for both species. If you do not see Western Grasswren at either Monkey Mia or Little Lagoon, look for them around the Peron Homestead in Francois Peron National Park. The park has been the focus of intensive landscape restoration that has included a re-introduction program for Malleefowl, (occasionally seen around the homestead) and threatened mammals such as the Bilby, Woylie, Banded Hare-wallaby and Rufous Hare-wallaby (Mala).

Rufous Fieldwren (ssp. *rubiginosis*) occurs in saltmarsh on the Peron Peninsula. Good areas to look include saltmarsh along Shark Bay Rd, along the road to Eagle Bluff (accessed 19 km south of Denham), and along an unnamed track that leads north to a saline estuary 1.5 km before you reach Eagle Bluff. Once at Eagle Bluff, scan the coastal zone for Osprey, which nest on the rocks on Eagle Island, just offshore from the bluff. Eagle Bluff is also one of the best places in Shark Bay to view marine life – look in the shallow waters below the boardwalk for Manta Ray, Green and Loggerhead Turtle, sharks and Dugong (mainly summer, when they move inshore to feed on the seagrass). The northern-most population of Rock Parrot reside on a few offshore islands in Shark Bay and are occasionally recorded at Baba Head on the southern side of Shark Bay. To get there, drive west along Useless Loop Rd (which turns off Shark Bay Rd, 87 km south of Denham) for 46 km to Tamala Homestead turn-off. Baba Head is located 2 km directly north of the Tamala Station turn-off.

Dirk Hartog Island (620 km^2) is located off the WA coastline, within the Shark Bay World Heritage Area. Consisting of shrub-covered sand dunes, the island is home to the 'Dirk Hartog' White-winged Fairy-wren (ssp. *leucopterus*): a subspecies that is black with white-wings. Dirk Hartog Island also has unique subspecies of Rufous Fieldwren (ssp. *hartogi*) and Southern Emu-wren (ssp. *hartogi* – scarce in the coastal heath southern side of the island). Access to the island is by boat from Denham, light aircraft from Monkey Mia Airport and, on weekends, a landing barge transfers vehicles (4WD) from Steep Point to Dirk Hartog Island. There are several camping grounds on the island and accommodation at the Dirk Hartog Island Resort. See www.dirkhartogisland.com.

The 'Shark Bay Variegated Fairy-wren' (ssp. *bernieri*) is found only in the low heath and shrubs of Bernier and Dorre Islands, in the Shark Bay World Heritage Area. Classified as 'A' Class nature reserves, the islands also support the last naturally occurring population of four Australian mammals: Shark Bay Mouse, Banded and Rufous Hare-wallaby (Mala), and Western Barred Bandicoot. To protect wildlife, there is no overnight access to the islands. Day visits are permitted on Bernier Island. See www.sharkbay.org/bernier_dorre_islands.aspx.

Carnarvon region

Key species: Dusky Gerygone, Yellow White-eye, White-breasted Whistler, Mangrove Grey Fantail, Western Quail-thrush
Other species: Black-tailed Treecreeper, Bourke's Parrot, Spinifex Pigeon, Black-eared Cuckoo, shorebirds, terns

Carnarvon sits adjacent to the Indian Ocean, 900 km north of Perth. It is a superb birding area, with a diverse selection of habitats presenting excellent opportunities for migratory waders, seabirds and waterbirds. Much of the area is within the Shark Bay World Heritage Area. A variety of accommodation is available in the Carnarvon township. See www.carnarvon.wa.gov.au.

Around Carnarvon township, there are several important roost sites for shorebirds. Boat Harbour (also known as Pickles Point) is located at the end of Harbour Rd; access is via Douglas St, which after 800 m becomes Harbour Rd, with a carpark at the end. As a roost site, Boat Harbour is often preferred by larger waders such as Bar-tailed Godwit, Eastern Curlew, Whimbrel, Grey-tailed Tattler, Great and Red Knot, Common Greenshank, Grey Plover and Greater Sand Plover, particularly between September to May. There is also a chance of finding something rarer, because recent records have included Broad-billed Sandpiper, Asian Dowitcher and Common Redshank. The coastal fringe here supports important stands of mangroves, which hold Yellow White-eye and Dusky Gerygone. There are also records of White-breasted Whistler at this site, though they aren't detected routinely. Shorebirds and mangrove species also occur at Pelican Point Reserve, which is located at the southern end of Babbage Island, directly opposite (west) of Boat Harbour; access is via Babbage Island Rd and then Pelican Point Rd. Both Boat Harbour and Pelican Point are also good for terns, with regular sightings of Fairy, Lesser Crested, Crested and Caspian Tern, and less frequent sightings of Roseate and Gull-billed Tern. Franklin's Gull, a vagrant within Australia, has also been recorded here. The Babbage Island jetty (accessed from Babbage Island Rd) is another good spot to look for terns. The Carnarvon Sewage Treatment Works is located on Babbage Island Rd, 1 km from Robinson St. Look here for waterbirds and shorebirds such as Musk Duck, Australasian, Hoary-headed and Great Crested Grebe, Great and Little Egret, Common, Wood and Terek Sandpiper, and Grey-tailed Tattler. The site has also hosted several other rarer species, including Freckled Duck, Long-toed Stint, Pectoral Sandpiper, Ruff and Little Curlew. Chinamans Pool is a permanent waterbody on the lower reaches of the Gascoyne River. When the river has stopped flowing, this 2–3 km long pool provides an important refuge for waterbirds. Access is via Marmion St off Robinson St; drive to the northern end of Marmion St to park in the small dirt parking area on the southern banks of the river. The pool extends both up and downstream from this point, but the lower reaches are usually the best to explore. Black-winged Stilt, Red-kneed and Black-fronted Dotterel, as well as migratory shorebirds such as Common Greenshank, Wood, Common and Sharp-tailed Sandpiper, along with large numbers of common waterfowl, often congregate here. In February 2013 the site hosted Australia's first recorded Eurasian Wigeon.

East of Carnarvon, the McNeill Claypan (1800 ha) is a large ephemeral freshwater lake, accessed via the North West Coastal Hwy, 7 km east of Carnarvon. When the lake contains shallow water, it may support large numbers of shorebirds and waterbirds, including Bar-tailed Godwit, Grey-tailed Tattler, Great and Red Knot, Red-necked Stint, Black-winged Stilt, and Grey, Lesser Sand (uncommon) and Greater Sand Plover. The claypan has supported a good number of rarities, with recent records including Australian Painted Snipe, Little Curlew, Ruff, Long-toed Stint, Little Ringed Plover and Oriental Pratincole. Around the claypan, look for Chiming Wedgebill and White-winged Fairy-wren.

Rocky Pool is a pleasant permanent freshwater pool surrounded by riverine vegetation. It is located 55 km east of Carnarvon, on the Gascoyne Junction Rd. Along the water's edge, there is a chance of Black Bittern (rare, several records), while the woodland and scrublands around the pool may hold Blue-winged Kookaburra, Black-tailed Treecreeper ('Allied Treecreeper' ssp. *wellsi),* Mulga and Bourke Parrot, Spinifex Pigeon (uncommon), Black-eared Cuckoo and Southern Whiteface. Keep an eye open for Grey Falcon because this rarity has been reported here on several occasions. On the banks of Gregory River, at the pools, there is a basic camping area with no facilities. Further to the east, Gascoyne Junction

Western Quail-thrush habitat near Gascoyne Junction: look for low stony rises (especially quartzite) with much bare earth and a sparse cover of low shrubs.

marks the westerly limit for Western Quail-thrush; take the main road at least 5 km east of the town and search the quartzite stony rises, with low sparse shrubs on either side of the road. The species' distinctive thin, high-pitched call usually betrays their presence. Other species include Splendid, White-winged and Variegated Fairy-wren, Chestnut-rumped Thornbill and Redthroat.

South of Carnarvon, the coastal fringe supports extensive tracts of mangroves, with some areas extending 2 km inland from the open sea. There are several access points, though Bush Bay is probably the best for birding. It can be reached from the North West Coast Hwy, 37.4 km south of Carnarvon; from the turn-off ,it is another 9.8 km. Once there, search for mangrove species such as White-breasted Whistler (uncommon), Yellow White-eye, Dusky Gerygone and Mangrove Grey Fantail (the southern-most limit for all these species). The samphire flats bordering the mangrove areas can also provide some good birding, with a chance of seeing Slender-billed Thornbill, Rufous Fieldwren, Redthroat, Horsfield's Bushlark and Little Button-quail here.

Miaboolya Beach, 8 km north of Carnarvon, is another good area for mangrove birds. The sandy flats and beach also support numerous terns, including Fairy,

Lake MacLeod

The ephemeral saline Lake MacLeod (covering an area of ~1500 km^2) is the western-most lake in Australia. Approximately 130 km long and 40 km across, it is located between Carnarvon and Coral Bay, 10 km from the coast. Aside from two permanent waterholes located in the north, much of the lake's surface is usually dry, except after heavy rain. The northern ponds, named Cygnet Pond and Ibis Pond, are a unique wetland system maintained by underground seepage of seawater through vents or sinkholes.
The number of waterfowl and shorebirds counted at Lake MacLeod has exceeded 50 000, with 70 species recorded to date. Shorebirds include Curlew Sandpiper, Red-necked Stint, Red Knot, Bar-tailed Godwit, Banded and Black-winged Stilt, and Red-necked Avocet, while Fairy Tern also occur occasionally. A good range of rarities have been recorded including Broad-billed Sandpiper, Long-toed and Little Stint, Red-necked Phalarope, Oriental Plover and Inland Dotterel. Lake MacLeod is also one of the few places in the world where an inland community of mangroves occurs. This unique habitat has been isolated from the coast for ~6000 years and, remarkably, supports two specialist mangrove birds: Dusky Gerygone and Yellow White-eye. Other species found in the shrublands around the lake include Chiming Wedgebill, Little Button-quail, White-winged Fairy-wren, Crimson Chat, Brown and Rufous Songlark and Crested Bellbird. Note that access to this site is difficult and requires local knowledge.

Lesser Crested and Crested, Caspian, White-winged Black and Common (the latter two in summer). The Australian endemic Hooded Plover has been recorded here, the northern-most record for this species in Western Australian. To get there from Carnarvon, head west along the North West Coastal Hwy for 5.3 km and then turn north along Bibbawarra Rd for 5.8 km, before finally travelling east along Miaboolya Rd for 9.5 km. The Bibbawarra Bore is located on Bibbawarra Rd, 16 km north of Carnarvon. Star and Zebra Finch drink at the bore, while Redthroat, Chiming Wedgebill, Masked and Black-faced Woodswallow, and Splendid Fairy-wren all occur in the surrounding vegetation. Point Quobba and the Blowholes are located 70 km north-west of Carnarvon. A small lighthouse overlooks the ocean here, which is an excellent seabird watching area; scan the seas for Roseate, White-winged Black and Common Tern, and Hutton's, Wedge-tailed and, less commonly, Streaked Shearwater. Beach Stone-curlew has also been recorded in the area. Approximately 1 km south of the Blowholes, there is a camping area near a rock pool.

Jupiter Well

Key species: Princess Parrot, Grey Falcon,
Other species: Black-breasted Buzzard, Pied and Black Honeyeater

The very remote Jupiter Well area has in recent years been one of the more reliable places in Australia to find the enigmatic Princess Parrot. The well is located in the heart of the Gibson Desert, 260 km west of the WA–NT border, with best access from Alice Springs; from there, take the Stuart Hwy then the Tanami Rd. After 118 km on the Tanami Rd turn left on to the Gary Junction Rd, travel 552 km (passing Kintore) to Kiwirrkurra (considered one of Australia's most remote communities). Jupiter Well is another 124 km west along the Gary Junction Rd. The Gary Junction Rd is often in good condition for such a remote area, but 4WD is still essential. An access permit is required; contact both the Ngaanyatjarra Council (08) 8950 1741 and the Kiwirrkurra Aboriginal Community (08) 8956 8612. At Jupiter Well, there is a pleasant bush camping area shaded by Desert Oak. Water is generally available here (but not guaranteed), but there are no other facilities. Note that the campsite is not marked by signage, but rather a collection of 44-gallon drums marks the spot. Jupiter Well is located in a very remote area, so you will need to be self-sufficient in terms of food, water and, fuel. Also ensure you carry appropriate remote area vehicle spares and communication equipment.

Princess Parrot appear to be wide-ranging nomads, arriving in an area to breed when conditions are good, only to disappear again, sometimes for decades, when dry times return. In recent years, the two most consistent areas for the species in Western Australia have been in the Jupiter Well region and within Neale Junction Nature Reserve (p. 426).

Ningaloo Reef

Ningaloo Reef is included here not for its bird life, but because it is a superb wildlife environment. The reef is 260 km long and is Australia's largest fringing coral reef. It is also the only major reef situated very close to a land mass. Best-known as a place where Whale Shark feed (Mar–Jun), some of the other wildlife found along the reef include Humpback Whale (winter), Bottlenose and Indo-Pacific Humpback Dolphin, Dugong and Manta Ray. The beaches are also a breeding ground for Loggerhead, Green and Hawksbill Turtle. In June 2011, the World Heritage Committee inscribed the Ningaloo Coast on the World Heritage List. Visit www.environment.gov.au/heritage/places/world/ningaloo.

The main area to look for Princess Parrot is along Gary Junction Rd to the east of Jupiter Well. Recent sightings of these parrots have been made 2.7 km and 44 km to the east. A highly nomadic species, sightings are never guaranteed. It is also worth looking for the parrots west of the well, with recent reports along Gary Junction Rd, 30 km and 47 km west. Other birds to look for in the area include Australian Bustard, Red-backed Kingfisher, Budgerigar, Little Button-quail, Rainbow Bee-eater, White-winged Triller, Masked and Black-faced Woodswallow, Crested Bellbird, Crimson Chat, and Pied, Black, White-fronted, Black-chinned (golden-backed ssp. *laetior*), Grey-headed, Singing and Brown Honeyeaters. Sought-after birds of prey include Black-breasted Buzzard and the rare Grey Falcon.

Karijini National Park and Tom Price

Key species: Grey Honeyeater, Striated (Pilbara) Grasswren, Western Bowerbird, Spinifex Pigeon, Grey Falcon
Other species: Painted Finch, Pheasant Coucal, Black-eared Cuckoo, Western Bowerbird

Karijini (formerly Hamersley Ranges) National Park (627 442 ha) is the second largest national park in WA. The park is situated in the heart of the Pilbara, and consists of richly coloured Spinifex-clad plateaus with spectacular rugged river canyons. With a climate best described as tropical semi-desert, the park is located immediately north of the Tropic of Capricorn. The region supports several locally endemic subspecies. Late autumn, winter and early spring, when the temperatures are pleasant and humidity is low, are the best seasons to visit. You can reach the park via a sealed road and, once in the park, most gravel roads are well maintained. Camping options in the park include the Dales Campground and the Karijini Eco Retreat www.karijiniecoretreat.com.au. Camping is administered by WA DPaW see www.dec.wa.gov.au, with park information available from the park visitor centre, (08) 9189 8121. There is also accommodation in Tom Price, including the Tom Price Tourist Park (08) 9189 1515, Tom Price Motel (08) 9189 1101 and Windawarri Lodge (08) 9189 1110.

Karijini National Park contains a wide selection of birds, most notably Grey Honeyeater and Striated (Pilbara) Grasswren (ssp. *whitei*, a candidate for full species status). Search for Grey Honeyeater in the

Mulga woodlands along Karijini Drive: the road that leads into the park from Tom Price. At the western park entrance, turn right and drive to the carpark at Mount Bruce, which is the second highest peak in WA at 1235 m above sea level. Search the Mulga around the base of Mount Bruce, following the Honey Hakea Track (4.6 km). The complete walk to the summit is 9 km and takes ~6 h. Grey Honeyeater has also been seen around the Dale's Gorge Campground. It may also be worth asking the rangers if there have been any recent sightings.

Other worthwhile walks at Karijini include the Gorge Rim track (2 km), a pleasant walk that follows the rim of Dales Gorge between the Circular Pool Lookout and the Fortescue Falls Track (800 m). You can also walk along Dales Gorge, following the creek-side trail (2 km) between Circular Pool Trail and the Fortescue Falls. Fern Pool, a 10 min walk from Fortescue Falls, is also worth visiting for its picturesque setting. Along all of these walks, look for Blue-winged Kookaburra (Pilbara ssp. *cliftoni*), Spinifex Pigeon (red-bellied ssp. *ferruginea*), Diamond Dove, Painted Finch, Pheasant Coucal, Black-eared Cuckoo, Western Bowerbird (there is usually a tended bower near the Circular Pool carpark), Budgerigar, Australian Ringneck, Little Woodswallow and Grey-headed Honeyeater. Another good area for birdwatching is near Hancock and Weano Gorges. Both Striated (Pilbara) Grasswren and Spinifexbird have been recorded near Oxer and Junction Pool Lookouts; check areas of large mature Spinifex.

Tom Price, a township located on the western side of Karijini National Park, is the highest town in WA (747 m asl). Grey Honeyeater has been recorded feeding in the town's main streets (e.g. Cedar St), and it is also worth looking in the *Eucalyptus* and *Acacia* trees (especially those with mistletoe) along the small creek line that runs north-west of the sewage treatment plant. The treatment plant is located on the intersection of Mines Rd and Nameless Valley Drive. Other birds to look for along the creek and around the plant include Black-winged Stilt, Red-kneed Dotterel, Blue-winged Kookaburra (ssp. *cliftoni*),

A typical landscape in the Pilbara – red rocky hills and spinifex clad slopes and flats: Spinifex Pigeon, Rufous-crowned Emu-wren, Striated (Pilbara) Grasswren, Spinifexbird and Painted Finch are just some of the species that occupy this habitat.

The red-bellied form of the Spinifex Pigeon ssp. *ferruginea* is common through much of the Pilbara. Sightings are possible in and around many of the gorges, campgrounds and carparks within Karijini National Park.

Red-backed Kingfisher and Little Button-quail. The treatment plant can also be a good place for finches, including Star, Painted and Zebra. The most prominent feature of Tom Price is the peak known as Mount Nameless (1128 m), 2 km south-east of town. The summit can be accessed via a 4WD track that heads west of Mines Rd, 4 km south of Tom Price, or alternatively you can walk to the top from town. Striated (Pilbara) Grasswren occur here; search the plateau 200 m north-west of the collection of buildings near the summit. Other birds to look for on Mount Nameless include Spinifex Pigeon, Grey-headed Honeyeater and Little Woodswallow.

Note that, if you are unable to find the Striated (Pilbara) Grasswren at Junction Pool Lookouts or Mount Nameless, it is worth searching for them immediately south of the township of Newman (located on the Great Northern Hwy). A good spot to look

Lacepede Islands Nature Reserve

The Lacepede Islands (180 ha) are a spectacular group of sand cays on a large platform reef 120 km to the north-west of Broome (19 km off the Dampier Peninsula coastline). Declared an 'A' Class Nature Reserve and administered by the WA DPaW, in 2012, these islands were identified by Birdlife International as an Important Bird Area. They provide habitat for enormous breeding colonies of Brown Booby (the world's largest colony at ~18 000 breeding pairs) and Lesser Frigatebird (~2700 pairs), and large numbers of Roseate Tern (> 30 000 individuals). Black Rat was eradicated from the islands in 1986, enhancing conditions for breeding seabirds. You will almost certainly also observe Masked Booby, Eastern Reef Egret, Lesser Crested, Crested and Bridled Tern, Common Noddy, and Australian Pied and Sooty Oystercatcher, and migratory shorebirds such as Grey-tailed Tattler, Great Knot, Greater Sand Plover and Ruddy Turnstone. The Lacepede Islands are also a significant breeding ground for Green Turtle. This is a remote site with difficult access, but several commercial cruises that also visit the Kimberley coast call at the islands. See www.kimberleybirdwatching.com.au.

is on the stony Spinifex-clad hillsides to the right of the Newman Pistol Club, situated at the end of Winchester Close, on the far south of the town.

Broome and the Broome Bird Observatory

Key species: Vast numbers of shorebirds, including Common Redshank, Asian Dowitcher, Broad-billed Sandpiper, White-breasted and Mangrove Golden Whistler, Yellow Chat, rarities
Other species: Eastern Yellow Wagtail, Dusky Gerygone

Established in 1988, the Broome Bird Observatory overlooks Roebuck Bay: one of the most significant sites in Australia for shorebirds. Some 120 000 shorebirds visit the bay annually (though numbers

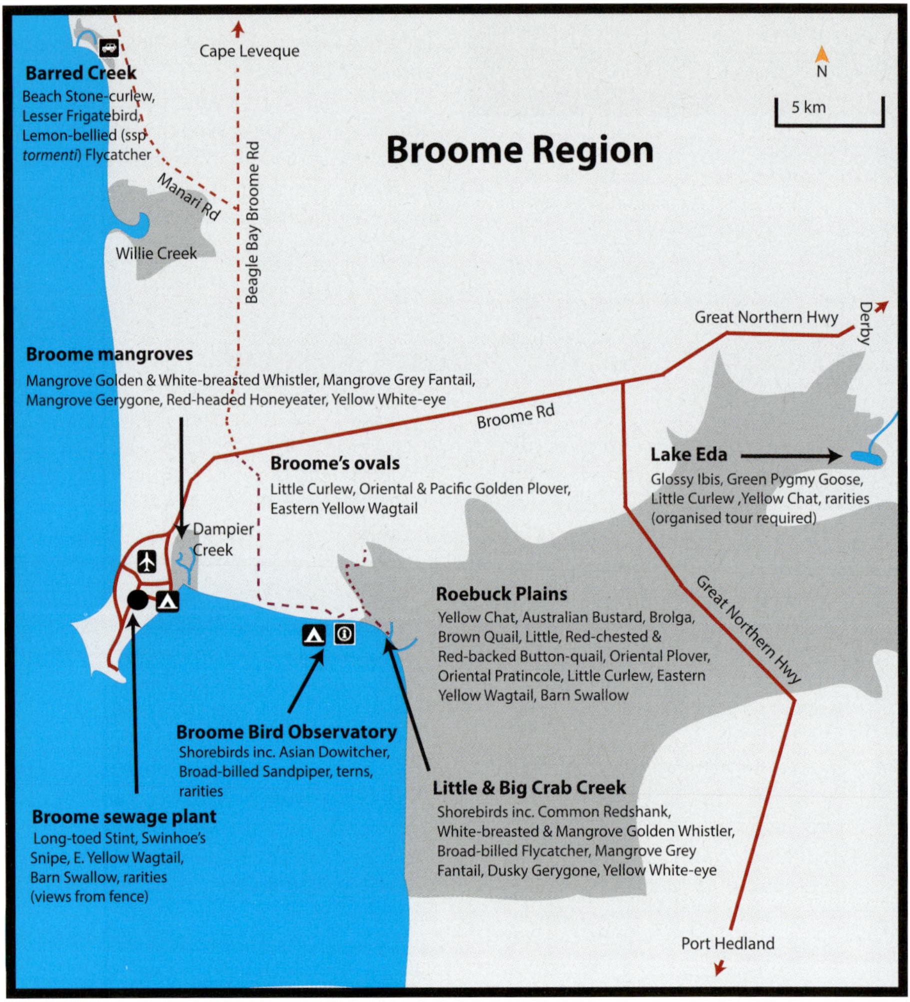

were once even higher) including standout species such as Asian Dowitcher and Common Redshank. The observatory is located 25 km east of Broome. To reach it, drive north from Broome along Broome Rd for 9 km, and then turn into Crab Creek Rd (unsealed); from there it is another 15 km. Check the road conditions during the wet season because the road can sometimes be closed. The observatory offers a range of accommodation, including a self-contained cottage, single, double, and family units and a campground. Facilities include a shade house, BBQ and amenities block. The observatory provides a range of birdwatching tours: www.broomebirdobservatory.com, (08) 9193 5600. There is a wide range of accommodation options in Broome www.broomevisitorcentre.com.au. Broome also has an international airport www.broomeair.com.au.

One of the main areas for viewing shorebirds is at the roost sites along the shores of the Roebuck Bay, located in the vicinity of the observatory. The best months for shorebirds are between September and April, although many birds overwinter. The best viewing times are during the high tide, when the shorebirds move off the vast expanses of mud flats as they are covered by the rising tide to rest at specific locations along the shoreline. The more common shorebirds seen include Terek and Curlew Sandpiper, Whimbrel, Eastern Curlew, Common Greenshank, Bar-tailed and Black-tailed Godwit, Great and Red Knot, Grey-tailed Tattler, Red-necked Stint, Ruddy Turnstone and Pacific Golden, Grey, and Lesser and Greater Sand Plover. The beach is also excellent for terns, including Lesser Crested, Little, Whiskered and White-winged Black Tern. Look carefully

The red sandy shoreline of Roebuck Bay bound the invertebrate-rich mud flats that extend for over 1 km out to the low tide mark. It is these mud flats that attract and support a large and exceptionally diverse concentration of migratory shorebirds.

at the Gull-billed Terns because both the large white Australian ssp. *macrotarsa* and the small greyer backed Asian ssp. *affinis* occur on the sand flats. Some of the notable species regularly recorded here include Asian Dowitcher, Broad-billed Sandpiper and Common Redshank. As a general guide, Common Redshank is most reliably seen on a rising tide by scanning the flats and especially the muddy drainage channels on the seaward side of the mangroves at the mouth of Little Crab Creek (ensure you leave enough time to 'sludge' back through the mangroves to the beach as the tide rises), while both Asian Dowitcher and Broad-billed Sandpiper are more easily found among the roosting shorebirds on the beaches at high tide; check with the Bird Observatory regarding the best current place to look for each species. Rarities seen here include Eurasian Curlew, Semipalmated Plover, Little Stint, and Franklin's and Lesser Black-backed Gull.

Little Crab Creek flows into Roebuck Bay 3 km east of the observatory. The mud flats at the creek mouth are a foraging site for shore birds, particularly during an approaching high tide, when they move onshore to roosting sites. In addition to Common Redshank (see above), look here for Striated Heron, Eastern Reef Egret, Bar-tailed Godwit, Grey-tailed Tattler, Common, Curlew and Terek Sandpiper, Eastern Curlew, Whimbrel, and Great and Red Knot. There is always a chance of rarer species such as Asian Dowitcher, Broad-billed Sandpiper and Oriental Plover. Among the mangroves bordering the creek, you may find White-breasted and Mangrove Golden Whistler, Broad-billed and Paperbark Flycatcher, Mangrove Grey Fantail, Dusky Gerygone and Yellow White-eye. As in most mangrove areas, mangrove birds respond well to 'pishing'.

The Roebuck Plains, which can be accessed 1 km north-east of the observatory via the Malurus Trail, are one of the most reliable places in Australia to see the rare Yellow Chat. At the end of the trail, there are several tracks following fence lines through open grassy plains and samphire flats; access is on foot only. Look here for Yellow Chat, Australian Bustard, Brolga, Brown Quail and Horsfield's Bushlark (ssp. *halli*). The grassy areas are at times particularly good for button-quail, with a chance of flushing Little, Red-chested and Red-backed Button-quail. Migratory species (from September to April) visiting the plains included Oriental Plover, Oriental Pratincole, Little Curlew, Eastern Yellow Wagtail and, occasionally, Barn Swallow.

Lake Eda is located 40 km east of Broome on the north-east side of the Roebuck Plains. It is an almost permanent body of freshwater and an important refuge site for birds in the southern Kimberley. At the lake, you may see Australian and Hoary headed Grebe, Brolga, all three Australian Ibis species, Green Pygmy Goose, Plumed and Wandering Whistling-Duck, Hardhead, Black Duck, Little Curlew, Long-toed stint (both summer), Whiskered Tern, Yellow Chat, Grey-crowned Babbler and Eastern Yellow Wagtail (the latter mostly Oct–Dec). A superb list of rarities has been recorded, including Australian Painted Snipe, Garganey, Freckled Duck, Spotted Redshank, Pectoral Sandpiper, Red necked Phalarope, Flock Bronzewing, Grey Wagtail and Rose-coloured Starling (Australia's first). The lake is on private property and access for visitors is only granted through pre-arranged tours; the best way to see the plains and wetland system is therefore via bird tours organised through the Bird Observatory.

The sewage works in Broome are always worth a look. They are accessed via a dirt track south of Clementson St, between Hunter St and Pembroke Rd. There is no direct access to the ponds, but they can be easily viewed from the vehicle track that circles the boundary fence. Regular species here include Australian Grebe, Plumed Whistling-Duck, Hardhead, Pacific Black Duck, Grey Teal, Black-winged Stilt, Black-fronted Dotterel, and Gull-billed, Whiskered and White-winged Black Tern (wet season). The ponds have outstanding potential for rarities; from October, Long-toed Stint, Eastern Yellow Wagtail and Barn Swallow start to appear, and, by November, both Swinhoe's and Pin-tailed Snipe can sometimes be found. Numbers of all these species generally decline in mid to late December after the first wet season rains arrive as they move out onto the flooded plains. Other less predictable rarities that have occurred here included Garganey, Little-ringed Plover, Semipalmated Plover, Ruff, Pectoral Sandpiper, Little Stint, Franklins Gull, White and Grey Wagtail, Red-rumped Swallow, nest-swiftlets and House Swift.

In Broome, during the wet season, the areas around the ovals are worth checking for Little Curlew, Eastern Yellow Wagtail and possible rarities. The town's mangroves hold Mangrove Golden and White-breasted Whistler, Mangrove Grey Fantail, Broad-billed Flycatcher, Dusky and Mangrove Gerygone, Red-headed Honeyeater and Yellow White-eye; the mangrove area behind the Roebuck Hotel, and especially Streeters Jetty accessed from Dampier Terrace just north of Short St, is a good place to start the search. Look here also for Whimbrel, Grey-tailed Tattler, Striated Heron, Eastern Reef Egret and Sacred Kingfisher (despite occasional reports, Collared Kingfisher is extremely rare in the Broome area). The port and public boat ramp area is another area within the Broome city limits that is worth visiting; check the beaches for shorebirds and terns, while Brown Booby often perch on the channel markers off the coast, and both Wilson's Storm-Petrel and Streaked Shearwater have been observed from the boat ramp carpark with the aid of a spotting scope. Lesser Frigatebirds are often seen overhead, especially in the late afternoon as they head north along the coastline (presumably to roost on the Lacepede Islands). In rough weather (i.e. immediately after a cyclone), the Port area can also be very productive for wind-blown seabirds. Gantheaume Point is another site worth visiting, because small numbers of shorebirds and terns regularly roost on the sandy beaches and rocky outcrops in the area. The whole region around Broome produces spectacular birding – in addition to those listed above, other rarities recorded include, Chinese Pond Heron, Spotted Redshank, Glossy Swiftlet, Arctic Warbler, Oriental Reed-Warbler, Blue and White Flycatcher, Red-throated Pipit and Eurasian Hoopoe.

The mangroves at Barred Creek (32 km north of Broome) have a sandy, rather than muddy, base, allowing relatively easy access on foot. It is a particularly good spot to find Beach Stone-curlew, Lesser Frigatebird (soaring overhead, especially late in the afternoon) and Lemon-bellied (Kimberley ssp. *tormenti*) Flycatcher – look

The striking Red-headed Honeyeater is a mangrove specialist distributed around the northern coastline. They are relatively easy to observe at mangrove sites around Broome, with Streeters Jetty off Damper Terrace being one of the best.

for the last in the outer foliage of the mangroves. Other birds that occur here include Brown Booby (out to sea), Eastern Reef Egret, Striated Heron, Black-necked Stork, Red-headed Honeyeater, White-breasted Whistler, Mangrove Gerygone, Red-winged Parrot, Little Bronze-Cuckoo, Broad-billed Flycatcher, Mangrove Grey Fantail and Northern Fantail. Birds of prey seen here include Osprey, White-bellied Sea-Eagle, Wedge-tailed Eagle and Collared Sparrowhawk. There is a camping area (an old pearler's camp), with no facilities. To get to Barred Creek, turn left off Broome Rd (9 km from Broome) on to the Beagle Bay–Broome Rd (the road to Cape Leveque). After 14.7 km, turn left again onto Manari Rd, then head north-west for 14.2 km. Here, there is a track that heads west to the banks of Barred Creek, a further 1.4 km. The roads and tracks are unsealed, and can be sandy, so a 4WD is required.

There are several significant freshwater wetlands dotted across the plains near Broome. Because they are on private property, a guided tour is the best option for access. Here, hundreds of Little Curlew congregate in the late dry season to drink and bath on the shores of Lake Eda through the heat of the day.

Mornington Station

Key species: Gouldian Finch, Purple-crowned Fairy-wren, Buff-sided Robin, White-quilled Rock-Pigeon, Sandstone Shrike-thrush
Other species: Pictorella Mannikin, Star Finch, raptors

Mornington Sanctuary is a vast conservation reserve covering 320 700 ha of the upper catchment of the Fitzroy river, as well as sections of the rugged King Leopold Ranges. The reserve is managed by the Australian Wildlife Conservancy: a non-government organisation that specialises in acquiring and managing significant habitats to ensure the ongoing protection of threatened plants and animals. A remarkable diversity of fauna occur within the reserve; besides the 202 species of bird, there are also 33 mammal, 76 reptile and 22 frog species that are known to occur on the property. Access is via the Gibb River Rd, with the turn-off to Mornington being 330 km from Derby or 565 km from Kununurra. From the turn-off, it is a further 88 km south to the camp area. Mornington is open to visitors between May and October, weather and road conditions permitting; 4WD is essential. Facilities include safari-style tents, each with en suite and a private balcony, a restaurant and bar and a small campground with unpowered sites. The reserve is probably best known for its large

Derby

Situated towards the head of King Sound, Derby has the most extreme tidal fluctuations of any town in Australia, with the differential between low and high tide nearing 12 m at peak times. The town is located 220 km from Broome. The mud flats around the jetty and boat ramp hold a selection of coastal species, including Grey-tailed Tattler, Marsh Sandpiper, Eastern Curlew, Whimbrel, Common Greenshank, Black-tailed Godwit, Grey Plover and Striated Heron, and there is a chance of Great-billed Heron in the mangroves bordering the jetty. These mangroves also hold Lemon-bellied (Kimberley) Flycatcher, Collared Kingfisher, Mangrove Robin, Little Bronze-Cuckoo, Mangrove Gerygone, White-breasted and Mangrove Golden Whistler, and Yellow White-eye. The sewage ponds (accessed via Conway St, off Derby Hwy) often support a good selection of waterbirds and shorebirds such as Plumed Whistling-Duck, Australasian Grebe, Royal Spoonbill, Common, Wood and Sharp-tailed Sandpiper, and Australian Spotted Crake, and there is a chance of wet season rarities such as Garganey, Little Ringed Plover, Long-toed Stint, Ruff and Oriental Pratincole, Swinhoe's Snipe, Barn Swallow and Eastern Yellow Wagtail. The ponds themselves often hold just a few ducks; it is the overflow ponds and marshland beyond the fenced ponds that are most productive. In Derby, it always worth being alert for something unusual, especially during the wet season. Vagrant species recorded in the area include Fairy Pitta (Australia's first record), Sabine's Gull and White Wagtail. There is a variety of accommodation in Derby. For further information, contact the Derby Visitor Centre www.derbytourism.com.au (08) 9191 1426.

population of Gouldian Finches, but it also hosts nine other species of finch including sought-after species such as Pictorella Mannikin, Painted Finch and Star Finch. The late dry season is often the best time to observe finches (and other seed-eaters) as they concentrate around the remaining waterholes. The thick riparian vegetation along Annie Creek provides excellent habitat for resident populations of Buff-sided Robin and Purple-crowned Fairy-wren, alongside Bar-breasted and White-gaped Honeyeater, Olive-backed Oriole, Crimson Finch and Northern Fantail. At night, Barking Owl frequently call in and around the camping area. Self-guided drives to Diamond Gorge and Sir John Gorge provide opportunities for sandstone escarpment species such as Sandstone Shrike-thrush and White-quilled Rock-Pigeon. Much of the reserve consists of tropical savannah woodlands; when key trees are in flower, these areas can be alive with nectar-feeding birds including Silver-crowned Friarbird,

Escarpment country and savannah woodland near Mornington Sanctuary in the Kimberley. Look in the woodlands for nectivores such as Varied Lorikeet, Silver-crowned Friarbird, and Banded and Black-chinned Honeyeater.

Banded and Black-chinned Honeyeater, Varied and Red-collared Lorikeet; Australian Bustard also occurs frequently in these open woodlands. Finally, the reserve supports an impressive number of raptors, including the rare and elusive Red Goshawk and Grey Falcon.

Mitchell Falls National Park

Key species: Black Grasswren, Kimberley Honeyeater, White-quilled Rock-Pigeon, Partridge Pigeon
Other species: Red Goshawk (rare), Sandstone Shrike-thrush, Rainbow Pitta, Green-backed Gerygone

Mitchell River National Park (115 300 ha), in the far north of the Kimberley, encompasses the Mitchell Plateau, and includes Mitchell Falls, Mertens Falls and Little Mertens Falls. The landscape around the plateau features superb escarpments alongside deep gorges and pandanus-lined creeks. The plateau is home to two Kimberley endemics: Black Grasswren and Kimberley Honeyeater. Largely because of the remoteness of the area, Black Grasswren is considered one of the hardest Australian birds to see; indicative of this, the nest and eggs of this species were not described until 1998. Access to the Mitchell Plateau is via the Gibb River Rd (295 km from Kununurra and 407 km from Derby), the Kalumburu Rd for a further 160 km, and then the Mitchell Plateau Track for the final 85 km to the Mitchell Falls (Punamii-unpuu) Campground. Access is strictly 4WD

The Buff-sided Robin is relatively common in the more dense riparian vegetation that lines many of the rivers in the Kimberley. In Kununurra, the species even frequents some of the caravan parks that front onto Lake Kununurra and the Ord River.

only; allow at least 2 days to drive in from Kununurra. The large campsite has toilets and fireplaces, with water available from the Mertens Creek. Park Rangers are based at Punamii-unpuu between May and October.

The most accessible site to see Black Grasswren is along the Punamii-unpuu (Mitchell Falls) Walk (8.6 km). The hike begins by the Mertens Creek crossing, travelling to the top of the Little Mertens Falls (15 min). Here you may see White-quilled Rock-Pigeon and Green-backed Gerygone. Look for Black Grasswren in areas of rocky escarpments vegetated with Spinifex on the right side of the creek, particularly north of the falls – though it also occurs on the south side of the falls. The best time to see Black Grasswren is early in the morning and in the late afternoon; listen for its chattering metallic call. The species can be quite a loud songster and the call sometimes echoes off rocky walls and gullies. If you miss seeing Black Grasswren near the Little Mertens Falls, it is also found in the areas of rocky escarpments and Spinifex along the walk to, and near, Mitchell Falls. White-quilled Rock-Pigeon can be common in this area, while Partridge Pigeon (yellow-faced ssp. *blaauwi*), is more common in the adjacent woodland (including the campground and the Mitchell Plateau access track). The other prime target in the Mitchell Falls National Park is Kimberley Honeyeater, only recently recognised as a full species (separated from White-lined Honeyeater). Once you leave Little Mertens Falls, you continue towards Mitchell Falls and the path leads through a shady patch of rainforest; the adjacent escarpment is a good area to look for Kimberley Honeyeater, while Buff-sided Robin and Rainbow Pitta sometimes frequent the rainforest thicket. Other birds on the Mitchell Plateau include Sandstone Shrike-thrush (common in escarpment country), Little Woodswallow, Grey Butcherbird (silver-backed ssp. *argenteus*), Variegated Fairy-wren (lavender-flanked ssp. *rogersi*), Banded, Bar-breasted, Yellow-tinted, White-gaped and Rufous-throated Honeyeater, and Northern Fantail. There is also a possibility of Red Goshawk in or over any of the wooded areas. While driving the Mitchell Plateau Track, stop and search large open grassy areas (particularly around burnt patches) with coarse gravelly substrate for Chestnut-backed Button-quail.

Drysdale River

Key species: Purple-crowned Fairy-wren, Crested (Northern) Shrike-tit, Buff-sided Robin
Other species: Pacific Baza, Black Bittern

The Drysdale River is a convenient rest site and camping spot on any trip into Mitchell Plateau, because it is situated on the Kalumburu Rd, 59 km north of the junction with the Gibb River Road, 360 km from Kununurra and 470 km from Derby. Camping is available on Drysdale River Station at either the homestead area or on the banks of the Drysdale River at the

A paperbark-lined billabong above Big Mertons Falls, Mitchell Plateau. This habitat supports Banded, Bar-breasted and White-gaped Honeyeater, and Northern Fantail, while the small rainforest patches along Mertons Creek provide opportunities for Rose-crowned Fruit-Dove, Rainbow Pitta and Buff-sided Robin.

Miners Pool. Both sites have toilets and hot showers, while the homestead area also offers motel-style accommodation, meals and a beer garden, fuel and a small general store.

The best area to visit is the area in the vicinity of Miners Pool (well signposted, 5 km from the homestead). Along the banks of this permanent waterhole, parties of Purple-crowned Fairy-wrens are reasonably common. Other birds that are common in the area include Azure Kingfisher, Blue-winged Kookaburra, Varied Lorikeet, Banded, Bar-breasted, Yellow-tinted, White-gaped and Rufous-throated Honeyeater, Silver-crowned Friarbird, Long-tailed and Crimson Finch, and Great Bowerbird. The other sought-after species that occurs in the Drysdale River crossing area is the Northern Shrike-tit (ssp.

The Black Grasswren is undoubtedly the star attraction at Mitchell Falls National Park. The species shows a preference for mature Spinifex on rocky 'jump ups' in proximity to water.

Bachsten Creek (Mount Elizabeth Station)

Black Grasswren inhabits the Spinifex-covered sandstone outcrops near Bachsten Creek, located on Mount Elizabeth Station. The station's main office is 30 km north off the Gibb River Rd – the turn-off is 340 km east of Derby and 365 km west of Kununurra www.mountelizabethstation.com, mt.elizabeth@bigpond.com, (08) 9191 4644. There is accommodation and camping at Mount Elizabeth Station. To search for Black Grasswren from Mount Elizabeth Station, travel 143 km along the Munja Track (an 8-h drive, and strictly 4WD only) to the Bachsten Creek campsite. With patience, Black Grasswren can be found in the 'lost city' rock formation, located next to the campsite, and it also occurs behind the manager's cottage. Facilities at the Bachsten Creek Campsite include cabins, campsites and showers. Note: the gate at the beginning of the Munja Track is usually locked – to open it, you need to collect the key from Mount Elizabeth Station. The station charges $100 per vehicle to travel the Munja Track, a fee that covers maintenance to re-open the track after each wet season.

On your way to the Bachsten Creek, Black Grasswren has also been recorded at the top of the first 'jump-up' (a step-rise along the Munja Track) only an hour's drive from Mount Elizabeth Station. Road conditions to this point are usually good; park your car near a creek line just before you reach the jump-up, and from here walk up the hill.

whitei). This species is considered particularly elusive by birdwatchers; just downstream of the crossing, search areas of open Ironbark woodland. The Northern Shrike-tit has a fairly soft, but quite distinctive, contact call, which is the best way to locate the species, though when ripping bark from trees in search of invertebrates they are also obvious and easily located!

Wyndham

Key species: Gouldian Finch, Mangrove Golden Whistler, White-breasted Whistler, Lemon-bellied Flycatcher, Chestnut-backed Button-quail, Arafura Fantail

Other species: Star Finch, Pictorella Mannikin, Broad-billed Flycatcher

Wyndham is WA's most northerly town and the oldest town in the Kimberley region. It is located on the Great Northern Hwy, 106 km from Kununurra. There is accommodation at the Three Mile Caravan Park (08) 9161 1064 on Baker St, at Parry Creek Farm in the Parry Lagoons Natures Reserve www.parrycreekfarm.com.au, and at the Wyndham Town Hotel, located on O'Donnell St, (08) 9161 1202.

Wyndham is one of Australia's most reliable places to see Gouldian Finch, particularly during the dry season when the species may venture into town to drink at any available water source (such as dripping taps). There are several good spots to look, including the Three Mile Caravan Park, the Wyndham District Hospital (located on Delamere St), and near the Aboriginal Dreamtime Statues (located in Warriu Park on Koolama St). Masked and Long-tailed Finch and, occasionally, Pictorella Mannikin also drink here. Early morning on the lawns in front of the council office can also be very good if the sprinklers are on.

There is a good area of mangroves along Foreshore Rd and Harbour Rd, near the Wyndham Wharf, that provides habitat for mangrove specialists such as Collared Kingfisher, Yellow White-eye, Broad-billed and Lemon-bellied (Kimberley ssp. *tormenti*) Flycatcher, Mangrove and Large-billed Gerygone, and Mangrove Grey Fantail. There is also a chance of Mangrove Golden (ssp. *melanura*) and White-breasted Whistler. Mangrove birds can respond well to 'pishing', which often attracts a multitude of species. Another rewarding area of mangroves is 16 km along King River Rd, which turns west from the Great Northern Hwy, 6 km south of Wyndham. In the mangroves that border the King River, you may see Mangrove Golden Whistler, Mangrove Gerygone, Lemon-bellied (Kimberley) and Shining Flycatcher, Mangrove Grey and Arafura Fantail, Yellow White-eye, Red-headed Honeyeater and Long-tailed and Star Finch. Chestnut-backed Button-quail has been recorded very close to Wyndham. Search the eucalypt woodland on the stony rises along Hugo Austla Drive, 800 m past the Wyndham District Hospital. Park 20 m past the Bastion sign (just north of a dirt track running east), and look for it in areas with patchy short grass and bare gravel. Alternatively, the species has also been reported on the road up to Five Rivers Lookout.

Parry Lagoons Nature Reserve and the Grotto

Key species: Huge numbers of water-birds, Star Finch, Pictorella Mannikin, Yellow-rumped Mannikin, Yellow Chat, Swinhoe's Snipe

The Ord River Floodplain includes some outstanding wetland systems. Parry Lagoons is one of the more accessible sites and includes vehicle access to a boardwalk and viewing platform.

Other species: Red-backed Button-quail, Lemon-bellied (Kimberley) Flycatcher, shorebirds including rarities

Parry Lagoons Nature Reserve (36 000 ha), located 20 km south of Wyndham, includes grasslands, open woodlands, sandstone country and an extensive series of wetlands. There is accommodation, camping, a cafe and restaurant at Parry Creek Farm (parrycreekfarm.com.au), and there is a range of accommodation in nearby Wyndham.

During the dry season, many of the lagoons in the reserve dry considerably, leading to large concentrations of waterbirds on the remaining waterbodies. The largest of the lagoons is the superb, Ramsar-listed, Marlgu Billabong, where you will find a boardwalk and bird hide. It is also worth walking from either side of the viewing areas, birdwatching in the grassy plains and along the creek lines. During the dry season, huge numbers of waterbirds can be present, feeding on the lagoon. You may see Brolga, Pied Heron, Black-necked Stork, Nankeen Night-Heron, Glossy Ibis, Royal Spoonbill, Comb-crested Jacana, Australasian Darter, and Caspian, Whiskered, White-winged Black (uncommon) and Gull-billed Tern, and waterfowl such as Plumed and Wandering Whistling-Duck, Green Pygmy-Goose, Magpie Goose, Radjah Shelduck, Pink-eared Duck and Hardhead. Rarer species include Garganey (several records, including occasional small flocks) and Northern Pintail (single record). The list of interesting species recorded around the edge of the lagoon is extensive: Star Finch, Pictorella and Yellow-rumped Mannikin, Yellow Chat, Brown Quail, Red-backed Button-quail, Horsfield's Bushlark, Golden-headed and Zitting (uncommon) Cisticola, Eastern Yellow Wagtail (wet) and Helmeted Guineafowl (feral). The lake fringes in this area are especially good for migratory shorebirds, with highlights including Australian Pratincole, Oriental Pratincole, Oriental Plover, Little Curlew, Swinhoe's Snipe, Australian Painted Snipe and Ruff. Other birds recorded in the reserve are Lemon-bellied (Kimberley ssp. *tormenti*) Flycatcher, Grey (Silver-backed ssp. *argenteus*) Butcherbird and Black-faced Woodswallow. Common raptors include Wedge-tailed Eagle, White-bellied Sea-Eagle, and Spotted and Swamp Harrier. Also keep an eye open for Black-breasted Buzzard circling overhead. At night, spotlight for Spotted Nightjar (dry season) and Barking Owl. The nearby Parry Creek Farm (50 ha) is a tourist resort on freehold land in the middle of Parry Lagoons Nature Reserve. A wide selection of finch species has been recorded on the farm, notably Gouldian and Star Finch, and you may also find Buff-sided Robin and Lemon-bellied Flycatcher.

The Grotto is a spectacular gorge and plunge pool that is a popular swimming hole 2 km off the Great Northern Hwy (39 km south of Wyndham). White-quilled Rock-Pigeon (ssp. *albipennis*), Spinifex Pigeon (red-plumed ssp. *ferrugenea*), Sandstone Shrike-thrush and Little Woodswallow are all reasonably common here. Birding in and around the Grotto may also produce sightings of Northern Fantail, Silver-crowned Friarbird, Varied Sittella, and Masked and Doubled-barred Finch. There is also a possibility of Gouldian Finch and Pictorella Mannikin. Nectivores are often seen feeding in flowering *Grevillea* and include Brown, White-gaped, Bar-breasted and Black-chinned (golden-backed ssp. *laetior*) Honeyeater, Silver-crowned Friarbird, and Varied and Red-collared Lorikeet.

Kununurra

Key species: Yellow-rumped Mannikin, Pictorella Mannikin, Star Finch, Buff-sided Robin, Black Bittern, Australian Little Bittern, White-browed Crake, Barn Swallow (wet season)

Other species: Oriental Cuckoo, Eastern Yellow Wagtail (both wet season)

Kununurra is located in the East Kimberley, 30 km west of the WA/NT border. By road, it is 1000 km from Broome, 800 km from Darwin and 100 km from Wyndham. There is a range of accommodation, including caravan parks and backpacker hostels. Information on accommodation and Lake Kununurra boat cruises are available from the Kununurra Visitor Centre www.visitkununurra.com, info@visitkununurra.com, (08) 9168 1177.

A Chestnut-breasted Mannikin perched on spear grass seed head. The irrigation area on the Weaber Plains, Kununurra is an excellent area for this and other species of finch.

Kununurra is an excellent birding location and one of Australia's best sites for a diversity of finches, in particular Yellow-rumped Mannikin. This species and, less commonly, Pictorella Mannikin and Star Finch occur in the Weaber Plains: an area of irrigated agricultural farmland immediately north of Kununurra. A good route to search for them is along Ivanhoe Rd, which runs north off the Victoria Hwy, 1 km west of the township. Drive north along the road and look for finches in the stands of native grass on the roadside verge and beside

Mirima (Hidden Valley) National Park

Likened to a small version of the Bungle Bungles, Mirima (Hidden Valley) National Park (2068 ha) is 2 km north-east of Kununurra. Access is via Hidden Valley Road, and the Hidden Valley Caravan Park sits adjacent to the park. Facilities include picnic tables, an information shelter and toilets. Lily Creek has eroded the sandstone massifs, creating spectacular sandstone domes and cliffs surrounded by grassy woodland. There are three short walks, all of which are worth doing: Gerliwany-gerring banan (1 km), Derdbe-gerring banan (800 m return) and Demboong banan Gap (500 m). Along these, look for sandstone specialists such as White-quilled Rock-Pigeon (ssp. *albipennis*) and Sandstone Shrike-thrush. Within the reserve, you may also encounter Northern Rosella, Northern Fantail, Little Woodswallow, Great Bowerbird, Rufous-throated and White-throated Honeyeater, Pacific Baza and Peregrine Falcon. In 2011, there were several reports of Dusky Honeyeater from within the reserve, though the species has still not been confirmed in WA. Mammals that frequent the park include Short-eared Rock-Wallaby and Euro.

irrigation channels, stopping whenever you see any bird activity. Other finches in the area include Crimson, Masked, Double-barred, Long-tailed and Zebra Finch, as well as Chestnut-breasted Mannikin. After 7.2 km, you come to River Farm Rd, which runs west for 3 km and is another good area for Yellow-rumped Mannikin. It is worth continuing north along Ivanhoe Rd until you reach Ivanhoe Crossing, again looking for Yellow-rumped Mannikin and other finches along the way. At the crossing, watch for Star Finch, Red-backed Kingfisher, Pheasant Coucal, Banded Honeyeater, White-throated Gerygone, Northern Fantail and Paperbark Flycatcher. Immediately to the left of the crossing, there is a good area of gallery forest, where you may see Buff-sided Robin, Shining Flycatcher and Yellow Oriole. If you miss seeing Star Finch at the above sites, they are fairly reliable further up the Ord River at Buttons Crossing, located on Parry Creek Rd, 8.6 km north of the Ivanhoe Crossing. Other birds to look for on the agricultural farmland north of Kununurra include Brolga, Pied Heron, Australian Bustard and Australian Pratincole. There is also the possibility of rarer species such as Grey Falcon or Eastern Grass Owl. In the wet season, Oriental Cuckoo and Barn Swallow occur here; check any swallows perched on powerlines carefully, because Red-rumped Swallow has also occurred. Finally, in January 2011, the second Australian record of a Javan Pond Heron was found in this area, just 19 km north of Kununurra.

Access to the Kununurra Sewage Treatment Works is via a track off Ivanhoe Rd, 800 m north of the Victoria Hwy. At the

Hidden Valley rock formations adjacent to Kununurra. These spectacular sandstone formations provide ideal habitat for the White-quilled Rock-Pigeon and Sandstone Shrike-thrush.

ponds, you may see waterbirds such as Pied Heron, Glossy Ibis, Australian Pratincole, Red-kneed Dotterel, Comb-crested Jacana, Black-winged Stilt, Plumed and Wandering Whistling-Duck, Radjah Shelduck and Magpie Goose. Between September and May, look for Common, Wood and Sharp-tailed Sandpiper. In the wet, search for Red-chested Button-quail in the rank grass adjacent to the ponds.

Interesting birds recorded at the Lake Kununurra Golf Course (located on Lake View Rd) include Buff-sided Robin (they inhabit the trees around the clubhouse), Star Finch, Yellow-rumped, Pictorella and Chestnut-breasted Mannikin, Red-browed Pardalote, Yellow Oriole, White-browed Crake and Banded Honeyeater. In the wet, there is a chance of migrants, such as Eastern Yellow Wagtail, and Barn and Red-rumped Swallow, though the last is a vagrant. Lake Kununurra (a section of the Ord River) and Lily Creek Lagoon are located on the southern edge of town. Both are good for waterbirds such as White-browed Crake, Black Bittern, Green Pygmy-Goose and Comb-crested Jacana. Oriental Reed-Warbler (wet season vagrant) has been recorded here; try the reeds at the boat ramp near Celebrity Tree Park (accessed via the Old Darwin Rd). Other birds that are possible include Star, Crimson and Double-barred Finch, Chestnut-breasted Mannikin, White-breasted and Masked Woodswallow, Pheasant Coucal and Paperbark Flycatcher. Several boat cruises operate on Lake Kununurra, some travelling up the Ord River to Lake Argyle. On these cruises, look for Black Bittern, Australian Little Bittern (uncommon), White-browed Crake, Comb-crested Jacana and Azure Kingfisher.

Lake Argyle

Key species: Yellow Chat, Purple-crowned Fairy-wren, White-quilled Rock-Pigeon, Gouldian Finch, waterbirds, shorebirds

Other species: Spinifex Pigeon, Black Bittern

Formed after construction of the Ord River Dam in 1971, the spectacular Lake Argyle is Australia's largest artificial lake. The turn off to Lake Argyle Village is 35 km east of Kununurra (7 km from the NT border) on the Victoria Hwy; from here, it is a further 35 km to the village which includes a store, fuel, visitor information centre, restaurant and caravan park, including cabin-style accommodation. Contact www.lakeargyle.com.au, resort@lakeargyle.com, (08) 9168 7777.

The birdwatching along scenic Lake Argyle Rd is excellent, with one especially good stop being Dead Horse Springs (24.5 km from the Victoria Hwy turn-off). The springs are a reliable site for White-quilled Rock-Pigeon (ssp. *albipennis*), Sandstone Shrike-thrush, Spinifex Pigeon (ssp. *ferrugenea*), Little Woodswallow, Red-browed Pardalote, and, occasionally, Gouldian and Painted Finch.

The Lake Argyle Tourist Village is a developed site that contains some good areas of riverine vegetation surrounded by high ridges. More common species that can be encountered around the village include Northern Rosella, Pheasant Coucal, Great Bowerbird, Yellow Oriole, Northern Fantail, and Masked and Long-tailed Finch, and a good selection of honeyeaters such as White-gaped, Brown, Yellow-tinted, White-throated and Blue-faced Honeyeater, Little and Silver-crowned Friarbird, and Yellow-throated Miner. Gouldian Finch occasionally feed around the caravan park, especially during the dry season. Look also for Spinifex Pigeon and White-quilled Rock-Pigeon on the bordering rocky ridges.

Regularly supporting around 150 000 waterbirds and shorebirds, Lake Argyle is listed wetland area under the Ramsar Convention. By far the best way to observe birds on and around the lake is by joining a specialised birdwatching boat charter

– contact the Lake Argyle resort for details. During such a boat cruise, you may see waterfowl such as Magpie Goose, Green Pygmy-Goose, Wandering Whistling-duck, Radjah Shelduck and Hardhead. Yellow Chat and Purple-crowned Fairy-wren occur in the floodplains and riverine forest at the southern end of Lake Argyle, near the outlet to the Ord River; the boat cruise targets both species, especially if you tell them of your interest. The shoreline and mud flats support Black-winged Stilt, White-browned Crake, Black Bittern, Nankeen Night Heron, Little, Intermediate and Great Egret, Glossy Ibis, Black-necked Stork and Brolga. The list of summer migratory shorebirds recorded from the cruise is extensive: Wood, Marsh, Curlew, Common and Sharp-tailed Sandpiper, Long-toed and Red-necked Stint, Common Greenshank, Bar-tailed Godwit, Oriental and Pacific Golden Plover, Oriental and Australian Pratincole, and Little Curlew have all been recorded. Other species you may encounter during a cruise include Sandstone Shrike-thrush, White-quilled Rock-Pigeon, Australian Bustard and Rainbow Bee-eater.

Purnululu (Bungle Bungles) National Park

Key species: White-quilled Rock-Pigeon, Sandstone Shrike-thrush, Painted Finch, Grey Falcon (rare)
Other species: Spinifex Pigeon, Spinifexbird

The spectacular (and World Heritage listed) Purnululu (Bungle Bungles) National Park (239 723 ha) contains an extraordinary array of banded sandstone domes (some towering to heights of over 300 m) covering a large portion of the park. It sits in a transitional zone, between the hot dry deserts of the WA arid zone to the south and the wetter monsoonal areas of the north. This results in a rich mix of species that frequent either tropical or arid environments. The park is accessed from Great Northern Hwy, via the Spring Creek Track (4WD and dry weather only), 252 km south of Kununurra (85 km north of Halls Creek). From the turn-off, it is another 53 km to the visitor centre. There are camping grounds at Walardi (12 km south of the visitor centre) or Kurrajonp (7 km north of the visitor centre), where facilities include toilets and water. There is also accommodation at the Bungle Bungles Wilderness Lodge. Depending on weather conditions, the park is usually open between April and December.

In the grassy woodlands on your way into the park, search along the roadsides for Australian Bustard, Emu and Brown Quail, stopping to look in any wooded areas for Crimson Chat, Black-tailed Treecreeper ('Allied Treecreeper' ssp. *wellsi*) and Red backed Kingfisher. The nationally rare Grey Falcon has occasionally been seen hunting over grassland areas.

Once in the park, there are several excellent walks. The Cathedral Gorge Walk (3 km) and Domes Walk (1 km circuit) both begin at the Piccaninny Creek carpark, taking you through the amazing beehive-shaped domes, along creek beds and through areas of Spinifex. These walks provide good opportunities for sightings of White-quilled Rock-Pigeon. Look also for Spinifex Pigeon, Spinifexbird, Painted and Long-tailed Finch, Red-browed Pardalote, Northern Fantail and Paperbark Flycatcher. If you have time (and inclination), it is worth considering the Piccaninny Gorge Walk (30 km): an overnight hike that passes through 7 km of relatively open terrain along the Piccaninny Creek before entering Piccaninny Gorge. There are two good walks on the northern side of the park: Mini Palms Gorge Walk (5 km return), which starts 9 km north of the visitor centre, and the Echidna Chasm Walk (2 km), which starts 19 km north of the visitor centre. Along both of these walks, look for Blue-winged Kookaburra, Great Bowerbird, Rainbow Bee-eater, Budgerigar, Northern

Rosella, Red-winged Parrot, Varied and Red-collared Lorikeet, Grey-headed, Rufous-throated, Blue-faced, Yellow-tinted, Grey-fronted, Brown and White-throated Honeyeater, and Silver-crowned and Little Friarbird. The park also supports a wide variety of mammals: Euro, Northern Nail-tail Wallaby, Short-eared Rock-wallaby and the world's smallest marsupial, the Long-tailed Planigale, all occur here.

Offshore islands and territories

There are eight offshore external Australian territories and more remote island areas that are of special interest to birdwatchers. Their appeal is that each supports either island endemic species or extra-limital species that don't typically occur elsewhere in Australia. Ashmore Reef, Christmas Island and the Cocos Islands are located in the tropical Indian Ocean – all are closer to Indonesia than the Australian mainland. Norfolk Island is located in the subtropical Pacific Ocean and is again situated closer to other nations, (in this case, both New Zealand and New Caledonia). Heard Island is one of the world's most isolated and difficult-to-access land masses, positioned both well south and approximately halfway between Australia and Africa in the subantarctic Indian Ocean. For completeness, Lord Howe Island (NSW), Macquarie Island (Tas) and Torres Strait Islands (Qld) are also dealt with in this section, given their geographic isolation. Christmas, Cocos, Norfolk and Lord Howe all support modern townships, have established tourist industries and are serviced by regular air services from the mainland. The northern islands of Torres Strait have small townships, no tourist infrastructure and require a permit to visit. Macquarie Island supports an established research base, with a permanent staff of 20–60 people, but is only accessible to birders who join one of the regular subantarctic cruises for tourists. Finally, both Ashmore Reef and Heard Island are uninhabited, have no infrastructure and can only be accessed on specifically tailored cruises.

Norfolk Island

Overview

Norfolk Island was discovered by Captain James Cook in 1774 and a British settlement was established in 1788, soon after Australia's first settlement at Sydney Cove, NSW. The island has a rather turbulent history. This included a period of occupation by Polynesians until perhaps 1000 years ago, occupation by British settlers (1788–1814), abandonment, then occupation as a penal institution for convicts (1825–55), before finally serving as the destination for the entire population of Pitcairn Island when the descendants of sailors who mutinied on the *Bounty* outgrew their tiny island home (1856). With some of the finest pre-colonial (<1850) examples of stone buildings in Australia, World Heritage listing and an abundance of interpretive information, it is no surprise that many visitors allocate a considerable portion of their time to the Kingston and Arthurs Vale historic area on the island. In addition to the historic precincts, Norfolk

Island supports a modern township, Burnt Pine, that includes numerous restaurants, cafes and duty free shops. Many of the ~60 accommodation options on the island are also close to the town, though there are also a range of more secluded options nestled against the national park or overlooking spectacular coastline.

Norfolk Island is situated 1610 km from Sydney, 1063 km north north-west of Auckland and 772 km south-east of Noumea. The main island has an area of 3460 ha, with about one-third of this being pasture. Large areas of native vegetation are now restricted to the National Park, though there are numerous smaller coastal reserves, some of which have been extensively replanted. Two small islands, Nepean and Phillip, lie to the south of the main island. Offshore, visits to Phillip Island and round island tours on chartered vessels are possible.

Because Norfolk Island was once part of the now submerged continent of Zealandia, biogeographically it shares many affinities with both New Caledonia and New Zealand. Pines belonging to the genus *Araucaria*, the Tasman Parakeet – an ally of the New Zealand Kakariki – and a dearth of native land mammals all bear witness to this relationship.

Most visitors to Norfolk stay for 4–7 days, though longer stays are of course possible. During a week-long stay several walks in the National Park and Botanic Gardens are recommended, as are visits to the various island headlands, including 100 Acres Reserve and Captain Cooks Monument, the grazed wetland areas of the Kingston Common and coastal reef sites

Views across Moo-oo Stone, Green Pool Stone and Bird Rock from Captain Cook Monument, Norfolk Island. An excellent vantage point from which to watch seabirds: look here for Black-winged Petrel, Red-tailed Tropicbird, Black Noddy, White Tern and Sooty Tern as well as the less common Brown Booby, Lesser Frigatebird and Great Frigatebird.

such as Slaughter Bay. A day trip to Phillip Island to observe the seasonally breeding seabirds would also be a highlight.

Norfolk Island has a mild subtropical climate. Temperature fluctuations are moderated by the surrounding ocean and average daily maximum temperatures for the 'summer' period (Oct–Mar) are 23–25°C, while the 'winter' period (Apr–Sep) averages 18–19°C. Minimum night time temperatures are ~6°C below the daily maximums. The wettest months are May–August and the driest November–January, with average monthly humidity varying little (74–79%). The summer period can be very rewarding, with many of the breeding seabirds and migrants (e.g. shorebirds) present. Winter has fewer birds, but this is when Little Shearwaters and the tiny colony of Providence Petrels breed. Visits in the 'shoulder' months of April and October have the potential to encounter both summer and winter species, but the total numbers of seabirds present are lower than at the height of the breeding season(s).

A hire car is necessary to get around with ease on the island and these are included in many accommodation packages. Even with an island speed limit of 50 km/h, at most, it is a 15 min drive from the township to the furthest destinations. Numerous steep hills mean that cycling is not as popular as it is on Lord Howe Island. The island is well serviced by a network of sealed roads, so a 2WD vehicle is perfectly adequate, even if some roads are a little bumpy. Daily scheduled flights operate from major eastern Australian capital cities. Because Norfolk is an external territory, a current Australian passport or a document of identity is required for Australian citizens, while others wishing to return to Australia will require a multiple entry Australian visa. No map is provided here because all the locations mentioned in the text are well represented on the many free tourist maps available on arrival at Norfolk.

Birding highlights

Norfolk Island supports three endemic species (Norfolk Island Gerygone, Slender-billed White-eye and Tasman Parakeet). Sadly, as with many island refuges, the island has lost numerous other species, including the White-chested White-eye, declared extinct as recently as 2011. The Pacific Robin (a close relative of the Scarlet Robin) is found nowhere else within Australia. The island also hosts several endemic subspecies, including unique forms of the Morepork (ssp. *undulata*), Golden Whistler (ssp. *xanthoprocta*), Grey Fantail (ssp. *pelzelni*) and Sacred Kingfisher (ssp. *norfolkiensis*). Norfolk also has the distinction of being the most reliable Australian site for two 'plastics': California Quail and feral chicken (aka Red Junglefowl). Seabirds are also impressive: Black-winged Petrel, Wedge-tailed Shearwater, Red-tailed Tropicbird, Black Noddy, and Sooty and White Tern all breed on the main island, while offshore islands support (among others) colonies of Little Shearwater, Masked Booby and Grey Ternlet. Following the eradication of rabbits and other pest mammals from Phillip Island, small colonies of Providence, White-necked and Kermadec Petrel are present there.

Norfolk Island National Park (main section) and Norfolk Island Botanic Gardens

Key species: Tasman Parakeet, Norfolk Morepork, Slender-billed White-eye, Pacific Robin, Long-tailed Cuckoo (slim chance, mostly spring)
Other species: Grey Fantail, Golden Whistler, Norfolk Island Gerygone

The Norfolk Island National Park is a large area of native vegetation in the northern part of the island. It incorporates the highest point on the island: Mount Bates (321 m).

The adjacent summit of Mount Pitt can easily be reached via a sealed road. There are several walking tracks within the park that link vehicle access at Mount Pitt, Palm Glen and Red Rd on the southern side and Captain Cook Monument on the northern side. Most of the walking tracks are well-formed park management tracks, though, because the park is quite hilly, the majority include a strenuous climb or descent in places. Mount Bates Track starting halfway up the road to Mount Pitt, has a particularly gentle grade, and the Palm Glen Circuit is also a pleasant stroll. Much of the park is infested by the invasive woody weed Cattley Guava, but park staff are making a concerted effort to remove this pest (and other weeds) and restore the vegetation to its natural state. All areas of native vegetation on any of these walks provide suitable habitat for the Slender-billed White-eye which, although Threatened, remains relatively common within the confines of the park. Usually feeding in association with the more abundant Silvereye, they are readily distinguished by their larger size and yellow-green bib. Pacific Robins are now all but restricted to the Park, preferring areas that support a dense cover of leaf litter under a canopy of native vegetation. The Critically Endangered Tasman Parakeet is often more easily encountered in and around the Botanic Gardens, but the bulk of the population (~240 birds) breeds within the park and some are always present here. Listen for their distinctive repetitive chatter '*kek-kek-kek-kek-kek*'. Try especially in and around the parking area at Palm Glen when the Guava are in fruit. Other species that can be expected here include the Norfolk Island Gerygone, Golden Whistler and extremely tame Grey Fantails, which may even land on you (or your tripod) if you make a soft squeak. Summer migrants include the New Zealand form of the Shining Bronze-Cuckoo. In spring, there is a chance of seeing Long-tailed Cuckoo; considered a rare vagrant to Australia, it is probably annual in small numbers at Norfolk on southward passage. Despite a loud and distinctive call, the species is notoriously difficult to see well in New Zealand, let-alone at Norfolk Island were most don't call – any observer that snags a sighting of this species should be pleased!

The population of boobooks on Norfolk Island now consists of hybrids between Norfolk Island Boobook and NZ Morepork. Most pairs hold territories in the National Park and adjacent Botanic Gardens.

At night, one can try for the Norfolk Island Morepork (ssp. *undulata*). Pairs of this extremely rare owl (perhaps just 30 individuals) occupy the Palm Glen, McLachlans Lane and Old Mountain Track areas and these can usually be heard calling to one another across the valleys on calm nights. The population now consists of hybrid individuals that resulted from a last ditch recovery effort that saw the sole remaining female of the subspecies pair with a male sourced from New Zealand in 1988. This pair raised four offspring over 2 years, forming the basis of the current population. Given the precarious state of the NI Morepork, and the fact that much of the ongoing management focusses on improving breeding success, call playback is to be avoided because most of this Critically Endangered population may well be within earshot from a single point!

The Botanic Gardens is a small (5.5 ha) reserve accessed from either the start of Mission Rd or Mount Pitt Rd. A series of short loop tracks traverse the slope, including the top-most track that is wheelchair accessible. The gardens incorporate extensive replanting on the upper slopes (including many fruiting plants that attract parakeets and white-eyes), while the lowest circuit passes through a dense rainforest gully festooned with giant vines. All tracks below the boardwalk include stairs, but none are particularly challenging and, if undertaken at a leisurely pace, are suitable for people of all fitness levels. Grey Fantail, Slender-billed White-eye, Silvereye and NI Gerygone are all reasonably common, and several pairs of Pacific Robin and Golden Whistler also occur. Pacific Emerald Dove can occasionally be found foraging in the garden beds. The Botanic Gardens are (usually!) the best place to locate the Tasman Parakeet. The species tends to be most active early and late in the day, retiring to the denser forest at other times, when they can be difficult to detect. Listen carefully for the distinctive call, check fruiting trees carefully and take note of the locations where the much more common (and introduced) Crimson Rosella feeds. For example, several peach trees are located in the small pastured valley below the Botanical Gardens. When in fruit (Dec–Feb), these trees provide foraging opportunities for the parakeet. This is private property (do not enter) but the fruit trees can be easily viewed from the gardens on the western side of Samson's Circuit or the steps on the Rainforest Gully Circuit below this.

Island Headlands, including Captain Cook Monument and 100 Acres Reserve

Key species: White Tern, Sooty Tern, Black Noddy, Masked Booby, Black-winged Petrel (summer)
Other species: Frigatebirds, Brown Booby, Common Noddy, NI Gerygone, California Quail

There are over 20 coastal lookouts accessible by vehicle or walking track around Norfolk Island. Many provide excellent opportunities for seabirds, especially if a spotting scope is used. Indeed, during a week-long April visit, one birder was able to observe all breeding seabirds, including individuals from tiny populations of Providence, White-necked and Kermadec Petrel from land. If extended periods of seawatching is not your thing, but the opportunity to observe some seabirds is still sought, then two lookouts in particular are worthy of a visit.

Captain Cook Monument is located on the northern side of the island. On the drive out from Burnt Pine, keep an eye out for the introduced California Quail – this species is quite common in rough pasture and farmland, but generally scarce in the well-vegetated National Park. The road leads right to the Monument and it is just a short walk to the lower of the two lookouts. Because the Monument is situated within the National Park, many of the more common bush birds can be found around the picnic area, including Norfolk Island Gerygone and Slender-billed White-eye, while a suite of introduced species, including California Quail and Crimson Rosella, frequent the lawned areas. The lower lookout provides views over several offshore rock stacks, including Moo-oo Stone, Green Pool Stone and Bird Rock. Here, binocular views of Masked Booby (nesting in summer) and Grey Ternlet, which roost on ledges on the steepest of the rock stacks, are possible. In summer, a steady stream of seabirds, including Black-winged Petrel, Red-tailed Tropicbird, Black Noddy, White Tern and Sooty Tern, fly over the point. The monument is also one of the best places to scan for visiting seabirds, including Brown Booby (check among the Masked Boobies perched on the rock stacks), Common Noddy (larger, more brown with a less distinct cap than the abundant Black Noddy), Little Black and Great Cormorant, and both Lesser and Great Frigatebird. In recent years, the frigatebirds have taken to roosting among the low shrubs on Moo-oo Stone and a careful check here may reveal individuals not immediately obvious to the naked eye.

Another wooded area and lookout worth a visit is the 100 Acres Reserve that includes Rocky Point (some maps refer to it as Rocky Point Reserve). Access is via New Farm Rd – on the drive from Burnt Pine, stop and scan both runways at the airport for migrant shorebirds. Pacific Golden Plover (summer), Double-banded Plover (winter) and other shorebirds roost on the runway and adjacent grass at high tide. Once you reach 100 Acres Reserve, enter the reserve via one of the two pedestrian gates. Much of the reserve is an arboretum. In the well-treed areas, colonies of the arboreal nesting Black Noddy and White Tern (both summer) occur. Bush birds include California Quail, Crimson Rosella, Pacific Emerald Dove, NI Gerygone and Silvereye. Slender-billed White-eye is still occasionally seen here, but the National Park and Botanic Gardens are much more reliable. In November 2010, a Long-tailed Cuckoo was also sighted here. The tracks from the two entrances form a loop and, at the head of this loop, a signposted track leads down to Rocky Point. It can be a bit slippery when wet, but the walk down (and back up!) is worth it for the views from Rocky Point. Rocky Point forms a 'corner' on Norfolk Island and thus, if the wind is strong, many seabirds pass near to the headland. Consequently, in the right conditions, it can be a good place to scan for shearwaters and those less common petrels.

The Tasman Parakeet is one of three endemic species on Norfolk Island. To find them, listen for their distinctive call and check fruiting trees around the Botanic Gardens, National Park and adjacent gardens.

Kingston Common, Slaughter and Emily Bay

Key species: Mallard, Wandering Tattler (summer), Pacific Golden Plover (summer), Double-banded Plover (winter), other visiting shorebirds and waterbirds
Other species: feral chicken, feral geese, California Quail

All visitors to Norfolk Island spend time in the Kingston and Arthurs Vale historic area, given the great historic importance of the site. The attraction for birders is that it is also the only large wetland site on the island. In dry times, the Kingston Common may consist of just a handful of pools, but, when rain has fallen, large areas of the pasture may be inundated. This means it is a magnet for regular and vagrant shorebirds and waterbirds alike. A scan of the wetlands will invariably produce sightings of White-faced Heron, Purple Swamphen and Mallard. True Mallard are essentially vagrants to mainland Australia (if one discounts the abundant farmyard hybrids found at many urban ponds) and thus Norfolk (along with Lord Howe and Macquarie Islands) are the only regular sites for wild-type Mallard. Sadly, even here, this introduced species greatly outnumbers (and hybridises with) the Pacific Black Duck. There is also a flock of feral geese here. Feral chickens, including some that show traits consistent with Red Junglefowl, are abundant. In the late afternoon, California Quail can usually be found feeding around the edges of the rougher pasture, especially up towards Watermill Dam. The grassy margins to the wetlands usually host Pacific Golden Plover (summer) and Double-banded Plover (winter), along with other visiting species. Pectoral Sandpiper appear as frequently as Sharp-tailed Sandpiper do, while there are usually a few Bar-tailed Godwit and Whimbrel (both summer) and small numbers of many other shorebirds have occurred. Check any Whimbrel carefully for the similar Bristle-thighed Curlew. There has been at least one previous report from Norfolk Island; situated well into the Pacific Ocean means the island is the most likely site within Australia for this potential mega-rarity. Other waterbirds also occur from time to time and thus it is not unusual to note one or two visitors – perhaps a cormorant, egret or ibis frequenting the Common or a Buff-banded Rail or Spotless

Kingston Common, Norfolk Island is a magnet for visiting and vagrant waterbirds and shorebirds. In addition to the resident Mallard, domestic geese and Purple Swamphen, and migrant Double-banded Plover (winter) and Pacific Golden Plover (summer), a few other species can usually be found on and around this freshwater wetland.

Crake at the margins to the reeds. Although there are a few other dams on the island (including Windmill Dam at the very 'top' of the Common), the only other large body of freshwater is located at Chapel Swamp, a dammed stream below St Barnabas Chapel on Douglas Rd. This too, is worth a careful scan each time it is passed, because Chapel Swamp has hosted several unusual waterbirds.

Just over the grassy bank from the Kingston Common is Slaughter Bay (so-named as a derivation of 'slack water' rather than some unspeakable horror). Owing to the expansive rock platform, the bay is perhaps the most reliable site on the island for Wandering Tattler (summer) and one or two can usually be found foraging on the rocks on falling or low tides. Offshore from Slaughter Bay and Emily Bay (the best swimming on the island) is Nepean Island. With binoculars (or, better, a spotting scope), views of the summer-breeding seabirds on Nepean Island can be had. Point Hunter, adjacent to the Lone Pine is the best vantage point for this. In addition to large numbers of Masked Booby and the occasional frigatebird, huge flocks of Wedge-tailed Shearwater can be seen wheeling over the sea – especially towards dusk.

Phillip Island

Key species: White-necked Petrel (summer), Kermadec Petrel (summer, chance), Providence Petrel (winter, chance), Little Shearwater (winter), Masked Booby, Grey Ternlet
Other species: Black-winged Petrel, Common Noddy, Sooty Tern

Phillip Island (197 ha) lies 7 km to the south of Norfolk and the entire island is part of Norfolk Island National Park. Once an island consisting solely of eroded red earth slopes, with the eradication of rabbits (1986) and other herbivores, the vegetation is slowly recovering. In the absence of feral predators, Phillip Island is also an Important Bird Area for seabirds, with at least 13 species breeding there. The island can only be visited as part of an organised tour with an accredited local guide. Generally, these run in the morning from the Kingston Jetty for a period of ~5 h. Currently only 'Charter Marine' http://trekking.nf/ can provide these services. Birders wishing to visit Phillip Island should contact Charter Marine soon after arrival because treks may only take place once a week and places are limited. Access is weather dependent and requires a reasonable degree of fitness. Here, the bulk of Norfolk's Black-winged Petrel population breed. It is also the only breeding site for the White-necked Petrel, and one of only two breeding sites for both the Providence and Kermadec Petrels, in Australia. All three species are, however, difficult to observe given daylight visiting hours and the fact that colonies are not right on the trekking route. Other species that breed on Phillip Island in summer include the Red-tailed Tropicbird, Masked Booby, a few pairs of Australian Gannets, Common Noddy, Sooty Tern and Grey Ternlet. In winter, most of the breeding seabirds have departed, but this is when both Providence Petrel and Little Shearwater come ashore to breed. There are few land birds on the Phillip Island and all are more easily observed on the main island, though there is always the chance of a vagrant; in May 2001, a Chaffinch in the company of Greenfinches was sighted near the peak.

Lord Howe Island

Overview

Lord Howe is one of Australia's most beautiful islands, with sandy beaches, crystal waters (especially within the extensive lagoon) and one of the world's most southerly coral reefs all contrasting with 'Jurassic park'-like southern peaks that are routinely shrouded in mist. Lord Howe supports a small township with a resident population of ~350 and tourist numbers are capped at 400 at any time. There are ~18 accommodation options that range from self-contained units to luxury resorts. Despite being settled since 1834, much of the island remains cloaked in native vegetation and almost 75% is designated as a Permanent Park Preserve (equivalent to a National Park).

Lord Howe Island is situated some 600 km east of the Australian mainland (off Port Macquarie, NSW). It is a relatively small island, being 11 km long and just 2 km wide, with a total land area of 1460 ha. The main island is dominated by the basalt peaks of Mount Lidgbird and Mount Gower at its southern end. The centre of the island is more low lying, while the northern section rises to the spectacular Malabar cliffs. From here, walkers are rewarded with dramatic views of the offshore rock stacks and islands, as well as picture-postcard views over the peaks to the south. The offshore islands provide numerous breeding refuges for seabirds, while further south the World Heritage-listed Balls Pyramid towers dramatically above the sea.

A week-long stay at Lord Howe is ideal. During such a stay, when not relaxing on the beach or snorkelling in the warm waters of the lagoon, most birders take several walks, including visits to the Malabar Cliffs in the north and Muttonbird Point in the south, spend time in the tern colonies behind Lagoon Beach and watch the display flights of petrels along with other nesting seabirds at Neds Beach. Night visits to a shearwater colony or two are a must. For the more energetic, there is also the trek to the summit of Mount Gower, which provides stunning views and opportunities to observe some of the other endemic birds that call Lord Howe home. If the weather is suitable, a boat trip to Balls Pyramid to observe the region's special seabirds would also be a highlight. Useful information on Lord Howe Island's birds can be found at www.lordhoweislandbirds.com Local ornithologist and naturalist Ian Hutton is also available for tours: www.lordhowe-tours.com.au.

Lord Howe Island's climate is mild all year round. Sea breezes prevent the summers from being too hot and the winters from being too cool. Maximum temperatures on the island are generally around 25°C throughout summer and 19°C in winter, while daily minimums range from 12 to 14°C in winter to 20°C in summer. Although Lord Howe Island is pleasant to visit at any time of year, the summer period is perhaps more rewarding for birders, with many of the summer-breeding seabirds and migrants, such as waders, present. Winter has fewer birds, but this is when Little Shearwaters and Providence Petrels breed. As with Norfolk Island, visits in the 'shoulder' months of April and October have the potential to encounter both summer and winter species, but total seabird numbers are generally lower than at the height of the breeding seasons.

Lord Howe Island Woodhens are a distinctive flightless rail. There are usually a few pairs within the township, though higher densities occur in areas south of the airport.

Because the island is small, and the southern half of the island is dominated by two mountain peaks, motor vehicles are limited to just a handful of sealed roads. As a consequence, the most popular form of transport is the bicycle. However, visitors are collected and returned to the airport as part of their accommodation package and there are regular shuttle services available to move between accommodation and eateries or to travel to the start of the Little Island and Mount Gower Trek towards the southern end of the island. A very limited number of hire cars is available, but these usually need to be booked well in advance. Regular scheduled flights operate from both Sydney and Brisbane.

Birding highlights

Lord Howe Island provides a spectacular (and relaxing!) backdrop to some quality birdwatching opportunities. The endemic Lord Howe (LH) Woodhen is a key target for most visitors, but so too are several seabirds that are difficult to observe elsewhere in Australia. Indeed, 14 species of seabird breed within the Lord Howe group. Almost the entire global population of the Providence Petrel breeds on the slopes of Mount Gower between April and November. Offshore islands and stacks provide the only Australian breeding site for the White-bellied Storm-Petrel and Ball's Pyramid supports the only substantial population of Kermadec Petrel in our region. Little Shearwaters, the most difficult of Australia's breeding shearwaters to observe, also breed (in winter) here. Land birds are rather depauperate, but there are three endemic subspecies: Silvereye (ssp. *tephropleurus*), Golden Whistler (ssp. *contempta*) and Pied Currawong (ssp. *crissalis*), that are generally easy to locate in wooded areas. Being an offshore island, there is always the chance of a vagrant too – Paradise Shelduck, South Island Pied Oystercatcher, Long-tailed Cuckoo, Yellowhammer, Chaffinch and Common Redpoll have all occurred. As is so true of many islands, the arrival of humans, and subsequently rats (in 1918), and other predators has seen the extinction of many species. Indeed, no fewer than nine unique birds have been lost from the island since human settlement.

Northern section

Key species: Black-winged Petrel, Wedge-tailed Shearwater, Flesh-footed Shearwater, Red-tailed Tropicbird, White Tern, Common Noddy (all spring–summer)
Other species: LH Woodhen, LH Pied Currawong, LH Golden Whistler, LH Silvereye

The northern section of Lord Howe Island contains the village, including most restaurant and accommodation options. Walking tracks in the far north provide access to North Bay, Mount Eliza, Kim's Lookout and the Malabar Cliffs. Most of these tracks have many stairs, so a slow steady pace is required. The Whistler, Silvereye and Pacific Emerald Dove are all regularly encountered along these tracks and there are usually a few pairs of the Endangered Pied Currawong. There are also some impressive opportunities to observe summer-breeding seabirds on these walks. Sooty Terns breed on many of the ledges around the Malabar Cliffs and at North Bay, while at North Bay also check the pines at the northern end for nesting Black Noddy. The larger Common Noddy nests among shrubs around the cliff slopes, including those in nearby Old Gulch. The Malabar Cliffs provide nesting opportunities for the bulk of the Red-tailed Tropicbird population at Lord Howe Island. Through the middle of the day, hundreds of tropicbirds perform spectacular flight displays (Nov–Jan). Kim's Lookout and Mount Eliza are

The Lord Howe Golden Whistler ssp. *contempta* is a relatively common species on the island and occurs in both forested habitats and the more lush gardens within the township.

good vantage points, with the added bonus that one of the resident Nankeen Kestrels may also be soaring in the updrafts. These cliffs have also hosted the occasional White-tailed Tropicbird and, in the summer of 2010–11, a Red-billed Tropicbird (the only Australian record) performed aerial displays and occupied a nest scrape here for several months. The '*pic-pic-pic*' calls of displaying Black-winged Petrels are also a summer feature, with numbers at their highest in the late afternoon. The cliffs here provide some opportunities for seabird-watching, though, being so elevated, a spotting scope is an advantage.

Back down around the village area, LH Whistler, LH Silvereye and Pacific Emerald Dove are generally common, as are Buff-banded Rail, and there are usually a few family groups of the Woodhen present. If a Woodhen family occupy the grounds of your accommodation, then you will enjoy daily sightings, but even if not, one would expect to encounter Woodhens around the village several times during a week-long stay. For those that enjoy snorkelling, in summer at Old Settlement Beach, relatively tame and approachable Green Turtles can also be found. Swampy areas behind Old Settlement Beach provide opportunities for Purple Swamphen, Masked Lapwing and Pacific Golden Plover (summer), while the denser forest in the Old Settlement area has at times harboured Masked Owl. The majority of the breeding White Terns (summer) can be found in the Norfolk Island Pines that line the lagoon. Neds Beach on the eastern side of the island is probably the best place to watch the displaying Black-winged Petrels (summer), harbours the most accessible Sooty Tern colony and has the added bonus of masses of tropical fish that congregate in the shallows for afternoon feeds.

Night forays to observe the summer-nesting shearwaters are an added bonus. Large numbers of Flesh-footed Shearwaters nest in the forest on the road to Neds Beach – just a five minute walk from the village centre, while the smaller Wedge-tailed Shearwater is most easily observed at Flagstaff Point or under the roadside vegetation between Pinetrees Resort and the start of the airfield and runway.

Southern section

Key species: Providence Petrel, LH Woodhen, Pied Currawong, Masked Booby
Other species: Black-winged Petrel, Red-tailed Tropicbird, LH Golden Whistler, LH Silvereye

The southern section of Lord Howe Island (defined here as south of the airport) is dominated by Mount Lidgbird (777 m) and Mount Gower (875 m). The airport runway and surrounding short grass can be good for shorebirds, including Pacific Golden Plover, Bar-tailed Godwit, Whimbrel and Ruddy Turnstone. At the eastern end of the runway, in the adjacent paddock, an area known as Mosley Park usually holds water. There is often a flock of Mallard here and a few White-faced Heron. It is also here that many of the rare waterbirds and shorebirds that appear from time to time are sighted. Magpie larks, which colonised the island almost 100 years ago, are common here, as they are in many other open areas on the island. At night, the introduced Masked Owl is also possible in this area. Almost adjacent to the turn into the airport buildings is the start of the Muttonbird Point walking track. Along the walk, you can expect to see the LH Golden Whistler and LH Silvereye, while at the point itself a lookout provides views of nesting Masked Boobies. This area can also be good for Common Noddy. You can return via the same track or continue onto Boat Harbour for more views along the coast before returning via Intermediate Hill. Woodhens are possible here, but are more reliable further south. Back on the road, the golf course area is worth a quick scan for shorebirds. Continuing over the hill past Capella Lodge, the paddocks adjacent to Salmon and Kings Beach are often a good location for the Woodhen – especially in the early morning or evening. Although the endemic form of the Pied Currawong occurs across the island, it is more common in the shadow of Mount Lidgbird. Here, one reaches the end of the road, the start of the Little Island walking track, and the trek to Mount Gower. The Kentia Palm forest section of the Little Island track (an easy 3 km return) is a good area for both the Woodhen and Pacific Emerald Dove and, from March to September, this area provides the best views of Providence Petrel for those not contemplating the steep Mount Gower trek. This trek continues on from the end of the Little Island track, but you must be accompanied by a licenced guide to proceed. This is the most strenuous

View to Mount Lidgbird near the start of the Little Island walking track, Lord Howe Island. Buff-banded Rail and Lord Howe Island Woodhen occasionally forage on the paddock margins here. This can also be a good vantage point to watch for Lord Howe Pied Currawongs if they are proving elusive elsewhere.

walk on Lord Howe Island, takes ~8 h and climbs from sea level to a height of 875 m. Many of the plants on the Mount Gower trek are endemic species that are difficult to see elsewhere and, at the summit, there is a unique moss forest and, on clear days, spectacular views. Woodhens are almost always encountered on the trek and, between March and September (April is best), it is possible to get up close and personal with the Providence Petrels that breed on the mountain slopes.

Balls Pyramid

Key species: Kermadec Petrel, White-bellied Storm-Petrel, White-necked Petrel (summer, chance), Little Shearwater (Apr–Nov), Providence Petrel (Mar–Nov)
Other species: Grey Ternlet, Masked Booby, Flesh-footed and Wedge-tailed Shearwater (summer)

Balls Pyramid is a spectacular 551 m volcanic stack that rises knife-like from the ocean some 23 km south of Lord Howe Island. Weather permitting, the area is accessible by charter boat for half- and full-day trips. The area is a popular fishing site and most local charter boats visit regularly. The best of these is Sea to Summit tours (www.lordhoweislandtours.net/) as the operator, Jack Shick, is a keen birder and photographer who understands the needs of birdwatchers. It is best to book on a trip as soon as you arrive on the island to avoid disappointment. Once clear of the reef, almost any subtropical seabird recorded from the region is possible (less common species have included Tahiti, Herald, Gould's and Cooks Petrel, and six species of albatross). Look for Red-tailed Tropicbird and Grey Ternlet on the cliffs and breeding Masked Booby on the inaccessible scree slopes below Mount Gower. Once out into deeper water, numbers of Flesh-footed and Wedge-tailed shearwater (summer) increase and you have an excellent chance of Kermadec Petrel and White-bellied Storm-Petrel – although both are summer breeders, they are present in local waters year round. From March to November, you will certainly see the winter-breeding Providence Petrel and you have a good chance of connecting with the winter-breeding Little Shearwater. Once down at the Pyramid, charter boats typically circumnavigate the 'rock' providing spectacular views. It is also here in the vicinity of the pyramid that Kermadec Petrel numbers are usually at their highest. It can be a spectacular sight surrounded by hundreds of seabirds with the Pyramid as a backdrop. In summer (especially Jan–Feb) there is also a chance of White-necked Petrel, which has been seen with increasing frequency (perhaps a pair or two have started calling Balls Pyramid home?). These charters operate no differently to the many 1-day pelagic trips that run off coastal ports on the mainland, so rug up, wear appropriate wet weather gear and take a seasickness pill if you are at all unsure of your sea legs!

Northern Torres Strait

Overview

Situated between 5 and 14 km off the southern coast of New Guinea, the islands of Boigu, Saibai and Dauan collectively form the north-western Torres Strait Island group. Despite their close proximity to the Papua New Guinea (PNG) coastline, these islands and their fringing coastal waters are Australian territory in the state of Qld. Saibai Island (population 380) is the largest of the group at 9625 ha. At its closest point, the open-water crossing between Saibai Island and PNG is just 3.7 km. Situated 40 km to the north-west of Saibai is the slightly smaller Boigu Island (population 340) covering some 7150 ha.

Because both islands were formed of alluvial soils from the rivers of PNG, they are low lying with similar topography. A coastal strip of mangroves is backed by a very narrow strip of dry vine forest, but the centre of each island is predominantly open grassy *Pandanus* and *Melaleuca* plains (more so on Saibai) and brackish swamps that support freshwater mangroves (more so on Boigu). The centre of each island is accessible via dirt tracks in the dry season, but, following wet season rains, much of each island becomes flooded and impassable. On the northern coast of each island, a township, associated airstrip, cemetery and refuse dump occupy small areas of the available land. In contrast, Dauan Island (population 165), located ~4 km west of Saibai is significantly smaller (325 ha), rocky and mountainous – at 297 m, the Devil's Marbles is the second highest peak in Torres Strait. The small township is situated along a narrow coastal strip, backed by steep hills on the north-east of the island and supports many large evergreen trees. Outside the township, there is a single large area of mangroves. The remainder of the coastline is mostly rocky, but here you will also find a couple of sandy beaches. The lower slopes of the Devil's Marbles are dominated by tall grass with emergent *Pandanus*, while boulder fields with a patchy cover of vine scrub dominate the upper slopes. On the south-facing slopes of the island, there are areas of denser vine thickets, but these areas are difficult to access and extremely difficult to traverse because of huge boulders covered in a matt of fallen leaves that conceal treacherous crevices.

Torres Strait is the most significant migration pathway for terrestrial bird species that migrate to and from Australia and many of the islands act as important stop-over points for birds in transit. In the north-western group, there are strong Papuan influences (more than 20 resident bird forms do not occur further south) and Australian influences, and thus the avifauna is both rich and unique. Coupled with the fact that vagrant PNG species occur from time to time, this destination has received increasing attention from visiting birdwatchers since the 1990s.

Although possible, visits to northern Torres Strait that are not part of an organised tour can be challenging. Because there is no infrastructure for tourists, all aspects of travel, including accommodation, transport and food, can be difficult to arrange. Additionally, before any trip, permission to visit each island must be obtained from each community in consultation with the Torres Strait Regional Authority. In recent years, groups have made day visits via charter flight from Horn Island, participated in voyages aboard charter vessels that have allowed several days ashore on each island or used scheduled air services to visit as small land-based groups. Birding on these islands can also be challenging because Estuarine Crocodiles are present in all waterways, mosquitos can be abundant and mosquito-borne Ross River Virus, Japanese Encephalitis, Dengue and Malaria have all occurred.

Like much of the tropical north, the islands of Torres Strait experience a low rainfall, low humidity 'dry season' (generally May–Nov) and a high rainfall, high humidity 'wet season' (generally Dec–Apr). It is typically slightly cooler in the middle of the dry (mean max temp of ~28°C in July) than it is in the late dry (mean max temp of ~32°C in Nov) but higher humidity in the period November–March means that it often feels much less pleasant during these months. Rainfall is highest in the months of January–April, but December can also be wet. Mosquito abundance is generally correlated with high rainfall months and atypical rain events (e.g. July 2007) can also lead to a rapid increase in these biting insects.

Birding is generally good all year round. Singing Starlings and Red-capped Flowerpeckers appear to be resident, though Collared Imperial-Pigeon numbers fluctuate in response to fruiting trees; although observed daily on many trips, they appear to be most consistently sighted during the wet season. Migration periods, especially September–November can be good for the passage of land birds. Although vagrant species have been found in all seasons, the highest incidence of vagrants occurs in the period November–February. This may be in part because trips at this time of year often encounter very windy conditions associated with cyclones or tropical lows that form to the south of the islands. That being said, cyclones directly over the northern-most islands of Torres Strait appear to be rare.

Access to the Saibai and Boigu islands can be gained by one of the daily scheduled services or charter of a light aircraft from Horn Island. Because Dauan Island does not have an airstrip, access here is via Saibai Island and must be arranged using the island 'ferry' service. This can usually be booked by talking to staff at the council offices on Saibai, though note that it does not necessarily operate on a daily basis. It is also possible to visit all three islands as part of an organised voyage aboard a live-aboard vessel, though these run infrequently. Regardless of how you get there, remember that none of the communities here has chosen to promote tourism and you are present as guests within the community – as such, it is especially important that you respect people's privacy, their property and property boundaries.

A mid-dry season view of Saibai Island. The wetland in the foreground usually retains some water year round. It can be accessed on foot by following tracks south-east from the southern end of the airstrip. The tropical savannah here supports Golden-headed Cisticola, Willie Wagtail (ssp. *melaleuca*) and occasional Peaceful Dove (ssp. *papua*) and Pheasant Coucal (ssp. *thierfelderi*).

It cannot be emphasised enough that these islands require permission to visit and that all logistics must be carefully planned before arrival. Once on the islands, access to birding sites is generally on foot. This means that on the larger islands of Boigu and Saibai a keen birder may end up walking significant distances.

Birding highlights

Despite the above difficulties, Northern Torres Strait provides some fantastic 'frontier' birding opportunities – for one, it is the only place in Australia where one can reasonably expect to observe Collared Imperial Pigeon, Coconut Lorikeet, Red-capped Flowerpecker and Singing Starling. Other regular PNG specialties include Spotted Whistling Duck and Uniform Swiftlet (both more frequent in the wet season). There is also a long list of 'novel' PNG subspecies that occur no further south than this island group. Many of these forms are common on the islands and the following list of unique subspecies (in an Australian context) typifies the range of species one might expect to encounter: Orange-footed Scrubfowl (ssp. *reinwardt*), Brown Goshawk (ssp. *dogwa*), Peaceful Dove (ssp. *papua*), Eclectus (ssp. *polychloros*) and Red-cheeked Parrots (ssp. *aruensis*), Barking Owl (ssp. *assimilis*), Little Kingfisher (ssp. *pusilla*), Mangrove (ssp. *pallida*) and Large-billed Gerygone (ssp. *brunneipectus*), Tawny-breasted (ssp. *saturatior*), Dusky (ssp. *fumata*) and Red-headed Honeyeaters (ssp. *erythrocephala*), Northern Fantail (ssp. *gularis*), Willie Wagtail (ssp. *melaleuca*), Spangled Drongo (ssp. *carbonarius*), Black Butcherbird (ssp. *alecto*) and Torresian Crow (ssp. *orru*). The northern islands have also hosted several vagrant species over the years, with some of the standouts being Gurney's Eagle (4+ records), Papuan Needletail (2 records), Black-headed Gull (2 records), Orange-bellied Fruit-Dove (1 record) and Lesser Paradise-Kingfisher (1 record).

Boigu Island

Key species: Collared Imperial-Pigeon, Coconut Lorikeet, Red-capped Flowerpecker, Singing Starling
Other species: Gurney's Eagle, Spotted Whistling-Duck, Eclectus Parrot, vagrant birds from PNG

On Boigu, the best areas to birdwatch are located in and around the township. Starting at the eastern end of town, there is an area adjacent to the beach where a raised gravel and earth wall runs back to the edge of the airstrip and then parallel to the airstrip to the rear of the school grounds. Near this walled area are several produce gardens and towards the taller mangroves (as far east as one can easily walk before entering true mangrove forest) there is a narrow strip of vine scrub. At the other end of town, beyond both the cemetery and the dump, there is another area of gardens that are associated with a similar narrow strip of vine thicket vegetation. Both areas support very similar bird communities. The vine thicket areas usually harbour Dusky and Red-headed Honeyeater (numbers fluctuate in response to flowering plants), Cicadabird, Varied Triller and Large-billed Gerygone. Olive-backed Sunbird is abundant and flycatchers can be especially well represented, with sightings of Spectacled, Black-winged and Black-faced Monarch, Leaden and Shining Flycatcher, Willie Wagtail, Northern and Rufous Fantail all possible. Willie Wagtail and Shining Flycatcher are both common residents, whereas the other flycatchers are mostly passage migrants with occasional wintering immature individuals. There are also huge Orange-footed Scrubfowl mounds in these vine thickets and the adults can occasionally be seen in the vicinity. These two vine thicket areas are also the most reliable areas for Red-capped Flowerpecker on Boigu; listen for a high pitched '*tzew*', which sounds remarkably similar to the familiar call of the Mistletoebird. Once located, it is often possible to watch this local speciality feeding on mistletoe berries among the mangroves. At night the *'chok-chok-chok'* call of the Large-tailed Nightjar resonates from these areas. The tall mangrove forest beyond these vine thickets are preferred habitat for Tawny-breasted Honeyeater, Black Butcherbird and Little Kingfisher; listen for the last's distinctive call as it dashes through the understorey.

The dump at the western end of the airstrip also provides good birding opportunities. Pied Heron are abundant here, as are Bar-shouldered Dove, White-breasted Woodswallow and House Sparrow (these colonised the northern islands in the 1990s), and Little Bronze-Cuckoo can usually be found in the surrounding mangrove thickets. The resident pair of Torresian Crow and their young are also often about. Being relatively open, the dump also provides a convenient vantage point from which to scan the surrounding grasslands and mangrove forest. In the early morning and late afternoon, this can be a good vantage point for flybys of Eclectus Parrot, Coconut Lorikeet and Collared Imperial-Pigeon. All three species appear to commute daily from the PNG mainland to both Saibai and Boigu Island (all seasons – but most commonly in the wet). The only record of an Orange-bellied Fruit-Dove was of one perched in a tall mangrove adjacent to the dump in November 2004, and Uniform Swiftlets (mostly wet season) may forage overhead. Dirt tracks across the saline grass flats provide access to the island interior. In the wet season, you can usually walk only a few hundred meters before it becomes impassable. Raptors include Whistling and Brahminy Kite, and White-bellied Sea-Eagle (all common), and Gurney's Eagle has been observed soaring over the area on several occasions (usually after 10 a.m. when thermal updrafts occur). Mangrove Robins (rare on the northern islands) have been observed towards the middle of the island in the dry and large numbers of waterbirds and shorebirds, including rarities such as Spotted Whistling-Duck and Black-headed Gull, have been observed foraging and roosting in the inundated clay pans in the wet.

Back in town, the fringes of the airstrip can be productive for roosting shorebirds and Red-backed Button-quail. At the rear of the airstrip is a freshwater pool of perhaps a hectare. In the dry season, this pool attracts large numbers of doves, honeyeaters (especially Brown-backed and Rufous-banded) and Spangled Drongo, which bathe and drink. It is also a good site for small numbers of waterfowl, including regular records of Spotted Whistling-Duck. Being close to town, the ducks are often quite wary, arriving after nightfall and departing soon after dawn, so an early morning check is best. Other waterbirds here include Striated Heron, Black Bittern, White-browed Crake and Purple Swamphen. In November 2004, a small group of Papuan Needletail foraged overhead and drank from the pool for several minutes before departing. The town water supply is also worth a look. Covered with a large white tarpaulin adjacent to the school and airstrip, it is a preferred high-tide shorebird roost; 100s of Lesser Sand Plover, Red-necked Stint and Little Tern, along with smaller numbers of other species, may occur. The surrounding cyclone fence is occasionally used as a roost site for swallows and martins – Barn, Welcome and Red-rumped Swallows, and Tree and Fairy Martins have all occurred. Claims of Pacific Swallow also exist, but confirmation of ID requires photographs that reveal structure and pattern of the spread tail (among other things).

The township area supports large numbers of Varied Honeyeater and House Sparrow, while Willie Wagtail, White-breasted Woodswallow and Olive-backed Sunbird are all common. Singing Starlings forage throughout the town in small parties, but are most easily observed in and around the church grounds; the starling breeds in the mangroves with an abundance of hollows visible from here and routinely feeds in the very large fig (partly broken and dying back in 2012). When in fruit, this tree can be a hive of activity, with Rose-crowned and Superb Fruit-Dove, Torresian Imperial-Pigeon, Olive-backed and Yellow (rare) Oriole and both Singing and Metallic Starling. Collared Imperial-Pigeon have been observed in this tree and other rare frugivores are likely to occur from time to time. In the heat of the day, the shade offered by the tree is a pleasant place to sit while watching the canopy for hidden birds and scanning the foreshore, strait and distant mangroves of PNG for birds flying to and from Boigu. Scanning the strait for flying birds in the mid to late afternoon is a fairly reliable method to locate Collared Imperial-Pigeons as they fly over the sea.

Away from the township, there are numerous other areas worth exploring but this must be done from a tender. The navigable channel that divides the island, the drier and more extensive vine thickets on the southern shoreline and several other areas where landings are possible would all be of interest if access in a boat can be arranged.

The Red-capped Flowerpecker is now known to be a breeding resident on both Boigu and Saibai Islands. The species frequents vine thickets and mangrove habitats. It's distinctive, Mistletoebird-like call is the best means of detection.

Saibai Island

Key species: Collared Imperial-Pigeon, Coconut Lorikeet, Red-capped Flowerpecker
Other species: Gurneys Eagle, Eclectus Parrot, Barking Owl, Brown Goshawk, vagrant birds from PNG

In many respects, the birding on Saibai can be quite similar to that on Boigu, but, having travelled so far, most visitors divide their time equally between the two islands to cover more ground and thus increase the probability of encountering more species. Notable differences between the avifauna of the two islands are that Silver Gull, Barking Owl and the Brown Goshawk (ssp. *dogwa*) are all regular on Saibai, but vagrant (or absent) on Boigu, while the reverse is true of Singing Starling.

As with Boigu, there are two narrow strips of vine thicket at either end of town that are easily accessible. At the eastern end of town (past the airstrip, past a few more houses), starting at the old dump (where there have been two tantalising reports of Rufous-bellied Kookaburra), work your way along the vine thicket habitat by walking along the narrow beach within the mangroves until you reach a small sandy clearing and an impassable creek. The only accessible pair of Collared Kingfishers on the northern islands can usually be found just across this creek; in the past, they have nested in an arboreal termite mound in sight of the clearing. At the other end of town, similar habitat can be accessed by walking through the cemetery beyond the oldest graves and then along the 'beach' through the mangroves for several hundred metres until you emerge into a small garden area next to a large tidal creek. All the species listed for these habitats on Boigu Island similarly occur here. Both of these areas, especially adjacent to the creeks, can be good for Red-capped Flowerpecker. Collared and Pied Imperial-Pigeon, and Eclectus Parrots have been observed here feeding on Wonga Plums when in season (Jan–Feb). Dead limbs on the tallest trees in the cemetery provide occasional perches for Red-capped Flowerpecker and, if you are having difficulty locating this species, then a patient wait here may be rewarding. This area can also be good for Coconut Lorikeet, though they are invariably seen in flight only. The dirt access track that links the cemetery to the township can be productive, with Little Kingfisher, Mangrove Robin and Mangrove Golden Whistler all recorded. At night, this is also a good area for Large-tailed Nightjar and Barking Owl.

The township itself has a fig tree adjacent to the church that is worth a check if in fruit, and directly behind the church in the saline grass flats Zitting Cisticola has occurred (but careful scrutiny is required to eliminate the much more abundant Golden-headed Cisticola). A dead Little Paradise-Kingfisher was found right out the front of the canteen and council offices in February 2010, highlighting the fact that rare species from PNG can turn up anywhere. The foreshore at Saibai is often productive, with a scatter of shorebirds to sort through, foraging Black-necked Storks at low tide and a resident flock of Silver Gull. This small gull flock (5–20 in recent years) is worth a careful check because a Black-headed Gull joined them in February 2010. Inland of the township are extensive grassland flats and *Pandanus* savannah. In the dry, an extended walk (or a much quicker drive!) on the main vehicle track takes the observer past the town water supply and out into the centre of the island where deeper freshwater wetlands usually still hold water. Here, Green Pygmy Goose and other waterfowl occur, while the grasslands support abundant Golden-headed Cisticola and occasional Peaceful Dove, Pheasant Coucal and Willie Wagtail. Areas of freshwater surrounded by large paperbarks and freshwater mangroves can be good for Brown-backed and Rufous-banded Honeyeater, while Oriental Cuckoo (wet season) occurs occasionally. After wet season rains, these tracks become impassable and it is usually only possible to walk a kilometre or so inland from the end of the airstrip. At this time, the wetlands can provide opportunities for thousands of waterbirds, including numerous egrets, herons and shorebirds. Most notable are the occasional flocks of many hundreds of Magpie Goose, Radjah Shelduck, Pacific Black Duck and Wandering Whistling-Duck; Spotted Whistling-Duck has also occurred here, but it seems more regular on Boigu. As with Boigu Island, there are other areas on the island that may prove to be productive birding sites, but, to visit these, it is necessary to arrange access in a boat.

Dauan Island

Key species: Eclectus Parrot, Noisy Pitta, Potential for vagrant birds from PNG
Other species: Frigatebirds, Pale White-eye (chance)

Most birders that visit Dauan Island concentrate their efforts in the township area, where species such as Olive-backed Sunbird are abundant, and the low-lying area around the dump on the north-western point, which supports small coastal vine thicket patches at the fringes of the single large area of mangroves. This is a productive area and can feature a range of cuckoos (Brush, Little Bronze and Oriental (wet season), Large-billed Gerygone, and Shining and Leaden Flycatcher. Noisy Pitta occurs here on passage (Oct–Nov) and there have been several sightings of Pale White-eye during the dry season.

For cultural reasons visitors, should not attempt to climb the Devils Marbles, but from the township and other nearby vantage points, these can be easily scanned for flights of larger birds – numbers of Eclectus Parrots, along with flocks of Pied Imperial-Pigeons, commute to Dauan from PNG when fruits are available. Being the only peak in northern Torres Strait, it is also worth keeping a careful eye on the sky here for swifts – Uniform Swiftlets have been reported on several occasions, and Fork-tailed Swift and White-throated Needletail are regular in the wet season.

Beyond the other end of the town, the eastern most point on the island projects out into much clearer seas than the sediment-rich waters off the townships on Boigu and Saibai. This can be a good vantage point from which to scan for seabirds in the strait – frigatebirds and several species of tern often soar overhead in favourable winds. Owing to the rocky shoreline, there are generally fewer waders present on Dauan when compared with the other nearby islands. Nevertheless, interesting species such as Wandering Tattler can turn up from time to time.

Eclectus Parrots visit all of the northern Torres Strait islands on a daily basis from nearby PNG. Because the females attend and guard nest sites year round on the adjacent 'mainland', the vast majority of sightings here involve males.

Ashmore Reef

Overview

The Australian external territory of Ashmore Reef is situated at the confluence of the Indian Ocean and Timor Sea, some 830 km west of Darwin and 630 km north of Broome. Significantly (from a migrant land bird perspective), it lies just 145 km south of the Indonesian island of Roti. The area has been designated a National Nature Reserve and the large fringing coral reef contains four lightly vegetated islands (total land area ~55 ha), and several additional sandbanks that rise above the high water mark. The reef supports a large population of seabirds, including some of the most important seabird rookeries on the North West Shelf. Many of these seabirds are breeding visitors and are thus present in large numbers on a seasonal basis. Ashmore Reef is also an important foraging area for migratory shorebirds that visit the region from the northern hemisphere. Numbers of shorebirds are highest between October and April. Owing to the exceptional numbers of birds, the reserve has been recognised as an Important Bird Area. Expeditions to Ashmore Reef offer some of the best opportunities in Australia to encounter Matsudaira's and Swinhoe's Storm-Petrel, Bulwer's and Jouanin's Petrel, the majority of Australia's tropical breeding seabirds and the likelihood of rare migrant land birds. Indeed, Ashmore Reef has a reputation as the 'frontier birding' site in Australia and generates more rare bird sightings per visitor-day than any other site in Australia.

Eight to 9-day expeditions generally take place in October, using a live-aboard vessel that accommodates small groups (<14 passengers), and are accompanied by expert guides. The vessel travels from Broome, WA, north over seabird-rich waters, past Scott Reef and onto Ashmore Reef where 3–4 days are then spent moored within the reef while birders undertake daily excursions to the islands. The return leg involves several more days at sea and, weather permitting, a morning ashore on the Lacepedes – a group of remarkable seabird-breeding islands off the Kimberley coast – before returning to Broome. Most participants make the most of being in Broome during the height of the migrant season and include at least a few days of birding in and around Broome (p. 448), while some add a side trip to the Kimberley (p. 452).

Ashmore Reef expeditions are timed to coincide with the peak period of southward migration of land birds, high counts of shorebirds and peak abundance of pelagic seabirds, while also providing excellent opportunities to observe breeding seabirds. This means annual trips usually take place during mid to late October. At this time of year, the sea-surface temperature is between 28 and 31°C and air temperatures are slightly higher. The voyage to and from Ashmore Reef typically encounters one or two completely glassy days, but invariably there are also one of two rougher days. Once at Ashmore, the vessel is stable because it is moored within the reef. On land, notably West Island, conditions can be hot (35+°C) because there is little cloud cover at this time of year. The expeditions work around this by making forays ashore in the cooler morning and late afternoon, while the middle of the day is reserved for boating excursions and snorkelling activities. As a side note, the snorkelling close to the mooring at Ashmore Reef is superb, with a myriad of tropical fish, outstanding coral formations and an excellent chance of an encounter with Green, Loggerhead or Hawksbill Turtle. Ashmore Reef can only be visited as part of an organised tour. Kimberley Birdwatching www.kimberleybirdwatching.com.au have been running tours there since the late 1990s. They hold all the necessary permits and operate the only regular expeditions.

Birding highlights

Expeditions to Ashmore Reef provide a unique opportunity to encounter many species of bird that are rarely observed elsewhere in Australia. The list is impressive and all first-time visitors can be almost assured of new sightings for their Australian list. The avifauna of Ashmore Reef and the surrounding waters had been little studied until the first birding expeditions visited in the late 1990s. These annual trips gather important data that have since been used for biodiversity monitoring and oil spill response planning. The sea voyage invariably provides some outstanding sightings, including regular Wilson's, Matsudaira's and Swinhoe's Storm-Petrel, Bulwer's, Tahiti and Jouanin's Petrel, Streaked, Hutton's and Wedge-tailed Shearwater, White-winged Black, Little and Lesser Crested Tern and all three Jaeger species. Roseate Tern can be particularly spectacular at the Lacepede Islands, with up to 34 000 sighted on a single morning there. Other more rarely encountered seabirds have included Arabian Shearwater, Leach's Storm-Petrel and Abbott's Booby. Incredibly, a total of 19 species of cetacean have been recorded since 2000 on these expeditions, with very good chances of Bottle-nosed, Pan-Tropical Spotted, Dwarf Long-snouted Spinner and Fraser's Dolphin and the possibility of 'mega' mammals such as Blue, Sperm and Humpback Whale.

Once at the reef, there are opportunities to observe the breeding seabirds when ashore on West Island and while circumnavigating East and Middle Island and the smaller Splittgerber Cay in tenders.

West Island is the most heavily vegetated of the four islands at Ashmore Reef. A ring of Octopus Bush fringes much of the island and provides shelter for many of the vagrant land birds that occur here.

Large colonies of Sooty Tern, Crested Tern, Common Noddy, Lesser Frigatebird and Brown Booby breed on these islands, while smaller breeding colonies of Wedge-tailed Shearwater, Masked and Red-footed Booby, Great Frigatebird, Eastern Reef and Little Egret, and Black and Lesser Noddy also occur. Significantly, Ashmore Reef is one of just three breeding sites for Lesser Noddy in Australian (the other accessible site being the Abrolhos Islands, WA).

Migrant shorebirds are also a feature of Ashmore Reef, with counts documenting up to 18 200 individuals using the reef. The extensive sand flats exposed at low tide provide foraging opportunities for internationally significant numbers of some species, including Grey Plover and Sanderling. Sought-after shorebirds such as Asian Dowitcher and Broad-billed Sandpiper are regularly observed, and rare shorebirds also appear from time to time, with recent records of Common Redshank, Little Stint and Red-necked Phalarope.

As if the seabirds and migrant shorebirds weren't enough, Ashmore Reef also has a reputation as one of Australia's most outstanding venues for migratory land birds and 'mega' rarities. Oriental, Brush, Horsfield's Bronze Cuckoo, Rainbow Bee-eater, Arctic Warbler, Yellow Wagtail and Barn Swallow are all regular migrants while every expedition encounters some exciting rarities. Since 2000, Ashmore Reef has hosted no less than 12 new species for Australia (Lesser Coucal, Island Monarch, Asian Brown Flycatcher, Grey-streaked Flycatcher, Pechora Pipit, Middendorff's Grasshopper Warbler, Yellow-browed Warbler, Eastern Crowned Warbler, Kamchatka Warbler, Asian Stubtail, Siberian Blue Robin and Chestnut-cheeked Starling). During the same period, other exciting rarities have included Red-legged

Lesser Noddy breeds at Ashmore Reef during the dry season and small numbers are present year round. An organised tour is the only way to access the key seabird islands that support this species.

Crake, Oriental Honey-Buzzard, Japanese Sparrowhawk, Dark-sided Flycatcher, Tiger and Brown Shrike, Grey Wagtail, Red-rumped Swallow, Oriental-Reed Warbler, Pallas's Grasshopper Warbler, Siberian Thrush and Pale White-eye. All the more remarkable is that some of these mega-rarities are now seen on an almost annual basis at Ashmore (e.g. Island Monarch, Asian Brown Flycatcher, Middendorff's Grasshopper Warbler and Oriental Reed-Warbler).

Christmas Island

Overview

Christmas Island is a natural paradise in the north-eastern Indian Ocean. The island is well known for a remarkable natural history event that occurs annually. Around the spring tide series that follows the first monsoon rains for the season (generally late Nov to mid-Dec), truly vast numbers of Christmas Island Red Crabs migrate from the rainforest plateau to the shoreline to spawn. A species of land crab, they spend the first month or so of life as aquatic larvae in the surrounding sea before coming ashore on the next high tide series as tiny crablings. The spectacle of the red crab migration is a natural history wonder that attracts large numbers of visitors every year. Christmas Island supports a series of small modern settlements and has a resident population of ~1500. There are a variety of accommodation options, including self-contained units, lodge and motel-style rooms and boutique hotels. Phosphate mining, a large Australian immigration detention centre and tourism are the major sources of employment for islanders. Although the island has an extensive history of mining, 63% of the island is designated as a national park.

Christmas Island is situated some 2600 km north-west of Perth, WA, 500 km south of Indonesia and 975 km east north-east of the Cocos (Keeling) Islands. The closest point to the Australian mainland is Exmouth, WA, 1560 km distant. The island is irregular in shape (sometimes likened to the outline of a dog) and is ~19 km in length and 14.5 km in breadth. The total land area is 13 500 ha, with 139 km of coastline. Christmas Island is unusual in that the interface between land and sea is mostly jagged limestone cliffs and the few beaches that occur are small, isolated and difficult to access. Above these cliffs, a somewhat depauperate forest grows on a lowland terrace of varying width and, behind this, the land rises abruptly again to a central plateau that supports tall rainforest. The highest point on Christmas Island is Murray Hill at 361 m. The plateau is the summit of a mostly submarine mountain that rises 4200 m from the adjacent sea bed, meaning that waters adjacent to the island provide foraging opportunities for truly pelagic seabirds. As an extinct volcano, there is some exposed basalt at places such as the Dales and Dolly Beach, but most of the surface is limestone accumulated from the growth of coral over millions of years. This process of limestone deposition continues today and the narrow fringing coral reef provides excellent diving and snorkelling opportunities.

A week-long stay at Christmas Island is ideal. During such a stay, birders can easily visit all of the major sites and make return visits to those that were most productive. Such a visit would include birding in and around the settlement areas, walks at Greta, Dolly, Lily and Ethel Beaches, several extended drives around the island to explore the national park, some time at the various lookouts watching tropical seabirds, including Christmas Island Frigatebird and White-tailed Tropicbird (Golden Bosunbird) soaring on the wind, and visits to less pleasant, but nonetheless very 'birdy', sites such as the dump and several marshy areas. Increasingly, visiting birdwatchers also include a half-day boat charter around the island to view marine life, including the tropical seabirds from a different perspective; if the timing is right (coinciding with crab spawning), these trips have the added potential bonus of snorkelling with Whale Sharks. Birding Tours Australia runs tailored trips for birders to Christmas Island at least twice per year: www.birdingtours.com.au.

Christmas Island has a tropical environment, where the heat and humidity are moderated by the surrounding ocean. Like Cocos, the island is strongly influenced by the ocean winds and experiences a SE trade wind season from March to October, and NW monsoon winds from November to April. Temperatures are fairly consistent through the year and maximum daily averages are in the range 26–28°C, with minimum night temperatures rarely dropping below 22°C. Average annual rainfall is 2110 mm with most falling in the

Rainforest-clad plateau and terraces visible from Margaret Knoll, Christmas Island. This is an excellent vantage point from which to observe soaring seabirds, including White-tailed Tropicbird (Golden Bosunbird) and Abbott's Booby. The endemic Christmas Island Fruit-bat may also be seen during the afternoon here soaring on updrafts.

period December–April. Christmas Island occasionally experiences cyclones during this period. International birders arrive year round, because most focus their efforts on the island endemic species, but Australian birders tend to visit in the period November–April in the hope that their visit will also coincide with the arrival of vagrant species from SE Asia.

A hire car is essential on Christmas Island and, for most birders, a small 4WD (e.g. RAV4) is recommended. The settlement areas and some of the steeper roads up onto the plateau are sealed, but, once away from the NE corner of the island, most roads are dirt. 4WD is required on some of the steeper tracks, especially in the wet season (e.g. Dolly Beach). Boat charters are available on Christmas Island, though none are tailored for birders and most visitors use these for snorkelling and diving on the coral reef, or swimming with Whale Sharks (Nov–Jan). Regular scheduled flights operate from Perth. Many of flights also stop at Cocos Island, so it can be convenient to include both islands in a single extended trip. The availability of flights, accommodation and hire cars for visitors can be strongly influenced by activity in the immigration detention centre; when at or above capacity, or during politically tumultuous times, bookings need to be made well in advance of any travel. Also note that, although the Red Crab migration is an outstanding spectacle, when underway, many of the islands roads are closed to reduce crab mortality and this can significantly reduce access to large areas of the island.

Birding highlights

Christmas Island has many highlights for the visiting birder. There are the five endemic species (Christmas Island (CI) Frigatebird, Abbotts Booby, CI Imperial-Pigeon, CI Hawk-Owl and CI White-eye) and a further four endemic subspecies (White-tailed Tropicbird (ssp. *fulvus*), CI Goshawk (ssp. *natilis*), CI Swiftlet (ssp. *natalis*), Common Emerald Dove (ssp. *natilis*) and Island Thrush (ssp. *erythropleurus*). The taxonomy of both the CI Goshawk and the CI Swiftlet are unstable: the goshawk has at times been recognised as a subspecies of both Variable Goshawk and Australian Brown Goshawk, while the swiftlet has been recognised as a subspecies of both Glossy Swiftlet and Linchi Swiftlet; for both forms, there may also be merit in considering them as island endemic species. Notwithstanding this uncertainty, here they are simply referred to as CI Goshawk and CI Swiftlet. There are also impressive numbers of other breeding seabirds, including some 20 000 Red-footed Boobies and the chance of overshooting or storm-driven rarities that see many Australian birders return repeatedly.

Flying Fish Cove, the settlements, gardens, airstrip and dump

Key species: Red-tailed Tropicbird, White-tailed Tropicbird, Java Sparrow, Island Thrush, Grey Wagtail
Other species: Frigatebirds, Brown Booby, Common Noddy, rarities, CI Fruit Bat

The town on Christmas Island is situated around the small industrial port at Flying Fish Cove. The cove provides reasonable swimming opportunities and occasionally hosts a shorebird or two. Common Sandpiper and Eastern Reef Egret are perhaps the two most frequent birds on the shoreline, but vagrant Kentish Plover and Common Kingfisher have also occurred. Out to sea, check the phosphate loader and jetty carefully for roosting terns and noddies. If the Red Crabs are spawning, the cove is also one of the best locations to observe this spectacle in the pre-dawn hours. Near the north-east point of the island, and a short drive from the airport, the township is rather sprawling and includes the 'suburbs' of Kampong and Settlement on the coastal margin and Silver City, Poon San and Drumsite on the plateau above. The lookout behind the information centre in Settlement provides good seabird-watching opportunities for most of the Christmas Island breeding species and, as an added bonus, there are sometimes Red-tailed Tropicbirds nesting at the base of the fig tree here. Another good vantage point for soaring seabirds (especially so for White-tailed Tropicbird (Golden Bosunbird), CI and Great Frigatebird and Red-footed Booby), is from the first 200 m of Gaze Rd, which links Settlement with Silver City and Poon San; take care because the road is narrow here and it is necessary to park further up and walk back down to stand behind the guard rail. As the many urban gardens provide different habitat opportunities to other areas on the island, short walks in these areas can be productive. CI Swiftlet and CI Imperial-Pigeon are often common, as are Tree Sparrows. The sought-after Java Sparrow is more difficult, but try any site where locals put out seed or breed (some of the restaurants and motels do this on an occasional basis). Urban areas have also hosted their share of rarities, including Daurian Starling in both Settlement and Silver City, Brown Shrike in Silver City, Oriental Honey Buzzard and Common Kingfisher in Drumsite. Another area to try for Java Sparrow is the nursery area adjacent to the Territory Day Park (behind the National Parks office) located in Drumsite. Small flocks occasionally occur here, especially at first light. The park can also be good for Island Thrush and CI Goshawk. Occasionally, CI Fruit Bat can be found here foraging within the larger fruit trees during daylight. Further along Murray Rd to the south are the Christmas Island Power Station and a phosphate mine works area. After rain, these areas often hold large puddles that are worth a quick check; Grey Wagtail and Chinese Pond Heron have both occurred here. A further 400–500 m down the road (after the intersection with Irvine Hill Rd), there is a market garden and banana plantation on the north side. This is private property and should not be entered, but a careful scan from the road is worthwhile because Grey Wagtail is regular here during the wet season. This area is also often good for CI Goshawk.

The island refuse dump is located off Hill Rd. Invariably unpleasant (household garbage in tropical heat), it is one of the best areas on the island for visiting and vagrant species. The list includes Javan and Chinese Pond Heron, Chinese Goshawk, Large Hawk Cuckoo, White Wagtail (occasional in Mar–Apr), Grey Wagtail (regular Nov–Mar) and Yellow Wagtail (occasional Nov–Mar), Barn Swallow (regular Nov–Mar), Red-rumped Swallow and Asian House Martin (occasional Dec–Mar). The dump is also an excellent site for CI Swiftlet and one of the more reliable sites for the now uncommon White-breasted Waterhen. Nearby, the northern end of the airstrip and adjacent cleared areas around Quarry Rd have hosted both Grey and Savannah

The Christmas Island Imperial-Pigeon is a large and very distinctive species that is easily encountered on the island.

Nightjar during the wet season. Finally, off Quarry Rd, there is a short walk to an historic cemetery and further on a look-out that provides superb views over the golf course and coastline below. This can be a good site to look for seabirds because the lookout sits above several frigatebird and Red-footed Booby colonies. Soaring Abbott's Booby can also occasionally be spotted from here.

Ethel and Lily Beach, CI Resort, golf course and cemeteries

Key species: CI Frigatebird, Great Frigatebird, CI Hawk-Owl, CI Swiftlet, vagrant waterbirds
Other species: Red-footed Booby, CI Imperial-Pigeon, CI White-eye, shorebirds

The coastal precinct beyond the Settlement on the north-eastern point of the island is another favoured area with visiting bird-watchers. There are good viewing opportunities for soaring seabirds around the various points and headlands, and large colonies of Red-footed Booby and CI and Great Frigatebird in the rainforest further up the terrace, especially between the golf course and the CI Resort. CI Swiftlet are often abundant along the roadside and foraging at low heights adjacent to the vegetation. Similarly CI Imperial-Pigeon and CI White-eye are usually easily found at various points along here. Check the golf course for open country birds: egrets and herons of various species are often present, and rarities such as Watercock, an unidentified Wheatear species, and Brown Shrike have all occurred. At night, the golf course is an excellent site for the diminutive CI Hawk-Owl. At the end of the golf course-Casino Rd is the CI Resort. Just before the resort gates, a natural spring often results in water flowing across the bitumen and into a small marshy area of grassland between the road and the coast. This site has a reputation as being one of the best for wetland birds on the island; as well as a chance of seeing the resident White-breasted Waterhen, records include Yellow and Cinnamon Bittern, Striated Heron, Watercock, Pin-tailed Snipe and Oriental Reed-Warbler. Look also for the freshwater-frequenting Blue Crabs in the drainage line here. The adjacent resort grounds are worth a look because CI Imperial-Pigeon, CI White-eye and Island Thrush are all common, while frigatebirds drink on the wing from the pool (when full) in hot weather, and there is always a chance of something unusual – the small cove just beyond the resort grounds has hosted Common Kingfisher.

Ethel and Lilly Beach are popular recreation sites for the Christmas Island locals. Red-footed Booby and Frigatebirds often nest in the trees behind the two carparks, while the area usually hosts a small mixed flock of shorebirds. A high-tide roost (mostly Pacific Golden Plover and Ruddy Turnstone, but anything could turn up) can sometimes be found on the point, which can be accessed via a short walking track immediately to the north of Lilly Beach.

Margaret Knoll, the Pink House, Greta and Dolly Beach, and South Point

Key species: Island Thrush, CI White-eye, CI Hawk-Owl, Common Emerald Dove, seabirds, rarities
Other species: feral chicken, CI Fruit Bat

The southern third of the island, accessed via the North–south Baseline Rd, includes large areas of coastal, terrace and plateau vegetation within the National Park, as well as several sizeable clearings created by mining activities. First stop along the road is Margaret Knoll Lookout (well signposted on the left). A short walk and flight of steps provides expansive views towards John D Point and South Point. Here, there is a good chance of passing seabirds, especially White-tailed Tropicbird (and occasional Abbott's Booby). The endemic CI Fruit-bat, which is partially diurnal (an anomalous behaviour for a fruit-bat), sometimes soars in the updrafts here, especially in the afternoon. Further along the North–south Baseline Rd, take the turn off to Greta and Dolly Beach. Just after the tight right-hand bend at the bottom of the steep decent off the plateau is Ross Hill Gardens. The Gardens are actually a series of freshwater springs that feed small creeks and rivulets. Where the water crosses the track, check for cryptic waterbirds: both Malayan Night-Heron and Schrenk's Bittern have occurred. To this point, access is usually suitable for 2WD vehicles, but after this 4WD is required. Continue onto first Dolly Beach and then Greta Beach for short walks to the coast and tiny white sandy beaches at the base of limestone cliffs. All of the usual forest birds occur along the vehicle and walking tacks; a vagrant Fairy Pitta has also been photographed in the forest here. Back on the North–south Baseline Rd, a short detour to the north via Grants Well Rd (before the Greta/Dolly Beach turn-off), or the East–west Baseline Rd and then the Pink House Rd, brings you to the Pink House. A now largely disused National Parks facility, the forest in this area is good for CI Hawk-Owl at night and provides opportunities for all of the other resident forest birds during the day (CI Goshawk, CI White-eye, CI Imperial-Pigeon, Common Emerald Dove and Island Thrush). Maintain a careful watch while driving these forest tracks as both Malayan Night-Heron and Grey Wagtail have been recorded on the forest edge during the wet season. An area known as LB3 on the East–west Baseline Rd, just off the North–south Baseline Rd often holds large puddles of water during the wet season. Check the boundaries of these for visiting waterbirds, including Yellow Bittern and Pond Herons. The forest edges here support numerous fruiting and seeding shrubs that may also provide opportunities for close views of CI Imperial-Pigeon. From here, head back to the North–south Baseline Rd and continue down to the South Point Temple for views to South Point. This area can be good for seabirds as they return to their breeding colonies in the late afternoon and, in the right conditions, can be a good vantage point to check for aerial feeding species such as CI Swiftlet, visiting swifts, swiftlets and possibly needletails, Barn Swallow and Asian House Martin; Nankeen Kestrel are also common here.

The Dales and Swiftlet Ally

Key species: Abbotts Booby, CI Swiftlet, CI Goshawk, vagrant swifts, swallows and martins
Other species: CI Thrush, CI White-eye, CI Imperial-Pigeon, visiting waterbirds

For the birder, there are three main areas of interest in the western third of the island. To get there, take Murray Rd from Drumsite, becoming North-west Point Rd, or travel via North–south Baseline Rd and then East–west Baseline Rd. The first of these two routes is often quite busy as traffic travels to and from the townships and the CI Detention Centre situated near North-west Point. The first area of interest is the tallest rainforest trees ~1 km before the CI Detention Centre on North-west Point Rd. This is one of the breeding sites for Abbott's Booby and adults atop their canopy nests can usually be found by slowly walking this section of road while scanning the tree tops. The area is also good for most of the other forest inhabiting endemics on the island. Before the gates to the Detention Centre, take the (public) road to the left. The road here passes through an

The Dales, Christmas Island, are a series of freshwater springs that have given rise to a unique wetland system. As freshwater is typically scarce on Christmas Island, it is not uncommon to find a visiting waterbird or two that has sought refuge here.

area of cleared vegetation that is recovering from mining activity interspersed with some remnant rainforest. This area, level with and just past the Detention Centre, has become popularly known as 'Swiftlet Ally' among visiting birdwatchers, because, in the right conditions (especially after tropical lows or cyclonic activity has occurred nearby in the period December–March), it can support huge numbers of aerial feeding birds. Not surprisingly, most individuals in any flock will be the local CI Swiftlet, but other exciting species reported here include Fork-tailed Swift, White-throated Needletail, Himalayan Swiftlet, Edible-nest Swiftlet, Barn Swallow, Red-rumped Swallow and Asian House Martin. After this stop, continue down the vehicle track to the Dales. From the carpark, take the walking track to the Dales, exploring each of the freshwater springs, the small water bodies that have formed at the head of many of these, and the fresh running streams below. In addition to all of the regular forest birds on Christmas Island, there is also the chance of an interesting waterbird or two: Malayan and Japanese Night-Heron, and Chinese Pond Heron have all occurred in the area. This area is also a great site for land crabs – not only are Red Crabs common, so too are the huge Robber Crabs and freshwater-frequenting Blue Crabs.

Cocos-Keeling Islands

Overview

The Cocos-Keeling group is the archetypical tropical coral atoll. The islands are low lying white sandy cays where Coconut Palms predominate and just about every ocean or lagoon view is a scene straight from a 'wish you were here' postcard. The Cocos group supports a population of ~600 who live in two settlements; residents in Home Island are mostly of Malay descent, while those on West Island are mostly of European descent. Tourist accommodation (motel and rental properties) is largely located on West Island, but there are also some (limited) options on Home Island.

Situated in the Indian Ocean, the nearest landfall is Christmas Island some 950 km to the north-west, but Indonesian Sumatra to the north is almost as close (1050 km). The closest point on the Australian mainland is on the Exmouth Peninsula, WA, some 2100 km to the south-east. The South Keeling atoll consists of 24 small islands (total land area 1310 ha) that form an incomplete ring around a large lagoon. In contrast, the atoll at North Keeling supports just one island situated 24 km to the north. North Keeling is a small (110 ha) C-shaped island with a narrow opening to the interior lagoon. The entire island and surrounding waters are designated as Pulu Keeling National Park.

Most of the southern islands were extensively cleared for coconut plantations in the 19th century and almost all vegetated areas now consist of rows upon rows of mature palms – because these plantations are no longer maintained, the understorey is often an impenetrable thicket of palm seedlings and fallen palm fronds. There are small pockets of native vegetation on most of the islands, especially on and behind the beaches. Of all the southern islands, Horsburgh retains the highest cover of native vegetation because it was reserved for recreational purposes during the period of early settlement. In contrast to the extensive palm plantations of the southern atoll, North Keeling Island was never extensively cleared and retains some of the best examples of native atoll vegetation in the Indian Ocean. Across the Cocos group, fresh water is mostly limited to water lenses on the larger islands – exceptions are small wetlands on West Island (Bechat Besar) and Horsburgh Island and an area that becomes swampy during heavy rain on Home Island. It is possible to see all of the regular species (see below) found nowhere else in Australia in a day, but most birders stay 4–7 days so that they can visit most of the larger islands. With some effort and a bit of luck, most groups also hope to locate a rarity or two.

Within the Cocos Keeling Island group, visitors can expect hot days and balmy nights. Being such an isolated coral atoll, the islands experience two main seasons that are dictated by the prevailing winds: the trade wind season from April to October, and the calmer doldrums season from November to April. Temperatures are fairly consistent no matter the season: 28–29°C during the day, with minimum night temperatures rarely below 24°C. Average annual rainfall is ~2000 mm but, unlike tropical areas on the mainland, expect higher rainfall in the period March–July, though low pressure systems at other times of the year can result in heavy falls. Because there are no island endemic species at Cocos Islands, and the specialties (Western Reef Egret, Green Junglefowl, White-breasted Waterhen and Saunders Tern) all occur elsewhere in the world, most visiting birders time their trip to also coincide with the potential for several vagrant species; as a consequence, almost all choose to visit in the period November–April. Indeed, unless otherwise stated, all the vagrant species mentioned in the site guide below have occurred in this period.

Most birders visiting the Cocos group hire a car on West Island because the island is ~11 km long (via road) and best results are achieved by visiting many of the sites on a daily basis. Access to both Horsburgh and South Island is by boat – suitable operators are indicated in the site accounts below. Access to Home Island is via the regular inter-island ferry service and, once here, it is possible to walk to the best birding sites.

Regular scheduled flights operate from Perth. These flights also stop at Christmas Island, so it can be convenient to include both islands in a single extended trip. Birding Tours Australia runs tailored trips here for birders twice a year: www.birding-tours.com.au. Pam Jones and Geoff Christie offer locally guided day tours, provide boating services to access Hosburgh Island and the coral reef and can provide packed lunches for independent travellers: www.cocosbirds.com.

Birding highlights

Although there are some good birding opportunities within the Cocos Islands, including several species that either occur nowhere else (or are very difficult elsewhere) in Australian territory, most bird-watchers visit the territory in the hope of also encountering vagrant species that have found themselves lost in the vast expanse of the tropical Indian Ocean. As a consequence, the following site guide is as much about the best areas to search for these vagrants as it is about seeking the handful of regular species. Green Junglefowl, Western Reef Egret and Saunders Tern (summer) are all relatively straightforward to observe within the Cocos group and occur nowhere else in Australia. White-breasted Waterhens are also abundant here and, as the species has declined in recent years at Christmas Island, this is now the most reliable site. The only regular passerines at Cocos are Christmas Island White-eyes (introduced on Horsburgh Island) and Barn Swallows, which visit most years in small numbers over the summer months. The vagrant list for the Cocos group is spectacular: Eurasian Teal, Greater Flamingo, Yellow and Cinnamon Bittern, Malayan Night-Heron, Japanese and Chinese Sparrowhawk, Eurasian Hobby, Indian Cuckoo, Large and Hodgson's Hawk-Cuckoo, European Roller, Blue and White, Asian Brown and Mugamaki Flycatcher, Blue Rock Thrush, Eye-browed Thrush, Tiger Shrike, and Rose-coloured and Duarian Starling are just some of the mega-rarities to have occurred there in recent years.

Perhaps the most striking of the 'special' birds resident on the Cocos-Keeling Islands is the introduced Green Junglefowl (abundant on West Island). Although feral chickens are also present, most individuals, including this cockerel, appear 'true to type'.

West Island

Key species: Green Junglefowl, White-breasted Waterhen, Western Reef Egret, White Tern, vagrant land and water birds

A walk around the settlement area on West Island (where most visitors are accommodated) is a good starting point for birders. There are no resident passerines or indeed non-passerine bush birds on West Island, so anything small and/or keeping to shrubs or trees is worth pursuing; Asian Koel, Oriental Cuckoo, Hodgson's Hawk-Cuckoo and Rose-coloured Starling have all been seen here. Adjacent to the settlement, it is possible to circumnavigate the airport and runway on the grassy vehicle track, so long as it's not too wet (also check for approaching aircraft before crossing the grass at either end and never drive on the tarmac!). There is usually a selection of larger species around the strip that include Green Junglefowl, White-breasted Waterhen, Eastern and Western Reef-Egret, and Nankeen Night Heron; Eastern Cattle and Intermediate Egret have also occurred. Shorebirds are occasionally found on the grassy verges and have included Pacific Golden Plover, Ruddy Turnstone and Oriental Pratincole. Aerial species are also attracted to the large grassy expanse – both Asian House Martin and White-throated Needletail have been recorded, and Barn Swallow occur annually in small numbers. During the period November–March, keep a sharp eye out for Watercock (rare). When present, they forage out from shrub cover into the rough grass and flush ahead of the vehicle with heavy flight and long trailing legs. Behind the weather station, there are some huge evergreen trees that occasionally provide refuge for scarce visitors; Japanese Goshawk, Oriental and Large Hawk Cuckoo have all been seen. If it has been raining, pools along this track may provide habitat for snipe (most are assumed to be Pin-tailed Snipe), though confirming identification can be difficult. The northern end the track can be a bit boggy. Stop here and scan across the North Lagoon because this can be a good area for Western Reef Egret.

Heading north out of the settlement, check the various cleared areas for birds. The town dump on the eastern side of the sealed road (waterhens), a brackish swampy area on the western side of the road (ducks and egrets, including records of both Javan and Chinese Pond Heron) and the fenced area that used to serve as a quarantine station (and more recently has been used

There are few views on the Cocos Islands that don't include at least some Coconut Palms, but stands of large trees (in this case, behind the Bureau of Meteorology Station on West Island) are more scarce and warrant a careful check for visiting and vagrant bird species.

as a detention centre for illegal immigrants) are all worth a short walk or careful scan. The quarantine area can be especially good for cuckoos if the broad-leafed trees are infested with caterpillars – Indian, Oriental, Hodgson's Hawk and Large Hawk Cuckoo have all been observed – and several predators including Chinese Sparrowhawk, Eurasian Hobby and Tiger Shrike have been sighted. Further north, the Rumah Baru Rd provides access to the new wharf area. There are several rectangular clearings ~220 and ~330 m down this road that warrant a quick stop because Chinese Sparrowhawk and Watercock have occurred. On the road itself, especially in the forested areas, Grey Wagtail has been seen. Check the wharf area for seabirds, notably terns, that occasionally rest here.

Back on the sealed road, and a further 400 m north, there is a track to the lagoon that leads through the old farm area. This clearing with storage sheds can be productive for birds – among others, the Asian subspecies of Dollarbird has been observed perched atop the tallest vantage points. Soon after passing the storage sheds (~100 m), there is a vehicle track that leads north through the palm forest for 300 m before a sharp turn to the west. Assuming it is not too overgrown or wet, drive to this bend then walk west for 20–30 m before striking out north through the palm forest for ~70 m to access the only large freshwater body on West Island. Known as Bechat Besar, this site has been a magnet for rarities. Expect to see some of the resident flock of Pacific Black Duck, egrets and perhaps a White-faced Heron. Sit or stand quietly at the southern end and carefully scan the wetland edges and overhanging vegetation. If there is exposed mud, shorebirds, including Pin-tailed Snipe, are possible. Rarities that have been sighted here include Australia's only Eurasian Teal (which stayed for several years) and Common (not Dusky) Moorhen, as well as Yellow and Black Bittern, Japanese and Chinese Sparrowhawk and Large Hawk-cuckoo. It is possible to access the northern end of Bechat Besar by returning to the Old Farm Rd and continuing onto the lagoon. Once at the lagoon, walk north along the beach, under huge trees that provide a nesting site for White Terns, for ~750 m to an old vehicle track that leads back into the palms – follow this to the first bend (~40 m) and the northern edge of the swamp is just visible from here to the south through dense vegetation.

Further north along the sealed road is the turn off to Trannies Beach. This is a popular swimming area and the best area for White Tern if they happen to be scarce during your visit (numbers fluctuate both seasonally and annually). The tall trees here may also harbour a visitor or two; previous records include Asian Brown Flycatcher and Chinese Sparrowhawk. The road also provides convenient access to view the open areas of the radar installation – additional access is available by taking the track south from Trannies Beach picnic area. Finally, the picnic area itself can be a good vantage point from which to scan the ocean for seabirds such as Masked and Red-footed Booby and Wedge-tailed Shearwater.

The southern half of West Island also has a few spots worth a check, including the 'bone yard' – a series of gravel scrapes accessible via a short track 200 m south of the grassed runway area – after heavy rains the area supports pools of water (Yellow Bittern and Barn Swallow), the edges of the radar installation (restricted access but viewable from the road at a couple of points), the Scout Park and finally the flats of South Lagoon. These flats are best viewed at low tide, from the very end of the road looking back towards the radar masts; here there is usually a selection of egrets, including the occasional Western Reef Egret. Chinese Pond Heron has also used this area to forage.

South Island

Key species: Saunders Tern, Western Reef Egret, shorebirds

South Island is most easily accessed by motorised outrigger canoe departing from the southern end of West Island. Such tours can be organised through Ash and Kylie James (www.cocosislandstourism.com.au), who can also ensure you are led to the right spot for congregating Saunders Terns and shorebirds. The roost sites are located on the lagoon side of the island and are accessed after a short walk through the palm forest. Visits need to be timed to occur on a rising tide for ease of access and the best chance of finding the birds. Take a spotting scope and, if the terns are not visible at the usual roost site, ensure the flats further to the east are also searched. Saunders Tern have not been recorded anywhere else in Australia, but are now known to be regular 'summer' visitors to South Island – closely related to the Little Tern, look for the strikingly black outer primaries. This roost site is also the best area for shorebirds in the Cocos group and up to 10 species might reasonably be expected here. There have been several intriguing sightings of possible Little Stint here, so it pays to scrutinise the shorebirds both carefully and with an eye for species not usually recorded in Australia!

Tiger Shrike is just one of a long (and growing) list of vagrants that have now been recorded on one or more of the Indian Ocean Territories: this particular species has now occurred on the Cocos-Keeling Islands, Christmas Island and Ashmore Reef.

Home Island

Key species: White-tailed Tropicbird, White Tern, vagrant land birds
Other species: White-breasted Waterhen

Home Island is easily accessed via the regular low-cost ferry service. Check the schedule on arrival – during weekdays, there are usually five or more return services, with fewer Saturday services (but access to Direction Island on this day) and no Sunday service. Once ashore on Home Island, most birders make a bee-line for the Clunies-Ross Mansion. This is a 300 m walk south along the road adjacent to the lagoon. On the way, maintain a lookout for White-tailed Tropicbird and White Tern overhead – both breed on Home Island. The large trees at the entrance to the mansion, the small park beside this and some of the foreshore areas are public access, but much of the grounds within the high stone walls that surround the mansion are private property. If you wish to bird these areas of the garden, you should seek permission from the owners. These sprawling English park-like gardens, with numerous mature trees and plenty of hiding spots for rare birds, have a deserved reputation as the best location for rarities on Home Island and possibly all of Cocos. Recent highlights include Japanese Sparrowhawk, Large Hawk and Hodgson's Hawk Cuckoo, Tiger Shrike, Dark-sided Flycatcher, Blue and White Flycatcher, Mugamaki Flycatcher, Eye-browed Thrush and Blue Rock-thrush. Asian Koel is also a regular wet season visitor and has occurred here and elsewhere on Home Island in recent years.

Most birders continue on beyond the Clunies-Ross Mansion, east along the foreshore to reach a small market garden and banana plantation area that forms the southern tip of the island. This area is encompassed by a 'ring' road that starts under the huge trees beyond the sports fields and adjacent to the extensive poultry sheds (500 m from the nearest wall of the mansion grounds) and loops around the southern point. The area within this 'ring' road is often quite boggy and thus has the potential to harbour waterbirds that have found their way to Home Island – make sure you seek permission before entering the actual gardens and be careful not to damage plantings, especially as some tropical food plants can appear somewhat weed-like (e.g. sago). In addition to the usual egrets, night herons and waterhens, birding groups have observed Watercock, Cinnamon and Yellow Bittern, and Pin-tailed Snipe here in recent years.

A careful search of the above areas usually provides a full day of birding on Home Island, but, if there is time, a walk through the settlement area, a check of the large trees both around the shops just inland from the jetty and further north towards the wind farm and pistol club may also be productive; for example, in December 2010, a Purple-backed Starling was photographed in the settlement.

Horsburgh Island

Key species: Christmas Island White-eye, feral chicken, vagrant land and waterbirds

Horsburgh Island is the northern-most of the southern group of islands and has an area of ~100 ha. Access is via charter operators. Geof's Glass Bottom Boats offers this service; contact on West Island can be made via the tavern or www.cocosbirds.com. Horsburgh is the only island in the Cocos group with a large breeding population of the introduced Christmas Island White-eye. The large trees behind the southern shore are worth a check, as is the large clearing with scattered shrubs through the centre of the island. Both Chinese Goshawk and Eurasian Hobby have been seen here and, in January 2011, Australia's first Eurasian Roller was photographed as it foraged in the centre of the island. On the northern side of the clearing, a narrow easily overlooked track (ask the boat operator for directions) leads through to the northern shore. A short walk, west along the beach, before heading back inland brings you to a small brackish wetland. There are several large trees here with dense shady foliage; in combination with the wetland, this location has much potential as a site for vagrant birds. As this is also the northern-most

Horsburgh Island supports the only resident passerine species in the entire Cocos-Keeling group – the introduced Christmas Island White-eye. This means that any other small bird found on the island group is likely to be a visitor from South-East Asia.

point of the southern island group a scan of the ocean here may also be productive – expect Red-footed Booby, Common Noddy and Wedge-tailed Shearwater, but keep your eyes peeled for rarities such as Bulwer's Petrel and Lesser Noddy.

North Keeling Island

Key species: Breeding seabirds, Buff-banded Rail

North Keeling Island is situated 25 km north of the Cocos group. It can only be accessed using one of the larger charter vessels based at the archipelago, and involves a sea journey of ~1.5 h. The entire island is a national park (Pulu Keeling) and much of the vegetation is in its original state. Access to the island can be hazardous, because it involves swimming from a boat across the reef to the shore, and is entirely weather dependent. Permits are required to access the park and visitors must be accompanied by either park staff or a permitted tour guide.

Pulu Keeling National Park remains one of the few examples in the Indian Ocean of a seabird island largely unaffected by feral animals. Seventeen bird species breed on the island, including one of the largest colonies of the Red-footed Booby in the world. Other breeding species include Masked Booby, Lesser and Greater Frigatebird, Wedge-tailed Shearwater, White-tailed Tropicbird and Common Noddy. The Endangered Cocos Buff-banded Rail (ssp. *andrewsi*) is found nowhere else. On the rarity front, Herald Petrels have been observed here and may breed and a Tropical Shearwater was found ashore at night in November 2009. Not to be outdone by the more southerly Cocos Islands, the only Greater Flamingo ever recorded in Australia frequented the lagoon at North Keeling Island in 1988.

Macquarie Island

Overview

Macquarie Island deservedly has a reputation as the Galapagos of the Southern Ocean – not only does it support a spectacular diversity of southern ocean vertebrates, but the sheer numbers involved can be staggering (try 100 000 King Penguins at Lusitania Bay, for example). To top it off, many species are tame, confiding and inquisitive. Sitting on a dark sandy beach, surrounded by an array of subantarctic wildlife while inquisitive King Penguins peck at your boots is something all nature lovers should experience once in a lifetime! Macquarie Island can only be visited as part of a subantarctic cruise, with most vessels travelling from Tasmania or southern NZ and including several NZ subantarctic islands in the same voyage. The entire island is a World Heritage-listed nature reserve and strict access requirements mean that visiting ships are granted just 2 days during any voyage when they can land passengers on the wildlife-rich shores.

Macquarie Island is situated ~1500 km south-south-east of Tasmania, about half way between Tasmania and Antarctica at latitude 55°S. The NZ subantarctic Auckland Island is the nearest landfall. 'Macca' as it is popularly known is ~34 km long and ~5 km wide at its widest point, with an area of 12 800 ha. Much of the island is dominated by a plateau at 150–200 m elevation, with the highest points being Mounts Hamilton and Fletcher (410 m) in the south. The west coast is rugged and difficult to access (especially by vessel), whereas the east coast is characterised by beaches of volcanic sands that provide some landing opportunities for visitors when the weather is calm. Behind most beaches is a narrow terrace vegetated in tussocks, before the land rises abruptly (cliffs or steep slopes) to the plateau above. Most wildlife congregate and breed on the sandy beaches and terrace of the east coast, though there are also significant penguin rookeries on the west coast. The northernmost point on the island is an isolated summit known as North Head and this is linked to the remainder of the island by a narrow, sandy isthmus. The Australian Antarctic Division research base is located on the isthmus adjacent to North Head.

Most vessels visit Macquarie island for 1–2 days and, weather permitting, tourist vessels generally make two landings; one at Sandy Bay and the other on the Isthmus adjacent to the Australian Antarctic Division base. A zodiac cruise off the vast King Penguin colony off Lusitania Bay is also usually included. While cruising off the island, and when at anchor, birders can expect to see the Macquarie Island Shag, Antarctic Tern and all four breeding penguin species, among other seabirds, around the vessel.

Macquarie Island weather can be summarised as cold, cloudy, wet and windy. During the summer season (Nov–Feb) when tourist vessels visit, maximum daily temperatures generally range between 6°C and 9°C. There are only a few days each year where neither rain nor snow falls and the island is one of the gloomiest places on Earth, with an average of just 856 h of sunshine per year! The best days for sunshine are in the period Nov– Feb, in part because at the height of summer there are only four hours of darkness! Macquarie Island has a remarkable annual average wind speed of 35 km h^{-1}! Given the above, to avoid disappointment, visitors should expect cold, wet and windy conditions while hoping for one of those rare calm sunny days!

Brothers Point, Macquarie Island. In addition to the scenic view, access to these grassy terraces via boardwalks from landing sites at Sandy Bay and on the Isthmus provide the best opportunity to encounter Common Redpoll.

Although yachts occasionally visit Macquarie Island, most birders would only be able to visit the island as part of a commercial cruise. Because these vessels visit the island specifically for nature-watching purposes, the trips are well run and quite well tailored to birders needs. For some trips, this may include the presence of experienced birdwatching guides. Access to the island itself is via rigid inflatable boats (RIBs) or Zodiacs, so a moderate level of fitness is required; cruise operators, some with decades of experience at Macquarie Island, can advise on requirements.

Birding highlights

In addition to the life changing natural history experience offered at Macquarie Island, a trip will provide sightings of King, Gentoo, Royal and Southern Rockhopper Penguin, Light-mantled Sooty Albatross, Macquarie Island Shag, both Giant-Petrel species, Brown Skua and Antarctic Tern. There is also a good chance of Wandering, Grey-headed and Black-browed Albatross, while a diverse array of southern ocean petrels and shearwaters is to be expected. Macquarie Island is also the only breeding location in Australia for the Common Redpoll: a self-introduced species that originated from the exotic New Zealand population.

Sandy Bay and Lusitania Bay

Key species: King Penguin, Royal Penguin, Light-mantled Sooty Albatross, Giant Petrels, Brown Skua, Common Redpoll and Southern Elephant Seal
Other species: Chinstrap Penguin (chance), Macquarie Island Shag

Sandy Bay is a popular landing site because of the large Royal Penguin colony on the slopes behind the beach and the large numbers of loafing seabirds and seals on the beach itself. There are strict protocols regarding approach distances to wildlife, but there are no restrictions on the wildlife approaching expeditioners! As a consequence, visitors can expect to get up close and personal with inquisitive King and Royal Penguins, Brown Skua and Southern Elephant Seal. Because of the sheer numbers involved, this is a sight to behold and an experience never to be forgotten. Both Giant Petrel species are also usually present on the beach and there is always a chance of something unusual. On average, station personal detect a Chinstrap Penguin per year making a brief visit or hauling out to moult on Macca – interestingly, many of these have been in Sandy Bay and tourist vessels have encountered them on several occasions. Rarer species also occur: in February 1997, when the *MV Kapitan Khlebnikov* landed, tourists at Sandy Bay, among the usual greeting party was an immature Emperor Penguin!

In addition to the loafing wildlife on the beach, a short distance to the north of the usual landing site there is a growing King Penguin colony. Because this species has an extended breeding cycle (just over 12 months), birds commence nesting during two periods in any year and thus the colony may contain a mix of adults on eggs and small chicks, while, at the same time, large fluffy brown chicks mill about in crèches. Behind the beach, a boardwalk takes visitors to a viewing platform that overlooks the Royal Penguin colony. This provides opportunities to get up close without causing undue disturbance; the sounds and smells of a Royal Penguin are quite an experience. This short boardwalk and the grassy slopes above Sandy Bay provide some opportunity for Common Redpoll: a species that is sought after by Australian listers. While reasonably common at 'Macca', short-stay visitors are most often rewarded with flight views; identification is

The abundant subantarctic wildlife at Macquarie Island is the major drawcard for visitors. Assuming the weather allows shore access, visitors are all but guaranteed close encounters with Royal Penguin (shown here), other loafing seabirds, including King and Gentoo Penguin, Brown Skua and Giant Petrels as well as Southern Elephant Seals.

straightforward because, barring vagrants, it is the smallest bird on the island (the next smallest being the introduced Starling).

The Isthmus and Australian Antarctic Division Station

Key species: Gentoo Penguin, Southern Rockhopper Penguin, Brown Skua, Antarctic Tern, Common Redpoll
Other species: Macquarie Island Shag

At the Isthmus, landings are usually made on the eastern shoreline, but occasionally, when atypical winds or seas prevail, the western shore in Buckles Bay may be used. This is an opportunity to visit a working subantarctic research station and many people head up to the 'mess hall' to meet research expeditioners and view the facilities, but the wildlife naturally still takes centre stage. The Isthmus is an important area for breeding Gentoo Penguins and visitors are guaranteed sightings of adults tending nests and chicks. There are also large numbers of Southern Elephant Seal here; depending on the timing of the visit, one might experience anything from dense harems of pupping females that are watched over by the huge male 'beachmasters' (Oct–Nov) to large numbers of 'weaners' – the rotund grey young of the year with doe-like eyes and comical faces (Nov–Jan). The tussock areas again provide opportunities for Common Redpoll, there are invariably some King Penguins hauled out on the beaches and to the north, in the rocky coves adjacent to the research base, are Southern Rockhopper Penguins. These penguins are not usually accessible on foot, but many expedition vessels provide zodiac cruises here or elsewhere on the island to ensure visitors obtain reasonable views of this species. The small rock stacks in this area are also used as breeding sites by the small resident population of Antarctic Terns. North Head presently supports the largest colony of the winter-breeding Grey Petrel; visitors that arrive very early (Oct) or late (Mar) in the season have a chance of seeing this species wheeling about the slopes if they are ashore in the late afternoon.

Waters around Macquarie Island

Key species: Northern and Southern Giant-Petrel, Southern Fulmar, White-headed, Soft-plumaged, Mottled, Blue and Grey Petrel, Antarctic Prion, Wanderer, Grey-headed and Light-mantled Sooty Albatross, Black-bellied Storm-Petrel

Other species: Antarctic Petrel (slim chance), Kerguelen Petrel, Broad-billed Prion, Little Shearwater (subantarctic form)

Ships that visit Macquarie Island typically use Hobart, Tas. and ports in the South Island of New Zealand to embark and disembark on subantarctic cruises. This means that visits to Macquarie Island include several days at sea in some exceptionally productive areas of ocean. Most vessels have experienced guides on board, but, even when they don't, there are usually a few keen and experienced seabirders present that can help the beginner with some of the trickier identifications. It is simply a matter of finding a suitable viewing platform on the vessel given the prevailing weather and sea conditions and putting in the hours watching the seabirds. There is a reasonable chance of most of the species listed above and this is especially so for trips earlier in the season (e.g. Nov) when cold waters extend further north. Whatever the timing, the further south you go, the higher the chance of encounters with cold-water species such as Southern Fulmar, Grey-headed Albatross and Blue Petrel. Finally, if the vessel is using Hobart as a port of call, check to see if it will pass over or near to the Soela Sea mounts 120–50 nautical miles south of Tasmania during daylight hours – if it does, it would pay to be especially vigilant here for Broad-billed Prion, because flocks of this Australian 'rarity' has been observed here in recent years.

Heard Island

Overview

The Australian external territory of Heard Island and McDonald Islands is a subantarctic island group located in the Southern Ocean, ~4000 km south-west of mainland Australia (and 4800 km south-east of Africa). It is the most remote and difficult-to-access location anywhere in Australia. But, for those lucky few that make it, it has its rewards. Heard Island sits below the Antarctic convergence (a marine front where very cold Antarctic and cold subantarctic waters meet); thus, it is permanently cold, with snow and ice to sea level year round. The islands are a contradiction of natural wonders in that they support the only active glaciers and the only active volcanoes in Australia!

Heard Island is ~40 km long from the tip of the Laurens Peninsula to the tip of Elephant Spit, and ~20 km wide from Saddle Point to Long Beach. It has an area of 36 800 ha and is dominated by Big Ben: a 2745 m active volcano. Just 5% of the island is covered in native vegetation, which is characterised by low grasses, mosses and lichens. Despite the ornithological riches (see below), an expedition to Heard Island is not for the faint-hearted because it is essentially a polar expedition to an extremely remote area of ocean with a reputation for some of the roughest seas on Earth.

Heard Island is typically cold and windy, with persistent low cloud cover and frequent precipitation; average maximum monthly temperatures range from just 0 to 4.2°C while average monthly wind speeds range from 26 to 33.5 km h^{-1}.

Birders wishing to visit Heard Island are best advised to contact expedition companies that specialise in subantarctic travel to enquire about upcoming voyages because they are few and far between. Indeed, to date, commercial expeditions have landed on Heard Island just four times. Potential participants would also be well advised to start saving – in 2012 most berths cost in excess of $20 000!

Birding highlights

Expeditions that visit Heard Island have typically allowed 2–3 days for potential landings before moving on. Sites at which landings have taken place are Atlas Cove and Spit Bay. Expeditions to Heard Island offer the only opportunity to observe the endemic Sheathbill and Shag and some of the best opportunities in Australia to encounter South Georgian Diving-Petrel and Fulmar Prion (summer breeders), Antarctic and Barau's Petrel (visitor) and a litany of other subantarctic seabirds.

The sea voyage invariably provides some excellent opportunities to observe southern ocean seabirds. Grey-backed and Black-bellied Storm-Petrel, White-headed, Soft-plumaged and Kerguelen Petrel, Salvin's and Antarctic Prion and the many species of southern Albatross can be expected, while Antarctic Petrel and Snow Petrel are both possible within Australian waters. Once at Heard Island, an impressive 19 breeding species may be present, including four penguin species (Macaroni, Gentoo, Southern Rockhopper and King). Other breeding species include the endemic Heard Island Sheathbill and Heard Island Shag, both usually easily seen from the ship if it is able to sail close to the island. Burrow-nesting petrels include South Georgian Diving-Petrel (which outnumbers the Common Diving-Petrel), Cape Petrel, Antarctic and Fulmar Prion, and Wilson's Storm-Petrel, while larger species such as Southern Giant-Petrel, Black-browed Albatross and Light-mantled Sooty Albatross nest on the surface. Brown Skua and Antarctic Tern also breed here. Being so remote and so far south, there are to date no records of land birds from Heard Island and any rarities are more likely to be visiting seabirds; for example, the first Australian claim of Kerguelen Tern was made here in April 2005.

Heard Island is the most remote and most southerly of Australian Territories. Notwithstanding the fact that Heard Island is an active volcano with glaciers that reach to the shoreline, seabirds are a key attraction: the long list of seabirds includes the highly sought-after Antarctic Petrel.

Annotated bird list of Australia and its territories

This section is a cross-referencing tool, linking a list of Australian birds with individual recommended sites referred to elsewhere in the book. In the list, broad site locations are given that mostly correspond with section headers. For example, if a bird is listed as occurring at Sturt NP, NSW, you can then turn to the site information for Sturt National Park to find more information on where you may find that bird at that site. As general rule, the places mentioned in the list below relate directly to a sites covered in the book. All sites have been indexed. For the more common and wide-ranging species, locations are not given, because you would expect to encounter them readily in their preferred habitat. Including vagrant and accidental records, ~900 species of bird have been recorded in Australia. Note that the abbreviated state name appears in brackets before all specific locations within that state.

Status

- **Common:** considerable numbers of individuals are usually present in their preferred habit and as such they are generally easily encountered.
- **Moderately common** (Mod common): moderate numbers of individuals are present in their preferred habit and individuals should be encountered on many, but not all, visits.
- **Uncommon:** small numbers of birds are present in their preferred habitat and generally the species is not readily located. Note that some species can be both uncommon yet encountered frequently at a small number of very specific sites.
- **Rare:** very small numbers of birds are present in their preferred habitat, and often requires a targeted effort to encounter the species. Again some species can be both rare yet encountered

frequently at a small number of very specific sites.

- **Introduced:** bird species living outside their native distributional range.
- **Vagrant:** accidental records for Australia or to particular states.

Abbreviations

ACT = Australian Capital Territory, Aust = Australia, BG = Botanical Garden, BR = Bushland Reserve, c = central, CA = Conservation Area, CP = Conservation Park (or Coastal Park in Vic), CR = Conservation Reserve, e = east, FFR = Flora and Fauna Reserve, FR = Flora Reserve, GR = Game Reserve, Is. = Island, Mt = Mount, n = north, NP = National Park, NR = Nature Reserve, NRA = Nature Recreation Area, NSW = New South Wales, NT = Northern Territory, OIT = Offshore island territory, Qld = Queensland, R = Reserve, RP = Regional Park, s = south, SA = South Australia, SF = State Forest, SP = State Park, ssp. = subspecies, Tas = Tasmania, Uni = University, Vic = Victoria, w = west, WA = Western Australia.

Residents and regular visitors

Ostrich (*Struthio camelus*). Introduced. Rare. Grassland and open woodland. Small feral population (SA) Mt Arden Station; (NSW) Moama–Barham Rd near Deniliquin. Long-term provenance of NSW population not established.

Southern Cassowary (*Casuarius casuarius*). Uncommon n Qld. Rainforest. (Qld) Cassowary House, Daintree, Mt Hypipamee NP, Mission Beach, Paluma Range NP.

Emu (*Dromaius novaehollandiae*). Common s Aust, rare to uncommon n Aust. Grassland and open woodland. (NT) Yellow Water, Caranbirini CR, Mereenie Loop Rd, Uluru-Kata Tjuta NP; (Qld) Lakefield NP, Mareeba Wetlands, Bladensburg NP, Bowra, Gulf Developmental Rd, Currawinya NP, Coolmunda Dam, Girraween NP; (NSW/ACT) Clarence River Estuary, Tidbinbilla NR, Kosciusko NP, Warrumbungle NP, Munghorn Gap, Round Hill CR, Mutawintji NP, Sturt NP, Kinchega NP; (Vic) Hattah-Kulkyne NP, Wyperfeld NP, Little Desert NP, Portland; (SA) Innes NP, Coorong NP, Gluepot; (WA) Purnululu (Bungle Bungles) NP, Shark Bay, Cue, Kalbarri NP, Sandford Rocks NR, Fitzgerald River NP, Wave Rock, Donnelly River, Lake McLarty NR.

Australian Brush-turkey (*Alectura lathami*). Common e Aust. Rainforest. Two distinct ssp.: Blue-wattled ssp. *purpureicollis* n Cape York, *lathami* remainder of range. (Qld) Mt Hypipamee NP, Lamington NP, Kutini-Payamu (Iron Range) NP; (NSW) Dharug NP, Ku-ring-gai Chase NP; (SA) Feral population Kangaroo Is.

Malleefowl (*Leipoa ocellata*). Rare Mallee woodland s Aust. (NSW) Round Hill NR, Nombinnie NR; (Vic) Wyperfeld NP, Hattah-Kulkyne NP, Murray-Sunset NP, Little Desert NP, Bronzewing FFR; (SA) Innes NP, Coorong NP, Gluepot R, Danggali CP, Billiatt CP, Lake Gilles CP, Yanerbie Beach, Yumbarra CP, Mamungari CP; (WA) Dragon Rocks NR, Fitzgerald River NP, Cape Arid NP, Sandford Rocks NR, Eyre Bird Observatory, Shark Bay, Kalbarri NP.

Orange-footed Scrubfowl (*Megapodius reinwardt*). Common n Aust. Rainforest, monsoon forest. (NT) Darwin BG, Charles Darwin Uni, Buffalo Creek, East Point, Holmes Jungle NR, Howard Springs, Fogg Dam, Tiwi Is.; (Qld) Kingfisher Park, Cape Hillsborough NP; (WA) Mitchell River NP; (OIT) Torres Strait Is.

California Quail (*Callipepla californica*). Introduced. Feral populations (Tas) King Is; (OIT) Norfolk Is.

Stubble Quail (*Coturnix pectoralis*). Common Farmland, grassland.

Brown Quail (*Coturnix ypsilophora*). Common Grassland, open woodland, wetlands. (NT) Pine Creek, Charles Darwin NP, Leanyer Sewage Ponds, Elsey NP, Abattoir Swamp; (Qld) Yungaburra, Lockyer Valley, Townsville Common, Kedron Brook Wetlands, Oxley Creek Common, Eagleby Wetlands, Lake Samsonvale, Lake Dyer, Inskip Point; (NSW) Ash Is., Crowdy Bay NP, Centennial Park, Royal NP, Barren Grounds NR, Fivebough Wetlands, Lake Cargelligo; (Vic) Cape Liptrap CP, Croajingolong NP, Wilsons Promontory NP, Lake Mokoan; (Tas) Peter Murrell R, Strahan, Waterhouse CA; (SA) Bool Lagoon, Mary Seymour CP, Messent CP, Robe; (WA) Parry Lagoons NR, Broome, Purnululu (Bungle Bungles) NP, Esperance, Lake Muir NR.

King Quail (*Excalfactoria chinensis*). Uncommon n Aust, rare s Aust. Grassland, heathland. (NT) Holmes Jungle NR, Pine Creek Treatment Plant; (Qld) Kutini-Payamu (Iron Range) NP, Great Sandy NP, Lake Samsonvale.

Indian Peafowl (*Pavo cristatus*). Introduced. Feral populations: (SA) Kangaroo Is.; (Tas) King Is., Flinders Is.; (WA) formerly Rottnest Is.

Red Junglefowl (*Gallus gallus*). Introduced. Feral populations Christmas Is. and Norfolk Is.

Green Junglefowl (*Gallus varius*). Introduced Cocos Keeling Is.

Common Pheasant (*Phasianus colchicus*). Introduced. (Tas) King Is., Flinders Is. (WA) Rottnest Is.

Wild Turkey (*Meleagris gallopavo*). Introduced Feral populations: (SA) Kangaroo Is.; (Tas) King Is., Flinders Is.

Magpie Goose (*Anseranas semipalmata*). Common n Aust, uncommon s Aust. Wetlands. (NT) Holmes Jungle NR, Howard Springs, Knuckey Lagoon, Fogg Dam, Mamukala Wetlands, Yellow Water; (Qld) Lotus Bird Lodge wetland, Centenary Lakes, Magnetic Is., Lake Mitchell, Hasties Swamp, Townsville Common, Tyto Wetlands, Horseshoe Lagoon, Tinchi Tamba Wetlands, Eagleby Wetlands, Lake Clarendon, Lake Dyer; (NSW/ACT) Hunter Wetland Centre, Macquarie Marshes, Fivebough Wetlands, Tidbinbilla NR; (Vic) Deen Maar, Reedy Lake, Serendip Sanctuary, Western Treatment Plant; (SA) Bool Lagoon; (WA) Lake Argyle, Kununurra, Parry Lagoons NR.

Spotted Whistling-Duck (*Dendrocygna guttata*). Rare Cape York. Wetlands. (Qld) First observed in Australian in 1995 at Weipa where now regular. Also Kutini-Payamu (Iron Range) NP and n Torres Strait.

Plumed Whistling-Duck (*Dendrocygna eytoni*). Common n and e Aust, uncommon s Aust. Wetlands, grasslands near water. (NT) Leanyer Sewage Ponds, Mamukala Wetlands, Yellow Water, Anbangbang Billabong, Murill Billabong; (Qld) Half Moon Bay Wetland, Cumberland Dam, Winton Treatment Works, Lake Bindegolly, Currawinya Lakes, Yorkeys Knob, Wolframite Dam, Lake Mitchell, Hasties Swamp, Lake Tinaroo, Tyto Wetlands, Townsville Common, Mackay Regional BG, Winton Treatment Works, Sawpits Waterhole, Lake Bindegolly, Currawinya NP, Eagleby Wetlands, Lockyer Valley; (NSW) Fivebough Wetlands, Macquarie Marshes, Carmel Lagoon, Fivebough Wetlands, South Myers Tank; (Vic) Lake Mokoan; (SA) Bool Lagoon, Koonchera Waterhole; (WA) Lake Argyle, Parry Lagoons NR, Broome, Derby, Kununurra.

Wandering Whistling-Duck (*Dendrocygna arcuata*) Common n Aust. Wetlands, grasslands near water. (NT) Leanyer Sewage Ponds, Howard Springs, Fogg Dam, Mamukala Wetlands, Yellow Water, Anbangbang Billabong, Angurugu Sewage Pond; (Qld) Lakefield NP, Lotus Bird Lodge, Weipa, Keating's Lagoon, Lake Moondarra, Centenary Lakes, Half Moon Bay Wetland, Cattana Wetlands, Yorkeys Knob, Wolframite Dam, Lake Mitchell, Mareeba Wetlands, Hasties Swamp, Tyto Wetlands, Townsville Common, Mackay Regional BG, Lake Moondarra, Eagleby Wetlands, Lockyer Valley; (NSW) Hunter Wetland Centre; (WA) Parry Lagoons NR, Kununurra.

Musk Duck (*Biziura lobata*). Common s Aust. Wetlands. (Qld) Seven Mile Lagoon, Lake Clarendon, Lake Atkinson, Coolmunda Dam; (NSW) Fivebough Wetlands, Lake Cargelligo, Centennial Park; (Vic) Hattah Lakes, Lake Boga, Hird Swamp, Chiltern Valley Number Dams, Lake Mokoan, Western Treatment Plant, Macleod Morass; (SA) Dry Creek Saltfields, Buckland Park, Goolwa, Murray Lagoon, Duck Lagoon, Bool Lagoon, Mungerannie Waterhole, Big Swamp; (WA) Carnarvon, Nallan Lake, Rowles Lagoon, Lake Warden, Vasse-Wonnerup Estuary, Lake McLarty NR, Herdsman Lake.

Freckled Duck (*Stictonetta naevosa*). Rare, nomadic. Wetlands. (NT) Alice Springs Waste Stabilisation Ponds; (Qld) Lake Moondarra, Lake Bindegolly, Lake Wyara, Wolframite Dam, Lake Atkinson; (NSW) Hunter Wetland Centre, Macquarie Marshes, Fivebough Wetlands; (Vic) Hattah Lakes, Round Lake, Lake Boga, Chiltern Valley Dams, Lake Lorne, Western Treatment Plant, Macleod Morass; (SA) Dry Creek Saltfields, Buckland Park, Goolwa Effluent Ponds, Murray Lagoon, Duck Lagoon, Bool Lagoon, Coongie Lakes, Koonchera Waterhole; (WA) Herdsman Lake, Lake McLarty, Lake Muir, Lake Warden, Rowles Lagoon.

Cape Barren Goose (*Cereopsis novaehollandiae*). Uncommon s Aust. Coastal wetlands, coastal grassland, islands: (Vic) Shallow Inlet, Corner Inlet, Phillip Is., Western Treatment Plant; (SA) Hindmarsh Is., Kangaroo Is., Bool Lagoon, Coorong NP, Big Swamp, Coffin Bay NP; (Tas) Pitt Water NR, Maria Is., Flinders Is., Waterhouse CA, Bass Strait Is.; (WA) Cape Arid NP, Woody Island NR, Lake Warden.

Black Swan (*Cygnus atratus*). Common. Wetlands.

Mute Swan (*Cygnus olor*). First introduced 1886. Feral population: (WA) Avon River, Northam.

Radjah Shelduck (*Tadorna radjah*). Mod common n Aust. Wetlands. (NT) Darwin BG, Leanyer Sewage Ponds, Howard Springs, Fogg Dam, Mamukala Wetlands, Yellow Water, Murill Billabong, Pine Creek Treatment Plant, Victoria River, Angurugu Sewage Pond; (Qld) Mungo Brush, Lakefield, Lotus Bird Lodge Wetland, Weipa, Normanton, Daintree River, Daintree Saltwater Barramundi Farm; (WA) Lake Argyle, Kununurra, Parry Lagoons NR.

Australian Shelduck (*Tadorna tadornoides*). Common s Aust. Wetlands.

Australian Wood Duck (*Chenonetta jubata*). Common. Wetlands, grassland, open woodland, farmland, golf courses.

Pink-eared Duck (*Malacorhynchus membranaceus*). Mod common, nomadic. Wetlands.

Cotton Pygmy-Goose (*Nettapus coromandelianus*). Uncommon Qld. Wetlands. (Qld) Lake Moondarra, Cumberland Dam, Yorkeys Knob, Skyrail Wetlands, Lake Mitchell, Keating's Lagoon, Mareeba, Nardellos Lagoon, Hasties Swamp, Lake Tinaroo, Tyto Wetlands, Townsville

Common, Mackay Regional BG, Lake Clarendon, Adare Rd Lagoon.

Green Pygmy-Goose (*Nettapus pulchellus*). Common n Aust. Wetlands. (NT) Fogg Dam, Mamukala, Yellow Water, Anbangbang Billabong; (Qld) Lakefield, Lotus Bird Lodge wetland, Lockhart River Treatment Plant, Weipa, Cumberland Dam, Lake Moondarra, Yorkeys Knob, Cattana Wetlands, Skyrail Wetlands, Wolframite Dam, Lake Mitchell, Mareeba, Nardellos Lagoon, Hasties Swamp, Tyto Wetlands, Townsville Common; (WA) Lake Argyle, Kununurra, Parry Lagoons NR, Roebuck Plains.

Garganey (*Anas querquedula*). Rare summer migrant to n Aust. Wetlands. Records from: (NT) Leanyer and Palmerston Sewage Ponds; (Qld) Weipa; (WA) Parry Lagoons NR, Derby.

Australasian Shoveler (*Anas rhynchotis*). Mod common s Aust. Wetlands. (Qld) Lake Bindegolly, Currawinya Lakes, Lake Atkinson, Coolmunda Dam; (NSW/ACT) Pitt Town Lagoon, Bushell's Lagoon, Ash Is., Carmel Lagoon, Fivebough Wetlands, Fyshwick Sewage Treatment Works; (Vic) Lake Boga, Chiltern Valley Dams, Western Treatment Plant; (SA) Buckland Park, Goolwa, Murray Lagoon, Bird Lake, Big Swamp; (WA) Albany, Vasse-Wonnerup Estuary, Lake McLarty NR, Thomsons Lake NR, Herdsman Lake, Lake Monger.

Grey Teal (*Anas gracilis*). Common Wetlands.

Chestnut Teal (*Anas castanea*). Common s Aust. Rare n Aust. Wetlands.

Northern Mallard (*Anas platyrhynchos*). Domesticated forms common on urban lakes and wetlands. Wild types rare except Norfolk Is., Lord Howe Is. and Macquarie Is. Hybridises with Pacific Black Duck.

Pacific Black Duck (*Anas superciliosa*). Common. Wetlands.

Hardhead (*Aythya australis*). Common Wetlands.

Blue-billed Duck (*Oxyura australis*). Uncommon s Aust. Wetlands. (NT) Alice Springs Waste Stabilisation Ponds; (Qld) Lake Bindegolly, Birdsville Lagoon, Lake Atkinson, Lake Galletly; (NSW/ACT) Lake Cargelligo Treatment Works, Macquarie Marshes, Fivebough Wetlands, Fyshwick Sewage Treatment Works; (Vic) Hattah Lakes, Round Lake, Lake Boga, Western Treatment Plant, Macleod Morass; (SA) Goolwa, Bool Lagoon; (WA) Herdsman Lake, Lake Monger, Mandurah region, Lake McLarty, Lake Muir, Vasse-Wonnerup Estuary, Rowles Lagoon, Lake Powell NR.

Red-tailed Tropicbird (*Phaethon rubricauda*). Tropical and subtropical ocean. Breeds Christmas Is., Cocos-Keeling, Lord Howe Is., Norfolk Is. and Ashmore Reef. Regularly reported in WA and Qld. In WA, formerly bred Rat Is., Rottnest Is and Sugarloaf Rock.

White-tailed Tropicbird (*Phaethon lepturus*). Tropical ocean. Golden morph Christmas Is. Two distinct breeding ssp.: *fulvus* is Critically Endangered; ssp. *lepturus* (Endangered in Aust. territories) breeds on Cocos-Keeling Is. and Ashmore Reefs. Indian Ocean. Ssp. *dorotheae* of the Pacific Ocean is rare non-breeding visitor off e Aust.

Australasian Grebe (*Tachybaptus novaehollandiae*). Common throughout Aust. Wetlands.

Hoary-headed Grebe (*Poliocephalus poliocephalus*). Common throughout Aust. Wetlands, coastal lagoons.

Great Crested Grebe (*Podiceps cristatus*). Uncommon. Wetlands. (Qld) Lake Moondarra, Lake Bindegolly, Currawinya Lakes, Wolframite Dam, Lake Barrine, Lake Samsonvale, Lake Clarendon; (NSW) Bushell's Lagoon, Deniliquin, Gulpa Creek, Kinchega NP; (Vic) Hattah Lakes, Round Lake, Lake Boga; (Tas) Pitt Water NR; (SA) Dry Creek Saltfields, Goolwa, Cooper Creek; (WA) Carnarvon, Rowles Lagoon, Fitzgerald River NP, Walpole-Nornalup NP, Herdsman Lake RP.

Rock Dove (*Columba livia*). Introduced. Common in urban and agricultural areas.

White-headed Pigeon (*Columba leucomela*). Mod common e coast. Range expanding in s-e Aust. Rainforest, woodlands. (Qld) Yungaburra, Wongabel SF, D'Aguilar NP, Lamington NP, Great Sandy NP; (NSW) Copeland Tops, Boorganna NR, Dorrigo NP, Nightcap NP, Bomaderry Creek RP, Mimosa Rocks NP; (Vic) Mallacoota (uncommon).

Barbary Dove (*Streptopelia roseogrisea*). Introduced. (SA) Rare. Urban areas of Adelaide.

Laughing Dove (*Streptopelia senegalensis*). Introduced. (WA) Common in urban areas of Perth.

Spotted Dove (*Streptopelia chinensis*). Introduced. Widespread, very common. Urban areas.

Brown Cuckoo-Dove (*Macropygia amboinensis*). Common in n-e Aust, uncommon in s-e Aust. Rainforest, woodlands. (Qld) Lockerbie Scrub, Kingfisher Park, Mt Lewis, Mt Hypipamee NP, Townsville Common, Lamington NP, D'Aguilar NP; (NSW) Royal NP, Barren Grounds NR, Morton NP, Nowra, Myall Lakes, Harrington, Copeland Tops, Dorrigo NP, Bruxner Park FR, Nightcap NP.

Emerald Dove (*Chalcophaps indica*). Common n and n-e Aust, uncommon s-e Aust. Rainforest, monsoon forest. (NT) Buffalo Creek, East Point, Howard Springs, Fogg Dam, Tiwi Is., Nourlangie; (Qld) Lockerbie Scrub, Mt Whitfield CP, Green Is., Kingfisher Park, Mt Lewis, Mt Hypipamee NP, Cape Hillsborough NP, Lamington NP, Great Sandy NP, Conondale Ranges; (NSW) Copeland Tops, Myall Lakes NP, Harrington, Wingham Brush NR, Iluka NR, Nightcap NP; (WA) occasional records Kimberley; (OIT) Christmas Is., Lord How Is., Norfolk Is.

Common Bronzewing (*Phaps chalcoptera*). Common. Forest, woodland, shrubland.

Brush Bronzewing (*Phaps elegans*). Uncommon s Aust. Heathland, forest, woodland. (Qld) Inskip Point; (NSW) Ku-ring-gai Chase NP, Barren Grounds NR, Booderee NP, Ben Boyd NP, Nadgee NR, Maxwell's FR; (Vic) Greater Bendigo NP, Great Otway NP, Portland, Cape Conran, Cape Liptrap CP, Croajingolong NP; (Tas) Mt Wellington, Bruny Is., Truganini Reserve, Meehan Range NRA, Strahan, King Is.; (SA) Para Wirra RP, Flinders Chase NP, Remarkable Rocks, Innes NP, Coorong NP, Robe, Lincoln NP, Coffin Bay NP; (WA) Dryandra Woodlands, Two Peoples Bay NR, Cheynes Beach, Fitzgerald River NP, Woody Island NR, Cape Arid NP, Eyre Bird Observatory.

Flock Bronzewing (*Phaps histrionica*). Nomadic and irruptive, fluctuating between common and rare. Mainly in open Mitchell Grass on black soil plains but widespread in arid regions. (NT) Elsey NP, Barkly Tableland, Connells Lagoon CR; (Qld) Bowra, Koolivoo Waterhole, Eyre Developmental Rd, Asterebla Downs NP, Diamantina NP; (NSW) Sturt NP; (SA) Coongie Lakes NP, Birdsville Track; (WA) (Broome) Roebuck Plains (rare).

Crested Pigeon (*Ocyphaps lophotes*). Common mainland Aust. Woodland, farmlands, gardens.

Spinifex Pigeon (*Geophaps plumifera*). Mod common. Rocky gorges. Three distinct ssp.: *plumifera* (white-bellied) n WA, NT; *leucogaster* (variable extent of white) NT, inland Qld; *ferruginea* (dark-bellied) c WA. (NT) Victoria River, Jasper Gorge, Caranbirini CR, Mt Gillen, Ormiston Gorge, Watarrka (Kings Canyon) NP; (Qld) Boodjamulla (Lawn Hill) NP, Riversleigh, Mt Isa, Bladensburg NP, Lark Quarry CP, Opalton; (WA) Lake Argyle, Parry Lagoons NR, Carnarvon, Tom Price, Purnululu (Bungle Bungles) NP, Karijini NP.

Partridge Pigeon (*Geophaps smithii*). Mod common Top End and Kimberley. Savannah woodland. Two distinct ssp.: *smithii* (red eye-ring) NT; *blaauwi* (yellow eye-ring) n WA. (NT) Yellow Water, Mardugal Billabong, Kakadu Hwy, Nourlangie, Murill Billabong, Pine Creek; (WA) Mitchell River NP.

Squatter Pigeon (*Geophaps scripta*). Mod common n-e Qld, rare s-e Qld. Savannah woodland. Two distinct ssp.: *peninsulae* (red eye-ring) Cape York; *scripta* (blue eye-ring) remainder of range. (Qld) Musgrave Roadhouse, Cumberland Dam, Currawinya NP, Mt Carbine, Mt Molloy, Mareeba Wetlands, Emerald Creek, Eungella Reservoir, Girraween NP, Durikai SF.

White-quilled Rock-Pigeon (*Petrophassa albipennis*). Mod common n-w Aust. Escarpment, sandstone gorges. Two distinct ssp.: *albipennis* (dark brown) Kimberley to Victoria River in n-w NT; *boothi* (light brown) n-w NT. (NT) Victoria River escarpment, Jasper Gorge; (WA) Lake Argyle, Mirima (Hidden Valley) NP, the Grotto, Mitchell River NP, Mornington Station, Purnululu (Bungle Bungles) NP.

Chestnut-quilled Rock-Pigeon (*Petrophassa rufipennis*). Mod common Top End. Arnhem escarpment, sandstone gorges. (NT) Nourlangie, Bardedjilidji, Gunlom, Katherine Gorge.

Diamond Dove (*Geopelia cuneata*). Common n and c Aust, less common s Aust. Grassy woodland.

Peaceful Dove (*Geopelia striata*). Common. Woodland, shrublands.

Bar-shouldered Dove (*Geopelia humeralis*). Common n Aust, uncommon s Aust. Woodland, forest.

Wonga Pigeon (*Leucosarcia picata*). Mod common e Aust. Forest, woodland. (Qld) Lamington NP; (NSW/ACT) Dharug NP, Royal NP, Morton NP, Nowra, Barrington Tops NP, Dorrigo NP, Nightcap NP, Mimosa Rocks NP, Namadgi NP; (Vic) Croajingolong NP, Fairy Dell NCP, Buchan Caves R.

Banded Fruit-Dove (*Ptilinopus cinctus*). Rare Top End. Monsoon forest in Arnhem escarpment. (NT) Nourlangie, Bardedjilidji, Gunlom, Yurmikmik.

Wompoo Fruit-Dove (*Ptilinopus magnificus*). Common n-e Aust, uncommon s-e Aust. Rainforest. (Qld) Lockerbie Scrub, Kutini-Payamu (Iron Range) NP, Jardine River NP, McIvor River Crossing, Mt Whitfield CP, Kuranda, Daintree, Kingfisher Park, Cooktown, Mt Hypipamee NP, Curtain Fig Tree, D'Aguilar NP, Mt Tamborine NP, Kuranda; (NSW) Copeland Tops, Werrikimbe NP, Dorrigo NP, Nightcap NP, Boorganna NR, Wongabel SF, Bruxner Park FR.

Superb Fruit-Dove (*Ptilinopus superbus*). Uncommon n-e Aust, rare s-e Aust. Rainforest and monsoon forest. (Qld) McIvor River Crossing, Kutini-Payamu (Iron Range) NP, Lockerbie Scrub, Mt

Whitfield CP, Kuranda, Kingfisher Park, Mt Hypipamee NP, Eungella Plateau; (NSW) Nightcap NP.

Rose-crowned Fruit-Dove (*Ptilinopus regina*). Common n Aust, uncommon s-e Aust. Rainforest, monsoon forest and mangroves. (NT) Darwin BG, East Point, Fogg Dam, Lee Point, Buffalo Creek, Holmes Jungle NR, Howard Springs, Tiwi Is., Groote Eylandt; (Qld) Lockerbie Scrub, Cooktown, Kutini-Payamu (Iron Range) NP, Green Is., Townsville Common, Cape Hillsborough NP, Eungella Plateau, D'Aguilar NP, Lamington NP, Inskip Point, Great Sandy NP, Conondale Ranges; (NSW) Crowdy Bay NP, Clarence River, Nightcap NP, Boorganna NR, Bruxner Park FR; (WA) Mitchell River NP.

Christmas Is. Imperial-Pigeon (*Ducula whartoni*). Common. Christmas Is. endemic.

Collared Imperial-Pigeon (*Ducula mullerii*). Uncommon. Mangroves n Torres Strait Is. (OIT) Boigu Is., Saibai Is.

Pied Imperial-Pigeon (*Ducula bicolor*). Common n Aust. Rainforest and mangroves. (NT) Darwin BG, Charles Darwin Uni, Lee Point, Holmes Jungle NR; (Qld) Kutini-Payamu (Iron Range) NP, Cairns Esplanade, Green Is., Daintree, Cape Hillsborough NP; (WA) Mitchell River NP.

Topknot Pigeon (*Lopholaimus antarcticus*). Common e Aust. Rainforest and eucalypt forest. (Qld) Kutini-Payamu (Iron Range) NP, Eungella Plateau, D'Aguilar NP; (NSW) Copeland Tops, Myall Lakes NP, Boorganna NR, Dorrigo NP, Nightcap NP, Royal NP, Budderoo NP, Morton NP, Mimosa Rocks NP, Ben Boyd NP, Maxwell's FR; (Vic) Croajingolong NP, Cabbage Tree Creek.

Tawny Frogmouth (*Podargus strigoides*). Common. Woodland and forest.

Papuan Frogmouth (*Podargus papuensis*). Mod common n Qld. Forest and rainforest. (Qld) Cooktown, McIvor River Crossing, Lakefield NP, Kutini-Payamu (Iron Range) NP, Uningan NR, Wenlock River, Lockerbie Scrub, Jardine River NP, Cairns BG, Daintree, Kingfisher Park, Paluma Range NP.

Marbled Frogmouth (*Podargus ocellatus*). Rare. Rainforest. Two distinct ssp.: *marmoratus* n Qld; *plumiferous* s-e Qld and n-e NSW. (Qld) Kutini-Payamu (Iron Range) NP, Lakefield NP, D'Aguilar NP, Lamington NP, Mt Tamborine NP, Great Sandy NP, Conondale Ranges; (NSW) Nightcap NP.

White-throated Nightjar (*Eurostopodus mystacalis*). Uncommon e Aust. Forest and woodland. (Qld) Kutini-Payamu (Iron Range) NP, Kuranda, Mareeba, Emerald Creek, Davies Creek NP, Great Sandy NP; (NSW) Crowdy Bay NP, Mt Coot-tha Forest, Capertee Valley, Dharug NP, Barren Grounds NR; (Vic) Chiltern-Mt Pilot NP, Warby Range SP, Croajingolong NP, Cape Conran, Bunyip SP.

Spotted Nightjar (*Eurostopodus argus*). Mod common. Open woodlands, shrublands. (NT) Royal Darwin Hospital, Fogg Dam, Gunlom, Elsey NP, Jasper Gorge, Kunoth Bore, Uluru-Kata Tjuta NP; (Qld) Weipa, Cumberland Dam, Boodjamulla (Lawn Hill) NP, Lake Moondarra, Bladensburg NP, Opalton, Bowra, Eulo Bore, Currawinya NP, Sundown NP; (NSW) Round Hill and Nombinnie NR, Cocoparra NP, Sturt NP; (Vic), Wyperfeld NP, Greater Bendigo NP; (SA) Gluepot R, Lake Gilles CP; (WA) Sandford Rocks NR, Neale Junction NR, Cocanarup Timber R, Parry Lagoons NR, Cue.

Large-tailed Nightjar (*Caprimulgus macrurus*). Common n Aust. Rainforest, monsoon forest. (NT) Buffalo Creek, East Point; (Qld) Cooktown, Lakefield NP, Kutini-Payamu (Iron Range) NP, Wenlock River, Lockerbie

Scrub, Abattoir Swamp, Mareeba Wetland, Paluma Range NP, Padaminka NR, Great Sandy NP; (OIT) Torres Strait Is.

Australian Owlet-Nightjar (*Aegotheles cristatus*). Common Woodland, forest.

Australian Swiftlet (*Aerodramus terraereginae*). Common n Qld. Aerial over forest, woodland, coastal range. (Qld) Kutini-Payamu (Iron Range) NP, Cairns Esplanade, Cairns BG, Newell Beach.

Christmas Island Swiftlet (*Collocalia esculenta natalis*). Common Christmas Is.

White-throated Needletail (*Hirundapus caudacutus*). Uncommon summer migrant e Aust. Aerial over forest, woodland, often ahead of storm fronts. (Qld) Kuranda, D'Aguilar NP, Great Sandy NP; (NSW) Myall Lakes NP, Crowdy Bay NP, Merriwindi SF, Fivebough Wetlands, Royal NP, Booderee NP, Mimosa Rocks NP; (Vic) Croajingolong NP; (Tas) Arthur River.

Fork-tailed Swift (*Apus pacificus*). Uncommon summer migrant. Aerial over open plains, shrublands, woodland, forest, often ahead of storm fronts. (NT) Barkly Tableland; (Qld) Kuranda, Newell Beach, Cumberland Dam; (NSW) Myall Lakes NP, Fivebough Wetlands, Royal NP; (Vic) Lake Mokoan; (SA) Bool Lagoon GR, Coorong NP; (WA) widespread.

Swinhoe's Storm-Petrel (*Hydrobates monorhis*). Uncommon. Tropical Indian Ocean. (OIT) Ashmore Reef.

Matsudaira's Storm-Petrel (*Hydrobates matsudairae*). Uncommon. Tropical Indian Ocean. (OIT) Ashmore Reef.

Wilson's Storm-Petrel (*Oceanites oceanicus*). Common Open ocean. (NSW) Port Stephens Pelagic, Sydney Pelagic, Wollongong Pelagic, Eden Pelagic; (Vic) Portland and Port Fairy Pelagic; (Tas) Pelagics; (SA) Port MacDonnell Pelagic; (WA) Albany Pelagic; (OIT) Heard Is., Macquarie Is., Ashmore Reef.

Grey-backed Storm-Petrel (*Garrodia nereis*). Common off s Aust, rare in NSW. Southern ocean. (Vic) Portland and Port Fairy Pelagic; (Tas) Pelagics; (SA) Port MacDonnell Pelagic; (OIT) Heard Is., Macquarie Is.

White-faced Storm-Petrel (*Pelagodroma marina*). Common off s Aust. Ocean. (NSW) Port Stephens Pelagic, Sydney Pelagic, Wollongong Pelagic, Eden Pelagic; (Vic) Mud Islands, Portland and Port Fairy Pelagic; (Tas) Pelagics; (SA) Port MacDonnell Pelagic; (WA) Albany Pelagic, Hillary's Boat Harbour Pelagic.

Black-bellied Storm-Petrel (*Fregetta tropica*). Generally rare, but mod common off Southport in winter. Open ocean. (Qld) Southport pelagic; (NSW) Port Stephens Pelagic, Sydney Pelagic, Wollongong Pelagic, Eden Pelagic; (Tas) Pelagics; (SA) Port MacDonnell Pelagic; (WA) Albany Pelagic; (OIT) Macquarie Is., Heard Is.

White-bellied Storm-Petrel (*Fregetta grallaria*). Rare except Lord Howe Is. where breeds. Sub-tropical Pacific Ocean. (NSW) Port Stephens Pelagic, Sydney Pelagic; (OIT) Lord Howe Is.

Wandering Albatross (*Diomedea exulans*). Uncommon Southern ocean. (NSW) Port Stephens Pelagic, Sydney Pelagic, Wollongong Pelagic, Eden Pelagic; (Vic) Portland and Port Fairy Pelagic; (Tas) Pelagics; (SA) Port MacDonnell Pelagic; (WA) Hillary's Boat Harbour Pelagic, Albany Pelagic; (OIT) Macquarie Is.

Antipodean Albatross (*Diomedea antipodensis*). Uncommon Southern ocean. Once considered conspecific with the

Wandering Albatross. (NSW) Port Stephens Pelagic, Wollongong Pelagic; (Vic) Portland and Port Fairy Pelagic; (Tas) Pelagics; (SA) Port MacDonnell Pelagic.

Southern Royal Albatross (*Diomedea epomophora*). Uncommon Southern ocean. Once considered conspecific with the Northern Royal Albatross, known as Royal Albatross. (NSW) Port Stephens Pelagic, Sydney Pelagic, Wollongong Pelagic, Eden Pelagic; (Vic) Portland and Port Fairy Pelagic; (Tas) Pelagics; (SA) Port MacDonnell Pelagic; (WA) Hillary's Boat Harbour Pelagic, Albany Pelagic. Best site Tas pelagic in spring.

Northern Royal Albatross (*Diomedea sanfordi*). Rare. Southern ocean. Once considered conspecific with the Southern Royal Albatross. (NSW) Port Stephens Pelagic, Sydney Pelagic, Wollongong Pelagic, Eden Pelagic; (Vic) Portland and Port Fairy Pelagic; (Tas) Pelagics; (SA) Port MacDonnell Pelagic.

Black-browed Albatross (*Thalassarche melanophris*). Common Southern ocean. (NSW) Port Stephens Pelagic, Sydney Pelagic, Wollongong Pelagic, Eden Pelagic; (Vic) Portland and Port Fairy Pelagic; (Tas) Pelagics; (SA) Port MacDonnell Pelagic; (WA) Albany Pelagic; (OIT) Macquarie Is., Heard Is.

Campbell Albatross (*Thalassarche impavida*). Common Southern ocean. Once considered conspecific with the Black-browed Albatross. (NSW) Port Stephens Pelagic, Sydney Pelagic, Wollongong Pelagic, Eden Pelagic; (Vic) Portland and Port Fairy Pelagic; (Tas) Pelagics; (SA) Port MacDonnell Pelagic.

Shy Albatross (*Thalassarche cauta*). Common Southern ocean. (NSW) Port Stephens Pelagic, Sydney Pelagic, Wollongong Pelagic, Eden Pelagic; (Vic) Portland and Port Fairy Pelagics; (Tas) Pelagics; (SA) Port MacDonnell Pelagic; (WA) Hillary's Boat Harbour Pelagic, Albany Pelagic.

Salvin's Albatross (*Thalassarche salvini*). Uncommon. Southern ocean. Once considered conspecific with the Black-browed Albatross. (NSW) Port Stephens Pelagic, Sydney Pelagic, Wollongong Pelagic, Eden Pelagic; (Vic) Portland and Port Fairy Pelagic; (Tas) Pelagics; (SA) Port MacDonnell Pelagic; (WA) Hillary's Boat Harbour Pelagic, Albany Pelagic.

Grey-headed Albatross (*Thalassarche chrysostoma*). Rare winter visitor. Southern ocean. (Vic) Portland and Port Fairy Pelagic; (Tas) Pelagics; (WA) Albany Pelagic; (OIT) Macquarie Is., Heard Is.

Indian Yellow-nosed Albatross (*Thalassarche carteri*). Common Southern and subtropical ocean. Once considered conspecific with the Atlantic Yellow-nosed Albatross. (NSW) Port Stephens Pelagic, Sydney Pelagic, Wollongong Pelagic, Eden Pelagic; (Vic) Portland and Port Fairy Pelagic; (Tas) Pelagics; (SA) Port MacDonnell Pelagic; (WA) Hillary's Boat Harbour Pelagic, Albany Pelagic.

Buller's Albatross (*Thalassarche bulleri*). Uncommon Pacific and southern ocean west to SA. (NSW) Port Stephens Pelagic, Sydney Pelagic, Wollongong Pelagic, Eden Pelagic; (Vic) Portland and Port Fairy Pelagic; (Tas) Pelagics; (SA) Port MacDonnell Pelagic. Best site Tas pelagic where often common.

Sooty Albatross (*Phoebetria fusca*). Rare. Southern ocean. (Vic) Portland and Port Fairy Pelagic; (SA) Port MacDonnell Pelagic; (Tas) pelagics; (WA) Albany Pelagic. Best sites Vic or SA in April and May.

Light-mantled Sooty Albatross (*Phoebetria palpebrata*). Rare. Southern ocean. (Tas) Pelagics; (OIT) Macquarie Is., Heard Is.

Southern Giant-Petrel (*Macronectes giganteus*). Uncommon winter-spring visitor. Southern ocean. (Qld) North Stradbroke Is., Long Reef Aquatic Reserve; (NSW) Sydney Pelagic, Wollongong Pelagic, Eden Pelagic; (Vic) Point Lonsdale Lighthouse, Cape Nelson Lighthouse, Portland and Port Fairy Pelagic; (Tas) Pelagics; (WA) Cape Naturaliste Lighthouse, Cape Leeuwin, Albany, Hillary's Boat Harbour Pelagic; (OIT) Macquarie Is., Heard Is.

Northern Giant-Petrel (*Macronectes halli*). Uncommon winter-spring visitor. Southern and subtropical ocean. (Qld) North Stradbroke Is., Southport Pelagic; (NSW) Sydney Pelagic, Wollongong Pelagic, Booderee NP, Ben Boyd NP; (Tas) Pelagics; (Vic) Cape Nelson Lighthouse, Portland and Port Fairy Pelagic; (WA) Cape Naturaliste Lighthouse, Hillary's Boat Harbour Pelagic; (OIT) Macquarie Is.

Southern Fulmar (*Fulmarus glacialoides*). Rare winter–spring visitor. Southern ocean. (Vic) Cape Nelson Lighthouse, Portland Pelagic; (Tas) Pelagics; (SA) Port MacDonnell Pelagic; (OIT) Macquarie Is.

Cape Petrel (*Daption capense*). Uncommon Southern ocean. (Qld) Southport Pelagic; (NSW) Wollongong Pelagic; (Vic) Cape Nelson Lighthouse, Portland and Port Fairy Pelagic; (Tas) Pelagics; (SA) Port MacDonnell Pelagic; (WA) Albany Pelagic; (OIT) Heard Is.

Blue Petrel (*Halobaena caerulea*). Rare winter–spring visitor. Southern ocean. (NSW) Wollongong Pelagic; (Vic) Portland and Port Fairy Pelagic; (Tas) Pelagics; (SA) Port MacDonnell Pelagic; (WA) Albany Pelagic; (OIT) Macquarie Is.

Salvin's Prion (*Pachyptila salvini*). Rare. Southern ocean. (Qld) Southport Pelagic; (Vic) Portland and Port Fairy Pelagic; (Tas) Pelagics; (SA) Port MacDonnell Pelagic; (WA) Albany Pelagic; (OIT) Heard Is.

Antarctic Prion (*Pachyptila desolata*). Rare. Southern ocean. (Qld) Southport Pelagic; (NSW) Maroubra, Eden Pelagic; (Vic) Portland and Port Fairy Pelagic; (WA) Albany Pelagic; (OIT) Macquarie Is., Heard Is.

Slender-billed Prion (*Pachyptila belcheri*). Uncommon Southern ocean. (Qld) Southport Pelagic; (NSW) Maroubra, Eden Pelagic; (Vic) Portland and Port Fairy Pelagic; (Tas) Pelagics; (SA) Port MacDonnell Pelagic; (WA) Albany Pelagic.

Fairy Prion (*Pachyptila turtur*). Common Southern ocean, occasionally inshore. (Qld) North Stradbroke Is., Southport Pelagic; (NSW) Maroubra, Eden Pelagic, Green Cape Lighthouse; (Vic) Wilsons Promontory, Point Lonsdale Lighthouse, Portland and Port Fairy Pelagic, Cape Nelson Lighthouse; (Tas) Pelagics, Maria Is., King Is., Freycinet NP; (SA) Port MacDonnell Pelagic.

Fulmar Prion (*Pachyptila crassirostris*). Southern ocean. Vagrant except Heard Is. where breeds.

White-chinned Petrel (*Procellaria aequinoctialis*). Common in s, rare in n, mostly spring–summer. Southern ocean. (Qld) Southport Pelagic; (NSW) Wollongong Pelagic; (Vic) Portland and Port Fairy Pelagic; (Tas) Pelagics; (SA) Port MacDonnell Pelagic; (WA) Albany Pelagic.

Black Petrel (*Procellaria parkinsoni*). Rare summer visitor. Subtropical Pacific ocean. (Qld) Southport Pelagic; (NSW) Port Stephens Pelagic, Sydney Pelagic, Eden Pelagic.

Grey Petrel (*Procellaria cinerea*). Rare to very rare winter visitor. Southern Ocean. (Vic) Portland and Port Fairy Pelagics; (Tas) Pelagics; (SA) Port MacDonnell Pelagic; (WA) Albany Pelagic; (OIT) Heard Is., Macquarie Is. Best chance is September Tas Pelagic.

Jouanin's Petrel (*Bulweria fallax*). Rare. Indian Ocean. Annual records of one or two from Ashmore Reef. Single record Cocos Keeling Is.

Bulwer's Petrel (*Bulweria bulwerii*). Tropical ocean. Common Ashmore Reef, vagrant elsewhere.

Wedge-tailed Shearwater (*Ardenna pacifica*). Common. Tropical and temperate ocean. (Qld) Lady Elliot Is., North Stradbroke Is., Southport pelagic; (NSW) Port Stephens Pelagic, Seal Rock Lighthouse, Long Reef Aquatic Reserve, Soldiers Point, Wollongong Pelagic, Cape St George Lighthouse, Eden Pelagic, Green Cape Lighthouse; (WA) Bathurst Point Lighthouse, Albany Pelagic, Hillary's Boat Harbour Pelagic, Robert Point, Cape Naturaliste Lighthouse; (OIT) Lord Howe Is., Norfolk Is., Ashmore Reef.

Buller's Shearwater (*Ardenna bulleri*). Rare to uncommon. Subtropical to southern Pacific Ocean. (Qld) Southport pelagic; (NSW) Port Stephens Pelagic, Sydney Pelagic, Wollongong Pelagic; (Tas) Pelagics.

Flesh-footed Shearwater (*Ardenna carneipes*). Common. Open ocean. (Qld) North Stradbroke Is., Southport Pelagic; (NSW) Port Stephens Pelagic, Seal Rock Lighthouse, Sydney Pelagic, Wollongong Pelagic, Eden Pelagic; (Vic) Portland and Port Fairy Pelagic; (SA) Port MacDonnell Pelagic; (WA) Penguin Is., Hillary's Boat Harbour Pelagic, Robert Point, Cape Naturaliste Lighthouse, Cape Leeuwin, The Gap, Albany Pelagic, Woody Island; (OIT) Norfolk Is., Lord Howe Is.

Sooty Shearwater (*Ardenna grisea*). Uncommon. Open ocean. (Qld) Southport Pelagic; (NSW) Long Reef Aquatic Reserve, Wollongong Pelagic, Eden Pelagic, Green Cape Lighthouse; (Vic) Portland and Port Fairy Pelagic, Cape Nelson Lighthouse; (Tas) Pelagics; (SA) Port MacDonnell Pelagic; (WA) Albany Pelagic.

Short-tailed Shearwater (*Ardenna tenuirostris*). Common. Mostly off s-e Aust but migrates north off NSW and Qld. (Qld) North Stradbroke Is., Southport Pelagic; (NSW) Port Stephens Pelagic, Seal Rock Lighthouse, Long Reef Aquatic Reserve, Sydney Pelagic, Soldiers Point, Cape St George Lighthouse, Eden Pelagic, Green Cape Lighthouse; (Vic) Cape Liptrap, Wilsons Promontory, Nobbies, Point Lonsdale Lighthouse, Point Addis Lookout, Portland and Port Fairy Pelagic, Cape Nelson Lighthouse; (Tas) Pelagics, Freycinet NP; (SA) Newland Head, Cape Du Couedic Lighthouse, Port MacDonnell Pelagic, Robe Lighthouse, Point Avoid Headland, Woody Island.

Streaked Shearwater (*Calonectris leucomelas*). Mostly rare. Tropical and subtropical ocean. (Qld) Southport pelagic; (NSW) Port Stephens Pelagic, Wollongong Pelagic; (OIT) Ashmore Reef (uncommon).

Fluttering Shearwater (*Puffinus gavia*). Uncommon to mod common. Eastern southern and subtropical ocean, harbours, occasionally inshore. (Qld) North Stradbroke Is., Southport Pelagic; (NSW) Seal Rock Lighthouse, Sydney Pelagic, Wollongong Pelagic, Cape St George Lighthouse, Eden Pelagic, Green Cape Lighthouse; (Vic) Cape Liptrap, Nobbies, Point Lonsdale Lighthouse, Point Addis Lookout, Portland and Port Fairy Pelagic, Cape Nelson Lighthouse; (Tas) Pelagics;

(SA) Newland Head, Port MacDonnell Pelagic, Robe Lighthouse.

Hutton's Shearwater (*Puffinus huttoni*). Uncommon to mod common. Southern to tropical ocean, harbours, occasionally inshore. (Qld) Southport Pelagic; (NSW) Port Stephens Pelagic, Sydney Pelagic, Wollongong Pelagic, Eden Pelagic, Cape Liptrap, Nobbies, Point Addis Lookout, Portland and Port Fairy Pelagic; (Tas) Pelagics; (SA) Port MacDonnell Pelagic; (WA) Hillary's Boat Harbour Pelagic, Cape Naturaliste Lighthouse, Cape Leeuwin, Albany Pelagic; (OIT) Ashmore Reef.

Little Shearwater (*Puffinus assimilis*). Rare, except s-w WA, Norfolk Is. and Lord Howe Is., where breeds. Southern to subtropical ocean. (NSW) Sydney Pelagic, Wollongong Pelagic; (Vic) Portland and Port Fairy Pelagic; (WA) Bathurst Point Lighthouse, Hillary's Boat Harbour Pelagic, Cape Naturaliste Lighthouse, Cape Leeuwin, The Gap, Albany Pelagic; (OIT) Norfolk Is., Lord Howe Is., Macquarie Is.

Tahiti Petrel (*Pseudobulweria rostrata*). Mostly rare. Subtropical to tropical ocean. (Qld) Southport Pelagic; (NSW) Port Stephens, Sydney Pelagic, Wollongong Pelagic; (OIT) Lord Howe Is., Ashmore Reef. Best sites Southport Pelagic in summer (common) and Ashmore Reef (uncommon).

Kerguelen Petrel (*Lugensa brevirostris*). Rare to very rare. Southern ocean. (Vic) Portland and Port Fairy Pelagic; (WA) Albany Pelagic; (OIT) Macquarie Is, Heard Is.

Kermadec Petrel (*Pterodroma neglecta*). Mostly rare. Temperate to subtropical Pacific ocean. (Qld) Southport Pelagic; (NSW) Port Stephens Pelagic, Sydney Pelagic, Wollongong Pelagic, Eden Pelagic; (OIT) Norfolk Is., Lord Howe Is. Best sites Southport Pelagic and Lord Howe Is.

Herald Petrel (*Pterodroma heraldica*). Tropical ocean. Breeds (Qld) Raine Island (not covered here), but otherwise vagrant.

Soft-plumaged Petrel (*Pterodroma mollis*). Rare to uncommon. Southern ocean. (NSW) Wollongong Pelagic; (Vic) Portland and Port Fairy Pelagic; (Tas) Pelagics; (WA) Hillary's Boat Harbour Pelagic; (OIT) Macquarie Is., Heard Is.

White-headed Petrel (*Pterodroma lessonii*). Uncommon in s, rare in NSW. Southern ocean. (NSW) Port Stephens Pelagic, Sydney Pelagic, Wollongong Pelagic, Eden Pelagic; (Vic) Portland and Port Fairy Pelagic; (Tas) Pelagics; (SA) Port MacDonnell Pelagic; (WA) Hillary's Boat Harbour Pelagic, Albany Pelagic; (OIT) Macquarie Is., Heard Is.

Great-winged Petrel (*Pterodroma macroptera*). Uncommon to common. Southern and subtropical ocean. (NSW) Port Stephens Pelagic, Sydney Pelagic, Wollongong Pelagic, Eden Pelagic; (Vic) Portland and Port Fairy Pelagic; (Tas) Pelagics; (SA) Port MacDonnell Pelagic; (WA) Hillary's Boat Harbour Pelagic, Albany Pelagic.

Providence Petrel (*Pterodroma solandri*). Uncommon to common. Southern to subtropical Pacific Ocean. (Qld) Southport Pelagic, North Stradbroke Is.; (NSW) Port Stephens Pelagic, Maroubra, Sydney Pelagic, Wollongong Pelagic, Eden Pelagic; (Tas) Pelagics; (OIT) Norfolk Is., Lord Howe Is.

Mottled Petrel (*Pterodroma inexpectata*). Rare. Pacific Ocean. (Qld) Southport Pelagic; (NSW) Sydney Pelagic, Wollongong Pelagic; (Tas) Pelagics; (OIT) Macquarie Is.

Gould's Petrel (*Pterodroma leucoptera*). Rare southern and Pacific Ocean. Ssp. *leucoptera*

Annotated bird list

breeds Cabbage Tree, Boondelbah and Little Broughton Is., off Port Stephens, NSW. (Qld) Southport Pelagic, North Stradbroke Is.; (NSW) Port Stephens Pelagic, Seal Rock Lighthouse Sydney Pelagic, Wollongong Pelagic; (Vic) Portland and Port Fairy Pelagic; (Tas) Pelagics; (OIT) Norfolk Is., Lord Howe Is.

White-necked Petrel (*Pterodroma cervicalis*). Rare. Tropical and subtropical Pacific Ocean. (Qld) Southport Pelagic, North Stradbroke Is.; (NSW) Port Stephens Pelagic, Sydney Pelagic, Wollongong Pelagic; (OIT) Norfolk Is., Lord Howe Is.

Black-winged Petrel (*Pterodroma nigripennis*). Tropical and subtropical Pacific Ocean. Rare on east coast pelagics. Breeds Norfolk Is., Lord Howe Is.

Common Diving-Petrel (*Pelecanoides urinatrix*). Mod common in s, rare in NSW. Oceans. (NSW) Maroubra, Green Cape Lighthouse; (Vic) Wilsons Promontory, Portland and Port Fairy Pelagic, Cape Nelson Lighthouse; (Tas) Maria Is. Pelagics.

South Georgian Diving-Petrel (*Pelecanoides georgicus*). Southern ocean. Breeds Heard Is., vagrant elsewhere.

Macaroni Penguin (*Eudyptes chrysolophus*). Resident Heard Is. Vagrant to mainland Aust.

Royal Penguin (*Eudyptes schlegeli*). Resident Macquarie Is. Vagrant to mainland Aust.

Gentoo Penguin (*Pygoscelis papua*). Resident Heard Is. and Macquarie Is. Vagrant to Tas.

Southern Rockhopper Penguin (*Eudyptes chrysocome*). Breeds Heard Is., Macquarie Is. Vagrant records SA, Tas, Vic and WA.

King Penguin (*Aptenodytes patagonicus*). Breeds Heard Is. and Macquarie Is. Vagrant records SA, Tas, Vic and WA.

Little Penguin (*Eudyptula minor*). Common. Coastal areas, oceans. (NSW) Long Reef Aquatic Reserve, Eden Pelagic Boat Trip, Ben Boyd NP; (Vic) Cape Liptrap CP, Phillip Is., Portland Harbour; (Tas) Pelagics, Bruny Is., Maria Is., The Nut SR, Burnie, King Is., Bicheno, Freycinet NP; (SA) Newland Head CP, Kangaroo Island, Port MacDonnell Pelagic; (WA) Penguin Is., Albany, Woody Island NR.

Lesser Frigatebird (*Fregata ariel*). Common. Tropical oceans. (NT) Darwin Harbour, Lee Point; (Qld) Weipa, Jardine River NP, Michaelmas Cay, Lady Elliot Is., North Stradbroke Is.; (WA) Barred Creek, Broome, Lacepede Islands NR; (OIT) Norfolk Is., Dauan Is.; Christmas Is.; North Keeling Island; Ashmore Reef.

Great Frigatebird (*Fregata minor*). Common tropical Pacific Ocean, rare to uncommon tropical Indian Ocean. (Qld) Weipa, Jardine River NP, Michaelmas Cay, Lady Elliot Is.; (OIT) Norfolk Is., Dauan Is., Ashmore Reef, Christmas Is., North Keeling Island.

Christmas Island Frigatebird (*Fregata andrewsi*). Tropical Indian Ocean. Breeds Christmas Is. Vagrant records NT, Cocos Keeling Is.

Abbott's Booby (*Papasula abbotti*). Tropical Indian Ocean. Breeds Christmas Is., rare in waters around Ashmore Reef.

Australasian Gannet (*Morus serrator*). Common. Coastal waters off southern half of Australia.

Masked Booby (*Sula dactylatra*). Rare to mod common. Tropical and subtropical ocean. (Qld) Michaelmas Cay, North

Stradbroke Is.; (WA) Lacepede Islands NR; (OIT) Norfolk Is., Lord Howe Is., Ashmore Reef, Christmas Is., Cocos-Keeling Is.

Red-footed Booby (*Sula sula*). Rare to mod common. Tropical ocean. (Qld) Michaelmas Cay, Lady Elliot Is.; (OIT) Ashmore Reef, Christmas Is., Cocos-Keeling Is.

Brown Booby (*Sula leucogaster*). Common. Tropical and subtropical oceans. (NT) Darwin Harbour, Lee Point; (NT) Portland Rd Township, Cape York, Michaelmas Cay, Green Is., Magnetic Is., Lady Elliot Is., North Stradbroke Is., Southport Pelagic Boat Trip; (WA) Broome, Lacepede Islands NR; (OIC) Lord Howe Is., Norfolk Is., Ashmore Reef, Christmas Is.

Australasian Darter (*Anhinga novaehollandiae*). Mod common. Wetlands.

Little Pied Cormorant (*Microcarbo melanoleucos*). Common. Wetlands and coastal areas.

Great Cormorant (*Phalacrocorax carbo*). Common. Wetlands and coastal areas.

Little Black Cormorant (*Phalacrocorax sulcirostris*). Commont. Wetlands and coastal areas.

Pied Cormorant (*Phalacrocorax varius*). Common. Wetlands and coastal areas.

Black-faced Cormorant (*Phalacrocorax fuscescens*). Mod common s Aust. Coastal areas. (NSW) Ben Boyd NP; (Vic) Cape Liptrap, Phillip Island, Point Lonsdale Lighthouse, Point Addis Lookout, Cape Nelson Lighthouse; (Tas) common coastal areas; (SA) Glenelg Beach, Newland Head CP, Kangaroo Is., Robe.

Heard Island Shag (*Leucocarbo nivalis*). Resident Heard Is.

Macquarie Island Shag (*Leucocarbo purpurascens*). Resident Macquarie Is.

Australian Pelican (*Pelecanus conspicillatus*). Common. Wetlands and coastal areas.

Black-necked Stork (*Ephippiorhynchus asiaticus*). Mod common in n, rare in NSW. Wetlands, floodplains. (NT) Fogg Dam, Mamukala Wetlands, Yellow Water, Anbangbang Billabong, Nitmiluk (Katherine Gorge) NP, Borroloola; (Qld) Cooktown, Marina Plains, Lotus Bird Lodge, Weipa, Cattana Wetlands, Daintree, Lake Mitchell, Mareeba Wetlands, Hasties Swamp, Townsville Common, Horseshoe Lagoon, Mackay, Karumba, Lake Moondarra, Tinchi Tamba Wetlands, Lake Samsonvale, Lake Atkinson, Great Sandy NP; (NSW) Ash Is., Hunter Wetland Centre, Crowdy Bay NP, Clarence River Estuary, Macquarie Marshes; (WA) Barred Creek, Lake Argyle, Parry Lagoons NR.

Australasian Bittern (*Botaurus poiciloptilus*). Rare to uncommon s Aust. with numbers decreasing. Wetlands with reedbeds/emergent vegetation. (Qld) Lake Samsonvale, Lake Clarendon; (NSW) Ash Is., Macquarie Marshes, Lake Cargelligo, Fivebough Wetlands, Deniliquin, Pitt Town Lagoon, Longneck Lagoon; (Vic) Hird Swamp, Johnson Swamp, Macleod Morass, Lake Connewarre, Fawthrop Lagoon, Deen Maar; (Tas) Waterhouse CA, Moulting Lagoon; (SA) Dry Creek Saltfields, Bool Lagoon GR; (WA) Thomsons Lake, Lake Muir, Lake Powell NR.

Australian Little Bittern (*Ixobrychus dubius*). Rare to uncommon. Wetlands with reedbeds/emergent vegetation. (Qld) Tyto Wetlands, Sherwood FP; (NSW) Macquarie Marshes, Lake Cargelligo, Fivebough Wetlands, Deniliquin, Bushell's Lagoon, Longneck Lagoon, Royal NP; (Vic) Hird Swamp, Johnson Swamp; (SA) Bool Lagoon

GR; (WA) Herdsman Lake, Lake Muir, Kununurra.

Black Bittern (*Ixobrychus flavicollis*). Uncommon. Tree-lined wetlands, mangroves. (NT) Holmes Jungle NR, Howard Springs, Fogg Dam, Yellow Water, Gunlom, Fergusson River, Nitmiluk (Katherine Gorge) NP, Elsey NP, Victoria River Crossing; (Qld) Kutini-Payamu (Iron Range) NP, Weipa, Half Moon Bay Wetland, Mossman River, Daintree, Tinaroo Creek, Tyto Wetlands, Boodjamulla (Lawn Hill) NP, Townsville Common, Lake Samsonvale; (NSW) Mill Creek, Mitchell Park, Royal NP; (Vic) Croajingolong NP (rare); (WA) Walpole-Nornalup NP (rare), Lake Argyle, Kununurra, Mitchell River NP.

White-necked Heron (*Ardea pacifica*). Mod common. Shallow fresh water, including farm dams and flooded pastures.

Great Egret (*Ardea modesta*). Mod common. Wetlands.

Intermediate Egret (*Ardea intermedia*). Mod common n Aust, uncommon s Aust. Wetlands, lakes, mudflats.

Great-billed Heron (*Ardea sumatrana*). Rare n Aust. Coastal rivers, creeks, waterways. (NT) Buffalo Creek, Leanyer Ponds, Yellow Water, Nitmiluk (Katherine Gorge) NP, Bitter Springs, Groote Eylandt; (Qld) Cooktown, Lakefield NP, Uningan NR, Wenlock and Pennefather River, Yorkeys Knob, Mossman River, Daintree River, Townsville Common, Karumba; (WA) Derby.

Cattle Egret (*Ardea ibis*). Common. Partial migrant. Paddocks, grasslands, usually near wetlands. First recorded in NT at Oenpelli 1948, now established widely.

Striated Heron (*Butorides striata*). Common n Aust and e Aust. Estuaries, mangroves, tidal rivers. (NT) Buffalo Creek, East Point, Mardugal; (Qld) Kutini-Payamu (Iron Range) NP, Cairns Esplanade, Mossman River, Daintree, Townsville Common, Magnetic Is., Cape Hillsborough NP, Eagleby Wetlands, Toorbul; (NSW) Ash Is., Crowdy Bay, Clarence River, Ballina, Boat Harbour, Bicentennial Park, Royal NP, Mimosa Rocks NP; (WA) Derby, Broome, Barred Creek.

Pied Heron (*Egretta picata*). Common n Aust. Coastal wetlands. (NT) Leanyer Sewage Ponds, Shoal Bay Waste Disposal Site, Holmes Jungle NR, Knuckey Lagoons CR, Fogg Dam, Mamukala Wetlands, Yellow Water; (Qld) Lakefield NP, Kutini-Payamu (Iron Range) NP, Weipa, Half Moon Bay Wetland, Lake Mitchell; (NSW) Macquarie Marshes (rare); (WA) Kununurra, Parry Lagoons NR.

White-faced Heron (*Egretta novaehollandiae*). Common. Wetlands, estuaries, mudflats.

Little Egret (*Egretta garzetta*). Common in n, moderately common in s. Wetlands, coastal areas.

Western Reef Egret (*Egretta gularis*). Uncommon resident on Cocos Keeling Is.

Eastern Reef Egret (*Egretta sacra*). Common n Aust, rare to uncommon s Aust. Coastline, estuaries, mudflats. (NT) Lee Point, East Point; (Qld) Cairns Esplanade, Green Is., Magnetic Is., Cape Hillsborough NP, Lady Elliot Is., Karumba; (NSW) Myall Lakes NP, Crowdy Bay NP, Ballina, Mimosa Rocks NP; (Vic) Croajingolong NP; (SA) Kangaroo Is., Innes NP, Coffin Bay NP; (WA) Rottnest Is., Leeuwin-Naturalist NP, Two Peoples Bay NR, Broome Bird Observatory, Barred Creek, Shark Bay.

Nankeen Night-Heron (*Nycticorax caledonicus*). Mod common, wherever there is permanent water. Uncommon in Tas (found The Nut SR, Stanley, King Is.)

Glossy Ibis (*Plegadis falcinellus*). Common n Aust, moderately common s Aust. Shallow wetlands, irrigated pastures (e.g. rice fields). (NT) Fogg Dam, Mamukala Wetlands, Yellow Water; (Qld) Lotus Bird Lodge wetland, Cattana Wetlands, Lake Mitchell, Mareeba Wetlands, Lake Tinaroo, Tyto Wetlands, Townsville Common, Lake Bindegolly, Currawinya Lakes, Lake Atkinson, Lake Dyer; (NSW) Ash Is., Macquarie Marshes, Lake Cargelligo Treatment Works, Fivebough Wetlands, Pitt Town Lagoon, Bushell's Lagoon; (Vic) Kerang Lakes, Lake Mokoan; (SA) Dry Creek Saltfields, Bool Lagoon, Mungerannie Waterhole, Whyalla Wetland; (WA) Lake Argyle, Broome Bird Observatory, Kununurra, Parry Lagoons NR.

Australian White Ibis (*Threskiornis molucca*). Common. Wetlands, moist grasslands, urban areas.

Straw-necked Ibis (*Threskiornis spinicollis*). Common. Wetlands, moist grasslands.

Royal Spoonbill (*Platalea regia*). Common, rare s WA. Wetlands, coastal lagoons.

Yellow-billed Spoonbill (*Platalea flavipes*). Common. Wetlands.

Eastern Osprey (*Pandion cristatus*). Common n Aust, rare to uncommon s Aust. Ocean beaches and estuaries. (NT) Charles Darwin Uni, Buffalo Creek, Yellow Water; (Qld) Marina Plains, Cairns Esplanade, Green Is., Mossman River Daintree River Lake Mitchell Mareeba Wetlands, Magnetic Is., Cape Hillsborough NP, Boondall Wetlands, Eagleby Wetland, Lake Samsonvale, Inskip Point; (NSW) Myall Lakes, Crowdy Bay NP, Ballina, Long Reef Aquatic, Mimosa Rocks NP; (SA) Kangaroo Is., Innes NP, Lincoln NP, Coffin Bay NP; (WA) Broome, Barred Creek, Houtman Abrolhos, Shark Bay, Kalbarri NP, Fitzgerald River NP, Lake McLarty NP, Rottnest Is., Penguin Is.

Black-shouldered Kite (*Elanus axillaris*). Common 'mainland' Aust, vagrant Tas. Open woodland, grasslands, farmlands.

Letter-winged Kite (*Elanus scriptus*). Rare, nomadic and irruptive in response to native rodent plagues. Arid and semi-arid inland Aust. Mostly nocturnal. (NT) Fogg Dam (rare), Barkly Tableland, Connells Lagoon CR, Old Andado Station; (Qld) Birdsville; (SA) Strzelecki Track, Bollards Lagoon Rd, Coongie Lakes, Birdsville Track.

Square-tailed Kite (*Lophoictinia isura*). Rare to uncommon. Eucalypt woodland and forest. (NT) Yellow Water, Elsey NP, Borroloola; (Qld) Mt Molloy, Big Mitchell Creek, Emerald Creek, Hasties Swamp, Mt Coot-tha Forest, D'Aguilar NP, Lake Samsonvale, Lake Atkinson; (NSW) Crowdy Bay NP, Baradine, Capertee Valley, Barren Grounds NR, Booderee NP, Nowra; (Vic) Chiltern-Mt Pilot NP, Warby Range SP, Cabbage Tree Creek; (WA) Wungong Dam, Bungendore Park, Dryandra Woodlands, Lake Muir, Stirling Range NP, Fitzgerald River NP, Cape Arid NP.

Black-breasted Buzzard (*Hamirostra melanosternon*). Uncommon arid and semi-arid inland Aust. (NT) Arnhem Hwy and Kakadu Hwy, Gunlom, Chinaman Creek, Barkly Tableland, Kunoth Bore, Barrow Creek, Ormiston Gorge NP, Newhaven, Old Andado Station; (Qld) Gulf Developmental Rd, Mt Isa, Bladensburg NP, Bowra, Eulo Bore, Currawinya NP, Birdsville; (NSW) Sturt NP; (SA) Strzelecki Track, Stuart Hwy; (WA) Cue, Jupiter Well, Broome, Parry Lagoons NR.

Pacific Baza (*Aviceda subcristata*). Uncommon n and n-e Aust. Forest and rainforest woodland. (NT) Charles Darwin Uni, Lee Point, Holmes Jungle NR, Fogg Dam, Yellow Water, Elsey NP; (Qld) Kutini-Payamu (Iron Range) NP, Cairns BG, Cattana Wetlands, Daintree River, Kingfisher Park, Mareeba Wetlands, Hasties Swamp, Padaminka NR, Eagleby Wetlands, Lake Samsonvale; (NSW) Dorrigo NP, Copeland Tops, Hawkesbury River Valley, Royal NP; (WA) Mirima (Hidden Valley) NP, Mitchell River NP, Drysdale River.

White-bellied Sea-Eagle (*Haliaeetus leucogaster*). Common coastal Aust. Coastlines, large watercourses.

Whistling Kite (*Haliastur sphenurus*). Common. Farmland, scrubland, watercourses, wetland.

Brahminy Kite (*Haliastur indus*). Common in n, uncommon in s. Estuaries, coastline, watercourses. (NT) Buffalo Creek, Yellow Water, Adelaide River, Charles Darwin Uni, Groote Eylandt; (Qld) Lakefield NP, Cairns Esplanade, Daintree River, Kingfisher Park, Hasties Swamp, Karumba, Boondall Wetlands, Eagleby Wetlands, Lake Samsonvale, Inskip Point; (NSW) Crowdy Bay NP, Clarence River Estuary, Ballina; (WA) Broome, Derby.

Black Kite (*Milvus migrans*). Common mainland Aust. Grassland, roadsides, rubbish dumps.

Brown Goshawk (*Accipiter fasciatus*). Common. Woodland, forest. Northern ssp. *didimus* (may be classified distinct sp.). Christmas Is. ssp. *natalis* currently lumped with Brown Goshawk but possibly a ssp. of Variable Goshawk *Accipiter hiogaster* or even a unique island species.

Collared Sparrowhawk (*Accipiter cirrocephalus*). Common. Woodland.

Grey Goshawk (*Accipiter novaehollandiae*). Uncommon. Temperate forest. Two distinct morphs, white and grey. (NT) Palmerston Sewage Works, Elsey NP; (Qld) Cooktown, Kutini-Payamu (Iron Range) NP, Lockerbie Scrub, Kingfisher Park, Mt Lewis, Big Mitchell Creek, Hasties Swamp, Cape Hillsborough NP, D'Aguilar NP, Lake Samsonvale, Great Sandy NP; (NSW) Hunter Wetland Centre, Copeland Tops, Dorrigo NP, Bruxner Park FR, Nightcap NP, Ku-ring-gai Chase NP, Scheyville NP, Royal NP Morton NP; (Vic) Croajingolong NP, Cabbage Tree Creek, Great Otway NP; (Tas) Arthur River, Mt Wellington.

Spotted Harrier (*Circus assimilis*). Uncommon to moderately common. Grasslands and open grassy woodland. (NT) Arnhem and Kakadu Hwys, Chinaman Creek, Barkly Tableland, Erldunda, Old Andado Station; (Qld) Emerald Creek, Hasties Swamp, Gulf Developmental Rd, Mt Isa, Bladensburg NP, Bowra, Lake Samsonvale, Lockyer Valley; (NSW) Capertee Valley, Round Hill NR, Cocoparra NP, Oolambeyan NP, Mutawintji NP, Sturt NP, Baker's Lagoon; (Vic) Wyperfeld NP, Little Desert NP, Terrick Terrick NP; (SA) Dry Creek Saltfields, Bool Lagoon, Telowie Gorge CP, Strzelecki Track, Vokes Hill Corner Rd; (WA) Lake McLarty NP, Stirling Range NP, Peak Charles NP, Cocklebiddy, Parry Lagoons NR.

Swamp Harrier (*Circus approximans*). Common. Wetlands, coastal heathland.

Red Goshawk (*Erythrotriorchis radiatus*). Rare. Forest. (NT) Fogg Dam, Tiwi Is., Mataranka, Yellow Water, Kakadu Hwy, Caranbirini CR; (Qld) Lake Moondarra, Lakefield NP, Musgrave Roadhouse, Kutini-Payamu (Iron Range) NP, Big Mitchell Creek; (WA) Mitchell River NP, Mornington Station.

Wedge-tailed Eagle (*Aquila audax*). Mod common. Rare Tas. Farmland, open plains, woodland, forest.

Little Eagle (*Hieraaetus morphnoides*). Uncommon to mod common, except Tas. Woodland, open plains.

Nankeen Kestrel (*Falco cenchroides*). Common. Farmland, open plains, woodland.

Brown Falcon (*Falco berigora*). Common. Woodland, opens plains, farmlands.

Australian Hobby (*Falco longipennis*). Mod common. Farmland, woodland, gardens.

Grey Falcon (*Falco hypoleucos*). Rare to very rare. Lightly timbered woodland and stony plains. (NT) Pine Creek, Nitmiluk (Katherine Gorge) NP, Elsey NP, Victoria River, Barkly Tableland, Connells Lagoon CR, Old Telegraph Station, N'Dhala Gorge, Newhaven Sanctuary, Mereenie Loop Rd, Old Andado Station; (Qld) Boodjamulla (Lawn Hill) NP, Lark Quarry CP, Mt Isa, Eulo Bore, Bowra, Currawinya NP; (NSW) Mutawintji NP, Sturt NP; (SA) Wilkawillina Gorge, Boolcoomatta R, Bollards Lagoon Rd, Coongie Lakes, Birdsville Track, Strzelecki Track; (WA) Jupiter Well, Carnarvon, Cue, Karijini NP, Mornington Station, Purnululu (Bungle Bungles) NP.

Black Falcon (*Falco subniger*). Uncommon. Semi-arid and arid interior, grassland, and open woodland. (NT) Victoria River, Barkly Tableland, Connells Lagoon CP, Alice Springs Waste Stabilisation Ponds, Mac Clark CR; (Qld) Lake Moondarra, Bowra, Lake Bindegolly, Eulo Bore, Currawinya NP, Birdsville, Lake Samsonvale, Lockyer Valley; (NSW) Round Hill NR; (Vic) Lake Mokoan, Terrick Terrick NP, Wyperfeld NP, Western Treatment Plant; (SA) Bool Lagoon, Strzelecki Track.

Peregrine Falcon (*Falco peregrinus*). Uncommon. Woodland, grassland, coastal cliffs, offshore islands, gardens. (NT) Caranbirini CR, Borroloola, Kunoth Bore, Mt Gillen, Ellery Creek Big Hole, Ormiston Gorge, N'Dhala Gorge, Uluru-Kata Tjuta NP; (Qld) Bowra, Eulo Bore, Eagleby Wetlands, Lake Samsonvale, Lockyer Valley, Girraween NP; (NSW) Hunter Wetland Centre, Warrumbungle NP, Mutawintji NP; (Vic) Hattah-Kulkyne NP, Wyperfeld NP, Chiltern-Mt Pilot NP, Lake Mokoan, Terrick Terrick NP, Phillip Is., Royal Park, You Yangs, Point Addis, Cape Nelson; (Tas) The Nut SR, Freycinet NP, Maria Is., Mt Field NP; (SA) Adelaide BG, Cleland CP, Belair NP, Dry Creek Saltfields, Newland Head CP, Bool Lagoon, Mt Remarkable NP, Flinders Ranges NP, Vulkathunha-Gammon Ranges, Coffin Bay NP; (WA) Mirima (Hidden Valley) NP, Esperance, Bungendore Park BR, Lake McLarty NR,

Sarus Crane (*Grus antigone*). Uncommon. Tropical wetlands. First recorded near Normanton (1966). (NT) Connells Lagoon CR; (Qld) Keating's Lagoon, Lakefield NP, Marina Plains, Weipa, Mareeba Wetlands, Hasties Swamp, Lake Tinaroo, Normanton, Karumba.

Brolga (*Grus rubicunda*). Common n Aust, rare to uncommon s Aust. Wetlands and wet grasslands. (NT) Holmes Jungle NR, Yellow Water, Arnhem Hwy, Connells Lagoon CR, Borroloola, Keating's Lagoon; (Qld) Marina Plains, Weipa, Mareeba Wetlands, Hasties Swamp, Lake Tinaroo, Townsville Common, Horseshoe Lagoon, Bladensburg NP, Currawinya NP, Tinchi Tamba Wetlands, Toorbul Shorebird Roost; (NSW/ACT) Clarence River Estuary, Macquarie Marshes, Fivebough Wetlands, Menindee Lakes, Tidbinbilla NR; (Vic) Terrick Terrick NP, Western Treatment Plant, Lake Connewarre; (SA) Bool Lagoon; (WA) Parry Lagoons NR, Roebuck Plains, Lake Eda, Kununurra, Lake Argyle.

Purple Swamphen (*Porphyrio porphyrio*). Common. Wetlands, waterways.

Chestnut Rail (*Eulabeornis castaneoventris*). Uncommon Top End and Kimberley. Mangrove forest. (NT) Charles Darwin Uni, Buffalo Creek, Charles Darwin NP, Palmerston Sewage Works, East Point, Tiwi Is., Groote Eylandt; (WA) Derby.

Red-necked Crake (*Rallina tricolor*). Uncommon far n Qld. Secretive. Tropical rainforest close to permanent water. (Qld) Lockerbie Scrub, Cassowary House, Kingfisher Park, Crater Lakes NP, Mission Beach, Paluma Range NP.

Lewin's Rail (*Lewinia pectoralis*). Uncommon e Aust. Highly secretive. Coastal marshland, heathland and moist lowland tropical forest. (Qld) Hasties Swamp, Kedron Brook Wetlands, Great Sandy NP, Conondale Ranges; (NSW/ACT) Maroubra, Sydney Olympic Park, Dharug NP, Barren Grounds NR, Jerrabomberra Wetlands NR; (Vic) Marlo CR, Western Treatment Plant, Fawthrop Lagoon, Deen Maar; (SA) Victor Harbour Treatment Ponds, Goolwa, Coorong NP.

Buff-banded Rail (*Gallirallus philippensis*). Mod common. Dense reeds and vegetation near wetlands. (NT) Fog Dam CR, Borroloola; (Qld) Yorkeys Knob, Green Is., Abattoir Swamp, Hasties Swamp, Lake Eacham, Tyto Wetlands, Townsville Common, Kedron Brook Wetlands, Eagleby Wetlands, Lake Samsonvale, Lake Dyer; (NSW) Ash Is., Crowdy Bay NP, Clarence River Estuary, Lake Cargelligo, Royal BG, Maroubra, Sydney Olympic Park, Royal NP; (Vic) Kerang Lakes, Macleod Morass, Western Treatment Plant, Mud Is., Belmont Common; (SA) Torrens Lake, Greenfields Wetland, Goolwa Effluent Ponds, Coorong NP, Coffin Bay NP; (WA) Herdsman Lake, Thomsons Lake, Penguin Is., Mandurah, Vasse-Wonnerup Estuary; (OIT) Norfolk Is., Lord Howe Is., Cocos-Keeling Is. ssp. *andrewsi*.

Lord Howe Woodhen (*Gallirallus sylvestris*). Resident Lord Howe Is.

Baillon's Crake (*Porzana pusilla*). Uncommon to mod common. Secretive. Dense reeds, vegetation near wetlands. (NT) Fogg Dam; (Qld) Abattoir Swamp, Lake Moondarra, Winton Treatment Works, Lake Bindegolly NP, Eagleby Wetlands, Lake Samsonvale; (NSW/ACT) Ash Is., Lake Cargelligo, Fivebough Wetlands, Sydney Olympic Park, Pitt Town Lagoon, Pugh's Lagoon, Baker's Lagoon, Jerrabomberra Wetlands NR; (Vic) Foster Swamp, Hird Swamp, Macleod Morass, Royal Park, Western Treatment Plant, Belmont Common, Deen Maar; (Tas) Goulds Lagoon; (SA) Dry Creek Saltfields, Victor Harbour Treatment Ponds, Goolwa Effluent Ponds, Bool Lagoon GR; (WA) Thomsons Lake.

Australian Spotted Crake (*Porzana fluminea*). Mod common. Secretive. Dense reeds and vegetation near wetlands. (NT) Alice Springs Waste Stabilisation Ponds; (Qld) Tyto Wetlands, Lake Moondarra, Winton Treatment Works, Gumholes Waterhole, Lake Bindegolly, Eagleby Wetlands; (NSW/ACT) Ash Is., Lake Cargelligo, Fivebough Wetlands, Pitt Town Lagoon, Royal NP, Jerrabomberra Wetlands NR; (Vic) Kerang Lakes, Western Treatment Plant, Belmont Common, Fawthrop Lagoon; (Tas) Goulds Lagoon; (SA) Dry Creek Saltfields, Goolwa, Bool Lagoon, Coorong NP; (WA) Thomsons Lake, Lake McLarty, Wagin Treatment Plant, Lake Muir, Rowles Lagoon, Derby.

Spotless Crake (*Porzana tabuensis*). Uncommon to mod common. Secretive. Dense reeds, vegetation near wetlands.

(NT) Alice Springs Waste Stabilisation Ponds; (Qld) Abattoir Swamp, Hasties Swamp, Kedron Brook Wetlands, Lake Samsonvale; (NSW/ACT) Ash Is., Lake Cargelligo, Fivebough Wetlands, Maroubra, Sydney Olympic Park, Pitt Town Lagoon, Jerrabomberra Wetlands NR; (Vic) Kerang Lakes, Goolwa, Bool Lagoon, Coorong NP, Western Treatment Plant, Belmont Common, Fawthrop Lagoon; (Tas) Goulds Lagoon; (SA) Dry Creek Saltfields, Goolwa, Bool Lagoon.

White-browed Crake (*Amaurornis cinerea*). Uncommon. Secretive. Dense reeds and vegetation near wetlands. (NT) Palmerston Sewage Works, Holmes Jungle NR, Fogg Dam; (Qld) Cairns BG, Yorkeys Knob, Cattana Wetlands, Cairns Crocodile Farm, Abattoir Swamp, Mareeba Wetlands, Hasties Swamp, Townsville Common; (WA) Kununurra, Lake Argyle.

Pale-vented Bush-hen (*Amaurornis moluccana*). Uncommon n Aust. Secretive. Dense reeds and vegetation near wetlands. (NT) Darwin BG, Leanyer Sewage Ponds, Palmerston Sewage Works, Victoria River; (Qld) Cairns BG, Cattana Wetlands, Daintree, Abattoir Swamp, Hasties Swamp, Mission Beach, Tyto Wetlands, Townsville Common, Eagleby Wetlands, Mackay, Lake Samsonvale, Conondale Ranges, Greater Brisbane, D'Aguilar NP, Canungra, Cooroy TP.

White-breasted Waterhen (*Amaurornis phoenicurus*). Common resident Cocos Keeling Is., rare resident Christmas Is., vagrant Ashmore Reef.

Black-tailed Native-hen (*Tribonyx ventralis*). Common near permanent or ephemeral inland and coastal wetlands. Dispersive.

Tasmanian Native-hen (*Tribonyx mortierii*). Common resident Tas.

Dusky Moorhen (*Gallinula tenebrosa*). Mostly common, rare n-w Aust and Tas. Wetlands, waterways.

Eurasian Coot (*Fulica atra*). Common. Wetlands, waterways.

Australian Bustard (*Ardeotis australis*). Uncommon. Dry plains, grasslands, open woodland. (NT) Arnhem Hwy, Fergusson River, Barkly Tableland, Connells Lagoon CR, Caranbirini CR, Borroloola, Newhaven Sanctuary, Uluru-Kata Tjuta NP; (Qld) Marina Plains, Peninsula Development Rd, Cumberland Dam, Gulf Developmental Rd, Karumba, Boodjamulla (Lawn Hill) NP, Bladensburg NP, Bowra, Currawinya NP, Diamantina Developmental Rd; (NSW) Sturt NP; (Vic) Wyperfeld NP (rare); (SA) Birdsville Track, Vokes Hill Corner Rd; (WA) Peak Charles NP, Cape Arid NP, Cocklebiddy-Rawlinna Rd, Neale Junction NR, Kununurra, Lake Argyle, Broome, Jupiter Well.

Black-faced Sheathbill (*Chionis minor*). Resident Heard Is.

Bush Stone-curlew (*Burhinus grallarius*). Mod common n Aust, rare to uncommon s Aust. Open woodland. Mainly nocturnal. (NT) Royal Darwin Hospital, Alyangula, Gunlom, Kunoth Bore; (Qld) Half Moon Bay Wetland, Kingfisher Park, Cairns BG, Magnetic Is., Tyto Wetlands, Cape Hillsborough NP; (NSW) Oolambeyan NP; (Vic) Glenlee FFR; (SA) Flinders Chase NP, Bordertown, Coongie Lakes, Lincoln NP; (WA) Cue.

Beach Stone-curlew (*Esacus magnirostris*). Rare to uncommon. Undisturbed open beaches, sand and mudflats. (NT) Buffalo Creek, East Point, Bing Bong, Alyangula; (Qld) Yorkeys Knob, Green Island, Mossman River, Yule Point, Coopers Creek, Magnetic Island, Karumba, Inskip

Point; (NSW) Manning Point, Sandon River, Wooli, Red Rock; (WA) Carnarvon, Barred Creek.

Australian Pied Oystercatcher (*Haematopus longirostris*). Common. Coastlines, estuaries, tidal flats.

Sooty Oystercatcher (*Haematopus fuliginosus*). Common. Coastlines, estuaries, tidal flats.

Black-winged Stilt (*Himantopus himantopus*). Common. Nomadic, dispersive. Coastlines, estuaries, wetlands, tidal flats.

Red-necked Avocet (*Recurvirostra novaehollandiae*). Common s Aust, uncommon n Aust. Nomadic, dispersive. Estuaries, wetlands, tidal flats.

Banded Stilt (*Cladorhynchus leucocephalus*). Uncommon. Nomadic, dispersive. Estuaries, wetlands. (NSW) Lake Clarendon; (Vic) Lake Tutchewop, Foster Swamp, Western Treatment Plant, Moolap Saltworks (SA) Dry Creek Saltfields, Murray Lagoon, Coorong NP, Bird Lake - Port Augusta, Whyalla Wetland, Big Swamp; (WA) Rottnest Is., Mandurah, Lake McLarty, Lake Muir, Vasse-Wonnerup Estuary, Lake Warden, Pink Lake, Lake MacLeod.

Pacific Golden Plover (*Pluvialis fulva*). Mod common summer migrant. Coastal wetlands, estuaries, tidal flats. (NT) Lee Point, Buffalo Creek's mouth, Leanyer Sewage Ponds, East Point, Mamukala Wetlands; (Qld) Cairns Esplanade, Hasties Swamp, Pioneer River, Lady Elliot Is., Boondall Wetlands, Lytton Wader Roost, Lake Atkinson, Pelican Point; (NSW) Ash Is., Stockton Sandspit, Harrington, Old Bar, Clarence River Estuary, Ballina, Long Reef Aquatic Reserve, Boat Harbour, Sydney Olympic Park, Comerong Island NR; (Vic) Shallow Inlet, Western Treatment Plant, Mud Islands, Lake Victoria; (Tas) Pitt Water NR; (SA) Dry Creek Saltfields, Goolwa, Little Bool Lagoon, Policeman's Point, Yanerbie Beach; (WA) Lake Argyle, Kununurra, Broome Bird Observatory; (OIT) Lord Howe Is., Norfolk Is.

Grey Plover (*Pluvialis squatarola*). Uncommon to mod common summer migrant. Coastal wetland, estuaries, tidal flats. (NT) Lee Point, Buffalo Creek's mouth, East Point; (Qld) Cairns Esplanade, Pioneer River, Lady Elliot Is., Karumba, Lytton Wader Roost, Inskip Point, Pelican Point; (NSW) Harrington, Boat Harbour, Towra Point NR, Comerong Island NR; (Vic) Foster Swamp, Hird Swamp, Western Treatment Plant, Mud Is.; (SA) Dry Creek Saltfields, Price Saltfields, Policeman's Point, Coffin Bay NP, Yanerbie Beach; (WA) Broome Bird Observatory, Derby, Shark Bay; (OIT) Ashmore Reef.

Little Ringed Plover (*Charadrius dubius*). Rare summer migrant n Aust. Coastal wetlands, estuaries, tidal flats. Recorded annually in NT, particularly Leanyer Sewage Ponds and Kimberley sector WA. Vagrant to other areas.

Red-capped Plover (*Charadrius ruficapillus*). Common resident. Coastal beaches, wetlands, estuaries, tidal flats.

Double-banded Plover (*Charadrius bicinctus*). Mod common winter migrant to e and s coastal Aust. Coastal wetlands, beaches, estuaries, tidal flats. (Qld) Inskip Point; (NSW) Harrington, Old Bar, Long Reef Aquatic R., Comerong Island NR, Mimosa Rocks NP; (Vic) Kerang Lakes, Hird Swamp, Western Treatment Plant, Mud Is., Lake Victoria, Killarney Beach; (Tas) Pitt Water NR; (OIT) Norfolk Is.

Lesser Sand Plover (*Charadrius mongolus*). Mod common summer migrant n

Aust, uncommon s Aust. Coastal wetlands, beaches, estuaries, tidal flats. (NT) Buffalo Creek, East Point; (Qld) Cooktown, Cairns Esplanade, Newell Beach, Townsville Common, Pioneer River, Lady Elliot Is., Karumba, Lytton Wader Roost, Inskip Point, Great Sandy Strait, Pelican Point; (NSW) Stockton Sandspit, Old Bar, Clarence River Estuary, Ballina; (Vic) Corner Inlet, Mud Is.; (SA) Thompson Beach; (WA) Broome Bird Observatory, Shark Bay.

Greater Sand Plover (*Charadrius leschenaultii*). Mod common summer migrant n Aust, uncommon s Aust. Coastal wetlands, beaches, estuaries, tidal flats. (NT) Buffalo Creek's mouth, East Point; (Qld) Cooktown, Cairns Esplanade, Newell Beach, Townsville Common, Pioneer River, Lady Elliot Is., Karumba, Lytton Wader Roost, Inskip Point Pelican Point; (NSW) Clarence River Estuary, Ballina; (Vic) Mud Is.; (SA) Thompson Beach; (WA) Broome Bird Observatory, Derby, Shark Bay.

Oriental Plover (*Charadrius veredus*). Uncommon to mod common summer migrant. Inland wetlands, grassy plains, farmland. (NT) Lee Point Rd, Leanyer Sewage Ponds, Knuckey Lagoons CR, Mamukala Wetlands, Timber Creek Airfield, Barkly Tableland, Connells Lagoon CR; (Qld) Karumba, Lake Moondarra; (WA) Lake Argyle, Parry Lagoons NR, Broome, Lake MacLeod, Eyre Bird Observatory (rare).

Inland Dotterel (*Charadrius australis*). Rare to uncommon inland Aust. Short grassland, stony desert, farmland. (NT) Erldunda, Mac Clark CR; (Qld) Bladensburg NP, Currawinya NP, Asterebla Downs NP, Diamantina NP, (NSW) Deniliquin, Sturt NP; (Vic) Neds Corner, Terrick Terrick NP (rare); (SA) Vulkathunha-Gammon Ranges, Boolcoomatta R, Strzelecki Track, Coongie Lakes, Birdsville Track, Stuart Hwy, Cook Airfield (Vokes Hill Corner Rd), Nullarbor NP; (WA) Cocklebiddy–Rawlinna Rd, Cue, Lake MacLeod.

Black-fronted Dotterel (*Elseyornis melanops*). Mod common. Coastlines, estuaries, wetlands, tidal flats.

Hooded Plover (*Thinornis rubricollis*). Rare to uncommon s Aust. Ocean beaches, coastal wetlands, estuaries, inland wetlands (WA only), tidal flats. (NSW) Mimosa Rocks NP, Ben Boyd NP; (Vic) Cape Conran Coastal Park, Cape Liptrap Coastal Park, Corner Inlet, Cape Woolamai, Lake Victoria, Point Roadknight, Great Ocean Rd; (Tas) Bruny Is., Maria Is., King Is., Bicheno, Freycinet NP; (SA) Waitpinga Beach, Stokes Bay (Kangaroo Is.), D'Estrees Bay (Kangaroo Is.), Parnka Point, Little Dip CP, Innes NP, Lincoln NP, Coffin Bay NP, Yanerbie Beach; (WA) Wagin Lakes, Fitzgerald River NP, Lake Warden, Pink Lake, Cape Arid NP, Carnarvon.

Red-kneed Dotterel (*Erythrogonys cinctus*). Mod common, except Tas. Inland wetlands, estuaries.

Banded Lapwing (*Vanellus tricolor*). Uncommon to mod common. Open plains, grasslands. (NT) Kunoth Bore, Erldunda; (Qld) Bowra; (NSW) Chat Alley (Round Hill NR), Deniliquin, Oolambeyan NP, Sturt NP, Pugh's Lagoon, Baker's Lagoon, Freeman's Reach Rd Turf Farm; (Vic) Neds Corner, Lake Tutchewop, Patho Plains, Western Treatment Plant; (SA) Hindmarsh Is., Strzelecki Track, Paniburra Bore, (WA) Rottnest Is., Lake McLarty NR, Rowles Lagoon.

Masked Lapwing (*Vanellus miles*). Common except s-e WA where rare. Two distinct ssp.: *miles* n Aust; *novaehollandiae* (Spur-winged Plover) s Aust. Coastal wetlands, inland wetlands, grasslands, parklands.

Plains-wanderer (*Pedionomus torquatus*). Rare. Grasslands, farmland. (NSW) Deniliquin, Oolambeyan NP; (Vic), Terrick Terrick NP; (SA) Boolcoomatta R.

Comb-crested Jacana (*Irediparra gallinacea*). Mod common to common n Aust. Wetlands. (NT) Fogg Dam, Mamukala, Yellow Water; (Qld) Lakefield NP, Cairns BG, Cattana Wetlands, Lake Mitchell, Mareeba Wetland, Hasties Swamp, Tyto Wetlands, Townsville Common, Horseshoe Lagoon, Mackay Regional BG, Cumberland Dam, Eagleby Wetlands, Lake Samsonvale, Seven Mile Lagoon, Lake Clarendon; (NSW) Rocky Creek Dam; (WA) Kununurra, Parry Lagoons NR.

Australian Painted Snipe (*Rostratula australis*). Rare. Wetlands. Nomadic. (Qld) Lake Atkinson, Lake Dyer; (NSW/ACT) Ash Is., Macquarie Marshes, Carmel Lagoon, Lake Cargelligo Treatment Works, Jerrabomberra Wetlands NR; (Vic) Hird Swamp, Johnson Swamp; (SA) Dry Creek Saltfields, Bool Lagoon; (WA) Parry Lagoons NR, Broome.

Lathams Snipe (*Gallinago hardwickii*). Uncommon to common summer migrant e Aust. Coastal and inland wetlands. (Qld) Lakefield NP, Lockhart River Treatment Plant, Cairns Esplanade, Abattoir Swamp, Wolframite Dam, Mareeba Wetlands, Hasties Swamp, Tyto Wetlands, Townsville Common, Kedron Brook Wetlands, Tinchi Tamba Wetlands, Oxley Creek Common, Lake Samsonvale, Lake Atkinson, Lake Dyer, Toorbul Shorebird Roost; (NSW/ACT) Ash Is., Hunter Wetlands Centre, Ballina, Fivebough Wetlands, Centennial Park, Pitt Town Lagoon, Longneck Lagoon, Blue Lake, Jerrabomberra Wetlands NR; (Vic) Hird Swamp, Chiltern Valley Number Dams, Lake Mokoan, Macleod Morass, Wilsons Promontory NP, Trin Warren Tam-boore wetlands, Belmont Common, Lake Lorne, Fawthrop Lagoon; (Tas) Goulds Lagoon, Pitt Water NR, Birchs Inlet, Waterhouse CA, Bicheno; (SA) Victor Harbour Treatment Ponds, Goolwa Barrage, Bool Lagoon, Coffin Bay NP.

Pin-tailed Snipe (*Gallinago stenura*). Rare summer migrant to coastal n-w Aust. Coastal and inland wetlands. (WA) Broome, Derby; (OIT) Christmas Is., Cocos-Keeling Is.

Swinhoes Snipe (*Gallinago megala*). Rare to uncommon summer migrant to coastal n and n-w Aust. Wetlands. (NT) Leanyer Sewage Ponds, Holmes Jungle NR, Knuckey Lagoons; (WA) Broome, Derby, Parry Lagoons NR.

Black-tailed Godwit (*Limosa limosa*). Mod common summer migrant. Coastal wetland, estuaries, tidal flats. (NT) Buffalo Creek, Leanyer Sewage Ponds, East Point, Holmes Jungle NR; (Qld) Cairns Esplanade, Mareeba Wetlands, Lake Moondarra, Lytton Wader Roost, Toorbul Shorebird Roost; (NSW) Ash Is., Stockton Sandspit, Fivebough Wetlands; (Vic) Western Treatment Plant, Mud Is.; (Tas) Pitt Water NR; (SA) Dry Creek Saltfields, Goolwa, Little Bool Lagoon, Coorong; (WA) Derby, Broome Bird Observatory, Vasse-Wonnerup Estuary, Lake McLarty NR, Thomsons Lake NR.

Bar-tailed Godwit (*Limosa lapponica*). Mod common to common summer migrant. Coastal wetlands, estuaries, tidal flats. (NT) Buffalo Creek, East Point, Mamukala; (Qld) Endeavour River, Cairns Esplanade, Mossman River, Townsville Common, Pioneer River, Boondall Wetlands, Lytton Wader Roost, Inskip Point, Esplanade, Toorbul Shorebird Roost; (NSW) Ash Is., Stockton Sandspit, Harrington, Old Bar, Hickey Is., Ballina, Towra Point, Sydney Olympic Park, Bicentennial Park, Comerong Is.; (Vic) Mallacoota, Corner Inlet, Western Treatment Plant, Mud Is., Reedy Lake; (Tas) Moulting Lagoon; (SA) Dry Creek Saltfields,

Goolwa, Little Bool Lagoon, Coorong; (WA) Lake Argyle, Derby, Broome Bird Observatory. Shark Bay, Carnarvon, Lake MacLeod, Houtman Abrolhos, Cape Arid NP, Albany, Lake McLarty NR, Mandurah; (OIT) Norfolk Is., Lord Howe Is.

Little Curlew (*Numenius minutus*). Mod common summer migrant n Aust, vagrant e and s Aust. Coastal and inland wetland, grasslands. (NT) Darwin BG, Lee Point Rd, Leanyer Sewage Ponds, East Point, Holmes Jungle NR, Knuckey Lagoons, Mamukala Wetlands, Connells Lagoon CR; (Qld) Marina Plains, Cairns Esplanade (uncommon), Lake Moondarra (uncommon), Karumba, Pioneer River (uncommon); (WA) Roebuck Plains, Parry Lagoons NR, Lake Argyle, Broome, Carnarvon, Nallan Lake.

Whimbrel (*Numenius phaeopus*). Common summer migrant n Aust, uncommon s Aust. Coastal wetland, estuaries, tidal flats.

Eastern Curlew (*Numenius madagascariensis*). Mod common summer migrant. Coastal, estuaries, tidal flats.

Terek Sandpiper (*Xenus cinereus*). Mod common summer migrant n Aust, uncommon s Aust. Coastal wetlands, estuaries, tidal flats. (NT) Lee Point, East Point; (Qld) Cairns Esplanade, Pioneer River, Karumba, Lytton Wader Roost, Great Sandy Strait, Toorbul Shorebird Roost; (NSW) Stockton Sandspit, Harrington lagoon, Clarence River, Ballina, Comerong Is.; (Vic) Mud Islands (uncommon), Western Treatment Plant (uncommon); (SA) Dry Creek Saltfields (uncommon); (WA) Derby, Broome Bird Observatory, Shark Bay, Carnarvon, Vasse-Wonnerup Estuary.

Common Sandpiper (*Actitis hypoleucos*). Mod common summer migrant n Aust, uncommon s Aust. Rocky coastlines, estuaries, wetlands.

Grey-tailed Tattler (*Tringa brevipes*). Common summer migrant n Aust, uncommon s Aust. Estuaries, reefs, mangroves.

Wandering Tattler (*Tringa incana*). Uncommon summer migrant. Reefs, estuaries. (Qld) Lady Elliot Is., Redcliffe, North Stradbroke Is.; (OIT) Lord Howe Is., Norfolk Is.

Common Greenshank (*Tringa nebularia*). Common summer migrant. Coastal wetlands, estuaries, wetlands, tidal flats.

Common Redshank (*Tringa totanus*). Rare. Estuaries, tidal flats. (WA) Broome (Crab Creek); (OIT) occasional records Ashmore Reef and Cocos Keeling Is.

Marsh Sandpiper (*Tringa stagnatilis*). Uncommon to mod common summer migrant. Coastal wetlands, estuaries, wetlands, tidal flats.

Wood Sandpiper (*Tringa glareola*). Common summer migrant n Aust, uncommon s Aust. Coastal wetland, estuaries, wetlands.

Ruddy Turnstone (*Arenaria interpres*). Uncommon to mod common summer migrant. Coastal wetlands, coastlines, estuaries, wetlands. (NT) Lee Point, East Point; (Qld) Cairns Esplanade, Green Is., Pioneer River, Lady Elliot Is., Redcliffe, Manly Boat Harbour breakwall; (NSW) Woody Head, Angels Flat Rock, Long Reef Aquatic Reserve; (Vic) Cape Conran, Mud Is., Western Treatment Plant, Point Lonsdale Lighthouse; (SA) D'Estrees Bay (Kangaroo Is.), Coffin Bay NP, Yanerbie Beach; (WA) Derby, Broome Bird Observatory, Shark Bay, Lacepede Islands NR, Houtman Abrolhos, Kalbarri NP, Eyre Bird Observatory, Albany, Mandurah, Rottnest Is., Penguin Is.

Asian Dowitcher (*Limnodromus semipalmatus*). Rare summer migrant. Coastal mudflats, estuaries. Regularly Roebuck Bay, WA and Ashmore Reef. Infrequent Cairns Esplanade, vagrant elsewhere.

Great Knot (*Calidris tenuirostris*). Uncommon to mod common summer migrant. Coastal wetland, estuaries, wetlands.

Red Knot (*Calidris canutus*). Mod common summer migrant. Coastal wetland, estuaries, wetlands.

Sanderling (*Calidris alba*). Mod common summer migrant. Coastal beaches. (NT) Buffalo Creek; (Qld) Cairns Esplanade, Inskip Point; (NSW) Old Bar, Ballina; (Vic) Corner Inlet, Edwards Point, Killarney Beach; (Tas) Pitt Water NR; (SA) Goolwa, Coorong, Coffin Bay NP; (WA) Derby, Broome Bird Observatory, Shark Bay, Houtman Abrolhos, Eyre Bird Observatory, Cape Arid NP, Fitzgerald River NP, Mandurah, Rottnest Is.; (OIT) Ashmore Reef.

Little Stint (*Calidris minuta*). Rare summer migrant. Coastlines, estuaries, wetlands.

Red-necked Stint (*Calidris ruficollis*). Common summer migrant. Coastlines, estuaries, wetlands.

Long-toed Stint (*Calidris subminuta*). Uncommon summer migrant. Coastal wetland, estuaries, wetlands. (NT) Leanyer Sewage Ponds; (Qld) Lake Moondarra; (NSW/ACT) Fivebough Wetlands,Jerrabomberra Wetlands NR; (Tas) Pitt Water NR; (Vic) Hird Swamp, Western Treatment Plant, Dry Creek Saltfields, Little Bool Lagoon; (WA) Derby, Broome, Lake Argyle, Lake MacLeod, Carnarvon, Nallan Lake, Albany, Lake McLarty NR.

Pectoral Sandpiper (*Calidris melanotos*). Rare to uncommon summer migrant. Coastal wetland, estuaries, wetlands. (NT) East Point; (Qld) Cairns Esplanade, Townsville Common, Eagleby Wetlands, Lake Clarendon, Toorbul Shorebird Roost; (NSW/ACT) Ash Is., Fivebough Wetlands, Pitt Town Lagoon, Comerong Is., Jerrabomberra Wetlands NR; (Vic) Western Treatment Plant, Lake Victoria; (SA) Dry Creek Saltfields, Price Saltfields, Little Bool Lagoon, Coorong; (WA) Derby, Broome, Shark Bay, Carnarvon, Lake McLarty NR, Mandurah, Thomsons Lake NR, Herdsman Lake RP.

Sharp-tailed Sandpiper (*Calidris acuminata*). Common summer migrant. Coastal wetland, estuaries, wetlands.

Curlew Sandpiper (*Calidris ferruginea*). Common summer migrant. Coastlines, estuaries, wetlands.

Broad-billed Sandpiper (*Limicola falcinellus*). Rare to uncommon summer migrant. Coastal wetland, estuaries, wetlands. (Qld) Karumba, Cairns Esplanade, Inskip Point, Lytton Wader Roost, Toorbul Shorebird Roost; (NSW) Ash Island, Comerong Is.; (Vic) Mud Is., Western Treatment Plant; (SA) Dry Creek Saltfields; (WA) Derby, Broome Bird Observatory, Shark Bay; (OIT) Ashmore Reef.

Ruff (*Philomachus pugnax*). Rare summer migrant. Coastal wetland, estuaries, wetlands. (Qld) Cairns Esplanade; (NSW) Ash Is., Pitt Town Lagoon; (Vic) Hird Swamp, Western Treatment Plant; (SA) Dry Creek Saltfields, Goolwa, Little Bool Lagoon, Coongie Lakes; (WA) Derby, Broome, Shark Bay.

Red-necked Phalarope (*Phalaropus lobatus*). Rare summer migrant in n-w, vagrant elsewhere. Coastal wetland, estuaries, wetlands. Recent records from (NT) Leanyer Sewage Ponds; (Qld) Lake Moondarra; (Vic) Western Treatment Plant; (SA) Dry Creek Saltfields; (WA) Rottnest Island, Lake MacLeod, Broome; (OIT) Ashmore Reef.

Red-backed Button-quail (*Turnix maculosus*). Uncommon to mod common n Aust. Grassy swampland, grassland. (NT) Tiwi Is., Holmes Jungle NR, Borroloola; (Qld) Townsville Common, Lake Samsonvale, Lake Dyer; (WA) Parry Lagoons NR, Broome.

Black-breasted Button-quail (*Turnix melanogaster*). Uncommon se Qld, ne NSW. Coastal rainforest, forest. (Qld) Lamington NP, Inskip Point, Fraser Is., Conondale Ranges.

Chestnut-backed Button-quail (*Turnix castanotus*). Uncommon to mod common Top End and Kimberley. Grassy woodland. (NT) Tiwi Is., Kakadu Hwy, Pine Creek, Chinaman Creek, Timber Creek; (WA) Wyndham, Mitchell River NP.

Buff-breasted Button-quail (*Turnix olivii*). Very rare far n Qld. Grassy woodland. (Qld) Cooktown, Mulligan Hwy near Lake Mitchell, Bakers Rd, Mareeba Wetlands.

Painted Button-quail (*Turnix varius*). Mod common. Open woodland, forest. (Qld) Mulligan Hwy near Lake Mitchell, Mareeba Wetlands, Davies Creek NP, Mt Coot-tha Forest, Lake Samsonvale, Lamington NP, Sheep Station Creek CP, Girraween NP, Sundown NP; (NSW) Capertee Valley, Ku-ring-gai Chase NP, Scheyville NP; (Vic) Chiltern-Mt Pilot NP, Greater Bendigo NP, You Yangs, Point Addis; (SA) Kaiserstuhl CP, Mary Seymour CP, Mt Remarkable; (WA) Dryandra Woodlands, Sandford Rocks NR, Houtman Abrolhos (ssp. *scintillans*).

Red-chested Button-quail (*Turnix pyrrhothorax*). Uncommon to mod common n Aust and e Aust. Grassland. (NT) Holmes Jungle NR, Connells Lagoon CR; (Qld) Townsville Common, Lake Samsonvale; (NSW) Deniliquin, Oolambeyan NP; (Vic) Terrick Terrick (rare), (WA) Kununurra.

Little Button-quail (*Turnix velox*). Mod common. Grassland, arid grassy woodland. (NT) Jasper Gorge, Kunoth Bore, Barrow Creek, Ellery Creek Big Hole, Ormiston Gorge, Tnorala CR; (Qld) McNamara's Rd, Bladensburg NP, Bowra, Currawinya NP; (NSW) Deniliquin, Oolambeyan NP; (Vic) Yarrara FFR, Wyperfeld, Glenlee FFR, Goschen BR, Tresco West BR, Terrick Terrick NP; (SA) Billiatt CP; (WA) Cocklebiddy, Tom Price, Jupiter Well, Lake MacLeod, Carnarvon, Cue.

Oriental Pratincole (*Glareola maldivarum*). Uncommon to common summer migrant in n, rare in s. Grassland, coastal wetlands. Regularly reported n Aust. (NT) Leanyer Sewage Ponds, Holmes Jungle NR, Knuckey Lagoons; (Qld) Boodjamulla NP, Lake Moondarra; (WA) Lake Argyle, Wyndham, Derby, Broome, Carnarvon, vagrant elsewhere; (OIT) Cocos Keeling Is.

Australian Pratincole (*Stiltia isabella*). Uncommon. Migratory/dispersive. Open plains, grasslands, farmland, wetlands. (NT) Mamukala Wetlands, Barkly Tableland, Connells Lagoon CR, Alice Springs Waste Stabilisation Ponds; (Qld) Marina Plains, Cairns Esplanade, Wolframite Dam, Karumba, Boodjamulla NP, Bladensburg NP, Currawinya NP, Lake Atkinson; (NSW) Oolambeyan NP, Sturt NP; (Vic) Neds Corner, Patho Plains; (SA) Coongie Lakes, Birdsville Track; (WA) Lake Argyle, Kununurra, Wyndham.

South Polar Skua (*Stercorarius maccormicki*). Rare. Passes offshore on n migration in autumn and s migration in spring.

Brown (Southern) Skua (*Stercorarius antarcticus*). Uncommon to mod common. Inshore, offshore waters.

Pomarine Jaeger (*Stercorarius pomarinus*). Uncommon. Inshore, offshore waters.

Arctic Jaeger (*Stercorarius parasiticus*). Mod common. Inshore, offshore waters.

Long-tailed Jaeger (*Stercorarius longicaudus*). Uncommon. Inshore, offshore waters.

Common Noddy (*Anous stolidus*). Uncommon to mod common. Inshore, offshore waters. (Qld) Kutini-Payamu (Iron Range) NP, Michaelmas Cay, Green Is., Lady Elliot Is., Southport Pelagic; (NSW) Myall Lakes NP, Ballina, Maroubra (rare), Sydney Pelagic; (WA) Houtman Abrolhos, Lacepede Is.; (OIT) Norfolk Is., Lord Howe Is., Ashmore Reef, Christmas Is., Cocos Keeling Is.

Black Noddy (*Anous minutus*). Uncommon to mod common n Aust. Inshore, offshore waters. (Qld) Michaelmas Cay, Lady Elliot Is., North Stradbroke Is.; (WA) Houtman Abrolhos; (OIT) Norfolk Is., Lord Howe Is., Ashmore Reef.

Lesser Noddy (*Anous tenuirostris*). Rare, localised. Tropical and subtropical Indian Ocean. (WA) Houtman Abrolhos; (OIT) Ashmore Reef.

White Tern (*Gygis alba*). Uncommon. Tropical and subtropical ocean. (Qld) Lady Elliot Is., North Stradbroke Is., Southport Pelagic; (NSW) Myall Lakes NP, Old Bar, Maroubra (rare), Sydney Pelagic; (OIT) breeds Norfolk Is., Lord Howe Is., Cocos Keeling Is.

Grey Ternlet (*Procelsterna cerulea*). Uncommon tropical and subtropical Pacific Ocean. (Qld) Lady Elliot Is., North Stradbroke Is., Southport Pelagic; (NSW) Maroubra (rare); (OIT) breeds Norfolk Is., Lord Howe Is.

Bridled Tern (*Onychoprion anaethetus*). Mod common. Tropical oceans. (NT) Groote Eylandt; (Qld) Kutini-Payamu (Iron Range) NP, Green Is., Lady Elliot Is., Southport Pelagic; (NSW) Ballina (rare) Sydney Pelagic; (WA) Rottnest Is., Penguin Is., Leeuwin-Naturalist NP, Lacepede Is.; (OIT) Ashmore Reef.

Sooty Tern (*Onychoprion fuscata*). Uncommon to common. Tropical oceans. (Qld) Kutini-Payamu (Iron Range) NP, Michaelmas Cay, Lady Elliot Is., North Stradbroke Is., Southport Pelagic; (NSW) Old Bar, Ballina, Maroubra (rare), Long Reef AR (rare), Sydney Pelagic; (WA) Houtman Abrolhos; (OIT) Norfolk Is., Lord Howe Is., Ashmore Reef.

Little Tern (*Sternula albifrons*). Uncommon to common. Coastlines. (NT) Buffalo Creek, East Point, Tiwi Is.; (Qld) Cairns, Mossman River, Mackay; (NSW) Ballina, Boat Harbour, Towra Point NR, Soldiers Point, Mimosa Rocks NP, Nadgee NR; (Vic) Croajingolong NP, Cape Conran CP, Marlo CR, Macleod Morass, Lake Tyers, Mud Is., Lake Victoria; (Tas) Waterhouse CA; (SA) Lincoln NP (WA), Broome; (OIT) Ashmore Reef.

Fairy Tern (*Sternula nereis*). Uncommon to common. Coastlines. Two distinct ssp.: *nereis* breeds in Aust; *exsul* breeds in New Caledonia, and is a vagrant. (NSW) Mimosa Rocks NP; (Vic) Croajingolong NP, Cape Conran CP, Macleod Morass, Lake Tyers, Mud Is., Lake Victoria; (Tas) Maria Is., King Is., Flinders Is., Waterhouse CA; (SA) Glenelg, Goolwa, Kangaroo Is., Coorong NP, Innes NP, Coffin Bay NP; (WA) Pelican Point, Penguin Is., Rottnest Is., Mandurah, Coodanup R, Leeuwin-Naturaliste NP, Albany, Lake Warden, Lake MacLeod, Houtman Abrolhos, Carnarvon, Lake MacLeod.

Saunder's Tern (*Sternula saundersi*). Summer migrant in small numbers to Cocos Keeling Is. Not recorded elsewhere.

Gull-billed Tern (*Gelochelidon nilotica*). Common in n, rare to uncommon in s. Coastlines, inland wetlands. Two distinct ssp.: Aust-breeding *macrotarsa* is widespread; smaller summer migrant *affinis* mostly n Australia.(NT) Alice Spring Stabilisation Ponds, Groote Eylandt, Mamukala, Fogg Dam CR, East Point R, Leanyer Sewage Ponds; (Qld) Lake Atkinson, Inskip Point, Southport pelagic, Lake Samsonvale, North Stradbroke Is., Manly Boat Harbour, Currawinya NP, Mackay, Horseshoe Lagoon, Michaelmas Cay, Cairns Esplanade; (NSW) Ballina, Harrington, Stockton Sandspit, Ash Island; (Vic) Winton Wetlands, Kerang; (SA) Coorong NP, Goolwa; (WA) Wagin TP, Nallan Lake, Shark Bay, Carnarvon, Broome, Roebuck Bay, Parry Lagoons NR.

Caspian Tern (*Hydroprogne caspia*). Mod common to common. Coastal and freshwater wetlands.

Whiskered Tern (*Chlidonias hybrida*). Mod common to common. Freshwater wetlands.

White-winged Black Tern (*Chlidonias leucopterus*). Uncommon to mod common. Freshwater and coastal wetlands. (NT) Leanyer Sewage Ponds, Fogg Dam, Mamukala; (Qld) Horseshoe Lagoon, Currawinya Lake; (NSW) Ballina, Ash Is., Fivebough Wetlands; (Vic) Kerang Lakes, Lake Mokoan, Western Treatment Plant; (WA) Lake McLarty, Broome, Carnarvon, Parry Lagoons NR; (OIT) Ashmore Reef.

Roseate Tern (*Sterna dougallii*). Uncommon. Coastlines and ocean. (NT) Buffalo Creek, Tiwi Is., Groote Eylandt; (QLD) Michaelmas Cay, Lady Elliot Is.; (WA) Rottnest Is., Penguin Is., Kalbarri NP, Houtman Abrolhos, Lacepede Is., Shark Bay, Carnarvon, Broome; (OIT) Ashmore Reef.

White-fronted Tern (*Sterna striata*). Uncommon winter migrant. Southern and eastern coastlines and ocean. (Qld) North Stradbroke Is.; (NSW) Ballina, Long Reef AR, Sydney Pelagic, Maroubra, Ben Boyd NP; (Vic) Cape Liptrap CP, Phillip Island, Point Addis, Portland and Port Fairy Pelagic, Portland; (Tas) Pelagics, Freycinet NP.

Black-naped Tern (*Sterna sumatrana*). Uncommon. Tropical coastlines and islands. (NT) Tiwi Is., Groote Eylandt; (Qld) Kutini-Payamu (Iron Range) NP, Michaelmas Cay, Green Is., Lady Elliot Is.

Common Tern (*Sterna hirundo*). Common to uncommon. Estuaries, coastlines and ocean. (NT) East Point; (Qld) North Stradbroke Is., Southport Pelagic, Redcliffe Point, Soldiers Point; (NSW) Maroubra; (Vic) Western Treatment Plant, Mud Is., Portland and Port Fairy Pelagic; (WA) Mandurah harbour, Carnarvon.

Arctic Tern (*Sterna paradisaea*). Rare. Estuaries, coastlines and ocean. (NSW) Ballina, Maroubra (rare); (Vic) Portland and Port Fairy Pelagic, Portland; (WA) Hillary's Boat Harbour Pelagic, Mandurah harbour.

Antarctic Tern (*Sterna vittata*). Subantarctic oceans. Resident Heard Is., Macquarie Is. Vagrant to mainland, with records from Kangaroo Is., SA. Also Vic, Tas, and WA.

Lesser Crested Tern (*Thalasseus bengalensis*). Uncommon to mod common. Tropical and subtropical coastlines. (NT) Buffalo Creek, East Point, Tiwi Is.; (Qld) Kutini-Payamu (Iron Range) NP, Michaelmas Cay, Green Is., Mossman River, Mackay, Lady Elliot Is., North Stradbroke Is.; (WA) Lacepede Is., Broome, Carnarvon.

Crested Tern (*Thalasseus bergii*). Common. Coastlines and inshore waters.

Pacific Gull (*Larus pacificus*). Uncommon to mod common s Aust. Coastline. (Vic) Croajingolong NP, Cape Liptrap CP, Wilsons Promontory NP, Portland, Killarney Beach; (Tas) Bicheno, Freycinet NP; (SA) Innes NP, Coffin Bay NP; (WA) Two Peoples Bay NR, Esperance, Cape Arid NP, Shark Bay.

Kelp Gull (*Larus dominicanus*). Uncommon to mod common s-e Aust esp. Tas. Coastlines. (NSW) Maroubra, Boat Harbour; (Vic) Cape Liptrap CP, Phillip Island, Bellarine Peninsula, Portland, Killarney Beach; (Tas) Bruny Is., Goulds Lagoon, Pitt Water NR, Maria Is., Bicheno, Freycinet NP.

Silver Gull (*Chroicocephalus novaehollandiae*). Common (often abundant). Widespread. Mostly coastal, but also inland wetlands.

Palm Cockatoo (*Probosciger aterrimus*). Mod common Cape York. Rainforest and woodland. (Qld) Kutini-Payamu (Iron Range) NP, Uningan NR, Lockerbie Scrub, Jardine River NP.

Red-tailed Black-Cockatoo (*Calyptorhynchus banksii*). Common n Aust, rare s Aust. Vic ssp. *graptogyne* Endangered. Forest, woodland, farmlands. (NT) Charles Darwin NP, Kakadu Hwy, Gunlom, Alice Springs Waste Stabilisation Ponds, Alice Springs Desert Park, Kunoth Bore, N'Dhala Gorge NP; (Qld) Peninsula Development Rd, Lake Mitchell, Cape Hillsborough NP, Karumba, Kinchega NP; (Vic) Lower Glenelg NP, Rennick SF; (WA) Wungong Dam, Bungendore Park, Serpentine NP, Walpole-Nornalup NP, Lake Muir NR, Boranup Karri Forest, Porongurup NP, Broome, Mitchell River NP, Parry Lagoons NR.

Glossy Black-Cockatoo (*Calyptorhynchus lathami*). Rare to uncommon in e Aust. Woodland dominated by *Allocasuarina*. Isolated Kangaroo Is. ssp. *halmaturinus* Endangered. (Qld) Lamington NP, Conondale Ranges, Girraween NP; (NSW) Barrington Tops NP, Werrikimbe NP, Crowdy Bay NP, Bruxner Park FR, Yuraygir NP, Pilliga Forest, Capertee Valley, Cocoparra NP, Dharug NP, Booderee NP, Bomaderry Creek RP, Mimosa Rocks NP, Ben Boyd NP, Maxwell's FR; (Vic) Croajingolong NP; (SA) Kangaroo Is.

Yellow-tailed Black-Cockatoo (*Calyptorhynchus funereus*). Mod common to common in s-e Aust. Woodland, pine plantations, farmlands. (Qld) Brisbane, Imbil SF, Fraser Is. (NSW) Sydney Royal BG, Centennial Park, Maroubra, Ku-ring-gai Chase NP, Royal NP; (Vic) Royal Park, Yarra Bend Park, Wilsons Promontory NP, Point Addis, Great Otway NP, Lower Glenelg NP; (Tas) Fern Glade R, King Is., Freycinet NP; (SA) Adelaide Hills, Mt Lofty Ranges, Eyre Peninsula.

Carnaby's Black-Cockatoo (*Calyptorhynchus latirostris*). Uncommon to mod common in s-w WA. Woodlands. (WA) Kings Park, Wungong Dam, Bungendore Park, Dryandra Woodlands, Walpole-Nornalup NP, Lake Muir NR, Boranup Karri Forest, Porongurup NP, Two Peoples Bay NR, Cheynes Beach, Stirling Range NP, Dragon Rocks NR, Fitzgerald River NP, Cocanarup Timber R, Cape Arid NP.

Baudin's Black-Cockatoo (*Calyptorhynchus baudinii*). Mod common in s-w WA. Woodland, forest. (WA) Bungendore Park, Serpentine NP, Walpole-Nornalup NP, Lake Muir NR, Boranup Karri Forest, Porongurup NP, Stirling Range NP.

Gang-gang Cockatoo (*Callocephalon fimbriatum*). Uncommon to mod common in s-e Aust. Forest, woodland. (NSW/ACT) Capertee Valley, Dharug NP, Budderoo NP, Morton NP, Booderee NP, Bomaderry Creek

RP, Kosciusko NP, Australian National BG, Tidbinbilla NR; (Vic) Buchan Caves R, Wilsons Promontory NP, Yarra Bend Park, Anglesea Heath, Point Addis, Sheoak Picnic Ground, Ralph Illidge Sanctuary, Lower Glenelg NP; (SA) Kangaroo Is. (introduced).

Major Mitchell's Cockatoo (*Lophocroa leadbeateri*). Uncommon. Open woodland, Mallee, open plains. Two distinct ssp.: *mollis* (scarlet crest) w and n Aust; *leadbeateri* (scarlet and yellow crest) s-e Aust. (NT) Alice Springs Waste Stabilisation Ponds, Simpson's Gap, Newhaven, Watarrka (Kings Canyon) NP, Mereenie Loop Rd, Erldunda, Uluru-Kata Tjuta NP; (Qld) Bowra, Lake Bindegolly, Eulo Bore; (NSW) Round Hill and Nombinnie NR, Cocoparra NP, Kinchega NP, Mutawintji NP; (Vic) Hattah-Kulkyne NP, Murray-Sunset NP, Wyperfeld NP; (SA) Danggali CP, Billiatt CP, Yumbarra CP, Mt Ive Station, Mamungari CP, Nullarbor NP; (WA) Eyre Bird Observatory.

Galah (*Eolophus roseicapillus*). Common. Woodland, forest, farmlands, urban parks.

Long-billed Corella (*Cacatua tenuirostris*). Common s-e Aust. Feral populations in many capital cities. Woodland, farmlands.

Western Corella (*Cacatua pastinator*). Uncommon to mod common in s-w WA. Open woodland, farmlands. Two distinct ssp.: *pastinator* found s of Perth, *derbyi* w of Perth. (WA) Wungong Dam, Lake Muir NR, Rocky Gully, Avon River.

Little Corella (*Cacatua sanguinea*). Common. Woodland, farmlands.

Sulphur-crested Cockatoo (*Cacatua galerita*). Common. Woodland, forest, farmlands, urban parks.

Cockatiel (*Nymphicus hollandicus*). Common to uncommon. Nomadic. Woodland, shrublands.

Rainbow Lorikeet (*Trichoglossus moluccanus*). Common. Forest, woodland, gardens. Introduced Perth.

Red-collared Lorikeet (*Trichoglossus rubritorquis*). Common. Forest, woodlands, gardens. n WA, n NT, n-w Qld.

Coconut Lorikeet (*Trichoglossus haematodus*). Mod common. Forest, mangroves. n Torres Strait; Boigu Is., Saibai Is.

Scaly-breasted Lorikeet (*Trichoglossus chlorolepidotus*). Common. Forest, woodland, gardens. Introduced Vic.

Varied Lorikeet (*Psitteuteles versicolor*). Mod common n Aust. Woodland, forest. (NT) Charles Darwin NP, Kakadu, Pine Creek, Fergusson River, Nitmiluk (Katherine Gorge) NP; (Qld) Cumberland Dam, Gregory River, Mt Isa, (WA) Derby, Mornington Station, Mitchell River NP, Drysdale River, Kununurra, Parry Lagoons NR, the Grotto, Purnululu (Bungle Bungles) NP.

Musk Lorikeet (*Glossopsitta concinna*). Common s-e Aust. Woodland, urban parks.

Little Lorikeet (*Glossopsitta pusilla*). Mod common e Aust. Woodland. (Qld) Big Mitchell Creek, Boondall WR, Sheep Station Creek CP, Durikai SF; (NSW) Border Ranges NP, Capertee Valley, Munghorn Gap NR, Londonderry; (Vic) Chiltern-Mt Pilot NP, Greater Bendigo NP, Yarra Bend.

Purple-crowned Lorikeet (*Glossopsitta porphyrocephala*). Uncommon s Aust. Woodland. (NSW) Deniliquin; (Vic) Little Desert NP, Glenlee FFR, Greater Bendigo NP, Dunach NCR, You Yangs, Belmont Common; (SA) Adelaide BG,

Kangaroo Is., Mary Seymour CP, Billiatt CP, Mt Remarkable NP, Telowie Gorge CP, Yumbarra CP; (WA) Kings Park, Rottnest Is., Dryandra Woodlands, Walpole-Nornalup NP, Lake Muir NR, Porongurup NP, Cheynes Beach, Stirling Range NP, Wave Rock, Fitzgerald River NP, Cape Arid NP, Eyre Bird Observatory.

Double-eyed Fig-Parrot (*Cyclopsitta diophthalma*). Mod common. Three distinct ssp.: *marshalli* (Marshall's Fig-Parrot) Cape York (mod common); *macleayana* (Macleay's' Fig-Parrot) Wet Tropics (mod common); *coxeni* (Coxen's Fig Parrot) s-e Qld, n NSW (possibly extinct). Rainforest. (Qld) Kutini-Payamu (Iron Range) NP, Cairns Esplanade, Cairns BG, Kuranda, Daintree, Kingfisher Park, Mt Lewis, Curtain Fig Tree, Mission Beach, Crater Lakes NP.

Tasman Parakeet (*Cyanoramphus cookii*). Uncommon resident Norfolk Is.

Eclectus Parrot (*Eclectus roratus*). Mod common Cape York and n Torres Strait. Rainforest, tropical woodland. (Qld) Kutini-Payamu (Iron Range) NP; (OIT) Boigu Is., Saibai Is., Dauan Is.

Red-cheeked Parrot (*Geoffroyus geoffroyi*). Mod common Cape York and n Torres Strait. Rainforest, tropical woodland. (Qld) Kutini-Payamu (Iron Range) NP; (OIT) Boigu Is., Saibai Is.

Australian King-Parrot (*Alisterus scapularis*). Mod common e Aust. Forest, woodland. (Qld) Mt Lewis, Yungaburra, D'Aguilar NP, Lamington NP, Sheep Station Creek CP; (NSW/ACT) Dorrigo NP, Pilliga Forest, Capertee Valley, Royal NP, Nowra, Mimosa Rocks NP, Kosciusko NP, Tidbinbilla NR; (Vic) Warby Range SP, Croajingolong NP, Buchan Caves R, Wilsons Promontory NP, Dandenong Ranges NP, Great Otway NP.

Red-winged Parrot (*Aprosmictus erythropterus*). Common e and n Aust. Nomadic. Open woodland.

Superb Parrot (*Polytelis swainsonii*). Uncommon s-e Aust. Open woodland, river edge forest. (NSW) Leeton, Deniliquin, Oolambeyan NP, Murray Valley NP; (Vic) Barmah region.

Regent Parrot (*Polytelis anthopeplus*). Mod common s Aust. Woodland, Mallee. (Vic) Hattah-Kulkyne NP, Murray-Sunset NP, Wyperfeld NP; (SA) Gluepot R, Danggali CP, Billiatt CP; (WA) Lake McLarty, Waychinicup NP, Stirling Range NP, Fitzgerald River NP, Cocanarup Timber R, Sandford Rocks NR, Dryandra Woodlands.

Princess Parrot (*Polytelis alexandrae*). Rare to very rare, generally elusive. Remote outback woodland (especially areas dominated by Marble Gum). (NT) Watarrka (Kings Canyon) NP, Mereenie Loop Rd, Uluru-Kata Tjuta NP; (SA) Mamungari CP; (WA) Neale Junction, Jupiter Well.

Green Rosella (*Platycercus caledonicus*). Common Tas. Forest, woodland, farmlands.

Crimson Rosella (*Platycercus elegans*). Common e Aust. Forest, woodland, farmlands. There are several distinct ssp.: crimson *elegans, nigrescens, melanoptera* Qld to s-e SA and Kangaroo Is.; orange *subadelaidae* (Adelaide Rosella); *fleurieuensis* SA; yellow *flaveolus* (Yellow Rosella) along Murray River and Murrumbidgee River.

Eastern Rosella (*Platycercus eximius*). Common e Aust, including Tas. Forest, woodland, farmlands. Three distinct ssp.: *elecica* (Golden-mantled Rosella) found n-e NSW, s-e Qld; *eximius* NSW, Vic, SA; *diemenensis* Tas.

Pale-headed Rosella (*Platycercus adscitus*). Common n-e Aust. Forest, woodland. Two distinct ssp.: *adscitus* (Pale-headed Rosella) n Qld, Cape York; *palliceps* (Blue-cheeked Rosella) Qld, n NSW.

Northern Rosella (*Platycercus venustus*). Uncommon to mod common n Aust. Woodland. (NT) Charles Darwin NP, Tiwi Is., Kakadu, Gunlom, Pine Creek, Copperfield Dam Fergusson River, Nitmiluk (Katherine Gorge) NP, Jasper Gorge, Caranbirini CR, Borroloola; (Qld) Boodjamulla (Lawn Hill) NP; (WA) Mornington Station, Mitchell River NP, Drysdale River, Kununurra, Mirima (Hidden Valley) NP, Lake Argyle, Purnululu (Bungle Bungles) NP.

Western Rosella (*Platycercus icterotis*). Mod common s-w WA. Woodland. (WA) Wungong Dam, Bungendore Park, Serpentine NP, Walpole-Nornalup NP, Lake Muir NR, Meelup RP, Boranup Karri Forest, Porongurup NP, Stirling Range NP, Wave Rock, Boorabbin NP, Goldfields Woodlands NP.

Australian Ringneck (*Barnardius zonarius*). Common. Woodland, Mallee, Mulga. Four distinct ssp.: *zonarius* (Port Lincoln Parrot) s NT, SA, WA; *semitorquatus* (Twenty-eight Parrot) s-w WA; *barnardi* (Mallee Ringneck) s-e SA, n Vic; *macgillivrayi* (Cloncurry Ringneck) w Qld.

Red-capped Parrot (*Purpureicephalus spurius*). Common s-w WA. Woodland. (WA) Thomsons Lake NR, Wungong Dam, Lake McLarty NR, Bungendore Park, Serpentine NP, Dryandra Woodlands, Walpole-Nornalup NP, Porongurup NP, Stirling Range NP, Cocanarup Timber R.

Blue Bonnet (*Northiella haematogaster*). Common. Woodland, farmlands. Four distinct ssp.: *haematogaster* (Yellow-vented Blue Bonnet) NSW, Vic, SA; *haematorrhous* (Red-vented Blue Bonnet) n NSW, s Qld; *narethae* (Naretha Blue Bonnet) w SA, s-e WA; *pallescens* n SA, s-w Qld.

Swift Parrot (*Lathamus discolor*). Rare to uncommon, nomadic. Forest, woodland. (Qld) Girraween (rare), Sheep Station Creek CP (rare); (NSW) Crowdy Bay NP (rare), Capertee Valley, Ku-ring-gai Chase NP, Londonderry, Castlereagh NR, Mimosa Rocks NP, (Vic) Chiltern-Mt Pilot NP, Warby Range SP, Greater Bendigo NP, Heathcote-Graytown NP, Royal Park, You Yangs RP; (Tas), Mt Wellington, Bruny Is., Maria Is., Meehan Range.

Red-rumped Parrot (*Psephotus haematonotus*). Common s-e Aust. Woodland, urban parks.

Mulga Parrot (*Psephotus varius*). Common arid s Aust. Mallee, Mulga, woodland.

Golden-shouldered Parrot (*Psephotus chrysopterygius*). Rare c Cape York. Tropical grassy woodland, in association with large termite mounds. (Qld) Musgrave, Artemis Station.

Hooded Parrot (*Psephotus dissimilis*). Uncommon s Top End. Savannah woodland in association with large termite mounds. (NT) Pine Creek, Copperfield Dam, Umbrawarra Gorge NP, Fergusson River, Nitmiluk (Katherine Gorge) NP, Chinaman Creek.

Budgerigar (*Melopsittacus undulatus*). Common arid Aust. Nomadic. Grasslands, open woodland.

Bourke's Parrot (*Neopsephotus bourkii*). Mod common to rare c Aust. Nomadic. Mulga woodland. (NT) Kunoth Bore, Erldunda; (Qld) Bowra, Eulo Bore, Kilcowera Station, Currawinya NP; (SA) Coongie Lakes; (WA) Goongarrie NP, Yeo Lake NR, Cue, Carnarvon.

Blue-winged Parrot (*Neophema chrysostoma*). Mod common s-e Aust. Coastal heathland, Mulga, saltbush, woodland. Uncommon to mod common. (Vic) Lake Tyrrell WR, Little Desert NP, Lake Tutchewop, Cape Liptrap CP, Shallow

Inlet, Corner Inlet, Phillip Is., Western Treatment Plant, Anglesea Heath, Point Addis Lookout, Great Otway NP, Cape Nelson Lighthouse; (Tas) Bruny Is., Meehan Range, Melaleuca, Narawntapu NP, King Is., Cradle Mountain-Lake St Clair NP, Waterhouse CA; (SA) Dry Creek Saltfields, Bool Lagoon, Coorong NP.

Elegant Parrot (*Neophema elegans*). Uncommon to mod common s and s-w Aust. Woodland, Mallee, coastal shrublands. (Vic) Little Desert NP; (SA) Dry Creek Saltfields, Price Saltfields, Port Gawler CP, Newland Head CP, Cox Scrub CR, Aldinga Scrub CP, Hindmarsh Island, Murray Mouth, Flinders Chase NP, D'Estrees Bay, Coorong NP, Messent CP, Mt Remarkable NP, Telowie Gorge CP, Flinders Ranges NP, Vulkathunha-Gammon Ranges NP, Australian Arid Land BG, Mt Ive Station; (WA) Bungendore Park, Dryandra Woodlands, Lake Muir, Porongurup NP, Stirling Range NP, Wave Rock, Fitzgerald River NP, Cocanarup Timber R, Monjingup Lake NR, Sandford Rocks NR.

Rock Parrot (*Neophema petrophila*). Uncommon to mod common s and s-w Aust. Coastal heathland, offshore islands. (SA) Price Saltfields, Newland Head CP, Hindmarsh Island Murray Mouth, Flinders Chase NP, Kangaroo Is., Innes NP, Lincoln NP, Coffin Bay NP; (WA) Rottnest Is., Leeuwin-Naturaliste NP, Two Peoples Bay NR, Betty's Beach, Woody Island NR, Cape Arid NP, Shark Bay.

Orange-bellied Parrot (*Neophema chrysogaster*). Rare. One of the world's rarest birds, estimated population of fewer than 100 birds. Migrates from Tas to coastal Vic and s-e SA in winter. Coastal heathland, sedgeland, wetlands. (Vic) Western Treatment Plant, Swan Bay; (Tas) Melaleuca, Birchs Inlet, King Is.; (SA) Little Dip CP.

Turquoise Parrot (*Neophema pulchella*). Uncommon to mod common s-e Aust. Woodland. (Qld) Durikai SF, Girraween NP, Sundown NP; (NSW) Barraba, Pilliga Forest, Warrumbungle NP, Capertee Valley, Munghorn Gap NR, Cocoparra NP; (Vic) Chiltern-Mt Pilot NP, Warby Range SP, Cape Conran, Canni Creek R.

Scarlet-chested Parrot (*Neophema splendida*). Rare. Nomadic. Open woodland, Mallee, Mulga. (SA) Danggali CP, Gluepot R, Yumbarra CP, Mamungari CP; (WA) Neale Junction, Anne Beadell Hwy, Connie Sue Hwy.

Eastern Ground Parrot (*Pezoporus wallicus*). Rare to uncommon s-e coast. Coastal heathland. (Qld) Great Sandy NP; (NSW) Yuraygir NP, Barren Grounds NR, Budderoo NP, Morton NP, Booderee NP, Ben Boyd NP, Nadgee NR; (Vic) Croajingolong NP, Cape Conran, Cape Liptrap CP; (Tas) Bruny Is., Melaleuca, Strahan.

Western Ground Parrot (*Pezoporus flaviventris*). Formerly considered conspecific with Eastern Ground Parrot, now a separate species. One of the world's rarest birds, estimated population of 110 individuals. Coastal heathland. (WA) Almost all occur within Cape Arid NP, also reported from Fitzgerald River NP and Waychinicup NP.

Night Parrot (*Pezoporus occidentalis*). Rare and elusive. The holy grail of Australian ornithology. Arid to semi-arid areas including Spinifex grasslands and chenopod shrublands. First-ever specimen collected near Coongie Lakes (Charles Sturt, 1845). No confirmed sightings between 1912 and 1979, leading to speculation of extinction. In May 1979, four birds reported near Lake Perigundi, SA. More recent records include: road-kill specimen on Diamantina Developmental Rd, 36 km n of Boulia, Qld (Oct 1990); several reports in the 1990s along the Dajarra Cloncurry Rd, Qld; a report from Camel Bore in Newhaven in the NT (Jan 1996); a report of three birds Minga Well near Fortescue Marshes, WA (Apr 2005); dead bird discovered Diamantina NP, Qld (Dec 2006); first photographed s-w Qld (2013).

Pheasant Coucal (*Centropus phasianinus*). Mod common to common. Woodland, roadside thickets. (NT) Darwin BG, Buffalo Creek, East Point, Yellow Water, Gunlom, Victoria River; (Qld) Green Is., Daintree, Kingfisher Park, Tinaroo Creek, Townsville Common, Cape Hillsborough NP, Karumba, Mt Coot-tha Forest, Lake Samsonvale, Sheep Station Creek CP, Great Sandy NP; (NSW) Myall Lakes NP, Crowdy Bay NP, Clarence River Estuary, Ku-ring-gai Chase NP, Royal NP; (WA) Kununurra, Lake Argyle, Karijini NP, Broome, Monington Station, Mitchell River NP, Kununurra.

Eastern Koel (*Eudynamys orientalis*). Common n and e Aust. Migratory. Forest, woodland, farmlands, gardens.

Channel-billed Cuckoo (*Scythrops novaehollandiae*). Common n and e Aust. Migratory. Forest, woodland.

Horsfield's Bronze-Cuckoo (*Chalcites basalis*). Common. Forest, woodland.

Black-eared Cuckoo (*Chalcites osculans*). Uncommon. Woodland, Mulga, Mallee. (NT) Uluru-Kata Tjuta NP; (Qld) Lake Moondarra, Bowra, Currawinya NP; (NSW) Pilliga Forest, Capertee Valley, Round Hill NR, Cocoparra NP, Oolambeyan NP, Sturt NP; (Vic) Hattah-Kulkyne NP, Wyperfeld NP, Yarrara FFR; (SA) Danggali CP, Gluepot R, Flinders Ranges NP, Australian Arid Land BG, Whyalla CP, Yumbarra CP, Mt Ive Station; (WA) Goongarrie NP, Sandford Rocks NR, Neale Junction NR, Karijini NP, Carnarvon, Shark Bay, Kalbarri NP.

Shining Bronze-Cuckoo (*Chalcites lucidus*). Common e and w Aust. Migratory. Forest, woodland.

Little Bronze-Cuckoo (*Chalcites minutillus*). Uncommon to common partial migrant n and n-e Aust. Rainforest, forest, mangroves. Two distinct ssp.: *minutillus* (Little Bronze-Cuckoo) NT, e Qld, n-e NSW; *russatus* (Gould's Bronze-Cuckoo) e Qld, n-e NSW. (NT) Darwin BG, Charles Darwin Uni., Leanyer Sewage Ponds, East Point, Yellow Water; (Qld) Wenlock River, Jardine River NP, Yorkeys Knob, Daintree, Hasties Swamp, Cape Hillsborough NP, Karumba, Lake Samsonvale, Adare Rd Lagoon.

Pallid Cuckoo (*Cacomantis pallidus*). Common. Migratory. Woodland, Mulga, open plains.

Chestnut-breasted Cuckoo (*Cacomantis castaneiventris*). Uncommon to mod common migrant or resident Cape York, rare Wet Tropics. Rainforest. (Qld) Cooktown, McIvor River Crossing, Kutini-Payamu (Iron Range) NP, Jardine River NP.

Fan-tailed Cuckoo (*Cacomantis flabelliformis*). Common migrant e and s Aust. Forest, woodland.

Brush Cuckoo (*Cacomantis variolosus*). Common e and n Aust. Migratory. Rainforest, forest, woodland.

Oriental Cuckoo (*Cuculus optatus*). Mod common summer migrant n Aust, rare s Qld and NSW. Rainforest, forest, mangroves. (NT) Darwin BG, East Point, Adelaide River; (Qld) Cooktown, Kutini-Payamu (Iron Range) NP, Daintree (WA) Broome, Derby, Kununurra.

Powerful Owl (*Ninox strenua*). Rare to uncommon e and s-e Aust. Forest.

(Qld) Mt Coot-tha Forest, Lamington NP, Fraser Is., Conondale Ranges; (NSW) Copeland Tops, Werrikimbe NP, Dorrigo NP, Capertee Valley, Royal BG, Dharug NP, Hawkesbury River Valley, Royal NP, Morton NP, Bomaderry Creek RP, Mimosa Rocks NP, Nadgee NR; (Vic) Croajingolong NP, Cape Conran, Fairy Dell NCR, Tarra-Bulga NP, Yellingbo NCR, Great Otway NP, Rennick SF.

Rufous Owl (*Ninox rufa*). Rare to uncommon n Aust. Rainforest, monsoon forest. (NT) Darwin BG, East Point, Holmes Jungle NR, Gunlom, Elsey NP; (Qld) Kutini-Payamu (Iron Range) NP, Cairns BG, Kuranda, Daintree, Emerald Creek, Curtain Fig Tree, Wongabel SF, Cape Hillsborough NP (rare); (WA) Mitchell River NP.

Barking Owl (*Ninox connivens*). Common in n., rare to uncommon in s. Woodland, open woodland. (NT) Darwin BG, Mamukala Wetlands, Yellow Water, Gunlom, Victoria River; (Qld) Lakefield NP, Lockerbie Scrub, Kingfisher Park, Karumba, Boodjamulla (Lawn Hill) NP; (NSW) Pilliga Forest, Capertee Valley, Barren Grounds NR; (Vic) Chiltern-Mt Pilot NP (rare), Warby Range SP (rare); (SA) Mary Seymour CP, Innamincka; (WA) Broome, Derby, Mornington Station, Mitchell River NP, Drysdale River, Kununurra, Parry Lagoons NR.; (OIT) Torres Strait Is.

Southern Boobook (*Ninox boobook*). Common Forest, woodland, farmlands. Distinct ssp. *lurida* (Red Boobook) Wet Tropics rainforest.

Morepork (*Ninox novaeseelandiae*). Formerly ssp. *leucopsis* of Southern Boobook. Mod common Tas. Forest, woodland. (Tas) Waterworks Re, Truganini Res, Bruny Is., Mt Field NP, Melaleuca, Strahan, Arthur River, Narawntapu NP, Flinders Is.; (OIT) Ssp. *undulata* (Norfolk Island Boobook) Norfolk Is. (very rare).

Christmas Island Hawk-Owl (*Ninox natalis*). Uncommon resident Christmas Is.

Sooty Owl (*Tyto tenebricosa*). Rare to uncommon. Tall forest. s-e Qld, e NSW, e Vic. (Qld) D'Aguilar NP, Lamington NP, Mt Tamborine NP, Conondale Ranges, Blackall Range; (NSW) Barrington Tops NP, Copeland Tops, Werrikimbe NP, Dorrigo NP, Nightcap NP, Border Ranges NP, Royal NP, Morton NP, Nadgee NR; (Vic) Cape Conran, Tarra-Bulga NP, Toolangi SF, Badgers Weir, Bunyip SP.

Lesser Sooty Owl (*Tyto multipunctata*). Formerly ssp. of Sooty Owl. Rare to uncommon Wet Tropics. Tall forest, rainforest. (Qld) Kuranda, Kingfisher Park, Mt Lewis, Davies Creek NP, Mt Hypipamee NP, Paluma Range NP.

Masked Owl (*Tyto novaehollandiae*). Rare to uncommon. Forest, woodland, farmlands edges. Distinct ssp. *castanops* Tas, sometimes considered a full species. (NT) Tiwi Is., Jasper Gorge; (Qld) Kingfisher Park, Emerald Creek, Great Sandy NP, Conondale Ranges; (NSW) Barrington Tops NP, Werrikimbe NP, Capertee Valley, Dharug NP, Bomaderry Creek RP, Mimosa Rocks NP, Nadgee NR; (Vic) Cape Conran Coastal Park; (Tas) Waterworks Res, Bruny Is., Truganini Res, Pitt Water NR, Fern Glade Res, Narawntapu NP; (OIT) introduced Lord Howe Is.

Eastern Barn Owl (*Tyto javanica*). Common Farmland, open plains, woodland.

Eastern Grass Owl (*Tyto longimembris*). Rare. Tall grassland, swamps, heathland. (NT) Holmes Jungle NR, Connells Lagoon CR; (Qld) Lakefield NP, Marina Plains, Tyto

Wetlands, Kedron Brook Wetlands, Lake Samsonvale, Great Sandy NP; (NSW) Ash Island, Crowdy Bay NP; (SA) Koonchera, Waterhole; (WA) Kununurra.

Azure Kingfisher (*Ceyx azureus*). Uncommon to mod common. Wetlands, waterways, estuaries, mangroves. (NT) Buffalo Creek, Howard Springs, Fogg Dam, Yellow Water, Elsey NP; (Qld) Mossman River, Daintree River, Lake Barrine, Tyto Wetlands, Boodjamulla (Lawn Hill) NP, Kedron Brook Wetlands; (NSW) Myall Lakes NP, Clarence River Estuary, Murray Valley NP, Hawkesbury River Valley, Royal NP, Bomaderry Creek RP, Mimosa Rocks NP; (Vic) Kerang Lakes, Ovens River, Double Creek Nature Walk, Lower Glenelg NP; (Tas) Strahan, Arthur River; (WA) Kununurra, Drysdale River.

Little Kingfisher (*Ceyx pusilla*). Uncommon n Aust. Wetlands, waterways, mangroves, estuaries. (NT) Buffalo Creek, Leanyer Sewage Ponds, Palmerston Sewage Work, Yellow Water; (Qld) Endeavour River, Uningan NR, Wenlock River, Pennefather River, Centenary Lakes, Jack Barnes Mangrove Boardwalk, Mossman River, Daintree River, Tyto Wetlands, Townsville Common; (OIT) Torres Strait Is.

Buff-breasted Paradise-Kingfisher (*Tanysiptera sylvia*). Mod common summer migrant to n Qld. Lowland rainforest. (Qld) Cooktown, McIvor River Crossing, Kutini-Payamu (Iron Range) NP, Mt Whitfield CP, Kuranda, Daintree, Kingfisher Park, Paluma Range NP, Cape Hillsborough NP.

Laughing Kookaburra (*Dacelo novaeguineae*). Common e Aust, introduced s-w WA, Tas and Kangaroo Is. Forest, woodland, farmlands.

Blue-winged Kookaburra (*Dacelo leachii*). Common n Aust. Forest, woodland, swampland. (NT) Fogg Dam, Mamukala Wetlands, Yellow Water, Anbangbang Billabong Walk Gunlom, Victoria River, (Qld) Cooktown, Uningan NR, Daintree, Mt Carbine, Tinaroo Creek, Townsville Common, Magnetic Is., Cape Hillsborough NP, Cumberland Dam, Boodjamulla (Lawn Hill) NP, Mt Isa, Lockyer Valley; (WA) Carnarvon, Karijini NP, Tom Price, Broome, Drysdale River, Purnululu (Bungle Bungles) NP.

Yellow-billed Kingfisher (*Syma torotoro*). Uncommon n Cape York. Rainforest. (Qld) Kutini-Payamu (Iron Range) NP, Uningan NR, Wenlock River, Lockerbie Scrub, Jardine River NP.

Forest Kingfisher (*Todiramphus macleayii*). Common e and n Aust. Forest, woodland, mangroves.

Red-backed Kingfisher (*Todiramphus pyrrhopygius*). Moderately common. Open woodland, arid shrublands. (NT) Kakadu, Pine Creek, Barkly Tableland, Alice Springs Desert Park, Kunoth Bore, Santa Teresa Rd, Ellery Creek Big Hole, Erldunda, Uluru-Kata Tjuta NP, Old Andado Station; (Qld) Tinaroo Creek, Gulf Developmental Rd, Karumba, Mica Creek, Bladensburg NP, Bowra, Eulo Bore, Kilcowera Station; (NSW) Cocoparra NP, Kinchega NP, Mutawintji NP, Sturt NP; (Vic) Tresco West Bushland Reserve, Wyperfeld NP, Hattah-Kulkyne NP; (SA) Danggali CP, Gluepot R, Brachina Gorge Wilkawillina Gorge, Vulkathunha-Gammon Ranges, Montecollina Bore, Cooper Creek, Australian Arid Land BG, Yumbarra CP; (WA) Goongarrie NP, Neale Junction NR, Cue, Kununurra, Tom Price, Jupiter Well.

Sacred Kingfisher (*Todiramphus sanctus*). Common. Summer migrant to s Aust. Forest, woodland, mangroves.

Annotated bird list

Collared Kingfisher (*Todiramphus chloris*). Moderately common n Aust. Mangroves, estuaries. (NT) Palmerston Sewage Works, East Point; (Qld) Kutini-Payamu (Iron Range) NP, Uningan NR, Cairns Esplanade, Jack Barnes Mangrove Boardwalk, Mossman River, Magnetic Is., Cape Hillsborough NP, Boondall Wetlands R, Kedron Brook Wetlands, Wynnum Mangrove Boardwalk, Eagleby Wetlands, Inskip Point, Toorbul Shorebird Roost; (WA) Wyndham, Derby.

Rainbow Bee-eater (*Merops ornatus*). Common. Summer migrant to s Aust. Open woodland, woodland.

Dollarbird (*Eurystomus orientalis*). Mod common summer migrant. Open woodland, forest edges. (NT) Leanyer Sewage Ponds, Yellow Water, Gunlom, Mataranka; (Qld) Kutini-Payamu (Iron Range) NP, Kingfisher Park, Lake Moondarra, Sundown NP; (NSW) Barraba, Capertee Valley, Munghorn Gap NR, Deniliquin, Dharug NP, Hawkesbury River Valley, Mimosa Rocks NP; (Vic) Ovens River (WA) Broome, Derby, Kununurra; (OIT) Cocos Keeling Is (Asian ssp. *orientalis*).

Red-bellied Pitta (*Pitta erythrogaster*). Uncommon summer migrant n Cape York. Rainforest. (Qld) Kutini-Payamu (Iron Range) NP, Lockerbie Scrub.

Noisy Pitta (*Pitta versicolor*). Mod common in e Aust. Tropical and subtropical rainforest. (Qld) Cooktown, McIvor River Crossing, Kutini-Payamu (Iron Range) NP, Lockerbie Scrub, Mt Whitfield CP, Jindalba Boardwalk, Kingfisher Park, Paluma Range NP, Magnetic Is., Eungella Plateau, D'Aguilar (Mt Glorious) NP, Lamington NP, Mt Tamborine NP, Great Sandy NP, Fraser Is., Conondale Ranges; (NSW) Copeland Tops, Crowdy Bay NP, Boorganna NR, Dorrigo NP, Bruxner Park FR, Iluka NR Nightcap NP, Border Ranges NP, Royal NP (rare); (OIT) Torres Strait Is.

Rainbow Pitta (*Pitta iris*). Mod common Top End and Kimberley. Monsoon rainforest. (NT) Buffalo Creek, East Point, Holmes Jungle NR, Howard Springs, Fogg Dam, Tiwi Is., Mardugal Park, Groote Eylandt; (WA) Mitchell River NP.

Albert's Lyrebird (*Menura alberti*). Uncommon to mod common s-e Qld and n-e NSW. Rainforest. (Qld) Lamington NP, Mt Tamborine NP; (NSW) Nightcap NP, Border Ranges NP.

Superb Lyrebird (*Menura novaehollandiae*). Uncommon to mod common s-e Aust. Forest, temperate rainforest. (Qld) Girraween NP; (NSW/ACT) Barrington Tops NP, Copeland Tops, Dorrigo NP, Capertee Valley, Munghorn Gap NR, Dharug NP, Royal NP, Barren Grounds NR, Budderoo NP, Morton NP, Mimosa Rocks NP, Ben Boyd NP, Kosciusko NP, Tidbinbilla NR, Namadgi NP; (Vic) Croajingolong NP, Cabbage Tree Creek, Buchan Caves R, Tarra-Bulga NP, Dandenong Ranges NP, Toolangi SF, Bunyip SP, Glen Nayook R; (Tas) Mt Field NP (introduced).

Rufous Scrub-bird (*Atrichornis rufescens*). Rare s-e Qld and n-e NSW. Temperate rainforest and forest with dense undergrowth. (Qld) Lamington NP; (NSW) Barrington Tops NP, Werrikimbe NP, Border Ranges NP.

Noisy Scrub-bird (*Atrichornis clamosus*). Rare. Coastal heathland s-w WA. (WA) Two Peoples Bay NR, Cheynes Beach.

White-throated Treecreeper (*Cormobates leucophaea*). Common e Aust. Forest, woodland, rainforest. Several distinct ssp. include *minor* (Little Treecreeper) n-e Aust.

White-browed Treecreeper (*Climacteris affinis*). Uncommon inland. Woodland, especially *Allocasuarina* and native pine. (Qld) Bowra, Eulo Bore, Currawinya

NP; (NSW) Binya SF; (Vic) Yarrara FFR, Timberoo FFR, Wyperfeld NP; (SA) Danggali CP, Gluepot R, Vulkathunha-Gammon Ranges, Mamungari CP; (WA) Cue, Kalgoorlie, Goongarrie NP.

Red-browed Treecreeper (*Climacteris erythrops*). Uncommon to mod common s-e Aust. Forest, alpine woodland. (Qld) D'Aguilar NP, Lamington NP; (NSW/ACT) Barrington Tops NP, Werrikimbe NP, Bruxner Park FR, Budderoo NP, Morton NP, Maxwell's FR, Kosciusko NP, Tidbinbilla NR, Namadgi NP; (Vic) Croajingolong NP, Dandenong Ranges NP, Toolangi SF, Glen Nayook R.

Brown Treecreeper (*Climacteris picumnus*). Common e Aust. Woodland. Two distinct ssp.: *picumnus* s-e Aust; *melanota* (Black Treecreeper) n-e Aust.

Black-tailed Treecreeper (*Climacteris melanura*). Uncommon to mod common n Aust. Woodland. (NT) Yellow Water, Arnhem Hwy, Bird Billabong, Nourlangie, Gunlom, Fergusson River, Nitmiluk (Katherine Gorge) NP, Chinaman Creek, Borroloola; (Qld) Mt Isa; (WA) Carnarvon, Karijini NP, Mornington Station, Purnululu (Bungle Bungles) NP.

Rufous Treecreeper (*Climacteris rufa*). Mod common to common s-w WA and w SA. Woodland, forest. (SA) Lake Gilles CP, Yumbarra CP, Mt Ive Station, Gawler Ranges; (WA) Bungendore Park, Dryandra Woodlands, Walpole-Nornalup NP, Boranup Karri Forest, Porongurup NP.

Spotted Catbird (*Ailuroedus melanotis*). Uncommon to mod common n-e Qld. Rainforest. (Qld) Kutini-Payamu (Iron Range) NP, Kuranda, Daintree, Kingfisher Park, Mt Lewis, Davies Creek NP, Mt Hypipamee NP, Curtain Fig Tree, Wongabel SF, Crater Lakes NP, Paluma Range NP.

Green Catbird (*Ailuroedus crassirostris*). Mod common e Aust. Temperate rainforest. (Qld) D'Aguilar NP, Lamington NP, Mt Tamborine NP, Great Sandy NP Conondale Ranges; (NSW) Barrington Tops NP, Copeland Tops, Myall Lakes NP, Werrikimbe NP, Boorganna NR, Dorrigo NP, Bruxner Park FR, Nightcap NP, Border Ranges NP, Dharug NP, Royal NP, Morton NP, Nowra.

Tooth-billed Bowerbird (*Scenopoeetes dentirostris*). Uncommon Wet Tropics. Upland rainforest (mostly 600-1400 m). (Qld) Mt Lewis, Mt Hypipamee NP, Wongabel SF, Crater Lakes NP, Paluma Range NP.

Golden Bowerbird (*Amblyornis newtonianus*). Rare Wet Tropics. Montane rainforest (above 900 m). (Qld) Mt Lewis, Davies Creek NP, Mt Hypipamee NP, Paluma Range NP.

Regent Bowerbird (*Sericulus chrysocephalus*). Uncommon to mod common s-e Qld, n-e NSW. Rainforest, forest. (Qld) Eungella Plateau, D'Aguilar NP (Mt Glorious), Lamington NP, Mt Tamborine NP, Great Sandy NP, Conondale Ranges; (NSW) Copeland Tops, Myall Lakes NP, Werrikimbe NP, Crowdy Bay NP, Boorganna NR, Wingham Brush NR, Dorrigo NP, Clarence River Estuary, Nightcap NP, Border Ranges NP.

Satin Bowerbird (*Ptilonorhynchus violaceus*). Mod common to common e Aust. Rainforest, forest.

Spotted Bowerbird (*Ptilonorhynchus maculatus*). Uncommon. Woodland. (Qld) Mt Isa, Bladensburg NP, Bowra, Lake Bindegolly, Eulo Bore, Currawinya NP, Sundown NP; (NSW) Macquarie Marshes, Pilliga Forest, Warrumbungle NP, Round Hill NR.

Western Bowerbird (*Ptilonorhynchus guttatus*). Mod common c and w Aust. Open woodland. (NT) Old Telegraph Station Historic R, Olive Pink BG, Alice Springs Desert Park, Simpson's Gap, Ormiston Gorge, Trephina Gorge, Watarrka (Kings Canyon) NP, Tnorala CR; (WA) Cue, Karijini NP.

Great Bowerbird (*Ptilonorhynchus nuchalis*). Common n Aust. Woodland. (NT) Charles Darwin NP, Howard Springs, Nourlangie, Gunlom, Nitmiluk (Katherine Gorge) NP; (Qld), Cooktown, +Kutini-Payamu (Iron Range) NP, Mt Carbine, Mt Molloy, Big Mitchell Creek, Cumberland Dam, Boodjamulla (Lawn Hill) NP, Lake Moondarra; (WA) Mirima (Hidden Valley) NP, Lake Argyle, Drysdale River, Purnululu (Bungle Bungles) NP.

Fawn-breasted Bowerbird (*Ptilonorhynchus cerviniventris*). Rare to uncommon n Cape York. Coastal rainforest, mangroves. (Qld) Kutini-Payamu (Iron Range) NP, Cowal Creek, Jardine River.

Superb Fairy-wren (*Malurus cyaneus*). Common s-e Aust and Tas. Forest, woodland, parks, gardens.

Splendid Fairy-wren (*Malurus splendens*). Common Woodland, Mulga, Mallee. Three distinct ssp.: *splendens* (Splendid Wren) s WA; *musgravi* (Turquoise Wren); *melonotus* (Black-backed Wren) arid e Aust.

Purple-crowned Fairy-wren (*Malurus coronatus*). Rare to uncommon n and n-w Aust. Cane grass and Pandanus thicket bordering water. (NT) Victoria River Crossing, Timber Creek, Jasper Gorge, Caranbirini CR, Borroloola; (Qld) Boodjamulla (Lawn Hill) NP, Gregory River; (WA) Lake Argyle, Drysdale River, Mornington Station.

Red-backed Fairy-wren (*Malurus melanocephalus*). Common Forest, woodland, coastal heath.

White-winged Fairy-wren (*Malurus leucopterus*). Mod common. Saltbush, open shrublands. Three distinct ssp.: *leuconotus* mainland Aust; *leucopterus* (Black and white Fairy-wren) Dirk Hartog Is.; *edouardi* (Black and white Fairy-wren) Barrow Is.

Variegated Fairy-wren (*Malurus lamberti*). Common Forest, woodland, open woodland. Four distinct ssp.: *lamberti* (Variegated Fairy-wren) coastal e Aust; *assimilis* (Purple-backed Fairy-wren) mainland Aust; *dulcis* (Lavender-flanked Fairy-wren) Top End and *rogersi* (also called Lavender-flanked Fairy-wren) Kimberley.

Lovely Fairy-wren (*Malurus amabilis*). Uncommon to mod common. Thickets, rainforest edge. (Qld) Cooktown, McIvor River Crossing, Kutini-Payamu (Iron Range) NP, Uningan NR, Wenlock River, Lockerbie Scrub, Jardine River NP, Mt Whitfield CP, Jack Barnes Mangrove Boardwalk, Cassowary House, Daintree, Kingfisher Park, Abattoir Swamp, Mt Lewis, Paluma Range NP.

Blue-breasted Fairy-wren (*Malurus pulcherrimus*). Mod common s-w SA and s WA. Open woodland, Mallee. (SA) Lake Gilles CP, Refuge Rocks, Lincoln NP, Kirton Point Caravan Park, Coffin Bay NP; (WA) Dryandra Woodlands, Stirling Range NP, Dragon Rocks NR, Fitzgerald River NP, Peak Charles NP, Cape Arid NP, Sandford Rocks NR, Eyre Bird Observatory, Shark Bay.

Red-winged Fairy-wren (*Malurus elegans*). Mod common to common s-w Aust. Karri forest, coastal heath. (WA) Wungong Dam, Serpentine NP, Walpole-Nornalup NP, Boranup Karri Forest, Porongurup NP, Waychinicup NP, Lake Powell NR, Stirling Range NP.

Southern Emu-wren (*Stipiturus malachurus*). Uncommon to mod common s-e and s-w Aust. Heathland, coastal heathland.

(Qld) Great Sandy NP; (NSW) Crowdy Bay NP, Maroubra, Royal NP, Barren Grounds NR, Budderoo NP, Morton NP, Booderee NP, Nowra, Ben Boyd NP, Nadgee NR; (Vic) Croajingolong NP, Cape Conran Coastal P, Marlo Coast R, Canni Creek R, Cape Liptrap Coastal P, Wilsons Promontory NP, Bunyip SP, Anglesea Heath, Great Otway NP; (Tas) Strahan, Melaleuca; (SA) Deep Creek CP, Cox Scrub CR, Flinders Chase NP, Mary Seymour CP, Coorong NP, Messent CP, Lincoln NP, Coffin Bay NP; (WA) Walpole-Nornalup NP, Lake Muir, Leeuwin-Naturaliste NP, Two Peoples Bay NR, Cheynes Beach, Betty's Beach, Waychinicup NP, Lake Powell NR, Stirling Range NP, Fitzgerald River NP, Cape Arid NP, Dirk Hartog Is., ssp. *hartogi.*

Mallee Emu-wren (*Stipiturus mallee*). Rare n-w Vic and e SA. Mallee in association with Spinifex. (Vic) Hattah-Kulkyne NP, Murray-Sunset NP; (SA) Ngarkat CP (very rare).

Rufous-crowned Emu-wren (*Stipiturus ruficeps*). Uncommon inland Aust. Spinifex. (NT) Santa Teresa Rd, Ellery Creek Big Hole, Ormiston Gorge, Newhaven Sanctuary; (Qld) Bladensburg NP, Lark Quarry CP, Opalton; (WA) Neale Junction NR.

Grey Grasswren (*Amytornis barbatus*). Uncommon e central Aust. Lignum, cane grass. Two distinct ssp.: *barbatus* (Bulloo) n-w NSW, s-w Qld; *diamantina* (Lake Eyre basin) n-e SA, s-w Qld. (Qld) Noccundra, Cooper Creek, Eyre Creek, Bulloo Overflow; (SA) Koolivoo Waterhole, Narriearra Station, Koonchera Dune.

Striated Grasswren (*Amytornis striatus*). Uncommon to mod common inland Aust. Four distinct ssp.: *whitei* (Pilbara Grasswren) Pilbara WA; *rowleyi* (Rusty Grasswren) c Qld; *striatus* (Striated Grasswren) NSW, n Vic, s-e SA; *oweni* (Sandhill Grasswren) c Aust. All four ssp. may be elevated to full species status in the future. Spinifex, Mallee woodland. (NT) Barrow Creek, Uluru-Kata Tjuta NP; (Qld) Bladensburg NP, Lark Quarry CP, Opalton; (Vic) Hattah-Kulkyne NP, Murray-Sunset NP; (SA) Danggali CP, Gluepot R, Billiatt CP; (WA) Neale Junction, Karijini NP.

Short-tailed Grasswren (*Amytornis merrotsyi*). Uncommon to mod common c SA. Recently elevated to full sp. status, separated from Striated Grasswren. Two distinct ssp.: *merrotsyi* (Flinders Ranges); *pedleri* (Gawler Ranges). Spinifex. (SA) Flinders Ranges NP, Arkaroola, Mt Ive Station.

White-throated Grasswren (*Amytornis woodwardi*). Rare Top End. Arnhem escarpment, spinifex. (NT) Gunlom, Yurmikmik, Plum Tree Creek.

Carpentarian Grasswren (*Amytornis dorotheae*). Rare n-w Qld, n-e NT. Spinifex, escarpment. (NT) Caranbirini CR (rare); (Qld) McNamara's Rd (Mt Isa), Barkly Hwy.

Thick-billed Grasswren (*Amytornis textilis*). Uncommon to mod common c Aust. Three distinct ssp.: *raglessi* (Flinders Ranges); *obscurior* (n-w NSW); *curnsamona* (Lake Frome basin). Cane grass, saltbush, rocky plains. (SA) Bimbowrie CP, Mt Lyndhurst Mine.

Western Grasswren (*Amytornis myall*). Mod common coastal central w WA, s-w SA. Two distinct ssp.: *myall* (Gawler Ranges); *textilis* (Shark Bay). Saltbush, rocky plains. Recently elevated to species, separated from Thick-billed Grasswren. (SA) Whyalla CP, Iron Knob Rd, Mt Ive Station; (WA) Shark Bay.

Dusky Grasswren (*Amytornis purnelli*). Mod common c Aust. Rocky grassland, spinifex, gorges. (NT) Santa Teresa Rd, Mt Gillen, Simpson's Gap, Ormiston

Gorge, Trephina Gorge, Newhaven Sanctuary, Watarrka (Kings Canyon) NP, Tnorala CR.

Kalkadoon Grasswren (*Amytornis ballarae*). Mod common Mt Isa region. Rocky plains, grassland. Recently elevated to full sp. status, separated from Dusky Grasswren. (Qld) Lake Moondarra, Mt Isa, Mica Creek (near Mt Isa).

Eyrean Grasswren (*Amytornis goyderi*). Uncommon c Aust. Sand dunes, cane grass. (NT) Old Andado Station; (Qld) Big Red; (SA) Strzelecki Track, Bollards Lagoon Rd, Coongie Lakes, Koonchera Dune, Lake Surprise Dune, Mungerannie Roadhouse.

Black Grasswren (*Amytornis housei*). Uncommon Kimberley. Spinifex, rocky gullies. (WA) Mitchell River NP, Bachsten Creek.

Eastern Bristlebird (*Dasyornis brachypterus*). Rare to uncommon e and s-e Aust. Coastal and mountain heathland. (Qld) Conondale Range NP (rare); (NSW) Border Range NP (rare), Barren Grounds NR, Budderoo NP, Morton NP, Booderee NP, Ben Boyd NP (rare), Nadgee NR; (Vic) Croajingolong NP (Howe Flat).

Western Bristlebird (*Dasyornis longirostris*) Rare to uncommon s-w WA. Coastal heathland. (WA) Two Peoples Bay NR, Cheynes Beach, Fitzgerald River NP.

Rufous Bristlebird (*Dasyornis broadbenti*). Uncommon to mod common s-e Aust. Three distinct ssp.: *broadbenti* s-w Vic, s-e SA; *caryochorus* s-w Vic; *littoralis* s-w WA (extinct). Coastal heathland, forest gullies. (Vic) Great Ocean Rd, Great Otway NP, Cape Nelson Lighthouse; (SA) Coorong NP, Robe.

Pilotbird (*Pycnoptilus floccosus*). Uncommon to mod common s-e Aust. Temperate rainforest, forest, woodland. (NSW/ACT) Dharug NP, Royal NP, Barren Grounds NR, Morton NP, Maxwell's FR, Kosciusko NP, Tidbinbilla NR; (Vic) Cabbage Tree Creek, Tarra-Bulga NP, Dandenong Ranges NP, Toolangi SF, Glen Nayook R.

Rockwarbler (*Origma solitaria*). Uncommon to mod common s-e NSW. Sandstone escarpment. (NSW) Royal NP, Morton NP, Bomaderry Creek RP, Ben's Walk (Nowra).

Fernwren (*Oreoscopus gutturalis*). Uncommon Wet Tropics. Montane rainforest (above 600 m). (Qld) Mt Lewis, Mt Hypipamee NP, Crater Lakes NP, Paluma Range NP.

Yellow-throated Scrubwren (*Sericornis citreogularis*). Common e Aust. Forest, rainforest. (Qld) Mt Lewis, Mt Hypipamee NP, Wongabel SF, Crater Lakes NP, Paluma Range NP, D'Aguilar NP, Lamington NP, Mt Tamborine NP; (NSW) Barrington Tops NP, Boorganna NR, Dorrigo NP, Nightcap NP, Royal NP, Budderoo NP, Morton NP, Bomaderry Creek RP.

White-browed Scrubwren (*Sericornis frontalis*). Common. Forest, temperate rainforest, woodland, parkland. Several distinct ssp. include: *maculatus; ashbyi, balstoni; rosinae* (Spotted Scrubwren) s Aust; *laevigaster* (Buff-breasted Scrubwren) coastal Qld; *frontalis* (White-browed Scrubwren) e Aust.

Tasmanian Scrubwren (*Sericornis humilis*). Common Tas. Forest, woodland. (Tas) Mt Wellington

Atherton Scrubwren (*Sericornis keri*). Uncommon Wet Tropics. Rainforest (above ~600 m). (Qld) Mt Lewis, Davies Creek NP, Mt Hypipamee NP, Wongabel SF, Crater Lakes NP, Paluma Range NP.

Large-billed Scrubwren (*Sericornis magnirostra*). Mod common to common e Aust. Forest, temperate rainforest.

Tropical Scrubwren (*Sericornis beccarii*). Mod common Cape York. Lowland rainforest. (Qld) Cooktown, McIvor River Crossing, Kutini-Payamu (Iron Range) NP, Wenlock and Pennefather River, Lockerbie Scrub, Jardine River NP.

Scrubtit (*Acanthornis magna*). Uncommon to mod common Tas. Forest. (Tas) Mt Wellington, Bruny Is., Snug Tiers NRA, Mt Field NP, Melaleuca, Birchs Inlet, King Is. (rare), Cradle Mountain-Lake St Clair NP.

Chestnut-rumped Heathwren (*Hylacola pyrrhopygia*). Uncommon s-e Aust. Forest, woodland, heathland. (Qld) Girraween NP; (NSW) Ku-ring-gai Chase NP, Royal NP, Barren Grounds NR, Morton NP, Ben Boyd NP; (Vic) Greater Bendigo NP, Heathcote-Graytown NP, Croajingolong NP, Anglesea Heath; (SA) Cleland CP, Morialta CP, Newland Head CP, Cox Scrub CR, Mt Remarkable, Telowie Gorge CP.

Shy Heathwren (*Hylacola cauta*). Uncommon to mod common s Aust. Mallee, coastal heathland. (NSW) Round Hill and Nombinnie NR, Cocoparra NP; (Vic) Hattah-Kulkyne NP, Bronzewing FFR, Wyperfeld NP, Little Desert NP, Greater Bendigo NP; (SA) Kangaroo Is., Coorong NP, Ngarkat CP, Innes NP, Gluepot R, Yumbarra CP; (WA) Fitzgerald River NP, Cape Arid NP, Peak Charles, Boorabbin NP, Goldfields Woodlands NP, Sandford Rocks NR, Eyre Bird Observatory.

Striated Fieldwren (*Calamanthus fuliginosus*). Mod common s-e Aust. Coastal heathland. (NSW) Morton NP, Ben Boyd NP; (Vic) Croajingolong NP, Marlo Coastal Cape Liptrap Coastal Park, Wilsons Promontory NP, Lake Victoria, Anglesea Heath, Deen Maar; (Tas) Strahan, Melaleuca, Narawntapu NP; (SA) Messent CP.

Rufous Fieldwren (*Calamanthus campestris*). Uncommon to mod common s Aust. Saltbush and bluebush shrublands. (Vic) Lake Tyrrell Wildlife R, Little Desert NP; (SA) Ngarkat CP, Bimbowrie CR, Mt Lyndhurst Mine, Montecollina Bore, Australian Arid Land BG, Whyalla CP, Lincoln NP, Vokes Hill Corner Rd, Nullarbor NP; (WA) Rawlinna Rd, Kalbarri NP, Cue, Shark Bay, Dirk Hartog Is., Carnarvon.

Western Fieldwren (*Calamanthus montanellus*). Uncommon to mod common w WA. Heathland, shrublands. Recently elevated to full sp. status, separated from Rufous Fieldwren. (WA) Stirling Range NP, Cheynes Beach, Two Peoples Bay NR, Lake Warden, Fitzgerald River NP, Cape Arid NP.

Redthroat (*Pyrrholaemus brunneus*). Uncommon to mod common. Mulga, Mallee. (NT) Old Telegraph Station Historic R, Kunoth Bore, Uluru-Kata Tjuta NP; (Qld) Bowra; (Vic) Wyperfeld NP; (SA) Mt Remarkable NP, Flinders Ranges NP, Vulkathunha-Gammon Ranges, Bimbowrie CR, Mt Lyndhurst Mine, Australian Arid Land BG, Whyalla CP; (WA) Wave Rock, Boorabbin NP, Goldfields Woodlands NP, Sandford Rocks NR, Neale Junction NR, Cape Arid NP, Cue, Kalbarri NP, Carnarvon.

Speckled Warbler (*Chthonicola sagittata*). Uncommon to mod common e Aust. Open woodland. (Qld) Lake Samsonvale, Durikai SF, Sundown NP; (NSW/ACT) Barraba, Pilliga Forest, Warrumbungle NP, Capertee Valley, Munghorn Gap NR, Cocoparra NP, Scheyville NP, Tidbinbilla NR; (Vic) Chiltern-Mt Pilot NP, Warby Range SP, Heathcote-Graytown NP, Dunach Nature CR, You Yangs.

Weebill (*Smicrornis brevirostris*). Common mainland Aust. Woodland, forest.

Brown Gerygone (*Gerygone mouki*). Common. Forest, rainforest. (Qld) Mt Hypipamee NP, Curtain Fig Tree, Wongabel SF, Great Sandy NP; (NSW) Barrington Tops NP, Royal NP, Budderoo NP, Nowra, Mimosa Rocks NP, Maxwell's FR; (Vic) Croajingolong NP, Cabbage Tree Creek, Fairy Dell Nature CR, Buchan Caves R, Dandenong Ranges NP, Toolangi SF, Glen Nayook R.

Norfolk Island Gerygone (*Gerygone modesta*). Common resident Norfolk Is.

Mangrove Gerygone (*Gerygone levigaster*). Mod common n and e Aust. Mangroves. (NT) Leanyer Sewage Ponds, East Point; (Qld) Townsville Common, Cape Hillsborough NP, Boondall Wetlands, Kedron Brook Wetlands, Tinchi Tamba Wetlands, Eagleby Wetlands, Great Sandy NP, Toorbul Shorebird Roost; (NSW) Ash Island, Crowdy Bay NP, Clarence River Estuary, Ballina; (WA) Wyndham, Derby, Broome Bird Observatory, Barred Creek.

Western Gerygone (*Gerygone fusca*). Mod common to common. Woodland, Mallee. (NT) Old Telegraph Station Historic R, Kunoth Bore, Mt Gillen, Simpson's Gap, Trephina Gorge, Newhaven Sanctuary, Erldunda, Uluru-Kata Tjuta NP; (Qld) Bowra, Durikai SF; (NSW) Barraba, Pilliga Forest, Round Hill NR, Cocoparra NP; (Vic) Murray Valley NP, Kerang Lakes, Warby Range SP, Terrick Terrick NP; (SA) Lincoln NP; (WA) Kings Park, Herdsman Lake, Rottnest Is., Wungong Dam, Bungendore Park, Dryandra Woodlands.

Dusky Gerygone (*Gerygone tenebrosa*). Common n-w WA. Mangroves, woodland. (WA) Lake MacLeod, Broome, Carnarvon.

Large-billed Gerygone (*Gerygone magnirostris*). Common n Aust. Mangroves, waterways, paperbark swamps. (NT) Charles Darwin Uni, Buffalo Creek, Bardedjilidji; (Qld) Cooktown, Lakefield NP, Uningan NR, Cairns Esplanade, Cairns BG, Jack Barnes Mangrove Boardwalk, Mossman River, Kingfisher Park, Wongabel SF, Townsville Common, Magnetic Island, Cape Hillsborough NP; (WA) Wyndham.

Green-backed Gerygone (*Gerygone chloronota*). Mod common n-w Aust. Forest, mangroves, paperbark swamps. (NT) Charles Darwin Uni, Buffalo Creek, East Point, Fogg Dam, Elsey NP, Groote Eylandt; (WA) Mitchell River NP.

Fairy Gerygone (*Gerygone palpebrosa*). Mod common e Qld. Rainforest, forest, mangroves. Two distinct ssp.: *personata* (Black-throated Gerygone) n Qld and Cape York; *flavida* (Fairy Gerygone) e Qld. (Qld) Kutini-Payamu (Iron Range) NP, Wenlock River, Lockerbie Scrub, Kingfisher Park, Big Mitchell Creek, Emerald Creek, Paluma Range NP, Cape Hillsborough NP, Great Sandy NP.

White-throated Gerygone (*Gerygone albogularis*). Mod common n, e and s-e Aust. Forest, woodland. (NT) Mamukala Wetlands, Bardedjilidji, Pine Creek, Elsey NP; (Qld) Musgrave, Big Mitchell Creek, Gulf Developmental Rd, Girraween NP, Sundown NP; (NSW) Crowdy Bay NP, Barraba, Pilliga Forest, Munghorn Gap NR, Hawkesbury River Valley, Bomaderry Creek RP, Maxwell's FR, Kosciusko NP; (Vic) Chiltern-Mt Pilot NP, Warby Range SP, Croajingolong NP, Fairy Dell Nature CR; (WA) Kununurra.

Slaty-backed Thornbill (*Acanthiza robustirostris*). Uncommon to mod common c and w Aust. Mulga shrublands. (NT) Kunoth Bore; (Qld) Eulo Bore, Kilcowera Station,

Currawinya NP; (SA) Mamungari CP; (WA) Neale Junction NR, Cue.

Striated Thornbill (*Acanthiza lineata*). Common e Aust. Forest, woodland.

Yellow Thornbill (*Acanthiza nana*). Common e Aust. Forest, *Acacia* woodland.

Yellow-rumped Thornbill (*Acanthiza chrysorrhoa*). Common s and inland Aust. Open woodland, wooded roadsides.

Chestnut-rumped Thornbill (*Acanthiza uropygialis*). Common s and inland Aust. Open woodland, Mallee, Mulga.

Buff-rumped Thornbill (*Acanthiza reguloides*). Mod common e and s-e Aust. Forest, woodland. (Qld) Durikai SF, Girraween NP; (NSW) Barraba, Pilliga Forest, Munghorn Gap NR, Scheyville NP; (Vic) Little Desert NP, Glenlee FFR, Chiltern-Mt Pilot NP, Greater Bendigo NP, Heathcote-Graytown NP, Dunach Nature CR, Anglesea Heath, Ironbark Basin Walk; (SA) Morialta CP, Kaiserstuhl CP.

Western Thornbill (*Acanthiza inornata*). Mod common s-w WA. Woodland, forest. (WA) Wungong Dam, Bungendore Park, Dryandra Woodlands, Stirling Range NP.

Slender-billed Thornbill (*Acanthiza iredalei*). Uncommon. Saltbush, samphire, heathland. Three distinct ssp.: *hedleyi* n-w Vic; *iredalei* s Aust (extinct n Aust); *rosinae* e SA. (Vic) Little Desert NP; (SA) Dry Creek Saltfields, Port Gawler CP, Ngarkat CP, Bimbowrie CP, Whyalla CP, Nullarbor NP; (WA) Goongarrie NP, Cocklebiddy-Rawlinna Rd. Carnarvon, Cue.

Tasmanian Thornbill (*Acanthiza ewingii*). Common Tas. Woodland, shrublands.

Inland Thornbill (*Acanthiza apicalis*). Common. Scrublands, Mallee, Mulga. (NT) Olive Pink BG, Kunoth Bore, Simpson's Gap, Watarrka (Kings Canyon) NP, Erldunda, Uluru-Kata Tjuta NP; (Qld) Mt Isa, Bladensburg NP, Opalton, Bowra; (NSW) Pilliga Forest, Warrumbungle NP, Round Hill NR, Cocoparra NP; (Vic) Hattah-Kulkyne NP, Yarrara FFR, Wyperfeld NP, Glenlee FFR, Greater Bendigo NP; (SA) Mt Remarkable NP, Gluepot R, Flinders Ranges NP, Vulkathunha-Gammon Ranges, Lake Gilles CP.

Brown Thornbill (*Acanthiza pusilla*). Common s and e Aust. Woodland, forest.

Mountain Thornbill (*Acanthiza katherina*). Mod common Wet Tropics. Montane rainforest (above 400 m). (Qld) Mt Lewis, Mt Hypipamee NP, Curtain Fig Tree, Wongabel SF, Crater Lakes NP, Paluma Range NP.

Southern Whiteface (*Aphelocephala leucopsis*). Common s Aust. Open plains, farmland, open woodland.

Chestnut-breasted Whiteface (*Aphelocephala pectoralis*). Rare c Aust. Gibber plains, arid lands. (SA) Old Mt Lyndhurst Mine, Stuart Hwy.

Banded Whiteface (*Aphelocephala nigricincta*). Uncommon c Aust. Arid lands, Mulga. (NT) Erldunda, Uluru-Kata Tjuta NP, Old Andado Station; (Qld) Birdsville; (SA) Strzelecki Track, Bollards Lagoon Rd, Birdsville Track; (WA) Cue.

Spotted Pardalote (*Pardalotus punctatus*). Common. Woodland, forest, Mallee, parkland. Two distinct ssp.: *punctatus* (Red-rumped Pardalote) s-e Aust; *xanthopygus* (Yellow-rumped Pardalote) s Aust.

Forty-spotted Pardalote (*Pardalotus quadragintus*). Rare Tas. Coastal forest. (Tas) Bruny Is., Partridge Is., Peter Murrell R, Maria Is., Flinders Is.

Red-browed Pardalote (*Pardalotus rubricatus*). Mod common n and c Aust. Woodland. (NT) Gunlom, Old Telegraph Station Historic R, Olive Pink BG, Ormiston Gorge, Watarrka (Kings Canyon) NP, Tnorala CR; (Qld) Gulf Developmental Rd, Boodjamulla (Lawn Hill) NP, Mt Isa, Bladensburg NP, Bowra; (NSW) Sturt NP; (SA) Vulkathunha-Gammon Ranges, Arkaroola, Strzelecki Track, Innamincka, Coongie Lakes; (WA) Lake Argyle, Kununurra, Purnululu (Bungle Bungles) NP.

Striated Pardalote (*Pardalotus striatus*). Common Woodland, forest. Several distinct ssp.: *striatus* (Yellow-tipped Pardalote) Tas, winter migrant to s-e Aust; *substriatus* (Striated Pardalote) s-e, c and w Aust; *ornatus* (Eastern Striated or Red-tipped Pardalote) s Aust; *melanocephalus* and *uropygialis* (Black-headed Pardalote) n Aust.

Eastern Spinebill (*Acanthorhynchus tenuirostris*). Common e Aust. Forest, heathland, woodland, gardens.

Western Spinebill (*Acanthorhynchus superciliosus*). Common s-w WA. Heathland, forest, woodland. (WA) Kings Park, Wungong Dam, Bungendore Park BR, Serpentine NP, Dryandra Woodlands, Lake Muir NR, Two Peoples Bay NR, Betty's Beach R, Cheynes Beach, Stirling Range NP, Fitzgerald River NP, Esperance, Peak Charles NP, Cape Arid NP.

Pied Honeyeater (*Certhionyx variegatus*). Rare to uncommon but nomadic and irruptive. Arid shrublands, arid woodland. (NT) Kunoth Bore, Santa Teresa Rd, Newhaven, Mereenie Loop Rd, Erldunda, Uluru-Kata Tjuta NP; (Qld) Bladensburg NP, Opalton, Eulo Bore, Kilcowera Station; (NSW) Sturt NP; (Vic) Goschen BR; (SA) Gluepot, Bimbowrie CR, Strzelecki Track, Australian Arid Land BG, Yumbarra CP; (WA) Yeo Lake NR, Jupiter Well, Cue, Kalbarri NP, Shark Bay.

Lewin's Honeyeater (*Meliphaga lewinii*). Common e Aust. Forest, rainforest.

Yellow-spotted Honeyeater (*Meliphaga notata*). Common n-e Qld. Forest, rainforest. (Qld) Cooktown, Kutini-Payamu (Iron Range) NP, Lockerbie Scrub, Cairns BG, Kuranda, Daintree, Kingfisher Park, Crater Lakes NP, Paluma Range NP, Townsville Common.

Graceful Honeyeater (*Meliphaga gracilis*). Common n-e Aust. Forest, rainforest. (Qld) Cooktown, Kutini-Payamu (Iron Range) NP, Uningan NR, Lockerbie Scrub, Cairns BG, Kuranda, Daintree, Kingfisher Park.

White-lined Honeyeater (*Meliphaga albilineata*). Uncommon Top End. Arnhem escarpment and gorges. (NT) Nourlangie, Bardedjilidji, Gunlom, Yurmikmik.

Kimberley Honeyeater (*Meliphaga fordiana*). Uncommon Kimberley. Sandstone escarpment and gorges. Recently elevated to full sp., separated from White-lined Honeyeater. (WA) Mitchell River NP.

Bridled Honeyeater (*Lichenostomus frenatus*). Mod common Wet Tropics. Rainforest. (Qld) Kingfisher Park (uncommon), Mt Lewis, Emerald Creek, Davies Creek NP, Wongabel SF, Crater Lakes NP, Paluma Range NP.

Eungella Honeyeater (*Lichenostomus hindwoodi*). Uncommon n-e Qld. Rainforest. Previously identified as Bridled Honeyeater, first recognised as a separate sp. 1983. (Qld) Eungella Plateau.

Yellow-faced Honeyeater (*Lichenostomus chrysops*). Common e Aust. Migratory (moving n winter). Woodland, forest.

Singing Honeyeater (*Lichenostomus virescens*). Common. Woodland, coastal heathland, arid shrublands.

Varied Honeyeater (*Lichenostomus versicolor*). Mod common n-e Qld. Mangroves, coastal woodland. (Qld) Portland Rd, Cairns Esplanade, Jack Barnes Mangrove Boardwalk, Townsville Common; (OIT) Torres Strait.

Mangrove Honeyeater (*Lichenostomus fasciogularis*). Mod common n-e and e Aust. Mangroves. (Qld) Ross River, Manly Boat Harbour, Great Sandy NP, Toorbul Shorebird Roost; (NSW) Clarence River Estuary.

White-gaped Honeyeater (*Lichenostomus unicolor*). Mod common n Aust. Riverine forest, mangroves. (NT) Charles Darwin Uni, Buffalo Creek, East Point, Howard Springs, Fogg Dam, Yellow Water, Gunlom, Groote Eylandt; (Qld) Townsville Common, Karumba, Boodjamulla (Lawn Hill) NP; (WA) Lake Argyle, the Grotto, Mitchell River NP, Drysdale River, Mornington Station.

Yellow Honeyeater (*Lichenostomus flavus*). Mod common n Qld. Riverine forest, mangroves. (Qld) Lakefield NP, Musgrave, Cairns BG, Cattana Wetlands, Abattoir Swamp, Emerald Creek, Mt Hypipamee NP, Townsville Common, Cape Hillsborough NP, Mackay Regional BG, Karumba.

White-eared Honeyeater (*Lichenostomus leucotis*). Common e Aust and s-w WA. Woodlands, forest, Mallee, coastal heathland.

Yellow-throated Honeyeater (*Lichenostomus flavicollis*). Common Tas. Woodland, forest.

Yellow-tufted Honeyeater (*Lichenostomus melanops*). Common s-e Aust. Forest, woodland. Distinct ssp. include Endangered *cassidix* (Helmeted Honeyeater). (Qld) Durikai SF, Girraween NP; (NSW) Hunter Valley Woodlands, Warrumbungle NP, Capertee Valley, Munghorn Gap NR, Ku-ring-gai Chase NP; (Vic) Bomaderry Creek RP, Chiltern-Mt Pilot NP, Warby Range SP, Greater Bendigo NP, Heathcote-Graytown NP, Yellingbo Nature CR (Helmeted Honeyeater), Dunach Nature CR.

Purple-gaped Honeyeater (*Lichenostomus cratitius*). Uncommon to mod common s Aust. Woodland, Mallee. (Vic) Bronzewing FFR, Wyperfeld NP, Little Desert NP, Greater Bendigo NP; (SA) Kangaroo Is., Coorong NP, Messent CP, Danggali CP, Billiatt CP, Innes NP, Lake Gilles CP, Lincoln NP, Coffin Bay NP; (WA) Stirling Range NP, Dragon Rocks NR, Fitzgerald River NP, Peak Charles NP, Cape Arid NP, Eyre Bird Observatory.

Grey-headed Honeyeater (*Lichenostomus keartlandi*). Mod common c and n-w Aust. Woodland, Mulga. (NT) Elsey NP, Caranbirini CR, Borroloola, Olive Pink BG, Mt Gillen, Simpson's Gap, Ellery Creek Big Hole, Ormiston Gorge, Trephina Gorge Nature Park, N'Dhala, Newhaven, Watarrka (Kings Canyon) NP, Uluru-Kata Tjuta NP; (Qld) Boodjamulla (Lawn Hill) NP, Mt Isa, Bladensburg NP, Opalton; (WA) Karijini NP, Tom Price, Jupiter Well, Purnululu (Bungle Bungles) NP.

Yellow-plumed Honeyeater (*Lichenostomus ornatus*). Mod common s Aust. Mallee woodland, arid woodland. (NSW) Round Hill NR, Nombinnie NR; (Vic) Hattah-Kulkyne NP, Murray-Sunset NP, Wyperfeld NP, Little Desert NP, Glenlee FFR, Greater Bendigo NP; (SA) Billiatt CR, Gluepot R, Flinders Ranges NP, Lake Gilles CP; (WA) Dryandra Woodlands, Stirling Range NP, Peak Charles NP, Cape Arid NP, Eyre Bird Observatory.

Grey-fronted Honeyeater (*Lichenostomus plumulus*). Uncommon to mod common. Woodland, Mallee, Mulga. (NT) Barkly

Tableland, Borroloola, Santa Teresa Rd, Trephina Gorge Nature Park, N'Dhala Gorge, Newhaven; (Qld) Cumberland Dam, Boodjamulla (Lawn Hill) NP, Mt Isa; (NSW) Round Hill NR, Nombinnie NR; (Vic) Murray-Sunset NP (rare); (SA) Danggali CP, Flinders Ranges NP, Telowie Gorge CP, Vulkathunha-Gammon Ranges NP, Mt Ive Station; (WA) Neale Junction NR, Kalbarri NP, Purnululu (Bungle Bungles) NP.

Fuscous Honeyeater (*Lichenostomus fuscus*). Mod common to common e Aust. Woodland, forest. (Qld) Lockyer Valley, Sheep Station Creek CP, Durikai SF; (NSW) Hunter Valley Woodlands, Barraba, Capertee Valley, Munghorn Gap NR, Ku-ring-gai Chase NP, Scheyville NP; (Vic) Chiltern-Mt Pilot NP, Warby Range SP, Greater Bendigo NP.

Yellow-tinted Honeyeater (*Lichenostomus flavescens*). Mod common to common n Aust. Woodland. (NT) Tiwi Is., Fergusson River, Elsey NP, Gregory NP, Jasper Gorge, Caranbirini CR, Borroloola; (Qld) Lakefield NP, Cumberland Dam, Karumba, Boodjamulla (Lawn Hill) NP, Mt Isa; (WA) Purnululu (Bungle Bungles) NP, Lake Argyle, Mitchell River NP, Drysdale River.

White-plumed Honeyeater (*Lichenostomus penicillatus*). Common. Woodland, forest, creek lines, parks, gardens.

White-fronted Honeyeater (*Purnella albifrons*). Uncommon to mod common inland Aust. Nomadic. Woodland, shrublands. (NT) Elsey NP, Kunoth Bore, Trephina Gorge, N'Dhala Gorge, Newhaven, Erldunda; (Qld) Bowra, Currawinya NP; (NSW) Round Hill NR, Nombinnie NR, Mutawintji NP; (Vic) Bronzewing FFR, Gama BR, Wyperfeld NP, Little Desert NP, Glenlee FFR, Goschen BR, Gama R, Greater Bendigo NP; (SA) Messent CP, Gluepot R, Billiatt CP, Flinders Ranges NP, Australian Arid Land BG, Lake Gilles CP; (WA) Boorabbin NP, Goldfields Woodlands NP, Eyre Bird Observatory, Neale Junction NR, Cue, Kalbarri NP, Jupiter Well, Shark Bay.

Bell Miner (*Manorina melanophrys*). Common s-e Aust. Forest, woodland, parkland. (Qld) Conondale Ranges, Mt Coot-tha Forest; (NSW) Border Ranges NP, Barrington Tops NP, Capertee Valley, Royal NP, Mimosa Rocks NP, (Vic) Croajingolong NP, Buchan Caves Reserve, Melbourne Zoo, CERES, Yarra Bend Park, Yellingbo Nature CR.

Noisy Miner (*Manorina melanocephala*). Common e Aust. Woodland, gardens, parkland.

Black-eared Miner (*Manorina melanotis*). Rare s-e Aust, Endangered. Mallee woodland. (Vic) Murray-Sunset NP, Bronzewing FFR; (SA) Danggali CP, Billiatt CP, Gluepot R.

Yellow-throated Miner (*Manorina flavigula*). Common Woodland, Mallee, Mulga.

Spiny-cheeked Honeyeater (*Acanthagenys rufogularis*). Common. Woodland, Mallee, coastal shrublands.

Western Wattlebird (*Anthochaera lunulata*). Uncommon to mod common s-w WA. Heathland, coastal woodland, gardens. (WA) Bungendore Park, Two Peoples Bay NR, Stirling Range NP, Fitzgerald River NP, Lake Warden, Monjingup Lake NR, Cape Arid NP.

Little Wattlebird (*Anthochaera chrysoptera*). Common s-e Aust. Woodland, coastal heathland, gardens.

Regent Honeyeater (*Anthochaera phrygia*). Rare to very rare s-e Aust, range contracting, now Critically Endangered. Forest, woodland. (Qld) Durikai SF; (NSW) Hunter Valley Woodlands, Barraba, Warrumbungle NP, Capertee Valley, Munghorn Gap NR, Londonderry; (Vic) Chiltern-Mt Pilot NP.

Red Wattlebird (*Anthochaera carunculata*). Common s Aust. Forest, woodland, parklands, gardens.

Yellow Wattlebird (*Anthochaera paradoxa*). Common Tas. Coastal heathland, woodland, gardens.

Brown-backed Honeyeater (*Ramsayornis modestus*). Common n-e Qld. Paperbark woodland, forest and rainforest edges, mangroves. (Qld) Cooktown, Lakefield NP, Kutini-Payamu (Iron Range) NP, Uningan NR, Lockerbie Scrub, Cairns BG, Abattoir Swamp, Townsville Common.

Bar-breasted Honeyeater (*Ramsayornis fasciatus*). Mod common to common n Aust. Paperbark woodland, creek lines. (NT) Charles Darwin Uni, East Point, Howard Spring, Fogg Dam, Yellow Water, Pine Creek, Victoria River, Nitmiluk (Katherine Gorge) NP, Elsey NP, Caranbirini CR, Borroloola, Groote Eylandt; (Qld) Musgrave, Mackay Regional BG, Boodjamulla (Lawn Hill) NP; (WA) the Grotto, Mitchell River NP, Drysdale River, Mornington Station.

Rufous-banded Honeyeater (*Conopophila albogularis*). Mod common far n Aust. Woodland, swampland, mangroves. (NT) Buffalo Creek, Howard Springs, Fogg Dam, Nitmiluk (Katherine Gorge) NP; (Qld) Lakefield NP.

Rufous-throated Honeyeater (*Conopophila rufogularis*). Mod common to common n Aust. Woodland, paperbark forest, mangroves. (NT) Yellow Water, Fergusson River, Nitmiluk (Katherine Gorge) NP, Chinaman Creek, Timber Creek, Caranbirini CR; (Qld) Cumberland Dam, Karumba, Boodjamulla, Mt Isa, Bladensburg NP; (WA) Mirima (Hidden Valley) NP, Mitchell River NP, Drysdale River, Purnululu (Bungle Bungles) NP.

Grey Honeyeater (*Conopophila whitei*). Rare c and w Aust. Mulga, arid shrublands. (NT) Olive Pink BG, Kunoth Bore; (WA) Cue, Paynes Find, Geraldton, Mt Magnet Rd, Great Northern Hwy, Wanjarri NR, Tom Price, Karijini NP.

Crimson Chat (*Epthianura tricolor*). Uncommon to mod common. Nomadic. Open woodland, arid shrublands. (NT) Jasper Gorge, Alice Springs Waste Stabilisation Ponds, Kunoth Bore, Santa Teresa Rd, Mt Gillen, Trephina Gorge, Newhaven Sanctuary, Watarrka (Kings Canyon) NP, Mereenie Loop Rd, Old Andado Station, Boodjamulla (Lawn Hill) NP; (Qld) Bladensburg NP, Bowra, Birdsville; (NSW) Cocoparra NP, Oolambeyan NP, Sturt NP; (Vic) Wyperfeld NP, Goschen BR; (SA) Mt Lyndhurst Mine, Bollards Lagoon Rd, Australian Arid Land BG, Mt Ive Station, Mamungari CP; (WA) Goongarrie NP, Eyre Bird Observatory, Cue, Lake MacLeod, Jupiter Well, Purnululu (Bungle Bungles) NP.

Orange Chat (*Epthianura aurifrons*). Uncommon to mod common. Nomadic. Shrublands, saltbush plains, salt lake edges, gibber plain. (NT) Borroloola, Alice Springs Waste Stabilisation Ponds, Newhaven Sanctuary, Erldunda, Mac Clark CR; (Qld) Bowra, Lake Bindegolly, Birdsville; (NSW) Chat Alley (Round Hill NR), Oolambeyan NP, Kinchega NP, Sturt NP; (Vic) Neds Corner, Lake Tyrrell, Lake Tutchewop, Hird Swamp; (SA) Gluepot R, Montecollina Bore, Strzelecki Track, Birdsville Track, Vokes Hill Corner Rd, Nullarbor NP; (WA) Cocklebiddy-Rawlinna Rd, Eyre Bird Observatory.

Yellow Chat (*Epthianura crocea*). Rare to uncommon, scattered sites n and c Aust.

Coastal swampland, inland wetland, bores. (NT) Knuckey Lagoons CR, Barkly Tableland, Connells Lagoon CR; (Qld) Mt Isa, Birdsville Bore, Marmor, Tortilla Plain (ssp. *Macgregor*); (WA) Lake Argyle, Parry Lagoons NR, Broome (Roebuck Plains).

White-fronted Chat (*Epthianura albifrons*). Common s Aust, uncommon and declining in coastal NSW. (NSW) Hunter River Estuary, Sydney Olympic Park (rare); (Vic) Kerang Lakes, Marlo Coastal, Wilsons Promontory NP, Western Treatment Plant, Bellarine Peninsula; (Tas) Strahan, Arthur River, Waterhouse CA, King Is.; (SA) Patawalonga Creek, Dry Creek Saltfields, Port Gawler CP, Murray Lagoon, Australian Arid Land BG, Whyalla CP; (WA) Rottnest Is., Mandurah, Eyre Bird Observatory.

Gibberbird (*Ashbyia lovensis*). Uncommon to mod common inland e Aust. Gibber and open plains. (NT) Old Andado Station; (Qld) Birdsville, Diamantina NP; (NSW) Sturt NP; (SA) Boolcoomatta R, Mt Lyndhurst Mine (rare), Strzelecki Track, Innamincka, Coongie Lakes, Birdsville Track.

Black Honeyeater (*Sugomel niger*). Uncommon to mod common. Nomadic. Open woodland, arid woodland. (NT) Ilparpa Swamp, Kunoth Bore, Ellery Creek Big Hole, Tnorala CR, Erldunda, Uluru-Kata Tjuta NP; (Qld) Bladensburg NP, Opalton, Bowra, Lake Bindegolly, Eulo Bore, Kilcowera Stations, Currawinya NP; (NSW) Round Hill NR, Nombinnie NR, Oolambeyan NP, Sturt NP; (Vic) Gama BR, Wyperfeld NP, Glenlee FFR, Goschen BR, Tresco West BR, Greater Bendigo NP; (SA) Danggali CP, Gluepot R, Coongie Lakes, Australian Arid Land BG; (WA) Yeo Lake NR, Jupiter Well, Shark Bay.

Dusky Honeyeater (*Myzomela obscura*). Common n and n-e Aust. Forest, rainforest, mangroves. (NT) Charles Darwin Uni., Howard Springs, Fogg Dam, Yellow Water, Nourlangie; (Qld) Cooktown, Lakefield NP, Musgrave, Cairns BG, Daintree, Kingfisher Park, Yungaburra, Crater Lakes NP, Cape Hillsborough NP, Eungella Plateau, Great Sandy NP, Conondale Ranges.

Red-headed Honeyeater (*Myzomela erythrocephala*). Uncommon to mod common far n Aust. Mangroves. (NT) Charles Darwin Uni, Buffalo Creek, Charles Darwin NP, Leanyer Sewage Ponds, Fogg Dam, Tiwi Is., Borroloola; (Qld) Kutini-Payamu (Iron Range) NP, Weipa, Jardine River, Karumba; (WA) Wyndham, Broome Bird Observatory, Barred Creek.

Scarlet Honeyeater (*Myzomela sanguinolenta*). Common e Aust. Woodland, forest, mangroves.

Green-backed Honeyeater (*Glycichaera fallax*). Uncommon n Cape York. Rainforest, monsoon forest. (Qld) Kutini-Payamu (Iron Range) NP.

Tawny-crowned Honeyeater (*Glyciphila melanops*). Mod common s Aust. Coastal and inland heathland, open woodland. (NSW) Yuraygir NP, Ku-ring-gai Chase NP, Royal NP, Budderoo NP, Booderee NP, Mimosa Rocks NP, Ben Boyd NP, Nadgee NR; (Vic) Wyperfeld NP, Little Desert NP, Greater Bendigo NP, Croajingolong NP, Cape Liptrap Coastal Park, Anglesea Heath; (Tas) Strahan, Waterhouse CA, Bruny Is.; (SA) Newland Head CP, Kangaroo Is., Messent CP, Billiatt CP, Innes NP, Flinders Ranges NP, Lincoln NP, Coffin Bay NP; (WA) Kings Park, Dryandra Woodlands, Two Peoples Bay NR, Cheynes Beach, Stirling Ranges NP, Fitzgerald River NP, Cape Arid NP, Goldfields Woodlands NP.

Banded Honeyeater (*Cissomela pectoralis*). Uncommon to mod common n Aust. Nomadic. Woodland, forest. (NT) Charles

Darwin NP, Mamukala Wetlands, Yellow Water, Nourlangie, Gunlom, Nitmiluk (Katherine Gorge) NP, Elsey NP, Victoria River, Jasper Gorge, Caranbirini CR; (Qld) Lakefield NP, Musgrave, Uningan NR, Jardine River, Mt Carbine, Big Mitchell Creek, Cumberland Dam, Boodjamulla (Lawn Hill) NP; (WA) Kununurra, Mornington Station, Mitchell River NP, Drysdale River.

Brown Honeyeater (*Lichmera indistincta*). Common (abundant) except s-e Aust. Woodland, forest, mangroves, gardens.

Crescent Honeyeater (*Phylidonyris pyrrhopterus*). Mod common s-e Aust. Forest. (NSW/ACT) Barrington Tops NP, Barren Grounds NR, Budderoo NP, Morton NP, Mimosa Rocks NP, Maxwell's FR, Kosciusko NP, Australian National BG, Tidbinbilla NR; (Vic) Croajingolong NP, Cabbage Tree Creek, Tarra-Bulga NP, Cape Liptrap CP, Wilsons Promontory NP, Dandenong Ranges NP, Toolangi SF, Bunyip SP, Great Otway NP; (SA) Cleland CP, Morialta CP, Kaiserstuhl CP, Kangaroo Is.; (Tas) Mt Wellington, Waterworks Res, Bruny Is., Truganini Res, Peter Murrell Res, Snug Tiers NRA, Meehan Range NRA, Mt Field NP, Melaleuca, Strahan, Cradle Mountain–Lake St Clair NP, Freycinet NP, King Is., Flinders Is.

New Holland Honeyeater (*Phylidonyris novaehollandiae*). Common s Aust. Woodland, coastal heathland, forest edges, parkland, gardens.

White-cheeked Honeyeater (*Phylidonyris niger*). Mod common to common e and w Aust. Woodland, coastal heathland, forest edges. (Qld) Abattoir Swamp, Davies Creek NP, Hasties Swamp, Wongabel SF, Paluma Range NP, Inskip Point, Great Sandy NP; (NSW) Hunter Wetlands Centre, Myall Lakes NP, Crowdy Bay NP, Ballina, Ku-ring-gai Chase NP, Royal NP, Barren Grounds NR, Booderee NP; (WA) Kings Park, Dryandra Woodlands, Bungendore Park, Two Peoples Bay NR, Cheynes Beach, Cape Arid NP, Eyre Bird Observatory.

White-streaked Honeyeater (*Trichodere cockerelli*). Uncommon Cape York. Woodland, rainforest edges, swampland. (Qld) Cooktown, McIvor River Crossing, Kutini-Payamu (Iron Range) NP, Lockerbie Scrub, Cowal Creek, Jardine River NP.

Black-chinned Honeyeater (*Melithreptus gularis*). Uncommon to mod common. Forest, woodland, open woodland. Two distinct ssp.: *gularis* e Aust; *laetior* (Golden-backed Honeyeater) NT, and n WA. (NT) Barkly Tableland, Pine Creek, Ormiston Gorge; (Qld) Boodjamulla (Lawn Hill) NP, Mt Isa, Lockyer Valley, Sheep Station Creek CP, Durikai SF, Sundown NP; (NSW) Hunter Valley, Barraba, Pilliga Forest, Warrumbungle NP, Capertee Valley, Munghorn Gap NP; (Vic) Chiltern-Mt Pilot NP, Warby Range SP, Terrick Terrick NP, Greater Bendigo NP, Heathcote-Graytown NP, Dunach Nature CR, You Yangs RP; (SA) Mary Seymour CP, Mt Remarkable NP; (WA) Jupiter Well, Mornington Station, the Grotto.

Strong-billed Honeyeater (*Melithreptus validirostris*). Common Tas. Forest, woodland. (Tas) Waterworks R, Bruny Is., Truganini R, Peter Murrell R, Snug Tiers NRA, Maria Is., Meehan Ranges, Mt Field NP, Melaleuca, Cradle Mountain-Lake St Clair NP, King Is., Flinders Is.

Brown-headed Honeyeater (*Melithreptus brevirostris*). Common s mainland Aust. Woodland, forest, Mallee.

White-throated Honeyeater (*Melithreptus albogularis*). Common n and n-e Aust. Woodland. (NT) Charles Darwin Uni.,

Howard Springs, Mamukala Wetlands, Yellow Water, Nourlangie, Gunlom, Elsey NP, Caranbirini CR; (Qld) Cooktown, Lakefield NP, Kutini-Payamu (Iron Range) NP, Cairns BG, Abattoir Swamp, Mt Carbine, Emerald Creek, Hasties Swamp, Townsville Common, Cape Hillsborough NP, Boodjamulla (Lawn Hill) NP, Eagleby Wetlands, Sheep Station Creek CP, Great Sandy NP; (WA) Mirima (Hidden Valley) NP, Lake Argyle, Purnululu (Bungle Bungles) NP.

White-naped Honeyeater (*Melithreptus lunatus*). Common e Aust. Woodland, forest.

Western White-naped Honeyeater (*Melithreptus chloropsis*). Also known as Swan River Honeyeater. Common, s-w WA. Woodland, forest. (WA) Wungong Dam, Lake McLarty, Bungendore Park, Serpentine NP, Dryandra Woodlands, Walpole-Nornalup NP, Lake Muir, Boranup Karri Forest, Cape Arid NP.

Black-headed Honeyeater (*Melithreptus affinis*). Common Tas. Forest, woodland, gardens. (Tas) Mt Wellington, Truganini R, Peter Murrell R, Snug Tiers NRA, Maria Is., Meehan Ranges, Mt Field NP, Melaleuca, Cradle Mountain-Lake St Clair NP, King Is., Flinders Is.

Blue-faced Honeyeater (*Entomyzon cyanotis*). Common n and e Aust. Forest, woodland, roadside vegetation.

Helmeted Friarbird (*Philemon buceroides*). Mod common Top End, NT. Woodland, forest, mangroves, rocky gorges. (NT) Tiwi Is., Charles Darwin NP, Palmerston Sewage Works, Leanyer Sewage Ponds, East Point, Gunlom, Nourlangie, Yurmikmik.

Hornbill Friarbird (*Philemon yorki*). Mod common n-e Qld. Woodland, forest, mangroves. (Qld) Cairns Esplanade, Jack Barnes Mangrove Boardwalk, Mossman River, Yungaburra, Magnetic Is. Recently elevated to full sp. status, separated from Helmeted Friarbird.

Silver-crowned Friarbird (*Philemon argenticeps*). Mod common n Aust. Woodland, forest, mangroves. (NT) Charles Darwin NP, East Point, Howard Springs, Tiwi Is., Yellow Water, Nourlangie, Gunlom, Pine Creek, Nitmiluk (Katherine Gorge) NP, Chinaman Creek, Victoria River, Caranbirini CR; (Qld) Kutini-Payamu (Iron Range) NP, Boodjamulla (Lawn Hill) NP, Mt Isa; (WA) Mornington Station, the Grotto, Derby, Drysdale River, Lake Argyle, Purnululu (Bungle Bungles) NP.

Noisy Friarbird (*Philemon corniculatus*). Common e mainland Aust. Woodland, forest.

Little Friarbird (*Philemon citreogularis*). Common n and e mainland Aust. Woodland, forest.

Macleay's Honeyeater (*Xanthotis macleayanus*). Mod common Wet Tropics. Rainforest, mangroves. (Qld) Kuranda, Daintree, Kingfisher Park, Curtain Fig Tree, Wongabel SF, Crater Lakes NP, Paluma Range NP.

Tawny-breasted Honeyeater (*Xanthotis flaviventer*). Mod common n Cape York and n Torres Strait. Rainforest, forest, mangroves. (Qld) Kutini-Payamu (Iron Range) NP, Uningan NR, Wenlock River, Lockerbie Scrub, Jardine River NP; (OIT) Boigu Is., Saibai Is.

Striped Honeyeater (*Plectorhyncha lanceolata*). Uncommon to mod common e Aust. Forest, woodland. (Qld) Bowra, Currawinya NP, Eagleby Wetlands, Sundown NP; (NSW) Crowdy Bay NP, Clarence River

Estuary, Barraba, Pilliga Forest, Capertee Valley, Munghorn Gap NR, Round Hill NR, Cocoparra NP, Kinchega NP, Mutawintji NP; (Vic) Hattah-Kulkyne NP, Neds Corner, Yarrara FFR, Wyperfeld NP; (SA) Gluepot R.

Painted Honeyeater (*Grantiella picta*). Rare to uncommon e Aust. Woodland, forest. (Qld) Bladensburg NP, Bowra; (NSW) Hunter Valley, Pilliga Forest, Capertee Valley, Binya SF; (Vic) Chiltern-Mt Pilot NP, Warby Range SP, Terrick Terrick NP, Dunach NCR.

Grey-crowned Babbler (*Pomatostomus temporalis*). Mod common n Aust, rare to uncommon s Aust. Woodland, farmland. (NT) Yellow Water, Bardedjilidji, Pine Creek, Jasper Gorge, Old Telegraph Station, Olive Pink BG, Alice Springs Desert Park, Ormiston Gorge; (Qld) Mt Carbine, Big Mitchell Creek, Mareeba Wetlands, Davies Creek NP, Cumberland Dam, Gregory River, Mt Isa, Bowra, Eulo Bore, Currawinya NP, Lockyer Valley, Sundown NP; (NSW) Hunter Valley, Pilliga Forest, Capertee Valley, Round Hill NR, Lake Cargelligo; (Vic) Loddon Weir, Back Swamp, Chiltern-Mt Pilot NP, Terrick Terrick NP.

Hall's Babbler (*Pomatostomus halli*). Uncommon to mod common inland s Qld and n NSW. *Acacia* shrublands, especially Mulga. (Qld) Bladensburg NP, Opalton, Bowra, Eulo Bore, Currawinya NP; (NSW) Mutawintji NP (rare).

White-browed Babbler (*Pomatostomus superciliosus*). Common mainland Aust. Woodland, forest, shrublands, Mulga.

Chestnut-crowned Babbler (*Pomatostomus ruficeps*). Uncommon to mod common inland s-e Aust. Mulga, Mallee, open woodland, saltbush plains. (Qld) Bowra, Lake Bindegolly, Eulo Bore, Currawinya NP; (NSW) Kinchega NP, Mutawintji NP, Sturt NP; (Vic) Hattah-Kulkyne NP, Terrick Terrick NP (rare); (SA) Gluepot R, Flinders Ranges NP, Mt Lyndhurst Mine, Bollards Lagoon Rd, Innamincka, Coongie Lakes.

Australian Logrunner (*Orthonyx temminckii*). Uncommon to mod common e Aust. Subtropical rainforest. (Qld) D'Aguilar NP, Lamington NP, Mt Tamborine NP, Conondale Ranges; (NSW) Copeland Tops, Werrikimbe NP, Boorganna NR, Dorrigo NP, Bruxner Park FR, Nightcap NP, Border Ranges NP, Nowra (Grotto Walk).

Chowchilla (*Orthonyx spaldingii*). Uncommon to mod common Wet Tropics. Tropical rainforest. (Qld) Kuranda, Mt Lewis, Davies Creek NP, Mt Hypipamee NP, Wongabel SF, Crater Lakes NP, Paluma Range NP.

Spotted Quail-thrush (*Cinclosoma punctatum*). Uncommon s-e Aust. Forest, woodland. (Qld) Mt Coot-tha Forest, Lamington NP, Girraween NP; (NSW/ACT) Pilliga Forest, Capertee Valley, Munghorn Gap NR, Dharug NP, Morton NP, Tidbinbilla NR, Namadgi NP; (Vic) Chiltern-Mt Pilot NP, Cabbage Tree Creek, Canni Creek R, Toolangi SF; (Tas) Mt Wellington, Bicheno region, Freycinet NP. Recently discovered isolated population near Ravenshoe, n Qld.

Chestnut Quail-thrush (*Cinclosoma castanotum*). Mod common s Aust. Mallee, Mulga, woodland. (NSW) Round Hill NR, Nombinnie NR; (Vic) Hattah-Kulkyne NP, Timberoo FFR, Bronzewing FFR, Wyperfeld NP; (SA) Danggali CP, Billiatt CP, Gluepot R, Lake Gilles CP, Lincoln NP, Yumbarra CP; (WA) Boorabbin NP, Goldfields Woodlands NP, Cape Arid NP, Eyre Bird Observatory.

Cinnamon Quail-thrush (*Cinclosoma cinnamomeum*). Mod common c Aust. Gibber plains, arid lands. (NT) Old Andado

Station; (NSW) Sturt NP; (SA) Bimbowrie CR, Boolcoomatta R, Mt Lyndhurst Mine, Strzelecki Track, Bollards Lagoon Rd, Coongie Lakes, Birdsville Track.

Nullarbor Quail-thrush (*Cinclosoma alisteri*). Mod common w SA, s-e WA. Found Nullarbor plains. (SA) Vokes Hill Corner Rd, Nullarbor NP; (WA) Cocklebiddy-Rawlinna Rd, Rawlinna, Naretha.

Chestnut-breasted Quail-thrush (*Cinclosoma castaneothorax*). Mod common inland s Qld and n NSW. Mulga, Mallee. (Qld) Bowra, Eulo Bore, Kilcowera Station, Currawinya NP.

Western Quail-thrush (*Cinclosoma marginatum*). Mod common c WA. Mulga. (WA) Goongarrie NP, Cue.

Eastern Whipbird (*Psophodes olivaceus*). Common e Aust. Forest, woodland.

Western Whipbird (*Psophodes nigrogularis*). Rare s SA and s-w WA. Dense coastal heath, Mallee. Four distinct ssp.: *nigrogularis* (Western Whipbird) s-w WA; *oberon* s-w WA; *leucogaster* (Mallee Whipbird) SA, n-e Vic (where probably extinct); *lashmari* Kangaroo Is. (SA) Kangaroo Is., Ngarkat CP, Innes NP, Lincoln NP; (WA) Cheynes Beach, Betty's Beach, Two Peoples Bay NR, Stirling Range NP, Fitzgerald River NP.

Chirruping Wedgebill (*Psophodes cristatus*). Uncommon to mod common, c Aust. Open woodland, shrublands. (Qld) Bowra, Eulo Bore; (NSW) Kinchega NP, Mutawintji NP, Sturt NP; (SA) Flinders Ranges NP, Vulkathunha-Gammon Ranges, Bimbowrie CR, Boolcoomatta R, Mt Lyndhurst Mine, Strzelecki Track, Innamincka, Coongie Lakes, Australian Arid Land BG.

Chiming Wedgebill (*Psophodes occidentalis*). Uncommon to mod common c and w Aust. Shrubland, open plains. (NT) Santa Teresa Rd, Erldunda, Uluru-Kata Tjuta NP; (WA) Yeo Lake NR, Shark Bay, Carnarvon, Lake MacLeod, Cue.

Varied Sittella (*Daphoenositta chrysoptera*). Mod common. Woodland, forest. Five distinct ssp.: *chrysoptera* (Orange-winged Sittella) s-e Aust; *pileata* (Black-capped Sittella) s-w Aust; *striata* (Streaked Sittella) n-e Aust; *leucocephala* (White-headed Sittella) e Aust; *leucoptera* (White-winged Sittella) n Aust.

Ground Cuckoo-shrike (*Coracina maxima*). Rare to uncommon. Open woodland, arid shrublands. (NT) Elsey NP, Victoria River, Barkly Tableland, Kunoth Bore, Barrow Creek, Watarrka (Kings Canyon) NP, Uluru-Kata Tjuta NP, Old Andado Station; (Qld) Gulf Developmental Rd, Boodjamulla (Lawn Hill) NP, Bowra, Lockyer Valley, Sundown NP; (NSW) Barraba, Sturt NP; (SA) Strzelecki Track, Bollards Lagoon Rd, Mt Ive Station; (WA) Neale Junction NR, Goongarrie NP, Eyre Bird Observatory, Cocklebiddy-Rawlinna Rd, Cue.

Black-faced Cuckoo-shrike (*Coracina novaehollandiae*). Common Woodland, forest.

White-bellied Cuckoo-shrike (*Coracina papuensis*). Common n Aust, uncommon to mod common s Aust. Woodland, forest.

Barred Cuckoo-shrike (*Coracina lineata*). Uncommon to mod common e and n-e Aust. Rainforest, forest. (Qld) Cooktown, Kutini-Payamu (Iron Range) NP, Kingfisher Park, Mt Lewis, Crater Lakes NP, Mission Beach, Paluma Range NP, Eungella Plateau, D'Aguilar NP (Mt Glorious), Lake Samsonvale, Conondale Ranges, Werrikimbe NP; (NSW) Crowdy Bay NP (rare), Bruxner Park FR (rare).

Cicadabird (*Coracina tenuirostris*). Mod common n Aust, uncommon s Aust. Woodland, forest. (NT) Darwin BG, Palmerston Sewage Works, Nourlangie; (Qld) Daintree, Kingfisher Park, Mareeba Wetlands, Eungella Plateau, Mt Coot-tha Forest, D'Aguilar NP, Conondale Ranges, Girraween NP; (NSW/ACT) Copeland Tops, Mungo Brush, Crowdy Bay NP, Border Ranges NP, Barraba, Munghorn Gap NR, Dharug NP, Royal NP, Bomaderry Creek RP Mimosa Rocks NP, Maxwell's FR, Tidbinbilla NR; (Vic) Croajingolong NP, Fairy Dell Nature CR, Bunyip SP.

White-winged Triller (*Lalage sueurii*). Common. Migrates n in winter. Woodland, grassy woodland.

Varied Triller (*Lalage leucomela*). Common. Forest, rainforest, woodland n and e Aust. (NT) Darwin BG, Charles Darwin Uni., Lee Point, Howard Springs, Fogg Dam, Mamukala Wetlands; (Qld) Cairns BG, Cattana Wetlands, Mossman River, Kingfisher Park, Mt Lewis, Great Sandy NP; (NSW) Myall Lakes NP, Crowdy Bay NP, Clarence River Estuary, Munghorn Gap NR.

Crested Shrike-tit (*Falcunculus frontatus*). Three distinct ssp. (that may reclassified as separate species): *whitei* (Northern Shrike-tit) rare in Top End and Kimberley; *frontatus* (Eastern Shrike-tit) mod common in e Aust; *leucogaster* (Western Shrike-tit) rare to uncommon in s-w WA. Forest, woodland. (NT) Fergusson River, Nitmiluk (Katherine Gorge) NP, Elsey NP, Chinaman Creek; (Qld) Hasties Swamp, Lockyer Valley, Sheep Station Creek CP, Girraween NP, Sundown NP; (NSW/ACT) Barrington Tops NP, Crowdy Bay NP, Capertee Valley, Deniliquin, Hawkesbury River Valley, Royal NP, Morton NP, Tidbinbilla NR; (Vic) Greater Bendigo NP, Dunach Nature CR, Canni Creek R, Tarra-Bulga NP, You Yangs; (SA) Cleland CP (rare), Aldinga Scrub CP, Mary Seymour CP; (WA) Stirling Range NP, Dryandra Woodlands, Walpole-Nornalup NP, Drysdale River.

Olive Whistler (*Pachycephala olivacea*). Uncommon to mod common s-e Aust. Forest, rainforest, coastal woodland; (Qld) Lamington NP; (NSW) Barrington Tops NP, Werrikimbe NP, Border Ranges NP, Kosciusko NP; (Vic) Croajingolong NP, Tarra-Bulga NP, Cape Liptrap CP, Dandenong Ranges NP, Toolangi SF, Bunyip SP, Great Otway NP, Cape Nelson Lighthouse, Lower Glenelg NP; (Tas) Mt Wellington, Bruny Is., Truganini R, Snug Tiers NRA, Mt Field NP, Melaleuca, Strahan, Fern Glade R, Cradle Mountain-Lake St Clair NP, King Is.

Red-lored Whistler (*Pachycephala rufogularis*). Rare, inland s-e Aust. Mallee. (NSW) Round Hill NR, Nombinnie NR; (Vic) Murray-Sunset NP; (SA) Danggali CP, Ngarkat CP, Billiatt CP, Gluepot R.

Gilbert's Whistler (*Pachycephala inornata*). Uncommon, s Aust. Woodland, Mallee. (NSW) Round Hill NR, Nombinnie NR, Murray Valley NP; (Vic) Yarrara FFR, Wyperfeld NP, Warby Range SP, Terrick Terrick NP; (SA) Billiatt CP, Gluepot R, Lake Gilles CP, Yumbarra CP; (WA) Boorabbin NP, Goldfields Woodlands NP, Victoria Rock NR.

Golden Whistler (*Pachycephala pectoralis*). Common. Forest, rainforest, woodland, Mallee.

Mangrove Golden Whistler (*Pachycephala melanura*). Uncommon to mod common n Aust. Mangroves, woodland along waterways. (NT) Darwin (rare), Adelaide River, Tiwi Islands, Borroloola; (Qld) Rare reports from s Qld, such as Cape Hillsborough NP; (WA) Wyndham, Derby, Broome Bird Observatory; (OIT) Torres Strait Is.

Grey Whistler (*Pachycephala simplex*). Mod common n Aust. Forest, rainforest, mangroves. Two distinct ssp. (that may be reclassified as a separate species): *peninsulae* (Grey or Grey-headed Whistler) n Qld; *simplex* (Brown Whistler) n NT.

Rufous Whistler (*Pachycephala rufiventris*). Common mainland Aust. Woodland, forest.

White-breasted Whistler (*Pachycephala lanioides*). Uncommon n and n-w Aust. Mangroves. (NT) Charles Darwin NP, Palmerston Sewage Works, Borroloola; (Qld) Karumba; (WA) Wyndham, Derby, Broome Bird Observatory, Barred Creek, Carnarvon.

Little Shrike-thrush (*Colluricincla megarhyncha*). Mod common e and n-e Aust. Rainforest, forest, woodland, mangroves. (NT) Howard Springs, Yellow Water, Gunlom, Nitmiluk (Katherine Gorge) NP, Caranbirini CR; (Qld), Cooktown, Kutini-Payamu (Iron Range) NP, Kingfisher Park, Big Mitchell Creek, Emerald Creek, Hasties Swamp, Curtain Fig Tree, Wongabel SF, Paluma Range NP, Townsville Common, Cape Hillsborough NP, Mt Coot-tha Forest, D'Aguilar NP, Inskip Point, Great Sandy NP; (NSW) Dorrigo NP, Clarence River Estuary, Nightcap NP.

Bower's Shrike-thrush (*Colluricincla boweri*). Mod common Wet Tropics. Rainforest (above 300 m). (Qld) Mt Lewis, Mt Hypipamee NP, Curtain Fig Tree, Crater Lakes NP, Paluma Range NP.

Sandstone Shrike-thrush (*Colluricincla woodwardi*). Mod common n Aust. Sandstone escarpment and gullies. (NT) Nourlangie, Nawurlandja Lookout, Bardedjilidji, Gunlom, Nitmiluk (Katherine Gorge) NP, Katherine Gorge, Victoria River, Jasper Gorge, Caranbirini CR; (Qld) Boodjamulla (Lawn Hill) NP; (WA) Mirima (Hidden Valley) NP, Purnululu (Bungle Bungles) NP, Lake Argyle, the Grotto, Mitchell River NP, Mornington Station.

Grey Shrike-thrush (*Colluricincla harmonica*). Common. Woodland, forest, Mallee.

Crested Bellbird (*Oreoica gutturalis*). Uncommon to mod common. Woodlands, Mallee, Mulga. (NT) Alice Springs Desert Park, Kunoth Bore, Newhaven Sanctuary, Uluru-Kata Tjuta NP; (Qld) Mt Isa, Bladensburg NP, Kilcowera Station, Currawinya NP; (NSW) Pilliga Forest, Round Hill NR, Cocoparra NP, Sturt NP; (Vic) Hattah-Kulkyne NP, Bronzewing FFRWyperfeld NP, Greater Bendigo NP, Heathcote-Graytown NP; (SA) Danggali CP, Gluepot R, Whyalla CP, Lake Gilles CP, Yumbarra CP; (WA) Fitzgerald River NP, Cape Arid NP, Sandford Rocks NR, Neale Junction NR, Cue, Jupiter Well, Lake Macleod, Shark Bay.

Australasian Figbird (*Sphecotheres vieilloti*). Common n and e Aust. Rainforest, forest, gardens, parks. Distinct ssp. include: *vieilloti* (Green Figbird) e Aust; *flaviventris* (Yellow Figbird) n Qld; *ashbyi* (also called Yellow Figbird) Top End and Kimberley.

Yellow Oriole (*Oriolus flavocinctus*). Common n Aust. Forest, mangroves, gardens. (NT) East Point, Howard Springs, Yellow Water Nitmiluk (Katherine Gorge) NP; (Qld) Cairns Esplanade, Cairns BG.

Olive-backed Oriole (*Oriolus sagittatus*). Common e and n Aust. Summer migrant in s. Woodland, forest.

White-breasted Woodswallow (*Artamus leucorhynchus*). Common. Woodland, waterways. Nomadic s Aust. (NT) Darwin Harbour, Leanyer Sewage Ponds, Pine Creek; (Qld) Lockhart River treatment plant, Cairns Esplanade, Bladensburg NP, Lockyer Valley, Inskip Point; (NSW) Crowdy Bay NP, Lake Cargelligo, Murray Valley NP, Kinchega NP; (Vic) Kerang Lakes, Chiltern-Mt Pilot NP; (WA) Broome, Derby, Kununurra.

Masked Woodswallow (*Artamus personatus*). Mod common to common and nomadic. Woodland, Mallee, Mulga, open plains. (NT) Pine Creek, Kunoth Bore, Ilparpa Swamp, Boodjamulla (Lawn Hill) NP; (Qld) Bladensburg NP, Bowra, Kilcowera Station; (NSW) Hunter Valley, Round Hill NR, Cocoparra NP, Murray Valley NP, Kinchega NP, Mutawintji NP; (Vic) Hattah-Kulkyne NP, Murray-Sunset NP, Wyperfeld NP, Glenlee FFR, Goschen BR, Terrick Terrick NP, Greater Bendigo NP, You Yangs; (SA) Gluepot; (WA) Yeo Lake NR, Kununurra, Jupiter Well, Carnarvon.

White-browed Woodswallow (*Artamus superciliosus*). Mod common to common and nomadic. Woodland, Mallee, Mulga, open plains. (NT) Boodjamulla (Lawn Hill) NP; (Qld) Bladensburg NP, Bowra, Kilcowera Station, Sundown NP; (NSW) Hunter Valley, Round Hill NR, Cocoparra NP, Murray Valley NP, Kinchega NP, Mutawintji NP; (Vic) Hattah-Kulkyne NP, Murray-Sunset NP, Wyperfeld NP, Glenlee FFR, Goschen BR, Terrick Terrick NP, Greater Bendigo NP, Canni Creek, You Yangs; (SA) Gluepot R, Arkaroola.

Black-faced Woodswallow (*Artamus cinereus*). Mod common to common. Woodland, salt lakes, open plains. (NT) Fogg Dam, Pine Creek, Nitmiluk (Katherine Gorge) NP, Elsey NP, Kunoth Bore, Santa Teresa Rd, Newhaven Sanctuary, Erldunda, Uluru-Kata Tjuta NP, Boodjamulla (Lawn Hill) NP; (Qld) Bladensburg NP, Bowra, Lake Bindegolly, Kilcowera Station, Currawinya NP, Birdsville; (NSW) Oolambeyan NP, Kinchega NP, Mutawintji NP, Sturt NP; (Vic) Lake Tyrrell, Kerang Lakes; (SA) Flinders Ranges NP, Strzelecki Track, Australian Arid Land BG; (WA) Thomsons Lake NR, Lake McLarty, Yeo Lake NR, Dryandra Woodlands, Kalbarri NP, Parry Lagoons NR, Jupiter Well, Carnarvon.

Dusky Woodswallow (*Artamus cyanopterus*). Common. Woodland, forest. Summer migrant s Aust, including Tas.

Little Woodswallow (*Artamus minor*). Uncommon to mod common. Nomadic. Woodland, gorges, escarpment. (NT) Nourlangie, Bardedjilidji, Gunlom, Pine Creek, Nitmiluk (Katherine Gorge) NP, Elsey NP, Caranbirini CR, Mt Gillen, Simpson's Gap, Ormiston Gorge, Trephina Gorge, Newhaven Sanctuary, Watarrka (Kings Canyon) NP, Tnorala CR, Boodjamulla (Lawn Hill) NP; (Qld) Mt Isa, Bladensburg NP, Bowra; (NSW) Mutawintji NP; (SA) Flinders Ranges NP, Arkaroola; (WA) Yeo Lake NR, Cue, Karijini NP, Lake Argyle, Mirima (Hidden Valley) NP, the Grotto, Mitchell River NP.

Black Butcherbird (*Cracticus quoyi*). Mod common n Aust. Rainforest edges, mangroves. (NT) Buffalo Creek, East Point; (Qld) Jardine River NP, Cairns BG, Jack Barnes Mangrove Boardwalk, Kuranda, Mossman River, Daintree, Townsville Common, Cape Hillsborough NP.

Grey Butcherbird (*Cracticus torquatus*). Mod common to common. Woodland, open forest, urban areas, farmland. Southern forms widespread and easily encountered. Distinct n Aust ssp. *argenteus* (Silver-backed Butcherbird), considered by some authorities a full sp. Selected sites for Silver-backed Butcherbird include: (NT) Gunlom, Murill Billabong, Pine Creek, Nitmiluk NP, Gregory NP; (WA) Parry Lagoons NR, Mitchell River NP.

Black-backed Butcherbird (*Cracticus mentalis*). Mod common Cape York. Woodlands. (Qld) Musgrave (along the Peninsula Development Rd), Weipa, Uningan NR, Jardine River.

Pied Butcherbird (*Cracticus nigrogularis*). Common. Woodlands, forest edges, farmland.

Australian Magpie (*Cracticus tibicen*). Mostly common, uncommon in far n Aust. Several distinct ssp.: *tibicen* and *terraereginae* (Black-backed Magpie) e Aust; *eylandtensis* (small) Groote Eylandt; *longirostris* (Long-billed Magpie) n WA; *tyrannica* and *telonocua* (White-backed Magpie) s Aust; *hypoleuca* (White-backed Magpie) Tas; *dorsalis* (Western Magpie) s-w WA. Woodland, forest, farmland, parkland, gardens.

Pied Currawong (*Strepera graculina*). Common e Aust. Forest, woodlands, mountain ranges, gardens.

Black Currawong (*Strepera fuliginosa*). Common Tas. Forests, woodland, gardens. (Tas) Mt Wellington, Mt Field NP, Cradle Mountain-Lake St Clair NP, Snug Tiers NRA, Freycinet NP.

Grey Currawong (*Strepera versicolor*). Common s Aust. Forest, woodland, Mallee, heathland. Several distinct ssp.: *versicolor* s-e Aust; *melanoptera* (Black-winged Currawong) w Vic, s-e SA; *intermedia* (Brown Currawong) SA; *arguta* (Clinking Currawong) Tas; *halmaturina* (Kangaroo Is.); *plumbea* WA.

Spangled Drongo (*Dicrurus bracteatus*). Common n Aust, uncommon in s. Forest, rainforest edges, gardens. (NT) Darwin BG, Buffalo Creek, East Point, Holmes Jungle NR, Howard Springs, Bardedjilidji; (Qld) Kutini-Payamu (Iron Range) NP, Cairns Esplanade, Mt Whitfield CP, Kingfisher Park, Mt Lewis, Hasties Swamp, Townsville Common, Cape Hillsborough NP, Great Sandy NP; (NSW) Hunter Wetlands Centre, Crowdy Bay NP, Clarence River Estuary, Nightcap NP.

Rufous Fantail (*Rhipidura rufifrons*). Mod common e Aust. Summer migrant to s-e Aust. Forests, rainforest. (Qld) Kutini-Payamu (Iron Range) NP, Uningan NR, Mt Lewis, Mt Hypipamee NP, Wongabel SF, D'Aguilar NP, Lamington NP, Great Sandy NP; (NSW) Copeland Tops, Mungo Brush, Crowdy Bay NP, Nightcap NP, Capertee Valley, Royal NP, Morton NP, Bomaderry Creek RP, Maxwell's FR, Kosciusko NP; (Vic) Croajingolong NP, Fairy Dell Nature CR, Tarra-Bulga NP, Kinglake NP, Great Otway NP, Ralph Illidge Sanctuary.

Arafura Fantail (*Rhipidura dryas*). Mod common Top End and Kimberley. Monsoon forest, woodland, mangroves. Formerly considered conspecific with Rufous Fantail. (NT) Buffalo Creek, Leanyer Sewage Ponds, Howard Springs, Adelaide River, Yellow Water, Borroloola, Groote Eylandt; (Qld) Karumba, Boodjamulla (Lawn Hill) NP, Gregory River; (WA) Wyndham, Mitchell River NP.

Grey Fantail (*Rhipidura albiscapa*). Common s and e Aust. Woodland, forest, coastal heathland.

Mangrove Grey Fantail (*Rhipidura phasiana*). Mod common. Mangroves. (NT) Buffalo Creek, Leanyer Sewage Ponds, Palmerston Sewage Works, Borroloola; (Qld) Karumba; (WA) Wyndham, Broome, Barred Creek, Shark Bay, Carnarvon.

Northern Fantail (*Rhipidura rufiventris*). Common n Aust. Forest, rainforest edges, mangroves. (NT) Charles Darwin Uni, Holmes Jungle NR, Howard Springs, Fogg Dam, Nitmiluk (Katherine Gorge)

NP, Caranbirini CR, Groote Eylandt; (Qld) Cooktown, Wenlock River, Abattoir Swamp, Mt Carbine, Big Mitchell Creek, Emerald Creek, Paluma Range NP, Boodjamulla (Lawn Hill) NP; (WA) Purnululu (Bungle Bungles) NP, Lake Argyle, Kununurra, Mirima (Hidden Valley) NP, the Grotto, Barred Creek, Mitchell River NP, Mornington Station.

Willie Wagtail (*Rhipidura leucophrys*). Common, except Tas. Woodland, forest, farmland, gardens.

Broad-billed Flycatcher (*Myiagra ruficollis*). Mod common n Aust. Monsoon forest, mangroves. (NT) Leanyer Sewage Ponds, Palmerston Sewage Works, Fogg Dam, Adelaide River, Yellow Water; (Qld) Kutini-Payamu (Iron Range) NP, Uningan NR, Cape Hillsborough NP, Karumba; (WA) Broome, Barred Creek, Wyndham.

Leaden Flycatcher (*Myiagra rubecula*). Common n Aust, mod common in e and n Aust. Summer migrant s-e Aust. Forest, woodland.

Satin Flycatcher (*Myiagra cyanoleuca*). Mod common e Aust. Summer migrant s-e Aust. Forest, rainforest edges.

Shining Flycatcher (*Myiagra alecto*). Mod common to common n and n-e Aust. Mangroves, creek lines. (NT) East Point, Howard Springs, Adelaide River, Tiwi Islands, Yellow Water, Groote Eylandt; (Qld) Cooktown, McIvor River Crossing, Marina Plains, Kutini-Payamu (Iron Range) NP, Uningan NR, Lockerbie Scrub, Cairns Esplanade, Jack Barnes Mangrove Boardwalk, Yorkeys Knob, Cattana Wetlands, Mossman River, Daintree, Townsville Common, Cape Hillsborough NP; (WA) Kununurra, Wyndham; (OIT) Torres Strait Is.

Restless Flycatcher (*Myiagra inquieta*). Common e and n Aust, s-w WA. Summer migrant s-e Aust. Woodland, forests.

Paperbark Flycatcher (*Myiagra nana*). Mod common n Aust. Woodland, forests, paperbark forests, mangroves. (NT) Mamukala, Adelaide River, Yellow Water, Nourlangie, Anbangbang Billabong Bitter Springs, Caranbirini CR; (Qld) Cumberland Dam, Karumba, Boodjamulla (Lawn Hill) NP, Gregory River, Mt Isa; (WA) Broome, Kununurra, Purnululu (Bungle Bungles) NP.

White-eared Monarch (*Carterornis leucotis*). Uncommon n-e Aust. Rainforest, forest, mangroves. (Qld) Kutini-Payamu (Iron Range) NP, Kuranda, Davies Creek NP, Crater Lakes NP, Mission Beach, Paluma Range NP, Townsville Common, Magnetic Island, Cape Hillsborough NP, D'Aguilar NP, Lamington NP, Great Sandy NP, Conondale Ranges; (NSW) Iluka NR, Nightcap NP.

Black-faced Monarch (*Monarcha melanopsis*). Mod common, e Aust, migrant s-e Aust. Forest, rainforest edges, creek lines. (Qld) Kuranda, Kingfisher Park, Curtain Fig Tree, Wongabel SF, Magnetic Island, Cape Hillsborough NP, Eungella Plateau, D'Aguilar NP, Lamington NP, Conondale Ranges; (NSW) Barrington Tops NP, Copeland Tops, Mungo Brush, Crowdy Bay NP, Nightcap NP, Royal NP, Barren Grounds NR, Morton NP, Bomaderry Creek RP, Mimosa Rocks NP, Maxwell's FR; (Vic) Croajingolong NP, Fairy Dell Nature CR, Glen Nayook R.

Black-winged Monarch (*Monarcha frater*). Mod common on Cape York, mostly in summer. Rainforest. (Qld) McIvor River Crossing, Kutini-Payamu (Iron Range) NP, Lockerbie Scrub.

Spectacled Monarch (*Symposiarchus trivirgatus*). Mod common n-e and e Aust. Forest, rainforest edges, mangroves. (Qld) McIvor River Crossing, Kutini-Payamu, Lockerbie Scrub, Kuranda, Daintree, Kingfisher Park, Mt Lewis, Davies Creek NP, Curtain Fig Tree, Wongabel SF, Paluma Range NP, Townsville Common, Magnetic Island, Cape Hillsborough NP, Eungella Plateau, D'Aguilar NP, Lamington NP. Great Sandy NP, Conondale Ranges; (NSW) Copeland Tops, Crowdy Bay NP, Boorganna NR, Bruxner Park FR, Iluka NR, Nightcap NP, Border Ranges NP.

Magpie-lark (*Grallina cyanoleuca*). Common. Woodland, farmland, parkland, gardens.

Frilled-necked Monarch (*Arses lorealis*). Mod common Cape York. Rainforest. Once considered ssp. of Frilled Monarch (*Arses telescophthalmus*). (Qld) Kutini-Payamu (Iron Range) NP, Wenlock River, Lockerbie Scrub.

Pied Monarch (*Arses kaupi*). Mod common Wet Tropics. Rainforest. (Qld) Kuranda, Kingfisher Park, Curtain Fig Tree, Wongabel SF, Crater Lakes NP.

Yellow-breasted Boatbill (*Machaerirhynchus flaviventer*). Uncommon to mod common n Qld. Rainforest. (Qld) Cooktown, Kutini-Payamu (Iron Range) NP, Wenlock River, Lockerbie Scrub, Jardine River NP, Kuranda, Daintree, Kingfisher Park, Mt Lewis, Mt Hypipamee NP, Curtain Fig Tree, Wongabel SF, Crater Lakes NP, Mission Beach, Paluma Range NP.

Australian Raven (*Corvus coronoides*). Common e mainland Aust and s-w WA. Woodland, forest, farmland, townships.

Forest Raven (*Corvus tasmanicus*). Mod common to common s-e Aust and Tas. Forest, woodland. Ssp. *boreus* (Relict Raven) confined to n NSW. (NSW) Myall Lakes NP, Crowdy Bay NP; (Vic) Cape Liptrap CP, Wilsons Promontory NP, Anglesea Heath, Great Otway NP, Ralph Illidge Sanctuary, Lower Glenelg NP; (Tas) common Tas; (SA) Bool Lagoon, Mary Seymour CP, Robe.

Little Raven (*Corvus mellori*). Common s-e mainland Aust. Woodland, forest, farmland, townships.

Little Crow (*Corvus bennetti*). Common arid Aust. Open plains, woodlands, arid areas, townships.

Torresian Crow (*Corvus orru*). Common n Aust. Woodland, open plains, arid areas, townships.

White-winged Chough (*Corcorax melanorhamphos*). Mod common e mainland Aust. Woodland, forest edges, Mallee.

Apostlebird (*Struthidea cinerea*). Mod common n and inland e Aust. Woodland, forest. (NT) Elsey NP; (Qld) Mt Carbine, Bowra, Coolmunda Dam; (NSW) Pilliga Forest, Round Hill NR, Cocoparra NP, Oolambeyan NP, Kinchega NP, Mutawintji NP, Sturt NP; (Vic) Hattah-Kulkyne NP; (SA) Danggali CP, Gluepot R, Flinders Ranges NP.

Trumpet Manucode (*Phonygammus keraudrenii*). Uncommon n Cape York. Rainforest. (Qld) Kutini-Payamu, Wenlock River, Lockerbie Scrub, Jardine River NP.

Paradise Riflebird (*Ptiloris paradiseus*). Uncommon s-e Qld and n-e NSW. Temperate rainforest. (Qld) D'Aguilar NP, Lamington NP, Conondale Ranges; (NSW) Barrington Tops NP, Copeland Tops, Werrikimbe NP, Dorrigo NP, Nightcap NP, Border Ranges NP.

Victoria's Riflebird (*Ptiloris victoriae*). Mod common Wet Tropics. Rainforest. (Qld) Cooktown, Kuranda, Daintree, Kingfisher

Park, Mt Lewis, Davies Creek NP, Mt Hypipamee NP.

Magnificent Riflebird (Ptiloris magnificus). Mod common Cape York. Rainforest. (Qld) Kutini-Payamu (Iron Range) NP, Wenlock River, Lockerbie Scrub, Jardine River NP.

Jacky Winter (*Microeca fascinans*). Mod common to common mainland Aust. Woodlands, farmland.

Lemon-bellied Flycatcher (*Microeca flavigaster*). Mod common n Aust. Woodland, mangroves. (NT) Charles Darwin Uni, East Point, Fogg Dam, Yellow Water, Elsey NP; (Qld) McIvor River Crossing, Musgrave, Kingfisher Park, Big Mitchell Creek Mareeba Wetlands, Emerald Creek, Paluma Range NP; (WA) Kimberly ssp. *tormenti*, Parry Lagoons NR, Wyndham, Derby, Barred Creek.

Yellow-legged Flycatcher (*Microeca griseoceps*). Uncommon Cape York. Rainforest. (Qld) Kutini-Payamu (Iron Range) NP, Lockerbie Scrub, Jardine River NP.

Pacific Robin (*Petroica multicolor*). Resident Norfolk Is. Uncommon and declining.

Scarlet Robin (*Petroica boodang*). Mod common. Forest, woodlands s-e Aust.

Red-capped Robin (*Petroica goodenovii*). Mod common mainland Aust. Woodland, Mallee, Mulga, shrublands.

Flame Robin (*Petroica phoenicea*). Common altitudinal migrant in s-e Aust: moves to lower elevations in winter. Forest, woodland, farmland.

Rose Robin (*Petroica rosea*). Uncommon to mod common e Aust. Forest, woodland. (Qld) Lamington NP; (NSW/ACT) Barrington Tops NP, Copeland Tops, Werrikimbe NP, Wingham Brush NR, Scheyville NP, Royal NP, Budderoo NP, Morton NP, Maxwell's FR, Kosciusko NP, Australian National BG, Tidbinbilla NR; (Vic) Warby Range SP, Croajingolong NP, Cape Conran, Cabbage Tree Creek, Fairy Dell Nature CR, Tarra-Bulga NP, Dandenong Ranges NP, Toolangi SF, Glen Nayook, Great Otway NP, Ralph Illidge Sanctuary.

Pink Robin (*Petroica rodinogaster*). Uncommon to mod common s-e Aust. Forest, temperate rainforest. (NSW) Kosciusko NP; (Vic) Tarra-Bulga NP, Toolangi SF, Glen Nayook, Great Otway NP; (Tas) Mt Wellington, Waterworks R., Bruny Is., Truganini R, Snug Tiers NRA, Mt Field NP, Melaleuca.

Hooded Robin (*Melanodryas cucullata*). Mod common and widespread mainland. Woodland, Mallee, Mulga, forest.

Dusky Robin (*Melanodryas vittata*). Mod common Tas. Forest, woodland. (Tas) Waterworks R, Bruny Island, Truganini R, Peter Murrell R, Meehan Range, Mt Field NP, Fern Glade R, Narawntapu NP, Melaleuca, Cradle Mountain-Lake St Clair NP, Waterhouse CA, Freycinet NP, King Is., Flinders Is.

Pale-yellow Robin (*Tregellasia capito*). Mod common e Aust. Rainforest. Two distinct ssp.: *nana* n Qld; *capito* se Qld and ne NSW. (Qld) Kingfisher Park, Mt Hypipamee NP, Curtain Fig Tree Wongabel SF Paluma Range NP, D'Aguilar NP, Lamington NP, Mt Tamborine NP, Great Sandy NP, Conondale Ranges; (NSW) Barrington Tops NP, Copeland Tops, Werrikimbe NP, Dorrigo NP, Bruxner Park FR, Nightcap NP, Border Ranges NP.

White-faced Robin (*Tregellasia leucops*). Mod common Cape York. Rainforest. (Qld)

Kutini-Payamu (Iron Range) NP, Lockerbie Scrub Jardine River NP.

Eastern Yellow Robin (*Eopsaltria australis*). Common e mainland Aust. Woodland, forest, coastal heathland.

Western Yellow Robin (*Eopsaltria griseogularis*). Uncommon to mod common s WA, w SA. Forest, woodland, Mallee. (SA) Lake Gilles CP, Refuge Rocks, Lincoln NP, Coffin NP, Yumbarra CP; (WA) Bungendore Park, Dryandra Woodlands, Lake Muir NR, Porongurup NP, Stirling Range NP, Dragon Rocks NR, Victoria Rock NR, Eyre Bird Observatory, Kalbarri NP.

White-breasted Robin (*Eopsaltria georgiana*). Uncommon to common s-w WA. Coastal heathland, coastal woodland. (WA) Wungong Dam, Walpole-Nornalup NP, Lake Muir NR, Leeuwin-Naturaliste NP, Porongurup NP, Two Peoples Bay NR, Cheynes Beach, Lake Powell NR, Stirling Range NP.

Mangrove Robin (*Peneonanthe pulverulenta*). Mod common n Aust. Mangroves. (NT) Palmerston Sewage Works; (Qld) Cooktown, Marina Plains, Kutini-Payamu (Iron Range) NP, Cairns Esplanade, Jack Barnes Mangrove Boardwalk, Yorkeys Knob, Mossman River, Townsville Common, Karumba; (WA) Derby; (OIT) Torres Strait Is.

Grey-headed Robin (*Heteromyias cinereifrons*). Mod common Wet Tropics. Rainforest (above 300 m). (Qld) Mt Lewis, Mt Hypipamee NP, Curtain Fig Tree, Wongabel SF, Crater Lakes NP, Paluma Range NP.

White-browed Robin (*Poecilodryas superciliosa*). Uncommon to mod common n-e Qld. Woodland, creek lines. (Qld) Shipton's Flat, McIvor River Crossing, Big Mitchell Creek, Emerald Creek, Paluma Range NP, Padaminka Nature Refuge.

Buff-sided Robin (*Poecilodryas cerviniventris*). Uncommon to mod common n Aust. Rainforest, woodland, creek lines. (NT) Mamukala Wetlands, Yellow Water, Gunlom, Pine Creek, Elsey NP, Timber Creek, Borroloola; (Qld) Boodjamulla (Lawn Hill) NP, Gregory River; (WA) Kununurra, Parry Lagoons NR, Drysdale River, Mornington Sanctuary, Mitchell River NP.

Northern Scrub-robin (*Drymodes superciliaris*). Uncommon n Cape York. Rainforest. (Qld) Kutini-Payamu (Iron Range) NP, Lockerbie Scrub.

Southern Scrub-robin (*Drymodes brunneopygia*). Uncommon to mod common s Aust. Mallee woodland, shrublands. (NSW) Round Hill NR; (Vic) Bronzewing FFR, Wyperfeld NP, Little Desert NP; (SA) Coorong NP, Messent CP, Danggali CP, Ngarkat CP, Telowie Gorge CP, Innes NP, Gluepot R, Flinders Ranges NP, Lake Gilles CP, Lincoln NP, Yumbarra CP; (WA) Stirling Range NP, Fitzgerald River NP, Peak Charles NP, Sandford Rocks NR, Eyre Bird Observatory, Kalbarri NP, Shark Bay.

Horsfield's Bushlark (*Mirafra javanica*). Uncommon to mod common. Grassland. (NT) Leanyer Sewage Ponds, Holmes Jungle NR, Connells Lagoon CR; (Qld) Townsville Common, Bladensburg NP; (NSW) Capertee Valley, Oolambeyan NP; (Vic) Terrick Terrick NP, Western Treatment Plant; (WA) Lake Argyle, Parry Lagoons, Broome, Carnarvon.

Eurasian Skylark (*Alauda arvensis*). Mod common s-e Aust. Introduced. Grassland.

Zitting Cisticola (*Cisticola juncidis*). Uncommon n Aust. Migratory. Coastal grassland, grassy swamps. (NT) Holmes Jungle NR, Knuckey Lagoons CR, Borroloola; (Qld) Lakefield NP, Karumba, Ayr, Marmor; (WA) Parry Lagoons NP; (OIT) Torres Strait Is.

Golden-headed Cisticola (*Cisticola exilis*). Common n and e Aust. Coastal grassland, reedbeds, grassy swamps.

Australian Reed-Warbler (*Acrocephalus australis*). Common. Migratory. Reedbeds, grassy swamps. Three distinct ssp.: *gouldi* s-w Aust; *carterae* n-w Aust; *australis* e and s-e Aust and Tas.

Tawny Grassbird (*Megalurus timoriensis*). Uncommon to mod common n and e Aust. Migratory and nomadic. Reedbeds, grassy swamps, coastal grassland. (NT) Holmes Jungle NR, Fogg Dam, Mamukala Wetlands; (Qld) Hasties Swamp, Lake Tinaroo, Tyto Wetlands, Townsville Common, Karumba, Kedron Brook Wetlands, Oxley Creek Common, Eagleby Wetlands, Lake Samsonvale, Great Sandy NP; (NSW) Ash Island; (WA) Lake Argyle, Parry Lagoons NR.

Little Grassbird (*Megalurus gramineus*). Mod common to common. Migratory. Reedbeds, grassy swamps. Three distinct Aust ssp.: *thomasi* s-w WA; *goulburni* e Aust; *gramineus* Tas.

Rufous Songlark (*Cincloramphus mathewsi*). Mod common. Migratory. Open woodland, grassland.

Brown Songlark (*Cincloramphus cruralis*). Mod common. Migratory. Grassland.

Spinifexbird (*Eremiornis carteri*). Uncommon c and w Aust. Spinifex grassland. (NT) Barrow Creek, Santa Teresa Rd, Ellery Creek Big Hole, Ormiston Gorge, Newhaven Sanctuary; (Qld) Riversleigh, Mt Isa, Bladensburg NP, Opalton, (WA) Karijini NP, Purnululu (Bungle Bungles) NP.

Christmas Island White-eye (*Zosterops natalis*). Common resident Christmas Island.

Pale White-eye (*Zosterops citrinella*). Locally common. Little Tuesday Is. (and other small islands), Torres Strait.

Yellow White-eye (*Zosterops luteus*). Mod common n Aust. Two distinct ssp.: *balstoni* coastal n-w WA; *luteus* coastal n Aust from far n WA (Kimberley) to Cape York and isolated population e Qld. Mangroves. (NT) Charles Darwin Uni, Buffalo Creek, Charles Darwin NP, Leanyer Sewage Ponds, East Point, Fogg Dam, Adelaide River, Tiwi Is.; (Qld) Marina Plains, Cape Hillsborough NP, Plantation Creek, Karumba; (WA) Wyndham, Derby, Broome Bird Observatory, Shark Bay, Carnarvon, Lake MacLeod.

Silvereye (*Zosterops lateralis*). Common. Woodland, forest, coastal heathland, parkland, gardens. The 'Capricorn Silvereye' (ssp. *chlorocephalus*), endemic to the southern Great Barrier Reef (see Lady Elliot Is) sometimes considered full sp., *chloronotus* (Western Silvereye) s-w WA, *cornwalli* s-e Aust, *vegetus* Qld, *pinarochrous* s-e Aust, *westernensis* s-e NSW, e Vic, *lateralis* (rufous-flanked) breeds Tas. Non-breeding migrant to s-e Aust, *tephropleurus* (Lord Howe Silvereye).

Slender-billed White-eye (*Zosterops tenuirostris*). Uncommon resident Norfolk Is.

White-backed Swallow (*Cheramoeca leucosterna*). Uncommon to mod common, inland Aust. Aerial over woodland, grassland. (NT) Kunoth Bore, Ormiston Gorge, Erldunda, Old Andado Station, Birdsville; (Qld) Lockyer Valley; (NSW) Hunter Valley,

Pilliga Forest, Capertee Valley, Round Hill NR, Cocoparra NP, Oolambeyan NP, Sturt NP; (Vic) Wyperfeld NP, Little Desert NP; (SA) Mt Lyndhurst Old Mine, Montecollina Bore, Bollards Lagoon Rd, Birdsville Track, Australian Arid Land BG; (WA) Neale Junction NR, Kalbarri NP.

Barn Swallow (*Hirundo rustica*). Rare to uncommon summer migrant n Aust. (NT) Lee Point; (Qld) Newell Beach; (WA) Kununurra, Derby, Broome; (OIT) Ashmore Reef.

Welcome Swallow (*Hirundo neoxena*). Common s Aust, uncommon far n Aust. Woodland, forest, parkland, farmland, gardens.

Fairy Martin (*Petrochelidon ariel*). Common. Migratory. Open woodland, farmland.

Tree Martin (*Petrochelidon nigricans*). Common. Migratory. Woodland, open woodland, farmland.

Red-whiskered Bulbul (*Pycnonotus jocosus*). Common in Sydney region, rare Melbourne, also Mackay. Introduced. Gardens, parkland. (Qld) Mackay; (NSW) Sydney, Sydney Olympic Park, Scheyville NP.

Bassian Thrush (*Zoothera lunulata*). Uncommon to mod common e Aust. Forest, rainforest. (Qld) Mt Lewis, Mt Hypipamee NP, Lamington NP, Conondale Ranges; (NSW/ACT) Barrington Tops NP, Copeland Tops, Boorganna NR, Royal NP, Budderoo NP, Morton NP, Booderee NP, Bomaderry Creek RP, Maxwell's FR, Kosciusko NP, Australian National BG; (Vic) Cabbage Tree Creek, Tarra-Bulga NP, Wilsons Promontory NP, Dandenong Ranges NP, Toolangi SF, Great Otway NP; (Tas) Mt Wellington, Bruny Is., Snug Tiers NRA, Mt Field NP, Cradle Mountain-Lake St Clair NP, Freycinet National Park; (SA) Mt Lofty BG, Kangaroo Is.

Russet-tailed Thrush (*Zoothera heinei*). Uncommon to mod common e Aust. Lowland rainforest, forest. (Qld) Mt Lewis, Paluma Range NP, D'Aguilar NP, Lamington NP, Mt Tamborine NP, Fraser Island, Conondale Ranges; (NSW) Barrington Tops NP, Copeland Tops, Wingham Brush NR, Dorrigo NP, Nightcap NP, Border Ranges NP.

Island Thrush (*Turdus poliocephalus*). Common Christmas Is.

Common Blackbird (*Turdus merula*). Common s-e Aust. Introduced. Gardens, parkland, woodland, forest.

Song Thrush (*Turdus philomelos*). Uncommon. Introduced. Resident in gardens in Vic., such as Melbourne and Geelong. Also common Lord Howe Is. and Norfolk Is.

Singing Starling (*Aplornis cantoroides*). Resident on Boigu Is. and visitor to Saibai Is., Torres Strait. Mangrove edges, gardens.

Metallic Starling (*Aplornis metallica*). Common Forest, rainforest edges. (Qld) Chilli Beach, Cairns Esplanade, Cairns BG, Kingfisher Park, Curtain Fig Tree.

Common Starling (*Sturnus vulgaris*). Common e Aust. Introduced. Gardens, farmland, woodland.

Common Myna (*Sturnus tristis*). Common e Aust, and expanding. Introduced. Gardens, roadsides.

Red-capped Flowerpecker (*Dicaeum geelvinkianum*). Uncommon resident on Boigu Is. and Saibai Is., Torres Strait. Mangrove edges, gardens.

Mistletoebird (*Dicaeum hirundinaceum*). Common, nomadic. Woodland, forest.

Olive-backed Sunbird (*Nectarinia jugularis*). Common n-e Qld. Rainforest, gardens, mangrove. (Qld) Portland Rd, Cairns BG, Kingfisher Park, Magnetic Is., Cape Hillsborough NP; (OIT) Torres Strait Is.

Zebra Finch (*Taeniopygia guttata*). Common mainland Aust. Grasslands, open woodland.

Double-barred Finch (*Taeniopygia bichenovii*). Common n and e Aust. Woodland, forest, grassland.

Long-tailed Finch (*Poephila acuticauda*). Common n and n-w Aust. Open woodland, grassy creek lines. Two forms red-billed form e of range, yellow-billed form w of range. (NT) Charles Darwin Uni., Leanyer Sewage Ponds, Howard Springs, Tiwi Is., Mamukala Wetlands, Arnhem Hwy, Gunlom, Pine Creek Cemetery, Fergusson River, Nitmiluk (Katherine Gorge) NP, Elsey NP, Victoria River, Timber Creek, Barkly Tableland, Borroloola; (Qld) Boodjamulla (Lawn Hill) NP, Lake Moondarra; (WA) Lake Argyle, Kununurra, Wyndham, Drysdale River, Purnululu (Bungle Bungles) NP.

Black-throated Finch (*Poephila cincta*). Uncommon to mod common n-e Aust. Woodland, forest. Three distinct ssp.: *cincta* (white-rumped) n Qld; *atropygialis* and *nigrotecta*, Cape York (both dark-rumped). (Qld) Lakefield NP, Musgrave, Mt Carbine, Big Mitchell Creek, Mareeba Wetlands, Tinaroo Creek Rd, Townsville Common, Cumberland Dam.

Masked Finch (*Poephila personata*). Mod common n Aust. Open woodland, grassy woodland. Two distinct ssp. *personata* n WA, n NT, n-w Qld; *leucotis* (White-eared Finch) n Qld. (NT) Charles Darwin Uni, Howard Springs, Tiwi Is., Mamukala Wetlands, Arnhem Hwy, Gunlom, Pine Creek Cemetery, Fergusson River, Nitmiluk (Katherine Gorge) NP, Elsey NP, Chinaman Creek, Gregory NP, Barkly Tableland, Borroloola; (Qld) Lakefield NP, Musgrave, Cumberland Dam; (WA) the Grotto, Lake Argyle, Kununurra, Wyndham.

Crimson Finch (*Neochmia phaeton*). Mod common n Aust. Tall grassland, cane fields. Two distinct forms white-bellied in the e, black-bellied in w. (NT) Leanyer Sewage Ponds, Howard Springs, Fogg Dam, Tiwi Is., Mamukala Wetlands, Arnhem Hwy, Gunlom, Fergusson River, Elsey NP, Victoria River, Timber Creek; (Qld) Lakefield NP, Yorkeys Knob, Cairns Crocodile Farm, Tyto Wetlands, Townsville Common, Horseshoe Lagoon, Boodjamulla (Lawn Hill) NP; (WA) Mornington Station, Mitchell River NP, Kununurra.

Star Finch (*Neochmia ruficauda*). Rare to uncommon n and n-w Aust. Grassland, grassy woodland, grassy swamps. Ssp. *ruficauda* probably extinct. (NT) Elsey NP, Victoria River, Buchanan Hwy, Timber Creek, Jasper Gorge, Borroloola; (Qld) Lakefield NP, Karumba; (WA) Carnarvon, Tom Price, Mornington Station, Kununurra, Parry Lagoons NR, Wyndham.

Plum-headed Finch (*Neochmia modesta*). Uncommon e Aust. Woodland, creek lines. (Qld) Townsville Common, Cumberland Dam, Bladensburg NP, Winton Treatment Works, Bowra, Oxley Creek Common (Brisbane), Lake Samsonvale, Lockyer Valley, Lake Dyer, Durikai SF, Coolmunda Dam, Girraween NP, Sundown NP; (NSW) Barraba, Pilliga Forest, Capertee Valley.

Red-browed Finch (*Neochmia temporalis*). Common e mainland Aust. Woodland, grassland, forest edges.

Diamond Firetail (*Stagonopleura guttata*). Mod common e Aust. Open woodland, grassy woodland. (Qld) Bowra, Durikai

SF, Girraween NP, Sundown NP; (NSW) Hunter Valley, Barraba, Pilliga Forest, Capertee Valley, Munghorn Gap NR, Lake Cargelligo Weir, Cocoparra NP, Murray Valley NP; (Vic) Little Desert NP, Glenlee FFR, Warby Range SP, Terrick Terrick NP, Greater Bendigo NP, Heathcote-Graytown NP, Canni Creek, Buchan Caves, You Yangs; (SA) Kaiserstuhl CP, Mary Seymour CP, Mt Remarkable NP, Telowie Gorge CP, Lincoln NP, Coffin Bay NP.

Beautiful Firetail (*Stagonopleura bella*). Rare to uncommon s-e Aust. Heathland, forest edges. (NSW) Royal NP, Barren Grounds NR, Budderoo NP, Morton NP, Booderee NP, Ben Boyd NP; (Vic) Croajingolong NP, Cape Conran, Marlo CR, Cape Liptrap CP, Wilsons Promontory NP, Yellingbo NCR, Bunyip SP, Great Otway NP, Cape Nelson SP; (Tas) Waterworks R, Bruny Is., Truganini R, Meehan Range, Melaleuca, Strahan, Douglas-Apsley NP, Freycinet NP; (SA) Newland Head CP, Deep Creek CP, Flinders Chase NP, Cape Gantheaume CP, Mary Seymour CP, Coorong NP.

Red-eared Firetail (*Stagonopleura oculata*). Uncommon s-w Aust. Coastal heathland, coastal forest. (WA) Wungong Dam, Serpentine NP, Walpole-Nornalup NP, Leeuwin-Naturaliste NP, Two Peoples Bay NR, Betty's Beach, Cheynes Beach, Lake Powell NR, Stirling Range NP, Woody Island NR, Monjingup Lake NR, Cape Arid NP.

Painted Finch (*Emblema pictum*). Uncommon to mod common c and w Aust. Rocky gorges, Spinifex grassland. (NT) Jasper Gorge, Santa Teresa Rd, Mt Gillen, Simpson's Gap, Ormiston Gorge, Trephina Gorge, N'Dhala Gorge NP, Newhaven Sanctuary, Watarrka (Kings Canyon) NP; (Qld) Boodjamulla (Lawn Hill) NP, Lake Moondarra, Bladensburg NP, Opalton; (SA) Mt Ives (rare); (WA) Lake Argyle, Mornington Station, Purnululu (Bungle Bungles) NP.

Blue-faced Parrot-Finch (*Erythrura trichroa*). Rare Wet Tropics. Rainforest edges. (Qld) Mt Lewis, Rose Gums Wilderness Retreat, Wooroonooran NP.

Gouldian Finch (*Erythrura gouldiae*). Rare to uncommon n Aust. Grassy woodland, open woodland, roadside grassland. (NT) Arnhem Hwy, Pine Creek Cemetery, Pine Creek, Fergusson River, Nitmiluk (Katherine Gorge) NP, Elsey NP, Chinaman Creek, Victoria River, Buchanan Hwy, Timber Creek, Jasper Gorge, Caranbirini CR; (Qld) Mareeba Wetlands (re-introduction program), Cumberland Dam (rare), Boodjamulla (Lawn Hill) NP (rare); (WA) Mornington Station, the Grotto, Lake Argyle, Wyndham.

Nutmeg Mannikin (*Lonchura punctulata*). Common e Aust. Introduced. Grassland, parkland.

Yellow-rumped Mannikin (*Lonchura flaviprymna*). Rare to uncommon Top End and Kimberley. Tall grassland, reedbeds. (NT) Leanyer Sewage Ponds, Palmerston Sewage Works, Knuckey Lagoons CR (rare at last three sites), Fogg Dam (uncommon), Chinaman Creek, Victoria River, Timber Creek, Jasper Gorge; (WA) Kununurra, Parry Lagoons NR.

Chestnut-breasted Mannikin (*Lonchura castaneothorax*). Mod common n and e Aust. Grassland, reedbeds, grassy woodland. (NT) Charles Darwin Uni, Leanyer Sewage Ponds, Holmes Jungle NR, Howard Springs, Fogg Dam, Elsey NP, Victoria River, Timber Creek; (Qld) Yorkeys Knob, Cairns Crocodile Farm, Mareeba Wetlands, Townsville Common, Horseshoe Lagoon, Cumberland Dam, Lake Samsonvale, Kedron Brook Wetlands (Brisbane), Oxley

Creek Common (Brisbane), Eagleby Wetlands, Lake Dyer; (NSW) Hunter Wetlands Centre, Baker's Lagoon; (WA) Kununurra.

Java Sparrow (*Lonchura oryzivora*). Introduced to Christmas Is., where uncommon and declining.

Pictorella Mannikin (*Heteromunia pectoralis*). Uncommon to mod common n Aust. Grassy woodland, grassland. (NT) Elsey NP, Victoria River, Buchanan Hwy, Jasper Gorge, Barkly Tableland, Connells Lagoon CR, Borroloola; (Qld) Cumberland Dam, Gregory River, Lake Moondarra, Bladensburg NP, Winton Treatment Works; (WA) Mornington Station, Kununurra, Parry Lagoons NR, Wyndham.

House Sparrow (*Passer domesticus*). Common e and c Aust. Introduced. Parkland, urban areas, gardens, farmland.

Eurasian Tree Sparrow (*Passer montanus*). Uncommon to mod common in gardens in Vic. Introduced. (Vic) Melbourne Zoo, CERES (Brunswick). Common on Christmas Is., naturally occurring vagrant to Ashmore Reef.

Australasian Pipit (*Anthus novaeseelandiae*). Common Grassland, grassy woodland.

Eastern Yellow Wagtail (*Motacilla tschutschensis*). Rare. Regular records from NT, WA (Kununurra, Parry Lagoons, Derby, Broome), Christmas Is., NSW (regular Ash Is). Vagrant other states. Green-headed Yellow Wagtail (ssp. *taivana*) occasional in Broome in late November and December, otherwise vagrant to n NT and WA.

European Goldfinch (*Carduelis carduelis*). Common s-e Aust. Introduced. Grassland, farmland, urban parkland.

Common Redpoll (*Acanthis flammea*). Common on Macquarie Is., where apparently self-introduced from naturalised population in NZ.

Common Greenfinch (*Chloris chloris*). Mod common s-e Aust. Introduced. Grassland, farmland, parkland.

Vagrants

Canada Goose (*Branta canadensis*). Vagrants from a feral population in New Zealand. Records from NSW, Lord Howe Is.

Paradise Shelduck (*Tadorna variegata*). Record from Lord Howe Is.

Northern Shoveler (*Anas clypeata*). Records from NSW, Vic, SA, WA.

Eurasian (Common) Teal (*Anas crecca*). Records from Cocos Keeling Is.

Northern Pintail (*Anas acuta*). Records from NT, NSW, WA.

Red-billed Tropicbird (*Phaethon rubricauda*). Record from Lord Howe Is.

Little Grebe (*Tachybaptus ruficollis*). Several records from Leanyer Sewage Ponds, NT.

Greater Flamingo (*Phoenicopterus ruber*). Single record from North Keeling Is.

Red collared Dove (*Streptopelia tranquebarica*). Records from Christmas Is.

Orange-bellied Fruit-Dove (*Ptilinopus iozonus*). Single record Boigu Is., Torres Strait.

Elegant Imperial-Pigeon (*Ducula concinna*). Records from Darwin, NT.

Savanna Nightjar (*Caprimulgus affinis*). Records from Christmas Is.

Grey Nightjar (*Caprimulgus indicus*). Records from Ashmore Reef, Christmas Is.

Glossy Swiftlet (*Collocalia esculenta*). A few records from Qld, WA.

Uniform Swiftlet (*Aerodramus vanikorensis*). Several records from Torres Strait Is., Cape York, Qld.

Black-nest Swiftlet (*Aerodramus maximus*). Unconfirmed records from Broome, WA, Christmas Is.

Edible-nest Swiftlet (*Aerodramus fuciphagus*). Records from Broome, WA, Ashmore Reef, Christmas Is.

Papuan Needletail Spine-tailed Swift (*Mearnsia novaeguineae*). Records from Boigu Is., Torres Strait, Qld.

House Swift (*Apus affinis*). Records from WA, NT, Qld, Ashmore Reef, Christmas Is.

Leach's Storm-Petrel (*Hydrobates leucorhoa*). Records from WA, Vic, Ashmore Reef.

Tristram's Storm-Petrel (*Hydrobates tristrami*). Single record from Sydney pelagic, NSW.

Polynesian Storm-Petrel (*Nesofregetta fuliginosa*). Record from near Norfolk Is. and Southport pelagic, Qld.

New Zealand Storm-Petrel (*Oceanites maorianus*). Increasing records off e Aust seaboard.

Laysan Albatross (*Phoebastria immutabilis*). Record from Norfolk Is.

Chatham Albatross (*Thalassarche eremita*). Records from NSW, Vic and Tas.

Atlantic Yellow-nosed Albatross (*Thalassarche chlororhynchos*). Records NSW, Vic, WA.

Antarctic Petrel (*Thalassoica antarctica*). Occasional records of beach-washed birds on southern mainland. Also Macquarie Is., Heard Is.

Snow Petrel (*Pagodroma nivea*). Records from Heard Is., Macquarie Is.

Broad-billed Prion (*Pachyptila vittata*). Records from WA, SA, Vic, Tas, NSW, Qld.

Westland Petrel (*Procellaria westlandica*). Records from e Aust seaboard.

Arabian Shearwater (*Puffinus persicus*). Records near Ashmore Reef.

Pink-footed Shearwater (*Ardenna creatopus*). Single record from Wollongong pelagic, NSW.

Great Shearwater (*Ardenna gravis*). Several southern ocean records with influx in 2011 and 2012.

Manx Shearwater (*Puffinus puffinus*). Records from NSW, SA.

Newell's Shearwater (*Puffinus newelli*). Record from Norfolk Is.

Audubon's (Tropical) Shearwater (*Puffinus lherminieri*). Records from Sydney and Wollongong pelagics, NSW.

Barau's Petrel (*Pterodroma baraui*). Records from Vic, WA, Christmas Is., Heard Is.

Juan Fernandez Petrel (*Pterodroma externa*). Records from NSW.

Collared Petrel (*Pterodroma brevipes*). Records from near Norfolk Is.

Cook's Petrel (*Pterodroma cookii*). Records from e Aust seaboard and SA.

Vanuatu Petrel (*Pterodroma occulta*). Single record from NSW.

Atlantic Petrel (*Pterodroma incerta*). Single record from Port Fairy pelagic, Vic.

Emperor Penguin (*Aptenodytes forsteri*). Records from Heard Is., Macquarie Is.

Adelie Penguin (*Pygoscelis adeliae*). Records from Tas, Heard Is., Macquarie Is.

Chinstrap Penguin (*Pygoscelis antarcticus*). Records from Vic, Tas, Heard Is., Macquarie Is.

Fiordland Penguin (*Eudyptes pachyrhynchus*). Records from s Aust seaboard.

Snares Penguin (*Eudyptes robustus*). Records from s Aust seaboard, Macquarie Is.

Erect-crested Penguin (*Eudyptes sclateri*). Records from s Aust seaboard., Macquarie Is.

Magellanic Penguin (*Spheniscus magellanicus*). Single record Phillip Is. Vic.

Northern Rockhopper Penguin (*Eudyptes moseleyi*). Vagrant records from WA.

Cape Gannet (*Morus capensis*). Records from Vic, WA.

Yellow Bittern (*Ixobrychus sinensis*). Records from Kalgoorlie, WA and several records Christmas Is., Cocos Keeling Is.

Schrenk's Bittern (*Ixobrychus eurhythmus*). Several records Christmas Is.

Cinnamon Bittern (*Ixobrychus cinnamomeus*). Record from Eighty-mile Beach, WA, several records Christmas Is., Cocos Keeling Is.

Grey Heron (*Ardea cinerea*). Several records from WA.

Chinese Pond Heron (*Ardeola bacchus*). Records from NT, WA, Christmas Is., Cocos Keeling Is.

Javan Pond Heron (*Ardeola speciosa*). Records from NT, WA, several records from Cocos Is.

Purple Heron (*Ardea purpurea*). Record from Perth WA.

Black-crowned Night-Heron (*Nycticorax nycticorax*). Records from Ashmore Reef, Cocos Keeling Is.

Japanese Night-Heron (*Gorsachius goisagi*). Single record Christmas Is.

Malayan Night-Heron (*Gorsachius melanolophus*). Several records from Christmas Is., Cocos Keeling Is.

Oriental Honey-Buzzard (*Pernis ptilorynchus*). Records from Kakadu, NT, Leinster, WA, Christmas Is., Ashmore Reef.

Gurney's Eagle (*Aquila gurneyi*). Several records from Boigu Is., Saibai Is., Torres Strait.

Eurasian Hobby (*Falco subbuteo*). Records from Christmas Is., Cocos Keeling Is.

Red-legged Crake (*Rallina fasciata*). Records from Broome, Whim Creek WA, Ashmore Reef.

Corncrake (*Crex crex*). Historic records from Randwick, NSW (1893), Jurien Bay, WA (1944).

Ruddy-breasted Crake (*Porzana fusca*). Historic records from Christmas Is. (1887, 1940).

Watercock (*Gallicrex cinerea*). Records from Christmas Is., Cocos Keeling Is., Ashmore Reef.

South Island Pied Oystercatcher (*Haematopus finschi*). Records from NSW, Qld, Lord Howe Is., Norfolk Is., Macquarie Is.

American Golden Plover (*Pluvialis dominicus*). Records from NSW, SA.

Ringed Plover (*Charadrius hiaticula*). Records from NT, Qld, NSW, Vic, SA, WA.

Semipalmated Plover (*Charadrius semipalmatus*). Records from Qld, NSW, Vic, WA.

Kentish Plover (*Charadrius alexandrinus*). Records from Old Bar, NSW, Darwin, NT.

Caspian Plover (*Charadrius asiaticus*). Records from Pine Creek and Lake Finniss, NT, Cocos Keeling Is.

Grey-headed Lapwing (*Vanellus cinereus*). Single record from Burren Junction, NSW.

Pheasant-tailed Jacana (*Hydrophasianus chirurgus*). Single record from Paraburdoo, WA.

Hudsonian Godwit (*Limosa haemastica*). Records from SA, Vic, NSW, Tas, WA.

Eurasian Curlew (*Numenius arquata*). Records from Eighty Mile Beach and Broome, WA.

Upland Sandpiper (*Bartramia longicauda*). Historic record from Sydney, NSW (1848).

Green Sandpiper (*Tringa ochropus*). Record from Darwin, NT.

Spotted Redshank (*Tringa erythropus*). Records from NT, NSW, Vic, WA.

Nordmann's Greenshank (*Tringa guttifer*). Three records from Eighty-mile Beach, WA.

Lesser Yellowlegs (*Tringa flavipes*). Records from Qld, NSW, Vic.

Short-billed Dowitcher (*Limnodromus griseus*). Single record from Corner Inlet, Vic.

White-rumped Sandpiper (*Calidris fuscicollis*). Several records from NSW, Vic, SA, WA.

Baird's Sandpiper (*Calidris bairdii*). Records from NT, Vic, Tas, SA.

Stilt Sandpiper (*Calidris himantopus*). Records from NT, Vic.

Dunlin (*Calidris alpina*). Single records from Qld, NSW.

[**Cox's Sandpiper**. (*Calidris* hybrid). Records from NSW, Vic, SA. Natural hybrid between Pectoral and Curlew Sandpiper.]

Buff-breasted Sandpiper (*Tryngites subruficollis*). Records from Qld, NSW, Vic, SA, Tas, WA.

Wilson's Phalarope (*Steganopus tricolor*). Records from Tullakool Saltworks, NSW, Lake Murdeduke, Vic.

Grey Phalarope (*Phalaropus fulicarius*). Records from Lake Mitchell, Qld, Magrath Flat, SA, Lake Woorinen, Discovery Bay and Port Fairy, Vic.

Black Tern (*Chlidonias niger*). Several records NSW.

Slaty-backed Gull (*Larus schistisagus*). Single record from Cooktown, Qld.

Lesser Black-backed Gull (*Larus fuscus*). Records from Broome, WA, Cocos Keeling Is.

Mew Gull (*Larus canus*). Single record from Christmas Is.

Black-tailed Gull (*Larus crassirostris*). Records from NT, Qld, Vic, WA.

Laughing Gull (*Leucophaeus atricilla*). Records from Qld, Vic, Tas, WA.

Franklin's Gull (*Leucophaeus pipixcan*). Records from NT, Qld, NSW, Vic, SA, WA.

Black-headed Gull (*Larus ridibundus*). Almost annual Dec–Mar around Darwin (Buffalo Creek and Darwin Port), NT. Other records Broome, WA, Torres Strait.

Sabine's Gull (*Xema sabini*). Records from NT, NSW, Vic, SA, WA.

Lesser Coucal (*Centropus bengalensis*). Single record Ashmore Reef.

Asian Koel (*Eudynamys scolopaceus*). Several records from Christmas Is., Cocos Keeling Is.

Large Hawk-Cuckoo (*Hierococcyx sparverioides*). Several records from Christmas Is., Cocos Keeling Is.

Long-tailed Cuckoo (*Urodynamys taitensis*). Rare migrant Norfolk Is., vagrant to Lord Howe Is.

Brown Hawk-Owl (*Ninox scutulata*). Records from Goldsworthy and Exmouth, WA, Ashmore Reef, Timor Sea oil rigs.

Common Kingfisher (*Alcedo atthis*). Several records from Christmas Is.

Lesser (Little) Paradise-Kingfisher (*Tanysiptera hydrocharis*). Single record from Saibai Is., Torres Strait.

Black-capped Kingfisher (*Halcyon pileata*). Single record from North Burracoppin, WA.

Blue-winged Pitta (*Pitta moluccensis*). Several records from WA, one record from Christmas Is.

Fairy Pitta (*Pitta nympha*). Records from Derby, WA, Christmas Is.

Hooded Pitta (*Pitta sordida*). Record from Barrow Is., WA.

Brown Shrike (*Lanius cristatus*). Records from Christmas Is., Ashmore Reef.

Tiger Shrike (*Lanius tigrinus*). Records from Browse Is., WA, Ashmore Reef, Christmas Is., Cocos Keeling Is.

Black-billed Magpie (*Pica pica*) Single record from Port Hedland, WA.

House Crow (*Corvus splendens*). Presumably ship-assisted. Records from NSW, Vic, WA.

Island Monarch (*Monarcha cinerascens*). Records from islands off Kimberley coast, WA, Ashmore Reef.

Oriental Reed-Warbler (*Acrocephalus orientalis*). Records from NT, Qld, NSW, WA, Ashmore Reef, Christmas Is.

Pallas's Grasshopper Warbler (*Locustella ochotensis*). Records from Christmas Is., Ashmore Reef.

Middendorff's Warbler (*Locustella ochotensis*). Records from Ashmore Reef.

Arctic Warbler (*Phylloscopus borealis*). Several records from WA, Ashmore Reef.

Dusky Warbler (*Phylloscopus fuscatus*). Single record Christmas Is.

Red-rumped Swallow (*Cecropis daurica*). Records from NT, Qld, WA, Ashmore Reef, Christmas Is.

Asian House Martin (*Delichon dasypus*). Records from Christmas Is., Cocos Keeling Is.

Blue Rock Thrush (*Monticola solitarius*). Records from Noosa, Qld, Cocos Keeling Is.

Isabelline Wheatear (*Oenanthe isabellina*). Record from Mt Carbine, Qld.

Narcissus Flycatcher (*Ficedula narcissina*). Single record from Barrow Is., WA.

Blue-and-White Flycatcher (*Cyanoptila cyanomelana*). Several records from WA, Christmas Is.

Dark-sided Flycatcher (*Muscicapa sibirica*). Records from Shay Gap, WA, Cocos Keeling Is., Ashmore Reef.

Asian Brown Flycatcher (*Muscicapa dauurica*). Records from Browse Is., WA, Ashmore Reef, Cocos Keeling Is.

Purple-backed Starling (*Sturnus sturninus*). Several records from Christmas Is., Cocos Keeling Is.

Rosy Starling (*Sturnus roseus*). Records from Coff's Harbour, NSW, Lake Eda, WA, Cocos Keeling Is.

Red-throated Pipit (*Anthus cervinus*). Records from Broome, WA.

Pechora Pipit (*Anthus gustavi*). Records from Browse Is., WA, Ashmore Reef.

Forest Wagtail (*Dendromanthus indicus*). Records from Alice Springs, NT, Christmas Island.

Citrine Wagtail (*Motacilla citreola*). Records from Woolooware Bay, NSW, Goolwa, SA, Christmas Is.

Grey Wagtail (*Motacilla cinerea*). Regularly records on Christmas Is., occasional records Ashmore Reef, Cocos Keeling Is., rare records from NT, Qld, SA, WA.

White Wagtail (*Motacilla alba*). Records from NT, Qld, NSW, Vic, WA, Christmas Is.

Common Chaffinch (*Fringilla coelebs*). Records from Lord Howe Is., Macquarie Is., Phillip Is. (off Norfolk Is.)

Yellowhammer (*Emberiza citrinella*). Records from Lord Howe Is., Macquarie Is.

Pale-headed Munia (*Lonchura pallida*). Records from Ashmore Reef.

Further reading

Baxter C 1995. *An Annotated List of the Birds of Kangaroo Island.* South Australia National Parks and Wildlife Service, SA.

Collins P 1995. *The Birds of Broome: an annotated List.* Broome Bird Observatory, Broome, WA.

Dolby T, Johns P, Symonds S 2009. *Where to See Birds in Victoria.* Allen & Unwin, Crows Nest, NSW.

Donato D, Wilkins P, Smith G, Alford L 1997. *Finding Birds in Australia's Northern Territory.* CSIRO Publishing, Collingwood, Vic.

Goodfellow DL, Stott M 2005. *Birds of Australia's Top End.* New Holland Publishing, Frenchs Forest, NSW.

Higgins P, *et al.* 1990–2006. *Handbook of Australian, New Zealand and Antarctic Birds.* Oxford University Press, South Melbourne, Vic. [7 Volumes.]

Marr N 1986. *Where to Find Birds in Western Australia.* Kangaroo Press, Kenthurst, NSW.

McCrie N, Watson J 2006. *Finding Birds in Darwin, Kakadu and the Top End, Northern Territory, Australia.* Self published, Casuarina, NT.

Montgomery I 2012. *Christmas and Cocos Keeling Islands Birding Guide.* Dept of Sustainability, Environment, Water, Population and Communities, Canberra ACT, www.environment.gov.au/parks/publications/christmas/birds-brochure.html

Morcombe M 2003. *Field Guide to Australian Birds.* Steve Parish Publishing, Archerfield, Qld. [Also available as a phone application, which includes recordings of sound from 650+ Australian bird species by David Stewart.]

Nevill S 2008. *Birds of the Greater South West Western Australia.* Simon Nevill Publications, WA 2008.

Nielsen L 1996. *Birds of Queensland's Wet Tropics and Great Barrier Reef.* Gerard Industries, Bowden, SA.

Pizzey G, Knight F, Pizzey S 2013. *The Field Guide to the Birds of Australia 9th edn.* HarperCollins, Pymble, NSW.

Plowright H, Buckingham R, Jackson L 2001–2007. *Field Guide to Australian Birdsong* [sound recording]. Bird Observers Club of Australia, Nunawading, Vic.

Reid R, Shaw N, Wheeler R 2004. *Birds of South-eastern Australia.* Overthefence Press, Hurstbridge, Vic.

Roberts P 1993. *Birdwatcher's Guide to the Sydney Region.* Kangaroo Press, Kenthurst, NSW.

Saunders DA, de Rebeira CP 1985. *The Birdlife of Rottnest Island.* Guildford, WA.

Simpson K, Day N 2010. *Field Guide to the Birds of Australia.* Penguin Australia, Ringwood, Vic.

Slater P, Slater P, Slater R 2003. *The Slater Field Guide to Australian Birds.* New Holland, Sydney, NSW.

Taylor M, Day N 1993. *Field Guide to the Birds of the ACT.* National Parks Association of the Australian Capital Territory Inc., Woden, ACT.

Further reading

Tzaros C 2005. *Wildlife of the Box-ironbark Country*. CSIRO Publishing, Collingwood, Vic.

van Delft R 1997. *Birding Sites Around Perth*. University of Western Australia Press, Nedlands, WA.

Watts D 2002. *Field Guide to Tasmanian Birds*. Reed New Holland, Frenchs Forest, NSW.

Watts D, Plowman C 2008. *Where to See Wildlife in Tasmania*. Jacana, Crows Nest, NSW.

Wieneke J 1992. *Where to Find Birds in North East Queensland*. Self published, Belgian Gardens, Qld.

Useful resources, contacts and birdwatching groups

Useful birding websites

Australian Birdline Services and Eremaea
This database includes over 2 million bird records, and incorporates the Australian Birdline services that provide daily updates on interesting sightings around the nation. Records submitted to Eremaea are also incorporated into the Birdlife Australia national bird atlas.
www.eremaea.com

Birdata
Coordinated by Birdlife Australia, this national bird atlas comprises over 7 million bird records with an online searchable interface to assist with trip planning.
www.birdlife.org.au

Birding-Aus
The largest online birding community in Australia, where members post and discuss all aspects of Australian wild birds and birdwatching. Users can subscribe to send and receive emails from the list-server. There is an online searchable archive that extends back the early 1990s and can prove valuable for trip planning.
www.birding-aus.org

Birding Western Australia
A comprehensive online guide to the birds and birding sites of Western Australia. www.birdingwa.iinet.net.au.

Nature Sound
Australian bird recordings from David Stewart.
www.naturesound.com.au

Southern Birding Service
An excellent online guide to the birds and birding sites of South Australia. www.sabirding.com/sites.html

Wildlifeimages.com.au
The most comprehensive collection of Australian bird images available online. www.wildlifeimages.com.au

Birdwatching and bird study groups

Australasian Wader Studies Group (AWSG)
C/- Birdlife Australia National Office Suite 2–05, 60 Leicester St, Carlton Vic 3053, Australia (03) 9347 0757 www.awsg.org.au

Birding NSW
PO Box Q277, Queen Victoria Building, Sydney, NSW 1230, Australia Email: info@birdingnsw.org.au www.birdingnsw.org.au

BirdLife Australia
National Office Suite 2–05, 60 Leicester St, Carlton, Vic 3053, Australia (03) 9347 0757 Email: info@birdlife.org.au www.birdlife.org.au

Birds Queensland
PO Box 3784, South Brisbane, Qld 4101, Australia birdsqueensland.org.au

Birds SA
C/- South Australian Museum, North Terrace, Adelaide, SA 5000, Australia Email: general@birdssa.asn.au www.birdssa.asn.au

Canberra Ornithologists Group (ACT)
PO Box 301,Civic Square, ACT 2608, Australia
canberrabirds.org.au

Cumberland Bird Observers (NSW)
PO Box 550, Baulkham Hills, NSW 1755, Australia www.cboc.org.au

Geelong Field Naturalists Club
PO Box 1047, Geelong, Vic 3220, Australia E-mail: gfnc@vicnet.net.au home.vicnet.net.au/~gfnc

Hunter Bird Observers Club (NSW)
PO Box 24, New Lambton, NSW 2305, Australia E-mail: info@hboc.org.au www.hboc.org.au

Illawarra Bird Observers Club (NSW)
PO Box 56 Fairy Meadow, NSW 2519, Australia www.iboc.org.au

Northern Territory Field Naturalists Club
PO Box 39565, Winnellie, NT 0821, Australia www.ntfieldnaturalists.org.au

Southern Oceans Seabird Study Association
PO Box 142, Unanderra, NSW 2526, Australia (02) 4272 4626 Email: sossa@tpg.com.au www.sossa-international.org

Other useful resources

NSW Environment and Heritage
NSW National Parks www.nationalparks.nsw.gov.au

Northern Territory Natural Resources, Environment, the Arts and Sport
National Parks and Reserves www.nretas.nt.gov.au/national-parks-and-reserves

Parks Australia
Department of Sustainability, Environment, Water, Populations and Communities Level 2, Lovett Tower, 13 Keltie St, Woden ACT, 2611, Australia (02) 6274 1111 www.environment.gov.au/parks

Parks Victoria
parkweb.vic.gov.au

Queensland Department of Environment and Resource Management
National Parks, Marine Parks and Forests www.derm.qld.gov.au/parks_and_forests

South Australian Department of Environment and Resource Management
Parks SA www.environment.sa.gov.au/parks/home

Tasmanian Department of Primary Industries, Parks, Water and Environment

Parks and Wildlife Service www.parks.tas.gov.au

Western Australian Parks and Wildlife

www.parks.dpaw.wa.gov.au

Index of bird common names

Index of place names